FEEDBACK NETWORKS
THEORY AND CIRCUIT APPLICATIONS

ADVANCED SERIES IN CIRCUITS AND SYSTEMS

Editor-in-Charge: **Wai-Kai Chen** (Univ. Illinois, Chicago, USA)
Associate Editor: **Dieter A. Mlynski** (Univ. Karlsruhe, Germany)

Published

Vol. 1: Interval Methods for Circuit Analysis
by *L. V. Kolev*

Vol. 2: Network Scattering Parameters
by *R. Mavaddat*

Vol. 3: Principles of Artificial Neural Networks
by *D Graupe*

Vol. 4: Computer-Aided Design of Communication Networks
by *Y-S Zhu and W K Chen*

Vol. 5: Feedback Networks: Theory and Circuit Applications
by *J Choma and W K Chen*

Advanced Series in Circuits and Systems — Vol. 5

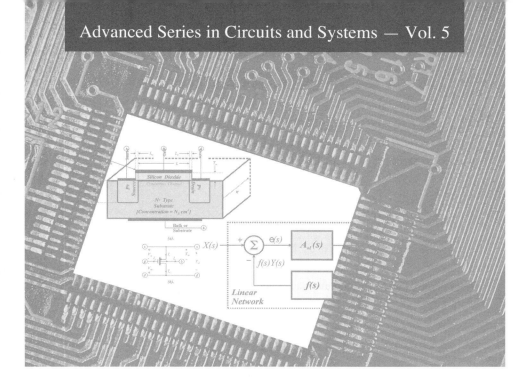

FEEDBACK NETWORKS
THEORY AND CIRCUIT APPLICATIONS

J Choma
University of Southern California

W K Chen
University of Illinois, Chicago

World Scientific

NEW JERSEY · LONDON · SINGAPORE · BEIJING · SHANGHAI · HONG KONG · TAIPEI · CHENNAI

Published by

World Scientific Publishing Co. Pte. Ltd.
5 Toh Tuck Link, Singapore 596224
USA office: 27 Warren Street, Suite 401-402, Hackensack, NJ 07601
UK office: 57 Shelton Street, Covent Garden, London WC2H 9HE

British Library Cataloguing-in-Publication Data
A catalogue record for this book is available from the British Library.

FEEDBACK NETWORKS: THEORY AND CIRCUIT APPLICATIONS
Advanced Series in Circuits and Systems — Vol. 5

Copyright © 2007 by World Scientific Publishing Co. Pte. Ltd.

All rights reserved. This book, or parts thereof, may not be reproduced in any form or by any means, electronic or mechanical, including photocopying, recording or any information storage and retrieval system now known or to be invented, without written permission from the Publisher.

For photocopying of material in this volume, please pay a copying fee through the Copyright Clearance Center, Inc., 222 Rosewood Drive, Danvers, MA 01923, USA. In this case permission to photocopy is not required from the publisher.

ISBN-13 978-981-02-2770-8
ISBN-10 981-02-2770-1

Printed in Singapore by World Scientific Printers (S) Pte Ltd

To our wives, Lorraine and Shiao-Ling,
for their constant encouragement, understanding, and patience.

Preface

Feedback Networks: Theory and Circuit Applications is not an explicit circuit design text, but it is very much a text that forges a foundation for the creative design of electronic circuits and systems. A circuit design initiative that culminates by responding positively to the demanding performance specifications of modern data processing, information transmission, and communication systems is a challenging and often daunting undertaking. The acquisition of design skills is complicated by the curious fact that computational precision is not the primary objective of design-oriented circuit analysis. Rather, the purpose of analyses undertaken in support of a design venture is to gain insights into the theoretic and phenomenological underpinnings of the mathematical solutions for the electrical responses of electronic networks. An insightful understanding is cultivated by solutions — albeit approximate solutions premised on clearly understood engineering presumptions — cast in forms that highlight circuit attributes, limitations, best case operating features, and worst case performance shortfalls. In short, design skills, methodologies, and guidelines are not necessarily nurtured by elegant and mathematically satisfying circuit solutions. They are more likely to derive from approximate circuit solutions that, when properly interpreted in light of given applications, paint an understandable engineering picture of pertinent circuit dynamics. A fundamental objective of this textbook is to paint compelling analytical images that support the innovative design of high frequency and high speed integrated electronics.

Two engineering reasons justify the obvious "feedback" focus of this textbook. The first of these reasons is that feedback signal flow paths are pervasive of all electronic networks. These feedback paths are either purposefully implemented or parasitically incurred and in many active networks destined for high frequency or high speed signal processing applications, both purposeful and parasitic feedback prevails. Included among the numerous advantages of feedback incorporated explicitly as a design vehicle is the realization of electronic circuits and systems exuding broadbanded steady state frequency responses, suitable impedance levels at input and output network ports, and acceptable desensitization of input-to-output transfer characteristics with respect to vagarious circuit elements or active device parameters. When feedback is not overtly adopted, it nonetheless prevails, generally in the form of undesirable capacitive coupling between device terminals or network node pairs, parasitic magnetic coupling between network branches, or unwanted conductive coupling between device terminals. For example, when an impedance is inserted in series with the source lead of a common source amplifier, the transistor channel resistance, which is not very large in deep submicron device technologies, is a bilateral, nonglobal feedback element interconnecting the drain and source terminals of the device. Rarely is parasitic feedback advantageous to circuit design objectives. Indeed, it commonly incurs degradation of the forward signal transmission characteristics, poor transient responses, difficulties in achieving maximum signal power gain, and in extreme cases, outright circuit instability.

A second reason for the expressed focus of this textbook is that the analysis of feedback networks is invariably cumbersome, particularly if attention is directed to high frequency circuit performance. As a result, definitive manual analyses are obviated in favor of only computer-based investigations that may not impart the circuit insights supportive of accurate, reliable, and reproducible circuit design. When manual analyses are executed, their value is often limited by unrealistic approximations and assumptions that either mask requisite insights or produce results that are incapable of mirroring engineering reality. In this textbook, a systematic analysis methodology for feedback electronics (which is actually a superfluous phrase, since feedback is pervasive of all electronics) is developed and ultimately applied to the design-oriented analysis of practical electronic networks. Aside from merely applying the new feedback analysis techniques to the standard cells of analog electronic systems, the new procedure actually facilitates circuit broadbanding innovations, as is demonstrated in the final chapter. In addition to rendering feedback network analysis less daunting and more practical than might be traditionally expected, the new methodology implicitly accounts for the facts that the feedback paths of interest may not be global and that purposefully implemented or parasitically encountered feedback is invariably bilateral in nature.

Feedback Networks: Theory and Circuit Applications is principally an advanced circuits and systems analysis text that forges a strong analytical foundation for more design-intensive analog and mixed signal integrated circuits and systems classes. It teaches students computationally efficient manual methods, complemented by meaningful computer-aided assessment and verification strategies, for analyzing the electrical dynamics of active networks destined for monolithic realization in silicon-germanium (SiGe) heterostructure bipolar, and principally complementary metal-oxide-semiconductor (CMOS) technologies. More than teaching mere analytical problem solving techniques, the text couches analyses in forms that foster the engineering insights underpinning a meaningful characterization and performance assessment of active circuits embedded in high frequency and/or high speed system applications. These insights are fundamental to consistently creative circuit and system design, for they enable realistic comparisons among candidate active devices and among plausible circuit architectures. They are also indispensable to the omnipresent design problem of mitigating the deleterious effects that parasitic energy storage and other high order device and circuit phenomena have on such performance metrics as bandwidth, signal delay in both time and frequency domains, gain and phase margins, phase noise, distortion, and transient step and impulse responses. In short, the formulation of insightful design-oriented analysis strategies commensurate with the realization of modern integrated circuits, and particularly analog high performance integrated circuits and systems, is the primary focus of this work.

This textbook is suitable for use in a senior elective circuits course whose students have successfully completed courses in basic circuit analysis, basic linear systems, and an introductory electronics course featuring exposure to linear electronics exploiting bipolar and MOS technology devices. It can also be used as a graduate level core course for the electronic circuits and systems arena. Students using this textbook should be comfortable with using one of the many available versions of SPICE computer-aided analysis software, such as HSPICE, PSPICE, TOPSPICE, Tanner SPICE, or the circuit analysis tools implicit to the CADENCE design suite.

Chapter 1 on *Circuit and System Fundamentals* offers the student an overview of the basic theoretic concepts traditionally addressed in sophomore and junior level circuits and linear systems classes. Although the issues covered in this chapter should not be foreign to senior and graduate level electrical engineering students, their interpretation in light of a variety of conventional circuit design requirements may comprise new material for neophyte electronic circuit design students. For example, the well-known theorems of Thévenin and Norton are reviewed and thence used directly to define the basic terminal characteristics of the four types of amplifiers encountered in modern electronic systems. Second order circuits are studied thoroughly in both the frequency and the time domains. In a representative graduate electrical engineering class, the instructor may elect to ignore Chapter 1 and proceed directly to Chapter 2. In such an event, the student is nonetheless strongly encouraged to read Chapter 1 and to attempt several of the Exercises at its conclusion to assess his/her ability to utilize fundamental theoretic detail in practical problems.

In Chapter 2 on *Two-Port Network Models and Analysis*, the models and analysis procedures surrounding the four fundamental types of two-port network characterizations are studied in detail. In the course of this study, the underpinnings of circuit feedback are introduced through the introduction of the concepts of open loop and loop gains. Computationally efficient methods of studying interconnections of two-port networks are developed, and power gain in active networks is defined and assessed. The concepts of potential network instability and unconditional network stability are introduced. Chapter 3 on *Scattering Parameters* embellishes the contents of Chapter 2 by considering the scattering parameter characterization of linear two-port networks. These scattering parameters are carefully interpreted and thence applied to the problem of designing lossless filters, which are commonly deployed in both narrowband and wideband electronics earmarked for high frequency communication systems.

Chapters 4, 5, and 6 collectively comprise a reasonably exhaustive treatise on the analysis of feedback networks. Chapter 4 on *Feedback Circuit and System Theory* addresses first order, second order, and multi-order feedback networks from the perspectives, of gain, transfer function sensitivity to critical parameters, bandwidth, gain-bandwidth product, and other performance metrics. Useful expressions for the gain and phase margins of second order networks are propounded, and the problems of overshoot, settling time, and delay time surrounding feedback network responses to transient step and impulsive inputs are thoroughly examined. Chapter 5, which considers *Signal Flow Methods of Feedback Network Analysis*, offers a general, tractable, and insightful method for analyzing both single loop and dual loop feedback amplifiers. The theoretic detail underpinning this analytical method, which embodies the analytical notions of null gain, return ratio, and null return ratio, is applied to the problem of definitively assessing the performance attributes and shortfalls indigenous to several types of traditionally encountered amplifiers. Chapter 6 on *Multiloop Feedback Amplifiers* complements the considerations of Chapter 5 by providing the reader with enhanced mathematical rigor. It also embellishes the dual loop discourse of Chapter 5 by developing a generalized and powerful mathematical technique for analyzing multiloop feedback circuits.

Chapter 7 on *Analog MOS Technology Circuits* exploits the theoretic disclosures proffered in preceding chapters by studying the canonic MOS technology analog cells at both low and high signal frequencies. In advance of these circuit studies, the circuit level models of MOS transistors, inclusive of deep submicron technologies, are studied in reasonably complete detail so that the circuit assessments developed in the chapter can be interpreted in terms device phenomenological issues and monolithic processing constraints. Chapter 8 on *MOS Technology Operational Amplifiers* builds on its predecessor chapter by studying, and developing design methodologies for, single stage and two-stage operational amplifiers destined for utilization in monolithic mixed signal applications.

In Chapter 9, *Broadband and Radio Frequency MOS Technology Amplifiers*, the feedback tools developed earlier are applied to the problem of optimizing amplifier performance at high signal frequencies. The circuit broadbanding schemes addressed include resistance-capacitance degeneration, shunt peaking, multi-order series peaking, and series-shunt peaking. In the course of these discussions, a new broadbanding scheme, premised on the realization of constant resistance compensation filters, is proposed. Impedance matching in tuned radio frequency amplifiers is discussed and critically assessed.

Almost all of the material included in this textbook has been used several times by Prof. Choma in a graduate level core course on electronic circuits at the University of Southern California Viterbi School of Engineering. To this end, the text benefits from the critique and constructive criticisms of many Viterbi School of Engineering graduate students and particularly, those unselfishly offered by Mr. Jonathan Roderick, who is scheduled to complete his electrical engineering doctorate in 2007. Prof. Choma also wishes to acknowledge the enormous benefits gleaned from numerous technical exchanges he enjoyed with his faculty colleague, Dr. Edward Maby, in the Ming Hsieh Department of Electrical Engineering of the USC Viterbi School of Engineering.

John Choma
San Dimas, California

Wai-Kai Chen
Fremont, California

Contents

Preface	**VII**
Chapter One: Circuit and System Fundamentals	**1**
1.1.0. Introduction	1
1.2.0. Thévenin's and Norton's Theorems	3
1.2.1. Thévenin and Norton Parameters	7
1.2.2. Engineering Observations	9
1.3.0. Dependent Sources and Amplifier Concepts	20
1.3.1. Voltage Amplifier	21
1.3.2. Transconductor	25
1.3.3. Transresistor	27
1.3.4. Current Amplifier	30
1.3.5. Buffers	32
1.3.5.1. Voltage Buffer	34
1.3.5.2. Current Buffer	40
1.3.6. Load Power Considerations	45
1.3.6.1. Maximum Power Transfer	48
1.3.6.2. The dBm Power Measure	50
1.3.6.3. Match Termination and Tuned Responses	51
1.4.0. Second Order Circuits and Systems	57
1.4.1. Second Order Filters	58
1.4.2. Frequency Response	63
1.4.2.1. Response Peaking	65
1.4.2.2. Bandwidth	66
1.4.2.3. Phase and Delay Responses	69

	1.4.3. Poles and Second Order System Parameters	75
	1.4.4. Time Domain Transient Responses	78
	1.4.4.1. Impulse Response	79
	1.4.4.2. Step Response	82
Exercises		89

Chapter Two: Two-Port Network Models and Analysis   111

2.1.0. Introduction		111
2.2.0. Two-Port Linearity Issues		112
2.3.0. Generalized Two-Port Parameters		117
	2.3.1. Hybrid h-Parameters	118
	2.3.1.1. Ideal Current Amplifier	122
	2.3.1.2. Parameter Measurement Issues	124
	2.3.2. Hybrid g-Parameters	128
	2.3.2.1. g- and h-Parameter Interrelationships	130
	2.3.2.2. Ideal Voltage Amplifier	132
	2.3.3. Short Circuit y-Parameters	135
	2.3.3.1. π-Type Network Model	137
	2.3.3.2. Ideal Transconductor	141
	2.3.3.3. Indefinite Admittance Matrix	142
	2.3.4. Open Circuit z-Parameters	146
	2.3.4.1. Tee-Type Network Model	148
	2.3.4.2. Ideal Transresistor	149
	2.3.5. Transmission Parameters	151
	2.3.5.1. Input and Output Impedances	153
	2.3.5.2. Voltage Transfer Function	154
	2.3.5.3. Cascade Interconnection	154
	2.3.5.4. Series and Shunt Branch Elements	156
2.4.0. Two-Port Methods of Circuit Analysis		158
	2.4.1. Circuit Analysis in Terms of h-Parameters	159
	2.4.1.1. Open Loop Gain and Loop Gain Concepts	160
	2.4.1.2. I/O Impedances	164

    2.4.2. Circuit Analysis Via g-Parameters 167
    2.4.3. Circuit Analysis in Terms of y-Parameters . . 169
    2.4.4. Circuit Analysis in Terms of z-Parameters . . 171
    2.4.5. Generalized Analytical Disclosures 176
  2.5.0. Systems of Interconnected Two Ports 177
    2.5.1. Series-Shunt Feedback Architecture 178
    2.5.2. Shunt-Series Feedback 186
    2.5.3. Shunt-Shunt Feedback 188
    2.5.4. Series-Series Feedback 189
  2.6.0. Power Flow and Transfer 191
    2.6.1. Power Gain Expressions 192
    2.6.2. Stability Considerations 195
    2.6.3. Maximum Transducer Gain 201
    2.6.4. Unilateralization 204
      2.6.4.1. Shunt-Antiphase Shunt
          Compensation 204
      2.6.4.2. Shunt-Antiphase Shunt Network
          Realization 206
  References 208
Exercises 208

Chapter Three: Scattering Parameters   225

  3.1.0. Introduction 225
  3.2.0. Reflection Coefficient 227
    3.2.1. Voltage Scattering 228
    3.2.2. Power Scattering 229
    3.2.3. Significance of the Reflection Coefficient . . 230
  3.3.0. Two-Port Scattering Parameters 239
    3.3.1. Parameters S_{11} and S_{21} 241
    3.3.2. Parameters S_{22} and S_{12} 243
    3.3.3. Port Voltage and Current Generalizations . . 244
    3.3.4. Scattering Analysis of a Generalized
       Two-Port 244
      3.3.4.1. Input and Output Reflection
          Coefficients 245
      3.3.4.2. Voltage Transfer Function 247
      3.3.4.3. Other Transfer Functions 249
    3.3.5. Scattering and Conventional Parameters . . . 250

3.3.5.1. S-Parameters in Terms of
    h-Parameters 250
3.3.5.2. h-Parameters in Terms of
    Scattering Parameters 252
3.4.0. Lossless Two-Port Networks 255
    3.4.1. Average Power Delivered to Complex
        Load 256
    3.4.2. Average Power Delivered to Two-Port
        Network 258
    3.4.3. Lossless, Passive Two-Port Network 260
    References 266
Exercises 266

Chapter Four: Feedback Circuit and System Theory — 277

4.1.0. Introduction 277
4.2.0. System Level Model of Feedback Circuit 277
4.3.0. Feedback Network Frequency Response 285
    4.3.1. Single Pole Open Loop Transfer
        Function 286
    4.3.2. Second Order Open Loop Transfer
        Function 288
    4.3.3. Stability Issues 291
        4.3.3.1. Phase Margin 292
        4.3.3.2. Gain Margin 295
        4.3.3.3. Alternative Damping and Undamped
            Frequency Expressions 297
    4.3.4. Compensation for Closed Loop Stability . . . 300
4.4.0. Time Domain Response 306
    4.4.1. Unit Step Response 306
    4.4.2. Settling Time 308
    References 312
Exercises 312

Chapter Five: Signal Flow Methods of Feedback Network Analysis — 321

5.1.0. Introduction 321
5.2.0. Feedback Network Analysis Fundamentals 322

            5.2.1. Calculation of Feedback Network
                   Parameters   326
                   5.2.1.1. Null Parameter Gain   327
                   5.2.1.2. Normalized Return Ratio   328
                   5.2.1.3. Normalized Null Return Ratio . . .   330
            5.2.2. Input and Output Impedances   337
                   5.2.2.1 Driving Point Input Impedance . . .   337
                   5.2.2.2 Driving Point Output Impedance . .   340
            5.2.3 Output Port-to-Local Port Feedback   345
    5.3.0   Special Case Feedback Network Examples   348
            5.3.1. Global Feedback   348
                   5.3.1.1. Transimpedance Feedback
                            Amplifier   348
                   5.3.1.2. Transadmittance Feedback
                            Amplifier   352
                   5.3.1.3. Voltage Feedback Amplifier   354
                   5.3.1.4. Current Feedback Amplifier   357
            5.3.2. Other Feedback Architectures   366
                   5.3.2.1. Feedback Branch Admittance   366
                   5.3.2.2. Feedback Branch Impedance   377
            5.3.3. Dual Loop Feedback   385
            5.3.4 Series-Series/Shunt-Shunt Feedback   388
                   5.3.4.1. Analysis of the Series-Series/
                            Shunt-Shunt Feedback Pair   390
                   5.3.4.2. Interpretation of Results and Design
                            Considerations   398
            5.3.5. Series-Shunt/Shunt-Series Feedback   405
                   5.3.5.1. Analysis of the Series-Shunt/
                            Shunt-Series Feedback Pair   405
                   5.3.5.2. Design Restrictions   416
        References   420
    Exercises   420

Chapter Six: Multiple Loop Feedback Amplifiers   441

    6.1.0. Introduction   441
    6.2.0. Indefinite Admittance Matrix   442
           6.2.1. Return Difference   448
           6.2.2. Null Return Difference   454
    6.3.0. Network Functions and Feedback   456

       6.3.1. Blackman's Formula 457
       6.3.2. Sensitivity Function 463
  6.4.0. Measurement of Return Difference 467
       6.4.1. Blecher's Procedure 469
       6.4.2. Impedance Measurements 472
  6.5.0. Multiloop Feedback 475
       6.5.1. Multiloop Feedback Theory 476
       6.5.2. Return Difference Matrix 480
       6.5.3. Null Return Difference Matrix 482
       6.5.4. Transfer Function Matrix 484
       6.5.5. Sensitivity Matrix 488
       6.5.6. Multi-Parameter Sensitivity 492
       References 495

Chapter Seven: Analog MOS Technology Circuits 497

  7.1.0. Introduction 497
  7.2.0. MOS Transistor Models 499
       7.2.1. Transistor Cross-Section and Electrical
              Symbol 500
       7.2.2. Static Volt-Ampere Relationships 504
              7.2.2.1. Cutoff 505
              7.2.2.2. Ohmic Electrical Regime 509
              7.2.2.3. Saturation Regime 512
       7.2.3. Small Signal Models 514
              7.2.3.1. Small Signal Model at High
                       Frequencies 518
              7.2.3.2. Unity Gain Frequency 522
  7.3.0. Common Source Amplifier 524
       7.3.1. Voltage Transfer Function 526
              7.3.1.1. Poles and Time Constants 528
              7.3.1.2. Miller-Limited Frequency
                       Response 532
              7.3.1.3. Output Port Time Constant
                       Dominance 540
       7.3.2. Input and Output Impedances 542
       7.3.3. Variants of the Common Source
              Topology 544
              7.3.3.1. NMOS Load 544
              7.3.3.2. PMOS Load 548

7.4.0. Common Drain Amplifier	552
7.4.1. Source Follower Transfer Function	555
7.4.2. Source Follower I/O Impedances	560
7.5.0. Common Gate Amplifier	568
7.5.1. Common Gate I/O Characteristics	571
7.5.2. Common Source-Common Gate Cascode . .	575
7.5.3. Enhanced Common Gate Cell	581
7.5.3.1. Low Frequency Circuit Properties	584
7.5.3.2. High Frequency Circuit Properties	585
7.5.3.3. Integrator Application	591
References	604
Exercises	605

Chapter Eight: MOS Technology Operational Amplifiers   629

8.1.0. Introduction	629
8.2.0. Op-Amp System Architectures	630
8.2.1. Single Stage Architecture	630
8.2.2. Two-Stage Architecture	632
8.2.3. Input Stage Transconductor	634
8.3.0. CMOS Input Stage Analysis	637
8.3.1. P-Channel Transconductor	638
8.3.1.1. Transconductor Small Signal Analysis	640
8.3.1.2. Transconductor Output Macromodel	642
8.3.1.3. First Stage Output Macromodel . . .	651
8.3.1.4. First Stage Static Analysis	654
8.3.2. N-Channel Transconductor	656
8.4.0. Phase Inverting Second Stage	657
8.4.1. Low Frequency Small Signal Analysis	658
8.4.2. Op-Amp Static Analysis	660
8.5.0. Frequency Compensation	661
8.5.1. Approximate High Frequency Analysis . . .	662
8.5.2. Miller Compensation	664
8.5.3. Improved Frequency Compensation	669
8.5.3.1. Buffered Capacitive Feedback	670

                    8.5.3.2. Passive Highpass Feedback 677
          8.6.0. Slew Rate Limitations 684
                    8.6.1. Fundamentals of Slew Rate Issues 685
                    8.6.2. Slew Rate Limiting due to Nonlinearity . . . 688
                    8.6.3. Full Power Bandwidth 690
          8.7.0. Biasing Subcircuits 691
                    8.7.1. Active Divider 692
                    8.7.2. Supply-Independent Biasing 702
                    References 710
          Exercises 711

Chapter Nine: Broadband and Radio Frequency MOS Technology Amplifiers   729

          9.1.0. Introduction 729
          9.2.0. Cascade of Dominant Pole Amplifiers 731
                    9.2.1. Bandwidth of N-Stage Cascade 732
                    9.2.2. Optimized Bandwidth of a Cascade 734
          9.3.0. Degenerative RC Broadbanding 736
                    9.3.1. Gain and Dominant Pole 737
                    9.3.2. Broadband Compensation 742
          9.4.0. Shunt Peaked Compensation 748
                    9.4.1. Common Source Stage Revisited 749
                    9.4.2. Shunt Peaked Amplifier Response 753
                              9.4.2.1. Maximally Flat Magnitude
                                          Response 756
                              9.4.2.2. Maximally Flat Delay Response . . 762
          9.5.0. Series Peaked Compensation 765
                    9.5.1. Second Order Compensated Response 767
                    9.5.2. Third Order Compensated Response 768
          9.6.0. Series-Shunt Peaked Compensation 771
                    9.6.1. Design Criteria 772
                    9.6.2. A Design Problem 774
                              9.6.2.1. Capacitance Bridged, Coupled
                                          Inductor Load 776
                              9.6.2.2. Constant Resistance Criteria 778
                              9.6.2.3. Transfer Relationship 781
          9.7.0. Broadbanding via Feedback 787
                    9.7.1. Low Frequency Characteristics 789

	9.7.2.	Amplifier Bandwidth	793
	9.7.3.	Input Impedance	795
	9.7.4.	Input Impedance Compensation	797
	9.7.5.	Output Impedance	806
9.8.0.	The f_T-Doubler		812
	9.8.1.	Small Signal Analysis	813
	9.8.2.	Realization of the f_T-Doubler	815
9.9.0.	Bandpass Feedback Amplifier		816
	9.9.1.	Common Source RF Amplifier	821
	9.9.2.	Impedance and Transfer Characteristics . . .	824
		9.9.2.1. Gate Impedance	825
		9.9.2.2. Voltage Transfer Function	832
	References		841
Exercises			842

Index   855

Chapter 1

Circuit and System Fundamentals

1.1.0. Introduction

The formulation of meaningful analytical procedures and design strategies for even the most advanced of electronic feedback circuits and systems relies on a thorough grasp of basic circuit and system concepts. Aside from abilities to apply and interpret the Kirchhoff voltage and current laws (KVL and KCL) in both the time and frequency domains, at least three issues underpin the mission of acquiring design-oriented analytical proficiency in the electronic circuits arena. The first of these is the theorems attributed to Thévenin and Norton. An ability to apply these theorems to the problems of exploring and understanding the electrical dynamics of electronic networks that couple specified signal sources to an arbitrary linear or nonlinear load is a virtual cornerstone of the electronic networks discipline. For example, Thévenin's and Norton's theorems might be gainfully applied to deduce the desired input/output (I/O) electrical characteristics of a preamplifier designed for insertion between the output terminals of a compact disc player and the input terminals of the power amplifier used to drive the audio speakers of a stereo system.

A second issue embraces transfer functions of linear networks. The capability of deducing the transfer characteristic and casting it into appropriate mathematical form serve a multitude of purposes. Included among these purposes are a delineation of the input to output gain of the network undergoing investigation, the determination of the network input and output impedances, an assessment of the relative stability of the system, and the determination of the time domain response of the subject circuit to specified transient and steady state input excitations. Phasor analyses in the sinusoidal steady state, which is fundamental to a stipulation of the manner in which the system gain and pertinent impedance levels depend on the frequency of the applied input signal, are intimately linked to network transfer functions. Phasors comprise the basis for deducing such electronic circuits and systems performance metrics as bandwidth, impedances, frequency response, and phase response. The bandwidth defines the frequency interval over which the I/O gain is maintained nominally constant. The impedance levels at the input and output terminals of an active network are instrumental in determining whether an amplifier is more suitable for voltage than for current amplification. The frequency response is essentially a mathematical snapshot of the manner in which the network under consideration performs over specified intervals of signal frequency. Finally, the phase response establishes the network delay, which defines the average time required by a system to process and ultimately deliver the desired steady state output response to a specified input signal.

The third issue is the intelligent use of the four types of dependent generators; namely, the voltage controlled current source (VCCS), the voltage controlled voltage source (VCVS), the current controlled current source (CCCS), and the current controlled voltage source (CCVS). Understanding the volt-ampere properties of these mathematical circuit branch elements is a prerequisite to formulating reasonably accurate, design-oriented, linearized circuit models for active devices, such as the metal-oxide-semiconductor field-effect transistor (MOSFET), the bipolar junction transistor (BJT), and the PN junction diode. Moreover, exploiting these properties prudently and creatively is fundamental to the intelligent application of Thévenin's and Norton's theorems and to the efficient deduction of the transfer characteristics of electronic systems.

In an attempt to vector the interested reader on a path toward ultimate electronic circuit design proficiency, the foregoing and a few related other concepts indigenous to basic circuit theory are reviewed and exemplified in this chapter. These reviews and illustrations serve to introduce the reader to an interesting and even perplexing paradox that underpins the genuinely difficult task (some might even argue art) of creative and innovative electronic circuit and system design. In particular, the fundamental purpose of circuit analysis is not the precise disclosure of either a circuit response or a specific circuit performance metric. Instead, analyses are conducted to gain an insightful understanding of the limitations and attributes of the time and frequency domain electrical dynamics pervasive of a circuit architecture deemed plausible for the design mission. As such, these design-oriented analyses respond to the time-honored adage that nothing should ever be built until that which is to be built is thoroughly understood.

Design-oriented engineering analysis is not a trivial undertaking because design itself is neither trivial nor straightforward. Design is a challenging undertaking because it is not the problem of finding the N solutions to a system of N equations in N unknowns. The most typical design problem is one in which there are more specifications that must be satisfied or more variables that need to be determined than there are independent equations that can be written. Basic algebra teaches that a problem for which the number of unknowns does not match the number of available independent equations has no unique solution. Since poorly structured mathematical problems are implicit to virtually all design environments, unique design solutions rarely prevail. Nevertheless, viable and even creative solutions can be determined. The best of these solutions, in the sense of yielding reliable, manufacturable, and cost effective electronic networks that meet operating specifications, are rarely forged by trial and error strategies. Instead, optimal solutions derive from fundamental phenomenological understanding. The task necessarily preceding such understanding is the conduct of thorough mathematical and computer-based analyses that insightfully highlight both the attributes and the limitations of the circuit and system architectures under consideration. The satisfying understanding that supports the completion of a design project ensues when analytical disclosures can be creatively interpreted and lucidly explained in terms of fundamental physical laws, basic circuit and system theories, and simple mathematical models.

1.2.0. Thévenin's and Norton's Theorems

Consider the system in Fig. 1.1(a), which abstracts two terminals of a generalized linear network coupled to a load branch. Since the subject network is stipulated as a linear entity, its intrinsic branch elements are exclusively linear resistors, linear capacitors, linear inductors, and linear controlled voltage and current sources. Although no sources of energy are presumed embedded in the structure, any number of independent energy sources can be applied. To this end and without loss of generality, two independent inputs — a voltage source, V_s, and a current source, I_s, — are depicted. It should be understood that the presumption of no intrinsic energy sources implies at least one of three possible operational circumstances. In particular, the internal capacitors and inductors may have zero initial voltages and currents, respectively, at the time at which the indicated input sources, V_s, and I_s, are applied. Alternatively, it may be that analytical interest focuses on only the steady state performance of the system. Accordingly, the effects

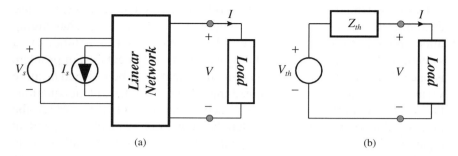

Figure 1.1. (a) A linear network driving an arbitrary load and excited by an independent voltage source, V_s, and an independent current source, I_s. (b) The system in (a) with the linear network supplanted by its Thévenin equivalent circuit consisting of a voltage source, V_{th}, connected in series with an impedance Z_{th}.

of initial capacitive and inductive energies have dissipated and no longer possess engineering significance. A third possibility is that initial capacitor voltages and inductor currents are treated as additional independently applied input excitations, similar to the signal sources, V_s, and I_s. In the present circumstance, it is tacitly assumed that analytical attention focuses exclusively on steady state electrical characteristics.

Thévenin's theorem states that the electrical characteristics at any port (or terminal pair) of a linear electrical network can be modeled by a voltage source in series with an impedance, as suggested by Fig. 1.1(b). The indicated voltage source, V_{th}, is termed the *Thévenin voltage* of the port undergoing scrutiny, while the subject series impedance, Z_{th}, is known as the *Thévenin impedance* of said port. If the port at which Thévenin's theorem is applied happens to be the output port of the network where signal responses to applied input excitations are to be delivered, the Thévenin impedance is also known as the network *output impedance*. When V_{th} and Z_{th} are correctly measured or calculated, the Thévenin equivalent circuit, or Thévenin model, "seen" by the load establishes a load voltage, V, and a load current, I, that are respectively identical to the load voltage and current supported by the original system in Fig. 1.1(a). It is important to underscore the fact that the foregoing assertions are independent of the nature of the load connected to the network port undergoing a Thévenin investigation. This is to say that the load at hand can be a passive, an active, a linear, or even a nonlinear electrical branch.

An alternative to Thévenin's theorem is Norton's theorem, which stipulates that any port of a linear electrical network can be represented as a

Circuit and System Fundamentals 5

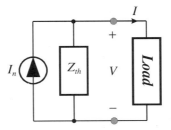

Figure 1.2. The Norton equivalent circuit for the system given in Fig. 1.1(a). As in the case of the Thévenin model in Fig. 1.1(b), the Norton circuit delivers a load voltage, V, and a load current, I, that are respectively identical to the load voltage and load current observed in the original system.

current source in shunt with an impedance, as suggested by Fig. 1.2. The current source, I_n, is termed the *Norton current* of the port undergoing scrutiny. The associated shunt impedance, which can be termed the *Norton impedance*, is, at risk of deflating Norton's ego, identical to the Thévenin impedance introduced in Fig. 1.1(b).

The topological simplicity of both the Thévenin and Norton models obscures their actual significance and engineering utility. An initial appreciation of these models can be garnered from the realization that the linear networks they represent can be large, intricate circuits comprised of hundreds of thousands of interconnected electrical branch elements. But architectural complexity notwithstanding, only two elements — voltage source and impedance in the case of Thévenin and current source and impedance in the case of Norton — are required for the unique determination of the voltage and corresponding current associated with an appended load. In short, it is likely easier to analyze the model in either Fig. 1.1(b) or in Fig. 1.2 than it is to analyze the entire electrical system abstracted in Fig. 1.1(a). The downside to this Thévenin or Norton analytical tack is that the replacement of the original system by either of the models shown in Fig. 1.1(b) or 1.2 leads to an irretrievable loss of branch voltage, branch current, and branch power information within the linear network. In most electronic circuit and system applications, this loss of information is an acceptable consequence of the expediency with which the load voltage, current, and power can be determined in terms of Thévenin or Norton parameters. In a few cases, such loss of information may prove unacceptable. For example, in some design environments, it may be essential to understand how nonzero network element tolerances or other manufacturing uncertainties deleteriously affect the ability of a linear network to establish and sustain required load voltage and current characteristics.

The Thévenin and Norton equivalent circuits are two distinctly different circuit topologies that serve to model any considered linear electrical system. Questions therefore naturally arise as to why two modeling approaches need be advanced when one model appears to suffice. To be sure, either a Thévenin or a Norton representation can be used to model a port of any linear network. Given the widespread analytical comfort levels associated with voltage sources, it is hardly surprising that the Thévenin equivalent circuit enjoys wider popularity than does its Norton counterpart. But in fact, some network ports are more amenable to Thévenin modeling, while others are more appropriate to Norton modeling. In idealized operating circumstances, it is even possible to encounter a network port for which Thévenin parameters can be calculated or measured, but Norton parameters are not deterministic, and vice versa. For example, and as is demonstrated in the following subsection of material, the Norton current is indeterminate for a network port whose Thévenin impedance is zero. In this case, only a Thévenin equivalent circuit can be meaningfully contrived and, as Fig. 1.3(a) illustrates, the subject network port emulates the volt-ampere characteristics of an ideal voltage source. On the other hand, the Thévenin voltage of a network port having infinitely large Thévenin impedance cannot be determined, which accordingly forces a Norton representation of said port. In this case, the network port at hand behaves as the ideal current source depicted in Fig. 1.3(b). A general extrapolation of the foregoing two statements is that a network port characterized by a small Thévenin impedance behaves as an approximate ideal voltage source and is therefore

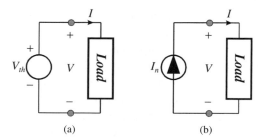

Figure 1.3. (a) The Thévenin equivalent circuit of a loaded linear network port whose Thévenin port impedance is zero. The port in question behaves as an ideal voltage source. (b) The Norton equivalent circuit of a loaded linear network port whose Thévenin port impedance is infinitely large. The port in question behaves as an ideal current source.

prudently modeled by a Thévenin equivalent circuit. On the other hand, a linear network port that emulates idealized current source characteristics by virtue of its large Thévenin impedance is best represented by a Norton equivalent circuit.

1.2.1. Thévenin and Norton Parameters

The experimental determination or analytical evaluation of the Thévenin parameters commences by noting in Fig. 1.1(b) that the load voltage, V, is

$$V = V_{th} - Z_{th}I, \qquad (1\text{-}1)$$

where I is obviously the current supplied by the network and conducted by the load connecting to, or terminating, the network port undergoing scrutiny. The fact that the Thévenin equivalent circuit is independent of the nature of the load encourages the exploitation of specific loads that facilitate the direct determination of the Thévenin voltage, V_{th}. To this end, if the actual load were to be supplanted by an open circuit, as shown in Fig. 1.4(a), current I is necessarily reduced to zero, which renders $V \equiv V_{th}$ in Eq. (1-1). In other words, the Thévenin voltage of a linear network port is the voltage established at that port when said port is open circuited. This observation explains the common reference to a Thévenin voltage as an *open circuit voltage*.

Consider now the case in which the original load is replaced by an ideal current source of value I_x, as is depicted in Fig. 1.4(b). The resultant network

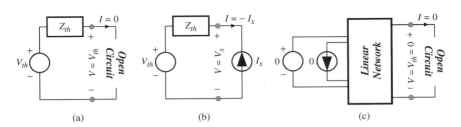

Figure 1.4. (a) The calculation of the Thévenin equivalent voltage at a port of a linear network. (b) The ohmmeter method of evaluating the Thévenin equivalent impedance. (c) The strategy for realizing zero Thévenin equivalent voltage. Zero Thévenin voltage expedites the ohmmeter method of computing the Thévenin equivalent impedance.

current is $I = -I_x$, and if V_x denotes the value of port voltage V corresponding to the applied current load, Eq. (1-1) delivers

$$\frac{V_x}{I_x} = Z_{th} + \frac{V_{th}}{I_x}, \qquad (1\text{-}2)$$

where it is important to note that voltage V_x is in disassociated polarity reference to current I_x. It would be delightful if V_{th} could be constrained to zero under the stipulated load current source constraint. In this event, the ratio of the current source voltage, V_x, -to- the current source current, I_x, is the Thévenin impedance, Z_{th}, in need of evaluation. The strategy for effecting null V_{th} in a physically sound sense derives from classic *superposition theory*, which applies to all linear networks. In particular, recall from Fig. 1.1(a) that the network undergoing study is excited by two sources of independent energy; namely, voltage V_s and current I_s. Superposition theory states that any branch voltage or any branch current of any linear electrical system, which certainly embraces a linear network under the condition of a linear load termination that happens to be a constant current source, is the algebraic superposition of the effects of all applied independent sources of voltage and current. It follows that

$$V_{th} = A_{st} V_s + Z_{st} I_s, \qquad (1\text{-}3)$$

where A_{st} and Z_{st} are understood to be constants (perhaps frequency dependent constants), independent of V_s and I_s. Obviously, V_{th} is zero if the applied signal energies are nulled, as is highlighted in the abstraction of Fig. 1.4(c). Thus, the Thévenin impedance at a network port can be determined as the ratio of a voltage, V_x, established in response to an applied load current, I_x, -to- I_x, under the special circumstance of all independently applied signal sources set to zero.

The procedure advanced for Thévenin impedance calculation effectively mirrors the operation of an ohmmeter used to measure the resistance between two electrical terminals. It might therefore be termed the *ohmmeter method* of impedance computation. At risk of inadvertently depressing the reader, there is no such beast as an ohmmeter. The ohmmeters commonly found in the laboratory are actually electronic systems that perform two functions when its leads are connected to a terminal pair of interest. The first function is the injection of a current (I_x) that is sufficiently small to preclude any significant electrical perturbation of the network undergoing characterization. In strictly linear networks, such as those considered in this discussion, superposition renders the actual value of I_x immaterial. In nonlinear structures, such as transistors or batteries, the value of I_x is so crucial

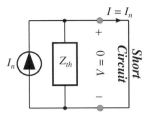

Figure 1.5. Calculation of the Norton equivalent current at a port of a linear network.

as to render a conventional ohmmeter ineffective for resistance evaluation. The second function performed by the ohmmeter is the monitoring of the resultant voltage (V_x) established at the port to which the current is applied. The reading observed on the ohmmeter is actually this voltage scaled to the applied current (V_x/I_x) and hence, it is the resistance evidenced at the port in question.

The determination of the Norton current, I_n, like the evaluation of the Thévenin voltage, relies on the fundamental fact that the parameters of the Norton equivalent circuit are independent of the load termination. If, therefore, the load appearing in Fig. 1.2 is replaced by an electrical short circuit, as indicated in Fig. 1.5, it is clear the that resultant current, I, flowing through the short-circuited load is identical to I_n. It follows that in general, the Norton current of a linear network port is the current supplied by that port to a short-circuited termination. Not surprisingly, the Norton current is often referred to as a *short circuit current*. And like V_{th}, I_n is the superposition of the effects of the applied input signal energies. With reference to Fig. 1.1(a),

$$I_n = Y_{sn}V_s + A_{sn}I_s, \qquad (1\text{-}4)$$

where Y_{sn} and A_{sn} are constants that are independent of applied signal voltage, V_s, and applied signal current, I_s.

1.2.2. Engineering Observations

Three useful and interesting sidebars ensue from the foregoing considerations. The first and most obvious of these is that the Thévenin equivalent circuit of the system in Fig. 1.1(a) can, by virtue of Fig. 1.1(b) and Eq. (1-3), be drawn in the topological format of Fig. 1.6(a). In this diagram, parameter A_{st} is a dimensionless parameter that represents the voltage transfer function, or voltage gain, from the port at which V_s is incident-to-the port that is terminated in the considered load. As such, A_{st} might logically be

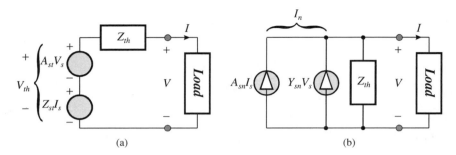

Figure 1.6. (a) Alternative form of the Thévenin equivalent circuit for the linear network of Fig. 1.1(a). The Thévenin voltage is cast explicitly in terms of the open circuit voltage gain, A_{st}, and the open circuit system transimpedance, Z_{st}. (b) Alternative form of the Norton equivalent circuit for Fig. 1.1(a). The Norton current is cast in terms of the short circuit system current gain, A_{sn}, and the short circuit system transadmittance, Y_{sn}.

termed the system *Thévenin voltage gain* or equivalently, the system *open circuit voltage gain*. On the other hand, parameter Z_{st} has units of ohms and is the *Thévenin transimpedance*, or *open circuit transimpedance*, evidenced between the port at which signal current I_s is applied and the load port. In other words, the transimpedance, like any impedance function, is a voltage-to-current ratio; it is literally the transfer impedance measured from the port of source current application-to-the load voltage response. A second, related observation is that because of Eq. (1-4), the Norton equivalent circuit in Fig. 1.2 can be delineated as the structure offered in Fig. 1.6(b). In this case, parameter A_{sn} is dimensionless and symbolizes the *Norton current gain*, or *short circuit current gain*, between the applied signal source current, I_s, and the load port at which voltage V is established. On the other hand, Y_{sn}, which has units of mhos, is the *Norton transadmittance*, or *short circuit transfer admittance*, from the port at which signal voltage V_s is applied and the load port that conducts current I. Collectively, both of the models in Fig. 1.6 underscore the fact that the Thévenin voltage and Norton current at a network port are respectively the superimposed effects of the energy sources applied to the network undergoing examination. They also highlight the various transfer relationships that link the Thévenin load voltage and the Norton load current to input signal energies.

The third observation derives from the explicit requirement that the Thévenin and Norton equivalent circuits applied to a given network port must each produce identical load voltage and current results under actual load termination conditions. In other words, one engineer using the

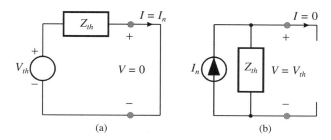

Figure 1.7. (a) The Thévenin equivalent circuit of a linear network port terminated in a short circuited load. (b) The Norton equivalent circuit of a linear network port that is open circuited.

Thévenin model and another using the Norton equivalent circuit must each compute the same load voltage and current responses. This necessity means that the Thévenin voltage, V_{th}, and the Norton current, I_n, are not independent variables. In order to arrive at the relationship between V_{th} and I_n, consider the model in Fig. 1.1(b) under the special circumstance of a short-circuited load, as depicted in Fig. 1.7(a). By definition, the resultant load current, I, is the short circuit, or Norton load current, I_n, which is

$$I \equiv I_n = \frac{V_{th}}{Z_{th}}. \tag{1-5}$$

This elegantly simple result shows that the Norton current at a port of a linear electrical network is nothing more than the ratio of the Thévenin voltage-to-the Thévenin impedance at said port. The application of the Norton model in Fig. 1.2 to the special case of an open-circuited load shown in Fig. 1.7(b) delivers a consistent result. Specifically, the open circuit load voltage V, which is now identical to the Thévenin voltage, V_{th}, "seen" by the load, is

$$V_{th} = Z_{th} I_n, \tag{1-6}$$

for which an understanding with respect to Eq. (1-5) assuredly instills pride in your high school algebra teachers.

Example 1.1. The circuit appearing in Fig. 1.8 is the linearized model of a bipolar junction transistor (BJT) voltage buffer, which is otherwise known as an emitter follower. The applied signal source is represented as a Thévenin equivalent circuit consisting of the series interconnection of a signal voltage, V_s, and a signal source resistance, R_s. The output signal voltage is V_o, which is taken as the voltage developed across a load capacitance of value C_l. Determine expressions for the Thévenin voltage, V_{th}, seen by the capacitive load, the Thévenin resistance, R_{th},

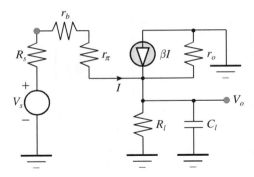

Figure 1.8. Linearized model of a bipolar junction transistor emitter follower.

facing this load, and the transfer function, $A_v(s) = V_o(s)/V_s(s)$. As a demonstration of the utility of the Thévenin analytical approach to evaluating the performance of an electronic network, examine the voltage transfer function from the perspective of determining the 3-dB bandwidth, ω_b, and plotting the frequency response of the amplifier. Numerically evaluate the Thévenin voltage gain, A_{st}, the Thévenin resistance, and the 3-dB bandwidth for transistor parameters of $r_b = 200\,\Omega$, $r_\pi = 2\,\text{K}\Omega$, $r_o = 50\,\text{K}\Omega$, and $\beta = 120\,\text{amps/amp}$. Additionally, take $R_s = 300\,\Omega$, $R_l = 3\,\text{K}\Omega$, and $C_l = 10\,\text{pF}$.

Solution 1.1.

(1) Figure 1.9(a) is the circuit diagram appropriate to the computation of the Thévenin voltage, V_{th}, established at the capacitive load port. Note in this diagram that the capacitive load branch has been removed; that is, the load has been open circuited. The resultant branch currents have been identified in this circuit diagram to appease Kirchhoff. Observe that the current, I, which controls the dependent current source, βI, is expressible as

$$I = \frac{V_s - V_{th}}{R_s + r_b + r_\pi}. \tag{E1-1}$$

A conventional nodal analysis then yields

$$\frac{V_{th}}{R_l} + \frac{V_{th}}{r_o} - (\beta + 1)I = \frac{V_{th}}{R_l} + \frac{V_{th}}{r_o} - \frac{(\beta + 1)(V_s - V_{th})}{R_s + r_b + r_\pi} = 0, \tag{E1-2}$$

from which the Thévenin voltage computes to be

$$V_{th} = \left[\frac{(\beta + 1)(r_o \| R_l)}{R_s + r_b + r_\pi + (\beta + 1)(r_o \| R_l)}\right] V_s. \tag{E1-3}$$

Circuit and System Fundamentals 13

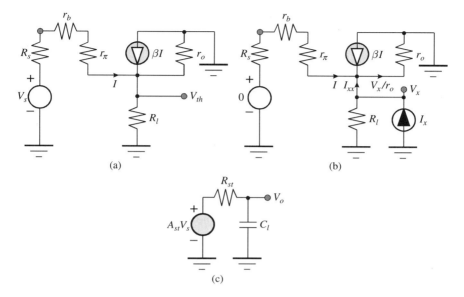

Figure 1.9. (a) Equivalent circuit used to evaluate the Thévenin voltage seen by the capacitance, C_l, in Fig. 1.8. (b) Equivalent circuit used to evaluate the Thévenin resistance seen by the capacitance, C_l, In Fig. 1.8. (c) Thévenin equivalent circuit pertinent to the output port of the amplifier in Fig. 1.8.

As a check on the propriety of Eq. (E1-3), note that $\beta = 0$ in Fig. 1.9(a) eliminates the current controlled current source, βI. Since resistance r_o is clearly in parallel with resistance R_l, it is hardly surprising that Eq. (E1-3) reduces to the simple voltage divider expression,

$$V_{th} = \left[\frac{(r_o \| R_l)}{R_s + r_b + r_\pi + (r_o \| R_l)}\right] V_s. \tag{E1-4}$$

(2) The circuit diagram used to determine the Thévenin resistance, R_{th}, seen by the load capacitance, C_l, in Fig. 1.8 is provided in Fig. 1.9(b), where the independent signal voltage, V_s, applied to the original circuit has been nulled. It is evident that the "ohmmeter" current, I_x, relates to the "ohmmeter voltage", V_x, in accordance with

$$I_x = \frac{V_x}{R_l} + I_{xx} = \frac{V_x}{R_l} + \frac{V_x}{r_o} - (\beta + 1)I, \tag{E1-5}$$

where current I is now

$$I = -\frac{V_x}{R_s + r_b + r_\pi}. \tag{E1-6}$$

Upon combining these two relationships, the pertinent Thévenin resistance is found to be

$$R_{th} = \frac{V_x}{I_x} = (R_l \| r_o) \left\| \left(\frac{r_\pi + r_b + R_s}{\beta + 1} \right) \right. . \qquad \text{(E1-7)}$$

For $\beta = 0$, this solution collapses to the expected result of a parallel combination of three effective circuit resistances; namely, the load resistance, R_l, the transistor model resistance, r_o, and the net resistance comprised of the series interconnection of resistances r_π, r_b, and R_s.

(3) From the solution for the Thévenin voltage in Step (1) above, the Thévenin voltage gain, A_{st}, is

$$A_{st} = \frac{V_{th}}{V_s} = \frac{(\beta + 1)(r_o \| R_l)}{R_s + r_b + r_\pi + (\beta + 1)(r_o \| R_l)}. \qquad \text{(E1-8)}$$

The resultant model for the evaluation of the overall voltage gain of the emitter follower is shown in Fig. 1.9(c). This simple model readily produces an overall gain expression of

$$A_v(s) = \frac{V_o(s)}{V_s(s)} = \frac{A_{st}(1/sC_l)}{R_{th} + 1/sC_l} = \frac{A_{st}}{1 + sR_{th}C_l}. \qquad \text{(E1-9)}$$

It is appropriate to interject that the product, $R_{th}C_l$, is the *time constant* attributed to the capacitance, C_l. In general, it can be stated that the time constant associated with a capacitor in a linear network is the product of said capacitance and the Thévenin resistance faced by the subject capacitor.

(4) In the laboratory, the amplifier at hand might very well be characterized under steady state sinusoidal operating conditions. With sinusoidal excitation, the steady state response derives from replacing the Laplace operator, s, in the preceding result by the imaginary frequency variable, $j\omega$. Thus,

$$A_v(j\omega) = \frac{V_o(j\omega)}{V_s(j\omega)} = \frac{A_{st}(1/j\omega C_l)}{R_{th} + 1/j\omega C_l} = \frac{A_{st}}{1 + j\omega R_{th}C_l}, \qquad \text{(E1-10)}$$

for which the magnitude of gain is

$$|A_v(j\omega)| = \left| \frac{V_o(j\omega)}{V_s(j\omega)} \right| = \left| \frac{A_{st}}{1 + j\omega R_{th}C_l} \right| = \frac{|A_{st}|}{\sqrt{1 + (\omega R_{th}C_l)^2}}. \qquad \text{(E1-11)}$$

Note that for very small radial signal frequencies, ω, the voltage transfer function is approximately constant, independent of frequency. On the other hand, large ω incurs a reduced magnitude of transfer function and thus, a degraded gain. Indeed, infinitely large ω results in zero gain magnitude. Such a transfer function characteristic is indicative of a *lowpass network*; that is, a network capable of passing with minimal gain reduction, or with minimal attenuation, low signal frequencies from its input to its output port, but incapable of processing very large frequencies without substantial signal amplitude attenuation. Of course, the reason for this lowpass characteristic is rendered transparent by the original circuit in Fig. 1.8. In particular, there is only one energy storage element — capacitor C_l — in the subject network. At very low signal frequencies, this capacitor emulates an open-circuited branch, thereby collapsing the network at hand to a purely resistive, so called *memoryless*, circuit. In a memoryless configuration, no branch element has an impedance that varies with signal frequency and accordingly, the gain of such a circuit is a constant, independent of signal frequency. At higher frequencies, the impedance of capacitor C_l decreases and in the limit of infinitely large frequency, the impedance of C_l approaches zero ohms. Since C_l is incident with the output port of the circuit, the magnitude of the output voltage, $V_o(j\omega)$, and thus the gain, $V_o(j\omega)/V_s(j\omega)$, decreases progressively toward zero for large signal frequencies.

(5) The gain expression deduced in the preceding computational step indicates that the zero frequency gain, say $A_v(0)$, is actually the Thévenin voltage gain, A_{st}. In the most general of circuit analyses, A_{st} is not identically equal to the zero frequency gain. It happens here that $A_{st} \equiv A_v(0)$ only because the load, which is removed from the otherwise memoryless network in the course of delineating the Thévenin gain, happens to be a capacitor. Thus, removal of the load in this example is tantamount to a consideration of zero signal frequency effects since a capacitive impedance at zero frequency is infinitely large.

In a lowpass circuit, the 3-dB frequency, ω_b, is the frequency at which the gain magnitude is a factor of the square root of two smaller than the magnitude of the zero frequency gain; that is,

$$|A_v(j\omega_b)| \triangleq \frac{|A_v(0)|}{\sqrt{2}} = \frac{|A_v(0)|}{|1 + j\omega_b R_{th} C_l|}. \tag{E1-12}$$

Evidently,

$$\omega_b = \frac{1}{R_{th} C_l}, \tag{E1-13}$$

which is little more than the inverse time constant associated with the lone capacitive element, C_l, in the original network.

A factor of *root two* gain degradation is equivalent to a gain magnitude deterioration of *three decibels* because of the definition of a decibel. In particular, the decibel value of any positive or a negative number, X, is $20 \log_{10}|X|$. If X is root two, its decibel (or dB) value is very close to 3. It follows that

$$A_v(j\omega_b) \text{ in db} = 20 \log_{10} |A_v(j\omega_b)| = 20 \log_{10} |A_v(0)| - 20 \log_{10} \sqrt{2}$$
$$\approx 20 \log_{10} |A_v(0)| - 3 \text{ dB};$$

that is, the decibel value of gain is reduced from its decibel value of zero frequency gain by an amount equal to 3 dB. Hence, the signal frequency effecting a root two gain magnitude reduction is termed the 3-dB frequency.

(6) The gain relationships in Step (4) can now be written in the forms

$$A_v(j\omega) = \frac{V_o(j\omega)}{V_s(j\omega)} = \frac{A_{st}}{1 + j\omega/\omega_b},\qquad \text{(E1-14)}$$

and

$$|A_v(j\omega)| = \left|\frac{V_o(j\omega)}{V_s(j\omega)}\right| = \frac{|A_{st}|}{\sqrt{1 + (\omega/\omega_b)^2}}.\qquad \text{(E1-15)}$$

The frequency response of an amplifier is simply the plot of its gain magnitude as a function of signal frequency. For the amplifier undergoing consideration herewith, this plot appears in Fig. 1.10, where the gain scale is in units of decibels and is normalized to the zero frequency gain, A_{st}. The frequency scale is normalized to the 3-dB bandwidth, ω_b.

The frequency response effectively pictures the ability of an amplifier to process applied input signals of varying frequencies. For example, the lowpass amplifier at hand is capable of providing an essentially constant I/O transmission over relatively low frequencies, but it is incapable of sustaining this transmission at high frequencies. To this end, the 3-dB frequency is a measure of amplifier effectiveness over frequency. To the extent that "essentially constant gain" can be viewed as a gain magnitude that is within *three decibels*

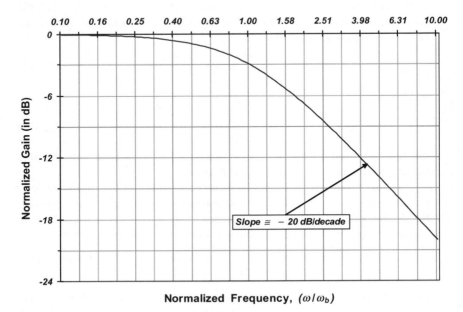

Figure 1.10. Frequency response of the amplifier studied in Example 1.1. The gain scale is normalized to the zero frequency value, A_{st}, of amplifier gain.

of its maximum (in this case, the low frequency) gain, the 3-dB bandwidth, ω_b, can be interpreted as the maximum frequency over which relatively constant gain is sustained. In this example, the maximum gain is actually less than one, which logically brings into question the utility of the considered amplifier. Despite this less than unity maximum gain, the buffer enjoys widespread popularity in electronic circuits and systems. More information about buffering applications is provided subsequently.

(7) For the stipulated numerical values of all device and circuit parameters, the Thévenin gain, the Thévenin voltage gain, Thévenin resistance, and 3-dB bandwidth are

$$A_{st} = 0.993 = -0.063 \text{ dB}$$
$$R_{th} = 20.51 \text{ } \Omega$$
$$\omega_b = 2\pi (775.9 \text{ MHz}).$$

In short, the buffer considered herewith establishes, within 3-dB error, almost unity voltage gain (0 dB) over a frequency passband extending from 0 Hz to slightly under 776 MHz, while providing a Thévenin resistance at its output port of slightly more than 20 Ω. Because the Thévenin resistance is indeed computed at the output port of the amplifier, this resistance metric is referred to as simply the amplifier *output resistance*.

Comments. The commentaries accompanying the preceding computational steps can be supplemented by the overarching observation of the profound simplicity of the Thévenin equivalent circuit. In particular, the original circuit in Fig. 1.8 contains eight (8) branch elements, while its Thévenin model in Fig. 1.9(c) contains only three (3) elements. This simplicity fosters design-oriented insights that are not rendered immediately transparent by the original configuration. For example, to the extent that the design objective is the realization of a buffer characterized by near unity low frequency gain and very low output resistance, the results highlighted by the Thévenin model suggest that large transistor β is essential. Note that the output resistance clearly satisfies

$$R_{th} < \frac{r_\pi + r_b + R_s}{\beta + 1},$$

which further dramatizes the importance of large β. Even the ability to achieve large 3-dB bandwidth is seen to be strongly dependent on transistor β, since

$$\omega_b = \frac{1}{R_{th} C_l} \approx \frac{\beta + 1}{(R_s + r_b + r_\pi) C_l}.$$

Example 1.2. Reconsider the circuit in Fig. 1.8 from the perspective of evaluating the Thévenin equivalent circuit presented to the signal source by the input port of the amplifier. Using the parameter values provided in the preceding example,

18 Feedback Networks: Theory and Circuit Applications

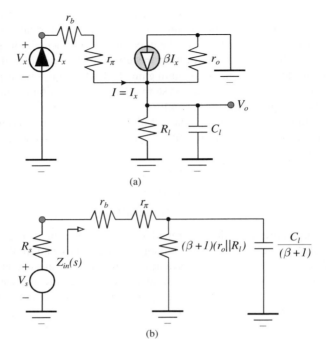

Figure 1.11. (a) Circuit model used to compute the Thévenin impedance presented to the signal source by the input port of the amplifier. (b) Topological depiction of the Thévenin input impedance determined for the equivalent circuit in (a).

evaluate the Thévenin input impedance at zero frequency, infinitely large signal frequency, and the previously computed 3-dB frequency of the amplifier.

Solution 1.2.

(1) Because the circuit capacitor in Fig. 1.8 is presumed to have zero initial charge and because no other energy sources appear within the network to the right of the signal source, the Thévenin equivalent circuit at the amplifier input port is comprised exclusively of an impedance, say $Z_{in}(s)$. The pertinent model for computing this impedance appears in Fig. 1.11(a) and reflects the fact that the "ohmmeter" current, I_x, is identical to the current, I, that controls the dependent current source, βI.

(2) Since the branch elements, R_l, r_o, and C_l are connected in parallel with one another and since the net current flowing through this shunt interconnection is $(\beta + 1)I = (\beta + 1)I_x$,

$$V_x = (r_b + r_\pi) I_x + \frac{(\beta + 1)(r_o \| R_l) I_x}{1 + s (r_o \| R_l) C_l}, \qquad \text{(E2-1)}$$

Circuit and System Fundamentals

whence

$$Z_{in}(s) = \frac{V_x}{I_x} = r_b + r_\pi + \frac{(\beta + 1)(r_o \| R_l)}{1 + s(r_o \| R_l) C_l}. \tag{E2-2}$$

This result can be rewritten in the form

$$Z_{in}(s) = r_b + r_\pi + \frac{(\beta + 1)(r_o \| R_l)}{1 + s[(\beta + 1)(r_o \| R_l)]\left[\dfrac{C_l}{(\beta + 1)}\right]}, \tag{E2-3}$$

which suggests representing the input amplifier port by the model offered in Fig. 1.11(b).

Because the Thévenin impedance found above pertains to the input port of the considered amplifier, it is often referred to as the *Thévenin input impedance* or, in abridged fashion, the *input impedance* of the amplifier. Note that this input impedance is computed with the capacitive load in tack; that is, the capacitive load is not removed from the circuit, as it is in the Thévenin voltage determination. In the jargon of circuit theory, the resultant input impedance, with the actual load connected, is sometimes called the *driving point input impedance*, as opposed to the open circuit input impedance, which would be $Z_{in}(s)$ under the condition of load removal.

(3) At zero signal frequency, the load capacitance behaves as an open circuit. More correctly, the admittance, sC_l, of the load capacitance at zero frequency is zero. From either the foregoing analytical disclosures or the representation in Fig. 1.11(b),

$$Z_{in}(0) = r_b + r_\pi + (\beta + 1)(r_o \| R_l) = 344.7 \, \text{K}\Omega.$$

(4) At infinitely large signal frequency, the load capacitance behaves as a short circuit. In particular, the impedance, $1/sC_l$, of the load capacitance is zero at infinitely large signal frequency. From either the foregoing analytical disclosures or the representation in Fig. 1.11(b),

$$Z_{in}(\infty) = r_b + r_\pi = 2.20 \, \text{K}\Omega.$$

(5) At the 3-dB bandwidth, ω_b, of the buffer, the driving point input impedance is

$$Z_{in}(j\omega_b) = r_b + r_\pi + \frac{(\beta + 1)(r_o \| R_l)}{1 + j\omega_b (r_o \| R_l) C_l}. \tag{E2-4}$$

From the preceding example, $\omega_b = 2\pi(775.9 \, \text{MHz})$ and accordingly,

$$Z_{in}(j\omega_b) = 2,200 + \frac{342.5(10^3)}{1 + j138.0} \approx (2{,}200 - j2{,}482) \, \Omega.$$

Since the imaginary part of this impedance function is negative, the input impedance at the 3-dB bandwidth of the amplifier is noted to be capacitive.

Comments. The use of Thévenin's theorem has served to highlight several important properties of a voltage buffer. The first of these properties, which derives from Example 1.1, is that the low frequency voltage transmission factor, or gain, is less than unity, but indeed, close to one. A second property is a low frequency input impedance that is significantly larger than the low frequency Thévenin output impedance. In particular, the I/O impedance transformation ratio is, from the present and preceding example, $344.7\,\mathrm{K}\Omega/20.51\,\Omega = 16.8(10^3)$. As is illustrated shortly, this dramatic ratio boasts utility in practical electronic systems. Third, the capacitive nature of the input impedance renders a significant reduction of this impedance over signal frequency. In this example, the difference between the low frequency and very high frequency input impedances is $344.7\,\mathrm{K}\Omega/2.20\,\mathrm{K}\Omega$, which is better than 156.

1.3.0. Dependent Sources and Amplifier Concepts

The Thévenin and Norton theorems and concepts addressed in the preceding section of material lay a foundation on which to build a fundamental understanding of general amplifiers and their respective properties. This understanding sets the stage for both open loop and closed loop electronic system design strategies by transforming the abstractness of dependent energy sources into topological tools that support design objectives. To this end, it is both instructive and interesting to be aware of the fact that there exist only four fundamental types of linear amplifiers and that these four amplifier configurations respectively emulate the four controlled sources that are an implicit part of basic circuit theory literature.

The most popular amplifying unit is the *voltage amplifier*, whose practical implementation delivers volt-ampere characteristics that emulate those of an ideal voltage controlled voltage source. The ubiquitous operational amplifier is an excellent example of a voltage amplifier. The second most common amplifier is the *transadmittance amplifier*, which is often referred to in the literature as simply a *transconductor*. The transconductor, which emulates the electrical characteristics of a voltage controlled current source, delivers an output port current that is directly proportional to applied input port voltage. It is the foundation of many broadband lowpass and tuned radio frequency (RF) amplifiers. It also enjoys utility as the gain cell implicit to wideband and ultra linear active resistance-capacitance (RC) filters. The *transimpedance amplifier*, or *transresistor*, is the dual of the transconductor. It converts an applied input current to an output voltage response and as such, its electrical dynamics approximate the ideal current controlled voltage source. Like the transconductor, the transresistor is often the core active element of broadband networks. It is often synthesized by appending appropriate feedback to a basic voltage amplifier. An operational amplifier operated with resistive feedback between its input and phase-inverted output ports is among the most common of transresistors. Finally, the volt-ampere characteristics of a *current amplifier* emulate the electrical properties of an ideal current controlled current source. The current amplifier is rarely used as a stand-alone circuit architecture. Instead, its impedance transformation attributes encourage its utilization in conjunction with transconductors to arrive at compensated circuits whose bandwidths are, under certain conditions, substantively larger than the bandwidth capabilities of transconductors operated without current amplifier compensation.

1.3.1. Voltage Amplifier

The circuit schematic symbol of a voltage amplifier is diagrammed in Fig. 1.12(a). Like all simple linear amplifiers, it is a *two-port* structure. Its input port, to which signal is applied to establish the differential input port voltage, V, which is ultimately amplified, is comprised of the two terminals labeled "+" and "−". The "+" terminal is called the *noninverting input terminal*, while the "−" terminal is termed the *inverting node*. The output port on the right of the circuit schematic symbol supports the Thévenin, or open circuit, voltage response, V_{th}, to applied input excitation. In the most general case, the open circuit, or Thévenin, response, V_{th}, is extracted differentially between the two terminals that comprise the amplifier output

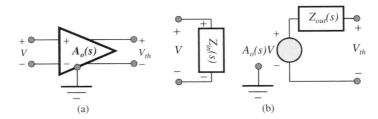

Figure 1.12. (a) Circuit schematic symbol of a voltage amplifier. The amplifier is depicted in its noninverting mode, since the controlling input voltage, V, is applied differentially from the noninverting input terminal-to-the inverting input terminal. (b) Circuit model of the amplifier in (a). Parameter $Z_{in}(s)$ is the driving point input impedance. To first order, this impedance is independent of the load termination. Parameter $Z_{out}(s)$ is the driving point output impedance of the amplifier. The parameter, $A_o(s)$, is the Thévenin voltage gain of the amplifier and is measured as the ratio of the differential open circuit output voltage, V_{th},-to the differential input port voltage, V.

port. If one of the two output port terminals is incident with the amplifier ground terminal, V_{th} is referred to as a *single ended output voltage*. If, as is diagrammed in the subject figure, neither of the two output port terminals is grounded, V_{th} is called a *differential output voltage*. Similarly, note that voltage V might be termed a *differential input voltage* because neither of the two input port terminals across which this voltage is established is indicated as common to the amplifier ground.

The indicated gain, $A_o(s)$, is a frequency dependent transfer function. For $V > 0$, which indicates that the noninverting input terminal lies at a signal potential that is larger than the potential established at its inverting counterpart, the Thévenin output voltage is $A_o(s)V$, thereby implying no phase inversion between the input and output ports. In other words, if V rises with time, the Thévenin output voltage is an amplified version of voltage V that likewise increases with time. On the other hand, for $V < 0$, which suggests that V is applied from the "−" input terminal-to-the "+" input terminal, as opposed to the polarity indicated in Fig. 1.12(a), the open circuit output voltage is $-A_o(s)V$, and 180 degree I/O phase inversion is evident.

From Thévenin's theorem, a viable equivalent circuit for the voltage amplifier abstracted in Fig. 1.12(a) is the model offered in Fig. 1.12(b). The input port model, which consists of a simple input impedance branch, $Z_{in}(s)$, reflects the presumption that no energy sources appear either within the active amplifier block or at the output port of the amplifier. Strictly speaking, $Z_{in}(s)$ is a driving point input impedance; that is, it is an impedance that is dependent on the load that terminates the amplifier output port. However, in this initial foray into the world of linear amplifiers, $Z_{in}(s)$ is presumed independent of the output port load. This presumption means that $Z_{in}(s)$ is unchanged whether the output port is terminated in a specific load or open circuited, as it is during the process of computing the Thévenin output port voltage. As the student will ultimately learn, this independence of the Thévenin input impedance on load termination is closely approximated if there is insignificant internal feedback implicit to the active amplifier cell. In turn, negligible internal feedback is generally a reasonable presumption at all but very high signal frequencies.

In the output port representation, $Z_{out}(s)$ is the usual Thévenin equivalent impedance seen looking into the output terminal. This impedance is, in fact, the driving point output impedance in that it is determined under the condition of the input port terminated in the internal impedance of the applied signal source. Like the nominal independence of $Z_{in}(s)$ on load termination, $Z_{out}(s)$ is also nominally independent of source impedance if negligible internal feedback prevails within the amplifier itself. The dependent voltage, $A_o(s)V$, is the Thévenin voltage established at the output port, while $A_o(s)$ is the Thévenin voltage gain measured from the differential amplifier input port, where voltage V prevails,-to-the open-circuited differential output port where the Thévenin voltage, V_{th}, is established.

In an actual linear application of the voltage amplifier, a signal voltage source having a Thévenin internal impedance of $Z_s(s)$ activates the input port, while a load impedance, $Z_l(s)$, terminates the output port, as is depicted in Fig. 1.13(a). Recalling the amplifier model hypothesized in Fig. 1.12(b), the system model pertinent to Fig. 1.13(a) is the topology appearing in Fig. 1.13(b). By inspection, the overall system voltage gain, say $A_v(s)$, is

$$A_v(s) = \frac{V_0}{V_s} = \frac{V_0}{V} \times \frac{V}{V_s} = A_0(s) \left[\frac{Z_l(s)}{Z_l(s) + Z_{out}(s)} \right] \left[\frac{Z_{in}(s)}{Z_{in}(s) + Z_s(s)} \right], \quad (1\text{-}7)$$

which shows that the overall voltage gain, compared to the Thévenin voltage gain, $A_o(s)$, is degraded by a factor equal to the product of input port and output port voltage dividers. The gain, $A_o(s)$, is the gain afforded by the amplifying device and is therefore the maximum possible gain achievable in a linear system in which this device is embedded. Accordingly, Eq. (1-7) underscores the fact that a linear system degrades the available device gain by the combined effects of nonzero amplifier driving point output impedance and finite amplifier driving point input impedance. This

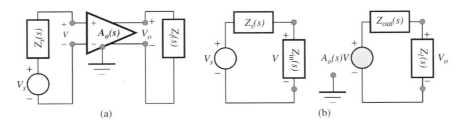

Figure 1.13. (a) System schematic depiction of a voltage amplifier terminated in a load impedance and driven at its input port by a voltage source. (b) Equivalent circuit of the system in (a). The input voltage, V, and the output response, V_o, are taken herewith as differential circuit branch voltages. However, and depending on the actual system architecture, either V or V_o, or both, can be single ended variables. If both V and V_o are extracted as single ended node voltages, the system in (a) is said to maintain a common ground between its input and output ports.

observation begets a stipulation of the electrical characteristics indigenous to an *ideal voltage amplifier*.

(1) *The driving point input impedance, $Z_{in}(s)$, is infinitely large for all signal frequencies and for all load terminations.* Note that the infinitely large input impedance property implies that zero current is drawn from the signal source by the amplifier input port. As a result, no voltage drop appears across the internal signal source impedance, thereby maximizing the transfer of applied Thévenin source voltage to the amplifier input port.

(2) *The driving point output impedance, $Z_{out}(s)$, is zero for all signal frequencies and for all signal source impedances.* This characteristic allows an output port voltage to be developed across any load impedance, inclusive (in principle only) of even a short-circuited load. More importantly, the voltage developed across the load termination is the Thévenin output port voltage, which is the maximum possible voltage that can be generated across the terminating load impedance.

(3) *In an ideal voltage amplifier, $A_o(s)$ is a constant, A_o, independent of signal frequency.* Properties 1 and 2 allow for a system gain that is identically equal to the voltage gain afforded by the amplifying device. Pragmatically, this gain, $A_o(s)$, is generally a suitably large, constant, real number, say A_o, at low frequencies. At high frequencies in the steady state, it attenuates at a minimum rate of 20 dB/decade because of unavoidable intrinsic energy storage parasitics. Observe that the idealized constant gain stipulation implies the unrealistic device capability of amplifying signals whose frequencies embody a range extending from "DC"-to-daylight.

Figure 1.14 summarizes the electrical properties of an ideal voltage amplifier.

Voltage amplifiers are often operated with differential input and single ended output ports. With reference to Fig. 1.12(a), the pertinent circuit schematic symbol is the structure shown in Fig. 1.15(a), and the applicable equivalent circuit appears in Fig. 1.15(b). In the interest of schematic simplicity, the diagram in Fig. 1.15(a) is generally cast in the form of Fig. 1.15(c), where it is understood that the output port voltage, V_{th}, is now referred to system ground. The diagrams in Figs. 1.13 and 1.14 remain applicable for single ended outputs, with the proviso that the system ground is now incident with the negative terminal of the output voltage response, V_o.

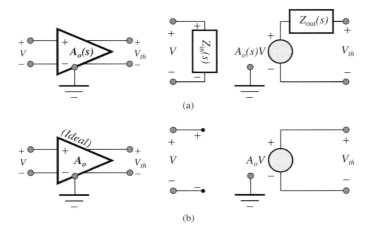

Figure 1.14. (a) System schematic diagram and circuit level model of a voltage amplifier. (b) System schematic diagram and circuit level model of an ideal voltage amplifier. The gain parameter, A_o, is a constant, independent of the frequencies of applied input signals.

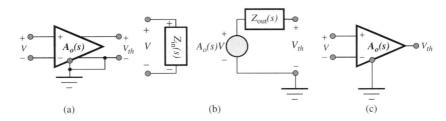

Figure 1.15. (a) Schematic portrayal of a voltage amplifier with single ended output. (b) Equivalent circuit of the single ended configuration abstracted in (a). (c) Simplified schematic symbol of a voltage amplifier with single ended output. In this depiction, the open circuit output voltage, V_{th}, is presumed measured with respect to the system ground.

1.3.2. Transconductor

The circuit schematic symbol of a transadmittance amplifier, or transconductor, appears in Fig. 1.16(a). Yet another name for this amplifier is *operational transconductor amplifier*, which is commonly abbreviated as "*OTA*." The differential input port voltage, V, which is established as a result of applied input signal, is a positive number when it is measured from the non-inverting input terminal ($+$) to the inverting terminal ($-$). This input port voltage is converted by the transconductor into a short circuit, or Norton, output current, I_n. The subject Norton current is proportional to

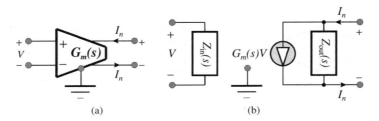

(a) (b)

Figure 1.16. (a) Circuit schematic symbol of a transconductance amplifier. (b) Circuit model of the amplifier in (a). Parameter $Z_{in}(s)$ is the driving point input impedance, while $Z_{out}(s)$ is the driving point output impedance of the amplifier. The parameter, $G_m(s)$, is the Norton transadmittance of the transconductor.

V with a proportionality constant, $G_m(s)$, whose dimension is mhos; that is, $I_n = G_m(s)V$. Note that the positive algebraic sense of I_n is a current flowing into the positive output terminal and flowing out of the negative output terminal of the transconductor when $V > 0$. The Thévenin and Norton concepts introduced earlier render the architecture of Fig. 1.16(b) a plausible two-port model of the linear transconductor.

Figure 1.17(a) offers a linear system application of the transconductor introduced in Fig. 1.16, while Fig. 1.17(b) depicts its corresponding circuit model. As usual, $Z_s(s)$ represents the source impedance of the applied voltage signal, and $Z_l(s)$ is the load impedance incident with the transconductor output port. By inspection, the I/O transadmittance, say $Y_f(s)$, is

$$Y_f(s) = \frac{I_o}{V_s} = \frac{I_o}{V} \times \frac{V}{V_s} = G_m(s)\left[\frac{Z_{out}(s)}{Z_{out}(s) + Z_l(s)}\right]\left[\frac{Z_{in}(s)}{Z_{in}(s) + Z_s(s)}\right].$$

(1-8)

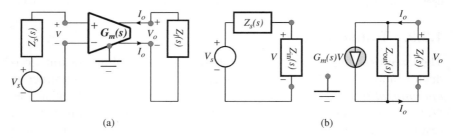

(a) (b)

Figure 1.17. (a) System schematic diagram of a transadmittance amplifier terminated in a load impedance and driven at its input port by a voltage source. (b) Equivalent circuit of the system in (a).

Analogous to the voltage gain expression in Eq. (1-8), this transadmittance function is the product of a maximum transfer function (in this case, a transadmittance function) and two dividers. The first of the two dividers on the right hand side of this relationship, which is a current divider for the system output port in Fig. 1.17(b), approaches unity as the output impedance, $Z_{out}(s)$, tends toward an open circuit. The second divider is an input port voltage divider, which approaches unity as the driving point input impedance, $Z_{in}(s)$, emulates the impedance of an open circuit. These observations lead forthwith to the electrical definitions implicit to an *ideal transadmittance amplifier*.

(1) *The driving point input impedance, $Z_{in}(s)$, is infinitely large for all signal frequencies and for all load terminations.* Infinitely large input impedance implies that zero current is drawn from the signal source by the amplifier input port. As a result, no voltage drop appears across the internal signal source impedance, thereby maximizing the transfer of applied Thévenin signal voltage to the amplifier input port.

(2) *The driving point output impedance, $Z_{out}(s)$, is infinitely large for all signal frequencies and for all signal source impedances.* This characteristic allows for an output current that is identical to the Norton output current and is therefore independent of load impedance.

(3) *In an ideal transconductor or transadmittance amplifier, $G_m(s)$ is a constant, independent of signal frequency.* Properties 1 and 2 allow for a system transadmittance that is identically equal to the transadmittance afforded by the amplifying device. Pragmatically, this forward transfer relationship is generally a suitably large, constant, real number, say g_m, at low frequencies. At high frequencies in the steady state, the effective forward transconductance attenuates owing to the unavoidable presence of internal energy storage parasitics.

Figure 1.18 reviews the foregoing electrical properties. Observe that the circuit model of an ideal transadmittance amplifier is identical to the schematic abstraction of a voltage controlled current source.

1.3.3. Transresistor

Figure 1.19(a) shows the circuit schematic symbol of a transimpedance amplifier, or more simply, a transresistor. This type of amplifier operates on applied input current, I, to generate an output port Thévenin voltage, $R_m(s)I$, that is proportional to current I. For a driving point input impedance

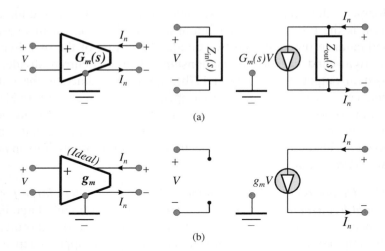

Figure 1.18. (a) System schematic diagram and circuit level model of a transadmittance amplifier, or transconductor. (b) System schematic diagram and circuit level model of an ideal transconductor. The transconductance parameter, g_m, is a constant, independent of frequency.

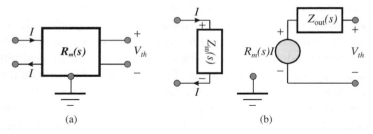

Figure 1.19. (a) Circuit schematic symbol of a transresistor. (b) Circuit model of the amplifier in (a). Parameter $Z_{in}(s)$ is the driving point input impedance, $Z_{out}(s)$ is the driving point output impedance, and $R_m(s)$, is the Thévenin transimpedance of the device.

of $Z_{in}(s)$ and a driving point output impedance of $Z_{out}(s)$, the electrical model is the topology offered in Fig. 1.19(b).

In system level applications of the transresistor, the input signal energy derives from a current source, I_s, whose presumably large Thévenin impedance is $Z_s(s)$, as depicted in Fig. 1.20(a). Also shown in this schematic diagram is a load impedance, $Z_l(s)$, that is incident with the transresistor output port and supports the resultant differential voltage response, V_o, to the input signal current source. The corresponding equivalent circuit in

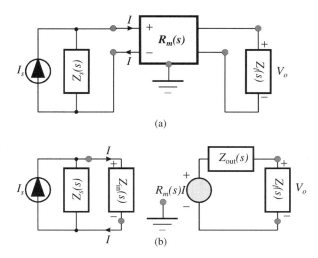

Figure 1.20. (a) System schematic diagram of a transimpedance amplifier terminated in a load impedance and driven at its input port by a current source. (b) Equivalent circuit of the system in (a).

Fig. 1.20(b) delivers an I/O transimpedance, $Z_f(s)$, given by

$$Z_f(s) = \frac{V_o}{I_s} = \frac{V_o}{I} \times \frac{I}{I_s} = R_m(s)\left[\frac{Z_l(s)}{Z_l(s)+Z_{out}(s)}\right]\left[\frac{Z_s(s)}{Z_s(s)+Z_{in}(s)}\right]. \tag{1-9}$$

An inspection of this relationship underscores the obvious fact that in the steady state, the magnitude, $|Z_f(j\omega)|$, of the overall transimpedance is less than the magnitude, $|R_m(j\omega)|$, of the Thévenin output port voltage-to-input port transimpedance. Accordingly, maximal forward transimpedance is afforded when both $Z_{in}(s)$ and $Z_{out}(s)$ approach the impedance of a short circuit. This observation readily leads to the definition of an *ideal transimpedance amplifier*.

(1) *The driving point input impedance, $Z_{in}(s)$, is zero for all signal frequencies and for all load terminations.* Zero input impedance means that no signal voltage can be sustained across the input port of a transresistor, which in turn suggests the impropriety of driving the input port of a transresistor with a voltage source.

(2) *The driving point output impedance, $Z_{out}(s)$, is zero for all signal frequencies and for all signal source impedances.* This characteristic

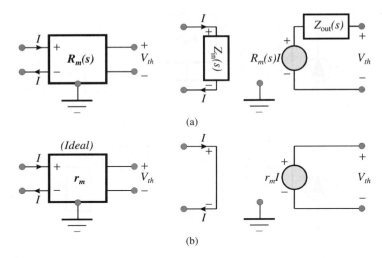

Figure 1.21. (a) System schematic diagram and circuit level model of a transimpedance amplifier, or transresistor. (b) System schematic diagram and circuit level model of an ideal transresistor. The transresistance parameter, r_m, is a constant, independent of frequency.

implies that the output voltage developed in response to applied input current is theoretically independent of all load terminations.

(3) *In an ideal transresistor or transimpedance amplifier, $R_m(s)$ is a constant, independent of signal frequency.* Properties 1 and 2 allow for a system transimpedance that is identical to the transimpedance of the amplifying device. Pragmatically, this forward transfer relationship is generally a large, constant, real number, say r_m, at low frequencies. At high frequencies in the steady state, the low frequency value of this transimpedance attenuates because of unavoidable intrinsic energy storage parasitics.

In Fig. 1.21, the foregoing electrical properties are reviewed and the electrical model of an ideal transimpedance amplifier is cast as a voltage controlled current source.

1.3.4. Current Amplifier

The circuit schematic symbol of a current amplifier appears in Fig. 1.22(a). This amplifier responds to applied input current, I, to establish a Norton output current, $I_n = B_o(s)I$. With a driving point input impedance of $Z_{in}(s)$

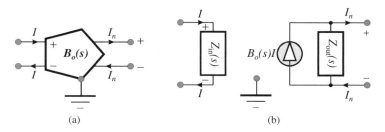

Figure 1.22. (a) Schematic symbol of a current amplifier. (b) Two-port model of the current amplifier in (a). Parameters $Z_{in}(s)$ and $Z_{out}(s)$ respectively denote the driving point input and output impedances, while $B_o(s)$ is the Norton current gain of the amplifier.

and a driving point output impedance of $Z_{out}(s)$, the electrical model of a current amplifier is the network in Fig. 1.22(b).

As in transresistor applications, the signal source applied to the input port of a current amplifier is a current source, I_s, having a relatively large source impedance, $Z_s(s)$. The resultant output response to this applied current is itself a current, I_o, conducted by load impedance $Z_l(s)$, which is connected across the amplifier output port. The system application at hand is abstracted in Fig. 1.23(a), for which the pertinent electrical model is the

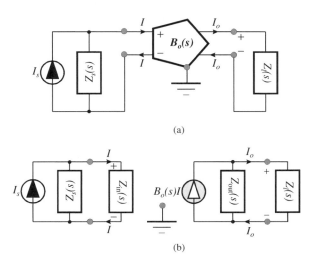

Figure 1.23. (a) System level application of a linear current amplifier. (b) Equivalent circuit of the current amplification system in (a).

circuit diagram shown in Fig. 1.23(b). This model generates a system current gain expression whose algebraic form is similar to that of the transfer relationships derived for the three previously studied amplification systems; namely,

$$A_i(s) = \frac{I_o}{I_s} = \frac{I_o}{I} \times \frac{I}{I_s} = B_o(s) \left[\frac{Z_{out}(s)}{Z_{out}(s) + Z_l(s)} \right] \left[\frac{Z_s(s)}{Z_s(s) + Z_{in}(s)} \right].$$

(1-10)

Clearly, $A_i(s)$ approximates $B_o(s)$, which is the maximum system current gain afforded by the utilized current amplification device, when $Z_{in}(s)$ is a very small impedance and $Z_{out}(s)$ is very large. It follows that an *ideal current amplifier* satisfies the requirements itemized herewith.

(1) *The driving point input impedance, $Z_{in}(s)$, is zero for all signal frequencies and for all load terminations.*
(2) *The driving point output impedance, $Z_{out}(s)$, is infinitely large for all signal frequencies and for all signal source impedances.* This characteristic implies that the output current developed in response to applied input current is theoretically independent of all load terminations.
(3) *In an ideal current amplifier, $B_o(s)$ is a constant, independent of signal frequency.* Properties 1 and 2 allow for a system current gain that is identical to the maximum current gain allowed by the amplifying device. This current gain is generally a large, constant, real number, say A_{io}, at low frequencies. At high frequencies in the steady state, the low frequency value of the current gain attenuates at a minimum rate of 20 dB/decade because of unavoidable intrinsic energy storage parasitics.

Figure 1.24 overviews the foregoing electrical properties and in the process, it depicts the electrical model of an ideal current amplifier as a current controlled current source.

1.3.5. Buffers

As might be suspected, the four types of amplifiers discussed in the preceding subsections of material are most commonly used to boost relatively anemic voltage or current signal amplitudes into more robust voltages and currents that can deliver required amounts of energy to specified loads. For example, consider the futility of connecting an audio speaker directly to the

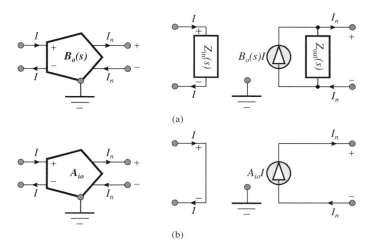

Figure 1.24. (a) System schematic diagram and circuit level model of a current amplifier. (b) System schematic diagram and circuit level model of an ideal current amplifier. The current gain, A_{io}, is a constant, independent of frequency.

output terminals of a compact disc (CD) player. Typical audio speakers have nominal input impedances in the range of eight-to-sixteen ohms and may require as many as tens of volts of excitation for proper performance and acceptable fidelity. In contrast, the Thévenin output impedance of representative CD units is 500 ohms or larger. Moreover, CD players rarely deliver open circuit output voltages larger than a few tens of millivolts. Since a 16 Ω speaker connected across a voltage source whose internal resistance is 500 Ω comprises a voltage divider of roughly 1/32, a CD unit having a 20 mV open circuit output voltage capability delivers only about 620 microvolts to the speaker terminals. This miniscule voltage is hardly sufficient to enjoy the Rolling Stones and thus, an appropriate amplifier (most likely a cascade of several amplifiers intertwined with requisite filters and signal processing subsystems) must be inserted between the CD player and the speaker.

If signal amplitude amplification is the dominant function of amplifiers, impedance buffering is the second most important application of amplifying networks. Buffers, which are ubiquitous in both analog and digital circuit technologies, perform impedance transformation between input and output ports so that the output voltage-to-signal source voltage transfer function or the output current-to-signal source current gain is maintained very close to unity for wide ranges of signal source and load impedances. Two types of buffers — the *voltage buffer* and the *current buffer* — are commonly found in electronic systems.

1.3.5.1. Voltage Buffer

With reference to the generalized ideal voltage amplifier diagrammed symbolically in Fig. 1.14(b), an *ideal voltage buffer* has a frequency invariant Thévenin voltage gain of unity ($A_o = 1$) in addition to infinitely large input impedance and zero output impedance for all load and source terminations, respectively. Since the Thévenin voltage gain is the largest possible voltage gain achievable in a system into which a voltage amplifier is embedded within the I/O signal path, it is only natural to question the pragmatism of an active device capable of only unity voltage gain.

A response to the foregoing inquiry begins by considering the simple voltage divider in Fig. 1.25(a). In this divider, the output voltage, V_o, is an attenuated version of the Thévenin signal source voltage, V_s, since

$$\frac{V_o}{V_s} = \frac{R_l}{R_l + R_s}. \quad (1\text{-}11)$$

If the hypothetical CD example considered in the preceding subsection is revisited herewith, the divider in question is $16/(16 + 500) = 1/32.25$, which suggests that only 3.1% of the Thévenin signal voltage is actually delivered to the load resistance, R_l. In other words, 96.9% of this signal

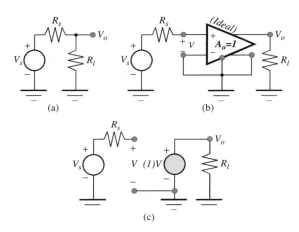

Figure 1.25. (a) A voltage divider for which the signal voltage of a source whose Thévenin resistance is R_s is to be delivered to load resistance R_l. (b) Ideal buffer inserted in I/O signal flow path. (c) Model of the buffered system in (b).

voltage is "lost" in the internal resistance, R_s, of the signal source. In an attempt to mitigate this signal loss, an ideal buffer is inserted between the source and the load, as suggested in Fig. 1.25(b). Since the subject buffer has infinitely large input impedance, no current is drawn from the signal source and as a result, no voltage is "lost" in the Thévenin resistance of the source. Moreover, the zero output resistance of the buffer allows an output voltage response to be established across a load resistance of any value. The propriety of these assertions is confirmed by the model in Fig. 1.25(c), which produces

$$\frac{V_o}{V_s} = \frac{V_o}{V} \times \frac{V}{V_s} = (1)(1) = 1. \qquad (1\text{-}12)$$

Thus, 100% of the Thévenin signal source voltage appears across the network output port as voltage V_o, independent of either load termination or source resistance.

Of course, no physically realizable voltage buffer is ideal. The practical buffer addressed in Examples 1.1 and 1.2, delivers a large, but nonetheless finite, input resistance of 344.7 KΩ, a small, but nonzero, output resistance of 20.51 Ω, and a nearly unity gain of 0.993. If this buffer supplants its idealized counterpart in Fig. 1.25, $R_s = 500\Omega$, $R_l = 16\,\Omega$, and Eq. (1-7) lead to a voltage transfer function of $V_o/V_s = 1/2.3$. This result is hardly the desired ideal unity value, but it is 14-times better than the nonbuffered value of 1/32.25.

Example 1.3. Operational amplifiers (op-amps) of reasonable quality can be gainfully exploited as voltage buffers in broadband electronic system applications. To this end, Fig. 1.26(a) depicts a voltage buffer realized with an op-amp having a single ended output port. For the purpose of this problem, assume that the op-amp has a Thévenin voltage gain (often referred to in the literature as the *open loop gain*) of $A_o = 80$ dB, an output resistance, r_o, of 35 Ω, and an input impedance that is purely capacitive. The net value of the input capacitance, which is plausibly attributed to the combined effects of the op-amp, incorporated compensation, and circuit parasitics, is $C_i = 300$ pF. The Thévenin resistance, R_s, of the signal source is 500 Ω, while the load resistance, R_l, driven by the buffer is 16 Ω. Derive general expressions for, and discuss the engineering significance of, the system voltage gain, $A_v(s) = V_o/V_s$, the output impedance, $Z_{out}(s)$, seen by the load resistance, R_l, and the input impedance, $Z_{in}(s)$, seen by the signal source.

Solution 1.3.

(1) Recalling Fig. 1.14(b) and using the information provided in this problem, Fig. 1.26(b) is the equivalent circuit of the buffer in Fig. 1.26(a). In terms of

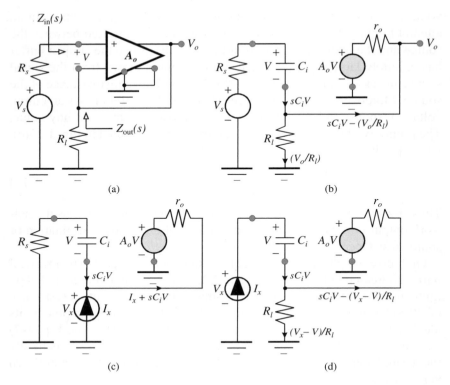

Figure 1.26. (a) Operational amplifier with single ended output configured as a voltage buffer. The signal source is represented as the series interconnection of voltage V_s and resistance R_s, while the load driven by the buffer is taken to be the resistance, R_l. (b) Equivalent circuit of the system in (a). The indicated branch currents are appropriate to a determination of the system voltage gain, $A_v(s) = V_o/V_s$. (c) Equivalent circuit for the determination of the driving point output impedance, $Z_{out}(s)$. (d) Equivalent circuit for evaluating the driving point input impedance, $Z_{in}(s)$.

the branch currents delineated in this diagram, KVL gives

$$A_o V = -r_o \left(sC_i V - \frac{V_o}{R} \right) + V_o,$$

$$V_s = R_s(sC_i V) + V + V_o.$$

Subsequent to elimination of variable V from these equilibrium relationships, a bit of algebra confirms an I/O voltage transfer function of the form,

$$A_v(s) = \frac{V_o}{V_s} = \left(\frac{kA_o}{1 + kA_o} \right) \left(\frac{1 + \dfrac{s}{z}}{1 + \dfrac{s}{p}} \right), \qquad \text{(E3-1)}$$

where

$$k = \frac{R_l}{R_l + r_o} \tag{E3-2}$$

is a voltage divider between the resistances, r_o and R_l. Moreover, the input port capacitance appears to generate a left half plane pole at frequency p, as well as a left half plane zero at frequency z. These critical frequencies are given by

$$p = \frac{1 + kA_o}{(R_s + kr_o) C_i} \tag{E3-3}$$

and

$$z = \frac{A_o}{r_o C_i}. \tag{E3-4}$$

(2) Several features of the voltage transfer function in Eq. (E3-1) warrant highlighting. First, observe that the system gain at zero frequency is

$$A_v(0) = \frac{kA_o}{1 + kA_o}, \tag{E3-5}$$

which is almost one by virtue of very large A_o. In the present case, $A_o = 80\,\text{dB} = 10{,}000$ and $k = 0.3137$, whence $A_v(0) = 0.9997$. It therefore appears that at least at low signal frequencies, the circuit in Fig. 1.26(a) very nearly satisfies the unity voltage gain objective of an ideal voltage buffer.

The locations of the pole and zero of the voltage transfer function define the frequency response of the buffer at hand. In the present case, $p = 2\pi(3.26\,\text{GHz})$, and $z = 2\pi(151.6\,\text{GHz})$. The frequency of the zero is better than 46-times larger than the frequency of the pole and is, in fact, so large as to render dubious its validity in light of the frequency response limitations implicit to the utilized simple model. Numerical validity notwithstanding, the frequency of the zero is so much larger than that of the pole as to justify its tacit neglect over a frequency passband extending from zero through, and somewhat beyond, the pole frequency. Accordingly,

$$A_v(s) = \frac{V_o}{V_s} \approx \frac{kA_o/(1 + kA_o)}{1 + \dfrac{s}{p}}, \tag{E3-6}$$

from which it is apparent that the 3-dB bandwidth, say ω_b, is

$$\omega_b \approx p = \frac{1 + kA_o}{(R_s + kr_o) C_i} \approx \frac{kA_o}{(R_s + R_l \| r_o) C_i} = 2\pi(3.26\,\text{GHz}). \tag{E3-7}$$

Thus, the buffer undergoing examination delivers very nearly unity gain, to within an error of 3-dB, from zero signal frequency to almost 3.3 GHz.

(3) The "ohmmeter" model pertinent to computing the driving point output impedance, $Z_{out}(s)$, seen by the load resistance, R_l, is depicted in Fig. 1.26(c). For the branch currents indicated in this diagram, KVL produces

$$0 = R_s(sC_i V) + V + V_x$$
$$V_x = r_o(I_x + sC_i V) + A_o V. \quad \text{(E3-8)}$$

Upon elimination of the voltage variable, V, in these two relationships, it is easily demonstrated that the output impedance, expressed in terms of steady state frequency variables, is

$$Z_{out}(j\omega) = \frac{V_x}{I_x} = \left(\frac{r_o}{1+A_o}\right)\left(\frac{1+j\omega\tau_z}{1+j\omega\tau_p}\right), \quad \text{(E3-9)}$$

where the time constant, τ_z, associated with the zero of the impedance function and the time constant, τ_p, attributed to the impedance function pole are respectively given by

$$\tau_z = R_s C_i \quad \text{(E3-10)}$$

and

$$\tau_p = \frac{(R_s + r_o) C_i}{1 + A_o}. \quad \text{(E3-11)}$$

(4) From Eq. (E3-9), the low frequency output impedance is

$$Z_{out}(0) = \frac{r_o}{1+A_o}, \quad \text{(E3-12)}$$

which is virtually zero because of the very large amplifier gain, A_o. Indeed, $Z_{out}(0)$ computes herewith to 0.0035 Ω, which assuredly emulates the zero output impedance indigenous to an ideal voltage buffer.

The time constant associated with the impedance zero is $\tau_z = 150$ nSEC, which corresponds to a frequency of $1/\tau_z = 2\pi(1.07\,\text{MHz})$. On the other hand, $\tau_p = 16.05$ pSEC, corresponding to a frequency, $1/\tau_p = 2\pi(9.92\,\text{GHz})$. Clearly, the pole frequency is significantly larger (over 9,000-times larger) than the zero frequency. It follows that for frequencies as large as an octave or two below the pole frequency,

$$Z_{out}(j\omega) \approx \left(\frac{r_o}{1+A_o}\right)(1+j\omega\tau_z), \quad \text{(E3-13)}$$

which suggests that the driving point output impedance is inductive. Specifically, this impedance reflects a resistance, say R_{eff}, connected in series with an inductance, say L_{eff}, such that

$$R_{\text{eff}} = \frac{r_o}{1+A_o}, \quad \text{(E3-14)}$$

and

$$L_{\text{eff}} = \left(\frac{r_o}{1+A_o}\right)\tau_z = \frac{r_o R_s C_i}{1+A_o}. \quad \text{(E3-15)}$$

The indicated effective resistance is, as anticipated, the previously determined zero frequency value of the output impedance. Although the effective series inductance is small, it can cause poor transient circuit responses and/or even resonant frequency responses when, as is commonly encountered, the buffer drives a strongly capacitive load.

(5) Figure 1.26(d) is the equivalent circuit pertinent to the evaluation of the driving point input impedance seen by the signal source. KCL and KVL applied to this circuit yield

$$I_x = sC_i V$$
$$0 = -A_o V + r_o \left(\frac{V_x - V}{R_l} - sC_i V\right) + V_x - V. \quad \text{(E3-16)}$$

Solving the second of these two equations for voltage V and substituting the solution into the first equation results in

$$Z_{in}(j\omega) = \frac{V_x}{I_x} = \left(\frac{1+kA_o}{j\omega C_i}\right)\left[1 + j\omega\left(\frac{kr_o C_i}{1+kA_o}\right)\right], \quad \text{(E3-17)}$$

which is clearly capacitive for most signal frequencies.

(6) Two interesting observations surface from an inspection of the result in Eq. (E3-17). First, note that at low frequencies, the effective input capacitance, say C_{ieff}, is very small and in particular, it is

$$C_{\text{ieff}} = \frac{C_i}{1+kA_o} = 95.6 \text{ fF}. \quad \text{(E3-18)}$$

This small capacitance contributes to the relatively broadband response of the buffer. To confirm this assertion, recall from preceding modeling exercises that the input port of an amplifier, where C_{ieff} is established in this exercise (at least at low signal frequencies), directly faces the signal source circuit. In this case, the signal source has an internal resistance of R_s, which implies that C_{ieff} establishes a time constant at the input port of $R_s C_{\text{ieff}} = 47.8$ pSEC. This time constant forges an input port pole at a frequency of $1/R_s C_{\text{ieff}} = 2\pi(3.33\,\text{GHz})$. Since C_{ieff} is the only capacitance in the buffering system, this pole frequency is necessarily the 3-dB bandwidth of the buffer. Indeed, the currently computed bandwidth of 3.33 GHz differs from that calculated previously in Eq. (E3-7) by only 2.1%. This miniscule computational difference is certainly understandable in light of the approximations invoked with respect to both Eqs. (E3-7) and (E3-18).

A second observation derives from analytical considerations at very high signal frequencies. Specifically, Eq. (E3-17) shows that

$$Z_{in}(j\infty) = kr_o = r_o \| R_l. \qquad \text{(E3-19)}$$

Not only is the input impedance purely resistive at high frequencies, its specific resistive value is obvious from an inspection of the model in Fig. 1.26(b). In particular, capacitance C_i becomes a short circuit at infinitely large frequencies. This short circuit constrains voltage V to zero, which nulls the dependent voltage, $A_o V$. When $A_o V$ is zero, resistance r_o is placed in parallel with load resistance R_l. It follows that with C_i effectively shorted, the signal source sees little more than the shunt interconnection of resistances r_o and R_l.

Comments. This example demonstrates how a generalized voltage amplifier model, which is itself predicated on basic Thévenin and Norton concepts, can be judiciously exploited for the purpose of assessing the performance of a simple voltage buffer. Specifically, the example shows that a commonly used operational amplifier topology can deliver a broadband frequency response having a low frequency voltage gain very near unity. The considered buffer also has an output resistance that is very low at low frequencies, a very large input impedance at low frequencies, and a driving point output impedance that is inductive at high signal frequencies. Although these performance metrics are generally evident in practical voltage buffers, caution must be exercised with respect to the numerical values gleaned for these metrics. Numerical errors accrue because the model invoked in this exercise is elementary in that it exploits but a single energy storage element (the input capacitance) in the system. In actual buffers, additional energy storage elements invariably prevail, as does a frequency dependence on the Thévenin gain parameter, $A_o(s)$.

1.3.5.2. Current Buffer

The electrical model pertinent to examining the electrical characteristics at the input and output ports of an ideal current amplifier is provided in Fig. 1.24(b). Note therein that the input impedance is zero for all signal frequencies and for all load terminations and that the output impedance is infinitely large for all frequencies and source impedances. An *ideal current buffer* boasts these two impedance signatures, in addition to a frequency invariant Norton current gain of unity ($A_{io} = 1$).

The utility of ideal current buffers can be rendered transparent with the help of the diagrams in Fig. 1.27. In Fig. 1.27(a), a current source having an

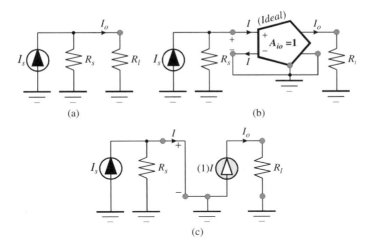

Figure 1.27. (a) A simple current divider for which the signal current of a source whose Thévenin resistance is R_s is to be delivered to a load resistance, R_l. (b) Ideal, single ended current buffer inserted in the signal flow path between the signal source and the load of the divider in (a). (c) Model of the buffered system in (b).

internal resistance of R_s is applied to a load resistance, R_l, with the result that the current, I_o, conducted by the load derives from

$$\frac{I_o}{I_s} = \frac{R_s}{R_s + R_l}. \qquad (1\text{-}13)$$

If R_l is comparable to, or even larger than, R_s, the current actually delivered to the load is an appreciably attenuated version of the current, I_s, available from the signal source. Obviously, negligible signal attenuation, or loss, occurs only if $R_l \ll R_s$. Rendering this desired inequality true is the fundamental purpose of a current buffer.

To the foregoing end, let a unity gain version of the current amplifier abstracted in Fig. 1.24(b) be inserted between the source and load, as indicated in Fig. 1.27(b). The corresponding model is the structure in Fig. 1.27(c), which offers

$$\frac{I_o}{I_s} = \frac{I_o}{I} \times \frac{I}{I_s} = (1)(1) = 1. \qquad (1\text{-}14)$$

As in the case of the voltage buffer considered previously, the current buffer provides an impedance transformation vehicle by which the load can be isolated from the source. Specifically, the signal source now drives a short circuit network input port, as opposed to the actual load resistance, R_l, thereby allowing the entire signal source current to be processed with unity gain. In turn, the load now faces an ideal current source, as opposed to the source resistance, R_s, wherein all of the processed signal current can be delivered to the load termination. The obvious keys to proper current buffering are a very low (ideally zero) input port impedance and a very high (ideally infinitely large) output impedance.

Example 1.4. Transconductors, which can be realized straightforwardly with either metal-oxide-semiconductor field-effect transistors (MOSFETs) or bipolar junction transistors (BJTs), can be configured to emulate ideal current buffers. To this end, consider the single ended current buffering configuration in Fig. 1.28(a), where the input signal is a current source comprised of current I_s and Thévenin

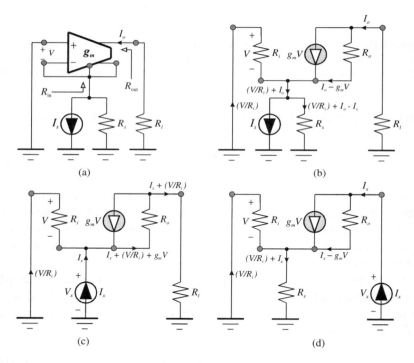

Figure 1.28. (a) Transconductor with single ended output configured as a current buffer. The signal source is represented as the shunt interconnection of current I_s and resistance R_s, while the load is taken to be the resistance, R_l. (b) Equivalent circuit of the system in (a). The indicated branch currents are appropriate to a determination of the system current gain, $A_i = I_o/I_s$. (c) Equivalent circuit for the determination of the driving point input resistance, R_{in}. (d) Equivalent circuit for evaluating the driving point output resistance, R_{out}.

resistance R_s, and the output response is taken as the current, I_o, conducted by the load resistance, R_l. Derive expressions for the current transfer ratio, I_o/I_s, the input resistance, R_{in}, seen by the signal source and the output resistance, R_{out}, seen by the load. Assume that only low signal frequencies are of interest, which allows the transconductor to be modeled by a shunt input resistance, R_i, shunt output resistance, R_o, and a frequency invariant transconductance of g_m.

Solution 1.4.

(1) Recalling the transconductor models in Fig. 1.18, the equivalent circuit of the current buffer in Fig. 1.28(a) is the circuit shown in Fig. 1.28(b). KVL applied to this model, in which branch currents have been delineated for analytical convenience, produces

$$0 = R_l I_o + R_o (I_o - g_m V) + R_s \left(\frac{V}{R_i} + I_o - I_s \right)$$

$$0 = V + R_s \left(\frac{V}{R_i} + I_o - I_s \right).$$

(E4-1)

A simultaneous solution of these two equilibrium relationships, followed by the obligatory algebra, results in a current transfer ratio of

$$A_i \triangleq \frac{I_o}{I_s} = \frac{(1 + g_m R_o)(R_i \| R_s)}{R_l + (R_i \| R_s) + [1 + g_m (R_i \| R_s)] R_o}.$$

(E4-2)

It is clear that this gain is smaller than one. However, $A_i \approx 1$ if

$$g_m R_o \gg 1,$$

(E4-3)

$$g_m (R_i \| R_s) \gg 1,$$

(E4-4)

and

$$R_o \gg \frac{R_l}{1 + g_m (R_i \| R_s)} + R_i \| R_s \left\| \frac{1}{g_m} \right..$$

(E4-5)

Observe that satisfying the foregoing three inequalities fundamentally requires suitably large R_o and sufficiently large g_m. Both of these parametric constraints are implicit to reasonably high performance transconductors.

(2) In the "ohmmeter" model of Fig. 1.28(c), which pertains to the evaluation of the driving point input resistance, R_{in}, note that $V \equiv -V_x$. Hence,

$$V_x = R_o \left(I_x - \frac{V_x}{R_i} - g_m V_x \right) + R_l \left(I_x - \frac{V_x}{R_i} \right),$$

(E4-6)

whence

$$R_{in} = \frac{V_x}{I_x} = \frac{(R_o + R_l) \| R_i}{1 + g_m R_o \left(\dfrac{R_i}{R_i + R_o + R_l} \right)}. \qquad \text{(E4-7)}$$

To the extent that $g_m R_o$ is a large number, the input resistance is relatively small and given approximately by

$$R_{in} \approx \frac{1}{g_m} \left(1 + \frac{R_l}{R_o} \right), \qquad \text{(E4-8)}$$

which collapses to $R_{in} \approx 1/g_m$ for the typically encountered circumstance of $R_o \gg R_l$.

(3) For the output resistance model in Fig. 1.29(d),

$$V = -R_s \left(\frac{V}{R_i} + I_x \right),$$

which implies

$$V = -(R_i \| R_s) I_x. \qquad \text{(E4-9)}$$

Since

$$V_x = R_o (I_x - g_m V) - V = R_o I_x + (1 + g_m R_o)(R_i \| R_s) I_x,$$
$$R_{out} = \frac{V_x}{I_x} = (R_i \| R_s) + [1 + g_m (R_i \| R_s)] R_o. \qquad \text{(E4-10)}$$

The driving point output resistance is seen to be a number larger than R_o, which is itself presumably large. Indeed, R_{out} can be substantially larger than R_o, since a transconductor is routinely designed to ensure relatively large g_m and large R_i. Moreover, if the transconductor is driven by a current source, as indicated in Fig. 1.29(a), R_s is, like R_i and R_o, a large resistance.

Comments. The foregoing analyses confirm that the transconductor configuration in Fig. 1.28(a) is a reasonable approximation of an ideal current buffer. The approximation is good only if the subject transconductance element is designed to offer large input resistance (R_i), large output resistance (R_o), and reasonably large forward transconductance (g_m). For these design constraints, the resultant input resistance of the buffer is small and roughly equal to the inverse of the transconductance of the transconductor, the output resistance is larger (and possibly significantly larger) than the transconductor output resistance, and the realized current transfer ratio is very nearly unity.

1.3.6. Load Power Considerations

The principle purpose of a voltage buffer is to ensure the transfer of maximum voltage between a signal source and the load imposed on this source. On the other hand, a current buffer functions to effect maximum current transfer between source and load. Neither of these two buffers serves to transfer maximum power from signal-to-load. For example, the power, which is fundamentally the product of voltage and current, delivered by a signal source to the input port of an ideal voltage buffer is zero because the infinitely large input impedance of this buffer precludes the flow of an input port current. Similarly, the power delivered to the input port of an ideal current buffer is zero by virtue of the fact that the zero input impedance indigenous to the current buffer precludes the establishment of a nonzero input port voltage. Obviously, practical buffers sustain nonzero input port power levels because their driving point input impedances are neither zero nor infinitely large. But well-designed voltage and current buffers certainly do not support input power levels that mirror the maximum power levels that applied signal source circuits are capable of generating.

Most digital electronic circuits and low-to-moderate frequency analog circuits and systems operate on applied signals whose implicit power levels are robust. As a result, the inherent inefficiency accompanying subcircuit, circuit, and subsystem interconnects that do not achieve the transfer of maximum signal power between a source and a load is of little, if any, concern in such applications. In other systems, such as high frequency and/or broadband communication electronics, signal power transfer is a major design issue because the available signal power levels are anemic. For example, consider the ubiquitous cellular telephone. The signal energy available at the antenna output terminals of a cell phone are typically so small as to be in danger of being masked by interference, or noise, generated either parasitically within the environment in which the cell phone operates or by the actual electronics used to detect, amplify, and otherwise process the antenna responses. Substantive signal power loss in this and analogous other applications must therefore be mitigated to maximize the likelihood of faithfully capturing and processing a low level signal in an unavoidably noisy electrical environment. To this end, care must be exercised to ensure that maximum signal power is indeed transferred to prescribed load terminations over the frequency response passband of interest. Stated quite simply, this means that for low level signal processing applications, maximum voltage transfer and maximum current transfer assume second place

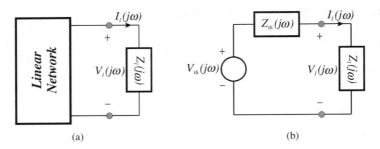

Figure 1.29. (a) A linear network terminated in a two terminal load impedance, $Z_l(j\omega)$. The system is presumed to operate in the sinusoidal steady state. (b) Thévenin equivalent circuit of the system in (a).

status to the fundamental design objective of assuring maximum signal power transfer.

The development of design criteria underlying the realization of maximum power transfer between a signal source and its imposed load commences with a study of the linear electrical system abstracted in Fig. 1.29(a). Assuming that the indicated linear network is excited by a sinusoid at radial frequency ω and that analytical attention focuses herewith on only steady state system performance, the pertinent Thévenin equivalent circuit is the model in Fig. 1.29(b). In the later diagram, $Z_l(j\omega)$ is the load impedance imposed on the linear network, $Z_{th}(j\omega)$ is the Thévenin impedance seen by this load, and $V_{th}(j\omega)$ is the phasor representation of the Thévenin voltage that drives the two terminal load impedance branch. Without loss of generality, the phase angle of $V_{th}(j\omega)$ can be taken as zero, so that $V_{th}(j\omega)$ is the real number,

$$V_{th}(j\omega) = \sqrt{2}V_{trms}e^{j0} = \sqrt{2}V_{trms}, \qquad (1\text{-}15)$$

where V_{trms} symbolizes the root mean square (RMS) value of the sinusoidal Thévenin voltage. For clarity, it should be understood that the time domain value, say $v_{th}(t)$, of this Thévenin voltage is

$$v_{th}(t) = \sqrt{2}V_{trms}\cos(\omega t). \qquad (1\text{-}16)$$

In response to the prevailing Thévenin voltage, a load voltage, $v_l(t)$, and a load current, $i_l(t)$, is established. Because of system linearity, both of these load variables are, like the applied Thévenin voltage, sinusoids at frequency ω. But depending on the nature of the load and Thévenin impedances in the circuit, the phase angles of these voltage and current variables are likely to be nonzero and nonidentical. Thus,

$$v_l(t) = \sqrt{2} V_{lrms} \cos(\omega t + \theta_v) \tag{1-17}$$

and

$$i_l(t) = \sqrt{2} I_{lrms} \cos(\omega t + \theta_i), \tag{1-18}$$

where V_{lrms} and I_{lrms} are the RMS values of the load voltage and load current, respectively, while θ_v and θ_i denote the respective phase angles, measured with respect to the presumed zero phase angle of the input signal, of these variables. In phasor notation,

$$V_l(j\omega) = \sqrt{2} V_{lrms} e^{j\theta_v} \tag{1-19}$$

and

$$I_l(j\omega) = \sqrt{2} I_{lrms} e^{j\theta_i}. \tag{1-20}$$

The instantaneous power, $p_l(t)$, delivered to, and dissipated by, the load impedance is, using Eqs. (1-17) and (1-18), and a good old fashioned trigonometric identity, is

$$p_l(t) = v_l(t) i_l(t) = V_{lrms} I_{lrms} [\cos(\theta_v - \theta_i) + \cos(2\omega t + \theta_v + \theta_i)]. \tag{1-21}$$

The average power, say P_l, delivered to the load is found by integrating this instantaneous power over one complete period, $T = 2\pi/\omega$, of the sinusoidal load voltage or current and then dividing this integrated value by T. In particular

$$P_l = \frac{1}{T} \int_0^T p_l(t)\, dt = V_{lrms} I_{lrms} \cos(\theta_v - \theta_i). \tag{1-22}$$

Several insightful observations pertain to this result. First, a purely resistive load termination sustains a terminal voltage and branch current that are in phase with one another. Accordingly, $(\theta_v - \theta_i) = 0$, and the average power dissipated by the load collapses to simply the product of the RMS values of load voltage and load current. Second, it is well known that capacitors store the energy delivered to them but do not dissipate any average power. Since a strictly capacitive load supports a load voltage that lags its branch current by exactly 90°, Eq. (1-22) confirms this zero average power fact, since $(\theta_v - \theta_i) = -\pi/2$ radians, whence $\cos(\theta_v - \theta_i) = 0$. Similarly, for a load comprised of a pure inductance, which also dissipates no average power, $(\theta_v - \theta_i) = +\pi/2$, whence $P_l = 0$.

1.3.6.1. *Maximum Power Transfer*

As a prelude to determining the criteria for maximum power transfer between load and source, Eq. (1-22) must be related to the power made available by the Thévenin signal source circuit that drives the load impedance. To this end, it is initially convenient to express the average dissipated power defined by Eq. (1-22) in terms of phasor load variables. Note in Eq. (1-20) that the complex conjugate of the phasor load current is

$$I_l(-j\omega) = \sqrt{2} I_{lrms} e^{-j\theta_i}, \quad (1\text{-}23)$$

whereupon from Eq. (1-19),

$$V_l(j\omega) I_l(-j\omega) = 2 V_{lrms} I_{lrms} e^{j(\theta_v - \theta_i)}$$
$$= 2 V_{lrms} I_{lrms} [\cos(\theta_v - \theta_i) + j\sin(\theta_v - \theta_i)]. \quad (1\text{-}24)$$

It follows from Eq. (1-22) that

$$P_l = \frac{1}{2} Re[V_l(j\omega) I_l(-j\omega)]. \quad (1\text{-}25)$$

An inspection of the circuit in Fig. 1.29(b) reveals a phasor load voltage of

$$V_l(j\omega) = \left[\frac{Z_l(j\omega)}{Z_l(j\omega) + Z_{th}(j\omega)}\right] V_{th}(j\omega) \quad (1\text{-}26)$$

and a phasor load current given by

$$I_l(j\omega) = \frac{V_{th}(j\omega)}{Z_l(j\omega) + Z_{th}(j\omega)}. \quad (1\text{-}27)$$

Decomposing the load and Thévenin impedances into their real (resistive) and imaginary (reactive) components,

$$Z_l(j\omega) = R_l + jX_l \quad (1\text{-}28)$$

and

$$Z_{th}(j\omega) = R_{th} + jX_{th}, \quad (1\text{-}29)$$

where it is understood that reactances X_l and X_{th} can be positive, negative, or zero, corresponding respectively to inductive, capacitive, or purely resistive impedances. On the other hand, R_{th} must be either a zero or a positive resistance for a strictly linear network, and R_l must be non-negative if $Z_l(j\omega)$ is a physically realizable passive load impedance. Inserting Eqs. (1-28) and (1-29) into Eqs. (1-26) and (1-27), and then substituting the resultant latter two expressions into Eq. (1-25), the average load power is found to be

$$P_l = \frac{1}{2} \frac{R_l |V_{th}(j\omega)|^2}{(R_l + R_{th})^2 + (X_l + X_{th})^2}. \tag{1-30}$$

The preceding result comprises a useful engineering disclosure in that expresses the average steady state power dissipated in a complex load impedance in terms of load parameters and the Thévenin parameters of the linear network that drives the load. Interestingly enough, this power is reduced by load and Thévenin reactances despite the fact that such reactances are incapable of dissipating power. Fortunately, reactances can be positive or negative and thus, a first step toward maximizing the load power entails choosing a load impedance having $X_l = -X_{th}$. Thus, an inductive load requires a capacitive Thévenin impedance, and vice versa, for load power maximization. Under this power maximization constraint, Eq. (1-30) reduces to

$$P_{l|X_l=-X_{th}|} = \frac{R_l |V_{th}(j\omega)|^2}{2(R_l + R_{th})^2}. \tag{1-31}$$

Clearly, the resultant load power displays a maximum with respect to the resistive component, R_l, of the load impedance, since the subject power is never negative and vanishes at both $R_l = 0$ and $R_l = \infty$. Remember that $V_{th}(j\omega)$ is discerned under an open circuit load condition and is therefore independent of all load parameters. The desired maximum can be determined by setting to zero the derivative of power with respect to R_l. When this analysis is executed, it is found that $R_l = R_{th}$ maximizes the power expression in Eq. (1-31), whence the maximum load power, say P_{lmax}, is

$$P_{lmax} = \frac{|V_{th}(j\omega)|^2}{8R_{th}} = \frac{V_{trms}^2}{4R_{th}}, \tag{1-32}$$

where Eq. (1-15) is used. Observe that the combined constraints, $R_l = R_{th}$ and $X_l = -X_{th}$, that lead to Eq. (1-32) imply a load impedance that is the complex conjugate of the Thévenin impedance, $Z_{th}(j\omega)$, of the linear network; that is,

$$Z_l(j\omega) \equiv Z_{th}(-j\omega). \tag{1-33}$$

Equation (1-33) defines the design condition commensurate with the transfer of maximum power between signal and load. When it is satisfied, the terminating load impedance is said to be *match terminated* to the source, and the equation itself is often referred to as the *match terminated design condition.* The resultant maximum average power delivered to the load termination is given by Eq. (1-32), which, in effect, also stipulates the maximum power capability of the signal source.

1.3.6.2. The dBm Power Measure

The maximum signal power levels routinely encountered in such low level electronics as audio preamplifiers, video amplifiers, and radio frequency (RF) communication networks are rarely larger than a few milliwatts. In many communication circuits, such as those exploited as first stages in cellular telephones, these power levels can be as small as only hundreds of picowatts. To illustrate, consider the simplified system level diagram of the front end of a high frequency communications receiver shown in Fig. 1.30(a). In this diagram, the antenna is the medium by which the signal earmarked for ultimate signal processing is captured. This antenna is coupled electrically to the input port of the first stage, or front end, amplifier of the communication cell by a cable or some other form of distributed transmission line that, under certain designable conditions, is characterized by a 50 Ω, purely resistive impedance. As suggested in Fig. 1.30(b), the input port of the amplifier is driven by a simple Thévenin equivalent circuit consisting of the Thévenin antenna voltage, $V_{th}(j\omega)$, and a Thévenin 50 Ω resistance, R_{th}, established by the transmission line interconnect. A match termination to this signal source medium therefore mandates an amplifier input impedance, Z_{in}, that is purely real and identical to 50 Ω. Assuming

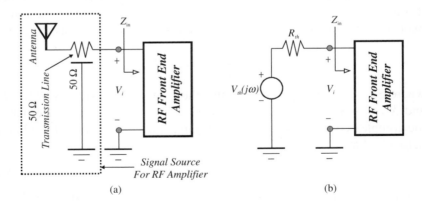

Figure 1.30. (a) Simplified system level diagram of the antenna signal source and front end (first stage) amplifier of a communication network. (b) The system in (a) with the antenna signal source replaced by its Thévenin equivalent network.

a RMS Thévenin signal of 300 μV, the available antenna signal power, and the signal power actually delivered to the amplifier when its input port displays a driving point impedance of $Z_{in} = 50\,\Omega$, is, by Eq. (1-32), $P_{lmax} = (300\,\mu V)^2/(4)(50) = 450\,\text{pW}$.

Signal power levels that are so low as the level just computed are commonly expressed as a normalized power in units of *decibels referred to a milliwatt*, or simply, *dBm*. For a signal power of P, the *dBm* value of P is

$$P(\text{in dBm}) = 10\log_{10}\left(\frac{P}{0.001}\right). \qquad (1\text{-}34)$$

A logarithmic multiplier of 10 is used in this definition, as opposed to the multiplier of 20 invoked in Example 1.1, because power is proportional to the square of either voltage or current. Thus, Eq. (1-34) is equivalent to a 20-times logarithm of voltage or current response. For the previously computed P_{lmax} of 450 pW, $P_{lmax} = -63.47\,\text{dBm}$, and it is therefore inferred that the indicated maximum load power is about 63.5 dB below a milliwatt. Note that 0 dBm corresponds to a signal power of one milliwatt, which is equivalent to 223.6 mV of RMS voltage established across 50 ohms of resistance.

1.3.6.3. *Match Termination and Tuned Responses*

Matched terminated loads comprise an effective vehicle for capturing the maximum signal power that an energy source is capable of delivering. But in the absence of design heroics that entail the incorporation of complex filters in the signal flow path of an electronic system, impedance matches can generally be effected at only a single frequency and at best, only over a restricted frequency passband that is geometrically centered about this so called *center frequency*. This operational circumstance provides the engineering backdrop for *tuned amplifiers* or in general, for electronic systems exhibiting *bandpass frequency responses*. *Bandpass amplifiers* provide zero gain at both low and high frequencies and offer nonzero gain in only the immediate neighborhood of the center frequency to which the system is *tuned*.

To illustrate the limitations and attributes of match terminated amplifier design, consider the system in Fig. 1.31(a). The diagram at hand depicts a signal source represented by its Thévenin equivalent circuit consisting of the series interconnection of the voltage phasor, V_s, and the resistance, R_s, applied to the input port of a transimpedance amplifier. For the purpose

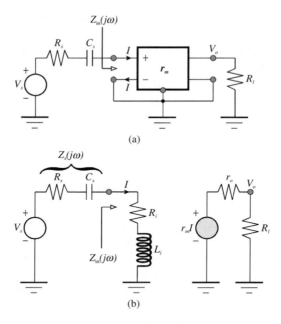

Figure 1.31. (a) System level diagram of a tuned transimpedance amplifier. (b) Electrical model of the system in (a).

of this discussion, the subject amplifier is terminated at its output port in resistance R_l, has a frequency invariant transresistance, r_m, and a purely real output impedance, r_o. Moreover, the transimpedance unit, as is the case with many practical realizations of such cells, has a driving point input impedance that can be represented as a series connection of a small resistance, R_i, and an inductance, L_i. Because of the desire to match terminate the front end of the indicated system, a capacitor, C_s, is inserted in series with the Thévenin source resistance, R_s. The electrical model corresponding to the foregoing stipulations is the topology depicted in Fig. 1.31(b).

An inspection of Fig. 1.31(b) suggests an effective source impedance of $Z_s(j\omega) = R_s - j/\omega C_s$, and an effective driving point amplifier input impedance of $Z_{in}(j\omega) = R_i + j\omega L_i$. Match terminated operation at the front end therefore requires $R_i = R_s$ and $(1/\omega C_s) = \omega L_i$. The latter of these two mandates can be satisfied at only one frequency, the center frequency, ω_o, such that

$$\omega_o = \frac{1}{\sqrt{L_i C_s}}. \tag{1-35}$$

Armed with this result and the requirement, $R_i = R_s$, KVL applied to the input loop of the model in Fig. 1.31(b) yields

$$I = \frac{V_s}{R_s + \dfrac{1}{j\omega C_s} + R_i + j\omega L_i} = \frac{V_s}{2R_s + j\omega_o L_i \left(\dfrac{\omega}{\omega_o} - \dfrac{\omega_o}{\omega}\right)}. \qquad (1\text{-}36)$$

At the center frequency, current I is simply $V_s/2R_s$, which is as expected since at $\omega = \omega_o$, $(1/j\omega_o C_s) + j\omega_o L_i = 0$; that is, the inductive impedance is precisely the negative of the capacitive impedance at the center frequency of the input port. Moreover, Eq. (1-36) confirms that I reduces to zero at both very low and very high frequencies. This observation mirrors engineering expectations since at very low frequencies, the capacitor is effectively an open circuit, while at very high signal frequencies, the inductor behaves as an open circuit.

The voltage gain, $A_v(j\omega)$, of the network follows forthwith from Fig. 1.31(b) and Eq. (1-36) as

$$A_v(j\omega) = \frac{V_o}{V_s} = \frac{V_o}{I} \times \frac{I}{V_s} = \left(\frac{R_l}{R_l + r_o}\right)\left(\frac{\dfrac{r_m}{2R_s}}{1 + jQ_o\left(\dfrac{\omega}{\omega_o} - \dfrac{\omega_o}{\omega}\right)}\right), \quad (1\text{-}37)$$

where

$$Q_o = \frac{\omega_o L_i}{2R_s} = \frac{\sqrt{L_i/C_s}}{2R_s} \qquad (1\text{-}38)$$

is known as the *quality factor*, or simply the "Q" of the input port circuit. Note that the quality factor is little more than a comparison of the inductive reactance at the center frequency-to-the net series resistance in the electrical loop in which the inductance is embedded. In practice, this "net series resistance" also includes the parasitic resistance unavoidably implicit to the conductive coil or metallization winding that comprises the inductor. Because this undesirable resistance can be lumped into the effective source resistance, R_s, Q_o in Eq. (1-38) diminishes and therefore, the reactive impact, or "quality" of the inductive coil is impaired.

Equation (1-37) confirms that the voltage gain at the tuned center frequency of the system in Fig. 1.31 is

$$A_v(j\omega_o) = \left(\frac{R_l}{R_l + r_o}\right)\frac{r_m}{2R_s}. \qquad (1\text{-}39)$$

This gain is the maximum available voltage gain since the magnitude of the gain in Eq. (1-37) decreases for both $\omega > \omega_o$ and $\omega < \omega_o$. It is important

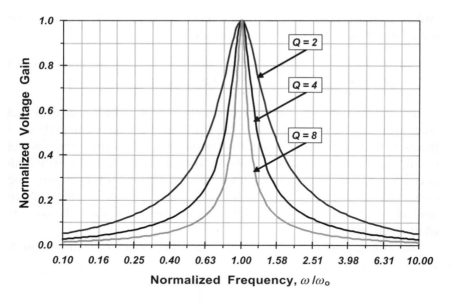

Figure 1.32. Frequency response of the tuned transimpedance amplifier in Fig. 1.31 for various values of the quality factor for the source/input port circuit.

to understand that high frequency gain deterioration is the result of the limited input port current caused by high inductive impedance. On the other hand, low frequency gain degradation is attributed to the high capacitive impedance in the input port of the transimpedance amplifier.

The resultant bandpass frequency response is illustrated in Fig. 1.32, which plots the normalized voltage gain magnitude,

$$|A_n(j\omega)| \triangleq \left|\frac{A_v(j\omega)}{A_v(j\omega_o)}\right| = \left|\frac{1}{1+jQ_o\left(\dfrac{\omega}{\omega_o}-\dfrac{\omega_o}{\omega}\right)}\right|$$

$$= \frac{1}{\sqrt{1+Q_o^2\left(\dfrac{\omega}{\omega_o}-\dfrac{\omega_o}{\omega}\right)^2}}, \quad (1\text{-}40)$$

versus the normalized signal frequency,

$$x = \omega/\omega_o \equiv f/f_o, \quad (1\text{-}41)$$

for various values of quality factor Q_o. Independent of Q_o, the normalized gain magnitude is one at the tuned frequency, ω_o, and falls off with frequency both above and below ω_o. The rate at which the magnitude rolls off with signal frequency is strongly influenced by Q_o. In particular, progressively larger quality factors result in increased rates of frequency response roll off and give less nebulously defined, crisper tuning at the center frequency.

At first blush, the limited frequency range over which substantive gains are possible in a tuned amplifier appear disadvantageous. Indeed, this narrowband response is a shortfall in applications that mandate circuit processing over broad frequency passbands. In other applications, and notably in commercial communications, narrow banding offers distinct operational advantages. One such advantage is that while a tuned electronic network assuredly amplifies signals whose frequencies are in the immediate neighborhood of the center frequency, ω_o, it implicitly rejects proximate frequencies, thereby minimizing potential interference incurred by undesirable, but nontheless unavoidable, signal energies at frequencies proximate to the tuned center frequency. This property is essential in commercial radio or television, wherein tuning to one broadcast station should preclude reception of another station that is broadcasting at a frequency near to that which is being received. For example, with $Q_o = 8$, a signal at a frequency that is one octave above ω_o (meaning twice as large as ω_o) has a normalized gain of 0.083. In comparison to the gain available at ω_o, the signal at frequency $2\omega_o$ is therefore attenuated by about 21.6 dB, or by a factor of roughly 12. Accordingly, this higher frequency signal may not pose a significant interference problem with respect to the signal at ω_o. A second advantage embraces electrical noise, to which the reader is exposed in due time. For the present, suffice it to say that the amount of random, spurious, electrical noise indigenous to an amplifier determines the minimum signal that can be faithfully detected, captured, and amplified by the utilized electronics. For progressively larger levels of electrical noise, applied input signals must be commensurately larger to ensure their accurate detection. As it materializes, such noise levels are directly proportional to the passband of the network frequency response. It follows that narrowband electronic systems are generally more capable of processing low-level signals than are otherwise comparable broadband systems.

Like the passband of a lowpass network, the passband of a tuned, or bandpass, system is defined as the frequency range over which the observable gain is within three decibels of the maximum available gain. In bandpass electronics, however, two 3-dB frequencies are evident, as is illustrated in Fig. 1.33. In the subject diagram, which plots normalized gain magnitude as a function of normalized frequency, the 3-dB frequencies are defined as those frequencies for which the gain magnitude is the inverse of root two,

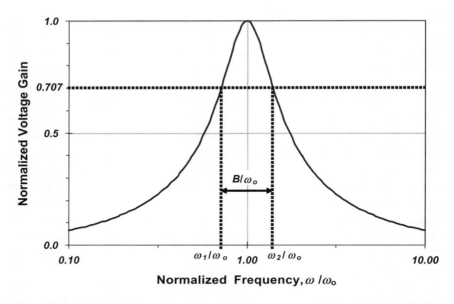

Figure 1.33. Generalized version of the frequency response for the tuned amplifier in Fig. 1.31. The plot conceptually illustrates the calculation of the circuit 3-dB bandwidth.

or 0.707. To this end, a higher than ω_o frequency, ω_2, and a lower than ω_o frequency, ω_1, are evidenced at the 3-dB down points. Since the normalized gain magnitude is the inverse of root two when the imaginary term coefficient on the right hand side of Eq. (1-40) is either plus one or negative one,

$$Q_o \left(\frac{\omega_2}{\omega_o} - \frac{\omega_o}{\omega_2} \right) = 1, \qquad (1\text{-}42)$$

and

$$Q_o \left(\frac{\omega_1}{\omega_o} - \frac{\omega_o}{\omega_1} \right) = -1. \qquad (1\text{-}43)$$

Solving these two equations for ω_2 and ω_1, respectively (both of which must be positive numbers),

$$\omega_2 = \frac{\omega_o}{2Q_o} \left(\sqrt{1 + 4Q_o^2} + 1 \right), \qquad (1\text{-}44)$$

$$\omega_1 = \frac{\omega_o}{2Q_o} \left(\sqrt{1 + 4Q_o^2} - 1 \right). \qquad (1\text{-}45)$$

It follows that the 3-dB bandwidth, B, is

$$B = \omega_2 - \omega_1 = \frac{\omega_o}{Q_o}, \qquad (1\text{-}46)$$

which supports an earlier contention of relative tuning sharpness with increasing circuit quality factor.

Interestingly enough, these results also suggest that

$$\omega_o = \sqrt{\omega_1 \omega_2}; \qquad (1\text{-}47)$$

that is, the center frequency is the geometric mean of the two 3-dB frequencies. However for very large Q_o, Eqs. (1-44) and (1-45) give

$$\begin{aligned}\omega_2 &\approx \omega_o + \frac{\omega_o}{2Q_o} = \omega_o + \frac{B}{2} \\ \omega_1 &\approx \omega_o - \frac{\omega_o}{2Q_o} = \omega_o - \frac{B}{2}\end{aligned}, \qquad (1\text{-}48)$$

which collectively depict ω_o as an approximate arithmetic mean of the upper and lower 3-dB frequencies.

1.4.0. Second Order Circuits and Systems

Although first order circuits containing but a single energy storage element, such as those addressed in Examples 1.1, 1.2 and 1.3, are relatively straightforward to analyze and assess, most electronic circuits and systems contain a multiplicity of energy storage elements and are therefore multi order in nature. Circuits whose transfer functions exhibit several poles and zeros are cumbersome to analyze and as a result, an engineering evaluation of their performance attributes and limitations can be a daunting undertaking. Fortunately, the salient properties of these high order circuits and systems can often be represented by second order mathematical models. Although these second order approximations are mathematically and topologically more intricate than are their first order counterparts, they do produce response estimates that, when carefully interpreted in light of all invoked approximations, track satisfactorily with the observable behavior of the circuits and systems they model.

The disclosure cited above comprises a sufficient reason to establish an adequate comfort level with the frequency and time domain electrical characteristics of second order networks. An additional justification for second order studies is that a broad class of programmable and reconfigurable electronic filters, known as *biquadratic filters*, are implemented as cascade interconnections of second order structures. These filters can be synthesized for virtually any type of requisite frequency response. For example, they can be bandpass units, such as the network in Fig. 1.31, which attenuate all signal frequencies except those that lie in the immediate neighborhood of a desired tuned center frequency. They can exhibit lowpass response properties, wherein low signal frequencies are processed with relatively constant gain, but high frequency signals are attenuated. The result is a reduction of potential high frequency interference threats imposed on an otherwise low-to-moderate frequency signal processor. Highpass biquadratic filters are the converse of lowpass architectures; that is, highpass units attenuate low frequencies while processing high frequencies with nominally constant gain. Finally, biquadratic filters can be realized as notch filters, which are the converse of bandpass units. They process all signal frequencies except those in the neighborhood of a center frequency. A common application of a notch filter entails the mitigation of the annoying 60-cycle "hum" evidenced in sensitive electronic units that are energized by conventional 60-Hz sinusoidal electric power.

1.4.1. Second Order Filters

Although filters are not the dominant focus of this discussion, they do provide a convenient vehicle for demonstrating the practical realization of a second order circuit. To this end, consider the lowpass active filter depicted in Fig. 1.34, which utilizes four (4) single ended operational transconductor amplifiers, or OTAs, and two (2) capacitors. In the interest of analytical

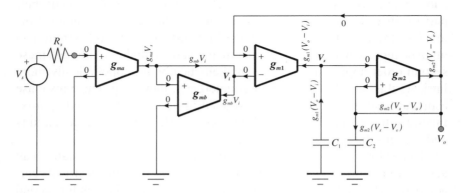

Figure 1.34. Schematic diagram of a second order lowpass filter realized with single ended operational transconductor amplifiers (OTAs).

simplicity, the four transconductors are taken herewith to be ideal; that is, each has infinitely large input impedance, infinitely large output impedance, and constant, frequency independent transconductance. The initial objective herewith is a delineation of the transfer function, $H(s) = V_o/V_s$, of the filter. To this end, the concepts set forth by the idealized transconductor model in Fig. 1.18(b) allow for the stipulation of the key circuit branch currents indicated in the subject schematic diagram. For example, since no current flows into the input port of an ideal transconductor, the input port voltage, measured from noninverting-to-inverting terminals, of the transconductor on the far left of the diagram is the signal source voltage, V_s. Accordingly, the current flowing into the output port of this active element is simply $g_{ma}V_s$. If the output port voltage of the second transconductor (transconductance of g_{mb}) is denoted as V_i, the feedback around this element forces the input port voltage to mirror V_i, whence an output port current flowing into the transconductor of $g_{mb}V_i$. The transconductor symbolized as g_{m1} has a resultant input port voltage of $(V_o - V_i)$, which produces an output port current of $g_{m1}(V_o - V_i)$. This output current is constrained to flow through the capacitance, C_1, because the subsequent transconductor (labeled g_{m2}) conducts zero input current. Finally, if the voltage developed across capacitance C_1 is symbolized as V_x, the input port voltage to the transconductor labeled g_{m2} is $(V_x - V_o)$, polarized from the inverting terminal-to-the noninverting terminal. Consequently, an output port current of $g_{m2}(V_x - V_o)$ flows out of the output port of the transconductance element on the far right of the subject schematic diagram. This current is forced to flow through the capacitance, C_2, because zero current is drawn by the noninverting input terminals of the third and fourth transconductors.

Nodal analysis applied to the output terminal of the g_{ma}-transconductor stipulates $g_{ma}V_s + g_{mb}V_i = 0$, whence

$$\frac{V_i}{V_s} = -\frac{g_{ma}}{g_{mb}}. \tag{1-49}$$

Before proceeding with the analysis, it is instructive to understand that the g_{mb}-transconductor functions as a load that presents an effective resistance of $(1/g_{mb})$ to a phase inverting transconductor amplifier whose transconductance is g_{ma}. In support of this contention, Fig. 1.35(a) is submitted to posture the second transconductor of the filter in Fig. 1.34 in a form appropriate to determining the effective resistance, V_i/I_i, established across the terminals of this subcircuit. Using Fig. 1.18(b), the pertinent model is the structure depicted in Fig. 1.35(b), which verifies that $I_i = g_{mb}V_i$ and hence, $V_i/I_i = 1/g_{mb}$. It follows that the cascade of the first two transconductors

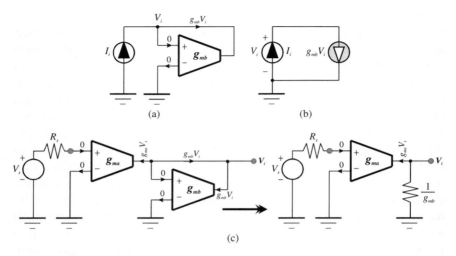

Figure 1.35. (a) Subcircuit consisting of the second transconductance unit in the filter of Fig. 1.34. (b) Electrical model for determining the terminal resistance of the subcircuit in (a). (c) Subcircuit consisting of the first two transconductance element stages in the filter of Fig. 1.34. The representation suggests that the second transconductance unit functions as an equivalent, two terminal resistance.

in the filter at hand is electrically equivalent to the macromodel offered in Fig. 1.35(c), whose voltage gain clearly subscribes to Eq. (1-49).

Having established that the g_{mb}-transconductor in Fig. 1.34 merely emulates a two-terminal resistance, questions naturally arise as to the propriety of using this active subcircuit when, in fact, a simple resistor whose resistance value is numerically equal to $(1/g_{mb})$ ostensibly suffices. To be sure, the simple resistance approach may be preferred in numerous applications because of noise, power dissipation, linearity, and other considerations. But one advantage to actively realizing the required terminating resistance is the ability to adjust actual resistance value electronically. Specifically, the value of OTA transconductance, g_{mb}, can be varied over at least a small range of values by adjusting the biasing voltages applied to the OTA. In general, *biasing* (not shown in the schematic diagrams) consists of one or more constant (or static) voltages appropriately applied to the transconductor amplifier to ensure its reasonably linear signal processing performance over the requisite range of signal amplitudes and frequencies. In effect, the g_{mb}-transconductor functions as a kind of electronic potentiometer, thereby allowing the design engineer to fine tune, or "tweak", the nominal design to achieve desired performance in the face of parametric device, circuit, or system uncertainties.

Returning to the analysis problem, the current conducted by capacitance C_1 is

$$sC_1 V_x = -g_{m1}(V_o - V_i), \quad (1\text{-}50)$$

while capacitor C_2 conducts

$$sC_2 V_o = g_{m2}(V_x - V_o). \quad (1\text{-}51)$$

Upon elimination of the voltage variable, V_x, from these two relationships, the voltage ratio, V_o/V_i, is found to be

$$\frac{V_o}{V_i} = \frac{1}{1 + \dfrac{sC_1}{g_{m2}} + \dfrac{s^2 C_1 C_2}{g_{m1} g_{m2}}}, \quad (1\text{-}52)$$

which is clearly a second order transfer function. Recalling Eq. (1-49), the desired transfer function is

$$H(s) = \frac{V_o}{V_s} = \frac{V_o}{V_i} \times \frac{V_i}{V_s} = -\frac{g_{ma}/g_{mb}}{1 + \dfrac{sC_1}{g_{m2}} + \dfrac{s^2 C_1 C_2}{g_{m1} g_{m2}}}. \quad (1\text{-}53)$$

Equation (1-53) renders immediately transparent the fact that the zero frequency gain, $H(0)$, of the active filter in Fig. 1.34 is

$$H(0) = -\frac{g_{ma}}{g_{mb}}. \quad (1\text{-}54)$$

The negative algebraic sign in this result indicates phase inversion between the input and output ports. This is to say that a rising input signal over time results in an amplified, but decreasing, output signal in the steady state. Conversely, a decreasing input is accompanied by an increasing steady state response. Since capacitors behave as open circuits for steady state, zero frequency inputs, the current conducted by C_1 at zero frequency is necessarily zero. By Eq. (1-50), this fact forces $V_o \equiv V_i$, whence Eq. (1-49) is seen to corroborate with Eq. (1-54).

In an attempt to garner insights about the responses evidenced by second order networks, it is expedient to write the second order transfer relationship of Eq. (1-53) in one of the two traditional generalized forms,

$$H(s) = \frac{H(0)}{1 + \dfrac{2\zeta s}{\omega_n} + \dfrac{s^2}{\omega_n^2}}, \tag{1-55}$$

or

$$H(s) = \frac{H(0)}{1 + \dfrac{s}{Q\omega_n} + \dfrac{s^2}{\omega_n^2}}, \tag{1-56}$$

where $H(0)$ symbolizes the circuit gain at zero signal frequency, which in this case is given by Eq. (1-54). Parameter ω_n is termed the *undamped natural frequency of oscillation*, or the *undamped self-resonant frequency*, of the system under consideration. A comparison of Eq. (1-55) or Eq. (1-56) with Eq. (1-53) suggests that for the filter at hand, ω_n (in units of radians-per-second) is

$$\omega_n = \sqrt{\frac{g_{m1} g_{m2}}{C_1 C_2}}. \tag{1-57}$$

Moreover, ζ is called the *damping factor* of the system, while Q is termed the system *quality factor*. From Eqs. (1-55), (1-56) and (1-53),

$$\frac{2\zeta}{\omega_n} = \frac{1}{Q\omega_n} = \frac{C_1}{g_{m1}}, \tag{1-58}$$

whence, by Eq. (1-57),

$$\zeta = \frac{1}{2Q} = \frac{1}{2}\sqrt{\frac{g_{m2} C_1}{g_{m1} C_2}}. \tag{1-59}$$

The engineering significance of damping factor ζ and of undamped self-resonant frequency ω_n is clarified by the subsections that follow. For the moment, suffice it to say that for nonzero damping factor, ω_n is a measure of the circuit 3-dB bandwidth. This is to say that for $\zeta \neq 0$, large ω_n produces large bandwidth, while small ω_n results in small circuit bandwidth. On the other hand, damping factor ζ is a measure of the *stability* of the circuit undergoing investigation. A stable linear circuit can be interpreted herewith as implying a circuit that is capable of establishing a steady state output response that is exclusively determined by, and linearly proportional to, the steady state input signal. A stereo amplifier is presumably stable since it delivers an electrical response to its connected speakers that is linearly related to the electrical signal established at the output terminals of a compact disc player, despite any interference caused by minor disc imperfections, local fluorescent lighting, or proximately operated household appliances. It is shown shortly that a negative damping factor is disastrous from a stability perspective. On the other hand, large ζ ensures consummate stability at the expense of a system inability to achieve steady state operation quickly. Damping factors slightly less than one offer the best compromise between adequate stability margins and expeditious response speeds.

1.4.2. Frequency Response

The frequency response is a traditional metric for evaluating the steady state performance of a linear circuit or system. This graphical tool effectively provides a snapshot of the manner in which the gain magnitude varies in the steady state with the frequency of an applied input sinusoid. An even cursory inspection of the frequency response therefore conveys information as to whether the circuit or system gain is too small or too large at certain signal frequencies, whether the gain is increasing too fast or too slowly over a range of frequencies, and whether the rate at which the gain magnitude diminishes with frequency is too dramatic.

Analytically, a frequency response study of a generalized lowpass second order network begins by supplanting the Laplace variable, s, in Eq. (1-55) by $j\omega$, since steady state responses to applied sinusoids are the order of the business at hand. In order to minimize algebra and forge analytical efficiency, it is convenient both to normalize the transfer function to its zero frequency gain and to normalize the signal frequency to the undamped self-resonant frequency. Accordingly, let the normalized transfer function, $H_n(j\omega)$, be

$$H_n(j\omega) \triangleq \frac{H(j\omega)}{H(0)}, \qquad (1\text{-}60)$$

and the normalized frequency, x, be

$$x \triangleq \frac{\omega}{\omega_n}. \qquad (1\text{-}61)$$

Then Eq. (1-55) can be recast as

$$H_n(jx) = \frac{1}{1 - x^2 + j2\zeta x}, \qquad (1\text{-}62)$$

whose magnitude is

$$|H_n(jx)| = \frac{1}{\sqrt{(1-x^2)^2 + (2\zeta x)^2}}. \tag{1-63}$$

Figure 1.36 plots the decibel value of the normalized gain magnitude delineated in Eq. (1-63) (20-times the base 10 logarithm of the magnitude) as a function of the normalized signal frequency for several values of the damping factor, ζ. Ideally, the normalized frequency response is a constant 0 dB, which is indicative of a constant gain in the amount of the "DC" gain, $H(0)$, over a frequency passband stretching from essentially zero frequency through to the 3-dB bandwidth of the circuit undergoing scrutiny. Obviously, the responses depicted in Fig. 1.36 are far from ideal. One observable problem is that considerable response peaking is evidenced for small damping factors. For example, $\zeta = 0.05$ results in a 20-dB peak, which implies that the network gain magnitude at some relatively high signal frequency is 10-times larger than the zero frequency gain. A slightly larger than 8-dB peak materializes for $\zeta = 0.2$.

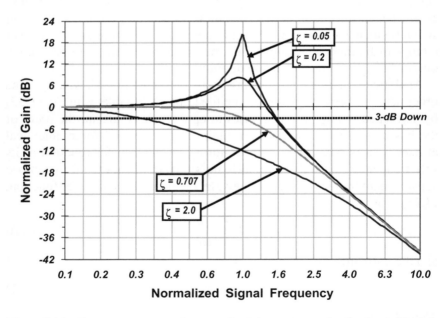

Figure 1.36. Frequency response of a generalized, lowpass second order circuit. The gain scale is normalized to the zero frequency gain of the circuit. The frequency scale is normalized to the undamped self-resonant frequency of the network.

Circuit and System Fundamentals

Excessive response peaking in lowpass configurations is undesirable for at least two reasons. First, such peaking accents high frequency signal amplitudes, while amplifying low frequencies with relatively constant, and often considerably smaller, gain. In a stereo system, the indicated peaking would result in shrill treble responses and anemic base responses. A second, and more alarming, consequence of excessive response peaking is the potential system instability it implies. For example, if ζ were to be nulled, Eq. (1-63) shows infinitely large gain magnitude at $x = 1$, which is equivalent to asserting infinitely large gain at the undamped (meaning $\zeta = 0$) self-resonant frequency. Infinitely large gain magnitude in the presence of finite output responses means that the network is curiously generating a response without a driving forcing function. As astonishing as this circumstance appears to be, it can happen in poorly designed electronics for which the interaction of high order energy storage parasitics with network gain elements reduce the effective circuit damping factor to zero. The network for which $\zeta = 0$ is said to oscillate, and since the infinitely large gain that results in a finite output response for zero inputs occurs at only the frequency, ω_n, the subject oscillatory response is a sinusoid of frequency ω_n.

1.4.2.1. Response Peaking

Because excessive peaking of the frequency response is undesirable and generally indicative of potential instability problems, exploring design-oriented means to avoid such peaking is a prudent undertaking. Steady state frequency response peaking is accompanied by a magnitude response that projects zero slope in the frequency domain. Accordingly, return to Eq. (1-63) and determine the value, say x_p, of the normalized frequency variable, x, where the derivative of the magnitude response with respect to x is zero; that is,

$$\left. \frac{d|H_n(jx)|}{dx} \right|_{x=x_p} = \left. \frac{d}{dx} \left[\frac{1}{\sqrt{(1-x^2)^2 + (2\zeta x)^2}} \right] \right|_{x=x_p} = 0. \quad (1\text{-}64)$$

The execution of this admittedly sloppy task results in two solutions for x_p; namely, $x_p = 0$ and

$$x_p \triangleq \frac{\omega_p}{\omega_n} = \sqrt{1 - 2\zeta^2} = \sqrt{1 - \frac{1}{2Q^2}}, \quad (1\text{-}65)$$

where ω_p symbolizes the radial frequency corresponding to zero frequency domain slope of the magnitude characteristic. The solution, $x_p = 0$, bodes no particular significance other than reaffirming the expectation of nominally constant gain in the immediate neighborhood of zero signal frequency. Equation (1-65) is the interesting solution in that it implies a second frequency at which zero slope is evidenced. The existence of this second solution implies a frequency response that does not diminish monotonically as the signal frequency increases from zero. However, note that for $\zeta > 1/\sqrt{2}$, or equivalently, $Q < 1/\sqrt{2}$, x_p in Eq. (1-65) is an imaginary number, which suggests that no real second frequency of zero magnitude response slope exists; in other words, no response peaking can be observed. In support of this disclosure, observe further that $x_p \equiv 0$ if ζ or Q is precisely $1/\sqrt{2}$. The constraint, $\zeta = 1/\sqrt{2}$, is therefore understandably referred to as the condition for *maximally flat magnitude* (*MFM*) frequency response in that it produces zero slope in the magnitude response at only a single frequency which is, in fact, zero frequency.

The solution in Eq. (1-65) can be plugged into Eq. (1-63) to ascertain the peak value, say M_p, of the frequency response magnitude at nonzero frequency. Biting this proverbial algebraic bullet results in

$$M_p = \frac{1}{2\zeta\sqrt{1-\zeta^2}} = \frac{Q}{\sqrt{1-\left(\frac{1}{2Q}\right)^2}}. \quad (1\text{-}66)$$

It should be understood that Eq. (1-66) is meaningful only for $0 < \zeta \leq 1/\sqrt{2}$ or equivalently, for $1/\sqrt{2} \geq Q < \infty$, which ensure that x_p in Eq. (1-65) is a real number. Figure 1.37 graphically depicts both the dependence of response peak and the frequency corresponding to response peaking on circuit quality factor.

1.4.2.2. Bandwidth

A common misperception of the maximally flat magnitude condition is that the MFM response yields maximal 3-dB bandwidth. Figure 1.36 confirms otherwise and shows that substantive bandwidth increases can be attained through a reduction in circuit damping factor. The generally unacceptable price paid for this bandwidth enhancement is progressively more pronounced response peaking.

The dependence of the 3-dB bandwidth on network damping factor can be discerned through another return to Eq. (1-63). In this case, the analytical

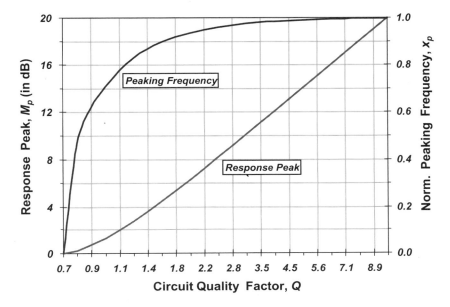

Figure 1.37. Peak transfer function magnitude and the normalized frequency at which peaking is evidenced for a second order network as a function of the network quality factor.

objective is the value, say x_b, of the normalized frequency that results in a normalized transfer function magnitude equal to the inverse of root two. After a few pages of annoying algebra, it can be shown that

$$x_b \triangleq \frac{\omega_b}{\omega_n} = \sqrt{(1 - 2\zeta^2) + \sqrt{1 + (1 - 2\zeta^2)^2}}, \quad (1\text{-}67)$$

where ω_b symbolizes the radial 3-dB bandwidth. Obviously, ω_b is directly proportional to the self-resonant frequency, ω_n. Indeed, $\omega_b \equiv \omega_n$ when the damping factor is $\zeta = 1/\sqrt{2}$. But the damping factor also impacts the achievable 3-dB bandwidth in the form that is depicted graphically in Fig. 1.38.

The last plot motivates a few useful observations and related considerations. First, note that the bandwidth falls dramatically in the damping range, $0 < \zeta < 1$. For example, at $\zeta = 1/\sqrt{2}$, the normalized bandwidth, x_b, is one, while at $\zeta = 1$, $x_b = 0.64$, which suggests a significant 36% bandwidth degradation with respect to the bandwidth evidenced under maximally flat operating conditions. For $\zeta > 1$, the bandwidth falls monotonically with

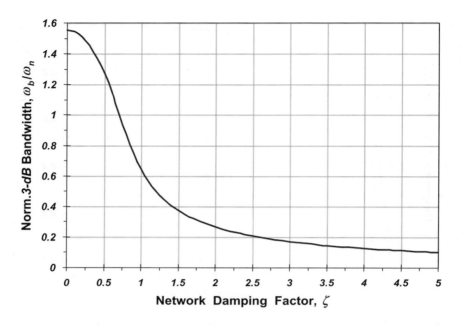

Figure 1.38. The dependence of network 3-dB bandwidth on the damping factor.

damping factor, degrading to slightly more than 20% of the MFM bandwidth at $\zeta = 2.5$. The lesson to be learned here is that while large ζ is comforting in the sense of achieving network stability, the price paid for unconditional stability is reduced network bandwidth; that is a progressive inability of the subject network to process faithfully high frequency signals. This observation justifies the common design objective of ensuring a damping factor that nominally satisfies the inequality, $1/\sqrt{2} < \zeta < 1$. A damping factor significantly smaller than $1/\sqrt{2}$ risks unacceptable response peaking, which hints at potential system stability problems, while a damping factor that is significantly larger than unity results in unacceptable bandwidth degradation.

Second, the case of a damping factor sufficiently larger than unity gives rise to a useful bandwidth approximation. With $\zeta > 1$, $(1 - 2\zeta^2)$ in Eq. (1-67) is a negative number. Accordingly, Eq. (1-67) can be recast as

$$\left(\frac{\omega_b}{\omega_n}\right)^2 = -(2\zeta^2 - 1) + \sqrt{1 + (2\zeta^2 - 1)^2}$$

$$= -(2\zeta^2 - 1) + (2\zeta^2 - 1)\sqrt{1 + \frac{1}{(2\zeta^2 - 1)^2}}.$$

For $(2\zeta^2 - 1)^2 \gg 1$, the radical on the right hand side of this relationship can be approximated by a two-term power series, with the result that

$$\left(\frac{\omega_b}{\omega_n}\right)^2 \approx -(2\zeta^2 - 1) + (2\zeta^2 - 1)\left[1 + \frac{1}{2(2\zeta^2 - 1)^2}\right],$$

and for $2\zeta^2 \gg 1$,

$$\left(\frac{\omega_b}{\omega_n}\right)^2 \approx \frac{1}{4\zeta^2}.$$

It follows that for sufficiently large ζ, the radial 3-dB bandwidth can be approximated by the simple relationship,

$$\omega_b \approx \frac{\omega_n}{2\zeta}. \tag{1-68}$$

A numerical comparison of Eq. (1-68) with Eq. (1-67) readily demonstrates that Eq. (1-68) incurs a bandwidth error of less than 10.9% for all $\zeta \geq 1.5$. Moreover, this error is always negative; that is, the approximated bandwidth is always smaller than the true 3-dB bandwidth. In integrated circuit design situations that are routinely plagued by a plethora of uncertainties surrounding the modeling of active devices, energy storage elements engendered by the physical layout of the circuit, and nonzero component tolerances, it is comforting to be afforded the opportunity of using a simple bandwidth expression that is guaranteed to yield slightly pessimistic bandwidth results. Computational simplicity aside, it is especially interesting to note that the bandwidth approximation in Eq. (1-68) is exactly the inverse of the s-term coefficient in the denominator of the second order transfer function delineated in Eq. (1-55). In other words,

$$H(s)|_{\zeta \geq 1.5} \approx \frac{H(0)}{1 + \dfrac{s}{\omega_b} + \dfrac{s^2}{\omega_n^2}}, \tag{1-69}$$

which suggests an approximate bandwidth evaluation deriving merely through discovery of the coefficient of the linear frequency term in the characteristic polynomial of the network transfer function.

1.4.2.3. Phase and Delay Responses

If the sinusoid,

$$v_s(t) = V_{sp} \cos \omega t, \tag{1-70}$$

is applied to a second order network whose transfer function is given by Eq. (1-55), the resultant steady state response is of the form,

$$v_o(t) = |H(0)|V_{sp} \cos[\omega t + \theta(\omega)]. \tag{1-71}$$

For a linear system, the frequency, ω, of the steady state output response is identical to the frequency of the applied, single frequency excitation. Equation (1-71) reaffirms the anticipated result that the amplitude of the steady state response is the amplitude, V_{sp}, of the input signal excitation, amplified (or multiplied) by the magnitude of the zero frequency gain, $|H(0)|$, of the network transfer function. Additionally, the steady state response is phase displaced by an amount, $\theta(\omega)$. Thus, for example, if $\theta(\omega) = -\pi/3$ radians, the output voltage is said to lag the input signal by 60°. The visible impact of this example phase angle is shown in Fig. 1.39, which plots, versus the normalized time, $\omega t/\pi$, the input signal of Eq. (1-70) and the output response of Eq. (1-71) for the case of a gain magnitude of $|H(0)| = 3$.

In general, the phase shift, $\theta(\omega)$, between the steady state input and output responses is a function of the input signal frequency. This generality is tacitly disturbing because, as is suggested by the plots in Fig. 1.39, the phase

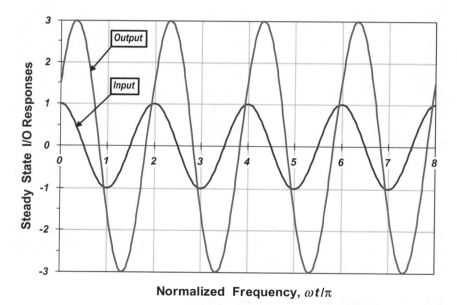

Figure 1.39. Steady state input and output waveforms for a linear second order network having a zero frequency gain magnitude of 3. The applied input excitation is a single frequency sinusoid. At this frequency, the phase angle attributed to the linear network is $-60°$.

shift of a linear network is indicative of steady state delay incurred in the signal processing between applied input and the resultant output response. It just takes a bit of time for all those electrons to navigate the interconnected electrical and electronic maze that comprises the signal flow path between input and output network ports. If the phase shift, and hence the signal delay, is dependent on frequency, it is conceivable that the processed low frequency components of a nonsinusoidal input signal waveform do not arrive at the output port at the same times that do the high frequency components of said waveform. This disparity comprises *phase distortion*, which can be acutely troublesome in certain applications. For example, if one is sitting in the front row of a concert hall listening to a live rock concert, one hears the bass guitar (low frequencies) accompanying a singing voice (higher frequencies) at nominally the same time that the voice is heard. But in a CD recording of the same interlude, a pronounced frequency dependence of the phase angle indigenous to the utilized stereo equipment can result in the speakers receiving the bass guitar input at a time that is appreciably delayed with respect to the voice response. In low cost stereo systems, this effect is typified by the "hollow" or "tunnel" sound, which is the bane of audiophiles.

Ideally, the foregoing delay issue can be mitigated by a phase angle that depends linearly on signal frequency. To wit, if $\theta(\omega)$ in Eq. (1-71) is given by

$$\theta(\omega) = -T_d\omega, \qquad (1\text{-}72)$$

where T_d is a constant, independent of frequency, the steady state sinusoidal output response in Eq. (1-71) becomes

$$v_o(t) = |H(0)|V_{sp} \cos\left[\omega(t - T_d)\right]. \qquad (1\text{-}73)$$

The last result is interesting in that it projects the impact of linear phase as a constant time delay in the amount of the proportionality constant, T_d, in Eq. (1-72). This is to say, that although the input signal is not processed instantaneously by the linear network, all input signal frequency components arrive in the steady state at precisely the times dictated by the applied input excitation. Referring to the preceding hypothetical rock concert, a stereo system boasting a linear phase response means that Keith Richard's guitar superimposed with Mick Jagger's voice would be heard through the speaker at the same times that they would be heard in a live concert setting.

Unfortunately, no physically realizable network can produce a linear phase response over frequency. But linear phase can be emulated over restricted frequency passbands. To this end, the *envelope delay*, $D(\omega)$, of a linear network or system is introduced in accordance with the definition,

$$D(\omega) = -\frac{d\theta(\omega)}{d\omega}. \tag{1-74}$$

Note that if $\theta(\omega)$ is the linear frequency relationship of Eq. (1-72), $D(\omega)$ is the constant delay, T_d, discussed in conjunction with Eq. (1-73). For any other phase function, the passband over which $D(\omega)$ is a reasonable approximation of a constant defines the signal frequency range over which nominally constant I/O delay is evidenced in the steady state.

For the second order network whose normalized, frequency domain transfer function is the expression in Eq. (1-62), the phase response, in terms of the normalized frequency variable, x, in Eq. (1-61) is

$$\theta(x) = -\tan^{-1}\left(\frac{2\zeta x}{1-x^2}\right). \tag{1-75}$$

Recalling Eqs. (1-74) and (1-61), the envelope delay, $D(x)$, as a function of x, is

$$D(x) = -\frac{d\theta(x)}{\omega_n dx},$$

whence a normalized envelope delay, $D_n(x)$, of

$$D_n(x) \triangleq \omega_n D(x) = -\frac{d\theta(x)}{dx}. \tag{1-76}$$

It follows that Eqs. (1-75) and (1-76) combine to produce

$$D_n(x) = \frac{2\zeta(1+x^2)}{1 + 2(2\zeta^2 - 1)x^2 + x^4} \tag{1-77}$$

as the normalized envelope delay of the second order network undergoing investigation. Figure 1.40 depicts Eq. (1-77) graphically in normalized format.

Several features of Eq. (1-77) warrant attention. First, observe a zero frequency normalized envelope delay of $D_n(0) = 2\zeta$ and hence, a zero frequency envelope delay of $D(0) = 2\zeta/\omega_n$. This delay is actually observable at not only zero frequency, but also at all low signal frequencies that conform to $x \ll 1$. Recall that $2\zeta/\omega_n$ is exactly the s-term coefficient in the denominator of the network transfer characteristic in Eq. (1-55). Moreover, and to the extent that the damping factor is at least as large as 1.5, $2\zeta/\omega_n$ approximates the inverse 3-dB bandwidth of the subject second order system. Thus, the low frequency delay of a second order system is precisely the

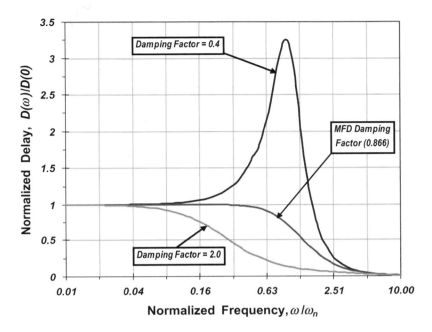

Figure 1.40. The frequency response of the envelope delay for a second order network for various values of the damping factor, ζ. Note that the maximally flat delay (MFD) condition yields a flat delay response over a reasonably wide range of signal frequencies. For a damping factor that is larger than that of the MFD value, flat delay also results, but over a more restricted frequency interval.

s-term coefficient in the network characteristic polynomial, and for $\zeta \geq 1.5$, this envelope delay is nominally the inverse of the 3-dB bandwidth, ω_b.

The foregoing observations are rendered transparent by Eq. (1-75). In particular, for small x, the argument of the arctangent function is small. Since the arctangent of a small numerical argument is approximately the argument itself,

$$\theta(x)|_{x \ll 1} = -\tan^{-1}\left(\frac{2\zeta x}{1 - x^2}\right)\bigg|_{x \ll 1} \approx -2\zeta x. \qquad (1\text{-}78)$$

It follows that

$$D_n(x)|_{x \ll 1} \approx -2\zeta, \qquad (1\text{-}79)$$

as hypothesized in the preceding paragraph.

A second observation is the fact that the envelope delay approaches zero at infinitely large frequencies. Thus, low frequency input signal components are delayed more than are high frequency components. Unfortunately, the envelope delay does not necessarily decay monotonically with increasing frequency, principally because the factor, $(2\zeta^2 - 1)$, in the denominator on the right hand side of Eq. (1-77) can be a negative number. Since nominally constant delay is a desirable performance metric, it is of interest to determine the operating condition that ensures a monotonically decreasing delay response. This constraint, which defines the so-called *maximally flat delay* (*MFD*) condition, is determined by ensuring that the first derivative, with respect to x, of the normalized delay function, $D_n(x)$, is zero at no real normalized frequency other than zero. Upon execution of this analytical task, the MFD condition is found to be $\zeta = \sqrt{3}/2$. Smaller damping factors incur nonmonotonicity, and hence peaking, in the delay response, while larger damping factors ensure a monotonically decreasing delay with signal frequency at the expense of a reduced frequency passband over which nominally constant delay is projected.

Example 1.5. Assume that the voltage transfer function of a preamplifier of one channel of a stereo system is the second order relationship given by Eq. (1-55). This preamplifier is to be designed for a maximally flat delay response that delivers constant delay to within 5% for signal frequencies extending to the upper limit of the audio spectrum namely 20 KHz. Determine the minimum 3-dB bandwidth that the preamplifier must deliver, as well as its self-resonant frequency. What is the low frequency delay of the designed amplifier?

Solution 1.5.

(1) Maximally flat delay (MFD) in a second order circuit requires a damping factor, ζ, of

$$\zeta = \frac{\sqrt{3}}{2}. \quad \text{(E5-1)}$$

An examination of the numerical computations precipitating the plots in Fig. 1.40 reveals that the resultant envelope delay remains within 5% of its zero, or low, frequency value through a normalized signal frequency that satisfies the inequality,

$$\frac{\omega}{\omega_n} = \frac{f}{f_n} < 0.5. \quad \text{(E5-2)}$$

If Eq. (E5-2) is to be satisfied for a frequency as large as $f = 20\,\text{KHz}$, it is clear that the required self-resonant frequency of the preamplifier must satisfy

$$f_n \geq \frac{f}{0.5} = \frac{20\,\text{KHz}}{0.5} = 40\,\text{KHz}.$$

(2) For a damping factor chosen in accordance with Eq. (E5-1), Eq. (1-67) stipulates the 3-dB bandwidth as

$$\frac{f_b}{f_n} = 0.786, \qquad \text{(E5-3)}$$

whence a bandwidth requirement of

$$f_b \geq 0.786(40\,\text{KHz}) = 31.5\,\text{KHz}.$$

(3) From Eqs. (1-76) and (1-77), the zero, and approximate low frequency, envelope delay, $D(0)$, evaluates as

$$D(0) = \frac{2\zeta}{\omega_n} = \frac{\zeta}{\pi f_n} = 6.89\,\mu\text{SEC}.$$

Comments. It is interesting that a maximally flat delay response in an audio amplifier requires a bandwidth that exceeds the 20 KHz upper frequency limit of the audio spectrum.

1.4.3. Poles and Second Order System Parameters

The preceding subsections of material underscore the significance of the damping factor, ζ, the quality factor, Q, and the radial undamped natural frequency, ω_n, as metrics that define the steady state frequency, phase, and delay responses of second order networks. It is often illuminating to cast these system parameters in terms of the network pole positions in the complex frequency plane. The most straightforward way of implementing this alternative characterization strategy is to relate the pole frequencies directly to the damping factor and self-resonant frequency.

The second order nature of the transfer function in Eq. (1-55) suggests the existence of two *critical frequencies*, or *poles*, say p_1 and p_2, such that

$$H(s) = \frac{H(0)}{1 + \dfrac{2\zeta s}{\omega_n} + \dfrac{s^2}{\omega_n^2}} = \frac{H(0)}{\left(1 + \dfrac{s}{p_1}\right)\left(1 + \dfrac{s}{p_2}\right)}. \qquad (1\text{-}80)$$

From a purely algebraic perspective, the poles define little more than the roots of the network characteristic polynomial, or denominator, of the transfer function. In particular, the roots herewith lie at $s = -p_1$ and at $s = -p_2$. On the presumption that p_1 and p_2 are real numbers, a necessary condition for network stability is that both p_1 and p_2 be positive. This requirement ensures that the two pole frequencies are negative and that the subject poles resultantly lie in the left half complex frequency plane. If p_1 and p_2 are complex numbers, physical realizability with lumped passive and active circuit elements demands that the two poles be complex conjugates. Additionally, network stability in the case of complex conjugate poles, like the stability constraint associated with real poles, mandates that complex poles also lie in the left half s-plane.

If the denominator on the far right hand side of Eq. (1-80) is expanded, $H(s)$ is expressible as

$$H(s) = \frac{H(0)}{1 + \frac{2\zeta s}{\omega_n} + \frac{s^2}{\omega_n^2}} = \frac{H(0)}{\left(1 + \frac{s}{p_1}\right)\left(1 + \frac{s}{p_2}\right)}$$

$$= \frac{H(0)}{1 + \left(\frac{1}{p_1} + \frac{1}{p_2}\right)s + \frac{s^2}{p_1 p_2}}. \qquad (1\text{-}81)$$

A simple comparison of like coefficients in the Laplace variable, s, produces

$$\omega_n = \sqrt{p_1 p_2} \qquad (1\text{-}82)$$

and

$$\frac{2\zeta}{\omega_n} = \frac{1}{p_1} + \frac{1}{p_2},$$

whence

$$\zeta = \frac{1}{2}\left(\sqrt{\frac{p_2}{p_1}} + \sqrt{\frac{p_1}{p_2}}\right). \qquad (1\text{-}83)$$

Observe that the undamped natural frequency of oscillation is exposed herewith as little more than the geometric mean of the two pole frequencies of a second order network. Moreover, the damping factor of the network is intimately related to the ratio of pole frequencies. Since negative damping factor ζ in Eq. (1-81) guarantees at least one right half plane pole (a characteristic polynomial having at least one root with a positive real part), a necessary condition for network stability is that the real solution of Eq. (1-83) must be a positive number.

Three special cases are of interest. The first of these is the *underdamped* case, wherein p_1 and p_2 are complex conjugate poles. From Eq. (1-82), underdamping necessarily implies that the pole frequencies satisfy

$$\left.\begin{array}{l} p_1 = \omega_n e^{j\varphi} = \omega_n \cos\varphi + j\omega_n \sin\varphi \\ p_2 = \omega_n e^{-j\varphi} = \omega_n \cos\varphi - j\omega_n \sin\varphi \end{array}\right\}, \qquad (1\text{-}84)$$

where angle ϕ must be larger than (and not equal to) $\pi/2$ radians and smaller than (but not equal to) $3\pi/2$ radians to guarantee network stability. Note that

Circuit and System Fundamentals

ω_n is observed to be the magnitude of either pole frequency. If Eq. (1-84) is substituted into Eq. (1-83), the damping factor for the underdamped network condition is found to be smaller than one, since

$$\zeta = \frac{1}{2}\left(e^{-j\varphi} + e^{j\varphi}\right) = -\cos\varphi, \qquad (1\text{-}85)$$

and $\pi/2 < \phi < 3\pi/2$. Thus, both the MFM and MFD cases considered earlier correspond to underdamped operating conditions. In particular, MFM requires $\zeta = 1/\sqrt{2}$, which corresponds to a pole angle, ϕ, of $3\pi/4$ radians. On the other hand, MFD stipulates $\zeta = \sqrt{3}/2$, or $\phi = 5\pi/6$ radians. Observe that the pole angle difference between MFM and MFD responses is a mere $\pi/12$ radians, or $15°$.

A special case of underdamping is an angle, ϕ, of $\pi/2$ radians, for which $\zeta = 0$ in Eq. (1-83). From Eq. (1-84), the poles corresponding to this zero damping case lie exclusively on the $j\omega$-axis of the complex frequency plane since $p_1 = +j\omega_n$ and $p_2 = -j\omega_n$; that is, the pole frequencies have null real parts. As is demonstrated subsequently, a network having zero damping responds to an impulsive input with a free running sinusoidal oscillation. This is to say that the output response is an eternal sinusoid even though the input excitation reduces ultimately to zero. Such an operating condition is certainly undesirable in linear amplification networks. For example, it would be annoying to hear a single frequency tone at the speakers of a stereo system as a background to the music recorded on a CD. However, zero damping is an essential design constraint of sinusoidal oscillators, which are exploited extensively in radios, television receivers, and numerous other communication media.

A second special interest is the *overdamped* case, for which the network poles are real numbers. If poles p_1 and p_2 are real, such that $p_2/p_1 = k$, a positive number, the damping factor in Eq. (1-83) becomes

$$\zeta = \frac{1}{2}\left(\sqrt{k} + \frac{1}{\sqrt{k}}\right), \qquad (1\text{-}86)$$

which can be demonstrated to yield $\zeta > 1$ for all positive values of k. For $k \gg 1$, the pole at frequency p_1 is said to be *dominant*, for the frequency, p_1, effectively determines the 3-dB bandwidth of the overdamped system. To demonstrate this contention, observe in Eq. (1-86) that

$$\zeta|_{k\gg 1} \approx \frac{\sqrt{k}}{2}. \qquad (1\text{-}87)$$

From Eqs. (1-87) and (1-68), the resultant 3-dB bandwidth computes as

$$\omega_b|_{k\gg 1} \approx \frac{\omega_n}{2\zeta} \approx p_1. \qquad (1\text{-}88)$$

While p_1 is said to be the dominant pole of an overdamped system having

$$p_2 = kp_1 \tag{1-89}$$

and $k \gg 1$, p_2 is commonly termed the *nondominant network pole*. The implication of this jargon is that since the 3-dB bandwidth is almost entirely determined by p_1, the dominant pole, p_2 is relatively unimportant or "nondominant."

The third special case is *critical damping*, for which p_1 and p_2 are real, positive, identical numbers. With $p_2 \equiv p_1$, k in Eq. (1-89) is one, whence a damping factor, from Eq. (1-86), of unity. Moreover, the undamped natural frequency, from Eq. (1-82), is now $\omega_n \equiv p_1$. Although a critically damped circuit is a stable structure, it is nonetheless undesirable for at least two reasons. The first of these reasons, which is explored later, is the fact that certain commonly encountered feedback signal paths around a critically damped circuit can incur unstable responses to bounded input excitations. The second reason is a deterioration of 3-dB bandwidth, which is the bane of design engineers tasked to realize broadbanded frequency responses. To wit, $\zeta = 1$ in Eq. (1-67) results in a 3-dB bandwidth, ω_b, of $\omega_b = 0.644 \omega_n = 0.644 p_1$, which is an almost 36% reduction from the bandwidth indigenous to a dominant pole response.

1.4.4. Time Domain Transient Responses

The frequency, phase, and delay responses of linear networks are steady state performance indices that quantify such commonly invoked metrics as I/O gain at zero or any other frequency, 3-dB bandwidth, and envelope delay at zero or another frequency of interest. Because these steady state performance barometers are relatively easy to deduce both analytically and experimentally, there is an industry-wide tendency to forego more intricate time domain characterizations of electronic networks. Unfortunately, a steady state performance characterization in the absence of companion transient response investigations documents an incomplete picture of network functionality and utility. Such a characterization is akin to walking into a theater at the concluding moments of a film; the viewer sees the ending scene without comprehending the vagaries of a compelling plot that leads to the concluding scene.

The steady state response is the output voltage or current produced after a sufficiently long time has elapsed subsequent to the time at which the input signal excitation is applied. It is literally the "concluding scene" of the circuit response. A circuit assessment limited to only the steady state fails to establish the length of time required by the circuit to achieve the steady state. It also fails to reveal the time domain nature of the *transient response*, which is a picture of the electrical response waveforms that prevail between the instant of time at which the input is applied and the time at which nominal steady state behavior is produced. These waveforms may be very slowly varying functions of time that imply an inordinately long time for the realization of steady state outputs. Or, the waveforms may significantly overshoot or undershoot the steady state response before actual steady state is achieved, thereby leading to unacceptable, potentially unstable, or even damaging electrical transients.

The impulse response and the step response are two commonly invoked tools for assessing the transient behavior of linear networks. In the subsections that follow, the impulse response of a second order oscillatory network is derived and scrutinized, as are the step responses to more general overdamped, critically damped, and underdamped second order circuits.

1.4.4.1. *Impulse Response*

The oscillatory nature of an undamped lowpass system can be established directly in the time domain through an investigation of the *impulse response* of the subject system. As is symbolically illustrated in Fig. 1.41, the impulse response, say $h(t)$, of a linear network having an *I/O* transfer function of $H(s)$ is the time domain output generated as a result of an applied impulse input, say $\delta(t)$. An impulsive source in the time domain is a pulse of infinitely large amplitude and zero time duration enclosing precisely unity area. This area is related to the energy said waveform delivers to the network port

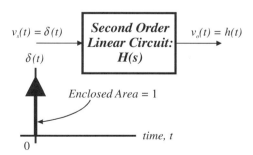

Figure 1.41. An abstraction of a second order circuit excited at its input port by a unit impulse of energy.

it drives. Idealized impulsive inputs are physically unrealizable, but mathematically, they forge a useful model for assessing the manner in which a linear system responds to abrupt excitation whose time duration is very short. For example, the laser tracking system within a compact disc player may encounter a speck of dust or a scratch on the disc media. This environmental parasitic causes a momentary undesirable input signal for which the system response hopefully abates quickly and inconsequentially. Communication systems also suffer from impulsive-like inputs when, for example, a fraction of the energy released by local lightening electromagnetically couples to the system antenna. In short, impulse responses are an ideal, but mathematically effective, way of gauging the impact exerted on a linear system by undesirably large and abrupt input energies.

For the undamped second order situation implied by $\zeta = 0$, the transfer function in Eq. (1-55) collapses to

$$H(s)|_{\zeta=0} = \frac{V_o(s)}{V_s(s)} = \frac{H(0)}{1 + \frac{s^2}{\omega_n^2}}, \qquad (1\text{-}90)$$

whose poles lie on the $j\omega$-axis at $s = \pm j\omega_n$. Since the Laplace transform of a unit impulse function is unity, the transform of the unit impulse response is simply the applicable transfer function. Accordingly the undamped time domain impulse response, say $h_o(t)$, is the inverse transform of the function appearing on the right hand side of Eq. (1-90). In particular,

$$v_o(t) \stackrel{\Delta}{=} h_o(t) = H(0)\omega_n \sin(\omega_n t); \qquad (1\text{-}91)$$

that is, the impulse response of interest is a bounded sinusoid whose radial frequency is the undamped natural frequency, ω_n, of the considered system. The curiosity here is that for time $t > 0$, the input energy, as is depicted in Fig. 1.41, is zero. Accordingly, the steady state gain, which is the amplitude, $H(0)\omega_n$, of the output sinusoid at frequency ω_n, divided by the input signal (which is zero for all nonzero time) is infinitely large. This deduction is confirmed by Eq. (1-90), which verifies infinite gain in the steady state at frequency ω_n, where s can be equated to $j\omega_n$. If frequency ω_n lies within the audio spectrum, for example, a stereo amplifier regrettably characterized by zero damping under certain operating conditions produces a piercing whistling tone in its speakers in response to a single, sharp beat of a drum in a musical passage recorded on a compact disc. More generally, the determination of the steady state response of an undamped linear network to any type of input is a pointless undertaking. The reason underlying this contention is that in the immediate neighborhood of the instant of time at which input signal is applied, said input emulates an impulse that produces a sinusoidal background response whose amplitude is not proportional to the steady state input signal amplitude. In a word, the network ceases to emulate input-to-output linearity.

Two other points surrounding Eq. (1-91) are noteworthy. The first of these points, and the one easiest to understand, is the explanation of why parameter ω_n is commonly referred to as the "undamped natural frequency of oscillation" for a second order system. In particular, note that zero damping not only gives rise to a sinusoidal impulse response, it produces an output sinusoid whose radial frequency is exactly ω_n. As such, ω_n is a natural resonant frequency evidenced only when zero damping (hence, "undamped") prevails in the system undergoing study.

The second point is more abstract but nonetheless important conceptually. In particular, the impulse response in Eq. (1-91) is an eternal sinusoid or equivalently, a sinusoid whose amplitude never diminishes. This eternal oscillation prevails despite the fact that the input giving rise to this response is zero for all times immediately subsequent to the application of said input. In other words, the output immediately after input application requires no input. Moreover, ostensibly nothing within the network serves to diminish the energy implicit to the sinusoidal response, which means that the subject network behaves as an ideal lossless entity. It follows that the damping factor, ζ, in an electrical or electronic circuit is a measure of the losses incurred by the resistances embedded within the I/O signal flow path of the circuit. Zero damping corresponds to zero energy losses and thus, no effective resistances in the signal flow path. Conversely, damping factors larger than zero imply a progressively more lossy circuit.

In any practical circuit, losses are inevitable. Thus, eternal oscillations cannot be sustained in a simple inductor-capacitor tank circuit because practical inductors have parasitic series resistances and practical capacitors have unavoidable shunt resistances. Accordingly, circuits expressly designed to behave as sinusoidal oscillators must exploit electronic amplifiers that utilize such devices as bipolar or metal-oxide-semiconductor (MOS) transistors. To be sure, these amplifiers supply gain when appropriately biased for nominally linear operation. But they can also establish requisite negative resistances that effectively cancel the net positive resistance implicit to I/O signal paths. As a result, they serve to constrain the effective network damping factor to zero, thereby conducing sinusoidal responses to virtually any form of input excitation (such as the turn on transient associated with switching in the batteries that bias the electronic circuits or the electrostatic noise coupled to network input ports by Aunt Milly's kitchen mixer). Once generated, the sinusoidal output response continues until the biasing power required to linearize embedded amplifying networks is removed or otherwise switched off. In short, sinusoidal input responses can never be generated in passive circuits, but they can be supported in active architectures that are configured to produce appropriate amounts of effective negative resistances.

1.4.4.2. *Step Response*

For practical, nonimpulsive inputs that are applied suddenly to a linear network, the energy storage elements within said network (and implicit to the interconnected passive and active components embedded within the electrical network) prohibit an instantaneous realization of steady state output responses. As a result, the *settling time* of a circuit, which is the time (measured immediately after input energy application) required to reach and maintain steady state output behavior to within an acceptable error tolerance, bodes obvious design-oriented interest. It is futile to deduce settling times for all possible input voltage and current waveforms. To this end, the unit step has been adopted as the applicable standard test vehicle for settling time delineation. As is abstracted in Fig. 1.42, the unit step of applied voltage or current changes instantaneously from zero value to unit value at an arbitrary time which, for convenience, can be taken as time $t = 0$. The step input arguably establishes a worst case measure of settling time since any "real" excitation, which cannot slew instantaneously at its time point of

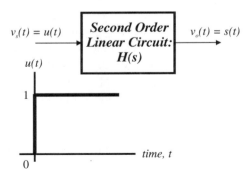

Figure 1.42. An abstraction of a second order circuit excited at its input port by a unit step of energy.

application, inherently provides the considered system with time to react. In other words, practical input waveforms offer the system a chance to track faithfully the applied excitation, thereby masking settling transients. Since the step input offers no such reaction opportunity at its point of application, an investigation of the step response and its associated settling time paints a picture of the inherent transient response limitations of the system undergoing study.

In general, the Laplace transform of the step response for the linear second order network in Fig. 1.42 is

$$V_o(s) = \mathcal{L}[s(t)] = \frac{H(0)}{s\left(1 + \frac{2\zeta s}{\omega_n} + \frac{s^2}{\omega_n^2}\right)} = \frac{H(0)}{s\left(1 + \frac{s}{p_1}\right)\left(1 + \frac{s}{p_2}\right)}, \quad (1\text{-}92)$$

where $\mathcal{L}[s(t)]$ denotes "Laplace transform of $s(t)$", Eq. (1-80) is recalled, and use is made of the fact that the Laplace transform of the unit step, $u(t)$, is $1/s$. Obviously, the time domain step response, $s(t)$, is little more than the inverse transform of either of the functional forms on the right hand side of Eq. (1-92). The consideration of three special cases expedites this inverse transformation task.

OVERDAMPED CASE

In an overdamped network, the damping factor, ζ, exceeds unity. Correspondingly, the pole frequencies, p_1 and p_2, are positive real numbers. If the pole ratio, k, in Eq. (1-89) is exploited, it can be demonstrated that the overdamped step response, say $s_o(t)$, normalized to the zero frequency gain, $H(0)$, is

$$\frac{s_o(t)}{H(0)} = 1 - \left(\frac{k}{k-1}\right)e^{-p_1 t} + \left(\frac{1}{k-1}\right)e^{-kp_1 t}, \quad t \geq 0+. \quad (1\text{-}93)$$

Figure 1.43 displays this step response as a function of the normalized time, $p_1 t$, for various values of the pole ratio, k. The plot at hand clearly shows a monotonically rising response over time. Moreover, it shows that progressively larger values of k result in faster responses and hence, reduced settling times. If the settling time, say t_s, is formally defined to be the time

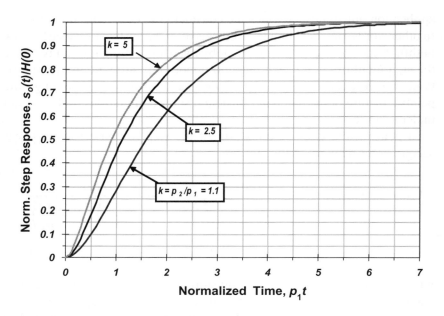

Figure 1.43. The normalized step response of an overdamped, linear, second order network for various values of the pole ratio, $k = p_2/p_1$.

required for the step response to rise to within 95% of its normalized steady state value of one,

$$\frac{s_o(t_s)}{H(0)} = 1 - \left(\frac{k}{k-1}\right)e^{-p_1 t_s} + \left(\frac{1}{k-1}\right)e^{-kp_1 t_s} \stackrel{\Delta}{=} 0.95. \quad (1\text{-}94)$$

For generalized k, this relationship requires an iterative numerical solution. To wit, $p_1 t_s = 4.53$ for $k = 1.1$, $p_1 t_s = 3.50$ for $k = 2.5$, and $p_1 t_s = 3.22$ for $k = 5$. Thus, the 95% settling time for $k = 5$ is about 41% smaller than the settling time with $k = 1.1$. The fact that it is only about 8.7% smaller than the $k = 2.5$ settling time suggests that a point of diminishing returns is reached as attempts are made to displace the less dominant pole to progressively higher frequencies.

For a network characterized by a dominant pole response, the pole at frequency $p_2 = kp_1$ is relatively inconsequential. The second term on the right hand side of Eq. (1-94) is resultantly negligible, thereby precipitating the approximate closed form solution,

$$p_1 t_s \approx 3 + \ln\left(\frac{k}{k-1}\right) \approx 3. \quad (1\text{-}95)$$

Note that the settling time for very large k differs from that of $k = 5$ by only 7.3%.

CRITICALLY DAMPED CASE

For critical damping, the damping factor, ζ, is unity, and the pole frequencies, p_1 and p_2, are positive, real, and identical numbers. The resultant step response, say $s_c(t)$, normalized to the zero frequency gain, $H(0)$, is

$$\frac{s_c(t)}{H(0)} = 1 - (1 + p_1 t) e^{-p_1 t}, \quad t \geq 0+. \tag{1-96}$$

Figure 1.44 plots this critically damped step response against the normalized time, $p_1 t$. An iterative numerical solution of Eq. (1-96) for the 95% settling time yields

$$p_1 t_s \approx 4.75, \tag{1-96}$$

which indicates a settling time that is better than 58% larger than the settling time indigenous to a dominant pole network. The significance of this larger settling time can be underscored through consideration of a hypothetical

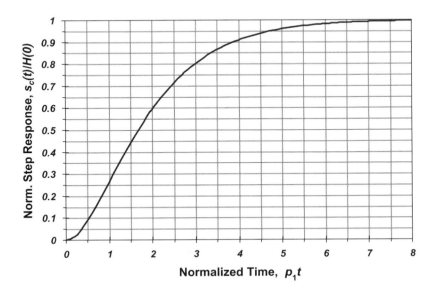

Figure 1.44. The normalized step response of a critically damped, linear, second order network plotted as a function of the normalized time variable, $p_1 t$.

circumstance in which a dominant pole network and a critically damped network are to produce identical 95% settling times. In this situation, the frequencies of the two identical poles in the critically damped configuration must be 58% larger than the frequency of the lone significant pole in the dominant pole system; that is, the critically damped network must be substantively broadbanded. As the reader ultimately learns, broadbanding a single pole, yet alone two poles, is rarely a trivial exercise, particularly since individual poles of a linear network are invariably dependent on the same, or parametrically related, circuit variables. Other reasons, such as stability issues that arise when global feedback is connected around a network, also discourage the exploitation of critical damping scenarios in electronic networks.

UNDERDAMPED CASE

An underdamped network having a bounded output step response has $0 \leq \zeta < 1$, which corresponds to complex conjugate poles having non-negative real parts. In this case, the step response, say $s_u(t)$, is the damped sinusoid,

$$\frac{s_u(t)}{H(0)} = 1 - \frac{e^{-\zeta \omega_n t}}{\sqrt{1-\zeta^2}} \sin\left[\sqrt{1-\zeta^2}\,(\omega_n t) + \cos^{-1}(\zeta)\right], \quad t \geq 0+. \tag{1-97}$$

The time domain nature of this function is dramatized in Fig. 1.45, which depicts step responses displaying potentially significant overshoot and undershoot of the steady state response value, depending on the value of the damping factor.

The nonmonotonic nature of the underdamped step response complicates the task of delineating the settling time. Before attempting to discern the time required for the response to achieve and maintain 95%, or any other percentage, of its steady state value, it is useful to note that the first term on the right hand side of Eq. (1-97) is indeed the normalized steady state output. Accordingly, the second term on the right hand side of the subject relationship can be viewed as an error response, $\varepsilon(t)$, such that

$$\frac{s_u(t)}{H(0)} = 1 + \varepsilon(t), \tag{1-98}$$

where

$$\varepsilon(t) = -\frac{e^{-\zeta \omega_n t}}{\sqrt{1-\zeta^2}} \sin\left[\sqrt{1-\zeta^2}\,(\omega_n t) + \cos^{-1}(\zeta)\right] \tag{1-99}$$

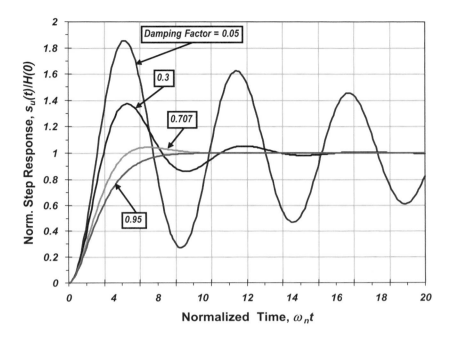

Figure 1.45. The normalized step response of an underdamped, linear, second order network plotted for various values of the damping factor, ζ, as a function of the normalized time variable, $\omega_n t$.

is plotted in Fig. 1.46. Note that for $\zeta > 0$, the amplitude of this error function diminishes with increasing time. It follows that a plausible analytical strategy for determining the settling time is to set the time slope, $d\varepsilon(t)/dt$, to zero in order to determine the time, say t_m, corresponding to the first error maximum beyond zero time. Ensuring that the maximum, or peak, error corresponding to this time lies below an acceptable value assures acceptably small response errors at any other time. The conduct of these messy tasks results in

$$\omega_n t_m = \frac{\pi}{\sqrt{1-\zeta^2}}, \tag{1-100}$$

and

$$\varepsilon_m = e^{-\zeta \omega_n t_m} = \exp\left(-\frac{\zeta \pi}{\sqrt{1-\zeta^2}}\right). \tag{1-101}$$

Figure 1.46 depicts $\omega_n t_m$ and ε_m for the case of a damping factor of $\zeta = 0.05$; in particular, $\omega_n t_m = 3.15$ and $\varepsilon_m = 0.85$. It follows that if ε_m is the tolerable

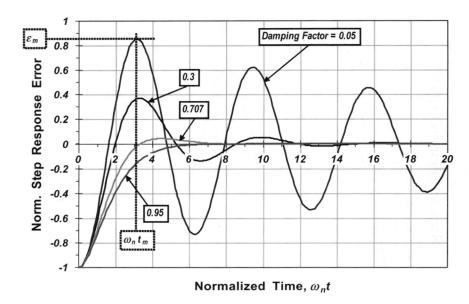

Figure 1.46. The normalized error response, with respect to the steady state output, of an underdamped, linear, second order network plotted for various values of the damping factor, ζ, as a function of the normalized time variable, $\omega_n t$. The peak error, ε_m, and the time, $\omega_n t_m$, corresponding to this peak are specifically delineated for the case of a damping factor of $\zeta = 0.05$.

maximum error associated with the settling time, t_s, Eq. (1-101) sets the requisite damping factor, which establishes the normalized settling time, $\omega_n t_s$, in Eq. (1-100).

Example 1.6. Assume that the voltage transfer function of a preamplifier of one channel of a stereo system is the second order relationship given by Eq. (1-55). This preamplifier is to be designed for 95% step response settling at a time that does not exceed the period associated with the theoretic upper frequency limit of the audio spectrum. What is the required 3-dB bandwidth of the amplifier?

Solution 1.6.

(1) If settling to within 95% of the steady state step response value is the required performance specification, ε_m in Eq. (1-101) must satisfy $\varepsilon_m \leq 0.05$. The corresponding damping requirement is therefore found to be $\zeta \geq 0.6901$.
(2) For a damping factor of 0.6901 Eq. (1-67) confirms a 3-dB bandwidth of

$$\frac{f_b}{f_n} = 1.024. \qquad (E6\text{-}1)$$

(3) The upper frequency limit of audio responses is 20 KHz. In accordance with the performance requirements of the amplifier at hand, the settling time must be no larger than $t_s = 1/2\pi\ (20\,\text{KHz}) = 7.958\ \mu\text{SEC}$. Using Eq. (1-100), the self-resonant frequency of the amplifier is

$$f_n = \frac{1}{2t_s\sqrt{1-\zeta^2}} \geq 86.82\,\text{KHz}. \tag{E6-2}$$

(4) Combining the foregoing two results, the requisite 3-dB bandwidth, f_b, must be at least as large as

$$f_b = (1.024)(86.82\,\text{KHz}) = 88.9\,\text{KHz}.$$

Comments. Superior performance in at least the senses of very short settling time and stringent settling error demands high bandwidth. It should be noted that the damping requirement herewith is not consistent with either maximally flat magnitude or maximally flat delay responses in the steady state. This observation underscores the necessity of investigating both the transient and the steady state responses in any electronic circuit and system design scenario. It also highlights the perennial need for design compromises. In this particular case, and in the absence of any compensation invoked on the second order transfer characteristic, decisions are mandated to ascertain whether settling time, maximally flat frequency response, maximally flat delay response, or some other performance metric comprises the dominantly important design theme.

Exercises

Problem 1.1
Under commonly encountered operating conditions, Fig. P1.1 is a valid linearized equivalent circuit of a voltage buffer realized in MOSFET device

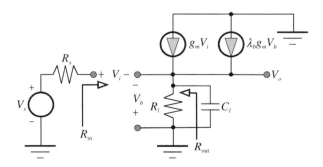

Figure P1.1.

technology. The input signal source is represented by its Thévenin equivalent circuit, which consists of voltage source V_s and resistance R_s. The response to this input signal is the indicated voltage, V_o, which is developed across the shunt interconnection of load resistance R_l and load capacitance C_l. The actual MOS transistor is modeled by the two voltage controlled current sources, $g_m V_i$ and $\lambda_b g_m V_b$, where g_m (typically of the order of hundreds of micromhos to a few millimhos) is the forward transconductance of the transistor, and λ_b (a dimensionless number often smaller than 0.1) emulates the impact exerted by the substrate on device forward transfer characteristics. Note that regardless of the nature of the transistor parameters, the model in Fig. P1.1 is a linear active circuit, not unlike the circuits addressed in this chapter.

(a) Determine, and express as a function of V_s, the Thévenin equivalent voltage, say V_{ot}, that drives the load capacitor, C_l. Simplify the expression for V_{ot} for the special case of an infinitely large load resistance, R_l. If R_l were to be omitted from the diagram in Fig. P1.1, would the resultant expression for V_{ot} be identical to the originally derived expression?

(b) What is the low frequency value of the voltage gain, V_o/V_s, of the circuit and how does this gain relate to the ratio, V_{ot}/V_s?

(c) Derive an expression for the Thévenin equivalent resistance, R_{out}, facing capacitance C_l.

(d) Derive an expression for the low frequency input resistance, R_{in}, "seen" by the signal source.

(e) What is the significance of the time constant, $R_{out}C_l$, to the frequency domain transfer function, $H(j\omega) = V_o(j\omega)/V_s(j\omega)$? Give an expression for this transfer relationship in terms of V_{ot}/V_s and the subject time constant.

(f) Give a simple expression for the 3-dB bandwidth of the circuit.

(g) Is there anything interesting about the gain bandwidth product, which is cleverly defined as the product of the magnitude of zero frequency gain and 3-dB bandwidth?

(h) Take $R_s = 300\,\Omega$, $R_l = 1,000\,\Omega$, $g_m = 5$ mmho, $\lambda_b = 0.08$, and $C_l = 8$ pF. Calculate the low frequency voltage gain, the output resistance, the time constant of the circuit, and the circuit 3-dB bandwidth.

Problem 1.2
Under commonly encountered operating conditions, Fig. P1.2 is a valid linearized equivalent circuit of a voltage amplifier realized in bipolar junction

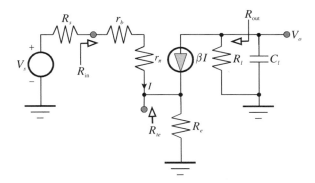

Figure P1.2.

transistor (BJT) device technology. The input signal source is represented by its Thévenin equivalent circuit, which consists of voltage source V_s and resistance R_s. The output, or response, to this input signal is the indicated voltage, V_o, which is developed across the shunt interconnection of load resistance R_l and load capacitance C_l. The actual BJT is modeled by the current controlled current source, βI, and the two resistances, r_b and r_π. Typically, β, which is dimensionless, is of the order of 100 or so, r_b can be as large as 200 Ω, and r_π is of the order of a few thousand ohms. The resistance, R_e, is a circuit element used for biasing and linearity purposes. It is generally chosen to be of the order of fifty to a few hundred ohms.

(a) Determine, and express as a function of V_s, the Thévenin equivalent voltage, say V_{ot}, that drives the load capacitor, C_l. Simplify the expression for V_{ot} for the special case of a very large current gain parameter, β.
(b) What is the low frequency value of the voltage gain, V_o/V_s, of the circuit and how does this gain relate to the ratio, V_{ot}/V_s?
(c) Derive an expression for the Thévenin equivalent resistance, R_{out}, facing capacitance C_l.
(d) Derive an expression for the low frequency input resistance, R_{in}, "seen" by the signal source.
(e) Derive an expression for the net effective resistance, say R_{te}, established across the terminals where resistance R_e is connected.
(f) What is the significance of the time constant, $R_{out}C_l$, to the frequency domain transfer function, $H(j\omega) = V_o(j\omega)/V_s(j\omega)$? Give an expression for this transfer relationship in terms of V_{ot}/V_s and the subject time constant.

(g) Give a simple expression for the 3-dB bandwidth of the circuit.
(h) Take $R_s = 300\,\Omega$, $R_l = 1,000\,\Omega$, $\beta = 120$, $r_b = 190\,\Omega$, $r_\pi = 1.5\,\text{K}\Omega$, $R_e = 100\,\Omega$, and $C_l = 8\,\text{pF}$. Calculate the low frequency voltage gain, the output resistance, the time constant of the circuit, the circuit 3-dB bandwidth, and the resistance parameter, R_{te}.

Problem 1.3

Consider the simple RLC circuit in Fig. P1.3, which can be viewed as a simplified model of the high frequency parasitics that underlie an interconnect between two integrated circuits on a circuit board. Interconnect lines have unavoidable distributed resistance, inductance, and capacitance that slow output responses to rapidly applied inputs. In extreme cases, these high frequency parasitics can incur undesirable oscillations or, depending on the electrical nature of the circuits they couple together, outright instability. You may unfortunately view this and the next problem as entailing significant mathematical "busy work", but the problems herewith are very practical and are commonly addressed by integrated circuit designers.

(a) The quality factor, Q of the circuit at hand is the ratio of the reactance of the inductor to the series resistance at the resonant frequency, say ω_o, of the circuit. Show that Q is given by

$$Q = \frac{1}{\omega_o RC} = \frac{1}{R}\sqrt{\frac{L}{C}}.$$

(b) Derive expressions for the transfer functions, $V_r(j\omega_o)/V_s(j\omega_o)$, $V_l(j\omega_o)/V_s(j\omega_o)$, and $V_c(j\omega_o)/V_s(j\omega_o)$. Use these functions to demonstrate that the magnitudes of the capacitor voltage, V_c, and the inductor voltage, V_l, are Q-times larger than the magnitude of the source voltage, V_s, at the resonant frequency of the circuit.

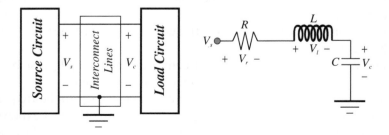

Figure P1.3.

(c) In terms of Q and ω_o, determine the 3-dB bandwidth, say ω_b, of the circuit transfer function, $V_c(j\omega_o)/V_s(j\omega_o)$. Using suitable software, plot the normalized bandwidth, ω_b/ω_o, versus Q for $0 < Q \le 6$.

(d) Show that in the steady state and at circuit resonance, the energy delivered to the inductor is the negative of the energy delivered to the capacitor. Give an engineering interpretation of this observation.

Problem 1.4

Reconsider the circuit of Fig. P1.3 under the condition that the source voltage, V_s, is an idealized unit step function. Moreover, take the capacitor voltage, V_c, as the response to this unit step excitation. In an ideal interconnect between two circuits, it is desirable that the output (V_c) respond instantaneously to the applied input. Clearly, this type of response is unrealizable because the capacitor prohibits instantaneous voltage changes. But in the steady state, the capacitor behaves as an open circuit and the inductor emulates a short circuit, thereby ultimately allowing the output to follow faithfully the applied input. This ability to follow the input is a desirable trait, but questions must be raised as to how much time elapses before steady state operating conditions are closely emulated.

(a) Show that the transfer function, say $H(s)$, of the circuit is of the form,

$$H(s) = \frac{V_c(s)}{V_s(s)} = \frac{H(0)}{1 + \left(\dfrac{2\zeta}{\omega_n}\right)s + \left(\dfrac{s}{\omega_n}\right)^2}.$$

Provide analytical expressions for $H(0)$, the damping factor, ζ, and the undamped natural frequency, ω_n, and give engineering interpretations of each of these parameters. Relate ζ and ω_n to Q and ω_o, respectively, as introduced in the preceding problem.

(b) What are the initial and steady state time domain values of the capacitor voltage response, $v_c(t)$?

(c) Assume that the circuit is underdamped; that is, $\zeta < 1$. Determine the time domain capacitor voltage, $v_c(t)$, and cast this voltage in the form,

$$v_c(t) = v_c(\infty) - v_e(t),$$

where $v_e(t)$ can be interpreted as an "error" signal between the steady state, or ultimately desired, response and the actual time domain response. Use suitable software to plot the error signal versus the normalized time, $\omega_n t$, for damping factor, ζ, values of $0.25, 0.5, 1/\sqrt{2}$, and 0.9.

(d) The one percent settling time, t_s, is the time required for the magnitude of the unit step response to achieve and forever maintain its steady state value to within $\pm 1\%$; that is,

$$|v_e(t_s)| \leq 0.01|v_c(\infty)|.$$

Derive a relationship for this settling time in terms of damping factor.

(e) What is the minimum damping factor commensurate with an error signal that is never any larger than one per cent of the steady state response? For a 1% settling time of 1 nSEC, what is the minimum tolerable circuit resonant frequency?

Problem 1.5

The circuit depicted in Fig. P1.5 utilizes three ideal transconductor amplifiers to realize a bandpass filter whose center frequency (in units of radians/sec) is ω_o and whose quality factor is Q. Note that two of the transconductors have identical transconductances, g_m, while the third unit has a transconductance of g_{m3}.

(a) Derive a generalized expression for the transfer function, $A_v(s) = V_o/V_s$.

(b) From the transfer function expression derived in Part (a), provide general expressions for the center frequency, ω_o, and the quality factor, Q, of the bandpass filter.

(c) What is the voltage gain at the center frequency of the filter?

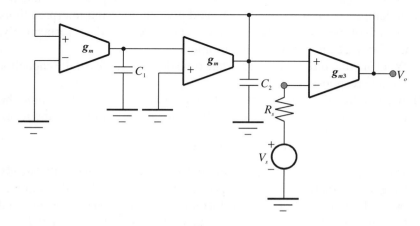

Figure P1.5.

(d) Assume that the transconductances, g_m and g_{m3}, are electronically adjustable. Can transconductor adjustments be made to control the center frequency and quality factor independently?
(e) Is it advantageous to control center frequency and quality factor independently in a commercial radio application of the filter? Explain briefly.

Problem 1.6
The circuit in Fig. P1.6 is a model of a commonly utilized amplifier that is compensated to ensure stable performance at high signal frequencies.

(a) Derive a generalized expression for the low frequency voltage gain, $A_v(0) \triangleq A_{vo} = V_o/V_s$.
(b) Derive an expression for the time constant attributed to the pole incurred by the indicated capacitance, C.
(c) If the transconductance parameter, g_m, can be varied at will, what is the maximum attainable gain-bandwidth product of the circuit?
(d) Derive an expression for the driving point input impedance seen by the signal source comprised of Thévenin voltage V_s and Thévenin resistance, R_s.
(e) Derive an expression for the driving point output impedance seen by the load resistance, R_l.

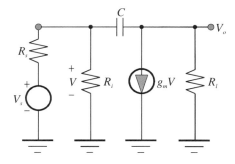

Figure P1.6.

Problem 1.7
Under very high frequency operating conditions, Fig. P1.7 is a reasonable approximation of the equivalent circuit of a tuned amplifier realized in submicron metal-oxide-semiconductor field-effect transistor (MOSFET) device technology. The indicated circuit architecture is a simplified version of a radio frequency (RF) amplifier commonly utilized in the front end of a

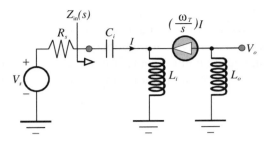

Figure P1.7.

radio receiver or cellular telephone. The input signal source is represented by its Thévenin equivalent circuit, which consists of voltage source V_s and resistance R_s. In an RF application, the Thévenin resistance, R_s, generally represents the characteristic impedance of the transmission line that couples the antenna signal source to the amplifier input port. The output, or response, to the input signal, V_s, is the indicated voltage, V_o, which is developed across the load inductance L_o. The actual MOSFET is modeled by the frequency dependent current controlled current source, $(\omega_T/s)I$, and the capacitance, C_i. Typically, ω_T is of the order of the mid-tens of gigaradians/sec, while C_i is typically in the range of the mid-tens of femtofarads. The inductance, L_i, is a circuit element that is exploited to achieve maximum power transfer between the applied input signal and the amplifier input port, whose input impedance is delineated as $Z_{in}(s)$. Note that regardless of the nature and numerical value of the transistor and circuit parameters, the model in Fig. P1.7 comprises a linear circuit.

(a) Show that the indicated input impedance, $Z_{in}(s)$, is expressible as,

$$Z_{in}(s) = R_{\text{eff}} + sL_{\text{eff}} + \frac{1}{sC_{\text{eff}}}.$$

Give, in terms of C_i, L_i, and ω_T, expressions for the effective input resistance, inductance, and capacitance, R_{eff}, L_{eff}, and C_{eff}, respectively.

(b) Let the resonant frequency of the input impedance be denoted as ω_i. What is ω_i in terms of inductance L_i and capacitance C_i? What design condition must be satisfied at the resonant frequency to achieve a match terminated input port; that is, $Z_{in}(j\omega_i) \equiv R_s$?

(c) Show that under steady state sinusoidal operating conditions and the match terminated constraint focused upon in the preceding part of this problem, the voltage gain of the RF amplifier can be written in the form,

$$A_v(j\omega) = \frac{V_0}{V_s} = -\frac{L_0/2L_i}{1 + jQ\left(\dfrac{\omega}{\omega_i} - \dfrac{\omega_i}{\omega}\right)},$$

where Q is the quality factor associated with the input amplifier port at the resonant frequency, ω_i.

(d) With a source resistance, R_s, of 50 Ω, a desired tuned center frequency, ω_i, of $2\pi(1200\,\text{MHz})$, and a transistor that has $\omega_T = 2\pi(20\,\text{GHz})$, compute the requisite values of output inductance, L_o, tuning inductance, L_i, and circuit quality factor, Q for a tuned center frequency gain magnitude of 20 dB.

Problem 1.8

The circuit in Fig. P1.8 uses two transconductors and two capacitors to realize a notch filter. Notch filters are often used in communication networks whenever an undesired input signal at a known frequency, say ω_o, must be sharply attenuated or even eliminated from the communication channel. Accordingly, an ideal notch filter delivers nonzero transfer function at both very low and very high signal frequencies and zero transfer function at the undesired frequency, ω_o.

(a) Derive a generalized expression for the transfer function, $A_v(s) = V_o/V_s$.
(b) What is the notch frequency of the filter?

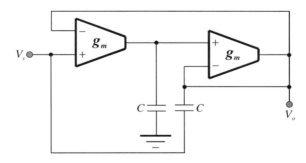

Figure P1.8.

(c) What is the value of the filter transfer function at both very low and very high frequencies?

Problem 1.9

The circuit in Fig. P1.9(a) is an equivalent circuit for a transconductance amplifier whose output port is terminated in a shunt interconnection of a load resistance, R_L, and a load capacitance, C_L. This equivalent circuit is to be reduced to the Norton architecture shown in Fig. P1.9(b), where the Norton transadmittance, $Y_n(s)$, is understood to be a function of frequency and pertinent circuit parameters.

(a) Derive an expression for the Norton transadmittance function, $Y_n(s)$.
(b) Derive an expression for the indicated Thévenin impedance, $Z_{th}(s)$.
(c) The capacitance, C_{ss}, creates both a left half plane pole and a left half plane zero in the voltage transfer function, V_o/V_s. If the time constant associated with the left half plane zero established by C_{ss} is selected to cancel, the time constant, $R_L C_L$, of the shunt load, give an expression for the resultant 3-dB bandwidth of the circuit.

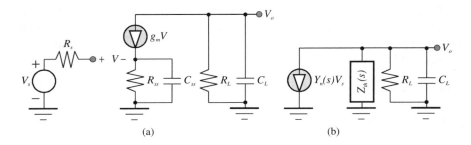

Figure P1.9.

Problem 1.10

Amplifiers are commonly exploited in monolithic analog technologies to synthesize effective resistances whose values can be controlled by suitable biasing voltages. A case in point is the equivalent circuit of such a structure offered in Fig. P1.10. Determine the effective resistance, say R_{Leff}, established by the circuit between Node ① and ground.

Circuit and System Fundamentals

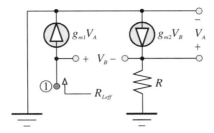

Figure P1.10.

Problem 1.11

Figure P1.11(a) is a valid linearized equivalent circuit of a voltage amplifier realized in bipolar junction transistor (BJT) device technology. The input signal source is represented by its Thévenin equivalent circuit, which consists of voltage source V_s and resistance R_s. The output response to this input signal is the indicated voltage, V_o, which is developed across the shunt interconnection of load resistance R_l and load capacitance C_l. The actual BJT is modeled by the current controlled current source, βI, and the three resistances, r_o, r_b, and r_π. The resistance, R_e, is a circuit element used for biasing and linearity purposes. It is generally chosen to be of the order of fifty to a few hundred ohms.

(a) Derive expressions for the Norton parameters, transconductance G_{sn} and resistance R_{out}, for the output port Norton equivalent circuit shown in Fig. P1.11(b).

(b) In terms of the aforementioned Norton parameters and the load variables, R_l and C_l, derive an expression for the overall voltage gain, $A_v(s) = V_o/V_s$.

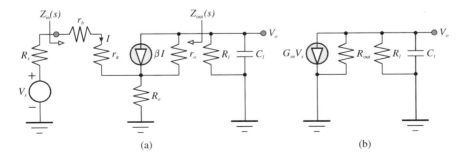

Figure P1.11.

(c) Using the results of the preceding two parts of this problem, find the low frequency value of the voltage gain, V_o/V_s, of the circuit. Simplify this gain expression for the case of large r_o and large β.
(d) In terms of R_{out} and R_l, give an expression for the 3-dB bandwidth, say ω_b, of the circuit. Approximate this result for the case of large r_o and large β.
(e) Derive an expression for the low frequency input resistance, $Z_{in}(0)$ ΔR_{in}, seen by the entire signal source circuit. Simplify this expression for the case of large r_o.
(f) Take $R_s = 300\,\Omega$, $R_l = 1\,\text{K}\Omega$, $\beta = 100$, $r_b = 190\,\Omega$, $r_\pi = 1.5\,\text{K}\Omega$, $r_o = 80\,\text{K}\Omega$, $R_e = 100\,\Omega$, and $C_l = 10\,\text{pF}$. Calculate the exact and the approximate values of the Norton transconductance, G_{sn}, the low frequency voltage gain, $A_v(0)$, the output resistance, R_{out}, the low frequency input resistance, R_{in}, and the circuit 3-dB bandwidth, ω_b. Compare respective exact and approximate computations by calculating percentage errors of the individual approximations.

Problem 1.12
The transfer function of the circuit studied in Problem 1.11 is expressible in the form,

$$A_v(s) = \frac{A_v(0)}{1 + s/\omega_b}.$$

(a) Determine the delay response $D(\omega)$ and the zero frequency value $D(0)$, of the input-to-output (I/O) delay in terms of the 3-dB bandwidth, ω_b.
(b) In terms of ω_b, what is the signal frequency, say ω_d, at which the delay is degraded from its zero frequency value by a factor of the square root of two?

Problem 1.13
The circuit model of the amplifier shown symbolically in Fig. P1.13(a) is the structure depicted in Fig. P1.13(b). The amplifier in question is utilized in the system offered in Fig. P1.13(c).

(a) Derive an expression for the Thévenin voltage gain seen by the terminating load resistance, R_l.
(b) Derive an expression for the Thévenin output resistance seen by the terminating load resistance.
(c) Simplify the expressions determined in the foregoing two parts of this problem for the case of small R_o and large A_o.

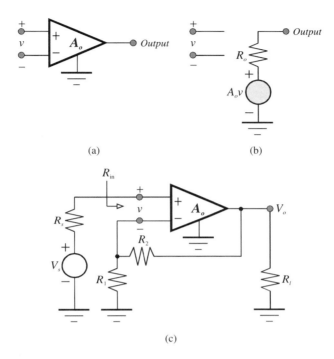

Figure P1.13.

(d) What is the driving point input resistance, R_{in}, "seen" by the applied signal source?
(e) Is the system in Fig. P1.13(c) better suited for voltage amplification, transimpedance amplification, or transconductor action?
(f) For small R_o and large A_o, how might R_2 be chosen to realize a nearly unity gain voltage buffer?

Problem 1.14
The amplifier addressed in Figs. 1.13(a) and 1.13(b) is utilized in the system offered in Fig. P1.14.

(a) Derive an expression for the Thévenin voltage gain seen by the terminating load resistance, R_l.
(b) Derive an expression for the Thévenin output resistance seen by the terminating load resistance.
(c) Simplify the expressions determined in the foregoing two parts of this problem for the case of small R_o and large A_o.

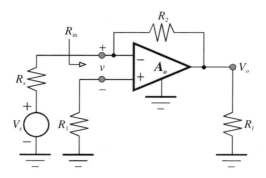

Figure P1.14.

(d) What is the driving point input resistance, R_{in}, "seen" by the applied signal source?
(e) Is the system in Fig. P1.14 better suited for voltage amplification, transimpedance amplification, or transconductor action?

Problem 1.15
Figure P1.15 depicts the schematic diagram of a two-pole lowpass filter. The two amplifiers indicated in the subject schematic representation can be viewed as ideal in the senses of delivering infinitely large input and zero output impedances. Observe that the amplifier providing a voltage gain of A_1 is a nonphase-inverting structure, while the amplifier that delivers a voltage gain magnitude of A_2 is a phase inverting unit.

(a) Derive an expression for the voltage transfer function, V_o/V_s, and cast this function in the form,

$$\frac{V_o}{V_s} = \frac{A(0)}{1 + \dfrac{s}{Q_o \omega_o} + \left(\dfrac{s}{\omega_o}\right)^2},$$

where $A(0)$ symbolizes the zero frequency gain of the circuit, Q_o is the circuit quality factor, and ω_o represents the undamped natural frequency of the circuit. Provide, in terms of R, C, A_1, and A_2, analytical expressions for $A(0)$, Q_o, and ω_o.

(b) Under what condition does the circuit become a sinusoidal oscillator? State this condition and give the corresponding oscillation frequency.
(c) What condition must be satisfied to ensure the unconditional stability of the circuit?

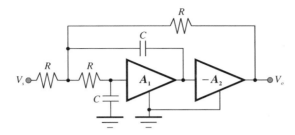

Figure P1.15.

(d) If Q_o is the inverse of *root two*, what is the 3-dB bandwidth of the resultant lowpass filter?

(e) For the condition in (d), give the pole locations of the filter.

Problem 1.16

Wideband analog and high-speed digital integrated circuits necessarily use minimal geometry transistors whose small breakdown voltages preclude their capability to sustain large collector-emitter (or drain-source) voltages over even relatively small time periods. To protect these devices from transient voltage overstress, a second order *LC* filter of the form shown in Fig. P1.16 is often inserted between the *ON/OFF* power line switch and the power supply pad of the integrated circuit. In this circuit, R_l represents the steady state load to which power is to be supplied and is nominally the ratio of the steady state load voltage to the steady state load current. Thus, if the desired quiescent pad voltage of an integrated circuit is 3.3 volts and if this circuit is to draw a quiescent current of 12 mA, $R_l = 3.3/12 \text{ mA} = 275 \, \Omega$. The filter itself consists of the inductance, L_s, which includes any parasitic

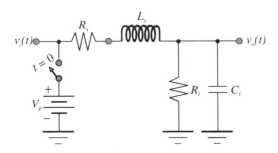

Figure P1.16.

inductance associated with the power supply bus routing on chip, and the capacitance, C_l, which includes parasitic power supply pad capacitance. The resistance, R_s is generally small and includes the effects of power bus losses and finite inductance quality factor (Q). By the way, the rubberized or plastic-coated "bump" you see in the power line that connects your laptop computer to an energy source is the inductance in Fig. P1.16. The indicated voltage, V_p is the Thévenin energizing voltage for the chip, while the switch, which is closed at *time* $t = 0$, allows the filter input voltage, $v_i(t)$, to emulate the step function, $V_p u(t)$. It is to be understood that the fundamental purpose of the filter is to slow the rate of power delivery from the input port, where $v_i(t)$ is measured, to the output port, where voltage $v_o(t)$ is established, so that $v_o(t)$ rises monotonically with time toward its steady state value with little or no voltage overshoot.

(a) The filter in Fig. P1.16 is clearly a second order circuit. In view of the discussion provided above, should the circuit poles, whose frequencies might be labeled, p_1 and p_2, be real numbers or complex conjugates? Briefly explain your rationale.

(b) Derive an expression for the transfer function, $H(s) = V_o(s)/V_i(s)$ and in the process, show that the pole frequencies satisfy the relationships,

$$\frac{1}{p_1} + \frac{1}{p_2} = \frac{L_s}{R_l + R_s} + (R_s \| R_l) C_l$$

and

$$\frac{1}{p_1 p_2} = \left(\frac{R_l}{R_l + R_s}\right) L_s C_l = H(0) L_s C_l.$$

(c) Assume that the poles are real and that their frequencies relate as $p_2 = k p_1$, where k is understood to be greater than or equal to one. For $k > 1$, show that the time domain response, normalized to the steady state value of the response, is

$$v_{on}(t) = \frac{v_o(t)}{H(0) V_p} = 1 - \left(\frac{k}{k-1}\right) e^{-p_1 t} + \left(\frac{1}{k-1}\right) e^{-k p_1 t},$$

while for $k = 1$, confirm that

$$v_{on}(t) = \frac{v_o(t)}{H(0) V_p} = 1 - (1 + p_1 t) e^{-p_1 t}.$$

(d) Plot the normalized responses determined in Part (c) versus the normalized time parameter, $t_n = p_1 t$ for $k = 1, 1.5, 3,$ and 10. What value of k might be desired to ensure the realization of the slowest possible step response for any given real number value of p_1?

(e) Let T_R represent the rise time of the filter; that is, T_R is the time required after the switch is closed for the output response to achieve 90% of its steady state value. For the optimal value of k (in the sense of a maximally slowed response) determined in Part (d), confirm that $p_1 T_R \approx 3.9$.

(f) Assume now that $R_l \gg R_s$ and $L_s \gg R_s R_l C_l$. For the optimal operating condition stipulated in Part (e), show that a rise time of T_R is achieved if

$$L_s \approx \frac{T_R(R_l + R_s)}{1.95}$$

and

$$C_l \approx \frac{T_R}{7.8 R_l}.$$

(g) Assume that a certain integrated circuit is to be energized by a 3.3-volt battery that is switched on at time $t = 0$. Assume further that the net effective Thévenin source resistance (R_s) is 15 Ω and that the effective steady state load resistance (R_l) is 1020 Ω. The latter resistance corresponds nominally to 3.3 volts delivered to a load drawing 3.23 mA. A 0 to 90% rise time (T_R) of at least 200 μ SEC is desired to protect the active devices in the given circuit. Design the protection filter and simulate it on SPICE to confirm the stipulated rise time objective.

Problem 1.17

The amplifier depicted in Fig. P1.17 has infinitely large shunt input resistance, zero Thévenin output port resistance, and a finite open loop voltage gain, A_o. The capacitance, C_i, represents the effective shunt input port

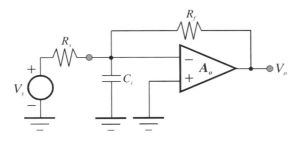

Figure P1.17.

capacitance and since no other amplifier capacitances are delineated, this capacitance is presumably the dominant energy storage element in the overall circuit. The amplifier is set up to function as an inverting buffer and accordingly, R_f is selected to equal the effective source resistance, R_s.

(a) Derive an expression for the closed loop voltage gain, $A_v(s) = V_o(s)/V_s(s)$.
(b) Derive an expression for the 3-dB bandwidth, say B, of the circuit.
(c) Derive an expression for the low frequency signal voltage, say V_i, developed across the amplifier input port and show that this voltage tends toward zero as the gain parameter, A_o, tends toward infinity.
(d) Since infinitely large open loop amplifier gains are observed only in academic environments, it is of engineering interest to investigate the response error precipitated by finite gain. To this end, define the error, ε, to be the difference between the magnitude of the input source signal voltage and the magnitude of the resultant response, V_o, under the simplifying condition of $V_s = 1$ volt. At low signal frequencies, what general condition must be satisfied by the gain parameter, A_o, if the design requirement is $\varepsilon \leq 2\%$?

Problem 1.18
Numerous signal processing applications, such as transconductance amplifiers, phase detectors, and oscillators, demand current sources and sinks characterized by extremely high resistances at their current output ports. This design requirement is a daunting challenge when frequency response objectives mandate the use of deep submicron MOS technology transistors, which are plagued by relatively small drain-source channel resistances. The circuit in Fig. P1.18(a) responds to the foregoing requirement by incorporating a feedback voltage amplifier into a traditional cascode current sink. In this exercise, assume that the amplifier is ideal in the senses of infinitely large input resistance, zero output resistance, and frequency-invariant opened loop voltage gain, A_o. The indicated voltage, V_{bias}, is constant in that it derives from a bandgap reference subcircuit or some other form of temperature stable supply.

(a) Describe qualitatively how the use of the presumably ideal amplifier encourages ideal (constant output current) current sink action.
(b) Use the small signal model of Fig. P1.18(b) to derive an expression for the indicated output resistance, R_{out}. Do not assume that the model

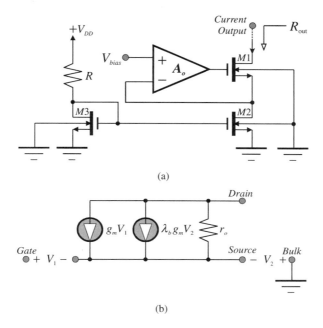

Figure P1.18.

parameters, λ_b, g_m, and r_o, are respectively identical for transistors $M1$ and $M2$.

Problem 1.19
The circuit abstracted in Fig. P1.19 is a gyrator, which has the capability of transforming capacitive load impedances, Z_l, to inductive input impedances, Z_{in}. Conversely, it can also transform inductive loads to driving point capacitive input impedances.

(a) Assuming ideal transconductors, derive a general expression for the driving point input impedance, Z_{in}.
(b) If the load impedance, Z_l, is the impedance of an inductance, say L, derive an expression for the resultant effective input capacitance, C_{in}.

Problem 1.20
An active realization of a biquadratic filter architecture is offered in Fig. P1.20, where all of the utilized transconductor amplifiers are ideal,

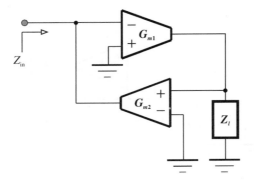

Figure P1.19.

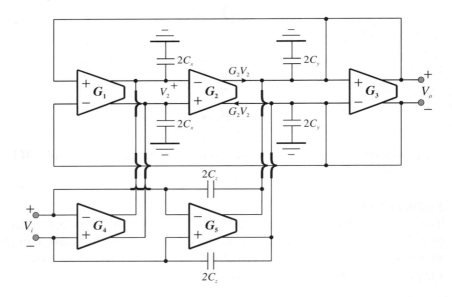

Figure P1.20.

balanced differential structures. Analyze the filter circuit to confirm that its voltage transfer function, $H(s) = V_o/V_s$, is

$$H(s) = \frac{c + bs + as^2}{1 + \dfrac{s}{Q\omega_n} + \left(\dfrac{s}{\omega_n}\right)^2}.$$

In particular, confirm that

$$a = \frac{C_x C_z}{G_1 G_2}, \quad b = \frac{G_5 C_x}{G_1 G_2}, \quad c = \frac{G_4}{\omega_n C_x},$$

$$\omega_n = \sqrt{\frac{G_1 G_2}{C_x (C_y + C_z)}},$$

and

$$Q = \sqrt{\left(\frac{G_1 G_2}{G_3^2}\right)\left(\frac{C_y + C_z}{C_x}\right)}.$$

Problem 1.21
The biquadratic filter whose topological structure is abstracted in Problem 1.20 is to be designed to realize a maximally flat, lowpass frequency response exhibiting unity gain at low signal frequencies and a 3-dB bandwidth of 800 MHz.

(a) Which transconductor(s) and which capacitor(s) can be removed from the given architecture?
(b) Design the circuit by calculating appropriate values of the remaining transconductances and capacitances. When possible, transconductance values can be equated to simplify the design methodology.
(c) Use SPICE to simulate the steady state frequency response and the time domain unit step response of the designed filter. Examine the resultant 3-dB bandwidth and compare with the design requirement. Investigate whether any overshoot observed in the step response is in agreement with theoretic predictions.

Problem 1.22
The biquadratic filter whose topological structure appears in Fig. P1.20 is to be designed to realize a bandpass frequency response exhibiting unity maximum gain at a center frequency of 800 MHz. The 3-dB bandwidth of the filter is to be 150 MHz.

(a) Which transconductance(s) and capacitance(s) can be removed from the architecture?
(b) Design the circuit by calculating appropriate values of the remaining transconductances and capacitances. When possible, transconductance values can be equated to simplify the design methodology.

(c) Use SPICE to simulate the steady state frequency response and the time domain unit step response of the designed filter. Examine the resultant 3-dB bandwidth and compare with the design requirement. Investigate the step response and provide engineering commentary on its time domain form.

Problem 1.23

The biquadratic filter whose topological structure appears in Fig. P1.20 is to be designed to realize a notch at a frequency of 800 MHz. The quality factor of the notch filter is to be at least five ($Q = 5$), and the filter is to provide unity gain magnitude at both very low and very high signal frequencies.

(a) Which transconductance(s) and capacitance(s) can be removed from the architecture?
(b) Design the circuit by calculating appropriate values of the remaining transconductances and capacitances. When possible, transconductance values can be equated to simplify the design methodology.
(c) Use SPICE to simulate the steady state frequency response and the time domain unit step response of the designed filter. Examine the resultant 3-dB bandwidth and compare with the design requirement. Investigate the step response and provide engineering commentary on its time domain form.

Problem 1.24

Reconsider the lowpass filter designed in Problem 1.21.

(a) Use SPICE to simulate the envelope delay response. What is the very low frequency value of this delay? Does this observation agree with theoretic predictions? Explain any disparity.
(b) Modify the circuit to ensure a maximally flat delay response at a value equal to the low frequency delay value observed in Part (a) of this problem.
(c) Use SPICE to simulate the steady state delay and frequency responses of the modified filter. Do the observed low frequency delay and 3-dB bandwidth (bandwidth of magnitude response) agree with theoretic predictions? Explain any disparities.

Chapter 2

Two-Port Network Models and Analysis

2.1.0. Introduction

Most electronic systems can be viewed as *two-port networks*. A network port consists of two electrical terminals, thereby suggesting that a two-port network has at least two accessible pairs of electrical terminals. One of these terminal pairs is the *input port*, to which a known signal voltage or current is applied. The network of interest processes this input excitation to produce a response at the second of the two network terminal pairs, which is called the *output port*. Basic circuit theories and analytical techniques allow the output response of a two-port network to be related to its applied input energy as a function of the volt-ampere characteristics of the individual electrical branches embedded in the network. Although this direct formulation of the output response to input signal relationship is systematic, three engineering issues limit the utility of this analytical strategy.

One engineering issue derives from the topological complexity of practical circuits. A useful electronic circuit may contain hundreds or even thousands of branch elements. The immediate effect of the ubiquity of electrical elements is an explosion of the required number of equilibrium equations that must be solved simultaneously to forge the desired input to output (I/O) transfer relationship. A manual solution is therefore daunting, if not impossible, thereby encouraging the use of the omnipresent computer. But neither cumbersome manual analyses nor computationally efficient computer-based numerical solutions are likely to inspire the engineering insights that underpin the realization of innovative and creative circuit topologies that support circuit design objectives.

A second shortcoming of direct circuit analyses is that the electrical characteristics of many internal network branches may not be defined well or understood clearly. This issue is nontrivial when the network in question contains active elements, such as bipolar junction transistors (BJTs) and metal-oxide-semiconductor field effect transistors (MOSFETs) that are incorporated for high performance processing of electrical energy. When such processing is mandated over broad frequency passbands, electrical responses are rendered vulnerable to stray capacitances, parasitic lead inductances, device processing uncertainties, and undesirable electromagnetic coupling from proximately located other electronics. To be sure, mathematical models of active elements and the aforementioned second order electrical phenomena can be constructed. But in the interest of acquiring the phenomenological understanding encouraged by analytical tractability, these macromodels are necessarily simplified to an extent that virtually ensures inaccuracies in, or even miscomprehension of, the ultimate fruits of analysis.

A final point of contention is the superfluous technical information generated as an implicit byproduct of conventional mesh, loop, and nodal analyses. The application of the Kirchhoff voltage law (KVL) and the Kirchhoff current law (KCL) produces, in addition to the desired I/O relationships, all node and branch voltages, and all mesh and loop currents intrinsic to the network undergoing examination. But in many applications, such as active filters that incorporate commercially available operational amplifiers or application specific high performance amplifiers, the primary engineering concern is the electrical relationship established between the I/O ports of the network. This relationship can be forged and exploited without an explicit awareness of the voltages and currents established within the utilized amplifiers. Indeed, the overall I/O transfer function and other performance metrics can be deduced as a function of only electrical properties gleaned from measurements executed at the I/O ports of these amplifiers. These electrical properties, or *two-port parameters*, are admittedly nonphysical entities in that they generally cannot be cast conveniently in terms of the physically sound phenomenology that underlie the internal network branches whose electrical interrelationships coalesce to produce observable I/O characteristics. Nonetheless, the parameters in question are useful because they do derive from reproducible and relatively simple port measurements, and they do inspire the formulation of analytically unambiguous overall network responses.

2.2.0. Two-Port Linearity Issues

The circuit analyses to which this chapter focuses pertains to linear two-port networks; that is, they apply to circuit configurations for which the response established at the output port is a linear function of the signal exciting the input port. This seemingly innocent backdrop is suspect because the transistors and other active devices embedded within electronic two-port networks are inherently nonlinear, which precludes the establishment of strictly linear I/O relationships. Fortunately, electronic circuits earmarked for such linear signal processing applications as amplification, impedance transformation, filtering, and buffering can be conditioned through appropriate biasing to deliver approximately linear driving point and I/O transfer characteristics. The key to this requisite biasing is the availability of one or more network ports, which may or may not be coincident with the signal input and signal output ports, to which static voltages can be applied to force the nominally linear device operation that is a prerequisite of linear signal processing.

In the electronic network abstraction of Fig. 2.1(a), biasing assumes the form of the two indicated power supply voltages, V_{CC} and V_{EE}, through which currents $i_{cc}(t)$ and $i_{ee}(t)$ respectively flow. These static voltage sources are presumed regulated so that their Thévenin resistances, which are not delineated in the diagram, are negligibly small. A signal voltage, $v_s(t)$, whose Thévenin impedance is Z_s and whose average value is zero, excites the network input port comprised of terminal pair 1-2. In response to the applied signal and biasing voltages, an output voltage, $v_o(t)$, is established across a load impedance, Z_l, which is incident with the output port defined by terminal pair 3-4. Corresponding to this output port voltage, a current, $i_o(t)$, flows through the load.

Although the responses, inclusive of $v_o(t)$ and $i_o(t)$, of unconditionally stable electronic networks are time variant, the steady state responses to the indicated biasing supplies are time invariant, or static voltages and currents that collectively comprise the *standby* or *quiescent* operating conditions of the network. Thus, with signal $v_s(t)$ nulled, which leaves only static excitations V_{CC} and V_{EE} as energy sources in the considered network, the I/O port voltages and currents are clamped to their quiescent values, V_{iQ}, V_{oQ}, I_{iQ}, and I_{oQ}. Moreover, and as is indicated in Fig. 2.1(b), the power supply currents assume their respective standby values, I_{ccQ} and I_{eeQ}, respectfully. In any electronic network, the numerical values of all of these variables depend not only on V_{CC} and V_{EE}, but also on the parameters indigenous to the nonlinear static volt-ampere characteristics of the active devices deployed in the network. The inherently nonlinear nature of the interrelationships of these static variables generally forces their numerical determination through approximate nonlinear manual analyses, circuit simulation, or other computer-assisted means.

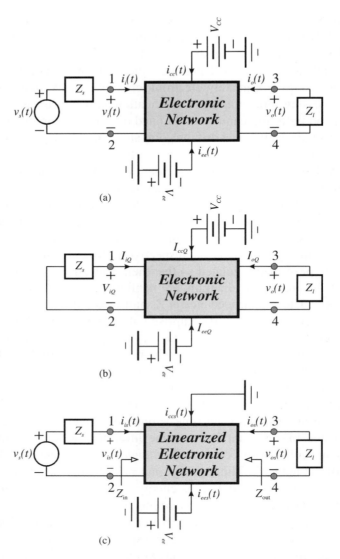

Figure 2.1. (a) An abstraction of a two-port electronic network for which two additional ports are provided for biasing. (b) The network of (a), with the input signal source nulled. The two power supplies establish the quiescent operating conditions for the network. (c) The linearized model of the network in (a). The biasing sources are nulled, and the input signal source is applied to compute the signal-induced perturbations in the port quiescent voltages and currents. The metrics, Z_{in} and Z_{out}, respectively symbolize the driving point input and output impedances.

With V_{CC} and V_{EE} sustained at their required biasing levels, the impact of an applied input signal, $v_s(t)$, can be rationalized as a perturbation in the quiescent values of all branch and node variables. A desirable attribute of the network biasing implementation is that the actual zero signal (quiescent) values of all network branch and node variables be rendered independent of these signal-induced perturbations so that classical superposition theory can be invoked with respect to the static and signal-induced, dynamic responses of the network. In other words, if quiescent responses are independent of dynamic responses, the variables indicated in Fig. 2.1(a) can be expressed in terms of those delineated in Fig. 2.1(b), which reflects null input signal energy, in accordance with

$$\left. \begin{aligned} v_o(t) &= V_{oQ} + v_{os}(t) \\ i_o(t) &= I_{oQ} + i_{os}(t) \end{aligned} \right\}, \tag{2-1}$$

$$\left. \begin{aligned} v_i(t) &= V_{iQ} + v_{is}(t) \\ i_i(t) &= I_{iQ} + i_{is}(t) \end{aligned} \right\}, \tag{2-2}$$

and

$$\left. \begin{aligned} i_{cc}(t) &= I_{ccQ} + i_{ccs}(t) \\ i_{ee}(t) &= I_{eeQ} + i_{ees}(t) \end{aligned} \right\}, \tag{2-3}$$

where the appended subscript, "s", denotes voltage or current perturbations, about a corresponding quiescent level, incurred exclusively by the signal source applied at the input port. In other words, $i_{os}(t)$ in Eq. (2-1), for example, is the change, $[i_o(t) - I_{oQ}]$, in net output port current about the quiescent value of this port current. For proper biasing, this current change does not affect the quiescent current, I_{oQ}, (or any other quiescent current or voltage in the considered network) and is identically zero when the input signal, $v_s(t)$, is set to zero.

In addition to ensuring that quiescent network variables are independent of the signal components of these variables, the biasing subcircuit must also be implemented to ensure nominally linear interrelationships among all such perturbed variables. Again by way of example, $v_o(t)$ in Eq. (2-1) cannot be expected to be a linear function of $v_i(t)$ in Eq. (2-2), principally because V_{oQ} is likely to be nonlinearly related to V_{iQ} by virtue of the fact that these two quiescent variables are nonlinear functions of the two applied network supply voltages. But the ability of the network in Fig. 2.1 to process input signals linearly does mandate that the adopted biasing scheme ensure the linear dependence of $v_{os}(t)$ on $v_{is}(t)$ (and on every other signal branch and node network variable).

The foregoing biasing assignment is a nontrivial exercise requiring that one or both of two conditions be satisfied. First, the biasing must ensure that for all signal and environmental conditions, every active device embedded in the subject network operates in a reasonably linear region of its static volt-ampere characteristics. This requirement translates into the plausibility of meaningfully approximating the Taylor series expansion, about a quiescent operating level, of the volt-ampere characteristics of an active device by strictly linear relationships. Second, all signal induced perturbations must be sufficiently small to render reasonable the linearization of the Taylor series expansions of device volt-ampere functions. Since these perturbations are linearly interrelated, the latter stipulation implies a sufficiently small input signal. Note, therefore, that low gain electronic networks, such as voltage buffers, may operate linearly despite a reasonably large input signal, while high gain circuits capable of theoretically large voltage changes at the output port mandate proportionately smaller input signal drives.

In effect, suitably designed biasing serves to linearize an electronic network in the immediate neighborhood of its quiescent operating points. This linearization allows the exploitation of classic superposition theory with respect to the problems of calculating both quiescent and dynamic branch voltages, branch currents, and node voltages. Figure 2.1(b) dramatizes this contention by depicting quiescent network variables as deriving from the conditions of zero applied signal and, of course, nonzero power supply voltages. Implicit to this circuit structure is the presumption that the ultimately applied input signal does not impact any of the standby electrical variables of the network.

Figure 2.1(c) is the second part of the superposition strategy in that it nulls the power supply voltages and applies the input signal to determine the signal-induced changes in all quiescent operating levels. Since these changes are linearly interrelated, the indicated "linearized" network is a circuit containing only linear resistors, linear capacitors, linear inductors, and/or linear current controlled or voltage controlled dependent sources. The topology of this linearized network is likely to differ from that of the original electronic network in Fig. 2.1(a) because its pertinence is confined only to a determination of signal-induced changes about standby operating points. In other words, the linearized structure is "equivalent" to the original electronic configuration only insofar as the accurate delineation of signal-induced changes in I/O port variables is concerned.

2.3.0. Generalized Two-Port Parameters

Figure 2.2 is a simplified version of the configurations offered in Fig. 2.1 in that it reflects only a generalized linear (or linearized) two-port electrical network. Because of an exclusive focus on linear I/O driving point impedance and transfer properties, the power supply ports shown in Fig. 2.1 are omitted. Moreover, the time domain notation with respect to signal source and all network electrical variables has been dropped in favor of more convenient peak, root mean square, or phasor designations. As in Fig. 2.1, the input port is formed by the terminal pair 1-2, while the output port is terminal pair 3-4. No energy sources lie within the two-port network, which implies that if any energy storage elements are embedded therein, zero state conditions apply. Energy is therefore applied to the two-port system at only its input port. In Fig. 2.2(a), this energy is represented by a Thévenin equivalent circuit comprised of the signal source voltage, V_s, and its internal series impedance, Z_s. Alternatively, the applied energy can be modeled as the Norton topology depicted in Fig. 2.2(b), where the Norton, or short circuit, equivalent input current, I_s, is

$$I_s = V_s / Z_s. \tag{2-4}$$

Because of the input signal excitation, a voltage, V_1, is established across the input port, a current, I_1, flows into this port, a current, I_2 flows into the output port, and a voltage, V_2, is developed across the output port.

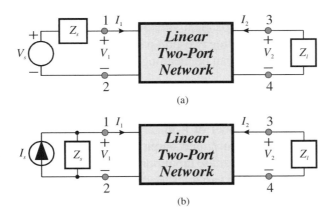

Figure 2.2. (a) A linear two-port network excited at its input port by a signal source whose Thévenin voltage is V_s and whose Thévenin impedance is Z_s. (b) A linear two-port network excited at its input port by a signal source whose Norton equivalent current is I_s and whose Thévenin impedance is Z_s.

Obviously, current I_2 and voltage V_2 are constrained by the Ohm's law relationship,

$$I_2 = -V_2/Z_l. \tag{2-5}$$

In any linear two-port system, two types of driving point specifications (input and output impedance or admittance) and four types of transfer properties (voltage gain, current gain, transimpedance, and transadmittance) can derive from the measurable input and output port voltages (V_1 and V_2) and the input and output port currents (I_1 and I_2). Moreover, these impedance, admittance, and transfer metrics can be quantified analytically even if the circuit architecture implicit to the two-port configuration of Fig. 2.2 is unknown, inaccessible, or simply too intricate for traditional circuit analyses. The upshot of the matter is that if "jumping into the two port box" to execute appropriate analyses is impossible or impractical, only two equilibrium equations can be written. One of these equations focuses on the input port, where the source energy, source impedance, and the input port variables, V_1 and I_1, reside, while the other addresses the output port, where the load impedance and the output port variables, V_2 and I_2, prevail. Since only two equations in the four variables, V_1, I_1, V_2, and I_2, can be written, the formulation of a unique network solution mandates that two of these variables be viewed as independent, and the remaining two be interpreted as dependent, variables. A viable solution also requires that V_2 and I_2 abide by the branch properties of the load termination, while V_1 and I_1 must be uniquely constrained by the source excitation and source impedance. The selection of the independent and dependent variable sets is arbitrary, subject to the proviso that the corresponding two-port parameters that define the electrical properties of the network can be meaningfully defined and measured.

2.3.1. Hybrid h-Parameters

A hybrid h-parameter model of a linear two-port network derives from choosing input port current I_1 and output port voltage V_2 as independent electrical variables. Resultantly, input port voltage V_1 and output port current I_2 are dependent signal variables. Because the two-port network undergoing scrutiny is linear, each dependent variable is necessarily a linear superposition of the effects of each independent variable. This observation gives rise to the symbolic volt-ampere relationships,

$$\left. \begin{array}{l} V_1 = h_{11}I_1 + h_{12}V_2 \\ I_2 = h_{21}I_1 + h_{22}V_2 \end{array} \right\}, \tag{2-6}$$

or in matrix format,

$$\begin{bmatrix} V_1 \\ I_2 \end{bmatrix} = \begin{bmatrix} h_{11} & h_{12} \\ h_{21} & h_{22} \end{bmatrix} \begin{bmatrix} I_1 \\ V_2 \end{bmatrix}. \tag{2-7}$$

The h-parameters, h_{ij}, are viewed as "hybrid" because of their differing dimensional units. To wit, note in Eqs. (2-6) and (2-7) that dimensional consistency requires that h_{11} be an impedance, h_{22} be an admittance, and h_{12} and h_{21} be dimensionless parameters.

A fact of linear circuit theory is that an analysis of a linear circuit produces a set of linear characteristic equations. It follows that a set of linear equations, such as is postulated by Eqs. (2-6) and (2-7), corresponds to a linear circuit, which becomes known as an electrical *model* reflective of the equilibrium state defined by the equation set. Circuit modeling is therefore the inverse of circuit analysis; that is, meaningful equations derive from circuit analysis, while ostensibly useful electrical models are premised on pertinent equilibrium equations.

In the case of Eqs. (2-6) and (2-7), the so called *h-parameter model,* or *h-parameter equivalent circuit*, of the linear two-port network in Fig. 2.2 is the topological structure shown in Fig. 2.3. This model mirrors basic linear circuit theory, which stipulates that any port of a linear circuit can be modeled by either a Thévenin or a Norton equivalent circuit. To this end, recall that the Thévenin and Norton models for a simple one-port configuration are cast in terms of the independent electrical variables that excite that port. In Fig. 2.3, observe that the Norton equivalent output port current,

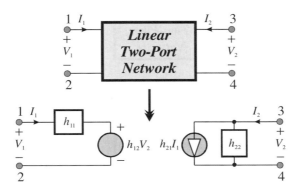

Figure 2.3. The h-parameter equivalent circuit of a linear two-port network. Parameter h_{11} is in units of ohms, h_{22} is in mhos, and both h_{12} and h_{21} are dimensionless parameters.

$h_{21}I_1$, and the Thévenin equivalent input port voltage, $h_{12}V_2$, are respectively proportional to the variables, I_1 and V_2, which have been selected as independent electrical variables in the formulation of the two-port characteristic equations.

Measurement procedures for the hybrid h-parameters derive directly from Eqs. (2-6) and (2-7). For example, if the output port voltage, V_2, is clamped to zero, which corresponds to the short-circuited output port depicted in Fig. 2.4(a),

$$h_{11} = \left.\frac{V_1}{I_1}\right|_{V_2=0},$$
$$h_{21} = \left.\frac{I_2}{I_1}\right|_{V_2=0}, \tag{2-8}$$

It follows that h_{11} is the short circuit (meaning that the output port is a short circuit) input impedance of the two-port undergoing examination, while h_{21} designates the forward short circuit current gain. Thus, h_{11} is a particular value of the driving point input impedance; specifically, h_{11} is the driving point input impedance under the special case of a short-circuited termination at the network output port. On the other hand, h_{21} is a measure of the forward gain of the network since it defines a value for the output port current corresponding to a given input port current. More specifically, h_{21} defines the maximum possible forward current gain in view of the fact

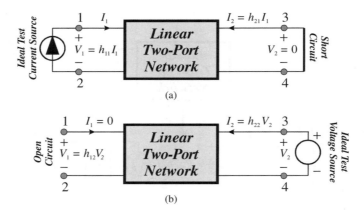

Figure 2.4. (a) Measurement of the short circuit hybrid h-parameters. (b) Measurement of the open circuit h-parameters.

that a short-circuited load termination is certainly conducive to maximal output port current.

With $I_1 = 0$, which reflects the open-circuited input port diagramed in Fig. 2.4(b), Eq. (2-6) or (2-7) yield

$$\left. \begin{array}{l} h_{12} = \left. \dfrac{V_1}{V_2} \right|_{I_1=0} \\ h_{22} = \left. \dfrac{I_2}{V_2} \right|_{I_1=0} \end{array} \right\}. \tag{2-9}$$

The parameter, h_{12}, is termed the reverse voltage gain, or the h-parameter feedback factor, of a two-port network. It is literally the maximum possible reverse voltage gain since the condition, $I_1 = 0$, is identical to a no load condition at the input port. A two-port network, and particularly an active two-port network, is naturally thought of as a system capable of delivering very large h_{21} so that maximal output signal is generated in response to input port excitation. But a portion of the output response can be returned, or fed back, to the input port because of the electrical nature of the devices implicit to the linear network and the manner in which the elements of said network are interconnected and laid out. Feedback can also be manifested by the electrical nature of the package in which the electronic circuit is embedded. Feedback can be an undesirable phenomenon, as is the case with packaging anomalies and when bipolar and MOS technology transistors are operated at high signal frequencies. It can also be a specific design objective, as when feedback paths are appended around active subcircuits to optimize overall circuit response. Regardless of the source of network feedback, parameter h_{12} is its measure in an h-parameter emulation of linear network I/O performance.

An alternative interpretation of h_{12} is that of an isolation factor between output and input ports. To this end, $h_{12} = 0$ reflects perfect isolation, while large h_{12} infers poor isolation, or good coupling, from the output port to the input port. In an attempt to clarify these assertions, return to Eq. (2-7) to solve for the ratio, V_1/I_1, which is literally the driving point input impedance of the considered network. In particular,

$$\dfrac{V_1}{I_1} = h_{11} + h_{12} \left(\dfrac{V_2}{I_1} \right). \tag{2-10}$$

In Eq. (2-10), the ratio, V_2/I_1, is the forward transimpedance of the circuit undergoing examination. For a fixed input current, I_1, this transimpedance function is certainly influenced by the load termination, across which the output port voltage, V_2, is established. For example, a short-circuited load necessarily renders $V_2 = 0$, whereby Eq. (2-10) confirms an input impedance that is identical to h_{11}, which is independent of any output port electrical variables. This result is certainly synergistic with the definition of h_{11} in Eq. (2-8). But the same impedance result, $V_1/I_1 = h_{11}$, is obtained if $h_{12} = 0$. Evidently $h_{12} = 0$ decouples, or isolates, the output port from its input counterpart in the sense that the input port does not respond to any output port voltage perturbation induced by load fluctuations, parasitic signal coupling, or other phenomena.

In an electronic system, it is generally advantageous to ensure $|h_{21}| \gg |h_{12}|$; that is, the magnitude of the maximum possible forward current gain is desirably much larger than the magnitude of the maximum possible reverse voltage gain. This design requirement is clearly satisfied when the I/O ports are perfectly isolated, in which case the subject two-port becomes known as a *unilateral network*. The term, "unilateral", refers to an ability of a network to transmit signal between I/O ports in only one direction; in this case, from the input port to the output port. A passive network can be shown to have $h_{21} = -h_{12}$, which implies $|h_{21}| \equiv |h_{12}|$. Any linear network for which $h_{21} = -h_{12}$ is said to be *bilateral*, which means that signal can be propagated equally well from input to output and from output to input ports.

Finally, h_{22} in Eq. (2-9) is the open circuit (meaning that the input port is open circuited) output admittance. Parameter h_{22} is the value of the driving point output admittance for the special case of an infinitely large source impedance. For $h_{22} = 0$, the output port in Fig. 2.3 emulates an ideal current source.

2.3.1.1. Ideal Current Amplifier

In an attempt to convey a sense of utility to the hybrid h-parameters, consider the prospect of using them to define the I/O port characteristics of an ideal current amplifier. Idealized current amplification is emulated by such commonly used analog canonic cells as the common gate and common base amplifiers and in general, they are routinely exploited in the conceptual phases of a circuit design undertaking. An ideal current amplifier, such as the abstraction offered in Fig. 2.5(a), provides an output port current that is directly proportional to the applied signal source current, independent of the source and load impedances. In the subject diagram, the current gain, which is the proportionality constant linking the output current to the

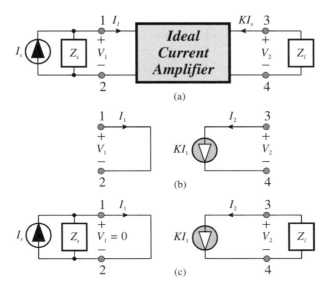

Figure 2.5. (a) Symbolic diagram of an ideal current amplifier coupling a signal source current to an impedance load at the output port. (b) The h-parameter model of the ideal amplifier in (a). (c) The h-parameter equivalent circuit of the system in (a).

source current is indicated as the frequency invariant constant, K. An ideal current amplifier provides zero driving point input impedance for all terminating load impedances. Additionally, its output port mirrors the electrical characteristics of an ideal current source, thereby implying an infinitely large driving point output impedance, independent of the Thévenin source impedance. Recalling Eq. (2-10), the driving point input impedance is rendered independent of load port variables if and only if $h_{12} = 0$. With zero internal feedback, zero driving point input impedance requires $h_{11} = 0$. From the h-parameter model in Fig. 2.3, the output port behaves as the requisite ideal current source if and only if $h_{22} = 0$. It follows that the h-parameter equivalent circuit of an ideal current amplifier is the topology shown in Fig. 2.5(b), where the pertinent short circuit current gain, h_{21}, is K. Accordingly, the h-parameter equations of the ideal current amplifier are given by the simple matric,

$$\begin{bmatrix} V_1 \\ I_2 \end{bmatrix} = \begin{bmatrix} 0 & 0 \\ K & 0 \end{bmatrix} \begin{bmatrix} I_1 \\ V_2 \end{bmatrix}, \qquad (2\text{-}11)$$

and the model of the current amplifier in Fig. 2.5(a) is the structure depicted in Fig. 2.5(c). In this model, the short-circuited input port precludes current conduction through the Thévenin source impedance, Z_s, thereby forcing the input port current, I_1, to be identical to the signal source current, I_s. At the output port in Fig. 2.5(c), the current, I_2, conducted by the load impedance, Z_l, is necessarily KI_1, whence the current transfer function, I_2/I_s, is identically K, independent of Z_s and Z_l.

2.3.1.2. Parameter Measurement Issues

A parametric measurement complication arises when the two-port network undergoing investigation is an electronic system, which inherently requires biasing to ensure reasonably linear driving point and transfer characteristics. Unfortunately, a short-circuited output port, which is required in the measurement of parameters h_{11} and h_{21}, and an open-circuited input port, which underpins the determination of parameters h_{12} and h_{22}, are likely to upset the required biasing conditions. But on the assumption that the test sources appearing in Fig. 2.4 are sinusoids having zero average value, this biasing dilemma can be circumvented by approximating a short circuit with a sufficiently large capacitance and an open circuit with a sufficiently large inductance. Figure 2.6 portrays these topological modifications to the h-parameter measurement strategy. Because the source and load impedances, Z_s and Z_l, may play a role in the biasing of the network, observe in this figure that the source impedance connection at the input port and the load impedance connection at the output port are sustained. It is to be understood that at the lowest test frequency of interest in Fig. 2.6(a), the capacitance serving as a "dummy" output port short circuit must present a branch impedance that is significantly smaller than the net resistance it effectively shunts. Since capacitors behave as open circuits at zero frequency, this dummy shunt is transparent to any output port biasing requirements. Similarly, the inductance serving as an input port open circuit in Fig. 2.6(b) must establish a branch impedance that is substantially larger than the resistance with which it is effectively placed in series. If the academic liberty of presuming infinitely large inductor quality factor is taken, this inductance is a short circuit at zero frequency and like the foregoing capacitance, it is transparent to static operating conditions.

Unfortunately, the simple measurement methodology described in the preceding paragraph proves effective only when the two-port network under test is earmarked for low frequency signal processing, such as that typified by audio and at best, video applications. For broadband systems spanning frequencies through several gigahertz, and for all other circuit and system

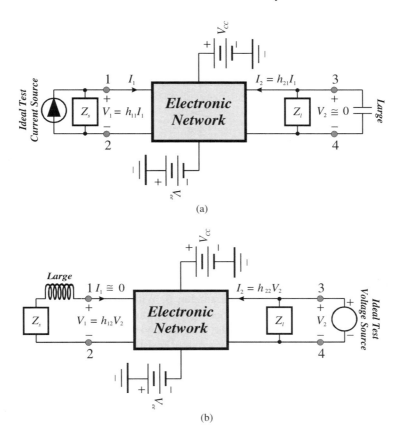

Figure 2.6. (a) Measurement of the short circuit hybrid h-parameters of an electronic network whose electrical characteristics are linearized by suitably applied biasing. (b) Measurement of the open circuit hybrid h-parameters of a linearized electronic network.

functions that require match terminated source and/or load impedances, the electronic subcircuits embedded within the two-port network are annoyed by open or short-circuited port terminations. The manifestations incurred by inappropriately terminating the I/O ports of broadbanded networks are severe I/O nonlinearities, poor transient responses in the senses of severe underdamping and/or long settling times, or even free running oscillations. The seriousness of the problem at hand forces an abandonment of a direct h-parameter characterization of such networks in favor of indirect measurement methods that do not require short-circuited or open-circuited network ports. More information about these indirect characterizations is

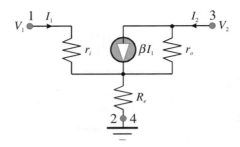

Figure 2.7 The small signal equivalent circuit of the amplifier addressed in Example 2.1.

provided later when the scattering parameters of linear two-port networks are discussed.

Example 2.1. Figure 2.7 depicts the low frequency, small signal equivalent circuit of a common emitter bipolar junction transistor amplifier that utilizes an emitter degeneration resistance, R_e, to achieve a forward gain that is nominally independent of the transistor current gain parameter, β. With reference to Fig. 2.1, note that an electrical ground is common between the input and output ports so that the port terminals, 2 and 4, are one and the same ground node. Let the amplifier input resistance, r_i, be 2.2 KΩ, the transistor output resistance, r_o, is 25 KΩ, $\beta = 90$, and $R_e = 80\,\Omega$. Derive general expressions for, and numerically evaluate, the four h-parameters, h_{ij}, of the emitter degenerated configuration.

Solution 2.1.

(1) Figure 2.8(a) is the equivalent circuit appropriate for evaluating the short circuit h-parameters, h_{11} and h_{21}. In particular, the subject figure is the diagram of Fig. 2.7 drawn under the condition of a short-circuited output port. A KVL equation written for the output port delivers

$$0 = r_o (I_2 - \beta I_1) + R_e (I_1 + I_2), \qquad \text{(E1-1)}$$

whence

$$h_{21} = \frac{I_2}{I_1} = \beta \left(\frac{1 - R_e/\beta r_o}{1 + R_e/r_o} \right) = 89.71 \text{ amps/amp}. \qquad \text{(E1-2)}$$

KVL around the input loop yields

$$V_1 = r_i I_1 + R_e (I_1 + I_2) = (r_i + R_e) I_1 + \beta R_e \left(\frac{1 - R_e/\beta r_o}{1 + R_e/r_o} \right) I_1, \qquad \text{(E1-3)}$$

and armed with Eq. (E1-2),

$$h_{11} = \frac{V_1}{I_1} = r_i + (\beta + 1)(r_o \| R_e) = 9.46 \text{ K}\Omega. \qquad \text{(E1-3)}$$

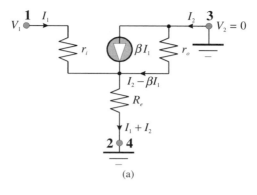

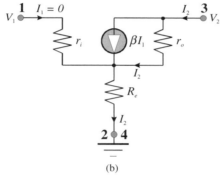

Figure 2.8. (a) Model used to calculate h-parameters h_{11} and h_{21} in the circuit of Fig. 2.7. (b) Equivalent circuit appropriate for the calculation of h-parameters h_{12} and h_{22} in Fig. 2.7.

Note that for $R_e \ll r_o$ and with $\beta > 1$, $h_{21} \approx \beta$, and $h_{11} \approx r_i + (\beta + 1)R_e$, for which the respective percentage numerical errors are 0.32% and 0.21%.

(2) Figure 2.8(b) is the diagram appropriate for the computation of h-parameters h_{12} and h_{22}. With $I_1 = 0$, it is clear that

$$\left. \begin{array}{l} V_2 = (r_o + R_e)I_2 \\ V_1 = R_e I_2 \end{array} \right\}, \tag{E1-4}$$

and thus,

$$h_{22} = \frac{I_2}{V_2} = \frac{1}{r_o + R_e} = 39.87\,\mu\text{S},$$

$$h_{12} = \frac{V_1}{V_2} = \frac{R_e}{r_o + R_e} = 3.19\,\text{mvolts/volt}. \tag{E1-5}$$

For the typically encountered case of $R_e \ll r_o$, $h_{22} \approx 1/r_o$, while $h_{12} \approx R_e/r_o$.

Comments. In addition to expediting amplifier modeling and circuit level analysis, the h-parameters are seen herewith as highlighting the electrical effects of an appended circuit element (in this case, the emitter degeneration resistance, R_e). For example, by inspecting the $R_e = 0$ (no emitter degeneration) values of the four h-parameters, the following conclusions are drawn. First, emitter degeneration minimally impacts the forward short circuit current gain, h_{21}, and the open circuit output port conductance, h_{22}, provided $R_e \ll r_o$. On the other hand, the emitter degeneration resistance substantially increases the open circuit input resistance, h_{11}, and is the cause of nonzero isolation, h_{12}, between the input and output amplifier ports.

2.3.2. Hybrid g-Parameters

The independent and dependent electrical variable sets used to define the *hybrid g-parameters* are the converse of those used in conjunction with the hybrid h-parameters. Specifically, the independent variables for g-parameter modeling are the input port voltage, V_1, and the output port current, I_2, thereby rendering the input port current, I_1, and the output port voltage, V_2, dependent electrical quantities. From superposition theory,

$$\begin{bmatrix} I_1 \\ V_2 \end{bmatrix} = \begin{bmatrix} g_{11} & g_{12} \\ g_{21} & g_{22} \end{bmatrix} \begin{bmatrix} V_1 \\ I_2 \end{bmatrix}, \quad (2\text{-}12)$$

which gives rise to the g-parameter equivalent circuit depicted in Fig. 2.9. Dimensional consistency requires g_{11} to be an admittance, g_{22} an impedance, and g_{12} and g_{21} both dimensionless.

As with the hybrid h-parameters, the measurement strategy for the hybrid g-parameters derives directly from the defining volt-ampere relationships in Eq. (2-12). In particular, parameters g_{11} and g_{21} are measured or evaluated under an open-circuited output port condition, while g_{12} and g_{22} evolve from a short-circuited input port constraint. Specifically,

$$\left. \begin{aligned} g_{11} &= \left. \frac{I_1}{V_1} \right|_{I_2=0} \\ g_{21} &= \left. \frac{V_2}{V_1} \right|_{I_2=0} \end{aligned} \right\}, \quad (2\text{-}13)$$

which casts g_{11} as an open circuit (meaning that the output port is open circuited) input port admittance and g_{21} as an open circuit forward voltage gain (also often referred to as the *open loop voltage gain*) of the two-port network. With reference to Fig. 2.1(c), g_{11} is the inverse of the input

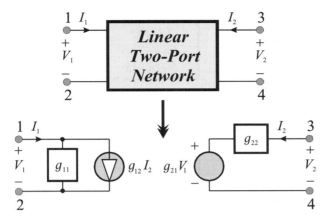

Figure 2.9. The g-parameter equivalent circuit of a linear two-port network. Parameter g_{11} is in units of mhos, g_{22} is in ohms, and both g_{12} and g_{21} are dimensionless parameters.

impedance, Z_{in}, under the special case of an infinitely large load impedance, while g_{21} is the I/O Thévenin voltage gain. Continuing in Eq. (2-12),

$$g_{12} = \left. \frac{I_1}{I_2} \right|_{V_1=0},$$
$$g_{22} = \left. \frac{V_2}{I_2} \right|_{V_1=0},$$
(2-14)

Thus, g_{22} is the short circuit (meaning that the input port is short circuited) output impedance of the two port. It is, in fact, the output impedance, Z_{out}, in Fig. 2.1(c) under the special case of a short-circuited Thévenin source impedance termination. Parameter g_{12}, like h-parameter h_{12}, is a measure of feedback from the output port to the input port. Equivalently, g_{12} measures the degree of isolation between input and output network ports. The salient implications of Eqs. (2-13) and (2-14) are summarized topologically by Fig. 2.10. Ignoring the previously discussed shortfalls surrounding the pragmatic implementation of short and open circuits in broadband electronic systems,, the requisite open circuit in Fig. 2.10(a) can be approximated by a sufficiently large inductance, while the short circuit in Fig. 2.10(b) can be achieved to first order by a large capacitance.

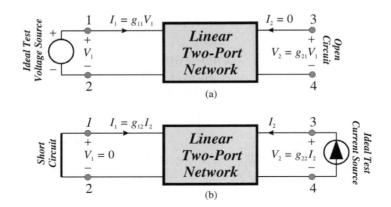

Figure 2.10. (a) Measurement of the open circuit hybrid g-parameters. (b) Measurement of the short circuit g-parameters.

2.3.2.1. g- and h-Parameter Interrelationships

At this juncture, two alternatives for the modeling and ultimate analysis of a linear two-port network have been presented. If h-parameters are selected as the vehicle for modeling a linear two-port network, the applicable equivalent circuit is the structure appearing in Fig. 2.3, while if g-parameters are adopted, the applicable model is the topology of Fig. 2.9. When either modeling approach is exploited for a given two-port system, both must necessarily predict identical driving point, forward transfer, and feedback properties. This homogeneity constraint implies that the h- and g-parameters are not independent network metrics.

A comparison of Eqs. (2-7) and (2-12) renders transparent the interrelationship between the h- and g-parameters of a linear two-port network. In particular, the g-parameter matrix on the right hand side of Eq. (2-12) is the inverse of the h-parameter matrix on the right hand side of Eq. (2-7); that is,

$$\begin{bmatrix} g_{11} & g_{12} \\ g_{21} & g_{22} \end{bmatrix} = \begin{bmatrix} h_{11} & h_{12} \\ h_{21} & h_{22} \end{bmatrix}^{-1}, \qquad (2\text{-}15)$$

and conversely,

$$\begin{bmatrix} h_{11} & h_{12} \\ h_{21} & h_{22} \end{bmatrix} = \begin{bmatrix} g_{11} & g_{12} \\ g_{21} & g_{22} \end{bmatrix}^{-1}. \qquad (2\text{-}16)$$

With

$$\Delta_h = h_{11}h_{22} - h_{12}h_{21}, \qquad (2\text{-}17)$$

representing the determinant of the h-parameter matrix on the right hand side of Eq. (2-7), Eq. (2-15) requires

$$\left. \begin{array}{l} g_{11} = h_{22}/\Delta_h \\ g_{22} = h_{11}/\Delta_h \\ g_{12} = -h_{12}/\Delta_h \\ g_{21} = -h_{21}/\Delta_h \end{array} \right\}, \tag{2-18}$$

Moreover, Eq. (2-16) is satisfied if and only if

$$\left. \begin{array}{l} h_{11} = g_{22}/\Delta_g \\ h_{22} = g_{11}/\Delta_g \\ h_{12} = -g_{12}/\Delta_g \\ h_{21} = -g_{21}/\Delta_g \end{array} \right\}, \tag{2-19}$$

where

$$\Delta_g = g_{11}g_{22} - g_{12}g_{21} \tag{2-20}$$

is the determinant of the g-parameter matrix on the right hand side of Eq. (2-12). Since the h-parameter and the g-parameter matrices are inverses of one another,

$$\Delta_h \Delta_g \equiv 1. \tag{2-21}$$

Although the selection of the type of parameters (h-parameters or g-parameters at this level of analytical development) exploited to model a two-port network is arbitrary, the actual choice may be dictated by circuit pragmatics. For example, observe in Eq. (2-18) that the g-parameters are not finite, and thus are not numerically deterministic, if the determinant of the h-parameter matrix is zero. A null determinant defines *matrix singularity* and thus, the g-parameters do not exist if the corresponding h-parameter matrix is singular. Moreover, Eq. (2-19) shows that h-parameters cannot be evaluated if the g-parameter matrix is singular; that is if $\Delta_g = 0$. It follows that some two-port networks, and particularly active two ports whose modeling has been simplified to reflect somewhat idealized volt-ampere characteristics, may have an h-parameter modeling representation in lieu of a g-parameter equivalent circuit and vice versa. For example the ideal current amplifier h-parameter matrix in Eq. (2-11) is obviously singular, thereby precluding a g-parameter model of such an idealized unit. In certain operating circumstances, it is possible that neither an h- nor a g-parameter model exists, whence other two-port parameter sets must be exploited.

It is interesting to note that the g-parameter feedback factor, g_{12}, is zero if its h-parameter counterpart, h_{12}, is likewise zero. Similarly, the feedforward parameter, g_{21}, is zero if $h_{21} = 0$. This observation suggests that both the g-parameter and the h-parameter equivalent circuits are analytically consistent in the senses of disclosing the degree of extant feedback, or isolation, and the degree of observable feedforward transmission.

2.3.2.2. Ideal Voltage Amplifier

Although the g-parameters are seemingly used far less often than are the h-parameters when arbitrary linear two-port networks are earmarked for characterization, they are, in fact, well suited for the modeling of an ideal voltage amplifier, for which the ubiquitous operational amplifier utilized at low signal frequencies is a pragmatic approximation. Formally, an ideal voltage amplifier, which is illustrated symbolically in Fig. 2.11(a), emulates the volt-ampere characteristics of a voltage controlled voltage source. It exudes infinitely large driving point input impedance for all load terminations, zero driving point output impedance for all Thévenin source impedances, and a frequency invariant I/O port voltage gain, say K, for all source and load impedances. Since g_{12}, like h_{12}, is a measure of the degree of isolation evidenced between network input and output ports, a driving point input impedance that is independent of the terminating load impedance requires $g_{12} = 0$. With $g_{12} = 0$, an infinitely large input port impedance additionally mandates $g_{11} = 0$. The zero I/O port isolation ensured by null g_{12} and the voltage amplifier requirement of zero driving point output port impedance stipulates $g_{22} = 0$. Given an I/O port voltage gain of K, $g_{21} = K$, and it follows that the g-parameter equilibrium equations of an ideal voltage amplifier are given by the matrix relationship,

$$\begin{bmatrix} I_1 \\ V_2 \end{bmatrix} = \begin{bmatrix} 0 & 0 \\ K & 0 \end{bmatrix} \begin{bmatrix} V_1 \\ I_2 \end{bmatrix}, \qquad (2\text{-}22)$$

for which the corresponding equivalent circuit is the topology offered in Fig. 2.11(b). Since the subject g-parameter matrix is singular, an ideal voltage amplifier does not have an h-parameter representation.

Figure 2.11(c) depicts the resultant electrical model of the system in Fig. 2.11(a). Since the input port is an open circuit, the input port current, I_1, is zero. This zero input port current ensures that the input port voltage, V_1, is identical to the signal source voltage, V_s, for all Thévenin source impedances, Z_s, and all load impedances, Z_l. It follows that

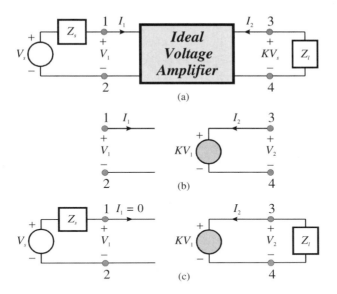

Figure 2.11. (a) Symbolic diagram of an ideal voltage amplifier coupling a signal source voltage to an impedance load incident with the output port. (b) The g-parameter model of the ideal amplifier in (a). (c) The g-parameter equivalent circuit of the system in (a).

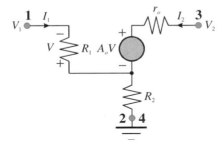

Figure 2.12. The small signal equivalent circuit of the amplifier studied in Example 2.2.

$KV_1 \equiv KV_s$ at the output port. With $g_{22} = 0$, no impedance appears in series with this dependent generator and the load impedance, which means that the output port voltage, V_2, developed across impedance Z_l is KV_s. It follows that the overall system voltage gain is $V_2/V_s = K$.

Example 2.2. Figure 2.12 is a simplified low frequency, small signal equivalent circuit of a voltage amplifier that exploits feedback via the resistance, R_2. Let $R_1 = 9 \text{ K}\Omega$, and $R_2 = 1 \text{ K}\Omega$. Assume an amplifier output resistance, r_o, of $100 \, \Omega$,

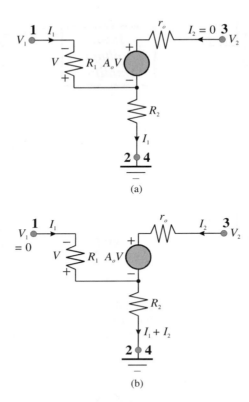

Figure 2.13. (a) Model used to calculate g-parameters g_{11} and g_{21} in the circuit of Fig. 2.12. (b) Model appropriate for the calculation of g-parameters g_{12} and g_{22}.

and an amplifier open loop gain (gain in the absence of feedback), A_o, of 100. Derive general expressions for, and numerically evaluate, the four g-parameters, g_{ij}, of the feedback amplifier.

Solution 2.2.

(1) Figure 2.13(a), which is the circuit diagram in Fig. 2.12 with an open-circuited output port, is the equivalent circuit for evaluating the g-parameters, g_{11} and g_{21}. By inspection of the input port mesh,

$$V_1 = (R_1 + R_2) I_1, \qquad \text{(E2-1)}$$

from which

$$g_{11} = \frac{I_1}{V_1} = \frac{1}{R_1 + R_2} = 100\,\mu S. \qquad \text{(E2-2)}$$

At the output port of Fig. 2.13(a),

$$V_2 = A_o V + R_2 I_1 = -A_o R_1 I_1 + R_2 I_1 = -(A_o R_1 - R_2)\frac{V_1}{R_1 + R_2}. \quad \text{(E2-3)}$$

It follows that

$$g_{21} = \frac{V_2}{V_1} = -A_o\left(\frac{1 - R_2/A_o R_1}{1 + R_2/R_1}\right) = -89.90 \text{ volts/volt}. \quad \text{(E2-4)}$$

(2) Figure 2.13(b) is the equivalent circuit pertinent to computing g-parameters g_{12} and g_{22}. With $V_1 = 0$, the I/O port currents are constrained by

$$0 = R_1 I_1 + R_2 (I_1 + I_2), \quad \text{(E2-5)}$$

whence,

$$g_{12} = \frac{I_1}{I_2} = -\frac{R_2}{R_1 + R_2} = -0.10 \text{ amps/amp}. \quad \text{(E2-6)}$$

Moreover,

$$V_2 = r_o I_2 - A_o R_1 I_1 + R_2 (I_1 + I_2). \quad \text{(E2-7)}$$

Using the preceding result for the current ratio, I_1/I_2, it is a simple matter to show that

$$g_{22} = \frac{V_2}{I_2} = r_o + (A_o + 1)(R_1 \| R_2) = 91 \text{ K}\Omega. \quad \text{(E2-8)}$$

Comments. The g-parameter model underscores the fact that for the amplifier at hand, feedback resistance R_2 enormously boosts the short circuit output resistance. It also lowers the open circuit input conductance. As expected, a feedback resistance of zero incurs no short circuit feedback within the amplifier.

2.3.3. Short Circuit y-Parameters

When the input port voltage, V_1, and the output port voltage, V_2, are chosen as independent variables in the course of analyzing a linear two-port network, the relevant volt-ampere relationships assume the form,

$$\begin{bmatrix} I_1 \\ I_2 \end{bmatrix} = \begin{bmatrix} y_{11} & y_{12} \\ y_{21} & y_{22} \end{bmatrix}\begin{bmatrix} V_1 \\ V_2 \end{bmatrix}. \quad (2\text{-}23)$$

In Eq. (2-23), for which the circuit-level interpretation is Fig. 2.14, the y_{ij} are called y-parameters of the two-port system undergoing investigation. The y_{ij} are also called *short circuit admittance parameters* because each element in the matrix on the right hand side of Eq. (2-23) is an admittance

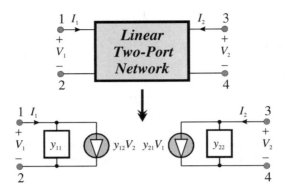

Figure 2.14. The y-parameter equivalent circuit of a linear two-port network. All of the short circuit admittance parameters, y_{ij}, are in units of mhos.

function measured under the condition of a short-circuited network port. Specifically,

$$\left. \begin{array}{l} y_{11} = \left. \dfrac{I_1}{V_1} \right|_{V_2=0} \\[6pt] y_{21} = \left. \dfrac{I_2}{V_1} \right|_{V_2=0} \end{array} \right\rangle, \tag{2-24}$$

and

$$\left. \begin{array}{l} y_{12} = \left. \dfrac{I_1}{V_2} \right|_{V_1=0} \\[6pt] y_{22} = \left. \dfrac{I_2}{V_2} \right|_{V_1=0} \end{array} \right\rangle. \tag{2-25}$$

It follows that parameter y_{11} is the short circuit (meaning that the output port is short circuited) input admittance of the two-port network, while y_{21}, is the short circuit forward transadmittance. Parameter y_{11} is therefore related to the driving point input admittance or impedance of the network. It is, in fact, the driving point input admittance observed under the condition of a short-circuited load terminating the output port. On the other hand, y_{21} is a measure of the forward gain. In Eq. (2-25) y_{12} is seen as a feedback factor, measured herewith as a reverse short circuit (meaning that the input port is short circuited) transadmittance, and y_{22} is the short circuit output admittance.

Because a two-port analysis predicated on y-parameters must produce results that are consistent with investigations based on either h- or

g-parameters, the y_{ij} of a linear network are necessarily related to both the h_{ij} and the g_{ij} of the same network. For example, compare Eqs. (2-7), (2-12), and (2-19), while recalling Eqs. (2-18) and (2-19), to deduce that with a short circuit imposed at the network output port ($V_2 = 0$),

$$\left. \begin{array}{l} y_{11} = \dfrac{I_1}{V_1}\bigg|_{V_2=0} = \dfrac{1}{h_{11}} = g_{11} - \dfrac{g_{12}g_{21}}{g_{22}} \\[2ex] y_{21} = \dfrac{I_2}{V_1}\bigg|_{V_2=0} = \dfrac{h_{21}}{h_{11}} = -\dfrac{g_{21}}{g_{22}} \end{array} \right\}. \qquad (2\text{-}26)$$

As expected, y_{21}, which is the y-parameter measure of forward signal transmission through a linear network, is proportional to both h_{21} and g_{21}, which are respectively the h-parameter and g-parameter counterparts to this network I/O gain metric. Moreover, recall that h_{11} is the short circuit input impedance of a linear two-port network. It is therefore only fitting that y_{11}, which represents the short circuit input admittance, is the inverse of h-parameter h_{11}. Continuing the comparison among Eqs. (2-7), (2-12), and (2-19),

$$\left. \begin{array}{l} y_{12} = \dfrac{I_1}{V_2}\bigg|_{V_1=0} = -\dfrac{h_{12}}{h_{11}} = \dfrac{g_{12}}{g_{22}} \\[2ex] y_{22} = \dfrac{I_2}{V_2}\bigg|_{V_1=0} = h_{22} - \dfrac{h_{12}h_{21}}{h_{11}} = \dfrac{1}{g_{22}} \end{array} \right\}. \qquad (2\text{-}27)$$

Also as expected, the y-parameter feedback parameter, y_{12}, is directly proportional to both the h-parameter feedback measure, h_{12}, and the corresponding g-parameter measure, g_{12}. Recalling that $h_{12} = -h_{21}$ for a bilateral two-port network, Eqs. (2-26) and (2-27) imply that bilateralness requires $y_{12} = y_{21}$ and $g_{12} = -g_{21}$. Despite the fact that h_{22} and y_{22} individually measure network output port admittance, y_{22} does not equal h_{22}. The reason underlying $y_{22} \neq h_{22}$ is that h_{22} monitors the output port admittance with the input port open circuited, while y_{22} is the output admittance under the condition of a short-circuited input port. Finally, since g_{22} is the output port impedance when the input port is short circuited, y_{22} is naturally the inverse of $1/g_{22}$.

2.3.3.1. π-Type Network Model

An interesting, and often utilitarian alternative to the basic y-parameter equivalent circuit in Fig. 2.14 results when Eq. (2-23) is rewritten in the form,

$$\left.\begin{aligned} I_1 &= y_{11}V_1 + y_{12}V_2 = (y_{11} + y_{12})V_1 - y_{12}(V_1 - V_2) \\ I_2 &= y_{21}V_1 + y_{22}V_2 = (y_{21} - y_{12})V_1 + (y_{22} + y_{12})V_2 - y_{12}(V_2 - V_1) \end{aligned}\right\}.$$

(2-28)

Upon introduction of the subsidiary admittance parameters,

$$\left.\begin{aligned} y_i &\stackrel{\Delta}{=} y_{11} + y_{12} \\ y_r &\stackrel{\Delta}{=} -y_{12} \\ y_f &\stackrel{\Delta}{=} y_{21} - y_{12} \\ y_o &\stackrel{\Delta}{=} y_{22} + y_{12} \end{aligned}\right\},$$

(2-29)

Eq. (2-28) becomes

$$\left.\begin{aligned} I_1 &= y_i V_1 + y_r (V_1 - V_2) \\ I_2 &= y_f V_1 + y_o V_2 + y_r (V_2 - V_1) \end{aligned}\right\},$$

(2-30)

which forges the alternative y-parameter model topology offered in Fig. 2.15(a). This π-type topology requires that the original linear two-port network operate with a common terminal between its input and output ports. This is to say that the original configuration is effectively a three terminal, two-port system for which the I/O terminal voltages, V_1 and V_2, are referenced to the lead that is common to the I/O ports. The alternative structure requires only one dependent generator, portrays y_i as a shunting input port branch admittance, y_o as a shunting output port branch admittance, y_r as an I/O port coupling admittance, and y_f as a barometer of forward transimpedance. For the special case of a bilateral, three terminal, two-port network, for which $y_{12} = y_{21}$, y_f is zero, and the model in Fig. 2.15(a) collapses to the true π-type architecture depicted in Fig. 2.15(b).

Example 2.3. Figure 2.16 is an approximate small signal model of a metal-oxide-semiconductor field-effect transistor (MOSFET) configured to operate ultimately as a common source amplifier. Terminal 1 represents the gate of the transistor, terminal 2 is the transistor drain terminal, and terminal 3 is the source terminal of the device. In this model, C_{gs} represents the gate to source capacitance of the transistor, C_{gd} is the gate to drain capacitance, C_{db} is the drain to bulk substrate capacitance (the substrate terminal is presumed grounded for small signals), r_o designates the source-drain channel resistance, and finally, g_m is the forward transconductance of the device. Find general expressions for each of the four common source y-parameters, y_{ij}. Also, derive a general expression for the short circuit (meaning

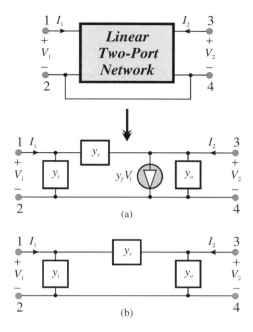

Figure 2.15. (a) Alternative form of a y-parameter equivalent circuit for a three terminal, linear two port network. (b) The model of (a) for the special case of a bilateral, three terminal linear network.

the drain terminal is short circuited to the source terminal) current gain and the frequency at which the magnitude of this short circuit current gain degrades to unity. If $g_m = 8\,\text{mS}$, $r_o = 8\,\text{K}\Omega$, $C_{gs} = 15\,\text{fF}$, $C_{gd} = 3\,\text{fF}$, and $C_{db} = 12\,\text{fF}$, numerically evaluate this unity gain frequency, which is commonly symbolized as f_T.

Solution 2.3.

(1) The good news about this problem is that the model in Fig. 2.16 is identical to that of the generalized topology shown in Fig. 2.15(a). By inspection, therefore, and with g_o signifying the conductance value of the resistance, r_o,

$$\left. \begin{aligned} y_i &= sC_{gs} \\ y_r &= sC_{gd} \\ y_f &= g_m \\ y_o &= g_o + sC_{db} \end{aligned} \right\}. \tag{E3-1}$$

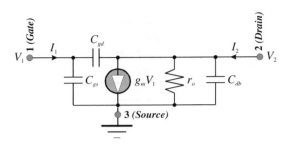

Figure 2.16. Small signal model of a MOSFET configured to operate as a common source amplifier.

From Eq. (2-29), it follows that

$$\left. \begin{array}{l} y_{12} = -y_r = -sC_{gd} \\ y_{11} = y_i - y_{12} = s\left(C_{gs} + C_{gd}\right) \\ y_{22} = y_o - y_{12} = g_o + s\left(C_{db} + C_{gd}\right) \\ y_{21} = y_f + y_{12} = g_m - sC_{gd} \end{array} \right\}. \qquad (E3\text{-}2)$$

(2) Extensive circuit analysis is not required for an evaluation of the short circuit current gain. Instead, it is necessary only that h-parameter h_{21} be recalled as the short circuit current gain of a linear two-port network. From Eq. (2-26),

$$h_{21} \equiv \frac{y_{21}}{y_{11}}, \qquad (E3\text{-}3)$$

and using the results of the preceding analytical step,

$$h_{21} = \frac{g_m - sC_{gd}}{s\left(C_{gs} + C_{gd}\right)}. \qquad (E3\text{-}4)$$

(3) The short circuit current gain, h_{21}, computed in the preceding solution step indicates the presence of a right half plane zero at a frequency of g_m/C_{gd}, which is 2π (424.4 GHz). This extremely large critical frequency is likely beyond the frequency range of validity of the model in Fig. 2.16 and is potentially much larger than the unity gain frequency, which is to be computed. Accordingly, if the unity gain frequency is indeed significantly smaller than the frequency of the right half plane zero,

$$h_{21} = \frac{g_m - sC_{gd}}{s\left(C_{gs} + C_{gd}\right)} \approx \frac{g_m}{s\left(C_{gs} + C_{gd}\right)}. \qquad (E3\text{-}5)$$

Under steady state sinusoidal signal conditions, s is replaced by $j\omega$ in the preceding disclosure, with the result that

$$h_{21} \approx \frac{g_m}{j\omega\left(C_{gs} + C_{gd}\right)}. \qquad (E3\text{-}6)$$

Clearly, the magnitude of h_{21} progressively decreases with increasing signal frequency and becomes one at

$$f_T \approx \frac{g_m}{2\pi \left(C_{gs} + C_{gd}\right)} = 70.7 \, \text{GHz}. \quad \text{(E3-7)}$$

Comments. In this example, two-port parameter theory leads expediently to the analytical and numerical definitions of a commonly invoked figure of merit for MOSFETs; namely, the unity gain frequency, f_T. The computed f_T — value of better than 70 GHz is quite large and indicative of deep submicron technology devices. However, it is worthwhile interjecting that although f_T commonly serves as a reasonably useful metric for comparing the high frequency attributes of one transistor against another, it is a less than meaningful metric for MOSFET circuits and systems. The unity gain frequency of an actual circuit that exploits MOSFET technology is invariably much smaller than the transistor f_T rating. A principle reason for this disparity is that the short circuit basis for enumerating f_T inherently ignores drain-bulk capacitance. It is also oblivious to the potentially dominant energy storage elements — parasitic or otherwise — embedded within the circuit topology in which the considered MOSFET is embedded.

2.3.3.2. Ideal Transconductor

An ideal *transconductance amplifier*, or *transconductor*, emulates the volt-ampere characteristics of an ideal voltage controlled current source, which is characterized by an infinitely large input impedance for all load terminations, and an infinitely large driving point output impedance for all source terminations. As is suggested by Fig. 2.17(a), the ideal transconductor provides an output port current that is linearly dependent on the signal source voltage, V_s, applied to the input port for all source and load impedances. The proportionality constant linking this output port current to the signal voltage is G_m, which is presumed independent of signal frequency. A comparison of the resultant model shown in Fig. 2.17(b) with the generalized y-parameter equivalent circuit in Fig. 2.14 confirms that an ideal transconductor has $y_{11} = y_{22} = y_{12} = 0$ and $y_{21} = G_m$. Accordingly, its y-parameter equilibrium equations subscribe to the matric relationship,

$$\begin{bmatrix} I_1 \\ I_2 \end{bmatrix} = \begin{bmatrix} 0 & 0 \\ G_m & 0 \end{bmatrix} \begin{bmatrix} V_1 \\ V_2 \end{bmatrix}. \quad (2\text{-}31)$$

In the corresponding system model of Fig. 2.17(c), note that the input port open circuit afforded by $y_{11} = y_{12} = 0$ forces $V_1 = V_s$, independent of the Thévenin source impedance, Z_s. Additionally, the idealized current source

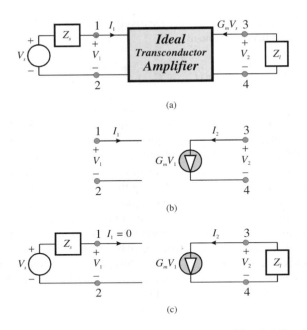

Figure 2.17. (a) Abstraction of an ideal transconductance amplifier coupling a signal source voltage to an impedance load. (b) The y-parameter model of the ideal transconductor in (a). (c) The y-parameter equivalent circuit of the system in (a).

nature of the output port ensures a load current, I_2, that is identical to $G_m V_1$ for all load impedances, Z_l. Since $V_1 \equiv V_s$, it follows that $I_2/V_s \equiv G_m$ for all Z_s and Z_l, as is suggested by the abstraction in Fig. 2.17(a).

2.3.3.3. Indefinite Admittance Matrix

An interesting and useful extension to the two-port y-parameter concept entails the consideration of the three terminal linear network given in Fig. 2.18. The network at hand is operated as a three port configuration, with each of the three ports supporting a voltage referenced to system ground. At first blush, the subject analytical extension is seemingly trivial since superposition theory allows the matric stipulation,

$$\begin{bmatrix} I_1 \\ I_2 \\ I_3 \end{bmatrix} = \begin{bmatrix} y_{11} & y_{12} & y_{13} \\ y_{21} & y_{22} & y_{23} \\ y_{31} & y_{32} & y_{33} \end{bmatrix} \begin{bmatrix} V_1 \\ V_2 \\ V_3 \end{bmatrix}. \qquad (2\text{-}32)$$

To be sure, Eq. (2-32) is a valid generality, but it incorrectly implies that each of the admittance parameters, y_{ij}, is independent of one another. In

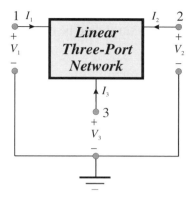

Figure 2.18. Schematic abstraction of a linear three-port network. Note that the reference terminal of all three-port voltages is system ground.

truth, only four of the nine indicated admittance metrics can be measured or computed independently.

A clarification of the preceding assertion begins by observing that the subject three-port network functions essentially as one giant, happy node in the sense that the three-port currents, I_1, I_2, and I_3, must sum to zero for any and all port voltages, V_1, V_2, and V_3. From Eq. (2-32), this constraint requires

$$\left. \begin{array}{l} y_{11} + y_{21} + y_{31} = 0 \\ y_{12} + y_{22} + y_{32} = 0 \\ y_{13} + y_{23} + y_{33} = 0 \end{array} \right\}, \tag{2-33}$$

which is tantamount to mandating that the sum of the elements in each of the three columns in the 3×3 matrix on the right hand side of Eq. (2-32) be zero. An additional parametric constraint is manifested through the consideration of the special circumstance in which all three-network-port voltages are identical. In this case, no potential difference exists, and thus no energy is applied, between any two of the three network terminals. Under the condition of null differential applied energy implied by $V_1 = V_2 = V_3$, $I_1 = I_2 = I_3 = 0$ if and only if

$$\left. \begin{array}{l} y_{11} + y_{12} + y_{13} = 0 \\ y_{21} + y_{22} + y_{23} = 0 \\ y_{31} + y_{32} + y_{33} = 0 \end{array} \right\}. \tag{2-34}$$

Accordingly, the admittance elements in each of the three rows of the 3×3 matrix in Eq. (2-32) must, like the elements in each column, sum to zero.

The immediate ramification of this disclosure is that the generalized volt-ampere relationship in Eq. (2-32) should be recast as,

$$\begin{bmatrix} I_1 \\ I_2 \\ I_3 \end{bmatrix} = \begin{bmatrix} y_{11} & y_{12} & -(y_{11}+y_{12}) \\ y_{21} & y_{22} & -(y_{21}+y_{22}) \\ -(y_{11}+y_{21}) & -(y_{12}+y_{22}) & (y_{11}+y_{12}+y_{21}+y_{22}) \end{bmatrix} \begin{bmatrix} V_1 \\ V_2 \\ V_3 \end{bmatrix}.$$

(2-35)

The 3×3 matrix on the right hand side of this relationship is termed the *indefinite admittance matrix* of the three-port shown in Fig. 2.18. For an n-terminal linear network, an $n \times n$ indefinite admittance matrix can be constructed straightforwardly in terms of the admittance parameters of the reduced $(n-1)$-terminal configuration simply by invoking the constraints of zero row and column sums in the resultant $n \times n$ matrix.

In an attempt to garner appreciation for the indefinite admittance matrix concept, consider the case in which the four admittance parameters of the two-port network resulting from $V_3 = 0$ have been measured or otherwise evaluated. The resultant volt-ampere expression, is Eq. (2-23), where it is tacitly assumed that the port voltages, V_1 and V_2, are referenced to electrical ground. It follows that the volt-ampere characteristics when V_3 is nonzero derive from Eq. (2-35), where again, all port voltages are presumed referenced to electrical ground. Suppose now that the network in Fig. 2.18 is operated with terminal 2 grounded and with nonzero V_1 and V_3. Since $V_2 = 0$, the second column of matrix elements on the right hand side of Eq. (2-35) can be discarded without loss of any information, since each of these elements multiply null voltage V_2 in the analytical course of establishing the three network port currents. Moreover, since $(I_1 + I_2 + I_3) = 0$ is guaranteed by the construction of the indefinite admittance matrix, the analytical information surrounding current I_2, which flows into the grounded second port, is superfluous; that is, once I_1 and I_3 are determined, current I_2 derives as simply the negative sum of these two currents. It follows that the second row of matrix elements in Eq. (2-35) can also be discarded, thereby giving rise to

$$\begin{bmatrix} I_1 \\ I_3 \end{bmatrix} = \begin{bmatrix} y_{11} & -(y_{11}+y_{12}) \\ -(y_{11}+y_{21}) & (y_{11}+y_{12}+y_{21}+y_{22}) \end{bmatrix} \begin{bmatrix} V_1 \\ V_3 \end{bmatrix} \quad (2\text{-}36)$$

as the applicable system of volt-ampere equations. Note that this system of equilibrium equations results exclusively from the characterization of the original system operated with terminal 3 grounded and does not require the execution of any substantive new analyses or alternative measurement procedures.

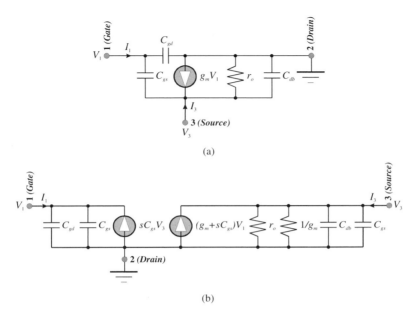

Figure 2.19. Small signal equivalent circuit of a MOSFET configured for common drain small signal processing. (b) The resultant y-parameter model of the common drain circuit in (a).

Example 2.4. Recall Fig. 2.16 as a small signal model of a MOSFET operated as a common source amplifier. Consider now the schematic diagram of Fig. 2.19(a), which depicts the same MOSFET operated in a common drain (also referred to as a source follower) configuration. Determine general expressions for each of the four common drain y-parameters, y_{11}, y_{13}, y_{31}, and y_{33}. Use these disclosures to draw the resultant source follower small signal equivalent circuit. The MOSFET model parameters remain $g_m = 8$ mS, $r_o = 8$ KΩ, $C_{gs} = 15$ fF, $C_{gd} = 3$ fF, and $C_{db} = 12$ fF.

Solution 2.4.

(1) The parameters, y_{11}, y_{12}, y_{21}, and y_{22}, for the common source circuit have been determined in Example 2.3. For convenience, they are repeated herewith, where it is recalled that $g_o = 1/r_o$. In particular,

$$\left. \begin{array}{l} y_{11} = s\left(C_{gs} + C_{gd}\right) \\ y_{12} = -sC_{gd} \\ y_{21} = g_m - sC_{gd} \\ y_{22} = g_o + s\left(C_{db} + C_{gd}\right) \end{array} \right\}. \quad \text{(E4-1)}$$

(2) The parameters for the common drain circuit in Fig. 2.19(a) derive directly from the foregoing results and Eq. (2-36). Specifically,

$$\left. \begin{aligned} y_{11} &= s\left(C_{gs} + C_{gd}\right) \\ y_{13} &= -(y_{11} + y_{12}) = -sC_{gs} \\ y_{31} &= -(y_{11} + y_{21}) = -\left(g_m + sC_{gs}\right) \\ y_{33} &= (y_{11} + y_{12} + y_{21} + y_{22}) = g_m + g_o + s\left(C_{gs} + C_{db}\right) \end{aligned} \right\}. \tag{E4-2}$$

Comments. The resultant common drain y-parameter equivalent circuit is offered in Fig. 2.19(b). This model shows that the output impedance of the considered source follower is small, even at low signal frequencies, because of the resistance, $1/g_m$, that appears directly across the output port. Moreover, the exclusively capacitive nature of the input port renders the driving point input impedance infinitely large at zero frequency, but small for very high signal frequencies.

2.3.4. Open Circuit z-Parameters

The use of open circuit impedance parameters is premised on selecting the input port current, I_1, and the output port current, I_2, as independent electrical variables in the two-port system of Fig. 2.2. The upshot of this selection is the volt-ampere matrix expression,

$$\begin{bmatrix} V_1 \\ V_2 \end{bmatrix} = \begin{bmatrix} z_{11} & z_{12} \\ z_{21} & z_{22} \end{bmatrix} \begin{bmatrix} I_1 \\ I_2 \end{bmatrix}, \tag{2-37}$$

where the z_{ij} are the network *open circuit impedance parameters*, or *z-parameters*. The equivalent circuit corresponding to Eq. (2-37) is shown in Fig. 2.20 and is seen to exploit Thévenin's theorem at both the input and output ports.

The designation of the z-parameters as open circuit impedances follows from the parametric measurement strategy implied by Eq. (2-37). In particular,

$$\left. \begin{aligned} z_{11} &= \left. \frac{V_1}{I_1} \right|_{I_2=0} \\ z_{21} &= \left. \frac{V_2}{I_1} \right|_{I_2=0} \end{aligned} \right\}, \tag{2-38}$$

Two-Port Network Models and Analysis

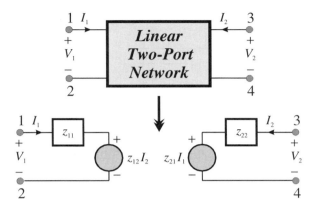

Figure 2.20. The z-parameter equivalent circuit of a linear two-port network. All of the open circuit impedance parameters, z_{ij}, are in units of ohms.

whence z_{11} is the open circuit (meaning an open-circuited output port) input impedance of the linear two-port network, while z_{21} is seen to be the open circuit transimpedance of the two-port system. Thus, like h_{11}, g_{11}, and y_{11}, z_{11} influences the input impedance of the subject two-port network, and like h_{21}, g_{21}, and y_{21}, z_{21} is a measure of the achievable forward gain. Equation (2-37) also confirms that

$$\left. \begin{array}{c} z_{12} = \dfrac{V_1}{I_2}\bigg|_{I_1=0} \\ z_{22} = \dfrac{V_2}{I_2}\bigg|_{I_1=0} \end{array} \right\}. \tag{2-39}$$

Accordingly, z_{12} is the feedback transimpedance under the condition of an open-circuited input port, and z_{22} is the open circuit output impedance measured under the same open-circuited input port.

A comparison of Eq. (2-37) with Eq. (2-23) suggests immediately that the z-parameter matrix of a two-port network is the inverse of the y-parameter matrix for the same network. Assuming that the y-parameter matrix is nonsingular and letting

$$\Delta_y = y_{11}y_{22} - y_{12}y_{21} \neq 0, \tag{2-40}$$

the z-parameters relate to their y-parameter counterparts in accordance with

$$\left.\begin{aligned} z_{11} &= \frac{y_{22}}{\Delta_y} = \frac{1}{y_{11} - \frac{y_{12}y_{21}}{y_{22}}} \\ z_{22} &= \frac{y_{11}}{\Delta_y} = \frac{1}{y_{22} - \frac{y_{12}y_{21}}{y_{11}}} \\ z_{12} &= -\frac{y_{12}}{\Delta_y} = -\frac{y_{12}}{y_{11}y_{22} - y_{12}y_{21}} \\ z_{21} &= -\frac{y_{21}}{\Delta_y} = -\frac{y_{21}}{y_{11}y_{22} - y_{12}y_{21}} \end{aligned}\right\}. \quad (2\text{-}41)$$

Note that for a bilateral two-port, which has $y_{12} = y_{21}$, z_{12} is identical to z_{21}. Note further that z-parameters cannot be determined for an ideal transconductance amplifier since, as is recalled from Eq. (2-31), the y-parameter matrix of an ideal transconductor is singular.

2.3.4.1. Tee-Type Network Model

Just as the generalized y-parameter model of a three terminal, bilateral two-port network can be converted to a π-type configuration, such as that depicted in Fig. 2.15(b), the z-parameter equivalent circuit of an analogous bilateral structure can be transformed into a tee-type topology. Returning to Eq. (2-37),

$$\left.\begin{aligned} V_1 &= z_{11}I_1 + z_{12}I_2 = (z_{11} - z_{12})\,I_1 + z_{12}\,(I_1 + I_2) \\ V_2 &= z_{21}I_1 + z_{22}I_2 = (z_{21} - z_{12})\,I_1 + (z_{22} - z_{12})\,I_2 + z_{12}\,(I_1 + I_2) \end{aligned}\right\}. \quad (2\text{-}42)$$

With

$$\left.\begin{aligned} z_i &\triangleq z_{11} - z_{12} \\ z_o &\triangleq z_{22} - z_{12} \\ z_f &\triangleq z_{21} - z_{12} \\ z_r &\triangleq z_{12} \end{aligned}\right\}, \quad (2\text{-}43)$$

Eq. (2-42) becomes expressible as

$$\left.\begin{aligned} V_1 &= z_i I_1 + z_r\,(I_1 + I_2) \\ V_2 &= z_f I_1 + z_o I_2 + z_r\,(I_1 + I_2) \end{aligned}\right\}. \quad (2\text{-}44)$$

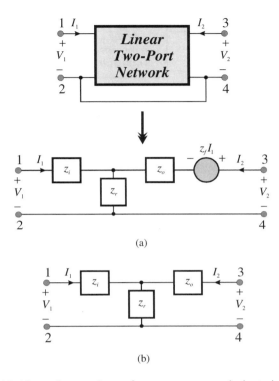

Figure 2.21. (a) Alternative topology of a z-parameter equivalent circuit for a three-terminal, linear two-port network. (b) The model of (a) for the special case of a bilateral, three-terminal linear network.

The equivalent circuit shown in Fig. 2.21(a) follows immediately from Eq. (2-44). For the bilateral network case, which entails $z_{21} = z_{12}$, whence $z_f = 0$, Fig. 2.21(b) is the applicable model.

2.3.4.2. Ideal Transresistor

An ideal *transresistance amplifier*, or *transresistor*, functions effectively as a current controlled voltage source. The ideal transresistance amplifier boasts zero input impedance for all load terminations, as well as zero output impedance for all signal source terminations. The input port is therefore most suitable for signal current, as opposed to signal voltage, excitation, and the output port behaves as an ideal voltage source. As is illustrated in Fig. 2.22(a), the ideal transresistor provides an output port voltage that

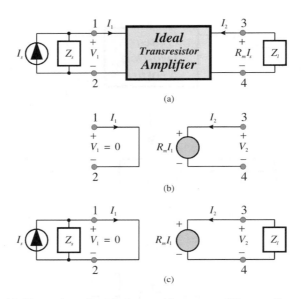

Figure 2.22. (a) Abstraction of an ideal transresistance amplifier coupling a signal source current to an impedance load. (b) The z-parameter model of the ideal transresistor in (a). (c) The z-parameter equivalent circuit of the system in (a).

is linearly dependent on the signal source current, I_s. The proportionality constant linking this output port voltage to the signal current is R_m, which is presumed independent of signal frequency. A comparison of the resultant model shown in Fig. 2.22(b) with the generalized z-parameter equivalent circuit in Fig. 2.20 leads to the conclusion that an ideal transresistor has $z_{11} = z_{22} = z_{12} = 0$ and $z_{21} = R_m$. Accordingly, its z-parameter equilibrium equations are

$$\begin{bmatrix} V_1 \\ V_2 \end{bmatrix} = \begin{bmatrix} 0 & 0 \\ R_m & 0 \end{bmatrix} \begin{bmatrix} I_1 \\ I_2 \end{bmatrix}. \tag{2-45}$$

In the corresponding system model of Fig. 2.22(c), note that the input port short circuit afforded by $z_{11} = z_{12} = 0$ forces $I_1 = I_s$, independent of the Thévenin source impedance, Z_s. Additionally, the idealized voltage source nature of the output port ensures a load voltage, V_2, that is identical to $R_m I_1$ for all load impedances, Z_l. Since $I_1 \equiv I_s$, it follows that $V_2/I_s \equiv R_m$ for all Z_s and Z_l, as is suggested in Fig. 2.22(a). Observe that the z-parameter matrix pertinent to an ideal transresistance amplifier is singular.

2.3.5. Transmission Parameters

The *transmission parameters*, which are often referred to as the *chain parameters*, c_{ij}, of the linear two-port network in Fig. 2.2 are implicitly defined by the matrix volt-ampere description,

$$\begin{bmatrix} V_1 \\ I_1 \end{bmatrix} = \begin{bmatrix} c_{11} & c_{12} \\ c_{21} & c_{22} \end{bmatrix} \begin{bmatrix} V_2 \\ -I_2 \end{bmatrix}. \qquad (2\text{-}46)$$

In the c-parameter representation of network volt-ampere characteristics, the selected independent variables are output port voltage and the negative of output port current, thereby remanding input port voltage and input port current to dependent variable status. The dimensions of the chain parameters are nonhomogeneous in that c_{11} and c_{22} are dimensionless, c_{12} has units of impedance, and c_{21} has units of admittance. In contrast to the four traditional two-port parameter representations, the chain parameters are rarely used for circuit modeling purposes. Instead, these parameters are used directly in the analysis and design of networks and are especially useful in the design of passive filters comprised of cascaded passive subcircuits.

As usual, the strategy underlying the measurement of the transmission parameters follows directly from the defining volt-ampere characteristics. To this end, let the output port of the considered system be open circuited so that the output port current, I_2, is nulled. Simultaneously, let the input port be voltage-driven to establish an input port voltage, V_1, and an input port current, I_1. The resultant test fixturing, shown in Fig. 2.23(a), is identical to that of Fig. 2.10(a), which is used to measure the open circuit hybrid g-parameters, g_{11} and g_{21}. From Eq. (2-46), $I_2 = 0$ produces

$$\left. \begin{array}{l} \dfrac{1}{c_{11}} = \dfrac{V_2}{V_1}\bigg|_{I_2=0} \\ \dfrac{1}{c_{21}} = \dfrac{V_2}{I_1}\bigg|_{I_2=0} \end{array} \right\rangle. \qquad (2\text{-}47)$$

Thus, the inverse of c-parameter c_{11} is the Thévenin voltage gain of the network undergoing test, while the inverse of c_{21} is the open circuit (meaning that the output port is open circuited), or Thévenin, forward transimpedance of the subject network. With the output port short circuited, as indicated in

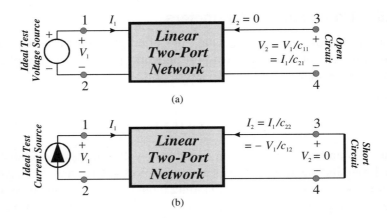

Figure 2.23. (a) Circuit for the measurement of the open circuit transmission, or chain, parameters, c_{11} and c_{21}. (b) Measurement of the short circuit transmission parameters, c_{12} and c_{22}.

Fig. 2.23(b), $V_2 = 0$ in Eq. (2-46) gives

$$\left. \begin{aligned} \frac{1}{c_{12}} &= -\frac{I_2}{V_1}\bigg|_{V_2=0} \\ \frac{1}{c_{22}} &= -\frac{I_2}{I_1}\bigg|_{V_2=0} \end{aligned} \right\rangle. \tag{2-48}$$

Clearly, the negative inverse of parameter c_{12} is the Norton forward transadmittance, and the negative inverse of parameter c_{22} is the Norton current gain of the network.

As is the case with any other set of two-port parameters, the chain parameters can be related to the h-, g-, y-, or z-parameters. Since the open circuit impedance parameters are the last parameters examined and thus still efficiently accessible in personal memory, compare Eq. (2-46) with Eq. (2-37). Specifically, observe that

$$\left. \begin{aligned} c_{11} &= \frac{V_1}{V_2}\bigg|_{I_2=0} = \frac{z_{11}}{z_{21}} \\ c_{21} &= \frac{I_1}{V_2}\bigg|_{I_2=0} = \frac{1}{z_{21}} \end{aligned} \right\rangle. \tag{2-49}$$

Moreover,

$$\left. \begin{array}{l} c_{22} = -\dfrac{I_1}{I_2}\bigg|_{V_2=0} = \dfrac{z_{22}}{z_{21}} \\ \\ c_{12} = -\dfrac{V_1}{I_2}\bigg|_{V_2=0} = \dfrac{z_{11}z_{22} - z_{12}z_{21}}{z_{21}} = \dfrac{\Delta_z}{z_{21}} \end{array} \right\}, \quad (2\text{-}50)$$

where $\Delta_z = z_{11}z_{22} - z_{12}z_{21}$ symbolizes the determinant of the z-parameter matrix. Returning to Eqs. (2-49) and (2-50), the determinant, Δ_c, of the c-parameter matrix is seen to be

$$\Delta_c = c_{11}c_{22} - c_{12}c_{21} = \dfrac{z_{12}}{z_{21}}. \quad (2\text{-}51)$$

It follows that the determinant of the transmission parameter matrix is unity if the network undergoing study is bilateral; that is if $z_{12} = z_{21}$. Interestingly enough, the c-matrix is singular if the network it represents is unilateral, which corresponds to a network divorced of feedback ($z_{12} = 0$).

2.3.5.1. Input and Output Impedances

The engineering attributes of transmission parameters begin to be revealed by applying the chain matrix concept to the problem of determining the driving point input and output impedances of the original two-port system in Fig. 2.2. The input impedance, Z_{in}, derives from Eq. (2-46) as

$$Z_{in} = \dfrac{V_1}{I_1} = \dfrac{c_{11}V_2 - c_{12}I_2}{c_{21}V_2 - c_{22}I_2}. \quad (2\text{-}52)$$

Since $V_2 = -Z_l I_2$, this relationship becomes expressible as

$$Z_{in} = \dfrac{V_1}{I_1} = \dfrac{c_{11}Z_l + c_{12}}{c_{21}Z_l + c_{22}}. \quad (2\text{-}53)$$

Equation (2-53) confirms that the driving point input impedance reduces to c_{11}/c_{21} for an open circuit load port. It should be noted that the consideration of the special case of an open-circuited load ($Z_l = \infty$) is tantamount to circuit analyses predicated on absorbing the extrinsic load impedance into the actual two-port network. Of course, such an analytical tack mandates that the network c-parameters, c_{ij}, be evaluated with the load impedance physically incorporated into the two-port network topology. In Eq. (2-52), observe that a short-circuited load termination delivers $Z_{in} = c_{12}/c_{22}$.

In order to derive an expression for the driving point output impedance, Z_{out}, the signal source voltage, V_s, in Fig. 2.2 is reduced to zero, thereby forcing $V_1 = -Z_s I_1$. This fact and Eq. (2-46) confirm that

$$Z_{out} = \frac{V_2}{I_2} = \frac{c_{22} Z_s + c_{12}}{c_{21} Z_s + c_{11}}. \tag{2-54}$$

In the limit of an ideal current source signal drive ($Z_s \to \infty$), the output impedance approaches c_{22}/c_{21}, while for an ideal voltage source signal excitation ($Z_s \to 0$), Z_{out} collapses to c_{12}/c_{11}.

2.3.5.2. Voltage Transfer Function

The voltage transfer function, V_2/V_1, from the input to the output port can be found by using Eq. (2-46) to write

$$V_1 = c_{11} V_2 - c_{12} I_2 = c_{11} V_2 + c_{12} \left(\frac{V_2}{Z_l} \right), \tag{2-55}$$

whence an I/O port voltage gain of

$$\frac{V_2}{V_1} = \frac{Z_l}{c_{11} Z_l + c_{12}}. \tag{2-56}$$

Note that Eq. (2-56) corroborates with Eq. (2-47) in the sense that an infinitely large load impedance delivers a port to port voltage gain of $1/c_{11}$. Since the input port voltage, V_1, relates to the signal source voltage, V_s, as the simple voltage divider,

$$\frac{V_1}{V_s} = \frac{Z_{in}}{Z_{in} + Z_s}, \tag{2-57}$$

this relationship, Eqs. (2-56), and (2-53) combine to deliver an overall system voltage gain of

$$\frac{V_2}{V_s} = \frac{Z_l}{(c_{11} Z_l + c_{12}) + (c_{21} Z_l + c_{22}) Z_s}. \tag{2-58}$$

As expected, Eq. (2-58) reduces to Eq. (2-56) when $Z_s = 0$, which constrains V_1 to V_s.

2.3.5.3. Cascade Interconnection

The transmission parameters and its associated c-parameter matrix are particularly useful analytical tools for ascertaining the transfer and driving point characteristics of networks whose architecture is comprised of a cascade of circuit sections. To this end, consider Fig. 2.24, which depicts a

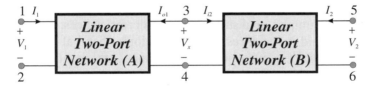

Figure 2.24. A cascade interconnection of two linear two-port networks. network A is presumed to have a transmission matrix, C_A, while C_B is the transmission matrix of network B.

two-stage cascade interconnection of two linear two-port networks. In terms of the indicated port voltages and port currents, the volt-ampere characteristics of network A are given by the matrix relationship,

$$\begin{bmatrix} V_1 \\ I_1 \end{bmatrix} = C_A \begin{bmatrix} V_x \\ -I_{o1} \end{bmatrix}, \qquad (2\text{-}59)$$

where C_A is the 2×2 chain matrix for Network A. Similarly for Network B,

$$\begin{bmatrix} V_x \\ I_{i2} \end{bmatrix} = C_B \begin{bmatrix} V_2 \\ -I_2 \end{bmatrix}, \qquad (2\text{-}60)$$

with C_B denoting the 2×2 chain matrix for network B. But since the input port current, I_{i2}, to network B is obviously the negative of the output port current, I_{o1}, of network A, Eqs. (2-59) and (2-60) combine to deliver

$$\begin{bmatrix} V_1 \\ I_1 \end{bmatrix} = C_A \begin{bmatrix} V_x \\ -I_{o1} \end{bmatrix} = C_A \begin{bmatrix} V_x \\ I_{i2} \end{bmatrix} = C_A C_B \begin{bmatrix} V_2 \\ -I_2 \end{bmatrix} \triangleq C \begin{bmatrix} V_2 \\ -I_2 \end{bmatrix}, \qquad (2\text{-}61)$$

where matrix C, which is

$$C = C_A C_B, \qquad (2\text{-}62)$$

is properly interpreted as the 2×2 transmission matrix for the overall cascade interconnection. Obviously, Eq. (2-62) can be extended to embrace a cascade of n network sections, wherein the overall transmission matrix becomes the ordered product of the chain matrices of the individual subcircuit members of the cascade. Once the overall chain matrix is evaluated, its elements can be used in conjunction with Eqs. (2-52), (2-54), and (2-58) to determine the I/O impedances and transfer function of the cascade.

Since matrix multiplication is noncommutative ($C_A C_B \neq C_B C_A$), the chain matrix, and thus the performance attributes, of a cascade interconnection of circuits is dependent on the ordering of the individual sections of the cascade. This observation mathematically confirms the transparent engineering fact that the observable network performance depends on the manner in which the individual elements and subsections of the network are interconnected. To the extent that on chip interconnect metallization is plagued by lumped or distributed parasitic resistances, conductances, capacitances, and inductances, the observation at hand also explains the sensitivity of high frequency circuit performance to the manner in which the individual components and subcircuits of a network topology are laid out.

2.3.5.4. *Series and Shunt Branch Elements*

Equation (2-62) offers an analytical foundation for the systematic evaluation of the transfer and driving point impedance functions of networks comprised of cascade interconnections of passive subcircuits. To wit, consider the impedance, Z, connected as shown in Fig. 2.25(a) as a branch element in series with an input port having port voltage V_1 and port current I_1 and an output port, where the pertinent voltage and current variables are V_2 and I_2, respectively. An application of Eqs. (2-47) and (2-48) to the impedance structure at hand gives a transmission matrix, say C_z, of

$$C_z = \begin{bmatrix} 1 & Z \\ 0 & 1 \end{bmatrix}, \tag{2-63}$$

where the determinant of C_z is seen to be unity, which is as expected in light of the presumed passivity of impedance Z. For Fig. 2.25(b), which depicts a branch admittance, Y, connected as an element that simultaneously shunts both input and output ports,

$$C_y = \begin{bmatrix} 1 & 0 \\ Y & 1 \end{bmatrix}. \tag{2-64}$$

It follows from Eq. (2-62) that for the configuration of Fig. 2.25(c), whose structure is a series impedance, Z_1, connected in cascade with a shunt admittance, Y_2, which in turn is cascaded with a second series impedance, Z_3,

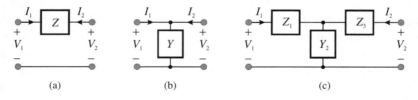

Figure 2.25. (a) Impedance branch placed in series between input and output ports. (b) Admittance branch connected in shunt with the input and output ports. (c) Tee-type network comprised of cascade interconnections of subcircuits having the topological forms in (a) and (b).

the network transmission matrix, say C_n, is

$$C_n = \begin{bmatrix} 1 & Z_1 \\ 0 & 1 \end{bmatrix} \begin{bmatrix} 1 & 0 \\ Y_2 & 1 \end{bmatrix} \begin{bmatrix} 1 & Z_3 \\ 0 & 1 \end{bmatrix} = \begin{bmatrix} 1 + Y_2 Z_1 & \{Z_3 + Z_1(1 + Y_2 Z_3)\} \\ Y_2 & 1 + Y_2 Z_3 \end{bmatrix}.$$
(2-65)

Example 2.5. The lowpass filter in Fig. 2.26 is terminated at its output port in a resistance, R_l. The Thévenin source resistance can either be presumed to be zero or otherwise absorbed into the resistance, R. Use transmission parameters to derive expressions for the voltage transfer function, $A_v(s) = V_2/V_1$, the driving point input impedance, $Z_{in}(s)$, and the driving point output impedance, $Z_{out}(s)$.

Solution 2.5.

(1) The transmission matrix of the filter follows directly from Eq. (2-65), where $Z_1 = R$, $Y_2 = sC$, and $Z_3 = sL$; namely,

$$C_n = \begin{bmatrix} 1 + sRC & sL + (1 + s^2 LC) R \\ sC & 1 + s^2 LC \end{bmatrix}.$$
(E5-1)

(2) With $Z_l = R_l$, Eq. (2-56) or alternatively, Eq. (2-58), wherein $Z_s = 0$, the voltage gain evaluates as

$$A_v(s) = \frac{V_2}{V_1} = \frac{\dfrac{R_l}{R_l + R}}{1 + s \left[\dfrac{L}{R_l + R} + (R_l \| R) C \right] + s^2 \left(\dfrac{R}{R_l + R} \right) LC}.$$
(E5-2)

(3) With $Z_l = R_l$, Eq. (2-53) yields a driving point input impedance of

$$Z_{in}(s) = (R_l + R) \left\{ \dfrac{1 + s \left[\dfrac{L}{R_l + R} + (R_l \| R) C \right] + s^2 \left(\dfrac{R}{R_l + R} \right) LC}{1 + s R_l C + s^2 LC} \right\}.$$
(E5-3)

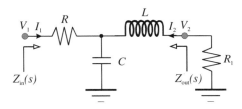

Figure 2.26. The lowpass *RLC* filter addressed in Example 2.5.

(4) From Eq. (2-54) and recalling that $Z_s = 0$, the driving point output impedance is found to be

$$Z_{out}(s) = R \left(\frac{1 + s\dfrac{L}{R} + s^2 LC}{1 + sRC} \right). \tag{E5-4}$$

Comments. It is notable that the transmission parameters and matrix allow for the formulation of the gain and I/O impedance relationships without need of laborious circuit analysis. As a check on the propriety of the results, observe that as expected, the zero frequency voltage gain is the resistive divider, $(R_l/R_l + R)$, the zero frequency value of the driving point input impedance is $(R_l + R)$, and the zero frequency output impedance is simply R.

It is also somewhat enlightening to observe that for the purely resistive load termination considered in this example, the poles of the voltage gain function are precisely the zeros of the driving point input impedance relationship. This discovery, which looms critical in the design of certain types of passive filters, can be generalized by returning to Eqs. (2-53) and (2-58) to write

$$Z_{in}(s) = \frac{c_{11} R_l + c_{12}}{c_{21} R_l + c_{22}}, \tag{E5-5}$$

and

$$A_v(s) = \frac{R_l}{c_{11} R_l + c_{12}}. \tag{E5-6}$$

It follows that

$$Z_{in}(s) = \frac{R_l}{A_v(s)\,(c_{21} R_l + c_{22})}, \tag{E5-7}$$

which clearly confirms that the poles of the voltage gain function are identical to the zeros of the input impedance expression. It should also be noted that for an infinitely large load resistance, which effectively materializes when said load is absorbed into the filter architecture,

$$A_v(s) Z_{in}(s)\big|_{R_l = \infty} = 1/c_{21}; \tag{E5-8}$$

that is, the product of forward voltage gain and driving point input impedance is simply the inverse of the transmission parameter, c_{21}. Since the poles of $A_v(s)$ are identical to the zeros of $Z_{in}(s)$, this result implies that the poles of $Z_{in}(s)$ are the zeros of parameter c_{21}, while the zeros of $A_v(s)$ are the poles of c_{21}.

2.4.0. Two-Port Methods of Circuit Analysis

The two-port parameter sets and models introduced in the preceding section of material allow for an unambiguous volt-ampere characterization of any linear two-port network in terms of the electrical properties that are observable and measurable at the input and output ports of the network. The equivalent circuits respectively corresponding to the hybrid h-, hybrid g-, short circuit y-, and open circuit z-parameters exploit superposition theory and comprise a modest extension of the classic one-port formulation of the Thévenin and Norton theorems. These models are the fundamental tools that expedite a computationally efficient analysis of the transfer and driving point characteristics of terminated electrical and electronic systems whose intrinsic circuit topologies are either unknown or too complicated for straightforward circuit analyses predicated on the Kirchhoff laws.

2.4.1. Circuit Analysis in Terms of h-Parameters

Recalling Fig. 2.3, the h-parameter equivalent circuit of the original, voltage-driven, terminated two-port system in Fig. 2.2 is the topology provided in Fig. 2.27. Although either version of the system diagrammed in Fig. 2.2 can be selected for analysis, the voltage-driven version is chosen because the Thévenin nature of the h-parameter input port model is synergistic with the application of the Kirchhoff voltage law to this port. In particular,

$$V_s = (Z_s + h_{11}) I_1 + h_{12} V_2, \quad (2\text{-}66)$$

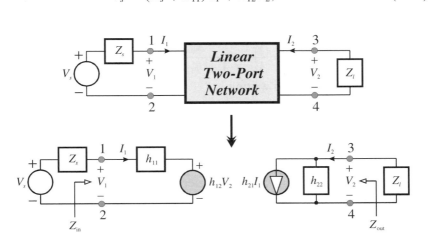

Figure 2.27. The h-parameter equivalent circuit of the linear two-port network in Fig. 2.2. The voltage-driven version of Fig. 2.2 is selected because of the Thévenin architectural nature of the h-parameter input port model.

while at the output port,

$$V_2 = -\frac{h_{21} I_1}{h_{22} + Y_l}, \qquad (2\text{-}67)$$

where Y_l is the admittance corresponding to the terminating load impedance, Z_l. If Eq. (2-67) is inserted into Eq. (2-66),

$$\frac{V_s}{I_1} = Z_s + h_{11} - \frac{h_{12} h_{21}}{h_{22} + Y_l}. \qquad (2\text{-}68)$$

Equations (2-68) and (2-67) resultantly yield a system voltage gain, A_v, of

$$A_v = \frac{V_2}{V_s} = \frac{-\dfrac{h_{21}}{(h_{11} + Z_s)(h_{22} + Y_l)}}{1 - \dfrac{h_{12} h_{21}}{(h_{11} + Z_s)(h_{22} + Y_l)}}. \qquad (2\text{-}69)$$

2.4.1.1. Open Loop Gain and Loop Gain Concepts

The cumbersome algebraic form of the voltage gain expression in Eq. (2-69) inspires neither engineering confidence nor design insights about the dynamic behavior of the system in Fig. 2.27. A more illuminating format begins to emerge from the substitution,

$$A_{vo} \triangleq -\frac{h_{21}}{(h_{11} + Z_s)(h_{22} + Y_l)}, \qquad (2\text{-}70)$$

whereupon Eq. (2-69) becomes

$$A_v = \frac{V_2}{V_s} = \frac{A_{vo}}{1 + h_{12} A_{vo}}. \qquad (2\text{-}71)$$

This result is equivalent to writing

$$V_2 = A_{vo}(V_s - h_{12} V_2), \qquad (2\text{-}72)$$

which can be mapped by the block diagram shown in Fig. 2.28. This diagram underscores h-parameter h_{12} as the factor of the output voltage response that is fed back to the system input port. Since feedback is such a critically important phenomenon in virtually all electrical and electronic circuits and systems, the parameters implicit to the diagram in Fig. 2.28 are worthy of further exploration.

To begin this complementary exploration, note in Fig. 2.28 that the signal path from the summing unit through the block labeled, A_{vo}, and thence through the h_{12} block, constitutes a closed loop, which is doubtlessly the reason that the gain expression in Eq. (2-71) is traditionally termed the

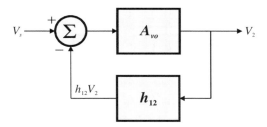

Figure 2.28. Block diagram representation of the circuit in Fig. 2.27. The diagram underscores the feedback, as manifested by parameter h_{12}, inherent to the subject circuit.

closed loop gain of the system undergoing study. Since setting h_{12} to zero reduces A_v in Eq. (2-71) to A_{vo}, and since $h_{12} = 0$ effectively opens the aforementioned closed loop, A_{vo}, in Eq. (2-70) is logically termed the *open loop gain*. Note that the circuit in Fig. 2.27 also verifies that $A_v = A_{vo}$ when $h_{12} = 0$ since by inspection,

$$I_1|_{h_{12}=0} = \frac{V_s}{Z_s + h_{11}}, \qquad (2\text{-}73)$$

whence by Eq. (2-67), the resultant gain is A_{vo}, as defined by Eq. (2-70).

The product, $h_{12}A_{vo}$, appearing in the denominator on the right hand side of Eq. (2-71) is seen in Fig. 1.28 as the multiplication of the open loop gain and the feedback factor. It is effectively the gain around the closed loop and is often termed the *loop gain* of the considered system. This h-parameter loop gain, say $T_h(Z_s, Y_l)$, is functionally dependent on the source impedance, Z_s, and the load admittance, Y_l. Although not delineated explicitly, this metric is also dependent on signal frequency owing to the likely dependence of the h-parameters on frequency, as well as presence of energy storage elements in either the source or load impedances or both. Formally, the loop gain is defined by

$$T_h(Z_s, Z_l) = h_{12}A_{vo} = -\left(\frac{h_{12}}{h_{11} + Z_s}\right)\left(\frac{h_{21}}{h_{22} + Y_l}\right), \qquad (2\text{-}74)$$

whereby Eq. (2-71) becomes expressible as

$$A_v = \frac{V_2}{V_s} = \frac{A_{vo}}{1 + T_h(Z_s, Z_l)}. \qquad (2\text{-}75)$$

It is interesting to note that the second parenthesized factor on the right hand side of Eq. (2-74) is the negative ratio of the output port voltage, V_2, to the input port current, I_1, where V_2 and I_1 are recalled as the independent variables implicit to h-parameter modeling. In other words, for an assumed

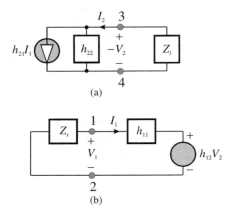

Figure 2.29. (a) Output port model of the equivalent circuit in Fig. 2.27, which is used to compute the negative input port to output port forward transimpedance, $-V_2/I_1$, component of the network loop gain. (b) Input port model of the equivalent circuit in Fig. 2.27, which is used to compute the output port to input port reverse transadmittance, I_1/V_2, component of the circuit loop gain function, $T_h(Z_s, Z_l)$.

input port current, I_1, the subject second factor is the negative forward transimpedance, $-V_2/I_1$, as is suggested in Fig. 2.29(a). On the other hand, and as is portrayed by Fig. 2.29(b), the first factor comprising the loop gain in Eq. (2-74), inclusive of the negative sign, is the reverse transadmittance ratio, I_1/V_2, for the case of zero signal source excitation, ($V_s = 0$). In other words, the negative first factor on the right hand side of Eq. (2-74) defines the input port current resulting exclusively from a presumed output port voltage. The loop gain concept is now apparent. In particular, the loop gain, which is always dimensionless, can be viewed conceptually as

$$T_h(Z_s, Z_l) = -\left(\frac{h_{12}}{h_{11} + Z_s}\right)\left(\frac{h_{21}}{h_{22} + Y_l}\right) = \left(\frac{I_1}{V_2}\right)_{|V_s=0}\left(\frac{-V_2}{I_1}\right), \quad (2\text{-}76)$$

which is a product of the forward and reverse gains formed of the independent modeling variables. In particular, it is the product of the phase-inverted forward gain (transimpedance) from input port to output port and the zero signal value of the reverse gain (transadmittance) from output port to input port. It is to be clearly understood that in Eq. (2-76), the first ratio, I_1/V_2, of independent variables is not the inverse of the second ratio, V_2/I_1, since said second ratio requires nonzero signal source voltage to establish an input port current I_1. Once this input port current is established, it gives rise to an output port voltage, V_2, whence the negative forward transimpedance, $-V_2/I_1$.

The loop gain, $T_h(Z_s, Z_l)$, then measures the amount of this generated output response that is fed back to the input port by shutting down the signal source and monitoring the resultant reverse transadmittance, I_1/V_2.

The loop gain of a linear circuit or system is a critically important metric for several reasons, two of which can be comprehended immediately. The first of these reasons derives from the presumption that over a stipulated range of signal frequencies, feedback is purposefully incorporated to realize a loop gain magnitude that is significantly larger than unity. Under this presumption, Eqs. (2-74) and (2-75) combine to deliver a closed loop, or actual, voltage gain of

$$A_v = \frac{V_2}{V_s} = \frac{A_{vo}}{1 + T_h(Z_s, Z_l)} \approx \frac{1}{h_{12}}, \tag{2-77}$$

which is dependent on only the feedback h-parameter, h_{12}, as opposed to being dependent on four h-parameters, a source impedance, and a load admittance. To the extent that h_{12} can be physically realized as a network function related to only a ratio of controllable passive components, the achievable voltage gain is rendered both reliable and analytically predictable. This laudable attribute is particularly germane to integrated electronics whose active components are invariably plagued by model parameters that are either ill-defined or poorly controlled during monolithic processing. In other words, a sufficiently large loop gain magnitude delivers an actual gain that is rendered relatively insensitive to both the h-parameters, h_{11}, h_{21}, and h_{22}, and the source and load terminations, Z_s, and Y_l, whose precise values are somewhat clouded by unavoidable parasitic losses and energy storage in practical systems. Note, however, that to the extent that closed loop gains larger than unity are generally desirable, h_{12} must necessarily be less than unity, which forces the open loop gain, A_{vo}, to be very large in magnitude. Unfortunately, large open loop gains, such as those that prevail in commercially available operational amplifiers, portend of potentially serious bandwidth limitations. The upshot of the matter is that while large loop gains are assuredly advantageous in active electronics from the perspective of gain desensitization with respect to parametric uncertainties and tolerances, they are generally difficult to sustain in broadband systems.

A second reason underlying the importance of the loop gain is the picture that this metric conveys with respect to the stability of active networks. In Fig. 2.29(b), consider the case in which over a range of signal frequencies, the algebraic sign of $h_{12}V_2$ is such as to increase the input port current, I_1. This so-called *positive feedback* would mean that feedback alone — divorced of any source excitation — increases I_1, which results in an increase in the magnitude of the output port voltage, V_2, which results in further increases in I_1, and so forth. In other words, the circuit or system under investigation is potentially unstable and, pending the extent by which I_1 increases, may be outright unstable. Although a definitive stability analysis is a complicated frequency domain task that is reserved for a forthcoming chapter, a clue as to the loop gain circumstances surrounding potential instability derives from an elementary consideration of low frequencies, where all parameters and terminations are real numbers. Specifically, I_1 increases and is therefore a positive current with $V_s = 0$ if $h_{12}V_2$ is negative. Because of Eq. (2-67), $h_{12}V_2 < 0$ requires $h_{12}h_{21}/(h_{22} + Y_l) > 0$. Since Eq. (2-76) yields

$$0 < \frac{h_{12}h_{21}}{h_{22} + Y_l} \equiv -(h_{11} + Z_s) T_h (Z_s, Z_l), \qquad (2\text{-}78)$$

the undesirable instability situation requires that the loop gain, $T_h(Z_s, Z_l)$, be negative, assuming the routine case of positive real impedances, h_{11} and Z_s. In other words, a negative loop gain (at least at low signal frequencies) is tantamount to network potential instability. In contrast, positive loop gain assures network stability. Since positive loop gain requires $h_{12}V_2 > 0$, input current I_1 decreases, the output response remains forever bounded for bounded inputs, and the system under consideration is said to be characterized by *negative feedback*.

The phrase, *potential instability*, is used in the foregoing paragraph in the context of stipulating $T_h(Z_s, Y_l) < 0$ as a necessary, but not sufficient, condition for circuit instability. For example, if the loop gain is negative, but has a magnitude less than unity, the circuit remains stable. If on the other hand, the loop gain is negative with a magnitude exceeding one, there must exist at least one frequency where the loop gain is precisely negative one. At these frequencies, the considered circuit is unstable in the sense that its gain is boundless, which means that output voltages are established even when the signal source is nulled. If $T_h(Z_s, Y_l) = -1$ at precisely one value of frequency, say ω_o, the circuit is technically unstable but nonetheless useful in communications and other systems as a sinusoidal oscillator whose steady state radial frequency of oscillation is ω_o.

2.4.1.2. I/O Impedances

The loop gain is also instrumental with respect to defining the nature of network impedances established at input and output ports. To this end, consider

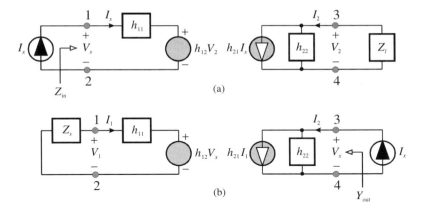

Figure 2.30. (a) Equivalent circuit for the evaluation of the driving point input impedance, Z_{in}, of the circuit modeled in Fig. 2.27. (b) Equivalent circuit for calculating the driving point output admittance of the circuit in Fig. 2.27.

Fig. 2.30(a), which relies on the model in Fig 2.27 to delineate the equivalent circuit pertinent to determining the driving point input impedance, Z_{in}. A straightforward circuit analysis confirms,

$$Z_{in} = \frac{V_x}{I_x} = h_{11} - \frac{h_{12}h_{21}}{h_{22} + Y_l} = h_{11}\left[1 - \frac{h_{12}h_{21}}{h_{11}(h_{22} + Y_l)}\right], \quad (2\text{-}79)$$

where I_x is a test independent input port current, V_x is the input port response to this current, and V_x/I_x exploits Ohm's law to determine the input impedance. Observe an input impedance of h_{11} in the absence of feedback ($h_{12} = 0$), which means that h_{11} can be thought of as an *open loop input impedance*. Moreover, Eq. (2-76) allows Eq. (2-79) to be expressed as

$$Z_{in} = h_{11} - \frac{h_{12}h_{21}}{h_{22} + Y_l} = h_{11}[1 + T_h(0, Z_l)], \quad (2\text{-}80)$$

where $T_h(0, Y_l)$ symbolizes the h-parameter loop gain under the condition of zero source impedance ($Z_s = 0$). Since the loop gain is proportional to the open loop gain, and since the magnitude of the open loop voltage gain for zero source impedance is surely larger than it is for nonzero source impedance, $|T_h(0, Y_l)| > |T_h(Z_s, Y_l)|$, which produces the inequality,

$$Z_{in} = h_{11} - \frac{h_{12}h_{21}}{h_{22} + Y_l} = h_{11}[1 + T_h(0, Z_l)] > h_{11}[1 + T_h(Z_s, Z_l)]. \quad (2\text{-}81)$$

Because the magnitude of the loop gain is desirably large for acceptable gain desensitization with respect to parametric vagaries, the last result indicates that the system in Fig. 2.27 is capable of a driving point input impedance that can be significantly larger than the h-parameter prediction of its counterpart open loop value.

Because of the Norton nature of the output port of the h-parameter network model, the driving point output admittance, Y_{out}, is more conveniently evaluated than is the corresponding output impedance. Figure 2.30(b) can be used to demonstrate straightforwardly that

$$Y_{out} = \frac{I_x}{V_x} = h_{22} - \frac{h_{12}h_{21}}{h_{11} + Z_s} = h_{22}[1 + T_h(Z_s, 0)], \qquad (2\text{-}82)$$

where Eq. (2-76) has been used once again, and $T_h(Z_s, 0)$ is the $Y_l = 0$ value of the h-parameter loop gain. Comments analogous to those made in regard to the input impedance can be proffered for the output admittance. In particular, Eq. (2-82) shows that the driving point output admittance can be considerably larger than its open loop value, which is h_{22}. It follows that the closed loop output impedance can be substantively smaller than the open loop output impedance predicted by h-parameters.

Example 2.6. Figure 2.31 is an extension of the model given in Fig. 2.7 in that it addresses an emitter-degenerated bipolar transistor amplifier whose input and output ports are terminated. The signal source applied to the amplifier input port has a Thévenin resistance, R_s, of 75 Ω, while the amplifier output port drives a resistive load, R_l, of 3 KΩ. The amplifier parameters are $r_i = 2.2$ KΩ, $r_o = 25$ KΩ, $\beta = 90$, and $R_e = 80\,\Omega$. The results of Example 2.1 are the h-parameters, $h_{11} = 9.46$ KΩ, $h_{21} = 89.71$ amps/amp, $h_{12} = 3.19$ mV/V, and $h_{22} = 39.87\,\mu$S. Determine the open loop voltage gain, A_{vo}, of the amplifier, the loop gain, $T_h(R_s, G_l)$, the closed

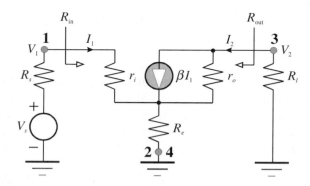

Figure 2.31. Small signal equivalent circuit of the emitter-degenerated common emitter amplifier considered in example 2.6. The model and circuit parameters are $r_i = 2.2\,\text{k}\Omega$, $r_o = 25\,\text{k}\Omega$, $\beta = 90$, $R_e = 80\,\Omega$, $R_s = 75\,\Omega$, and $R_l = 3\,\text{k}\Omega$.

loop gain, A_v, the driving point input resistance, R_{in}, and the driving point output resistance, R_{out}.

Solution 2.6.

(1) Using Eq. (2-70) with $Z_s = R_s = 75\,\Omega$ and $Y_l = 1/R_l = 1/(3\,\text{K}\Omega)$, the open loop voltage gain of the subject amplifier is found to be $A_{vo} = -25.21$, or a magnitude of 28.03 dB. From Eq. (2-74), the loop gain is $T_h(R_s, G_l) = -80.42(10^{-3})$, which is only -21.89 dB. It follows from Eq. (2-71) that the closed loop, or actual, voltage gain is $A_v = -27.42$, or a closed loop gain magnitude of 28.76 dB.

(2) The $R_s = 0$ value of the loop gain is, by Eq. (2-76), $T_h(0, G_l) = -81.06(10^{-3})$, while the $G_l = 0$ value of this metric is $T_h(R_s, 0) = -752.77(10^{-3})$. Using Eq. (2-81) and (2-82), the driving point input resistance is calculated to be $R_{in} = 8.69\,\text{K}\Omega$, while, the driving point output conductance is $G_{out} = 9.86\,\mu\text{S}$, whence an output resistance of $R_{out} = 1/G_{out} = 101.45\,\text{K}\Omega$.

Comments. The single stage emitter-degenerated bipolar amplifier is so ubiquitous in state of the art electronics that rarely is it perceived as a positive feedback architecture. Yet, the loop gain, as calculated above, is a negative number. Fortunately, it is a negative number whose magnitude is considerably less than one. Indeed, this loop gain magnitude is so much smaller than one that it is often tacitly ignored in gain and input impedance calculations, which probably explains why its positive feedback nature is typically unnoticed. But because of this small loop gain, the amplifier at hand is a terrible voltage signal processor, principally because its extremely high output impedance renders the closed loop voltage gain vulnerable to even modest load termination uncertainties.

2.4.2. Circuit Analysis Via *g*-Parameters

The circuit analyses executed in terms of a *g*-parameter model of a linear two-port network are similar to those conducted in conjunction with *h*-parameters, and indeed, with *y*- and *z*-parameters. In the case of *g*-parameters, the model in Fig. 2.9 allows the system of Fig. 2.2 to be represented by the equivalent circuit offered in Fig. 2.32. The shunt nature of the input port for the *g*-parameter model renders a Norton representation of the signal source more expedient than a Thévenin architecture. Moreover, the series nature of the *g*-parameter output port model encourages a view of the load termination as an impedance, as opposed to an admittance.

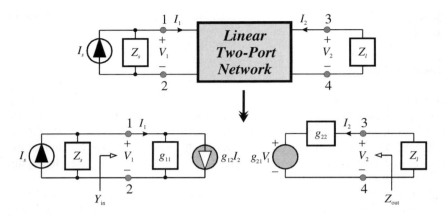

Figure 2.32. The g-parameter equivalent circuit of the linear two port network in Fig. 2.2. The current-driven version of Fig. 2.2 is selected because of the Norton nature of the g-parameter input port model.

A conventional circuit analysis of the equivalent circuit in Fig. 2.32 yields a current gain, A_i, of

$$A_i = \frac{I_2}{I_s} = \frac{-\dfrac{g_{21}}{(g_{11} + Y_s)(g_{22} + Z_l)}}{1 - \dfrac{g_{12}g_{21}}{(g_{11} + Y_s)(g_{22} + Z_l)}}, \tag{2-83}$$

which can be cast into the form,

$$A_i = \frac{I_2}{I_s} = \frac{A_{io}}{1 + T_g(Y_s, Z_l)}. \tag{2-84}$$

In the last expression,

$$A_{io} = -\frac{g_{21}}{(g_{11} + Y_s)(g_{22} + Z_l)} \tag{2-85}$$

is the open loop current gain; that is, it is the current gain of the two-port system in Fig. 2.2 under the condition that the feedback g-parameter, g_{12}, is nulled. Moreover,

$$T_g(Y_s, Z_l) \triangleq g_{12}A_{io} = -\left(\frac{g_{12}}{g_{11} + Y_s}\right)\left(\frac{g_{21}}{g_{22} + Z_l}\right) \tag{2-86}$$

is the g-parameter loop gain of the network. Note that this loop gain is expressed as an explicit function of the Thévenin source admittance, Y_s, and the terminating load impedance, Z_l. Moreover, observe the striking similarity in form of Eqs. (2-84) through (2-86) with their respective h-parameter counterpart relationships. The stability and closed loop gain desensitization commentaries proffered in conjunction with the loop gain premised on h-parameters apply equally well to the g-parameter loop gain defined by Eq. (2-86).

The aforementioned functional similarity with corresponding h-parameter expressions applies equally well insofar as the driving point input admittance, Y_{in}, and the driving point output impedance, Z_{out}, in Fig. 2.32 are concerned. Specifically, it can be demonstrated that

$$Y_{in} = g_{11} - \frac{g_{12}g_{21}}{g_{22} + Z_l} = g_{11}\left[1 + T_g(0, Z_l)\right], \quad (2\text{-}87)$$

and

$$Z_{out} = g_{22} - \frac{g_{12}g_{21}}{g_{11} + Y_s} = g_{22}\left[1 + T_g(Y_s, 0)\right]. \quad (2\text{-}88)$$

In Eq. (2-87), the source admittance (not the source impedance) is nulled in the course of evaluating the input admittance. In Eq. (2-88), the load impedance (not the load admittance) is set to zero for the output impedance computation. Observe that parameter g_{11} can now be interpreted as the open loop value of the driving point input admittance. On the other hand, g_{22} is the open loop value of the driving point output impedance.

2.4.3. Circuit Analysis in Terms of y-Parameters

If the two-port system of Fig. 2.2 is modeled by short circuit admittance, or y-, parameters, the model in Fig. 2.14 prevails, and the equivalent circuit in Fig. 2.33 results. The shunt input port nature of the y-parameter model suggests the prudence of a Norton representation of the signal source, while the similar Norton topology of the y-parameter output port encourages an admittance representation of the load admittance. With port voltage V_2 as the independent output variable and Norton signal current I_s as the known independent energy source, the closed loop gain is a transimpedance, Z_f, which can be shown to be

$$Z_f = \frac{V_2}{I_s} = \frac{Z_{fo}}{1 + T_y(Y_s, Y_l)}, \quad (2\text{-}89)$$

where

$$T_y(Y_s, Y_l) \triangleq y_{12} Z_{fo} = -\left(\frac{y_{12}}{y_{11} + Y_s}\right)\left(\frac{y_{21}}{y_{22} + Y_l}\right) \quad (2\text{-}90)$$

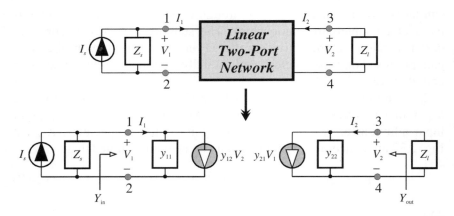

Figure 2.33. The y-parameter equivalent circuit of the indicated linear two-port network. The current-driven version of Fig. 2.2 is selected as the system architecture because of the Norton nature of the y-parameter input port model.

is the loop gain expressed in terms of y-parameters, and

$$Z_{fo} = -\frac{y_{21}}{(y_{11} + Y_s)(y_{22} + Y_l)} \tag{2-91}$$

is the open loop transimpedance; that is, the closed loop transimpedance under the zero feedback constraint, $y_{12} = 0$. A conventional Ohm's law type of circuit analysis confirms driving point input and output admittances, Y_{in} and Y_{out}, respectively, of

$$Y_{in} = y_{11} - \frac{y_{12}y_{21}}{y_{22} + Y_l} = y_{11}\big[1 + T_y(0, Y_l)\big] \tag{2-92}$$

and

$$Y_{out} = y_{22} - \frac{y_{12}y_{21}}{y_{11} + Y_s} = y_{22}\big[1 + T_y(Y_s, 0)\big]. \tag{2-93}$$

It is to be understood that the short circuit admittances, y_{11} and y_{22}, are respectively the open loop, or zero feedback value of, the driving point input and output admittances. Moreover, note that the signal source admittance (not the source impedance) is nulled in the loop gain term implicit to the input admittance expression of Eq. (2-92). On the other hand, the load admittance (not the load impedance) is set to zero in the y-parameter loop gain, $T_y(Y_s, Y_l)$, in the course of computing the driving point output admittance of the linear network undergoing study.

2.4.4. Circuit Analysis in Terms of z-Parameters

The open circuit impedance, or z-, parameter model of a linear two-port network is shown in Fig. 2.20. The use of this model in conjunction with the system of Fig. 2.2 delivers the equivalent circuit in Fig. 2.34. The series nature of both the input and output ports of this equivalent circuit encourages an impedance characterization of the load termination and a Thévenin representation of the applied signal source. Accordingly, the most expediently evaluated forward gain is the transadmittance, I_2/V_s, where I_2 is the output port current and, of course, V_s is the Thévenin signal voltage.

An analysis of the model in Fig. 2.34 stipulates the aforementioned closed loop transadmittance, Y_f, as

$$Y_f = \frac{I_2}{V_s} = \frac{Y_{fo}}{1 + T_z(Z_s, Z_l)}, \tag{2-94}$$

where the open loop value, Y_{fo}, of this closed loop transadmittance is

$$Y_{fo} = -\frac{z_{21}}{(z_{11} + Z_s)(z_{22} + Z_l)}. \tag{2-95}$$

Moreover, the z-parameter loop gain, $T_z(Z_s, Z_l)$, in Eq. (2-8) is given by

$$T_z(Z_s, Z_l) \triangleq z_{12} Y_{fo} = -\left(\frac{z_{12}}{z_{11} + Z_s}\right)\left(\frac{z_{21}}{z_{22} + Z_l}\right). \tag{2-96}$$

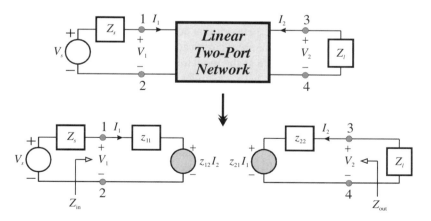

Figure 2.34. The z-parameter equivalent circuit of the indicated linear two-port network. The voltage-driven version of Fig. 2.2 is selected as the system architecture because of the Thévenin nature of the z-parameter input port model.

Finally, the indicated input and output impedances, Z_{in} and Z_{out}, respectively, can be shown to be

$$Z_{in} = z_{11} - \frac{z_{12}z_{21}}{z_{22} + Z_l} = z_{11}[1 + T_z(0, Z_l)] \qquad (2\text{-}97)$$

and

$$Z_{out} = z_{22} - \frac{z_{12}z_{21}}{z_{11} + Z_s} = z_{22}[1 + T_z(Z_s, 0)]. \qquad (2\text{-}98)$$

Example 2.7. The equivalent circuit given in Fig. 2.35 is an approximate, low frequency, common source model of a MOSFET amplifier that can be designed to realize match terminated resistance characteristics. With reference to the diagram in Fig. 2.35, "match termination" means that for a Thévenin source resistance, R_s, equal to the terminating load resistance, R_l, the driving point input resistance, R_{in}, and the driving point output resistance, R_{out}, are equal and, in fact, are equal to R_s. In other words, if R symbolizes the numerical value of the source and load resistances, $R_s = R_{in} = R_{out} = R_l = R$ under match terminated operating conditions. Use two-port parameter techniques to determine the values of the circuit resistances, R_x and R_f, commensurate with a match terminated condition. For this condition, derive an expression for the resultant voltage gain, $A_v = V_2/V_s$. For the purpose of this problem, assume that the resistance, R_i, which represents the static resistance seen between the gate and source terminals of the utilized MOSFET, is very large. No presumption should be made about the degeneration resistance, R_g, in the source terminal of the MOSFET circuit model.

Solution 2.7.

(1) The first step in the solution procedure is a determination of the appropriate two-port parameters of the network that couples the signal source to the load termination in Fig. 2.35. The short circuit admittance parameters are somewhat arbitrarily selected herewith as the modeling tool and to this end, Fig. 2.36(a), which depicts the subject network under the condition of a short-circuited output port, is pertinent to the evaluation of the y-parameters, y_{11} and y_{21}. Note in this figure that the current conducted by resistance R_g is

$$\frac{V}{R_i} + g_m V = \frac{V}{R_i}(1 + g_m R_i),$$

whence the input port voltage, V_1, is expressible as

$$V_1 = \frac{V}{R_i}[R_i + (1 + g_m R_i) R_g]. \qquad (E7\text{-}1)$$

It follows from Fig. 2.36(a) that

$$I_1 = \frac{V}{R_i} + \frac{V_1}{R_f},$$

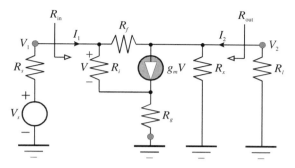

Figure 2.35. The low frequency, small signal equivalent circuit of the match terminated amplifier considered in Example 2.7.

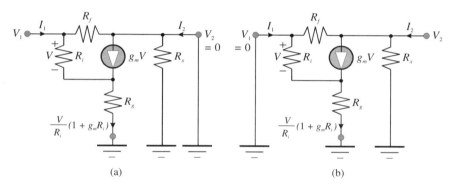

Figure 2.36. (a) Circuit model used to evaluate the y-parameters, y_{11} and y_{21}, in Example 2.7. (b) Equivalent circuit for calculating the y-parameters, y_{12} and y_{22}, in Example 2.7.

from which the short circuit input admittance, y_{11}, evaluates as

$$y_{11} = \left.\frac{I_1}{V_1}\right|_{V_2=0} = \frac{1}{R_f} + \frac{1}{R_i + (1 + g_m R_i) R_g} \approx \frac{1}{R_f}. \quad \text{(E7-2)}$$

Moreover, since

$$I_2 = g_m V - \frac{V_1}{R_f},$$

the short circuit forward transadmittance, y_{21}, is

$$y_{22} = \left.\frac{I_2}{V_1}\right|_{V_2=0} = \frac{g_m R_i}{R_i + (1 + g_m R_i) R_g} - \frac{1}{R_f} \approx g_{me} - \frac{1}{R_f}, \quad \text{(E7-3)}$$

where large R_i is presumed and

$$g_{me} \triangleq \frac{g_m}{1 + g_m R_g} \tag{E7-4}$$

defines an effective forward transconductance. The immediately observed impact of the source terminal degeneration resistance, R_g, is a potentially substantive degradation of the original value of the MOSFET forward transconductance, g_m.

(2) Figure 2.36(b) offers the equivalent circuit appropriate to the computation of the y-parameters, y_{12} and y_{22}. In this figure, the current conducted by resistance R_g, is identical to the current flowing through R_g in Fig. 2.36(a). Hence, the input port voltage, V_1, in Eq. (E7-1) has the same algebraic expression for both circuit models, which means that with $V_1 = 0$, voltage V is also zero. It follows immediately that

$$y_{12} = \left. \frac{I_1}{V_2} \right|_{V_1=0} = -\frac{1}{R_f}, \tag{E7-5}$$

and

$$y_{22} = \left. \frac{I_2}{V_2} \right|_{V_1=0} = \frac{1}{R_f} + \frac{1}{R_x}. \tag{E7-6}$$

(3) If the Thévenin source impedance and the terminating load impedance are purely resistive and each equal to the stipulated match terminated resistance, R, $Y_s = Y_l = 1/R$ in Eqs. (2-92) and (2-93). A further inspection of these two relationships indicates that the driving point input admittance, Y_{in} (a conductance in this case), and the driving point output admittance, Y_{out} (also presently a conductance), are equal to one another for $Y_s = Y_l = 1/R$ if and only if $y_{11} = y_{22}$. Recalling Eqs. (E7-2) and (E7-6), $y_{11} = y_{22}$ if the shunting output port resistance, R_x, is chosen to satisfy

$$R_x = R_g + \left(1 + g_m R_g\right) R_i, \tag{E7-7}$$

which, for large R_i, reduces to

$$R_x \approx \left(1 + g_m R_g\right) R_i. \tag{E7-8}$$

A large gate-source resistance, R_i, is seen to command a proportionately large shunt output port resistance, R_x.

(4) Returning to either Eq. (2-92) or Eq. (2-93),

$$\frac{1}{R} = y_{11} - \frac{y_{12}y_{21}}{y_{11} + \frac{1}{R}}, \tag{E7-9}$$

where, from the preceding computational step, use is made of the fact that the selection of resistance R_x in accordance with Eq. (E7-7) forces $y_{22} = y_{11}$. Using Eqs. (E7-2), (E7-5), and (E7-7), Eq. (E7-9) implies

$$R_f = \frac{(2 + g_{me}R_x)R_x}{(R_x/R)^2 - 1}. \tag{E7-10}$$

Since R_x, is very large by virtue of its direct proportionality to the very large model resistance, R_i, Eq. (E7-10) reduces to the approximate result,

$$R_f \approx g_{me}R^2 = \frac{g_m R^2}{1 + g_m R_g}, \tag{E7-11}$$

where Eq. (E7-4) is exploited with respect to the effective transconductance parameter, g_{me}.

(5) The transimpedance of the network in Fig. 2.35 is given collectively by Eqs. (2-89) through (2-91). The transimpedance, Z_f, relates to the network voltage gain, A_v, in accordance with

$$A_v = \frac{V_2}{V_s} = \frac{V_2}{I_s R} = \frac{Z_f}{R}, \tag{E7-12}$$

where I_s is the Norton signal current corresponding to the Thévenin signal voltage, V_s, and the source resistance, R_s, which is R in the present case. With $y_{11} = y_{22}$ and $Y_s = Y_l = R_s \equiv R$, Eq. (E7-12) and the aforementioned transimpedance equations produce

$$A_v = \frac{V_2}{V_s} = -\frac{y_{21}/R}{\left(y_{11} + \frac{1}{R}\right)^2 - y_{12}y_{21}}. \tag{E7-13}$$

The substitution of the pertinent computational results of Steps 1–4 into this expression, followed by a bit of algebraic gymnastics, yields

$$A_v = \frac{V_2}{V_s} \approx -\left(\frac{g_{me}R - 1}{2}\right). \tag{E7-14}$$

Comments. Match terminated amplifiers are pivotal to state of the art communication circuits. Circuits earmarked for communication system applications are plagued by signal strengths that are comparable to electrical noise floors, thereby rendering reliable signal detection problematic. Matched terminations at a given port assure maximum power transfer therein, which ensures minimal, if any, signal power loss between the effective source and the load driven by the source.

Although the traditional engineering design approaches that underpin the realization of a match terminated amplifier can prove to be a cumbersome undertaking, this example points out the relevant propriety of two-port parameters. In particular, the two-port parameters, and particularly the y-parameters exploited herein, support the design objective by highlighting the fundamental requirements supportive of, and streamlining the design-oriented computation steps that result in, match terminated circuit operation.

2.4.5. Generalized Analytical Disclosures

A review of the open loop gain, closed loop gain, loop gain, input impedance, and output impedance expressions disclosed in the preceding four subsections of material suggest an obvious commonality among the mathematical forms of these network performance metrics. In particular, a further inspection of these results suggests that the open loop gain, say G_{ol}, of any linear two-port network can always be written in the form,

$$G_{ol} = \frac{\eta_{21}}{(\eta_{11} + \Gamma_s)(\eta_{22} + \Gamma_l)}, \qquad (2\text{-}99)$$

while the corresponding closed loop gain, G_{cl}, is expressible as

$$G_{cl} = \frac{G_{ol}}{1 + T_\eta(\Gamma_s, \Gamma_l)}, \qquad (2\text{-}100)$$

where the loop gain, $T_h(\Gamma_s, \Gamma_l)$, is given by

$$T_\eta(\Gamma_s, \Gamma_l) = \eta_{12} G_{ol} = -\left(\frac{\eta_{12}}{\eta_{11} + \Gamma_s}\right)\left(\frac{\eta_{21}}{\eta_{22} + \Gamma_l}\right). \qquad (2\text{-}101)$$

Moreover, the driving point input or output impedance, say Ψ_{in}, which is often referenced generically as an input or output *immittance*, has the mathematical form,

$$\Psi_{in} = \eta_{11}\left[1 + T_\eta(0, \Gamma_l)\right], \qquad (2\text{-}102)$$

while the counterpart immittance, Ψ_{out}, for the output port is

$$\Psi_{out} = \eta_{22}\left[1 + T_\eta(\Gamma_s, 0)\right]. \qquad (2\text{-}103)$$

The profitable use of the foregoing general results is predicated on a prudent two-port parameter modeling vehicle. In turn, the choice of the two-port model parameters depends on the nature of the open loop and closed loop gains of interest. This selection also determines whether $\Gamma_s(\Gamma_l)$ is a signal source(load) impedance or admittance and whether $\Psi_{in}(\Psi_{out})$ is a driving point input(output) impedance or admittance. Table 2.1 itemizes

Table 2.1. Parameters used in the generic relationships that define the I/O performance of a linear two-port network.

GAIN	G_{ol}	G_{cl}	η_{ij}	Γ_s	Γ_l	Ψ_{in}	Ψ_{out}
Voltage	A_{vo}	A_v	h_{ij}	Z_s	Y_l	Z_{in}	Y_{out}
Current	A_{io}	A_i	g_{ij}	Y_s	Z_l	Y_{in}	Z_{out}
Transimpedance	Z_{fo}	Z_f	y_{ij}	Y_s	Y_l	Y_{in}	Y_{out}
Transadmittance	Y_{fo}	Y_f	z_{ij}	Z_s	Z_l	Z_{in}	Z_{out}

the nature of the parameters underlying the use of Eqs. (2-99) through (2-103). For example, when using g-parameters ($\eta = g$) to evaluate the performance of a linear two-port network, the open loop and closed loop gains in Eqs. (2-99) and (2-100) are current gains ($G_{ol} = A_{io}$ and $G_{cl} = A_i$), the Thévenin source immittance is represented as an admittance ($\Gamma_s = Y_s$), and the load termination is viewed as an impedance ($\Gamma_l = Z_l$). Moreover, the pertinent driving point input immittance is an admittance ($\Psi_{in} = Y_{in}$), while the driving point output immittance is an impedance ($\Psi_{out} = Z_{out}$).

2.5.0. Systems of Interconnected Two Ports

The closed loop gain (voltage gain, current gain, transimpedance, or transadmittance) of a linear two-port network is nominally independent of the source and load terminations and nominally inversely dependent on the two-port feedback parameter (h_{12}, g_{12}, y_{12}, or z_{12}) if the magnitude of the network loop gain is sufficiently large. For most active networks divorced of purposefully applied external feedback and all passive circuits, the internal feedback parameter is too small to deliver the loop gain magnitude required for this laudable parametric desensitization.

To ensure reliable and predictable closed loop performance, the internal feedback parameter of an active two-port network is commonly augmented by incorporating a second feedback network extrinsic to the active cell. This appended feedback loop is routinely, but not universally, formed of passive components for at least two reasons. First, the values of passive elements, and especially the ratios of passive element component values, are more predictable than are the parameters of active devices. Thus, to the extent that the incorporated external feedback network is the dominant vehicle for determining the effective feedback parameter of the overall closed loop system, the resultant closed loop gain becomes a predictable system performance metric. A second reason underpinning the use of passive external feedback structures stems from the fact that passive circuits are reciprocal architectures that are incapable of providing gain. They are thus stable structures that do not substantially alter the forward gain magnitude and phase characteristics of the active cells with which they interact. In particular, the magnitude of the forward transmission parameter of a passive feedback cell is likely to be significantly smaller than the magnitude of the corresponding forward transmission parameter of the active cell to which it is interconnected.

When the appended feedback loop is connected from the output port of the original open loop two-port network to its input port, the nature of the feedback is said to be *global*. There are four commonly exploited forms of global feedback: *series-shunt feedback, shunt-series feedback, shunt-shunt feedback*, and *series-series feedback*. Assuming that the two-port parameters of the original open loop structure and those of the appended feedback subcircuit are unaltered by their input and output port electrical interconnections, the analysis of the resultant global architecture can be straightforwardly formulated in terms of a mere superposition of their appropriate respective two-port parameters. That which constitutes "appropriate" two-port parameters in the analysis of global feedback systems is largely a matter of technical common sense, as the following subsections illustrate.

2.5.1. Series-Shunt Feedback Architecture

Series-shunt feedback materializes by interconnecting two networks in such a way as to ensure the satisfaction of two operational constraints. The first of these constraints is that the overall input port voltage of the global interconnection is the sum of the input port voltages of the individual two networks. Second, the net output port current is precisely the sum of the respective output port currents for the two interconnected structures. The situation at hand is abstracted in Fig. 2.37(a), which shows an "uncompensated" linear two-port network, Network A, "compensated" by a second linear two-port network, Network B, that interconnects with the first network as a series-shunt feedback structure. The series nature of the interconnected input ports of these two networks is confirmed by the observation that the net input port voltage, V_1, relates to the input port voltage, V_{1a}, of Network A and the

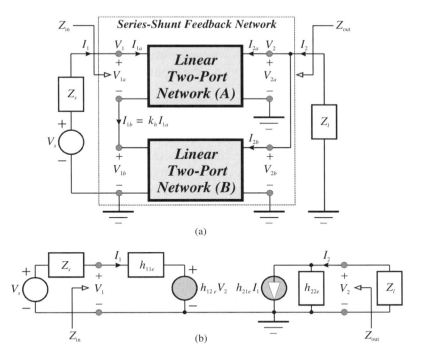

Figure 2.37. (a) Series-shunt feedback interconnection of two linear two-port networks. (b) The h-parameter equivalent circuit of the system in (a).

input port voltage, V_{1b}, of Network B as

$$V_1 = V_{1a} + V_{1b}. \tag{2-104}$$

Moreover, the net output port current, I_2, conducted by the extrinsic load impedance, Z_l, is seen to be the sum of the output currents, I_{2a} and I_{2b}, of Networks A and B, respectively; that is,

$$I_2 = I_{2a} + I_{2b}. \tag{2-105}$$

The shunt interconnected nature of the two network output ports is further confirmed by the observation that the overall output port voltage, V_2, is identical to the output port voltage V_{2a}, of Network A and the output port voltage, V_{2b}, of Network B. Specifically,

$$V_2 = V_{2a} = V_{2b}. \tag{2-106}$$

Although a strictly series input port architecture mandates that the net input current, I_1, is identical to both of the individual input port currents, I_{1a} and

I_{1b}, such current equality cannot generally be guaranteed, particularly when Network A is an active configuration. However, the linearity of the interconnected closed loop system allows conjecturing that the Network B input port current, I_{1b}, is proportional to I_{1a}, the input port current of Network A. Thus, let

$$I_{1b} = k_h I_{1a}, \qquad (2\text{-}107)$$

where k_h is determined by the electrical properties of Network A. It is a constant, and indeed possibly a frequency variant constant, that is independent of all network voltages and currents.

Because the h-parameter equivalent circuit is effectively a series-shunt architecture evoking input port voltage V_1 and output port current I_2, both of which are sums of respective port variables as per Eqs. (2-104) and (2-105), as dependent variables, h-parameters are used to model each of the two subsystems in Fig. 2.37(a). Consequently,

$$\begin{bmatrix} V_{1a} \\ I_{2a} \end{bmatrix} = \begin{bmatrix} h_{11a} & h_{12a} \\ h_{21a} & h_{22a} \end{bmatrix} \begin{bmatrix} I_{1a} \\ V_{2a} \end{bmatrix} = \begin{bmatrix} h_{11a} & h_{12a} \\ h_{21a} & h_{22a} \end{bmatrix} \begin{bmatrix} I_1 \\ V_2 \end{bmatrix}, \qquad (2\text{-}108)$$

where Eq. (2-106) and the fact that I_{1a} is the overall input port current, I_1, are exploited. For Network B,

$$\begin{bmatrix} V_{1b} \\ I_{2b} \end{bmatrix} = \begin{bmatrix} h_{11b} & h_{12b} \\ h_{21b} & h_{22b} \end{bmatrix} \begin{bmatrix} I_{1b} \\ V_{2b} \end{bmatrix} = \begin{bmatrix} h_{11b} & h_{12b} \\ h_{21b} & h_{22b} \end{bmatrix} \begin{bmatrix} k_h I_{1a} \\ V_{2b} \end{bmatrix}$$

$$= \begin{bmatrix} k_h h_{11b} & h_{12b} \\ k_h h_{21b} & h_{22b} \end{bmatrix} \begin{bmatrix} I_1 \\ V_2 \end{bmatrix}, \qquad (2\text{-}109)$$

where Eqs. (2-106), (2-107), and the rules of matrix algebra, have been applied. Noting that the input port current, I_1, and the output port voltage, V_2, appear as common independent variables in both of the preceding two relationships, Eqs. (2-104) and (2-105) combine with Eqs. (2-108) and (2-109) to deliver

$$\begin{bmatrix} V_1 \\ I_2 \end{bmatrix} = \begin{bmatrix} (h_{11a} + k_h h_{11b}) & (h_{12a} + h_{12b}) \\ (h_{21a} + k_h h_{21b}) & (h_{22a} + h_{22b}) \end{bmatrix} \begin{bmatrix} I_1 \\ V_2 \end{bmatrix}. \qquad (2\text{-}110)$$

This simple result straightforwardly delineates the h-parameter equivalent circuit for the series-shunt feedback network of Fig. 2.37(a). This model appears in Fig. 2.37(b), where from Eq. (2-110), it is understood that the effective h-parameters, h_{ije}, of the overall feedback system are

$$\begin{bmatrix} h_{11e} & h_{12e} \\ h_{21e} & h_{22e} \end{bmatrix} = \begin{bmatrix} (h_{11a} + k_h h_{11b}) & (h_{12a} + h_{12b}) \\ (h_{21a} + k_h h_{21b}) & (h_{22a} + h_{22b}) \end{bmatrix}. \quad (2\text{-}111)$$

The performance metrics of the series-shunt feedback structure now derive directly from the pertinent equations given in Section 2.4.1, subject to the proviso therein that, h_{ij} is replaced by the effective h-parameter, h_{ije}.

The engineering propriety of the analyses inferred by the composite h-parameters in Eq. (2-111) is demonstrated in the next example. Before such demonstration, it should be understood that feedback applied in series-shunt with an active two-port network results in closed loop performance appropriate to at least a first order approximation of the transfer characteristics of ideal voltage amplification. Recall that ideal voltage amplifiers have infinitely large input impedance (thus, allowing for zero attenuation of an applied voltage) and zero output impedance, which allows an output port voltage to be established across any terminating load impedance. In Fig. 2.37, the driving point input impedance, Z_{in}, is not necessarily infinitely large, but it can be rendered suitably large for intended applications. To wit, Z_{in}, which is simply the ratio, V_1/I_1, can be expressed as

$$Z_{in} = \frac{V_1}{I_1} = \frac{V_{1a} + V_{1b}}{I_{1a}} = \frac{V_{1a}}{I_{1a}} + k_h \frac{V_{1b}}{I_{1b}}, \quad (2\text{-}112)$$

where Eqs. (2-104) and (2-107) have been used. The first term on the far right hand side of Eq. (2-26) is clearly the input impedance, say Z_{ina}, of Network A, while the ratio, V_{1b}/I_{1b}, in this expression is the input impedance, Z_{inb}, of Network B. Thus, the overall, or closed loop, input impedance is

$$Z_{in} = \frac{V_1}{I_1} = Z_{ina} + k_h Z_{inb}. \quad (2\text{-}113)$$

To the extent that both Z_{ina} and Z_{inb} are positive real impedances and the constant, k_h, is large, Eq. (2-113) suggests that Z_{in} can be considerably larger than Z_{ina}, which effectively represents the network input impedance in the absence of applied series-shunt feedback.

For the output impedance of the structure in Fig. 2.37,

$$Z_{out} = \frac{V_2}{I_2} = \frac{V_2}{I_{2a} + I_{2b}} = \frac{1}{\dfrac{1}{Z_{outa}} + \dfrac{1}{Z_{outb}}} = Z_{outa} \| Z_{outb}, \quad (2\text{-}114)$$

where Z_{outa} and Z_{outb} respectively symbolize the output impedances of Networks A and B. For positive real Z_{outa} and Z_{outb}, the closed loop output impedance, Z_{out}, is clearly smaller than Z_{outa}. Depending on the value of Z_{outb}, it can be significantly smaller than Z_{outa}. Collectively, Eqs. (2-113) and (2-114) suggest a prudent implementation of series-shunt feedback around a linearized electronic two-port network can produce a reasonable engineering approximation of ideal voltage amplification.

Example 2.8. Figure 2.38(a) schematically depicts series-shunt feedback implemented in bipolar junction transistor (BJT) technology. As is suggested in this figure, active Network A is the two-stage transistor amplifier, while feedback Network B is the voltage divider comprised of the resistances, R_1 and R_2. Assume that both transistors are identical and are biased identically (biasing is not shown) and that Figure 2.38(b) is a sufficiently accurate low frequency model of each transistor. In this model, take $r_i = 2\,\text{K}\Omega$ and $\beta = 120$. Additionally, let $R_1 = 150\,\Omega$, and $R_2 = 900\,\Omega$. Assume that the signal source resistance, R_s, is $75\,\Omega$, while the terminating load resistance, R_l, is $1.2\,\text{K}\Omega$. Calculate the overall voltage gain, $A_v = V_2/V_s$, the driving point input resistance, R_{in}, and the driving point output resistance, R_{out}. Additionally, deduce appropriately simplified analytical relationships for each of these three performance indices.

Solution 2.8.

(1) The equivalent circuit of the overall feedback is depicted in Fig. 2.39(a), where use has been made of the bipolar model suggested in Fig. 2.38(b). Since feedback Network B is connected in series-shunt with active Network A, h-parameters are the appropriate modeling vehicle for Network B. To this end,

$$\left. \begin{array}{l} h_{11b} = \dfrac{V_{1b}}{I_{1b}}\bigg|_{V_{2b}=0} = R_1 \| R_2 \\[6pt] h_{22b} = \dfrac{I_{2b}}{V_{2b}}\bigg|_{I_{1b}=0} = \dfrac{1}{R_1 + R_2} \\[6pt] h_{12b} = \dfrac{V_{1b}}{V_{2b}}\bigg|_{I_{1b}=0} = \dfrac{R_1}{R_1 + R_2} = -h_{21b} \end{array} \right\} \qquad \text{(E8-1)}$$

This result promotes Fig. 2.39(b) as the h-parameter equivalent circuit of feedback Network B. If the feedback network in the amplifier of Fig. 2.39(a) is replaced by the h-parameter model of Fig. 2.39(b), the equivalent circuit of the overall amplifier can be configured as illustrated in Fig. 2.39(c). In the latter diagram, use is made of the fact that current I in Fig. 2.39(a) is $-\beta I_{1a} = -\beta I_1$, whence βI in the same figure becomes $-\beta^2 I_1$. Moreover, observe that the input current, I_{1b}, of feedback Network B is $(\beta + 1)I_{1a} = (\beta + 1)I_1$, which implies that the constant, k_h, introduced in Eq. (2-102) is $k_h = (\beta + 1)$. It follows that $h_{21b}I_{1b} = (\beta + 1)h_{21b}I_1$.

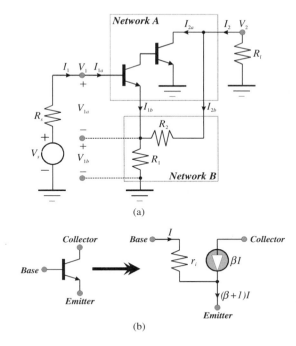

Figure 2.38. (a) Series-shunt feedback amplifier realized in bipolar junction transistor (BJT) technology. (b) Simplified small signal, low frequency model of a BJT.

(2) By comparing the model in Fig. 2.39(c) with the generalized h-parameter equivalent circuit in Fig. 2.37, the hybrid h-parameters of the overall feedback amplifier can be deduced by mere inspection. In particular,

$$\left.\begin{aligned} h_{11e} &= r_i + (\beta+1)(R_1 \| R_2) = 17.56 \, \text{K}\Omega \\ h_{21e} &= -\left[\beta^2 + (\beta+1)\left(\frac{R_1}{R_1+R_2}\right)\right] = -14.42 \, \text{KA/A} \\ h_{12e} &= \frac{R_1}{R_1+R_2} = 1/7 \, \text{V/V} \\ h_{22e} &= \frac{1}{R_1+R_2} = 952.4 \, \mu S \end{aligned}\right\} \quad \text{(E8-2)}$$

(3) The closed loop performance of the subject amplifier can now be determined directly from pertinent relationships in Section 2.4.1. From Eq. (2-70), the open loop voltage gain is $A_{vo} = 457.9 = 53.22 \, \text{dB}$, and from Eq. (2-71), the corresponding closed loop voltage gain is $A_v = 6.90 = 16.77 \, \text{dB}$. Note a loop gain of $T_h(R_s, G_l) = h_{12e} A_{vo} = 65.41 = 38.31 \, \text{dB}$.

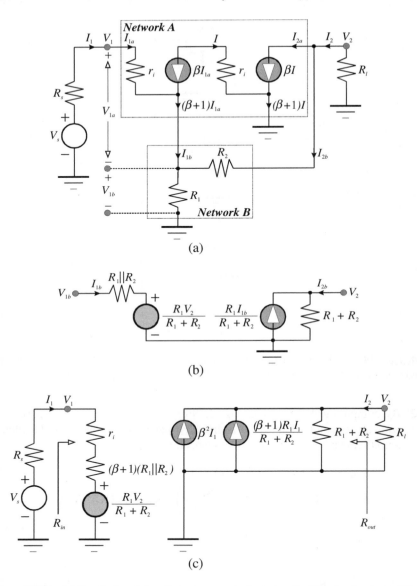

Figure 2.39. (a) Small signal, low frequency equivalent circuit of the feedback amplifier in Fig. 2.38(a). (b) The h-parameter model of feedback network B. (c) Simplified version of the equivalent circuit in (a). Note that the topology of this architecture reflects that of the classic h-parameter equivalent circuit of a linear network.

(4) Using Eq. (2-80), the closed loop input resistance is $R_{in} = 1.17$ Meg Ω, while Eq. (2-82) yields a closed loop output conductance of $G_{out} = 117.8$ mS, which equates to an output resistance of $R_{out} = 8.49\,\Omega$.

(5) Since the loop gain in this example is very large, several simplified analytical relationships can be forged. Recalling Eq. (2-71) once again, the closed loop gain is approximately

$$A_v \approx \frac{1}{h_{12e}} = 1 + \frac{R_2}{R_1} = 7.0, \qquad (E8\text{-}3)$$

which is only 1.45% higher than the precisely computed voltage gain.

(6) Because of large loop gain, the driving point input resistance, R_{in}, is expressible as

$$R_{in} \approx h_{11e} T_h(0, G_l),$$

which for large β is equivalent to

$$R_{in} \approx \beta^2 R_1 \left(\frac{R_l}{R_l + R_1 + R_2}\right) = 1.15\,\text{Meg}\Omega. \qquad (E8\text{-}4)$$

This result is lower than that computed in Step 4 by only 1.54%. Similarly, the output conductance, $G_{out} = 1/R_{out}$, is

$$G_{out} \approx h_{22e} T_h(R_s, 0),$$

or

$$R_{out} = \frac{1}{G_{out}} \approx \frac{R_2}{\beta} = 7.5\,\Omega. \qquad (E8\text{-}5)$$

When compared to the calculation in Step 4, this result is seen to be in error by -11.7%.

Comments. Very large input resistance and very small output resistance combine to make the series-shunt feedback amplifier an excellent voltage amplifier. In particular, its low output resistance allows small load resistances to be driven without rendering the voltage gain significantly dependent on the terminating load resistance. Similarly, its large input resistance effectively desensitizes the voltage gain to uncertainties in source resistances. Best of all, the resultant approximate voltage gain is dependent on a resistance ratio, which is relatively easy to control accurately during monolithic processing. It is important to underscore the fact that these design-oriented observations derive directly from the fruits of a straightforward two-port parameter analysis of the subject feedback configuration. In the absence of this two-port methodology, a conventional circuit analysis of the amplifier at hand would likely have generated the documented numerical results without the articulated insights that implicitly underpin circuit design innovation.

2.5.2. Shunt-Series Feedback

The shunt-series feedback amplifier, whose system level diagram is abstracted in Fig. 2.40(a), is suitable for current amplification because of the low input impedance afforded by the shunt feedback connection at its input port and the high output impedance that derives from the series feedback interconnection at its output port. The signal source is therefore represented as a Norton structure, and it is understood that the most appropriate output response variable of interest is the overall output port current, I_2. The shunt nature of the respective input port connections is underscored by the fact that the overall input port voltage, V_1, the input port voltage, V_{1a}, of Network A (which is generally the active component of the feedback system), and the input port voltage, V_{1b}, of Network B (whose electrical nature is generally passive) are identical. Moreover, the overall input port current, I_1, is the sum of currents I_{1a} and I_{1b}, which respectively symbolize the currents flowing into the input ports of Networks A and B. On the other hand,

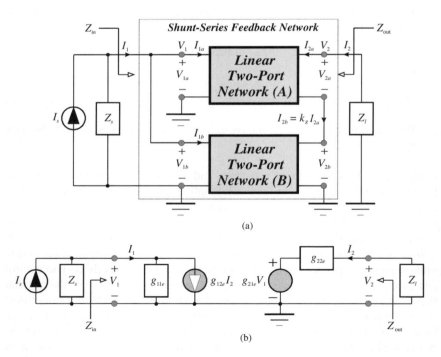

Figure 2.40. (a) Shunt-series feedback interconnection of two linear two-port networks. (b) The g-parameter equivalent circuit of the network in (a).

the overall output port voltage, V_2, is the sum of the output port voltage, V_{2a}, of Network A and V_{2b}, the voltage developed across the output port of Network B. Although this output port voltage relationship suggests a series topology, the current, I_{2b}, conducted by the output port of Network B is not necessarily the same as the output port current, I_{2a}, of Network A because of electrical constraints implicit to the circuit architecture of Network A. However, system linearity compels that I_{2b} be proportional to I_{2a} and to this end, a proportionality constant, k_g, is introduced, as is indicated in Fig. 2.40(a), so that

$$I_{2b} = k_g I_{2a}. \tag{2-115}$$

Because the g-parameter equivalent circuit is a shunt-series topology, g-parameters can be gainfully exploited for an efficient analysis of the shunt-series feedback amplifier. For Network A,

$$\begin{bmatrix} I_{1a} \\ V_{2a} \end{bmatrix} = \begin{bmatrix} g_{11a} & g_{12a} \\ g_{21a} & g_{22a} \end{bmatrix} \begin{bmatrix} V_{1a} \\ I_{2a} \end{bmatrix} = \begin{bmatrix} g_{11a} & g_{12a} \\ g_{21a} & g_{22a} \end{bmatrix} \begin{bmatrix} V_1 \\ I_2 \end{bmatrix}, \tag{2-116}$$

and for Network B,

$$\begin{bmatrix} I_{1b} \\ V_{2b} \end{bmatrix} = \begin{bmatrix} g_{11b} & g_{12b} \\ g_{21b} & g_{22b} \end{bmatrix} \begin{bmatrix} V_{1b} \\ I_{2b} \end{bmatrix} = \begin{bmatrix} g_{11b} & g_{12b} \\ g_{21b} & g_{22b} \end{bmatrix} \begin{bmatrix} V_{1b} \\ k_g I_{2a} \end{bmatrix} = \begin{bmatrix} g_{11b} & k_g g_{12b} \\ g_{21b} & k_g g_{22b} \end{bmatrix} \begin{bmatrix} V_1 \\ I_2 \end{bmatrix}. \tag{2-117}$$

Since $I_1 = I_{1a} + I_{1b}$ and $V_2 = V_{2a} + V_{2b}$, it follows that

$$\begin{bmatrix} I_1 \\ V_2 \end{bmatrix} = \begin{bmatrix} (g_{11a} + g_{11b}) & (g_{12a} + k_g g_{12b}) \\ (g_{21a} + g_{21b}) & (g_{22a} + k_g g_{22b}) \end{bmatrix} \begin{bmatrix} V_1 \\ I_2 \end{bmatrix}, \tag{2-118}$$

whence the effective g-parameters in the g-parameter equivalent circuit shown in Fig. 2.40(b) are implicitly given by

$$\begin{bmatrix} g_{11e} & g_{12e} \\ g_{21e} & g_{22e} \end{bmatrix} = \begin{bmatrix} (g_{11a} + g_{11b}) & (g_{12a} + k_g g_{12b}) \\ (g_{21a} + g_{21b}) & (g_{22a} + k_g g_{22b}) \end{bmatrix}. \tag{2-119}$$

The performance merits derived in Section 2.4.2 can now be applied directly to characterize the shunt-series feedback amplifier.

For the typical case in which Network A is an active structure and Network B is a passive unit, $|g_{21a}| \gg |g_{21b}|$ and $|g_{12a}| \ll |k_g g_{12b}|$. This is to say that the input to output, or forward, signal path is dominated by the gain afforded by the active subcircuit, while the feedback network is designed to ensure that it dominates the output to input, or feedback signal flow path. From Section 2.4.2, a large loop gain delivers a closed loop current gain, A_i, of

$$A_i = \frac{I_2}{I_s} \approx \frac{1}{g_{12e}} \approx \frac{1}{k_g g_{12b}}, \qquad (2\text{-}120)$$

which is nominally independent of the parametric vagaries implicit to active circuits. Moreover, large loop gain produces a driving point input admittance, Y_{in}, of

$$Y_{in} = \frac{1}{Z_{in}} \approx -\frac{k_g g_{12b} g_{21a}}{g_{22a} + k_g g_{22b} + Z_l} \qquad (2\text{-}121)$$

and a driving point output impedance, Z_{out}, of

$$Z_{out} \approx -\frac{k_g g_{12b} g_{21a}}{g_{11a} + g_{11b} + Y_s}, \qquad (2\text{-}122)$$

where, of course, Y_s is the admittance value of the source impedance, Z_s. The negative algebraic signs appearing on the right hand sides of the preceding two expressions are not a concern in unconditionally stable shunt-series systems, which have $k_g g_{12b} g_{21a} < 0$ at low signal frequencies.

2.5.3. Shunt-Shunt Feedback

In the shunt-shunt feedback architecture of Fig. 2.41(a), y-parameters comprise the most expeditious analytical tool. Because of the shunt feedback interconnections at both the input and the output ports of the closed loop system, the structure at hand produces low input impedance and low output impedance, thereby rendering it suitable for transimpedance applications. Armed with the experiences gleaned from the preceding two considered feedback architectures, it is easily demonstrated that the pertinent y-parameter equivalent circuit is the model in Fig. 2.41(b), where the effective y-parameters, y_{ije}, are

$$\begin{bmatrix} y_{11e} & y_{12e} \\ y_{21e} & y_{22e} \end{bmatrix} = \begin{bmatrix} (y_{11a} + y_{11b}) & (y_{12a} + y_{12b}) \\ (y_{21a} + y_{21b}) & (y_{22a} + y_{22b}) \end{bmatrix}. \qquad (2\text{-}123)$$

If Network A is an active circuit and Network B is a passive structure, $|y_{21a}| \gg |y_{21b}|$ and $|y_{12a}| \ll |y_{12b}|$ for reasons analogous to those proffered in conjunction with the shunt-series feedback amplifier. Under these conditions and for large loop gain, the closed loop transimpedance is

$$Z_f = \frac{V_2}{I_s} \approx \frac{1}{y_{12e}} \approx \frac{1}{y_{12b}}. \qquad (2\text{-}124)$$

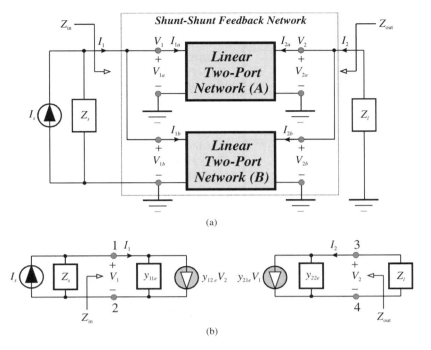

Figure 2.41. (a) Shunt-shunt feedback interconnection of two linear two-port networks. (b) The y-parameter equivalent circuit of the network in (a).

Moreover, the driving point input and output admittances, Y_{in} and Y_{out} respectively, are

$$Y_{in} \approx -\frac{y_{12b} y_{21a}}{y_{22a} + y_{22b} + Y_l}, \tag{2-125}$$

and

$$Y_{out} \approx -\frac{y_{12b} y_{21a}}{y_{11a} + y_{11b} + Y_s}. \tag{2-126}$$

2.5.4. Series-Series Feedback

In the series-series feedback system of Fig. 2.42(a), Thévenin equivalent I/O port models, and thus z-parameter modeling, are appropriate. The considered system at hand produces high input and output driving point closed loop impedances, whence the configuration functions best as a transadmittance amplifier. As is the case with the series components of

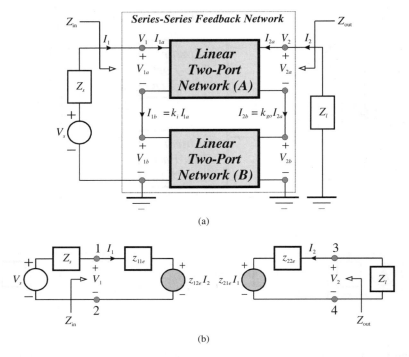

Figure 2.42. (a) Series-series feedback interconnection of two linear two-port networks. (b) The z-parameter equivalent circuit of the network in (a).

the series-shunt and shunt-series amplifiers, note that the input and output port currents of Network B are indicated as respectively proportional to the input and output port currents of the presumably active subcircuit, Network A. The applicable z-parameter equivalent circuit is provided in Fig. 2.42(b), where the effective z-parameters, z_{ije}, are defined by

$$\begin{bmatrix} z_{11e} & z_{12e} \\ z_{21e} & z_{22e} \end{bmatrix} = \begin{bmatrix} (z_{11a} + k_i z_{11b}) & (z_{12a} + k_o z_{12b}) \\ (z_{21a} + k_i z_{21b}) & (z_{22a} + k_o z_{22b}) \end{bmatrix}. \quad (2\text{-}127)$$

Assuming large loop gain, $|z_{21a}| \gg |k_i z_{21b}|$, and $|z_{12a}| \ll |k_o z_{12b}|$, the closed loop transadmittance is

$$Y_f = \frac{I_2}{V_s} \approx \frac{1}{z_{12e}} \approx \frac{1}{k_o z_{12b}}, \quad (2\text{-}128)$$

the closed loop input impedance, Z_{in}, is

$$Z_{in} \approx -\frac{k_o k_i z_{12b} z_{21a}}{z_{22a} + k_o z_{22b} + Z_l}, \quad (2\text{-}129)$$

and the closed loop output impedance, Z_{out}, is

$$Z_{out} \approx -\frac{k_o k_i z_{12b} z_{21a}}{z_{11a} + k_i z_{11b} + Z_s}. \qquad (2\text{-}130)$$

2.6.0. Power Flow and Transfer

For the generalized two-port network depicted in Fig. 2.2, the simplest measure of I/O power transfer is the *power gain*. The power gain, G_p, is defined as the ratio of average power P_l delivered to the terminating load impedance, Z_l, to the average power, P_i, presented to the input port. Thus,

$$G_p \triangleq \frac{P_l}{P_i}. \qquad (2\text{-}131)$$

Because G_p is a measure of signal transfer from the network input port to the network output port, it is obviously a function of the network two-port parameters. Moreover, since G_p is the signal power dissipated in the load impedance, normalized to the power evidenced at the network input port, it is a function of the load impedance, Z_l, and is independent of the source impedance, Z_s. For a passive network, G_p cannot exceed unity and for a lossless, passive configuration (meaning no elements within the two-port network are capable of dissipating power), G_p is identically unity.

A second measure of power transfer is the *available power gain*, G_{pa}, which is defined as the ratio of the average maximum, or available, power, P_{la}, delivered to load impedance Z_l to the average maximum power, P_{ia}, available at the signal source. Accordingly,

$$G_{pa} \triangleq \frac{P_{la}}{P_{ia}}. \qquad (2\text{-}132)$$

The maximum power available to the terminating load impedance is determined exclusively by the open circuit output voltage (or the short circuit output current) and the Thévenin output impedance of the network. On the other hand, the maximum power that can be generated by the applied signal source is a function of only the signal source voltage (or current) and the Thévenin impedance of said source. Like the power gain G_p, the available power gain, G_{pa}, is dependent on network two-port parameters, but unlike G_p, G_{pa} depends on the signal source impedance, Z_s, and is independent of the terminating load impedance, Z_l.

The third, and arguably most utilitarian, measure of power transfer is the *transducer power gain*, G_{pt}, which is defined as the ratio of average power delivered to the load-to-the average power maximally available from the signal source. In particular,

$$G_{pt} \triangleq \frac{P_l}{P_{ia}}. \tag{2-133}$$

Clearly, G_{pt} is functionally dependent on network two-port parameters and both the source impedance and the terminating load impedance. By comparing the average power delivered to the load impedance with the average power that the source is capable of supplying, G_{pt} measures the efficacy of the two-port network in the sense of characterizing the ability of the network to transfer the total average power available at the source to the terminating load.

It should be clearly understood that in the sinusoidal steady state, all three of the foregoing power definitions are real functions of radial signal frequency ω. The frequency dependence of these power gain metrics derives from the fact that in general, the source and load impedances, Z_s and Z_l, respectively, can be expected to be frequency dependent; that is Z_s is merely shorthand notation for $Z_s(j\omega)$, while Z_l analytically abridges $Z_l(j\omega)$. Moreover, each of the four two-port parameters of the network undergoing study is invariably frequency dependent. The power gains are real, as opposed to complex, frequency functions because average signal power levels depend on squares of absolute voltage and current levels and not mere voltages and currents.

2.6.1. Power Gain Expressions

Each of the three power gains defined in the preceding subsection can be related explicitly to the two-port parameters of the network in question. To this end, consider Fig. 2.43(a), which depicts a linear two-port network driven at its input port by a sinusoidal signal voltage, V_s, whose intrinsic Thévenin impedance is Z_s. The subject two-port configuration is terminated at its output port in a load impedance, Z_l. In response to the indicated signal excitation, there results an input port current, I_1, an output port current, I_2, an input port voltage, V_1, and an output port voltage, V_2, (all of which are frequency dependent). Although the two-port network at hand can be modeled by any convenient set of two-port parameters, open circuit impedance parameters are selected arbitrarily herewith, thereby resulting in the pertinent equivalent circuit offered in Fig. 2.43(b).

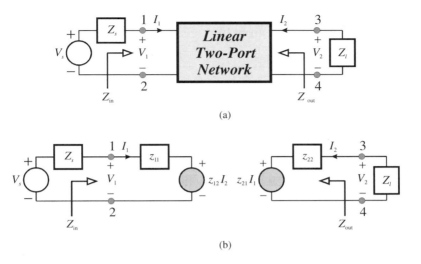

Figure 2.43. (a) A linear two-port network driven at its input port by a sinusoidal voltage, V_s, whose Thévenin impedance is Z_s. The load impedance, Z_l, like Z_s, is a complex function of radial frequency ω. (b) The z-parameter equivalent circuit of the network in (a). All z-parameters, z_{ij}, are frequency variant.

An inspection of the circuit in Fig. 2.43(b) reveals that the input port to output port current gain is

$$\frac{I_2}{I_1} = -\frac{z_{21}}{z_{22} + Z_l}, \tag{2-134}$$

while a recollection of the disclosures proffered in Section 1.3.6 permits writing

$$P_l = \frac{|I_2|^2}{2} Re\,(Z_l) \tag{2-135}$$

for the average power delivered to the terminating load impedance. It is understood that the notation, $Re(Z_l)$, signifies the real, or resistive, component of the complex load impedance, $Z_l = Z_l(j\omega)$. On the other hand, the average power, P_i, delivered by the signal source to the input port of the linear network is

$$P_i = \frac{|I_1|^2}{2} Re\,(Z_{in}) = \frac{|I_1|^2}{2} Re\left(z_{11} - \frac{z_{12}z_{21}}{z_{22} + Z_l}\right), \tag{2-136}$$

where Eq. (2-97) has been used. The last three expressions lead directly to a network power gain, G_p, given by

$$G_p \triangleq \frac{P_l}{P_i} = \left| \frac{z_{21}}{z_{22} + Z_l} \right|^2 \frac{Re(Z_l)}{Re\left(z_{11} - \dfrac{z_{12}z_{21}}{z_{22} + Z_l}\right)}. \tag{2-137}$$

Observe that the power gain is simply the product of the square magnitude of network current gain and the ratio of the real part components of load impedance-to-driving point input impedance.

The available power, P_{ia}, at the network input port is the maximum possible power that can be supplied by the applied signal source, V_s. This input power, which is actually observed if and only if the driving point input impedance, $Z_{in} = Z_{in}(j\omega)$, is a matched conjugate of the Thévenin source impedance, $Z_s = Z_s(j\omega)$, is, recalling the discourse of Section 1.3.6 once again,

$$P_{ia} = \frac{|V_s|^2}{8 Re(Z_s)}. \tag{2-138}$$

From Fig. 2.43(b),

$$\frac{I_2}{V_s} = \frac{I_2}{I_1} \times \frac{I_1}{V_s} = -\left(\frac{z_{21}}{z_{22} + Z_l}\right)\left(\frac{1}{Z_s + Z_{in}}\right)$$

$$= -\frac{z_{21}}{(z_{11} + Z_s)(z_{22} + Z_l) - z_{12}z_{21}}. \tag{2-139}$$

This result, together with Eqs. (2-138), (2-135), and (2-133), deliver a transducer power gain of

$$G_{pt} \triangleq \frac{P_l}{P_{ia}} = 4 \left| \frac{z_{21}}{(z_{11} + Z_s)(z_{22} + Z_l) - z_{12}z_{21}} \right|^2 Re(Z_s) Re(Z_l). \tag{2-140}$$

A useful alternative form of this rather cumbersome expression is

$$G_{pt} = \left| \frac{z_{21}}{z_{22} + Z_l} \right|^2 \frac{4 Re(Z_s) Re(Z_l)}{|Z_s + Z_{in}|^2}. \tag{2-141}$$

Note that if the driving point input impedance, Z_{in}, is indeed a conjugate match to the Thévenin source impedance, Z_s, $(Z_s + Z_{in})$ becomes $2Re(Z_s)$, whence

$$G_{pt}\big|_{Z_{in}=Z_s^*} = \left| \frac{z_{21}}{z_{22} + Z_l} \right|^2 \frac{Re(Z_l)}{Re(Z_s)}, \tag{2-142}$$

where the superscript asterisk (*) indicates the operation of complex conjugation; that is, $Z_{in}(j\omega) = Z_s(-j\omega)$. Since the denominator on the right hand side of Eq. (2-137) is the real part component of Z_{in}, which is identical to $Re(Z_s)$ if Z_{in} and Z_s are complex conjugates, the transducer power gain and the power gain are identical when the input port of the considered network is match terminated.

The available load power, P_{la}, in Eq. (2-132) is best evaluated by viewing the network output port in Fig. 2.43(b) as a Thévenin equivalent circuit. In particular, the Thévenin output voltage, say V_{2T}, which is necessarily evaluated under the open circuit load condition tantamount to zero output port current, I_2, is simply $V_{2T} = z_{21}I_1$. Moreover, $I_2 = 0$ implies that for the input port, $V_s = (Z_s + z_{11})I_1$. It follows that the open circuit, or Thévenin, voltage gain of the network under consideration is

$$\frac{V_{2T}}{V_s} = \frac{z_{21}}{z_{11} + Z_s}. \tag{2-143}$$

For a driving point output impedance, Z_{out}, the maximum power available for dissipation in the terminating load impedance is

$$P_{la} = \frac{|V_{2T}|^2}{8Re(Z_{out})} = \frac{|V_{2T}|^2}{8Re\left(z_{22} - \dfrac{z_{12}z_{21}}{z_{11} + Z_s}\right)}. \tag{2-144}$$

In view of Eqs. (2-138) and (2-143), the available power gain in Eq. (2-132) is seen to be

$$G_{pa} = \left|\frac{z_{21}}{z_{11} + Z_s}\right|^2 \frac{Re(Z_s)}{Re(Z_{out})} = \left|\frac{z_{21}}{z_{11} + Z_s}\right|^2 \frac{Re(Z_s)}{Re\left(z_{22} - \dfrac{z_{12}z_{21}}{z_{11} + Z_s}\right)}. \tag{2-145}$$

2.6.2. Stability Considerations

The three power gain metrics on which the preceding two subsections of material focus yield clues as to possible stability dilemmas in linear active two-port systems. To illustrate, return to Eqs. (2-136) and (2-137) to observe the possibility that power gain G_p can be negative or even infinitely large for passive load terminations. Because of the presence of the internal feedback dependent product, $(z_{12}z_{21})$, in these relationships, negative or null values of the real part of the driving point input impedance on the right hand sides of the aforementioned two equations cannot be precluded over all signal frequencies, even if impedance parameters z_{11} and z_{22} are positive real functions. If $Re(Z_{in}) < 0$ for at least one frequency, say ω_o, Eq. (2-136) confirms that the power delivered to the network input port at $\omega = \omega_o$ is negative, independent of the (positive) power available at the signal source. This situation is disturbing, for it implies that the network input port is itself generating signal power, independent of the source, which is presumably applied for the express purpose of delivering power to said input port. If $Re(Z_{in}) = 0$ at $\omega = \omega_o$, the power gain is infinitely large at ω_o, which gives rise to the prospects of output port signal responses in the absence of signal excitation applied to the input port. Such an intriguing circumstance is tantamount to conceding network oscillation at frequency ω_o. Accordingly, it is both natural and reasonable to expect an input port featuring a positive real driving point impedance over all frequencies of interest for otherwise, output responses can be produced magically without an applied input signal source. Equations (2-144) and (2-145) allow similar statements to be proffered in regard to the engineering improprieties of negative or null real part output impedances.

The potential stability problem associated with negative real driving point input impedance is even more dramatically underscored by the transducer gain expression in Eq. (2-141). If indeed $Re(Z_{in}) < 0$ for at least one frequency, ω_o, it is possible to implement a signal source having a positive real Thévenin impedance, Z_s, such that the impedance sum, $(Z_s + Z_{in})$, vanishes at $\omega = \omega_o$. It follows under this circumstance that the transducer power gain, G_{pt}, at frequency ω_o is infinitely large, thereby implying system oscillation at $\omega = \omega_o$.

It is important to note that the generally undesirable condition, $Re(Z_{in}) \leq 0$, does not guarantee system instability in the sense of the self-sustaining oscillations implied by $G_{pt} = \infty$. If $Re(Z_{in}) \leq 0$ at one or more frequencies, stable system operation is sustained as long as the Thévenin signal source impedance is characterized by a sufficiently large and positive real part component at the offending frequencies. On the other hand, $Re(Z_{in}) > 0$ precludes unstable responses in the sense of assuring that no positive real Thévenin source impedance can be implemented to produce either a negative or an infinitely large transducer power gain. Implicit to this assertion is the tacit presumption that Z_l is the impedance function of a passive load structure (and is therefore positive real) and that the open circuit network impedance parameters, z_{11} and z_{22}, are likewise positive real functions.

The preceding arguments can be formalized with reference to the generalized system of Fig. 2.43(a). To this end, assume that the Thévenin source impedance, Z_s, the terminating load impedance, Z_l, the open circuit two-port network input impedance, z_{11}, and the open circuit two-port network output impedance, z_{22}, are all positive real functions of frequency. This is to say that for all radial signal frequencies, ω,

$$\left. \begin{array}{l} Re(Z_s) > 0 \\ Re(Z_l) > 0 \\ Re(z_{11}) > 0 \\ Re(z_{22}) > 0 \end{array} \right\}, \qquad (2\text{-}146)$$

which is tantamount to stipulating passive source and load impedances and open circuit impedances that can be realized with positive resistors, capacitors, and inductors. The embedded two-port network, which can be a single device or an entire electronic signal processing unit, is said to be *unconditionally stable* if $Re(Z_{in}) > 0$ and $Re(Z_{out}) > 0$ for all signal frequencies of interest. Branding a two-port network as unconditionally stable means that no passive source or load terminations can be introduced to yield either negative or infinitely large power gain, available power gain, or transducer power gain. In contrast, an embedded two-port structure or amplifier is said to be *potentially unstable* if $Re(Z_{in}) \leq 0$ or $Re(Z_{out}) \leq 0$ for at least one signal frequency. Thus, a potentially unstable two-port network is a structure for which at least one passive load or source termination can be implemented to yield a power transfer metric that is either negative or infinitely large for at least one signal frequency.

Despite the facts that input impedance Z_{in} is dependent on terminating load impedance Z_l and output impedance Z_{out} is functionally dependent on Thévenin source impedance Z_s, the criterion assuring unconditional stability in a system, such as the one depicted in Fig. 2.43(a), is determined exclusively by the two-port parameters implicit to the embedded linear network. This circumstance is fortuitous for it allows a circuit designer to gauge the system propensity for oscillation by simply examining the two-port parameters of the linear network used to couple an arbitrary passive source to an equally arbitrary load. To wit, assume the conditions postured by Eq. (2-146) and write

$$\left. \begin{array}{l} Z_l = R_l + jX_l \\ Z_{in} = R_{in} + jX_{in} \\ z_{11} = r_{11} + jx_{11} \\ z_{22} = r_{22} + jx_{22} \\ z_{12}z_{21} = \alpha + j\beta \end{array} \right\}. \qquad (2\text{-}147)$$

Resultantly, the driving point input impedance is

$$Z_{in} = z_{11} - \frac{z_{12}z_{21}}{z_{22} + Z_l} = r_{11} + jx_{11} - \frac{\alpha + j\beta}{(r_{22} + R_l) + j(x_{22} + X_l)}. \quad (2\text{-}148)$$

With R_{in} symbolizing the real, or resistive (but invariably frequency dependent), component of Z_{in}, a rationalization of the last term in this relationship leads immediately to

$$\frac{R_{in}}{r_{11}} = \frac{(r_{22} + R_l)^2 - \dfrac{\alpha}{r_{11}}(r_{22} + R_l) + (x_{22} + X_l)^2 - \dfrac{\beta}{r_{11}}(x_{22} + X_l)}{(r_{22} + R_l)^2 + (x_{22} + X_l)^2}. \quad (2\text{-}149)$$

Alternatively,

$$\frac{R_{in}}{r_{11}} = \frac{\left[(r_{22} + R_l) - \dfrac{\alpha}{2r_{11}}\right]^2 + \left[(x_{22} + X_l) - \dfrac{\beta}{2r_{11}}\right]^2 - \dfrac{\alpha^2 + \beta^2}{(2r_{11})^2}}{(r_{22} + R_l)^2 + (x_{22} + X_l)^2}. \quad (2\text{-}150)$$

Since resistance parameter r_{11} is positive by virtue of Eq. (2-146), an unconditionally stable system mandates that the right hand side of Eq. (2-150) be positive for all frequencies and for all load terminations. The denominator of this right hand side is clearly always positive and thus, it suffices to ensure only that the numerator on the right hand side of Eq. (2-150) be positive. Observe that the second term in the subject numerator term is itself always non-negative. If steps are taken to ensure a positive numerator under the reactive load condition,

$$X_l = \frac{\beta}{2r_{11}} - x_{22}, \quad (2\text{-}151)$$

which is physically realizable for positive, negative, or null X_l, unconditional stability remains assured for any and all other values of the load reactance. The stability requirement now reduces to

$$\left[(r_{22} + R_l) - \frac{\alpha}{2r_{11}}\right]^2 > \frac{\alpha^2 + \beta^2}{(2r_{11})^2}, \quad (2\text{-}152)$$

for which the worst possible case is $R_l = 0$; that is, the satisfaction of Eq. (2-152) for $R_l = 0$ guarantees unconditional stability for any and all passive load terminations. Recalling Eq. (2-147), let

$$\lambda_c = \frac{2r_{11}r_{22} - \alpha}{\sqrt{\alpha^2 + \beta^2}} \equiv \frac{2\,Re(z_{11})\,Re(z_{22}) - Re(z_{12}z_{21})}{|z_{12}z_{21}|}, \quad (2\text{-}153)$$

which is commonly referenced in the literature as the *Llewellyn stability factor*. Then, subject to the constraints imposed by Eq. (2-146), the preceding two relationships combine to stipulate that a two-port system, such as the one abstracted in Fig. 2.43(a), is unconditionally stable for all passive load terminations if and only if the parameters of the embedded linear network satisfy the inequality, [6]–[8]

$$\lambda_c = \frac{2\text{Re}(z_{11})\,Re(z_{22}) - Re(z_{12}z_{21})}{|z_{12}z_{21}|} > 1. \qquad (2\text{-}154)$$

Several noteworthy points can be underscored in regard to the mathematically elegant stability constraint of Eq. (2-154). First, and as the reader is invited to confirm in Problem 2.25, $\lambda_c > 1$ ensures, in addition to $Re(Z_{in}) > 0$ for all passive load terminations, $Re(Z_{out}) > 0$ for all passive Thévenin source impedances. In other words, $\lambda_c > 1$ ensures both positive real driving point input and driving point output impedances for any passive load and source impedances. Second, although Eq. (2-154) derives from a z-parameter modeling premise, a similar expression prevails for all other traditional two-port parameters. In particular, for any consistent set of two-port parameters, η_{ij},

$$\lambda_c = \frac{2\text{Re}\,(\eta_{11})\,Re\,(\eta_{22}) - Re\,(\eta_{12}\eta_{21})}{|\eta_{12}\eta_{21}|}. \qquad (2\text{-}155)$$

Third, the unconditional stability requirement, $\lambda_c > 1$, must prevail for all signal frequencies, in addition to prevailing for all positive real load and source terminations. If but a single frequency exists for which $\lambda_c \leq 1$, the network is potentially unstable in the sense that positive real load or source terminations can be determined to induce self-sustaining system oscillations. Finally, and to the extent that positive real load and source terminations are presumed, the Llewellyn constraint is independent of the Thévenin signal source impedance and the load impedance incident with the system output port. As such, the Llewellyn stability factor is a signature of the two-port device or amplifier utilized in the two-port system, as opposed to reflecting a performance assessment metric for the global, terminated two-port system.

Example 2.9. A very simple small signal model of a MOSFET configured to operate as a common source amplifier is given in Example 2.3. Determine the Llewellyn stability factor of this device and assess its stability.

Solution 2.9.

(1) In Example 2.3, the short circuit admittance parameters for the common source MOSFET are determined to be

$$\left.\begin{aligned} y_{11} &= j\omega\left(C_{gs} + C_{gd}\right) \\ y_{12} &= -j\omega C_{gd} \\ y_{21} &= g_m - j\omega C_{gd} \\ y_{22} &= g_o + j\omega\left(C_{gd} + C_{db}\right) \end{aligned}\right\}, \qquad \text{(E9-1)}$$

where g_m is the forward transconductance, C_{gs} is the gate-source capacitance, C_{gd} is the gate-drain capacitance, g_o is the drain-source channel conductance, and C_{db} is the drain-bulk capacitance. Observe that as a result of Eq. (E9-1),

$$\left.\begin{aligned} Re(y_{11}) &= 0 \\ Re(y_{22}) &= g_o \end{aligned}\right\}, \qquad \text{(E9-2)}$$

and

$$y_{12}y_{21} = -\left(\omega C_{gd}\right)^2\left[1 - \frac{g_m}{j\omega C_{gd}}\right]. \qquad \text{(E9-3)}$$

(2) Using $\eta_{ij} = y_{ij}$ in Eq. (2-155), The Llewellyn stability factor for the common source amplifier is

$$\lambda_c = \frac{1}{\sqrt{1 + \left(\dfrac{g_m}{\omega C_{gd}}\right)^2}}. \qquad \text{(E9-4)}$$

Since $\lambda_c < 1$ for all signal frequencies, ω, the common source MOSFET amplifier, to the extent that the model invoked in Example 2.3 is an accurate representation of its small signal I/O characteristics, is a potentially unstable entity.

Comments. The stability news conveyed by Eq. (E9-4) is not as distressing as might be perceived at first blush. To be sure, the result suggests potential instability for all signal frequencies. But it should be remembered that the small signal model, shown in Fig. 2.16, exploited in the determination of the Llewellyn stability factor is an extremely elementary topology that ignores resistive losses in the drain and source leads and parasitic shunt resistance incident across the gate-source terminals. A more accurate approach to a determination of the Llewellyn stability factor entails HSPICE two-port parameter simulations executed on a suitably high level small signal model, followed by a numerical evaluation of λ_c over all relevant signal frequencies. Also, keep in mind that the mere fact that $\lambda_c < 1$ in the present case does not ensure instability. Instead, $\lambda_c < 1$ simply ensures potential instability in the sense that suitable source and load terminations can be implemented to induce self-sustaining oscillations of the network into which the considered transistor is embedded. Were it not for $\lambda_c < 1$ over at least certain frequencies, the common practice of exploiting common source amplifiers in the design of a wide variety of sinusoidal oscillators would comprise a classic exercise in futility.

2.6.3. Maximum Transducer Gain

Maximum possible transducer gain is realized when the driving point input impedance, Z_{in}, in the network of Fig. 2.43(a) is a conjugate match to the Thévenin source impedance, Z_s, and simultaneously, the driving point output impedance, Z_{out}, is a conjugate match to the terminating load impedance, Z_l. This dual conjugate matching constraint is a daunting exercise because the presence of intrinsic feedback within the linear two-port network renders Z_{in} dependent on Z_l, which in turn must be a conjugate match to an output impedance influenced as a result of internal feedback by the value of Z_s chosen to match terminating Z_{in}. As daunting as this "chase one's tail" problem is, it is soluble. With reference to Fig. 2.43(b), let the source impedance commensurate with a conjugate match to Z_{in} under the condition that Z_l is a conjugate match to Z_{out} be written symbolically as

$$Z_s = Z_s(j\omega) = R_{so} + jX_{so} \equiv Z_{in}(-j\omega), \qquad (2\text{-}156)$$

where it is understood that the requisite resistive and reactive components, R_{so} and X_{so}, respectively, are functions of signal frequency. Correspondingly, let the load impedance that match terminates the network output port when the input port is match terminated be written as

$$Z_l = Z_l(j\omega) = R_{lo} + jX_{lo} \equiv Z_{out}(-j\omega). \qquad (2\text{-}157)$$

It can be shown that[9]

$$\left. \begin{array}{l} X_{so} = -Im(z_{11}) + \dfrac{Im(z_{12}z_{21})}{2Re(z_{22})} \\[2mm] X_{lo} = -Im(z_{22}) + \dfrac{Im(z_{12}z_{21})}{2Re(z_{11})} \end{array} \right\}, \qquad (2\text{-}158)$$

which are easy enough to apply in a design environment. The expressions for R_{so} and R_{lo} are considerably more interesting and may even explain why circuit designers charged with the responsibility of achieving conjugate impedance matches at the input and output ports of their linear active networks tend to age and/or grow bald prematurely. The results are

$$\begin{aligned} R_{so} &= \left[Re(z_{11}) - \frac{Re(z_{12}z_{21})}{2Re(z_{22})} \right] \sqrt{1 - \left(\frac{1}{\lambda_c}\right)^2} \\ R_{lo} &= \left[Re(z_{22}) - \frac{Re(z_{12}z_{21})}{2Re(z_{11})} \right] \sqrt{1 - \left(\frac{1}{\lambda_c}\right)^2} \end{aligned} \quad (2\text{-}159)$$

The aforementioned aging phenomena is experienced by circuit designers who are tacitly unaware that it is impossible to achieve simultaneous conjugate impedance matches at the input and output ports of a linear two-port network unless said two-port network is unconditionally stable. Specifically, Eq. (2-159) confirms that R_{so} and R_{lo} are physically realizable if and only if the Llewellyn stability factor, λ_c, for the utilized active network exceeds unity. Frustrations are exacerbated when the linear two-port network embodies uncompensated submicron MOS technology devices, which are rarely unconditionally stable throughout the frequency spectra of design interest.

It should be noted that Eqs. (2-158) and (2-159) collapse to anticipated results if no internal feedback ($z_{12} = 0$) prevails. In this case λ_c becomes infinitely large, assuming the reasonable circumstance of nonzero and finite real parts to impedance parameters z_{11} and z_{22}. With $z_{12} = 0$, the preceding two equations show, as expected, that the requisite source impedance is simply the conjugate of the open circuit input impedance, while the required load impedance is little more than the conjugate of the open circuit output impedance. It is worthwhile interjecting that Eqs. (2-158) and (2-159) can be cast in terms of any consistent set of two-port network parameters, η_{ij}, provided that the results are judiciously interpreted. For example, if short circuit admittance parameters are exploited, Eq. (2-158) defines the necessary source and load susceptances, while Eq. (2-159) defines requisite source and load conductances.

Assuming unconditional stability, Eqs. (2-158) and (2-159) can be substituted into Eq. (2-141) to produce the maximum possible transducer power gain, G_{ptm}. This exercise, which accelerates aging in educators and authors, gives

$$G_{ptm} = \frac{|z_{21}|^2}{[2Re(z_{11})Re(z_{22}) - Re(z_{12}z_{21})]\left(1 + \sqrt{1 - \left(\frac{1}{\lambda_c}\right)^2}\right)}. \quad (2\text{-}160)$$

With the help of Eq. (2-154), Eq. (2-160) can be expressed in the slightly more compact form,

$$G_{ptm} = \frac{|z_{21}/z_{12}|}{\lambda_c \left(1 + \sqrt{1 - \left(\frac{1}{\lambda_c}\right)^2}\right)}. \qquad (2\text{-}161)$$

Since λ_c must exceed unity if Eq. (2-161) is to be meaningful, and since the denominator on the right hand side of Eq. (2-161) increases monotonically with λ_c, the impedance magnitude ratio, $|z_{21}/z_{12}|$, is the largest possible value of the maximum transducer power gain. This particular transducer gain is achieved for $\lambda_c = 1$, which corresponds to a utilized two-port network that is on the boundary of potential instability and unconditional stability. These disclosures uncover the interesting and unsurprising fact that progressively smaller maximum possible transducer power gains are afforded when the utilized active two-port is robustly unconditionally stable in the sense of large λ_c.

For the special case of zero internal feedback, as reflected by the constraints, $z_{12} = 0$ and positive real impedance parameters, z_{11} and z_{22}, Eqs. (2-158) and (2-159) give match terminated source and load impedances of

$$\left. \begin{array}{l} Z_s|_{z_{12}=0} = Z_s(j\omega) = Re(z_{11}) - j\,Im(z_{11}) \equiv z_{11}(-j\omega) \\ Z_l|_{z_{12}=0} = Z_l(j\omega) = Re(z_{22}) - j\,Im(z_{22}) \equiv z_{22}(-j\omega) \end{array} \right\}. \qquad (2\text{-}162)$$

Moreover, λ_c in Eq. (2-74) becomes infinitely large. It follows that the $z_{12} = 0$ value, say G_{ptu}, of the maximum transducer gain in Eq. (2-74) is

$$G_{ptu} \stackrel{\Delta}{=} G_{ptm}\big|_{z_{12}=0} = \frac{|z_{21}|^2}{4Re(z_{11})Re(z_{22})}. \qquad (2\text{-}163)$$

Because Eq. (2-163) is premised on the source and load terminations defined by Eq. (2-76), which in turn rely on the surreal circumstance of null feedback within the embedded two-port network, G_{ptu} is not an especially useful metric for circuit and system design. Nonetheless, the frequency response of G_{ptu} is ubiquitously exploited by the semiconductor industry for purposes of assessing the high frequency signal processing capabilities of both discrete component and monolithic active devices. An apparent reason for this widespread utility is the relative ease of measuring, albeit indirectly, parameters z_{21}, $Re(z_{11})$, and $Re(z_{22})$. Of particular interest to the semiconductor industry is the frequency, ω_{max}, at which G_{ptu} degrades to unity. This frequency is labeled as the unity power gain frequency, which is somewhat misleading in that G_{ptu} is a true measure of maximum possible transducer power gain only if null internal feedback prevails within the linear two-port network. Nevertheless, ω_{max} does serve as a useful barometer of high frequency device performance attributes and indeed, it may even be a more reasonable high frequency assessment than are other commonly invoked measures, such as the short circuit unity gain frequency.

2.6.4. Unilateralization

The foregoing stability and power gain testimony attest to the troublesome nature of two-port network intrinsic feedback, as measured by the generalized parameter, η_{12}. All active devices and most electronic networks that might be exploited as the linear two-port network in Fig. 2.43(a) are plagued with internal feedback. When the two-port network in Fig. 2.43(a) is a MOS technology common source amplifier or a buffered version of same, such feedback is dominantly manifested by gate-drain capacitance in the common source input stage. In addition to affecting deleteriously the attainable 3-dB system bandwidth, the presence of this feedback possibly precludes unconditional system stability, which in turn prohibits the maximization of the system power gain. Strategies aimed toward minimizing this feedback capacitance, if not outright eliminating it, are therefore a laudable design objective. If indeed the net internal feedback between output and input ports of an active two-port network is entirely annihilated, the resultant configuration becomes known as a *unilateral amplifier* in that no responses to applied input signals feed back from the amplifier output port-to-the input port.

2.6.4.1. *Shunt-Antiphase Shunt Compensation*

The system depicted in Fig. 2.43(a) is unilateral if its internal feedback parameter, z_{12} (or η_{12} in general) is zero. To the extent that this intrinsic feedback arises from phenomena that produce a parasitic impedance or admittance connected between the input and output ports, the neutralization of η_{12} might be accomplished by connecting the negative of this phenomenological admittance directly between the input and output ports. Unfortunately, such a shunt-shunt interconnection is productive only if the appended admittance is the algebraic negative of the intrinsic shunt admittance. But a shunt-antiphase shunt interconnection of a suitable admittance may serve effectively to trick the linear two-port network into thinking that

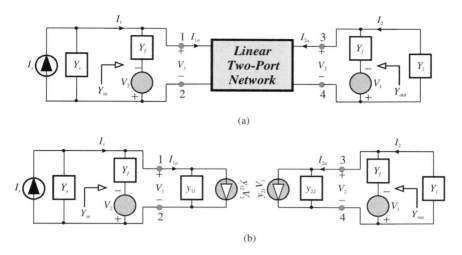

Figure 2.44. (a) The linear two-port network of Fig. 2.43(a) with appended shunt-antiphase shunt compensation. The appended network consisting of admittance Y_f and the two controlled voltage sources are inserted for the purpose of achieving a unilateral two-port system. (b) The y-parameter equivalent circuit of the network in (a).

the negative of its internal I/O feedback admittance is incident between the system input and output ports.

To the foregoing end, Fig. 2.44(a) is submitted as a system level schematic diagram of a shunt-antiphase shunt feedback amplifier. Two compensating admittances, Y_f, are deployed. One of these admittances is connected between the amplifier input port and the phase inverted voltage prevailing at the output port, while the other is incident between the output port and the phase inverted signal voltage established across the input port. The small signal model of the structure in Fig 2.44(a) is provided in Fig. 2.44(b) where, because of the shunting nature of appended compensation, short circuit admittance, or y-parameters are chosen as the modeling vehicle. If the linear two-port network is characterized by y-parameters y_{ij}, the currents, I_{1a} and I_{2a}, flowing into the amplifier input port and output port, respectively, are given by

$$\begin{bmatrix} I_{1a} \\ I_{2a} \end{bmatrix} = \begin{bmatrix} y_{11} & y_{12} \\ y_{21} & y_{22} \end{bmatrix} \begin{bmatrix} V_1 \\ V_2 \end{bmatrix}. \qquad (2\text{-}164)$$

Since

$$\left. \begin{array}{l} I_1 = Y_f(V_1 + V_2) + I_{1a} \\ I_2 = Y_f(V_2 + V_1) + I_{2a} \end{array} \right\}, \qquad (2\text{-}165)$$

the y-parameter matric relationship for the overall system is

$$\begin{bmatrix} I_1 \\ I_2 \end{bmatrix} = \begin{bmatrix} (y_{11} + Y_f) & (y_{12} + Y_f) \\ (y_{21} + Y_f) & (y_{22} + Y_f) \end{bmatrix} \begin{bmatrix} V_1 \\ V_2 \end{bmatrix}. \tag{2-166}$$

An inspection of the last expression indicates that the effective short circuit feedback admittance is $(Y_f + y_{12})$. If Y_f is chosen, such that

$$Y_f = -y_{12}, \tag{2-167}$$

the system under consideration is unilateral, and the short circuit admittance parameters, y_{ijc}, of the resultantly compensated network are

$$\left. \begin{aligned} y_{11c} &= y_{11} + Y_f = y_{11} - y_{12} \\ y_{12c} &= y_{12} + Y_f = 0 \\ y_{21c} &= y_{21} + Y_f = y_{21} - y_{12} \\ y_{22c} &= y_{22} + Y_f = y_{22} - y_{12} \end{aligned} \right\}. \tag{2-168}$$

In view of the unilateral nature of the compensated system, the power gain metric, G_{ptu} in Eq. (2-163) is now a meaningful network performance barometer. For the compensated values of the y-parameters in Eq. (2-168), this power gain is

$$G_{ptu} = \frac{|y_{21} + Y_f|^2}{4Re(y_{11} + Y_f)Re(y_{22} + Y_f)} = \frac{|y_{21} - y_{12}|^2}{4Re(y_{11} - y_{12})Re(y_{22} - y_{12})}. \tag{2-169}$$

2.6.4.2. Shunt-Antiphase Shunt Network Realization

The shunt-antiphase shunt compensation scheme discussed in the preceding section is easily realized by balanced differential amplifiers, which are discussed in depth in a subsequent chapter of this text. For the moment, the design strategy is best illustrated by the system diagram of the differential pair offered in Fig. 2.45. In this depiction, the dual active two-port networks must be identical and must also be biased identically to ensure their linear operation. Each of these two active two-port structures can be viewed as a replica of the linear two-port network embedded in the system of Fig. 2.44(a), provided that said two-port network operates as a three terminal circuit by connecting together Terminals 2 and 4. The current source, I_k, which can be realized with transistors, is required to provide a path to system ground for the static biasing currents conducted by each

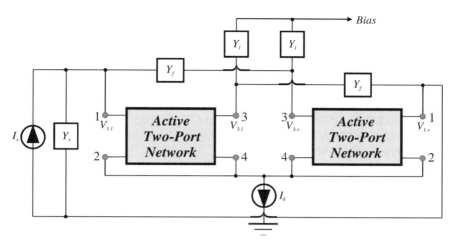

Figure 2.45. Balanced differential realization of shunt-antiphase shunt compensation of a linear active two-port network. Output responses can be extracted with respect to ground at either of the indicated terminals 3 or differentially between the two terminals labeled 3.

active two-port network. The input signal, shown herewith as the signal current, I_s, whose Thévenin admittance is Y_s, is not applied between ground and any of the two input ports for the interconnected network. Rather, this signal is applied differentially between the two input ports of the active structure. In this balanced differential configuration and to the extent that current source I_k emulates an ideal element, the voltage response, V_{3r}, at Terminal 3 (with respect to ground) of the right active network is precisely 180° phase displaced with respect to the voltage response, V_{3l}, produced between Terminal 3 of the left active network and system ground. Similarly, voltage V_{1l} at Terminal 1 of the left active network (with respect to ground) is precisely the negative of voltage V_{1r} developed from Terminal 1-to-ground of the right active two-port network. It can therefore be seen that the controlled voltage sources exploited in the model of Fig. 2.44(b) are realized as a result of the balanced operation of the differential system at hand. In particular, one of the two compensation admittances, Y_f, is connected between left Terminal 1, where the signal voltage is V_{1l}, and right Terminal 3, which supports voltage V_{3r}, which in turn is the negative of voltage V_{3l}; that is, $V_{3r} = -V_{3l}$. Analogously, the other admittance, Y_f, is incident at left Terminal 3, which supports voltage V_{3l}, and right Terminal 1, at which voltage V_{1r}, the negative of V_{1l}, prevails.

References

1. W. K. Chen, *Active Network and Feedback Amplifier Theory* (McGraw-Hill Book Company, New York, 1980), Chapter 3.
2. W. K. Chen, *Active Network Analysis* (World Scientific Publishing Co., New Jersey, 1991).
3. E. F. Bolinder, Survey of some properties of linear networks, *IRE Trans. Circuit Theory* **CT-4** (1957) 70–78 (correction in **CT-5** (1958) 139).
4. E. S. Kuh and R. A. Rohrer, *Theory of Linear Active Networks* (Holden-Day, San Francisco, California, 1967).
5. S. S. Haykin, *Active Network Theory* (Addison-Wesley, Reading, Massachusetts 1970).
6. J. G. Linvill and J. F. Gibbons, *Transistors and Active Circuits* (McGraw-Hill Book Company, Inc., New York, 1961), Chapter 11.
7. J. M. Rollett, Stability and power-gain invariants of linear two-ports, *IRE Trans. Circuit Theory* **CT-9** (1962) 29–32.
8. J. O. Scanlan and J. S. Singleton, The gain and stability of linear two-port amplifiers, *IRE Trans. Circuit Theory* **CT-9** (1962) 29–32.
9. J. R. Miller (ed.), *Solid-State Communications* (McGraw-Hill Book Company, New York, 1966), pp. 50–58.

Exercises

Problem 2.1
The model shown within the symbolic network "box" in Fig. P2.1 is a linearized equivalent circuit of an emitter-degenerated, common emitter amplifier. The amplifier is driven by a voltage source whose Thévenin resistance is R_s, and its load termination consists of a resistance, R_l in shunt with a capacitance, C_l. In this problem, $R_s = 100\,\Omega$, $R_l = 1.2\,\text{K}\Omega$, $R_e = 100\,\Omega$, and $C_l = 150\,\text{fF}$. The transistor model parameters are $r_b = 90\,\Omega$, $r_\pi = 2.2\,\text{K}\Omega$, $r_o = 30\,\text{K}\Omega$, and $\beta = 150$.

(a) Derive analytical expressions for each of the four h-parameters of the linearized common emitter amplifier.
(b) Using the h-parameter expressions deduced in Part (a), derive analytical expressions for the low frequency, open loop values of voltage gain, V_2/V_s, input resistance, R_{in}, and output resistance, R_{out}.
(c) Using the h-parameter expressions deduced in Part (a), derive an analytical expression for the low frequency loop gain of the amplifier. Discuss

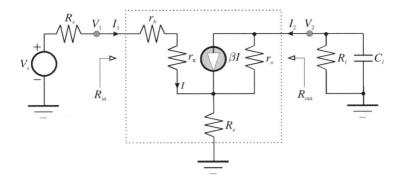

Figure P2.1.

the influence that the emitter degeneration resistance, R_e, has on this loop gain.

(d) Using appropriate preceding results, compute numerical, low frequency values for:

   i. open loop input resistance;
   ii. driving point input resistance;
   iii. open loop output resistance;
   iv. driving point output resistance;
   v. loop gain;
   vi. open loop voltage gain;
   vii. closed loop voltage gain.

(e) Can the analytical expressions for the closed loop voltage gain deduced in Part (b) and the loop gain deduced in Part (c) be adequately approximated by relatively simple algebraic expressions? If said expressions can be approximated, explain the conditions that render the approximations valid and give the resultant analytical approximations.

(f) In terms of the h-parameters and suitable other circuit variables, give a general expression for the 3-dB bandwidth of the circuit. Numerically evaluate this bandwidth.

Problem 2.2

The model shown within the symbolic network "box" in Fig. P2.2 is an approximate small signal equivalent circuit of a MOSFET configured to operate as a common source amplifier. The amplifier is driven by a voltage source whose Thévenin resistance is R_s, and its load termination consists

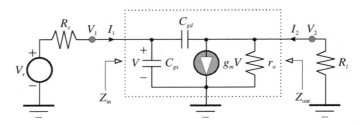

Figure P2.2.

exclusively of the resistance, R_l. Take $R_s = 100\,\Omega$, $R_l = 5\,\text{K}\Omega$, $g_m = 25\,\text{mmho}$, $r_o = 20\,\text{K}\Omega$, and $C_{gs} = 30\,\text{fF}$, and $C_{gd} = 5\,\text{fF}$.

(a) Derive analytical expressions for each of the four short circuit admittance parameters of the linearized transistor.
(b) Using the y-parameter expressions deduced in Part (a), derive analytical expressions for the open loop values of voltage gain V_2/V_s, input impedance Z_{in}, and output impedance Z_{out}.
(c) Using the y-parameter expressions deduced in Part (a), derive an analytical expression for the loop gain of the amplifier. Discuss the influence that the gate-drain capacitance, C_{gd}, has on this loop gain.
(d) Using appropriate preceding results, compute numerical values for the following circuit metrics at a signal frequency of 1 GHz:

    i. open loop driving point input impedance;
    ii. closed loop driving point input impedance;
    iii. open loop driving point output impedance;
    iv. closed loop driving point output impedance;
    v. loop gain;
    vi. open loop voltage gain;
    vii. closed loop voltage gain.

Problem 2.3

The twin tee structure shown in Fig. P2.3 is commonly used as a notch filter; that is, it ideally affords zero transmission between its input and output ports at a single frequency, say ω_n, which is known as the notch frequency.

Without evaluating the actual transfer function, V_o/V_i, determine an expression for notch frequency ω_n in terms of resistance R and capacitance C.

HINT: Deduce the y-parameters of the network by noting that the filter consists of two RC two-port networks connected in shunt-shunt architecture.

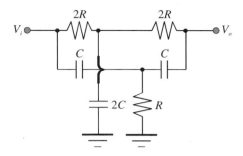

Figure P2.3.

This is to say that the input, output, and ground terminals of each filter section are respectively connected together. Recall the y-parameter properties of a generalized passive network, and exploit the fact that the gain of a network is necessarily proportional to its y-parameter, y_{21}.

Problem 2.4

Figure P2.4 depicts a simplified, high frequency, small signal model of an N-channel MOSFET, wherein bulk-induced modulation of threshold voltage is tacitly ignored.

(a) Derive general expressions for the steady state short circuit admittance parameters for the common source interconnection of the subject transistor.
(b) Use the results of Part (a) to deduce the indefinite short circuit admittance matrix of the MOSFET operated in small signal mode.
(c) Use the results of Part (b) to arrive at the steady state short circuit admittance parameters for a common gate interconnection of the MOSFET. Note that in a common gate configuration, I_3 and V_3 are the input variables, while I_2 and V_2 are the output variables.

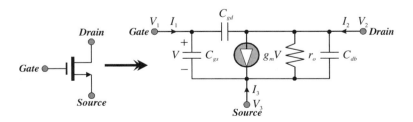

Figure P2.4.

(d) Use the results of Part (c) to develop the small signal common gate equivalent circuit of the MOSFET.

Problem 2.5
Repeat Parts (c) and (d) of the preceding problem for the case of a source follower interconnection of the MOSFET. In a source follower, or common drain configuration, I_1 and V_1 comprise the input variables, and I_3 and V_3 are the output quantities.

Problem 2.6
Use the pertinent results of Problem 2.4 to determine the common gate h-parameters of a MOSFET.

Problem 2.7
Consider the two-stage feedback amplifier of Fig. P2.7, where biasing is not shown in the interests of preserving topological simplicity. Each transistor has the simplified small signal equivalent circuit that is also shown in the figure.

(a) Is the amplifier best utilized for voltage gain, current gain, transimpedance, or transadmittance purposes? What kind of two-port parameter set is optimally suited for an analysis of this configuration? The following parts of this problem pertain to this most appropriate parameter set.

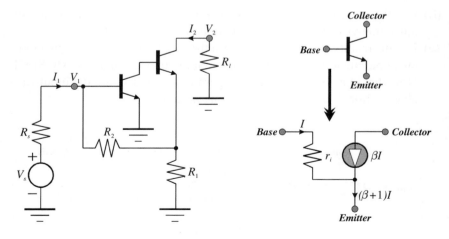

Figure P2.7.

(b) Determine expressions for the open loop values of gain, input resistance seen by the signal source $(V_s - R_s)$, and output resistance seen by the load resistance (R_l).

(c) Assuming very large loop gain, determine expressions for the closed loop values of gain, input resistance, and output resistance.

(d) What design criterion (or criteria) must be satisfied to ensure that feedforward through the feedback network is inconsequential?

(e) What design criterion (or criteria) must be satisfied to ensure sufficiently large loop gain?

(f) In the indicated amplifier, a capacitance, say C, is connected in shunt with resistance R_2. Assuming that the criteria determined in the preceding two parts of this problem are satisfied at even high signal frequencies, what are the radial frequencies of the gain function pole and zero established by capacitance C?

Problem 2.8

Convert the π-type resistive architecture of Fig. P2.8(a) into the tee-type configuration depicted in Fig. P2.8(b) by using any two-port parameters as the foundational tool for expressing the resistances, R_X, R_Y, and R_Z, as functions of the resistances, R_A, R_B, and R_C. The conversion must be such as to ensure that the electrical characteristics at terminals 1 and 2 with respect to ground are identical for both networks.

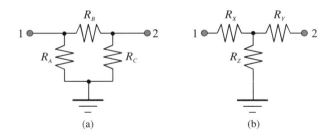

Figure P2.8.

Problem 2.9

Determine the transmission parameter matrix for each of the resistive structures offered in Fig. P2.8.

Problem 2.10

Figure P2.10 is a linearized equivalent circuit of a voltage amplifier.

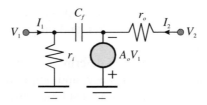

Figure P2.10.

(a) Determine the hybrid g-parameters for the amplifier at hand.
(b) What are the frequencies of the pole and zero indigenous to the open loop voltage gain of the amplifier?
(c) Give the approximate g-parameters for the special case of $r_i = \infty$, $r_o = 0$, and $C_f = 0$.

Problem 2.11
Figure P2.11 is a linearized equivalent circuit of a current amplifier.

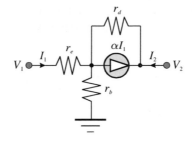

Figure P2.11.

(a) Determine the hybrid h-parameters for the amplifier at hand.
(b) Give the approximate h-parameters for the special case of $r_d = \infty$, $r_e = 0$, and $r_b = 0$.

Problem 2.12
Determine the open circuit z-parameters for, and draw the corresponding z-parameter equivalent circuit of, the resistive network shown in Fig. P2.12.

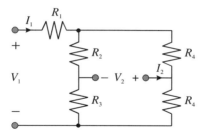

Figure P2.12.

Problem 2.13

The terminal volt-ampere characteristics of the linear network depicted in Fig. P2.13(a) subscribe to the matric,

$$\begin{bmatrix} I_1 \\ I_2 \\ I_3 \end{bmatrix} = \begin{bmatrix} 1 & -1 & -2 \\ -1 & 1 & -2 \\ -3 & 0 & 3 \end{bmatrix} \begin{bmatrix} V_1 \\ V_2 \\ V_3 \end{bmatrix},$$

where all elements in the 3×3 coefficient matrix are in units of siemens, and terminal voltages V_1, V_2, and V_3 are measured with respect to network terminal 4. Find the short circuit admittance matrix, **Y**, of the modified architecture offered in Fig. P2.13(b), such that

$$\begin{bmatrix} I_1 \\ I_3 \\ I_4 \end{bmatrix} = \mathbf{Y} \begin{bmatrix} V_1 \\ V_3 \\ V_4 \end{bmatrix},$$

with the understanding that terminal voltages V_1, V_3, and V_4 are measured with respect to network terminal 2.

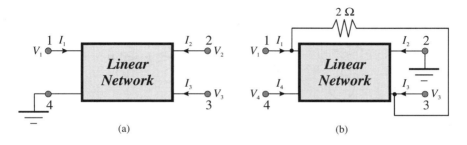

Figure P2.13.

Problem 2.14

The power supply voltages, V_{CC} and V_{EE}, which are incorporated to ensure reasonably linear operation of the transistors in the three-stage feedback amplifier shown in Fig. P2.14, can be presumed to be idealized voltage sources. All three bipolar junction transistors are identical, and their small signal terminal volt-ampere characteristics behave in accordance with the constraints imposed by the model given in Problem 2.7.

(a) What kind of feedback is established by the resistive subcircuit comprised of resistors R_A, R_B, and R_C?

(b) In concert with the type of feedback established in Part (a), for what type of signal processing gain is the closed loop amplifier best suited?

(c) For the type of feedback disclosed in Part (a), determine the most appropriate two-port parameter equivalent circuit for the subject resistive feedback subcircuit.

(d) Using the result deduced in Part (c), draw the small signal equivalent circuit of the entire feedback amplifier. For this overall amplifier model and using the results of Part (c), find the overall two-port parameters for the entire amplifier.

(e) Recalling the generalized performance metrics documented in Section 2.4.5, and assuming that the transistor parameter, β, is large, derive approximate expressions for the following performance indices:

  i. open loop driving point input resistance;
  ii. open loop driving point output resistance;

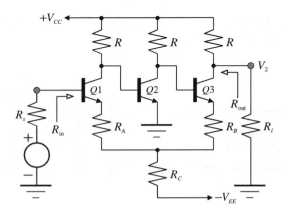

Figure P2.14.

iii. open loop gain;
iv. loop gain;
v. closed loop driving point input resistance, R_{in};
vi. closed loop driving point output resistance, R_{out};
vii. closed loop gain.

(f) Using the applicable results of Part (e), derive an approximate expression for the closed loop voltage gain, V_2/V_s.

Problem 2.15

The power supply voltages, V_{DD} and V_{SS}, which are incorporated to ensure reasonably linear operation of the MOSFETs in the three-stage feedback amplifier shown in Fig. P2.15 can be presumed to be idealized voltage sources. All three MOS transistors are identical, and their small signal terminal volt-ampere characteristics behave fundamentally in accordance with the constraints imposed by the model given in Problem 2.4. However, take r_o in this model as infinitely large and assume an analytical focus on only relatively low frequencies, which implies that all capacitances indigenous to the MOSFET model can be supplanted by open circuits.

(a) What kind of feedback is established by the feedback subcircuit comprised of resistance R_f and capacitance C_f?
(b) In concert with the type of feedback established in Part (a), for what type of signal processing gain is the closed loop amplifier best suited?

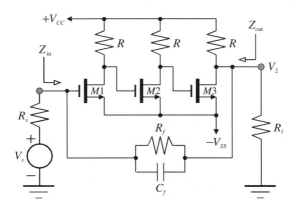

Figure P2.15.

(c) For the type of feedback disclosed in Part (a), determine the most appropriate two-port parameter equivalent circuit for the subject feedback subcircuit.
(d) Using the result deduced in Part (c), draw the small signal equivalent circuit of the entire feedback amplifier. For this overall amplifier model and using the results of Part (c), find the overall two-port parameters for the entire amplifier.
(e) Recalling the generalized performance metrics documented in Section 2.4.5, and assuming that the transistor parameter, g_m, is large, derive approximate expressions for the following performance indices:

  i. open loop driving point input resistance;
  ii. open loop driving point output resistance;
  iii. open loop gain;
  iv. loop gain;
  v. closed loop driving point input impedance, Z_{in};
  vi. closed loop driving point output impedance, Z_{out};
  vii. closed loop gain.

(f) Using the applicable results of Part (e), derive an approximate expression for the closed loop voltage gain, V_2/V_s.
(g) For the gain determined in Part (f), what is the approximate closed loop 3-dB bandwidth?

Problem 2.16
The linear two-port network abstracted in Fig. P2.16(a) is characterized by the terminal volt-ampere relationship,

$$\begin{bmatrix} V_1 \\ I_2 \end{bmatrix} = \begin{bmatrix} h_{11} & 0 \\ h_{21} & h_{22} \end{bmatrix} \begin{bmatrix} I_1 \\ V_2 \end{bmatrix},$$

where the pertinent h-parameters, h_{ij}, of the network indicate that zero internal feedback prevails. In terms of these h-parameters, derive expressions for the effective resistance, R_{eff}, indicated in the topological structures of both Figs. P2.16(b) and P2.16(c). Simplify these results for the case of very large h_{21} and very small h_{22}.

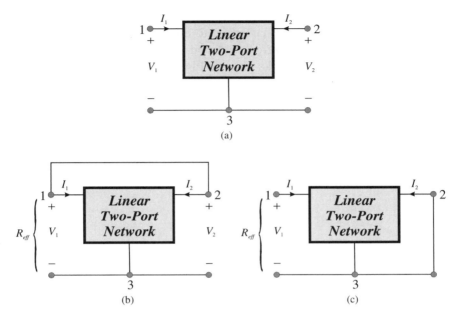

Figure P2.16.

Problem 2.17

The linear two-port network in Fig. P2.16(a) is now configured as the sub-circuit offered in Fig. P2.17. In terms of the h-parameters, h_{ij}, and the resistances, R_x and R_y, derive an expression for the indicated effective resistance, R_{eff}. Simplify this result for the case of very large h_{21} and very small h_{22}.

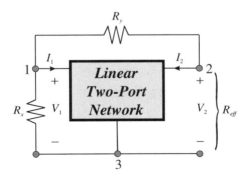

Figure P2.17.

Problem 2.18
Use the appropriate two-port parameter to determine expressions for the frequencies of the pole and zero associated with the short circuit current gain of each of the RC networks given in Fig. P2.18.

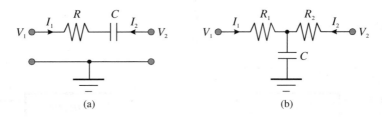

Figure P2.18.

Problem 2.19
For the two RC networks shown in Fig. P2.18, use the appropriate two-port parameters to determine expressions for the frequencies associated with the pole and zero of the forward transadmittance function.

Problem 2.20
The two-port network of Fig. P2.20 represents a circuit level model of a certain voltage amplifier. Determine the hybrid g-parameters of this unit and provide a g-parameter equivalent circuit for the amplifier.

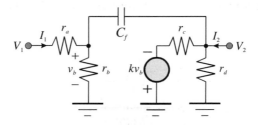

Figure P2.20.

Problem 2.21
Figure P2.21 depicts the schematic diagram of a modified common gate amplifier formed of nominally identical transistors biased at roughly identical quiescent operating points. Determine the short circuit admittance

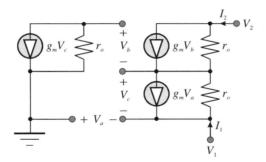

Figure P2.21.

parameters of this amplifier, and draw the corresponding y-parameter equivalent circuit.

Problem 2.22
Express the power gain, available power gain, and transducer power gain of the system shown in Fig. 2.43(a) in terms of the short circuit admittance parameters of the utilized two-port network.

Problem 2.23
Express the power gain, available power gain, and transducer power gain of the system shown in Fig. 2.43(a) in terms of the hybrid h-parameters of the utilized two-port network.

Problem 2.24
Express the power gain, available power gain, and transducer power gain of the system shown in Fig. 2.43(a) in terms of the hybrid g-parameters of the utilized two-port network.

Problem 2.25
The Llewellyn stability constraint articulated by Eq. (2-68) was derived by ensuring that the real part of the driving point input impedance of a two-port network is positive for all positive real load terminations. Show that the same stability requirement evolves when steps are taken to ensure that the real part component of the driving point output impedance of a linear two-port network is positive for all positive real source impedance terminations.

Problem 2.26

The linear active two-port in Fig. P2.26 has short circuit admittance parameters,

$$\left.\begin{array}{l} y_{11} = g_{22} + j\omega C_{11} \\ y_{12} = -g_{12} + j\omega C_{12} \\ y_{21} = g_{21} - jb_{21} \\ y_{22} = g_{22} + j\omega C_{22} \end{array}\right\},$$

where b_{21}, all g_{ij}, and all C_{ij} are non-negative numbers. The admittances, Y_f and Y_c, are incorporated as shown in an attempt to render unilateral the resultant two-port structure enclosed by the dashed box. If Y_f represents the admittance of a capacitance, C_f, and Y_c designates the admittance of an inductance, L_c, such unilateralization can be accomplished at a single frequency, say ω_o, to which the overall network might ultimately be tuned in a narrowband signal processing application.

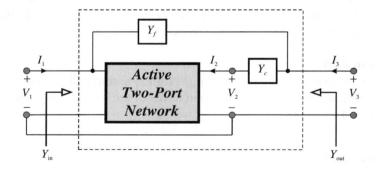

Figure P2.26.

(a) Show that in general, unilateralization requires an admittance, Y_f, of

$$Y_f = \left(\frac{Y_c}{Y_c + y_{22}}\right) y_{12}.$$

(b) With $Y_f = j\omega C_f$ and $Y_c = 1/j\omega L_c$, use the expression given in Part (a) to show that unilateralization is accomplished if

$$\omega_0^2 L_c C_f = \frac{g_{12}}{g_{22}}$$

$$C_f = \left(\frac{g_{12}}{g_{22}}\right) C_{22} - C_{12}.$$

(c) Observe a problem in the foregoing compensation strategy when parameter g_{12} is null. How might admittance Y_f be modified to circumvent this dilemma? Give unilateralization design equations similar to those offered in Part (b) for this modified feedback admittance.

Chapter 3

Scattering Parameters

3.1.0. Introduction

The short circuit admittance, open circuit impedance, hybrid h-, and hybrid g-parameters are commonly used to formulate two-port circuit models that globally interrelate the driving point input and output impedances, and forward and reverse transfer characteristics of linear networks. These models are simple architectures in that they embody only four electrical parameters whose measurement or calculation exploit the electrical implications of short or open circuits imposed at the input and output ports of the network undergoing study. Although the parameterization of these conventional two-port models can generally be executed straightforwardly and accurately at relatively low signal frequencies, high signal frequencies present at least two measurement challenges in broadband electronics. The first of these is the unavoidable parasitic inductance implicit to circuit interconnects that renders perfect short circuits an impossibility at very high signal frequencies. Moreover, very low impedance paths at either the input or the output port of electronic circuits may force embedded active devices to function nonlinearly or even to fail because of excessive current conduction. Second, the inherent potential instability of most high frequency or broadband electronic networks is exacerbated when these networks are constrained to operate with open circuited input or output ports. For example, attempts to measure the open circuit impedance parameters of a broadband electronic circuit are invariably thwarted by parasitic network oscillations incurred by the action of opening either an input or an output network port. In a word, circuits, and especially electronic circuits, designed to process very high signal frequencies are invariably finicky about the imposition of short or open circuits at their input or output port.

The daunting challenge posed by the measurement of conventional two-port parameters motivates the *scattering*, or *S-parameter* characterization of linear two-port systems. In contrast to the impedance, admittance, and hybrid parameters, the scattering parameters of linear electrical or electronic networks are measured without the need of short-circuiting or open-circuiting input and output ports. Instead, these ports are terminated in fixed and known reference impedances that can be similar or even identical to the terminating impedances incorporated in the design. Accordingly, the dynamic performance and operational integrity of a network under test are not compromised by test fixturing adopted for its scattering parameter characterization. Neither is the comfort level most engineers have with conventional two-port parameters compromised, for measured scattering parameters can always be converted to corresponding admittance, impedance, and hybrid parameters. While such conversion practices are ubiquitous, they are actually unnecessary since the driving point and transfer properties of linear passive or active two-port configurations can be determined directly in terms of the measured or computed network S-parameters.

The preceding paragraph implies that S-parameters are a measurement-friendly alternative to conventional two-port network parameters. Indeed, scattering parameters are conveniently extracted in the laboratory, but they are also explicitly useful in design. Because the S-parameters of a linear two-port network interrelate incident and reflected waves of energy at input and output ports, as opposed to intertwining input and output port voltages and currents, they are useful in the design of microwave amplifiers. The reason underlying this design utility is that delivered energies and average signal power levels are easier to quantify at microwave frequencies than are the voltages and currents associated with these energy waves. Additionally, the scattering parameters of lossless two-ports have a unique property that enables engineering expediency in the design of lossless matching networks that achieve maximum signal power transfer from a source to its terminating load.

This chapter develops the scattering parameter concept and defines the S-parameters of a generalized linear two-port network. Following a demonstration of the strategy used to convert S-parameters to conventional two-port parameters, it addresses the utility of S-parameters in the analysis of active networks. An interesting feedback perspective of S-parameters invoked in the analysis of high frequency circuits is offered prior to concluding with an exploration of the utility of S-parameters in the design of lossless matching filters for RF circuit and system applications.

3.2.0. Reflection Coefficient

The scattering parameter concept for two-port network analyses is best introduced by first considering the scattering characterization of the simple one-port network consisting of the load impedance, Z_l, shown in Fig. 3.1. The load impedance at hand consists of a real part, R_l, and a reactive component, X_l, which is positive for an inductive termination and negative for a capacitive load. The subject figure portrays a voltage source, V_s, whose Thévenin equivalent impedance is the real resistance, R_o. In the vernacular of scattering theory, this Thévenin impedance is often referred to as the *reference impedance* of the one-port network for in effect, the reflection coefficient quantifies the observable input impedance of the part as a quantity normalized to the measurement reference impedance.

Let the current, I, conducted by the load in Fig. 3.1 be decomposed into the algebraic superposition of an incident current, I_i, and a reflected, or "scattered" current, I_r, such that

$$I = I_i - I_r. \tag{3-1}$$

In the last relationship, I_i is defined as the value of current I for the special case of a load impedance, Z_l, matched to the one-port reference impedance, R_o, that is, $Z_l \equiv R_o$. Thus,

$$I_i = \frac{V_s}{2R_o}, \tag{3-2}$$

whence the reflected current, I_r, follows

$$I_r = I_i - I = \frac{V_s}{2R_o} - \frac{V_s}{R_o + Z_l} = \left(\frac{Z_l - R_o}{Z_l + R_o}\right)\frac{V_s}{2R_o} = \left(\frac{Z_l - R_o}{Z_l + R_o}\right)I_i. \tag{3-3}$$

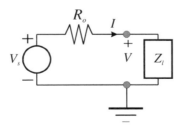

Figure 3.1. A one-port load impedance driven by a voltage source whose Thévenin equivalent resistance is R_o.

The parenthesized quantity on the far right hand side of Eq. (3-3) is known as the *scattering parameter*, or more simply, the *reflection coefficient*, ρ, of the considered one-port load impedance. Specifically,

$$\rho \triangleq \frac{Z_l - R_o}{Z_l + R_o} = \frac{Z_{ln} - 1}{Z_{ln} + 1}, \qquad (3\text{-}4)$$

where

$$Z_{ln} = \frac{Z_l}{R_o} \qquad (3\text{-}5)$$

is the load impedance normalized to the characterizing, or reference, source impedance. Note that Eq. (3-4) allows Eq. (3-3) to be expressed as

$$I_r = \rho I_i, \qquad (3\text{-}6)$$

which implies that the net load current in Eq. (3-1) can be written as

$$I = (1 - \rho) I_i. \qquad (3\text{-}7)$$

Equation (3-7) confirms that when $\rho = 0$, which corresponds to a terminating load impedance matched to the reference impedance, the only current supplied by the signal source is the incident current. Note that Eq. (3-7) also conveys a branch current model as an algebraic sum of an incident current component, which flows in the direction of the actual branch current, and a reflected current component, which flows opposite to the direction of the observed branch current.

3.2.1. Voltage Scattering

Like the load current, the load voltage, V, can be decomposed into an incident component, V_i, and a reflected component, V_r, in accordance with

$$V = V_i + V_r. \qquad (3\text{-}8)$$

The incident component of load voltage is the value of voltage V under the condition of a load impedance matched to the circuit reference impedance, that is,

$$V_i = \left(\frac{Z_l}{R_o + Z_l}\right) V_s \bigg|_{Z_l = R_o} = \frac{V_s}{2}. \qquad (3\text{-}9)$$

It follows that the scattered, or reflected, load voltage component is

$$V_r = \left(\frac{Z_l - R_o}{Z_l + R_o}\right) \frac{V_s}{2} = \rho V_i, \qquad (3\text{-}10)$$

Scattering Parameters

which allows the net load voltage to be expressed as

$$V = (1+\rho)V_i. \tag{3-11}$$

Observe from Eqs. (3-6), (3-9), and (3-10) that

$$\begin{bmatrix} V_i \\ V_r \end{bmatrix} \begin{bmatrix} R_o & 0 \\ 0 & R_o \end{bmatrix} \begin{bmatrix} I_i \\ I_r \end{bmatrix}; \tag{3-12}$$

that is, the incident component of load voltage, V_i, is directly proportional to the incident component of load current, I_i, while the reflected load voltage, V_r, is directly proportional to the reflected load current, I_r. Moreover, the incident load voltage is independent of the reflected load current, and analogously, the reflected load voltage is independent of the incident load current.

3.2.2. Power Scattering

The signal power, P, delivered to the load impedance can also be written as a superposition of incident (P_i) and reflected (P_r) powers, that is,

$$P = P_i - P_r. \tag{3-13}$$

If the load voltage, V, is a sinusoid of amplitude V_p, the average incident power delivered to the matched load is

$$P_i = \frac{\left(V_p/\sqrt{2}\right)^2}{R_o} = \frac{V_s^2}{8R_o}, \tag{3-14}$$

where V_s is understood to represent the amplitude of the applied sinusoidal input voltage. The average power delivered to the actual load impedance is limited to the power dissipated in the real part, say R_l, of the load impedance, and is

$$P = \frac{|I|^2 R_l}{2} = \frac{V_s^2 R_l}{2|Z_l + R_o|^2}. \tag{3-15}$$

A bit of algebra applied to the combination of Eqs. (3-13), (3-14), and (3-15) confirms that

$$P_r = |\rho|^2 P_i, \tag{3-16}$$

from which the average load power is seen to be

$$P = \left(1 - |\rho|^2\right) P_i = (1 + |\rho|)(1 - |\rho|) P_i. \tag{3-17}$$

Thus, if the load impedance is matched to the circuit reference impedance, ρ is zero and resultantly, no power is reflected to the signal source from the load, and maximum signal source power is delivered to said load.

3.2.3. Significance of the Reflection Coefficient

The reflection coefficient introduced in Eq. (3-4) is a bilinear transformation that encodes ohms in the Cartesian impedance plane to a dimensionless complex number in the reflection plane. Because of the bilinear nature of this transformation, the reflection coefficient of an impedance is unique for a given impedance function and conversely, a unique impedance derives from a stipulated reflection coefficient. To demonstrate the latter contention, Eq. (3-4) can be solved for load impedance Z_l to obtain

$$Z_l = \left(\frac{1+\rho}{1-\rho}\right) R_o, \qquad (3\text{-}18)$$

which confirms impedance uniqueness for any given value of reflection coefficient. The engineering implication of this uniqueness is that the measurement of the reflection coefficient for a one-port network, as is routinely accomplished by laboratory network analyzers, is equivalent to a determination of the impedance that terminates the characterized one-port.

Assuming that the real part of the load impedance is non-negative for all signal frequencies, which is tantamount to asserting that Z_l is a passive two-terminal impedance, Eq. (3-4) shows that the magnitude of the load reflection coefficient is at most unity. Note that for a short-circuited load impedance ($Z_l = 0$), $\rho = -1$, while an open-circuited load ($Z_l = \infty$) produces $\rho = 1$. In view of the fact that the magnitude of ρ can never exceed unity for passive loads, it follows that all possible passive impedances, whose values reflect a range extending from a short circuit to an open circuit, map to coordinates lying within, or directly on, the unit circle centered at the origin of the reflection plane. This observation is dramatized in Fig. 3.2, where the reflection plane is taken as a coordinate system whose horizontal axis is the real part of ρ and whose vertical axis is the imaginary part of ρ. If, for example, $\rho_r = 0.3$ and $\rho_i = 0.75$ in the subject figure, the normalized impedance is, from Eq. (3-18), $Z_l/R_o = 0.330 + j1.425$. If the reference impedance is 50 ohms and the signal frequency is 1 GHz, this normalized impedance consists of the series combination of a resistance of 16.51 ohms and an inductance of 11.34 nanohenries.

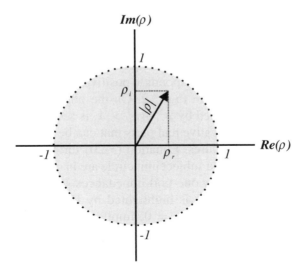

Figure 3.2. Cartesian coordinate system of the reflection coefficient plane. The shaded area enclosed by the unit circle centered at the origin of the plane is the loci of all possible passive impedances. The indicated vector symbolizes a reflection coefficient, ρ, of $\rho_r + j\rho_i$, which corresponds to the unique one-port impedance function, Z_l, defined by Eq. (3-18).

It is worthwhile elaborating on the unit circle boundary for the magnitude of the reflection coefficient. To this end, let the reflection coefficient, ρ, in Eq. (3-18) be written as $\rho = \rho_r + j\rho_i$, whence the normalized load impedance, Z_{ln}, becomes

$$Z_{ln} = \frac{(1 + \rho_r) + j\rho_i}{(1 - \rho_r) - j\rho_i} = \frac{[1 - (\rho_r^2 + \rho_i^2)] + j(2\rho_i)}{(1 - \rho_r)^2 + \rho_i^2}. \quad (3\text{-}19)$$

If the normalized impedance is viewed as a series interconnection of a normalized resistance, R_{ln}, and a normalized reactance, X_{ln}, Eq. (3-19) shows that

$$R_{ln} = \frac{1 - (\rho_r^2 + \rho_i^2)}{(1 - \rho_r)^2 + \rho_i^2} \quad (3\text{-}20)$$

and

$$X_{ln} = \frac{2\rho_i}{(1 - \rho_r)^2 + \rho_i^2}. \quad (3\text{-}21)$$

Equations (3-20) and (3-21) illuminate several reflection plane properties that comprise the foundation of the Smith chart,[1,2] a well-known graphical tool used in the analysis of distributed transmission lines and in the design of many linear communication and microwave circuits. The first of these important properties derives from the observation that the relationship, $\rho_r^2 + \rho_i^2 = 1$, defines the unit circuit centered at the origin, $(\rho_r, \rho_i) = (0, 0)$, of the reflection coefficient plane. Accordingly, Eq. (3-20) implies that the area enclosed by this unit circle corresponds to positive real part load impedances ($R_{ln} > 0$), as is suggested in Fig. 3.2. On the other hand, the area outside of the unit circle defined by $\rho_r^2 + \rho_i^2 > 1$, is seen as embracing load impedances having the negative real parts that can be generated by active, potentially unstable, structures. Equation (3-20) further confirms that all points lying precisely on the subject unit circle are in one to one correspondence with purely imaginary one-port impedances, that is, $R_{ln} = 0$.

A second set of properties is highlighted by Eq. (3-21). Specifically, purely real load impedances ($X_{ln} = 0$) imply $\rho_i = 0$. For a passive one-port, $\rho_i = 0$ corresponds to all points lying within the unit circle and directly on the horizontal axis of the reflection plane. On the other hand, inductive loads ($X_{ln} > 0$) in passive one-port networks map into the region of the unit circle lying above the horizontal axis, while capacitive passive loads ($X_{ln} < 0$) establish loci within the unit circle and below the horizontal axis of the reflection plane. These observations, as well as those of the preceding paragraph are summarized in Fig. 3.3.

A generalization of the foregoing stipulations follows from recasting Eq. (3-20) in the form,

$$\left(\rho_r - \frac{R_{ln}}{R_{ln} + 1}\right)^2 + \rho_i^2 = \left(\frac{1}{R_{ln} + 1}\right)^2. \qquad (3\text{-}22)$$

This relationship stipulates that the reflection plane loci of constant real part load impedance are circles centered on the horizontal axis at $\rho_r = R_{ln}/(R_{ln} + 1)$ and having radii of $1/(R_{ln} + 1)$. To this end, infinitely large real part load impedance is seen to be a point located at $(\rho_r, \rho_i) = (1, 0)$. In concert with a previous disclosure, observe in Eq. (3-22) that zero real part load impedance corresponds to the infinity of points lying on the unit circle centered at the origin of the reflection coefficient plane. Figure 3.4 provides a graphical overview of these observations.

It is also interesting to learn that constant load reactances also map to circles in the ρ-plane. In particular, Eq. (3-21) can be shown to be equivalent to

$$(\rho_r - 1)^2 + \left(\rho_i - \frac{1}{X_{ln}}\right)^2 = \left(\frac{1}{X_{ln}}\right)^2. \qquad (3\text{-}23)$$

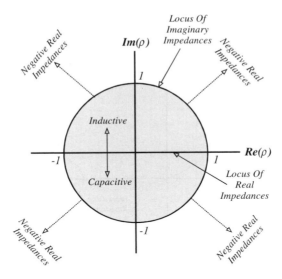

Figure 3.3. The impedance implications of reflection coefficient loci. The shaded region bounded by the unit circle corresponds to all impedances characterized by positive real part components. The area enclosed by the unit circle and lying above the horizontal axis corresponds to inductive positive real impedances. On the other hand, the area enclosed by the unit circle and lying below the horizontal axis corresponds to capacitive positive real impedances.

The subject constant reactance circles are centered at $(\rho_r, \rho_i) = (1, 1/X_{ln})$ and have radii of $1/X_{ln}$. Note therefore that the centers of these circles are located above the horizontal reflection plane axis for inductive loads and below the horizontal axis for capacitive loads.

The reflection coefficient in Eq. (3-4) can be partitioned into its explicit real and imaginary parts to allow for the plotting of the reflection coefficient contour into a Smith-type chart. To this end, write the normalized load impedance, Z_{ln}, in the form, $Z_{ln} = R_{ln} + jX_{ln}$, to obtain $\rho = \rho_r + j\rho_i$, where

$$\rho_r = \frac{R_{ln}^2 - 1}{(R_{ln} + 1)^2 + X_{ln}^2} \quad (3\text{-}24)$$

and

$$\rho_i = \frac{2X_{ln}}{(R_{ln} + 1)^2 + X_{ln}^2}. \quad (3\text{-}25)$$

As the signal frequency is varied, R_{ln} and X_{ln} can be computed, whence the real and imaginary parts of the reflection coefficient in Eqs. (3-24)

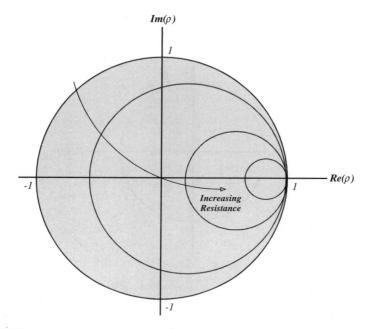

Figure 3.4. The contours of constant real part load impedance in the reflection coefficient plane.

and (3-25), respectively, can be evaluated for each frequency of interest. The resultant plot of ρ_i versus ρ_r comprises the resultant reflection contour, with signal frequency used as an implicit parametric variable.

Equation (3-25) suggests that the imaginary part of the reflection coefficient is a nonmonotonic function of the normalized reactive component of the port impedance undergoing characterization. Specifically, observe that ρ_i is zero at both $X_{ln} = 0$ and $X_{ln} = \infty$, which implies that ρ_i is maximized at an intermediate value of X_{ln}, say X_{lno}, that constrains the slope, $d\rho_i/dX_{ln}$, to zero. It is a simple matter to show that

$$X_{lno} = \pm(R_{ln} + 1), \tag{3-26}$$

for which the corresponding reflection coefficient, say ρ_o, is

$$\rho_o = \left(\frac{1}{R_{ln} + 1}\right)\left(\frac{R_{ln} - 1}{2} \pm j\right). \tag{3-27}$$

It is understood that the positive algebraic sign in Eqs. (3-26) and (3-27) applies for inductive port impedances, while the negative sign pertains to capacitive loads. As is demonstrated in the following example, Eq. (3-26)

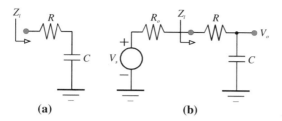

Figure 3.5. (a) Series RC lowpass filter addressed in Example 3.1. (b) The filter in (a) driven by a voltage source having a Thévenin internal resistance of R_o, the reference impedance used in the characterization of the impedance established by the RC subcircuit.

boasts at least tacit engineering significance when it is applied to lowpass dominant pole networks. For such networks, the frequency for which Eq. (3-26) is satisfied is the 3-dB bandwidth of the network formed by driving the subject impedance with a voltage source whose Thévenin resistance is the reference resistance used in the impedance characterization.

Example 3.1. Plot the real and imaginary parts of the reflection coefficient, as a function of signal frequency, for the impedance established by the lowpass RC network in Fig. 3.5(a). The indicated series circuit resistance, R, is 300 Ω, while the capacitance, C, is 50 pF. The normalizing reference impedance (R_o) can be taken to be 50 Ω.

Solution 3.1.

(1) The impedance of the RC network load in Fig. 3.5(a) is

$$Z_l = R + \frac{1}{j\omega C}, \tag{E1-1}$$

whence the normalized load resistance is

$$R_{ln} = \frac{R}{R_o}, \tag{E1-2}$$

and the normalized load reactance is

$$X_{ln} = -\frac{1}{\omega R_o C}. \tag{E1-3}$$

(2) Equations (E1-2) and (E1-3) for R_{ln} and X_{ln}, respectively, can be substituted into Eqs. (3-24) and (3-25) to deduce the frequency dependencies of the real and imaginary parts of the reflection coefficient in the stipulated 50 Ω reference environment. The pertinent plots are displayed in Fig. 3.6.

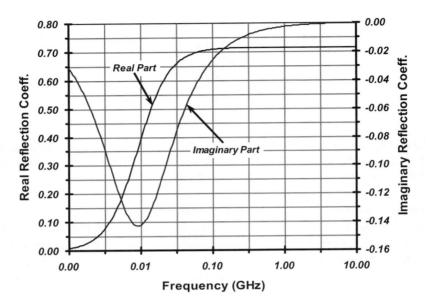

Figure 3.6. The real part and the imaginary part components of the reflection coefficient associated with the impedance established by the simple RC filter in Fig. 3.5. The circuit resistance, R, is $300\,\Omega$, the circuit capacitance, C, is $50\,\text{pF}$, and the reference impedance, R_o, is $50\,\Omega$.

Comments. From Eq. (3-26), the imaginary component of the load reflection coefficient is minimized at $X_{lno} = -(R_{ln} + 1) = -(R_l + R_o)/R_o = -7$. From Eq. (3-27), the corresponding reflection coefficient is $\rho_o = 0.357 - j0.142$, which is confirmed in Fig. 3.6.

The signal frequency, ω_o, corresponding to the minimization of the imaginary component of the reflection parameter is, with the help of Eq. (3-26),

$$\omega_o = \frac{1}{(R_l + R_o)C} = 2\pi(9.1\,\text{MHz}), \tag{E1-4}$$

which is indeed the 3-dB bandwidth of the circuit in Fig. 3.5(b). Specifically, the subject frequency is the 3-dB bandwidth of the lowpass circuit formed by driving the load impedance undergoing investigation with a signal voltage source whose Thévenin resistance is identical to the reference impedance used in the impedance characterization.

Example 3.2. Plot the real and imaginary parts of the reflection coefficient, as a function of signal frequency, for the impedance established by the RLC network in Fig. 3.7. In the course of plotting these reflection coefficient components, normalize

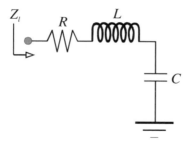

Figure 3.7. Series RLC bandpass filter addressed in Example 3.2.

the signal frequency to the center frequency, ω_o, of the impedance function. The indicated series circuit resistance, R, is 300 Ω, while the quality factor, Q, of the RLC impedance is 3. The reference impedance (R_o) is 50 Ω.

Solution 3.2.

(1) The impedance of the RLC network in Fig. 3.7 is

$$Z_l = R + j\omega L + \frac{1}{j\omega C}, \tag{E2-1}$$

which is equivalent to writing

$$Z_l = R + jQR\left(\frac{\omega}{\omega_o} - \frac{\omega_o}{\omega}\right). \tag{E2-2}$$

In the last expression,

$$\omega_o = \frac{1}{\sqrt{LC}} \tag{E2-3}$$

is the resonant frequency (in radians-per-second) of the load impedance, while

$$Q = \frac{\omega_o L}{R} = \frac{\sqrt{L/C}}{R} \tag{E2-4}$$

is quality factor of the load inductance at the circuit resonant frequency.

(2) The normalized load resistance, R_{ln}, follows

$$R_{ln} = \frac{R}{R_o}, \tag{E2-5}$$

and the normalized reactive component of this load impedance is

$$X_{ln} = Q\frac{R}{R_o}\left(\frac{\omega}{\omega_o} - \frac{\omega_o}{\omega}\right) = QR_{ln}\left(\frac{\omega}{\omega_o} - \frac{\omega_o}{\omega}\right). \tag{E2-6}$$

The foregoing two analytical results can now be substituted into Eqs. (3-24) and (3-25) to generate the plots provided in Fig. 3.8.

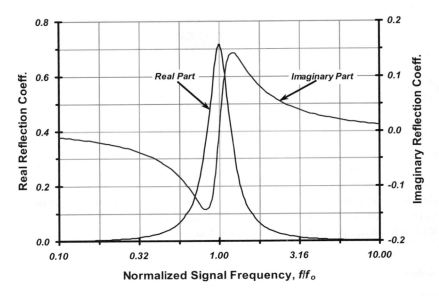

Figure 3.8. The real part and the imaginary part components of the reflection coefficient associated with the impedance established by the filter in Fig. 3.7. The circuit resistance, R, is 300 Ω, the impedance quality factor, Q, is 3, and the reference impedance, R_o, is 50 Ω.

Comments. As expected, the imaginary part of the reflection coefficient vanishes at the resonant frequency (at $\omega/\omega_o = f/f_o = 1$), since this imaginary reflection parameter is directly proportional to the effective reactance of the considered impedance. At signal frequencies that are smaller than the resonant frequency, where the impedance is dominantly capacitive, the imaginary reflection component is negative, analogous to the disclosures offered in the preceding example. On the other hand, ρ_i is understandably positive for $f > f_o$, where the impedance is dominated by circuit inductance. The indicated peaks in the ρ_i-profile occur at frequency values that satisfy Eq. (3-26). It can be shown that the frequency difference, say $\Delta\omega_o$, between these observed maxima is a measure of the impedance quality factor and specifically,

$$\Delta\omega_o = \frac{1}{Q}\left(1 + \frac{1}{R_{ln}}\right). \quad \text{(E2-7)}$$

Finally, note that ρ_r, the real part component of the reflection coefficient, is maximized at the resonant frequency, where $\rho_i = 0$. Since $\rho_i = 0$ is in one-to-one correspondence with $X_{ln} = 0$, this maximum real part reflection coefficient is, from Eq. (3-24),

$$\rho_{rmax} = \frac{R_{ln} - 1}{R_{ln} + 1}. \quad \text{(E2-8)}$$

In other words, for a given reference impedance, R_o, the measurement of ρ_{rmax} is tantamount to a measurement of R, the resistance appearing in series with the inductor. This observation is often exploited profitably to determine the parasitic resistance that unavoidably appears in series with circuit inductance.

3.3.0. Two-Port Scattering Parameters

The concept of the reflection coefficient, which is essentially the lone scattering parameter of a one-port linear network, serves as a springboard for the scattering parameter characterization and analysis of linear two-port networks. To this end, consider the abstraction in Fig. 3.9, which depicts a linear two-port network to which test sinusoidal voltages, V_{s1} and V_{s2}, are applied to the input and output ports, respectively. It is critical to observe that the series resistances associated with these two test voltages are identical. These resistances, which are symbolized as R_o, are the known reference impedances for which the network at hand is undergoing an input/output (I/O) port characterization. Invariably, R_o is 50 ohms and in such a case, the network under test is said to be undergoing a characterization in a 50 Ω test fixture or a 50 Ω test environment.

The indicated port voltages, V_1, and V_2, which are established in response to the applied test signals, V_{s1} and V_{s2}, can be viewed as respective superpositions of incident and reflected components. Following the lead of Eq. (3-8),

$$\left. \begin{array}{l} V_1 = V_{1i} + V_{1r} \\ V_2 = V_{2i} + V_{2r} \end{array} \right\}, \qquad (3\text{-}28)$$

where the incident component, V_{1i}, of input port voltage V_1 is simply $V_{s1}/2$, and V_{2i}, the incident component of output port voltage V_2, is $V_{s2}/2$.

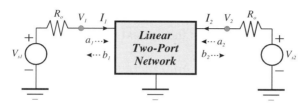

Figure 3.9. Abstraction of a linear two-port system undergoing a scattering parameter characterization in a test environment whose reference impedance is R_o. The a_i denote incident energy waves at the network ports, while the b_i are reflected energy waves at the system ports.

The immediate effect of test signal V_{s1} is the application of energy to the input port of the network. When the driving point input impedance of the network is matched to the reference impedance, R_o, maximum energy transfer from V_{s1} to the input port occurs and in this case, no energy is reflected from the input port to the applied signal source. This maximal energy is symbolized in Fig. 3.9 as the incident energy wave, a_1. Since the input impedance is generally not matched to the reference impedance, the actual energy transferred is smaller than a_1. The resultant difference between a_1 and the actual energy transferred to the input port is the reflected energy wave, which is denoted in the figure as b_1. Analogous statements can be proffered for the incident and reflected energy waves, a_2, and b_2, respectively, at the network output port. These a_i and b_i are written as

$$\left. \begin{array}{l} a_1 = \dfrac{V_{1i}}{\sqrt{R_o}} = \sqrt{R_o}\, I_{1i} \\ a_2 = \dfrac{V_{2i}}{\sqrt{R_o}} = \sqrt{R_o}\, I_{2i} \end{array} \right\rangle \qquad (3\text{-}29)$$

and

$$\left. \begin{array}{l} b_1 = \dfrac{V_{1r}}{\sqrt{R_o}} = \sqrt{R_o}\, I_{1r} \\ b_2 = \dfrac{V_{2r}}{\sqrt{R_o}} = \sqrt{R_o}\, I_{2r} \end{array} \right\rangle, \qquad (3\text{-}30)$$

where Eq. (3-12) has been exploited. These seemingly strange algebraic relationships are placed in engineering perspective by noting that the square magnitudes of the various a_i and b_i represent incident and reflected port power levels.

The scattering parameters, S_{ij}, of the two-port network under consideration can now be introduced in the context of the matrix relationship,

$$\begin{bmatrix} b_1 \\ b_2 \end{bmatrix} = \begin{bmatrix} S_{11} & S_{12} \\ S_{21} & S_{22} \end{bmatrix} \begin{bmatrix} a_1 \\ a_2 \end{bmatrix}. \qquad (3\text{-}31)$$

The definition and the measurement of these four scattering parameters proceed directly from this defining relationship. But unlike conventional two-port parameters, considerable engineering care must be exercised to ensure an accurate and meaningful interpretation of the scattering parameters.

3.3.1. Parameters S_{11} and S_{21}

To begin, note that under the condition of $a_2 = 0$,

$$\left. \begin{aligned} S_{11} &= \frac{b_1}{a_1}\bigg|_{a_2=0} \\ S_{21} &= \frac{b_2}{a_1}\bigg|_{a_2=0} \end{aligned} \right\}. \tag{3-32}$$

The constraint, $a_2 = 0$, implies zero signal incidence at the output port of the network under test. Since the output port in Fig. 3.9 is driven by a test signal, V_{s2}, placed in series with the measurement reference impedance, R_o, of the system, $a_2 = 0$ is equivalent to the constraint, $V_{s2} = 0$. This assertion derives from the general observation that energy incidence at the network output port materializes from only two sources. The first and most obvious of these two sources is purposefully applied output port energy in the form of the signal voltage, V_{s2}, which is nulled in this particular exercise. The second of these sources is signal reflection from the load impedance. But such reflection is zero when the load impedance is identical to the measurement reference impedance of the system. Accordingly, a_2 is held fast at zero precisely because V_{s2} is set to zero and the terminating load impedance at the output port is R_o.

From Eqs. (3-29) and (3-30), S_{11} in Eq. (3-32) is equivalent to

$$S_{11} = \frac{b_1}{a_1}\bigg|_{a_2=0} = \frac{V_{1r}}{V_{1i}}\bigg|_{a_2=0}. \tag{3-33}$$

Recalling Eq. (3-10) and the fact that $a_2 = 0$ implies an exclusively passive load termination identical to the reference impedance, R_o, S_{11} is seen as the input port reflection coefficient under the special circumstance of an output port terminated in the characteristic impedance of the measurement fixture. Letting $Z_{in}(R_o)$ denote this specialized input impedance, where the parenthesized R_o underscores the requisite output port load termination,

$$S_{11} = \frac{Z_{in}(R_o) - R_o}{Z_{in}(R_o) + R_o} = \frac{Z_{in}(R_o)/R_o - 1}{Z_{in}(R_o)/R_o + 1}. \tag{3-34}$$

It is imperative to understand that the input port impedance, $Z_{in}(R_o)$, in Eq. (3-34) is not the driving point input impedance of the network identified for characterization, that is, it is <u>not</u> the input impedance under actual load circumstances. Instead, it is the input impedance under the test condition of a load impedance set equal to the reference impedance of the test fixture. Figure 3.10(a) highlights this observation.

Returning to Eqs. (3-29), (3-30), and (3-32),

$$S_{21} = \left.\frac{b_2}{a_1}\right|_{a_2=0} = \left.\frac{V_{2r}}{V_{1i}}\right|_{a_2=0} = 2\frac{V_2}{V_{s1}}, \tag{3-35}$$

where use is made of the facts that zero energy incidence at the output port means that the output port voltage, V_2, is solely a reflected voltage component, V_{2r}, and $V_{1i} \equiv V_{s1}/2$. As is confirmed in Fig. 3.10(a), parameter S_{21} is seen as simply twice the voltage gain under the conditions of a Thévenin source impedance of R_o and an output load impedance likewise equal to R_o. It is interesting to note that the squared magnitude of parameter S_{21} is intimately related to the signal power gain of the subject network. In particular,

$$|S_{21}|^2 = \left|2\frac{V_2}{V_{s1}}\right|^2 = \frac{|V_2|^2/R_o}{|V_{s1}|^2/4R_o}. \tag{3-36}$$

The numerator on the right hand side of this relationship is clearly the power delivered by the network output port to the reference impedance load of R_o. On the other hand, the denominator on the right hand side of Eq. (3-36) represents the maximum possible power deliverable by the signal source applied at the input port. Thus, when the network at hand is terminated at

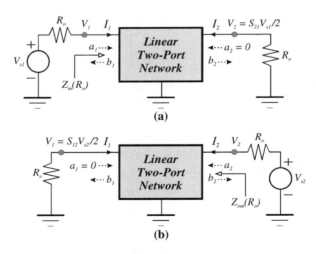

Figure 3.10. (a) Test fixturing for the measurement of the scattering parameters, S_{11} and S_{21}, of a linear two-port network. Observe that no energy is applied to the output port. (b) Test fixturing for the measurement of parameters S_{22} and S_{12}. In this case, zero energy is applied to the input port.

its output port in R_o and driven at its input port by a signal source whose Thévenin equivalent resistance is R_o, the squared magnitude of S_{21} is simply the ratio of delivered output power to maximum available source power. This ratio is termed the *transducer power gain*, G_T, of a linear network, that is,

$$|S_{21}|^2 = \frac{|V_2|^2/R_o}{|V_{s1}|^2/4R_o} = 4\left|\frac{V_2}{V_{s1}}\right|^2 \triangleq G_T. \tag{3-37}$$

It is worthwhile interjecting that the frequency response of the transducer power gain of an active network generally provides a more meaningful indication of attainable circuit frequency response than does either the current gain of a device or the voltage or current gain of a device. This statement derives from the simple fact that active networks are generally utilized for the explicit purpose of boosting signal power gain, as opposed to providing current or voltage amplification. In a lowpass system therefore, the signal frequency at which the transducer power gain degrades from its low frequency value by 6-dB is a more realistic measure of high frequency circuit properties than is either the traditional unity gain frequency of a device or the 3-dB bandwidth of test circuit current or voltage gain.

3.3.2. Parameters S_{22} and S_{12}

The analytical definitions of parameters S_{22} and S_{12} derive in a fashion that is analogous to the derivation of S_{11} and S_{21}. From Eq. (3-31),

$$\left. \begin{array}{c} S_{22} = \left.\dfrac{b_2}{a_2}\right|_{a_1=0} \\ S_{12} = \left.\dfrac{b_1}{a_2}\right|_{a_1=0} \end{array} \right\}. \tag{3-38}$$

The condition of zero input port energy incidence, $a_1 = 0$, is equivalent to the stipulations of a zero input signal source, $V_{s1} = 0$, and an input port terminated to signal ground through the reference impedance, R_o. Accordingly, S_{22} is the corresponding reflection coefficient of the output port,

$$S_{22} = \frac{Z_{\text{out}}(R_o) - R_o}{Z_{\text{out}}(R_o) + R_o} = \frac{Z_{\text{out}}(R_o)/R_o - 1}{Z_{\text{out}}(R_o)/R_o + 1}, \tag{3-39}$$

where, as is inferred by Fig. 3.10(b), $Z_{\text{out}}(R_o)$ specifies the output port impedance measured under the condition of an input port terminated to ground in the test fixturing reference impedance. Moreover, and under the same termination circumstances, S_{12} is related to the reverse, or feedback, voltage gain of the network. Specifically,

$$S_{12} = \left.\frac{b_1}{a_2}\right|_{a_1=0} = \left.\frac{V_{1r}}{V_{2i}}\right|_{a_1=0} = 2\frac{V_1}{V_{s2}}. \qquad (3\text{-}40)$$

3.3.3. Port Voltage and Current Generalizations

A final noteworthy observation derives from Eqs. (3-28) through (3-30). In particular, the input and output port voltages, V_1 and V_2, respectively, can be written as

$$\left.\begin{array}{l}V_1 = \sqrt{R_o}(a_1 + b_1)\\ V_2 = \sqrt{R_o}(a_2 + b_2)\end{array}\right\}. \qquad (3\text{-}41)$$

Since currents are expressed as a difference between incident and reflected components, Eqs. (3-29) and (3-30) also confirm that

$$\left.\begin{array}{l}I_1 = \dfrac{a_1 - b_1}{\sqrt{R_o}}\\[6pt] I_2 = \dfrac{a_2 - b_2}{\sqrt{R_o}}\end{array}\right\}. \qquad (3\text{-}42)$$

It follows that short-circuited linear network ports, which are required in the measurement of the admittance parameters, require $a_i = -b_i$. On the other hand, open-circuited ports, which are demanded of impedance parameter extraction, mandate $a_i = b_i$.

3.3.4. Scattering Analysis of a Generalized Two-Port

Figure 3.11 depicts the general case of a linear two-port network undergoing an I/O characterization in the sinusoidal steady state. The signal source applied at the input port of the network is represented as a Thévenin equivalent circuit whose voltage is V_s and whose impedance is Z_s. The output port of the subject linear network is terminated to ground through an impedance of Z_l. The two-port network itself is presumed to have scattering parameters, S_{ij}, which are extracted with respect to a known real reference impedance, R_o. Accordingly, the indicated I/O incident and reflected energy waves, a_1, b_1, a_2, and b_2, subscribe to Eq. (3-31). It is also assumed that the source and load impedances are measured indirectly as source and load reflection coefficients, ρ_s and ρ_l, referenced as well to R_o. Thus, the incident and

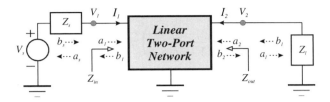

Figure 3.11. Linear network terminated at its input and output ports in generalized load and source impedances. The two-port network is presumed to have measured or otherwise known scattering parameters referenced to a characteristic impedance of R_o.

reflected energy waves, a_s and b_s, pertinent to the source termination satisfy the relationship,

$$\rho_s = \frac{b_s}{a_s} = \frac{Z_s - R_o}{Z_s + R_o}. \tag{3-43}$$

Similarly, for the output port,

$$\rho_l = \frac{b_l}{a_l} = \frac{Z_l - R_o}{Z_l + R_o}. \tag{3-44}$$

Since no signal energy is explicitly applied at the output port as an independent voltage or current source, the energy wave, a_2, incident at the output port of the linear network is necessarily equal to the energy reflected back to said port by the load impedance; that is, $a_2 = b_l$. Additionally, all energy reflected at the output port of the linear network is incident to the load impedance, thereby establishing the constraint, $b_2 = a_l$. It follows that the output port reflection coefficient, ρ_2, relates to the load reflection coefficient, ρ_l, as

$$\rho_2 = \frac{b_2}{a_2} = \frac{a_l}{b_l} = \frac{1}{\rho_l}. \tag{3-45}$$

3.3.4.1. Input and Output Reflection Coefficients

The determination of the reflection coefficient for the input port of the subject linear two-port network is tantamount to determining the driving point input impedance, Z_{in}; that is, the input impedance under actual load termination conditions. Since $b_2 = a_l$ and $a_2 = b_l = \rho_l a_l$ in Fig. 3.11, Eq. (3-31) yields

$$a_l = S_{21}a_1 + S_{22}b_l = S_{21}a_1 + \rho_l S_{22}a_l, \tag{3-46}$$

or

$$a_l = \left(\frac{S_{21}}{1 - \rho_l S_{22}}\right) a_1. \tag{3-47}$$

Returning to Eq. (3-31),

$$b_l = S_{11}a_1 + S_{12}a_2 = S_{11}a_1 + S_{12}b_l = S_{11}a_1 + \rho_l S_{12}a_l. \tag{3-48}$$

Upon substitution of Eq. (3-47) into Eq. (3-48), the reflection coefficient, ρ_1, of the input port is seen to be

$$\rho_1 = \frac{b_1}{a_1} = S_{11} + \frac{\rho_l S_{12} S_{21}}{1 - \rho_l S_{22}}. \tag{3-49}$$

Three noteworthy observations surface from the preceding result. First, note that $\rho_1 = S_{11}$ when $\rho_l = 0$, which defines a load impedance matched to the reference impedance ($Z_l = R_o$). This result is expected in view of the fact that S_{11} is, by definition, the input port reflection coefficient under the special case of a load termination matched to the reference impedance. Second, $\rho_1 = S_{11}$ even if $\rho_l \neq 0$, provided $S_{12} = 0$. The constraint, $S_{12} = 0$, equates to a zero feedback condition for the linear two-port. In turn, zero feedback isolates the input port from the output port, which means that the input impedance is independent of the terminating load impedance. Equivalently, the input port reflection coefficient is unaffected by load terminations, thereby rendering an input port reflection coefficient identical to S_{11}, which represents input port reflection for null reflected load energy. Finally, since

$$\rho_1 = \frac{b_1}{a_1} = \frac{Z_{\text{in}} - R_o}{Z_{\text{in}} + R_o}, \tag{3-50}$$

Eq. (3-49) uniquely stipulates the network driving point input impedance in accordance with

$$Z_{\text{in}} = \left(\frac{1+\rho_1}{1-\rho_1}\right) R_o = \left[\frac{(1-\rho_l S_{22})(1+S_{11}) + \rho_l S_{12} S_{21}}{(1-\rho_l S_{22})(1-S_{11}) - \rho_l S_{12} S_{21}}\right] R_o. \tag{3-51}$$

Analyses that mirror the foregoing thrusts lead forthwith to an expression for the output port reflection coefficient, ρ_2. In particular, it can be shown that

$$\rho_2 = \frac{b_2}{a_2} = \frac{Z_{\text{out}} - R_o}{Z_{\text{out}} + R_o} = S_{22} + \frac{\rho_s S_{12} S_{21}}{1 - \rho_s S_{11}}, \tag{3-52}$$

which defines the driving point output impedance, Z_{out}, as

$$Z_{out} = \left(\frac{1+\rho_2}{1-\rho_2}\right) R_o = \left[\frac{(1-\rho_s S_{11})(1+S_{22}) + \rho_s S_{12} S_{21}}{(1-\rho_s S_{11})(1-S_{22}) - \rho_s S_{12} S_{21}}\right] R_o. \quad (3\text{-}53)$$

For reasons analogous to those provided in conjunction with the input port reflection coefficient, $\rho_2 = S_{22}$ when either ρ_s or S_{12} (or both) are zero.

3.3.4.2. Voltage Transfer Function

In Fig. 3.11, the input port to output port voltage transfer function is the voltage ratio, V_2/V_1, which from Eq. (3-41) is

$$\frac{V_2}{V_1} = \frac{a_2 + b_2}{a_1 + b_1} = \frac{a_2}{a_1}\left(\frac{1+\rho_2}{1+\rho_1}\right). \quad (3\text{-}54)$$

The energy ratio, a_2/a_1, in this relationship can be expressed in terms of measurable parameters by returning to Eq. (3-31). In particular,

$$b_2 = S_{21} a_1 + S_{22} a_2 = \rho_2 a_2, \quad (3\text{-}55)$$

from which

$$\frac{a_2}{a_1} = \frac{S_{21}}{\rho_2 - S_{22}}. \quad (3\text{-}56)$$

Inserting Eq. (3-56) into Eq. (3-54) and using Eq. (3-45), the I/O port voltage transfer function becomes

$$\frac{V_2}{V_1} = \frac{S_{21}}{1 - \rho_l S_{22}}\left(\frac{1+\rho_l}{1+\rho_1}\right). \quad (3\text{-}57)$$

Now, the input port voltage, V_1, is related to the signal source voltage, V_s, through the simple voltage divider expression,

$$\frac{V_1}{V_s} = \frac{Z_{in}}{Z_{in} + Z_s}. \quad (3\text{-}58)$$

In terms of pertinent reflection coefficients, this divider is the equivalent relationship,

$$\frac{V_1}{V_s} = \frac{Z_{in}}{Z_{in} + Z_s} = \frac{\dfrac{1+\rho_1}{1-\rho_1}}{\dfrac{1+\rho_1}{1-\rho_1} + \dfrac{1+\rho_s}{1-\rho_s}} = \frac{(1+\rho_1)(1-\rho_s)}{2(1-\rho_1\rho_s)}. \quad (3\text{-}59)$$

Since the overall I/O voltage transfer function, A_v, is

$$A_v \triangleq \frac{V_2}{V_s} = \frac{V_2}{V_1} \times \frac{V_1}{V_s}, \tag{3-60}$$

Eqs. (3-57), (3-59), and (3-49), combine to deliver

$$A_v = \frac{V_2}{V_s} = \frac{S_{21}}{2} \left[\frac{(1-\rho_s)(1+\rho_l)}{(1-\rho_s S_{11})(1-\rho_l S_{22}) - \rho_s \rho_l S_{12} S_{21}} \right]. \tag{3-61}$$

In an attempt to make sense of the algebraic garble that defines the system voltage transfer function in terms of measured two-port scattering parameters and source and load reflection coefficients, note in Eq. (3-61) that when $\rho_s = \rho_l = 0$, A_v collapses to $S_{21}/2$. This result corroborates with an earlier disclosure that postures S_{21} as twice the forward voltage gain of a network terminated at its output port in the reference impedance and driven at its input port by a Thévenin equivalent source resistance that likewise equals the system reference impedance. Under such an operational circumstance, S_{21} as an illuminating measure of forward gain. Moreover, when the I/O ports are approximately terminated in reference impedances so that $Z_s \approx R_o$ and $Z_l \approx R_o$ to render $\rho_s \approx 0$ and $\rho_l \approx 0$, the frequency response of S_{21} becomes a meaningful measure of the achievable system frequency response.

A second observation is that for the case of zero internal feedback, the resultant voltage gain, which rightfully can be termed the open loop gain, say A_{vo}, of the system, is

$$A_{vo} = A_v|_{S_{12}=0} = \frac{S_{21}(1-\rho_s)(1+\rho_l)}{2(1-\rho_s S_{11})(1-\rho_l S_{22})}. \tag{3-62}$$

It is expedient to write this open loop gain in the form,

$$A_{vo} = (1-\rho_s) A_{ve} (1+\rho_l), \tag{3-63}$$

where

$$A_{ve} = \frac{S_{21}/2}{(1-\rho_s S_{11})(1-\rho_l S_{22})}. \tag{3-64}$$

Resultantly, Eq. (3-61) is expressible as

$$A_v = (1-\rho_s) \left(\frac{A_{ve}}{1 - 2\rho_s \rho_l S_{12} A_{ve}} \right) (1+\rho_l), \tag{3-65}$$

for which an effective loop gain, say T_s, of

$$T_s = -2\rho_s \rho_l S_{12} A_{ve} \tag{3-66}$$

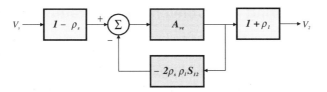

Figure 3.12. Block diagram model of the scattering parameter voltage transfer relationship for the generalized linear system offered in Fig. 3.11.

becomes transparent. Equation (3-65) gives rise to the block diagram representation offered in Fig. 3.12, where the factors, $(1-\rho_s)$ and $(1+\rho_l)$, in this diagram and in Eq. (3-65) account for the cognizant effects of impedance mismatches at the input and output ports, respectively.

3.3.4.3. Other Transfer Functions

The voltage transfer function in Eq. (3-61) serves as an analytical foundation for the determination of transadmittance, transimpedance, and current gain I/O relationships. For example, the linear system in Fig. 3.11 confirms $V_2 = -I_2 Z_l$, whence

$$A_v = \frac{V_2}{V_s} = -\frac{I_2 Z_l}{V_s}. \tag{3-67}$$

It follows that the forward transadmittance, Y_f, is

$$Y_f = \frac{I_2}{V_s} = -\frac{A_v}{Z_l}. \tag{3-68}$$

Substituting Eq. (3-44) for the load impedance, Z_l, and Eq. (3-61) for the voltage gain, A_v, into Eq. (3-68), the resultant normalized forward transadmittance, Y_f, can be shown to be

$$Y_f R_o = -\frac{S_{21}}{2}\left[\frac{(1-\rho_s)(1-\rho_l)}{(1-\rho_s S_{11})(1-\rho_l S_{22}) - \rho_s \rho_l S_{12} S_{21}}\right]. \tag{3-69}$$

Analogous expressions for current gain and transimpedance can be similarly constructed.

3.3.5. Scattering and Conventional Parameters

Since the scattering parameters of a linear two-port network serve to define its terminal I/O and transfer characteristics, these parameters must be consistent with conventional two-port parameters (hybrid h-, hybrid g-, admittance, and impedance), which also describe the I/O and transfer properties of a linear system. Consider, for example, the network in Fig. 3.11 modeled by the h-parameter equivalent circuit provided in Fig. 3.13. On the assumption that the desired S-parameters are to be referred to a characteristic impedance of R_o, care has been exercised in Fig. 3.13 to terminate both the input and output ports in R_o.

3.3.5.1. S-Parameters in Terms of h-Parameters

An inspection of Fig. 3.13(a) reveals a reference-terminated input impedance of

$$Z_{in}(R_o) = h_{11} - \frac{h_{12}h_{21}R_o}{1 + h_{22}R_o} \qquad (3\text{-}70)$$

and a reference-driven, reference-terminated voltage gain of

$$A_v(R_0) = \frac{V_2}{V_{s1}} = -\frac{h_{21}R_o}{(h_{11} + R_o)(1 + h_{22}R_o) - h_{12}h_{21}R_o}. \qquad (3\text{-}71)$$

Equations (3-70) and (3-34) combine to produce an input port scattering parameter, S_{11}, of

$$S_{11} = \frac{(h_{11} - R_o)(1 + h_{22}R_o) - h_{12}h_{21}R_o}{(h_{11} + R_o)(1 + h_{22}R_o) - h_{12}h_{21}R_o}, \qquad (3\text{-}72)$$

while Eqs. (3-71) and (3-35) yield

$$S_{21} = -\frac{2h_{21}R_o}{(h_{11} + R_o)(1 + h_{22}R_o) - h_{12}h_{21}R_o}. \qquad (3\text{-}73)$$

As expected, the scattering parameter, S_{21}, which is a measure of achievable forward gain, is nominally proportional to h-parameter h_{21}. If negligible

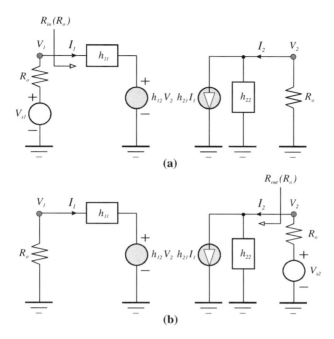

Figure 3.13. (a) The h-parameter model of the linear network in Fig. 3.11 used in the determination of the input impedance and forward voltage gain under the condition of the output port terminated in the characteristic impedance, R_o. (b) The model of (a) configured for the evaluation of the output impedance and reverse (or feedback) voltage gain, with the input port terminated in R_o.

feedback ($h_{12} \approx 0$) prevails within the network undergoing test and characterization, parameter S_{21} is simply twice the open loop voltage gain, while S_{11} in Eq. (3-72) reduces to

$$S_{11} \approx \frac{h_{11} - R_o}{h_{11} + R_o}. \qquad (3\text{-}74)$$

The last expression correctly infers a match-terminated, driving point input impedance, h_{11}, that is independent of load termination when negligible internal feedback prevails.

In Fig. 3.13(b), the output impedance under the condition of the input port terminated to ground through the reference impedance, R_o, derives from

$$\frac{1}{Z_{\text{out}}(R_o)} = h_{22} - \frac{h_{12} h_{21} R_o}{h_{11} + R_o}, \qquad (3\text{-}75)$$

whence by Eq. (3-39),

$$S_{22} = \frac{(h_{11} + R_o)(1 - h_{22}R_o) + h_{12}h_{21}R_o}{(h_{11} + R_o)(1 + h_{22}R_o) - h_{12}h_{21}R_o}. \qquad (3\text{-}76)$$

With negligible feedback ($h_{12} \approx 0$),

$$S_{22} \approx \frac{1 - h_{22}R_o}{1 + h_{22}R_o} = \frac{\dfrac{1}{h_{22}} - R_o}{\dfrac{1}{h_{22}} + R_o}, \qquad (3\text{-}77)$$

which properly casts $1/h_{22}$ as the driving point output impedance established under zero feedback conditions. A straightforward analysis of the same circuit model also confirms a reverse voltage gain of

$$\frac{V_1}{V_{s2}} = \frac{h_{12}R_o}{(h_{11} + R_o)(1 + h_{22}R_o) - h_{12}h_{21}R_o}, \qquad (3\text{-}78)$$

which, by Eq. (3-40), infers a feedback S-parameter, S_{12}, of

$$S_{12} = \frac{2h_{12}R_o}{(h_{11} + R_o)(1 + h_{22}R_o) - h_{12}h_{21}R_o}. \qquad (3\text{-}79)$$

Observe that S_{12} vanishes if the network provides no internal feedback. Additionally, Eqs. (3-79) and (3-73) show that

$$\frac{S_{21}}{S_{12}} = -\frac{h_{21}}{h_{12}}, \qquad (3\text{-}80)$$

which ensures $S_{21} \equiv S_{12}$ for a bilateral linear network.

3.3.5.2. *h*-Parameters in Terms of Scattering Parameters

Recall the previously noted difficulties with regard to maintaining reliable and accurate short-circuited and open-circuited network ports at high signal frequencies. As a result, the conventional two-port parameters of passive and active linear devices, circuits, and systems earmarked for high frequency signal processing applications are invariably deduced indirectly in terms of measured scattering parameters. Moreover, and as is underscored in the preceding subsection of material, the analytical expressions that define the electrical nature of the two-port driving point and transfer characteristics in terms of S-parameters are somewhat cumbersome and therefore challenging to exploit from a design perspective. It follows that relating the conventional two-port parameters explicitly to S-parameters is a warranted task, which is undertaken herewith for the hybrid h-parameters.

The h-parameters, h_{11} and h_{21}, for a linear two-port network such as the one abstracted in Fig. 3.13, are evaluated under the condition of a short-circuited output port, that is, $V_2 = 0$. From Eq. (3-41), $V_2 = 0$ implies $a_2 = -b_2$, which means that the energy variable incident at the output port is the negative of its reflected counterpart or equivalently, the output port reflection coefficient, ρ_2, is negative one. Recalling Eq. (3-31), $a_2 = -b_2$ implies

$$b_1 = S_{11}a_1 + S_{12}a_2 = S_{11}a_1 - \left(\frac{S_{12}S_{21}}{1+S_{22}}\right)a_1, \qquad (3\text{-}81)$$

and a resultant input port reflection coefficient,

$$\rho_1 = \left.\frac{b_1}{a_1}\right|_{a_2=-b_2} = S_{11} - \frac{S_{12}S_{21}}{1+S_{22}}. \qquad (3\text{-}82)$$

From Eqs. (3-41) and (3-42),

$$h_{11} \triangleq \left.\frac{V_1}{I_1}\right|_{V_2=0} = \left.\frac{R_o(a_1+b_1)}{a_1-b_1}\right|_{a_2=-b_2} = \left.R_o\left(\frac{1+\rho_1}{1-\rho_1}\right)\right|_{a_2=-b_2}. \qquad (3\text{-}83)$$

A combination of Eqs. (3-82) and (3-83) leads to the following expression for the open circuit input impedance:

$$h_{11} = \left[\frac{(1+S_{11})(1+S_{22}) - S_{12}S_{21}}{(1-S_{11})(1+S_{22}) + S_{12}S_{21}}\right]R_o. \qquad (3\text{-}84)$$

While maintaining $V_2 = 0$ or equivalently, $a_2 = -b_2$, Eq. (3-42) assists in casting the short circuit current gain, h_{21}, in the form

$$h_{21} = \left.\frac{I_2}{I_1}\right|_{V_2=0} = \left.\frac{a_2-b_2}{a_1-b_1}\right|_{a_2=-b_2} = \left.\frac{a_2}{a_1}\left(\frac{2}{1-\rho_1}\right)\right|_{a_2=-b_2}. \qquad (3\text{-}85)$$

Using Eqs. (3-81) and (3-82), Eq. (3-85) becomes

$$h_{21} = -\frac{2S_{21}}{(1-S_{11})(1+S_{22}) + S_{12}S_{21}}. \qquad (3\text{-}86)$$

As expected, the short circuit current gain is nominally proportional to S-parameter S_{21}, provided that negligible feedback ($S_{12} \approx 0$) prevails.

The h-parameters, h_{22} and h_{12}, are open circuit metrics of a linear two-port network. With reference to Fig. 3.13(b), these two parameters are evaluated with the input port current, I_1, nulled, which is equivalent to the constraint, $a_1 = b_1$, in Fig. 3.11. For this situation, Eq. (3-31) delivers

$$a_1 = b_1 = \left(\frac{S_{12}}{1 - S_{11}}\right) a_2 \qquad (3\text{-}87)$$

and

$$\rho_2 = \left.\frac{b_2}{a_2}\right|_{a_1=b_1} = S_{22} + \frac{S_{12}S_{21}}{1 - S_{11}}. \qquad (3\text{-}88)$$

It follows that the open circuit output admittance, h_{22}, is

$$h_{22} \triangleq \left.\frac{I_2}{V_2}\right|_{I_1=0} = \left.\frac{(a_2 - b_2)}{R_o(a_2 + b_2)}\right|_{a_1=b_1} = \left.\frac{1}{R_o}\left(\frac{1 - \rho_2}{1 + \rho_2}\right)\right|_{a_1=b_1}. \qquad (3\text{-}89)$$

Putting Eq. (3-88) into Eq. (1-89),

$$h_{22} = \left[\frac{(1 - S_{11})(1 - S_{22}) - S_{12}S_{21}}{(1 - S_{11})(1 + S_{22}) + S_{12}S_{21}}\right]\left(\frac{1}{R_o}\right). \qquad (3\text{-}90)$$

For the feedback h-parameter, h_{12},

$$h_{12} = \left.\frac{V_1}{V_2}\right|_{I_1=0} = \left.\frac{a_1 + b_1}{a_2 + b_2}\right|_{a_1=b_1} = \left.\frac{a_1}{a_2}\left(\frac{2}{1 + \rho_2}\right)\right|_{a_1=b_1}. \qquad (3\text{-}91)$$

Using Eqs. (3-87) and (3-88), h_{12} is expressible as

$$h_{12} = \frac{2S_{12}}{(1 - S_{11})(1 + S_{22}) + S_{12}S_{21}}. \qquad (3\text{-}92)$$

Clearly, both h_{12} and S_{12} are measures of intrinsic feedback in the sense that $S_{12} = 0$ nulls h_{12}.

The pertinent results for the foregoing four h-parameters are conveniently summarized by the hybrid h-parameter matrix,

$$h = \begin{bmatrix} \left\{\dfrac{(1+S_{11})(1+S_{22}) - S_{12}S_{21}}{(1-S_{11})(1+S_{22}) + S_{12}S_{21}}\right\} R_o & \dfrac{2S_{12}}{(1-S_{11})(1+S_{22}) + S_{12}S_{21}} \\ \dfrac{-2S_{21}}{(1-S_{11})(1+S_{22}) + S_{12}S_{21}} & \left\{\dfrac{(1-S_{11})(1-S_{22}) - S_{12}S_{21}}{(1-S_{11})(1+S_{22}) + S_{12}S_{21}}\right\}\left(\dfrac{1}{R_o}\right) \end{bmatrix}. \qquad (3\text{-}93)$$

The matrices for the hybrid-g, short circuit admittance, and open circuit impedance two-port parameters can be similarly formulated.

3.4.0. Lossless Two-Port Networks

Filters are ubiquitous in electronic circuits designed for communication system applications. For example, bandpass, lowpass, notch, and other types of frequency responses are implemented to ensure that undesired signals captured at antennas and other signal source media are adequately attenuated prior to ultimate signal processing. By limiting the bandwidth of filters incorporated in critical signal flow paths, the deleterious effects of noise generated by passive and active components are mitigated. Yet another filtering application is impedance matching aimed toward maximizing signal power transfer between critical source and load ports. Although an endless array of filter architectures is available to satisfy numerous engineering requirements, certain pragmatic guidelines underpin the design of filters for a broad variety of high frequency or broadband radio frequency (RF) systems.

The first of these requirements derives from the likely impropriety of adopting active network design strategies for many types of filtering applications. One concern is the frequency response shortfalls inherent to the electronic subcircuits implicit to active filter realizations. As a rule, the design of active filters for frequency operation above the mid-hundreds of megahertz comprises a daunting, if not impossible, engineering challenge. A second problem is that active filters tend to be considerably noisier than their passive counterparts. This noise issue poses a serious dilemma in low-level input stage signal processing where available signal strengths may be dangerously close to electrical noise floors. Another active filter red flag that surfaces, particularly in portable electronics, is the biasing power that active elements require. Finally, active filters are not as linear as passive architectures. This nonlinearity poses problems when input signal maxima are obscure and when signals of comparable amplitudes appear in proximate frequency bands to incur potentially unacceptable intermodulation responses.

Passive filters circumvent many of the foregoing limitations, but they too are hardly examples of unimpeachable perfection. Resistor, capacitor, and inductor component tolerances loom potentially troublesome when bandwidths or other frequency response metrics must be accurately achieved. Inductors have finite quality factors that exacerbate circuit noise problems, increase circuit power dissipation, and limit achievable narrow band responses in bandpass circuits. Resistors associated with practical inductors and those embedded within filters incur signal power losses as well. This loss problem is especially serious in low-level communications that feature minimal available signal input power. No leap of faith is required to understand that if an already anemic power level is processed by an inordinately lossy filter, the resultant response at the filter output port may be substantively masked by random noise phenomena, thereby precluding reliable signal capture and accurate signal processing.

The foregoing commentary warrants at least cursory attention paid to the realization of lossless, passive filters. To this end, the scattering parameter concept provides a vehicle for the development of a systematic design strategy for the realization of two-port filters whose only branch elements are either inductors or capacitors. To be sure, neither inductors nor capacitors are perfectly lossless elements. In all but the most exceptional of cases, the power losses incurred by the parasitic resistances associated with practical inductors and capacitors are far less significant than those of branch resistances purposefully embedded in filters.

3.4.1. Average Power Delivered to Complex Load

To begin a discussion addressing the design of lossless two-port filters, it is worthwhile reviewing from Chapter 1 the relationships pertinent to the steady state power dissipated in a complex load impedance. To this end, consider the simple branch impedance, $Z(j\omega)$, in Fig. 3.14. Let this impedance conduct a steady state sinusoidal current, $i(t)$, in response to a sinusoidal voltage, $v(t)$, established across the branch. If the subject branch voltage is

$$v(t) = \sqrt{2} V_m \cos(\omega t + \theta_v) \qquad (3\text{-}94)$$

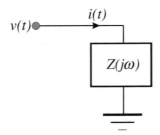

Figure 3.14. A complex load impedance to which a sinusoidal voltage, $v(t)$, is applied and a sinusoidal current, $i(t)$, resultantly flows in the steady state.

and the corresponding branch current is

$$i(t) = \sqrt{2} I_m \cos(\omega t + \theta_i), \qquad (3\text{-}95)$$

the resultant instantaneous power, $p(t)$, dissipated by, and delivered to, the load impedance is

$$p(t) = v(t)i(t) = 2V_m I_m \cos(\omega t + \theta_v) \cos(\omega t + \theta_i). \qquad (3\text{-}96)$$

In Eqs. (3-94) and (3-95), V_m and I_m respectively symbolize the root mean square (RMS) voltage and current amplitudes, while the phase difference, $(\theta_v - \theta_i)$, is the phase angle of the branch impedance. Equation (3-96) may be expanded to

$$p(t) = V_m I_m [\cos(\theta_v - \theta_i) + \cos(2\omega t + \theta_v + \theta_i)], \qquad (3\text{-}97)$$

which renders evident an average power delivery, and hence an average dissipated power, of

$$P_{\text{avg}} = \frac{1}{T} \int_0^T p(t)\, dt = V_m I_m \cos(\theta_v - \theta_i). \qquad (3\text{-}98)$$

A more useful expression for the average power dissipated in a complex load impedance is predicated on a phasor representation of the pertinent voltage and current. Specifically, write $v(t)$ and $i(t)$ in the phasor formats,

$$\left. \begin{array}{l} V(j\omega) = V_m e^{j\theta_v} \\ I(j\omega) = I_m e^{j\theta_i} \end{array} \right\}. \qquad (3\text{-}99)$$

It follows from Eq. (3-98) that

$$P_{\text{avg}} = \text{Re}\big[V(j\omega) I(-j\omega)\big]. \qquad (3\text{-}100)$$

It is to be understood that $I(-j\omega)$ in the last relationship is the conjugate of the phasor, $I(j\omega)$, in Eq. (3-100). If the impedance, $Z(j\omega)$ is a lossless entity, which forces its phase angle for all frequencies, ω, to be either positive ninety (inductive) or negative ninety (capacitive), no average power is dissipated by $Z(j\omega)$ and thus, $P_{\text{avg}} = 0$.

3.4.2. Average Power Delivered to Two-Port Network

The foregoing results are easily extended to embrace two-port networks, such as the one that appears in Fig. 3.11. Since the network in question has two accessible signal ports, the net power it dissipates is

$$P_{avg} = \text{Re}\big[V_1(j\omega)I_1(-j\omega) + V_2(j\omega)I_2(-j\omega)\big]. \qquad (3\text{-}101)$$

The development of design criteria for lossless two-port networks is facilitated through the introduction of the port voltage and port current vectors,

$$\left.\begin{array}{l}\mathbf{V}(j\omega) = \begin{bmatrix} V_1(j\omega) \\ V_2(j\omega) \end{bmatrix} = \sqrt{R_o}\begin{bmatrix} a_1(j\omega) + b_1(j\omega) \\ a_2(j\omega) + b_2(j\omega) \end{bmatrix} \\[1em] \mathbf{I}(j\omega) = \begin{bmatrix} I_1(j\omega) \\ I_2(j\omega) \end{bmatrix} = \dfrac{1}{\sqrt{R_o}}\begin{bmatrix} a_1(j\omega) - b_1(j\omega) \\ a_2(j\omega) - b_2(j\omega) \end{bmatrix}\end{array}\right\}, \qquad (3\text{-}102)$$

where Eqs. (3-41) and (3-42) are exploited and, of course, $a_i(j\omega)$ and $b_i(j\omega)$ respectively symbolize steady state incident and reflected energy variables at the ith port. If the energy vectors, $\mathbf{a}(j\omega)$ and $\mathbf{b}(j\omega)$, are introduced, such that

$$\left.\begin{array}{l}\mathbf{a}(j\omega) = \begin{bmatrix} a_1(j\omega) \\ a_2(j\omega) \end{bmatrix} \\[1em] \mathbf{b}(j\omega) = \begin{bmatrix} b_1(j\omega) \\ b_2(j\omega) \end{bmatrix}\end{array}\right\}, \qquad (3\text{-}103)$$

the average total power dissipated by the two-port network can be related explicitly to port incident and reflected variables in accordance with

$$\begin{aligned}P_{avg} &= \text{Re}\big[\mathbf{V}^T(j\omega)\mathbf{I}(-j\omega)\big] \\ &= \text{Re}\big[\{\mathbf{a}^T(j\omega) + \mathbf{b}^T(j\omega)\}\{\mathbf{a}(-j\omega) - \mathbf{b}(-j\omega)\}\big]. \end{aligned} \qquad (3\text{-}104)$$

The superscript, "T", in this relationship denotes the mathematical operation of matrix transposition, wherein matrix rows are interchanged with matrix columns.

A convenient reduction of the average power relationship in Eq. (3-104) results from an expansion of the right hand side of this equation.

Specifically,

$$P_{\text{avg}} = \text{Re}\left[\mathbf{a}^T(j\omega)\mathbf{a}(-j\omega) - \mathbf{b}^T(j\omega)\mathbf{b}(-j\omega)\right]$$
$$+ \text{Re}\left[\mathbf{b}^T(j\omega)\mathbf{a}(-j\omega) - \mathbf{a}^T(j\omega)\mathbf{b}(-j\omega)\right]. \quad (3\text{-}105)$$

Because both $\mathbf{a}(j\omega)$ and $\mathbf{b}(j\omega)$ are column vectors, the matrix product, $\mathbf{b}^T(j\omega)\mathbf{a}(-j\omega)$, is a scalar, and since the transpose of a scalar is certainly the same as the scalar quantity itself,

$$\mathbf{b}^T(j\omega)\mathbf{a}(-j\omega) = \left[\mathbf{b}^T(j\omega)\mathbf{a}(-j\omega)\right]^T = \mathbf{a}^T(-j\omega)\mathbf{b}(j\omega). \quad (3\text{-}106)$$

Accordingly,

$$P_{\text{avg}} = \text{Re}\left[\mathbf{a}^T(j\omega)\mathbf{a}(-j\omega) - \mathbf{b}^T(j\omega)\mathbf{b}(-j\omega)\right]$$
$$+ \text{Re}\left[\mathbf{a}^T(-j\omega)\mathbf{b}(j\omega) - \mathbf{a}^T(j\omega)\mathbf{b}(-j\omega)\right]. \quad (3\text{-}107)$$

The first term in the second real part quantity on the right hand side of Eq. (3-107) is simply the complex conjugate of the second term in this real part component. Since the difference of two complex conjugate terms is a purely imaginary number, the second real part term in question is identically zero. Thus, the net average power dissipated by a linear two-port network reduces to

$$P_{\text{avg}} = \text{Re}\left[\mathbf{a}^T(j\omega)\mathbf{a}(-j\omega) - \mathbf{b}^T(j\omega)\mathbf{b}(-j\omega)\right]. \quad (3\text{-}108)$$

Since each product term implicit to the real part operation in this equation is a scalar formed of a product of a number and its complex conjugate, each term within the brackets is a real number, thereby allowing the average power relationship to be simplified as

$$P_{\text{avg}} = \mathbf{a}^T(j\omega)\mathbf{a}(-j\omega) - \mathbf{b}^T(j\omega)\mathbf{b}(-j\omega). \quad (3\text{-}109)$$

An additional mathematical manipulation leads to forging the final form of a design strategy for lossless two-port filters. In particular, observe from Eq. (3-31) that the incident and reflected variables of a two-port network interrelate as the matrix relationship,

$$\mathbf{b}(j\omega) = \mathbf{S}(j\omega)\mathbf{a}(j\omega), \quad (3\text{-}110)$$

for which

$$\mathbf{b}(-j\omega) = \mathbf{S}(-j\omega)\mathbf{a}(-j\omega) \quad (3\text{-}111)$$

and

$$\mathbf{b}^T(j\omega) = [\mathbf{S}(j\omega)\mathbf{a}(j\omega)]^T = \mathbf{a}^T(j\omega)\mathbf{S}^T(j\omega). \quad (3\text{-}112)$$

Equation (3-109) now becomes, with the help of Eq. (3-112),

$$P_{\text{avg}} = \mathbf{a}^T(j\omega)\mathbf{a}(-j\omega) - \mathbf{a}^T(j\omega)\mathbf{S}^T(j\omega)\mathbf{S}(-j\omega)\mathbf{a}(-j\omega), \quad (3\text{-}113)$$

or more compactly,

$$P_{\text{avg}} = \mathbf{a}^T(j\omega)\big[\mathbf{U} - \mathbf{S}^T(j\omega)\mathbf{S}(-j\omega)\big]\mathbf{a}(-j\omega), \quad (3\text{-}114)$$

where **U** is the identity matrix. Although this relationship has been derived for the specific case of a two-port linear network, it is applicable to the more generalized environment of an m-port multiport. For an m-port, **U** is an identity matrix of order m, and **S** is a square matrix, also of order m. It is worth underscoring the fact that Eq. (3-114) is valid for any linear m-port network, regardless of whether said network is active, passive, unilateral, bilateral, lossless, or dissipative.

3.4.3. Lossless, Passive Two-Port Network

The lossless condition for a two-port network is rendered transparent by Eq. (3-114), specifically,

$$\mathbf{U} \equiv \mathbf{S}^T(j\omega)\mathbf{S}(-j\omega). \quad (3\text{-}115)$$

For two-port networks, Eq. (3-115) is equivalent to the stipulation,

$$\begin{bmatrix} 1 & 0 \\ 0 & 1 \end{bmatrix} = \begin{bmatrix} S_{11}(j\omega) & S_{21}(j\omega) \\ S_{12}(j\omega) & S_{22}(j\omega) \end{bmatrix} \begin{bmatrix} S_{11}(-j\omega) & S_{12}(-j\omega) \\ S_{21}(-j\omega) & S_{22}(-j\omega) \end{bmatrix}, \quad (3\text{-}116)$$

which implies

$$|S_{11}(j\omega)|^2 + |S_{21}(j\omega)|^2 = 1, \quad (3\text{-}117)$$

$$|S_{22}(j\omega)|^2 + |S_{12}(j\omega)|^2 = 1, \quad (3\text{-}118)$$

$$S_{11}(j\omega)S_{12}(-j\omega) + S_{21}(j\omega)S_{22}(-j\omega) = 0, \quad (3\text{-}119)$$

and

$$S_{12}(j\omega)S_{11}(-j\omega) + S_{22}(j\omega)S_{21}(-j\omega) = 0. \quad (3\text{-}120)$$

In light of Eq. (3-119), Eq. (3-120) is a superfluous relationship because each of its terms is the complex conjugate of its corresponding terms in Eq. (3-119). Since the transducer power gain, $G_T(j\omega)$, which is intimately related to the reference-terminated voltage gain in accordance with Eq. (3-37), is the squared magnitude of S-parameter S_{21}, Eq. (3-117) yields

$$|S_{11}(j\omega)|^2 = S_{11}(j\omega)S_{11}(-j\omega) = 1 - G_T(j\omega). \qquad (3\text{-}121)$$

As is demonstrated in a forthcoming example, this relationship is the foundation upon which rests the design strategy governing the realization of a lossless two-port filter satisfying any realistic frequency response requirement. In particular, the transducer power gain can be computed via Eqs. (3-37) and (3-35) for a given or desired input voltage to output voltage frequency response. For example, this frequency response may reflect a Butterworth, Tchebyschev, Bessel, or any other meaningful lowpass, bandpass, highpass, notch, or other frequency response form. Then the squared magnitude of parameter S_{11} can be factored into a product of a complex number and its conjugate, whence the input impedance under reference-terminated output port conditions can be evaluated. This input impedance function can then be expanded as a *continued fraction expansion* (essentially repeated long division) to arrive at the required filter topology. It is important to underscore the fact that since Eq. (3-121) applies to a lossless electrical circuit, the elemental nature of the input impedance commensurate with the foregoing design procedure is either inductive or capacitive. Therefore, this impedance is necessarily proportional to a ratio of a numerator polynomial in $(j\omega)$ to a denominator polynomial in $(j\omega)$, the orders of which differ precisely by one.

Example 3.3. A communication system application requires a lowpass, third order Butterworth filter that provides a 3-dB bandwidth of $B = 2\pi(1.2\,\text{GHz})$. The filter is to drive a 50 Ω resistive load (R_l) from an antenna whose coupling to the filter is a coaxial cable having a characteristic impedance that is likewise 50 Ω (R_s). Realize the filter as a lossless (LC) two-port topology.

Solution 3.3.

(1) The conceptual schematic diagram of the required lossless filter is offered in Fig. 3.15(a). The terminating load resistance is $R_l = 50\,\Omega$ and the effective Thévenin source resistance is $R_s = 50\,\Omega \equiv R_l$. Since the filter is lossless and therefore contains only presumably ideal inductances and capacitances as branch elements, and since the requisite two-port delivers a lowpass frequency response, the zero frequency voltage "gain" is simply the divider $R_l/(R_l+R_s)$, or 0.5.

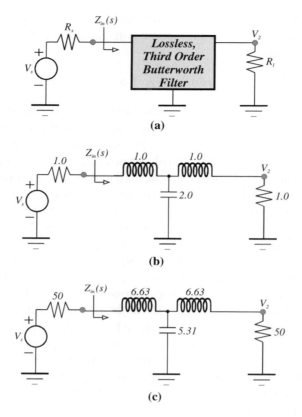

Figure 3.15. (a) System level abstraction of the filter design problem addressed in Example 3.3. (b) The normalized filter realization. Resistances are referred to 50 Ω, inductances to 6.6315 nH, and capacitances to 2.6526 pF. (c) Filter realizing a third order Butterworth response. All resistances are in ohms, inductances are in nanohenries, and the capacitance is in units of picofarads. The 3-dB frequency of the lowpass filter is 1.2 GHz.

(2) The third order Butterworth polynomial for a lowpass filter whose 3-dB bandwidth is normalized to 1 radian-per-second, in terms of the Laplace operator, "s" is $(s+1)(s^2+s+1)$. This declaration means that the voltage transfer function of the system abstracted in Fig. 3.15(a) is

$$H(s) = \frac{V_2}{V_s} = \frac{0.5}{(1+s)(1+s+s^2)}. \quad \text{(E3-1)}$$

From Eq. (3-37), the required transducer power gain, $G_T(s)$, is

$$G_T(s) = |S_{21}(s)|^2 = |2H(s)|^2 = 4H(s)H(-s) = \frac{1}{1-s^6}. \quad \text{(E3-2)}$$

(3) Using Eq. (3-121), the scattering parameter, S_{11}, satisfies

$$|S_{11}(s)|^2 = 1 - G_T(s) = \frac{-s^6}{1 - s^6}. \tag{E3-3}$$

Since the squared magnitude of $S_{11}(s)$ is the product of $S_{11}(s)$ and its conjugate, $S_{11}(-s)$,

$$S_{11}(s) = \frac{s^3}{(1+s)(1+s+s^2)}, \tag{E3-4}$$

where the factoring of the denominator of the squared magnitude of $S_{11}(s)$ is facilitated by the observation that the poles of $S_{11}(s)$ are necessarily the same as the poles of the desired voltage transfer function.

(4) Recalling Eq. (3-34), the required input impedance function is

$$Z_{in}(s) = R_o \left[\frac{1 + S_{11}(s)}{1 - S_{11}(s)} \right] = R_o \left(\frac{1 + 2s + 2s^2 + 2s^3}{1 + 2s + 2s^2} \right), \tag{E3-5}$$

with the understanding that the subject driving point input impedance is that which materializes when the output port of the filter is terminated in the requisite load resistance, R_l. Ordinarily, the constant multiplier, R_o, in this impedance relationship is taken to be the source resistance, R_s. In this case, the source and load resistances are matched to ensure unity transducer power gain at low frequencies. Thus, R_o in the present case is either R_s or R_l, both of which happen to be 50 Ω.

The frequency in all of the foregoing performance relationships has been normalized to the desired filter bandwidth of $B = 2\pi(1.2\,\text{GHz})$, that is, 1.0 Hz corresponds to 1.2 GHz. If circuit branch impedances are normalized to R_o, 1.0 Ω corresponds to 50 Ω. The synthesis procedure is expedited, at least in a numerical sense, by proceeding to normalize all inductive and capacitive impedances at the circuit bandwidth to the reference, or normalizing, resistance R_o.[3] Thus, if L_α symbolizes the *normalizing inductance*,

$$L_\alpha = \frac{R_o}{B} = 6.6315\,\text{nH}, \tag{E3-6}$$

meaning that 1.0 H in the circuit realization maps to 6.6315 nH in the finalized design. For branch capacitances,

$$C_\alpha = \frac{1}{B R_o} = 2.6526\,\text{pF}. \tag{E3-7}$$

Thus, 1.0 F in the prototypical realization is in one to one correspondence with an actual branch capacitance of 2.6526 pF. It follows that the prototype impedance to be synthesized derives from the normalized expression,

$$Z_{in}(s) = \frac{1 + 2s + 2s^2 + 2s^3}{1 + 2s + 2s^2}. \tag{E3-8}$$

(5) A continued fraction expansion of the right hand side of Eq. (E3-8) results in

$$Z_{in}(s) = s + \frac{s+1}{2s^2 + 2s + 1} = s + \frac{1}{2s + \dfrac{1}{s+1}}. \tag{E3-9}$$

The far right hand side of this equation infers the normalized prototype topology offered in Fig. 3.15(b). In particular, the circuit is a 1.0 H inductance connected in series with a topology consisting of a 2.0 F capacitance shunting a series combination of a 1.0 H inductance and a 1.0 Ω resistance. The subject 1.0 Ω resistance is, of course, the normalized load termination. Figure 3.15(c) portrays the de-normalized, or actual, lossless filter required by the given operating specifications.

Comments. The SPICE simulation of the magnitude response for the filter in Fig. 3.15(c) is shown in Fig. 3.16. Note a response magnitude that is flat to within nominally three decibels up to a frequency of 1.2 GHz, which is the 3-dB bandwidth objective of the design.

Although the filter design procedure exemplified herewith is most easily applied to the problem of realizing lowpass filter architectures, the resultant lowpass architecture can form the basis for other types of frequency response forms. In

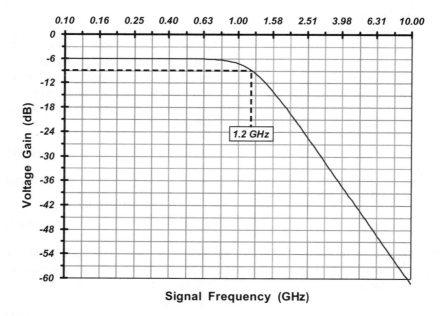

Figure 3.16. SPICE simulation of the frequency response for the third order, lowpass Butterworth filter given in Fig. 3.15.

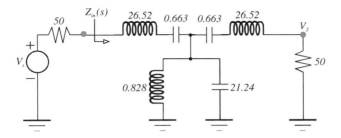

Figure 3.17. Butterworth bandpass filter realized by applying a lowpass to bandpass frequency transformation to the lowpass filter in Fig. 3.15(c). The center frequency of the filter is designed to be 1.2 GHz, and the 3-dB bandwidth is approximately 300 MHz, which implies an effective quality factor of $Q = 4$.

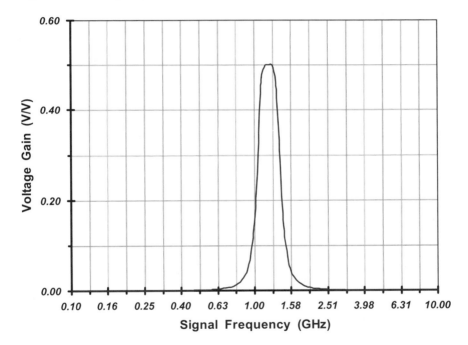

Figure 3.18. SPICE simulation of the frequency response for the third order, bandpass Butterworth filter given in Fig. 3.17.

particular, well-known frequency response transformation tools can transform a prototype lowpass filter into a bandpass, highpass, notch, or other type of physically realizable filter.[4] For example, Fig. 3.17 submits a bandpass filter architecture predicated on the lowpass structure designed in this example. In effect, observe that inductances appearing in the lowpass structure are supplanted by series-resonant LC circuits, while capacitances in the lowpass circuit are replaced by parallel-resonant LC branches. The indicated element values in Fig. 3.17 pertain to a tuned center frequency of 2π (1.2 GHz) and a 3-dB bandwidth of 300 MHz, which implies an effective quality factor of four. The SPICE frequency response plot in Fig. 3.18 confirms the functionality of the bandpass structure.

References

1. P. H. Smith, An improved transmission line calculator, *Electronics* **17** (1944) 130.
2. K. K. Clarke and D. T. Hess, *Communication Circuits*: *Analysis and Design* (Addison-Wesley Publishing Company, Reading, Massachusetts, 1978), pp. 444–447.
3. J. Choma, Jr., *Electrical Networks*: *Theory and Analysis* (Wiley-Interscience, New York, 1985), pp. 509–513.
4. W.-K. Chen, *Passive and Active Filters*: *Theory and Implementations* (John Wiley & Sons, New York, 1986), pp. 92–101.

Exercises

Problem 3.1
This problem addresses the interrelationships between the short circuit admittance (y-) parameters and the scattering (S-) parameters of a linear two port network. Assume that the S-parameters are measured with respect to a reference impedance that equals the resistance value, R_o.

(a) Derive generalized expressions for the y-parameters in terms of the network scattering parameters.
(b) Derive generalized expressions for the S-parameters in terms of the network short circuit admittance parameters.

Problem 3.2
This problem addresses the interrelationships between the open circuit impedance (z-) parameters and the scattering (S-) parameters of a linear two port network. Assume that the S-parameters are measured with respect to a reference impedance that equals the resistance value, R_o.

(a) Derive generalized expressions for the z-parameters in terms of the network scattering parameters.

(b) Derive generalized expressions for the S-parameters in terms of the network open circuit impedance parameters.

Problem 3.3
This problem addresses the interrelationships between the hybrid g-parameters and the scattering (S-) parameters of a linear two-port network. Assume that the S-parameters are measured with respect to a reference impedance that equals the resistance value, R_o.

(a) Derive generalized expressions for the g-parameters in terms of the network scattering parameters.
(b) Derive generalized expressions for the S-parameters in terms of the network g-parameters.

Problem 3.4
This problem addresses the interrelationships between the transmission (c-) parameters and the scattering (S-) parameters of a linear two-port network. Assume that the S-parameters are measured with respect to a reference impedance that equals the resistance value, R_o.

(a) Derive generalized expressions for the c-parameters in terms of the network scattering parameters.
(b) Derive generalized expressions for the S-parameters in terms of the network transmission parameters.

Problem 3.5
The linear network abstracted in Fig. P3.5 is known as a *gyrator* in that its two-port equations subscribe to
$$\begin{bmatrix} V_1 \\ V_2 \end{bmatrix} = \begin{bmatrix} 0 & -r \\ r & 0 \end{bmatrix} \begin{bmatrix} I_1 \\ I_2 \end{bmatrix},$$
where r is a frequency invariant constant whose dimension is ohms.

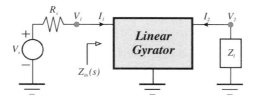

Figure P3.5.

(a) Derive the scattering parameters of a gyrator characterized in a reference impedance environment of R_o.
(b) For a terminating capacitive load, $Z_l = 1/sC_l$, what is the resultant input port reflection coefficient, ρ_{1c}?
(c) For a terminating inductive load, $Z_l = sL_l$, what is the resultant input port reflection coefficient, ρ_{1l}?
(d) Examine the reflection coefficients derived in Parts (b) and (c) to offer a conclusion as to the respective natures of the driving point input impedance, Z_{in}. Exploit this examination to project an opinion about the engineering function of a gyrator.

Problem 3.6
Determine the two-port scattering matrix for an ideal voltage amplifier whose I/O port voltage gain is A_{vo}.

Problem 3.7
Determine the two-port scattering matrix for an ideal transimpedance amplifier whose I/O port transimpedance is Z_{fo}.

Problem 3.8
Determine the two-port scattering matrix for an ideal transadmittance amplifier whose I/O port transadmittance is Y_{fo}.

Problem 3.9
Determine the two-port scattering matrix for an ideal current amplifier whose I/O port current gain is A_{io}.

Problem 3.10
The reflection characteristics of a one-port network comprised of the complex impedance, Z_l, are traditionally evaluated with respect to a real reference impedance, say R_o. However, reflection properties can also be determined when the reference impedance is a complex number of the form, $Z_o = R_o + jX_o$.

(a) Derive an expression for the current reflection coefficient, ρ_i, of a linear one-port network whose characteristic, or reference, impedance, is the complex number, Z_o.
(b) Repeat Part (a) but now, find the voltage reflection coefficient, ρ_v.
(c) What relationship exists between ρ_i in Part (a) and ρ_v in Part (b)? Confirm that $\rho_i \equiv \rho_v$ when $Z_o = R_o$, a real impedance.

(d) When the reference impedance is the complex number, Z_o, what relationship exists between reflected and incident powers, P_r and P_i, respectively? Show that this cumbersome expression collapses to Eq. (3-16) when $Z_o = R_o$, a real number.

Problem 3.11
Figure P3.11 depicts the system level diagram of four-port, lossless, reciprocal directional coupler. As suggested by the arrows within this diagram, a signal incident at the first port couples power into the second and fourth ports, but not into the third port. Likewise, power applied to the second port is coupled into the first and third ports, but not into the fourth port, and so forth. The coupler is designed to have $S_{12} = A$ and $S_{13} = jB$, where "A" and "B" are positive, real constants. Derive the simplest possible scattering matrix for the coupler in terms of "A" or "B," but not both.

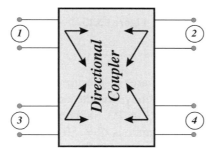

Figure P3.11.

Problem 3.12
Design a three-pole, lossless, lowpass Butterworth filter to couple a $50\,\Omega$ source to a $50\,\Omega$ load. The 3-dB bandwidth of this filter is to be 2.4 GHz.

(a) Show a finalized schematic diagram of your design and submit a SPICE frequency response simulation to confirm the propriety of your design.
(b) Revisit the simulation of the finalized design by allowing each inductor utilized therein to have a quality factor (Q) of 3.5 at 2.4 GHz. Do not attempt to execute closed form mathematical analyses to respond to this question, but submit SPICE plots that compare the idealized design with the practical, finite Q, design. Discuss observed behavior, particularly as regards the values of the zero frequency transfer function and the 3-dB bandwidth.

Problem 3.13

Plot the real and imaginary parts of the input port reflection coefficient, as a function of signal frequency, for the admittance established by the shunt RC network shown in Fig. P3.13. The indicated shunt resistance is $R = 300\,\Omega$, and the shunt capacitance is $C = 50\,\text{pF}$. The reference impedance (R_o) can be taken as $50\,\Omega$.

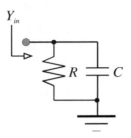

Figure P3.13.

Problem 3.14

Plot the real and imaginary parts of the input port reflection coefficient, as a function of signal frequency, for the admittance established by the shunt RLC network shown in Fig. P3.14. The indicated shunt resistance is $R = 300\,\Omega$, the shunt capacitance is $C = 50\,\text{pF}$, and the shunt inductance is $L = 2\,\text{nH}$. The reference impedance (R_o) can be taken as $50\,\Omega$.

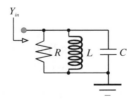

Figure P3.14.

Problem 3.15

For the generalized linear two-port system in Fig. 3.11, derive an expression for the I/O current gain, A_i, in terms of the system scattering parameters, S_{ij}. The subject scattering parameters are characterized with respect to a real reference impedance, R_o. Assume that the input current pertinent to the current gain evaluation is the Norton equivalent of the indicated Thévenin signal source circuit.

Problem 3.16
For the generalized linear two-port system in Fig. 3.11, derive an expression for the I/O transimpedance, Z_f, in terms of the system scattering parameters, S_{ij}. The subject scattering parameters are characterized with respect to a real reference impedance, R_o. Assume that the input current pertinent to the transimpedance evaluation is the Norton equivalent of the indicated Thévenin signal source circuit.

Problem 3.17
The generalized tee-type topology depicted in Fig. P3.17(a) gives rise to the specific lossless tee architecture of Fig. P3.17(b), which is proposed as a delay filter for high frequency signal processing applications.

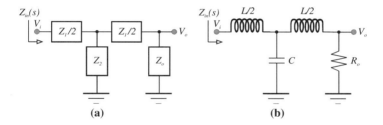

Figure P3.17.

(a) Let the generalized filter structure be terminated in an impedance, Z_o, as shown in Fig. P3.17(a). Show that when

$$Z_o = \sqrt{Z_1 Z_2 \left(1 + \frac{Z_1}{4Z_2}\right)},$$

the indicated input impedance, $Z_{in}(s)$, is identical to Z_o. Thus, when the filter is driven by a voltage source whose Thévenin impedance is Z_o, the driving point output impedance seen by the load impedance, which is also Z_o, is likewise Z_o. As a result, Z_o assumes the stature of a characteristic impedance for the indicated lumped filter circuit.
(b) Discuss the engineering advantages of satisfying the relationship given in conjunction with Part (a) of this problem when a filter application

calls for the implementation of a cascade of structures, each member of which is identical to the topology of Fig. P3.17(a).

(c) Demonstrate that for the filter given in Fig. P3.17(b), the characteristic impedance, as a function of complex frequency s, is

$$Z_o(s) = \sqrt{R_o\left[1 + \left(\frac{s}{\omega_c}\right)^2\right]},$$

where

$$R_o = \sqrt{\frac{L}{C}}$$

the low frequency resistance value of the characteristic network impedance and the frequency parameter, ω_c, is given by

$$\omega_c = \frac{2}{\sqrt{LC}}.$$

Provide an engineering interpretation of this result for the case in which the steady state utilization of the filter in Fig. P3.17(b) is confined to signal frequencies that are significantly smaller than ω_c.

(d) The network in Fig. P3.17(a) is implemented as the lossless structure in Fig. P3.17(b), in which it is noted that the load termination at the filter output port is realized as the aforementioned low frequency characteristic impedance, R_o. Plot, as a function of the normalized frequency, ω/ω_c, the real and imaginary components of the resultant reflection coefficient at the filter input port. Discuss the engineering implications of these plots.

(e) For the circuit addressed in Part (d) of this problem, show that the input port to output port voltage transfer function is expressible as

$$H(p) = \frac{V_o}{V_i} = \frac{1}{1 + 2p + 2p^2 + 2p^3},$$

where p is the normalized complex frequency, $p = s/\omega_c$.

(f) If the input port of the filter in Fig. P3.17(b) is driven by a signal source comprised of a voltage, V_s, and a Thévenin resistance identically equal to R_o, as defined in Part (c) of this problem, give an expression for the resultant overall voltage transfer function, $H(p) = V_o/V_s$.

(g) Demonstrate that the I/O envelope delay, say $D(y)$, corresponding to the port transfer function deduced in Part (e) is

$$D(y) = \frac{2(1 - y^2 + 2y^4)}{(1 - 2y^2)^2 + [2y(1 - y^2)]^2},$$

where y is the normalized real frequency, $y = \omega/\omega_c$.

(h) Design the filter in Fig. P3.17(b) for a zero frequency delay of 20 pSEC when the indicated load termination is 50 Ω. Use SPICE to investigate the magnitude, phase, and delay responses of the I/O port voltage transfer function, V_o/V_i. Scrutinize the simulated magnitude and phase responses by comparing them to the implications of the analytical results deduced in Parts (e) and (g) of this problem.

Problem 3.18

The lossless delay filter offered in Fig. P3.18 is an ostensibly improved version of the topology appearing in Fig. P3.17(b). The terminating load resistance, R_o, is the low frequency value of the frequency dependent characteristic impedance, $Z_o(s)$. Moreover, parameter m is a number in the range, $0 < m < 1$, while the inductance, L_1, is given by

$$L_1 = \frac{(1 - m^2)L}{4m}.$$

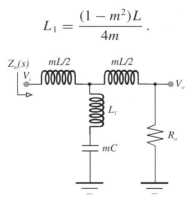

Figure P3.18.

(a) Derive an expression for the characteristic impedance, $Z_o(s)$. What is the low frequency value, $Z_o(0)$, of this impedance, and over what nominal range of signal frequencies is $Z_o \approx Z_o(0)$? Study this observation and submit an argument as to why the filter at hand may indeed be an improved version of the structure shown in Fig. P3.17(b).

(b) Design the filter for the same specifications as those proffered in conjunction with Part (h) of Problem 3.17. Use SPICE to investigate the magnitude, phase, and delay responses of the I/O port voltage transfer function, V_o/V_i. Compare these simulated results with those gleaned in the preceding problem for the filter in Fig. P3.17(b).

(c) Set up SPICE to simulate the frequency responses of the real and imaginary parts of the input port reflection coefficient of the filter designed in Part (b) of this problem. Assume a reference impedance of $R_o = 50\,\Omega$.

Problem 3.19

At a specified quiescent operating point, a certain type of monolithic bipolar junction transistor has the following measured scattering parameters at a signal frequency of 2 GHz and for I/O reference impedances of 50 Ω. The transistor is to process signals from a 75 Ω source to a 50 Ω load.

$$S_{11} = 0.692e^{-j43°} \quad S_{12} = 0.038e^{+j54°}$$
$$S_{21} = 6.459e^{+j151°} \quad S_{22} = 0.610e^{-j56°}$$

(a) Evaluate the voltage gain of this transistor under actual load and source conditions.
(b) Evaluate the power gain of this transistor under actual load and source conditions.
(c) Evaluate the y-parameters of the subject transistor at 2 GHz.
(d) Using the results of Part (c) of this problem, deduce a π-type small signal equivalent circuit for the transistor. When possible, numerically indicate resistance, capacitance, or inductance values for the branch elements of this model.

Problem 3.20

For the amplifier whose small signal model is provided in Fig. P3.20, determine the four scattering parameters, with the understanding that the reference impedance for both the input and the output ports is a resistance, R_o.

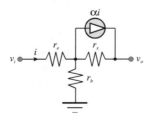

Figure P3.20.

Problem 3.21

Repeat Problem 3.20 for the amplifier whose small signal equivalent circuit appears in Fig. P3.21.

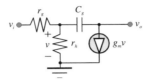

Figure P3.21.

Chapter 4

Feedback Circuit and System Theory

4.1.0. Introduction

The preceding two chapters provide the mathematical tools and models for investigating the I/O transfer and impedance properties of linear two-port networks for which feedback is applied purposefully or, in the case of high frequency energy storage effects and other high order phenomena, inadvertently. These tools and models, which derive explicitly from fundamental circuit theoretic concepts, establish straightforward analytical recipes for deducing forward gain, feedback transmission, and I/O impedance characteristics for relatively elementary and even topologically complex two-port networks. However, they do not directly reveal the designable circuit parameters that establish the relative quality of a network in the sense of its ability to deliver and sustain desired I/O performance specifications. These barometers of quality include, but are not limited to, the steady state network frequency and phase responses, the rise and fall times associated with the network step response, and the gain and phase margins commensurate with an assurance of network stability. This chapter addresses the foregoing issues in terms of the designable and controllable parameters of block diagram, or system level, models for generalized linear networks. At some risk of oversimplification, the fundamental purpose of this chapter is the deduction of meaningful stability metrics that ensure acceptable I/O network performance within the constraint of acceptable margins of stability.

4.2.0. System Level Model of Feedback Circuit

Figure 4.1 abstracts a generalized system level model of a linear network to which *global feedback* (feedback directly from output port to input port)

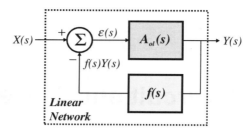

Figure 4.1. Block diagram representation of a linear network for which the transformed response to the applied transformed input, $X(s)$, is $Y(s)$.

is applied. The Laplace transform, $X(s)$, of the applied input signal can be a voltage or a current, as can the transform, $Y(s)$, of the resultant output response. If $X(s)$ and $Y(s)$ are both either voltage or current transforms, the delineated *open loop gain*, $A_{ol}(s)$, as a function of the complex frequency operator, "s", is a dimensionless voltage or current transfer relationship. On the other hand, if $X(s)$ is a current transform and $Y(s)$ is a voltage transform, the system operates as a transimpedance amplifier and $A_{ol}(s)$ represents a transimpedance having units of ohms. For a transadmittance amplifier, $X(s)$ is a voltage transform, $Y(s)$ is a current transform, and $A_{ol}(s)$ is necessarily a transadmittance whose units reflect admittance.

The function, $f(s)$, in Fig. 4.1 symbolizes the *feedback factor* of the system. Specifically, the product, $f(s)Y(s)$, stipulates the transformed amount of output signal that is fed back to the network input port and therefore made available for a comparison to the input signal transform. This comparison is manifested as a transformed *error signal*, $\varepsilon(s)$, where, as is suggested by the subject block diagram,

$$\varepsilon(s) = X(s) - f(s)Y(s). \qquad (4\text{-}1)$$

In accordance with the mathematical abstraction of Fig. 4.1, $\varepsilon(s)$ in a feedback system implementation represents the transform of the time domain signal applied at the actual input port of the subcircuit whose transfer function is $A_{ol}(s)$. Since the abstraction at hand also confirms

$$Y(s) = A_{ol}(s)\varepsilon(s), \qquad (4\text{-}2)$$

the transform of the error signal relates to the transformed input signal as

$$\varepsilon(s) = \frac{X(s)}{1 + f(s)A_{ol}(s)}. \qquad (4\text{-}3)$$

The product, $f(s)A_{ol}(s)$, in this relationship is dimensionless, which implies that the dimension of the feedback factor must be the inverse of the units indigenous to the open loop gain. This product, which is ordinarily symbolized as $T(s)$, is logically termed the *loop gain* of the network undergoing study; that is,

$$T(s) \triangleq f(s)A_{ol}(s) \tag{4-4}$$

is the gain associated with the signal path loop from the error signal node to the output node of the feedback factor block.

If Eq. (4-1) is inserted into Eq. (4-2) and Eq. (4-4) is exploited, the *closed loop gain*, $A_{cl}(s)$, which is the observable gain in an engineering application of the subject system, evolves as

$$A_{cl}(s) \triangleq \frac{Y(s)}{X(s)} = \frac{A_{ol}(s)}{1 + f(s)A_{ol}(s)} \equiv \frac{A_{ol}(s)}{1 + T(s)}. \tag{4-5}$$

Aside from reflecting the generalized feedback function forms deduced in conjunction with the two-port network parameters in Chapter 2, several other noteworthy issues are projected by this result. The simplest of these is the fact that since the loop gain, $T(s)$, is a unitless feedback network metric, the dimension of the closed loop gain, $A_{cl}(s)$, is identical to that of the open loop function, $A_{ol}(s)$. Thus, for example, if $A_{ol}(s)$ is an open loop transadmittance, the resultant closed loop gain is likewise a transadmittance function.

A situation that underscores the principle reason for utilizing appropriate feedback in electronic networks derives from a consideration of the special case of a large loop gain. Assume that for all real frequencies, ω, of interest, the loop gain satisfies the constraint,

$$|T(j\omega)| = |f(j\omega)A_{ol}(j\omega)| \gg 1. \tag{4-6}$$

By Eq. (4-5), the closed loop gain resultantly collapses to

$$A_{cl}(j\omega) \triangleq \frac{Y(j\omega)}{X(j\omega)} \approx \frac{A_{ol}(j\omega)}{f(j\omega)A_{ol}(j\omega)} = \frac{1}{f(j\omega)}; \tag{4-7}$$

that is, the closed loop gain is determined largely by the inverse of the feedback factor and is rendered essentially independent of the open loop gain. This desensitization of closed loop performance with respect to the open loop gain is the dominant attribute of prudently designed feedback networks. In monolithic electronic circuits, the open loop gain is vulnerable to the parametric vagaries of the active elements and even certain passive components whose collective topological interconnections define the equilibrium equations that establish the open loop gain function. These plausibly significant uncertainties derive from routine manufacturing tolerances or transistor model parameterization whose accurate and reproducible numerical delineation is often a daunting challenge. They may be the ramifications of high frequency energy storage effects or distributed phenomena whose meaningful quantification via conventional lumped circuit modeling is difficult or even impossible. Regardless of the sources of uncertainties, parametric vagaries in the open loop are precisely reflected by the closed loop system gain when zero feedback, in the sense of $f(j\omega) = 0$, prevails. But to the extent that feedback is incorporated in such a way as to ensure the satisfaction of Eq. (4-6), the closed loop gain in Eq. (4-7), becomes independent of the open loop gain and the unreliability attendant to its implicit parametric uncertainties.

Several important considerations surface from the foregoing feedback system signature. First, the alleged desensitization is meaningful only to the extent that the feedback factor, $f(j\omega)$, can be accurately predicted and reliably sustained in anticipated operating environments. This engineering necessity explains why active subcircuits are rarely incorporated into the subsystem that produces the desired feedback factor, for if active configurations conduce characteristic uncertainties in the open loop function, they affect the feedback factor similarly. Accordingly, the feedback function is generally realized with passive subcircuits and preferably by ratios of passive element values, since parametric ratios can be controlled more accurately than can absolute values of passive elements. One shortfall of this design strategy is that passive subcircuits are bilateral; that is, they are capable of transmitting signals from their input port to their output port as easily as transmitting energy in the reverse direction. However, the feedback block diagram in Fig. 4.1 tacitly presumes signal conduction through the feedback subsystem only in the direction of the indicated arrows. Fortunately, and particularly at low signal frequencies, the signal flow (in the direction of the arrows) through the open loop subsystem is invariably far more substantial than is the feedback flow against the indicated arrows through the feedback component.

A second point of interest involves the special cases of voltage and current amplification, for which $A_{ol}(s)$ and $A_{cl}(s)$ are dimensionless. In these signal processing applications, a gain greater than unity is presumably desired throughout the closed loop passband. If the closed loop is desensitized with respect to its open loop gain, Eq. (4-7) shows that greater than unity closed loop gain requires a feedback factor whose magnitude is smaller than one over the signal frequency range of interest. But if $|f(j\omega)| < 1$, the desensitization requirement of Eq. (4-6) mandates a commensurately larger open loop gain magnitude. Thus, for example, if a closed loop gain of 100 (40 dB) is the design objective, $|f(j\omega)|$ must be 0.01 (-40 dB), whence $|A_{ol}(j\omega)|$ must be of the order of at least 1,000 (60 dB) to ensure the reasonable satisfaction of Eq. (4-6). Unfortunately, large open loop gains incur closed loop bandwidth penalties and, as is ultimately demonstrated, potential instability problems. These bandwidth and stability shortfalls explain why operational amplifiers, which are used in a plethora of low frequency feedback applications, have enormous open loop gains and embarrassingly small 3-dB bandwidths. They also explain why feedback amplifiers exploited in radio frequency (RF) and broadband analog applications rarely deliver closed loop gains that exceed 10 to 20 dB.

It should also be noted that when feedback is designed to ensure very large loop gain, the transformed error signal defined by Eq. (4-3) approaches zero for finite input signal amplitude. If the error signal indeed approaches zero, Eq. (4-1) straightforwardly predicts a closed loop gain of $Y(s)/X(s) = 1/f(s)$, which concurs with the approximate result in Eq. (4-7). Since the error signal represents the energy delivered to the input port of the open loop subcircuit, these observations suggest a viable strategy for analyzing the closed loop behavior of relatively complex feedback networks. In particular, the analysis of these networks might be expedited by presuming at the outset that the input port of the open loop subcircuit operates as a *virtual short circuit* in the sense of simultaneously supporting zero voltage and conducting zero current. Despite the idealism implicit to the virtual ground approximation, the strategy expedites the circuit level analysis of many global feedback systems. Except for certain classes of high performance feedback electronics in which exceptionally large loop gains are impractical, it generally produces acceptably accurate performance estimates.

Aside from the foregoing analytical facilitation, the error signal boasts at least two other important engineering properties. The first of these addresses electronic feedback networks for which the open loop contains inherently nonlinear active elements. Generally, the linear domain of open loop operation can be determined by comparing the signal swing at the output port with signal responses predicted by strictly linear circuit models. Once this maximum possible signal excursion is determined, the maximum permissible error signal swing derives by mere division of the output swing by the open loop gain magnitude. Given a loop gain magnitude chosen in accordance with open loop parametric desensitization or other criteria, the maximum allowable input signal swing follows from Eq. (4-3).

The second error signal property of interest deals with feedback network stability, for which detail is provided subsequently. When the magnitude of the time domain error signal, $\varepsilon(t)$, is bounded by a suitably small number for all time "t", the implemented feedback is said to be *negative feedback*, and the corresponding system is *unconditionally stable*. On the other hand, if $\varepsilon(t)$ becomes boundless with time, *positive feedback* is said to prevail, and the feedback system at hand is *unstable*. A large percentage of electronic feedback networks are *potentially unstable*, which is to say that $\varepsilon(t)$ becomes boundless only for specific ranges of source and load terminations, certain biasing conditions pertinent to the open loop, and/or other designable factors. A relaxation or sinusoidal error signal oscillation, sustained despite a nulled input signal, corresponds to the unstable, but nonetheless useful, case of an *oscillator*, which is indigenous to virtually all communication system electronics. As might be surmised from a recollection of Chapter 1, these conditions are intimately related to the critical frequencies of the feedback network *characteristic polynomial*, $[1 + T(s)]$, which determines the poles of both the error transform and the closed loop gain.

Example 4.1. The operational amplifier (op-amp) in Fig. 4.2(a) has a large, but nonetheless finite, open loop gain, A_o, of 90,000 (99.1 dB), infinitely large input impedance at its inverting (−) and noninverting (+) input ports, and zero output port impedance. To first order, the static input/output (I/O) characteristic curve of the subject op-amp is as depicted in Fig. 4.2(b), where the applied biasing is $V_{pp} = 9$ volts. The op-amp in Fig. 4.2(a) is exploited in the feedback circuit shown in Fig. 4.2(c), where the feedback resistance, R_f, is 4.5 KΩ, the Thévenin signal source resistance, R_s, is 300 Ω, the biasing resistance, R, is 280 Ω, and the resistance, R_l, terminating the amplifier output port is 1 KΩ. In this diagram, the power supply voltages are omitted in the interest of topological simplicity and because the op-amp is tacitly presumed to operate in its linear I/O regime. Analyze the feedback amplifier at hand to determine its loop gain, feedback factor, closed loop gain, and the maximum input signal amplitude commensurate with nominal I/O linearity of the op-amp for low signal frequencies.

Solution 4.1.

(1) The small signal model, shown in Fig. 4.2(d) exploits the facts that the input impedances at both of the op-amp input ports are infinitely large and the output impedance seen by the terminating load resistance is zero. It also uses the static I/O characteristic diagrammed in Fig. 4.2(b), which projects that the output voltage, V_o, is clearly $-A_o v$. Since no current is conducted by the noninverting

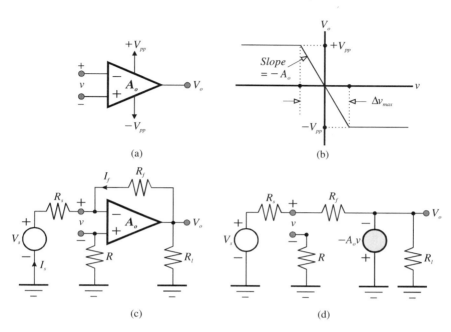

Figure 4.2. (a) Schematic diagram of an operational amplifier biased at voltages of $\pm V_{pp}$. The input port voltage, v, is applied from the inverting ($-$) terminal to the noninverting ($+$) terminal. The open loop gain of the op-amp is A_o. (b) The static I/O characteristic of the op-amp in (a). Observe that linearity prevails over the nominal output signal range, $-V_{pp} < V_o < +V_{pp}$. (c) Feedback amplifier formed with the op-amp in (a). (d) Small signal, low frequency equivalent circuit of the feedback configuration in (c).

input terminal, zero voltage drop prevails across resistance R, which means that the voltage, v, across the two op-amp input ports is effectively the voltage established at the inverting input node with respect to signal ground, It follows by the Kirchhoff laws that

$$\frac{v - V_s}{R_s} + \frac{v - (-A_o v)}{R_f} = 0, \tag{E1-1}$$

or

$$v = \frac{V_s}{1 + (A_o + 1)\left(\dfrac{R_s}{R_f}\right)}. \tag{E1-2}$$

Since $V_o = -A_o v$, the closed loop gain, A_{cl}, is seen to be

$$A_{cl} = \frac{V_o}{V_s} = -\frac{A_o}{1 + (A_o + 1)\left(\dfrac{R_s}{R_f}\right)} = -15.0 = 23.5 \text{ dB}. \tag{E1-3}$$

(2) Equation (E1-3) can be recast as

$$A_{cl} = \frac{-A_o}{1 + \left[\left(-\frac{A_o + 1}{A_o}\right)\left(\frac{R_s}{R_f}\right)\right](-A_o)}, \quad \text{(E1-4)}$$

which precisely reflects the form given in the generalized closed loop gain relationship of Eq. (4-5). It follows that the open loop gain of the entire amplifier is

$$A_{ol} = -A_o = -90{,}000, \quad \text{(E1-5)}$$

which happens to be the open loop gain of the utilized operational amplifier. Meanwhile, the feedback factor is

$$f = \left(-\frac{A_o + 1}{A_o}\right)\left(\frac{R_s}{R_f}\right) = -66.67\,(10^{-3}). \quad \text{(E1-6)}$$

Accordingly, the loop gain is, by Eq. (4-4),

$$T = fA_{ol} = 6{,}000, \quad \text{(E1-7)}$$

which is certainly more than sufficiently large to satisfy the guideline dictated by Eq. (4-6). Because the loop gain is extremely large and the open loop gain is likewise very large, the closed loop gain collapses to the familiar result,

$$A_{cl} = \frac{V_o}{V_s} \approx -\frac{R_f}{R_s} \approx \frac{1}{f} = -15.0 = 23.5\,\text{dB}, \quad \text{(E1-8)}$$

as predicted by Eq. (4-7).

(3) An inspection of Fig. 4.2(b) reveals that the maximum possible signal swing commensurate with nominal linearity at the output port of the amplifier is $2V_{pp} = 18$ volts. For the stipulated op-amp gain magnitude of 90,000, this output swing equates to a maximum input port swing of $\Delta v_{\max} = 18/90{,}000 = 200\,\mu\text{V}$. This voltage compliance is essentially the maximum permissible error swing ensuring amplifier linearity. Recalling Eq. (E1-2), the corresponding maximum permissible input signal excursion (a peak to peak swing) is

$$\Delta V_{s\,\max} = \left[1 + (A_o + 1)\left(\frac{R_s}{R_f}\right)\right]\Delta v_{\max} = 1.2\,\text{volts}. \quad \text{(E1-9)}$$

As expected, this swing, multiplied by the closed loop gain magnitude of 15.0, gives an output maximal swing response of 18 volts, which is exactly the net biasing voltage applied to the op-amp.

Comments. The delineation of the open loop gain, the feedback factor, and the loop gain in this simple example follows from straightforward algebraic manipulation of the results of the circuit equilibrium equations. For more complex feedback networks, analogous algebraic gymnastics are likely to prove impractical and annoying and to this end, a more elegant strategy underlying the determination of the foregoing performance metrics is forthcoming. At this juncture, however, the point made is that sufficient negative feedback serves to desensitize the closed loop with respect to open loop parameters in a practical, albeit simplified, circuit application. Moreover, the large open loop gain, acting in concert with the large loop gain, constrains the op-amp input port voltage to values that are sufficiently small to sustain circuit linearity, despite a relatively large applied input signal swing.

The presumptions of negative (stable) feedback and large loop gain also facilitate analyses without the need of resorting to elaborate circuit models. In particular, revisit the circuit diagram in Fig. 4.2(c) with no current conducted by either the inverting or the noninverting op-amp input terminals and with the amplifier input port voltage, v, which represents the error signal in this example, presumed nulled. Then, the signal source current, I_s, in Fig. 4.2(c) is simply V_s/R_s, while the current, I_f, flowing through the feedback resistance, R_f, is V_o/R_f. Because of the assumptions of infinitely large input impedance, the current sum, $(I_s + I_f)$, is constrained to zero, whence, the closed loop gain, $V_o/V_s = -R_f/R_s$, follows forthwith.

4.3.0. Feedback Network Frequency Response

As is illustrated in Chapter 1, the frequency response of a linear circuit or system benchmarks performance in the senses of gain, bandwidth, response peaking, envelope delay, and other frequency domain metrics. Through inverse Laplace transformation, the transfer function underlying the frequency response additionally supports an assessment of time domain responses. The core of these frequency and time domain evaluations is the delineation of the poles and zeros implicit to the transfer function of the network undergoing test. In multi-order circuits and systems, the invariably unpleasant experience accompanying a determination of critical network frequencies is muddled by prevailing feedback, which can significantly alter the frequency domain characteristics of the open loop and even cause the resultant closed loop to become unstable. Because the task of accurately discerning the frequencies of all poles and zeros implicit to practical active circuits and systems is genuinely daunting, two relatively simple frequency domain models have evolved. These are the *single pole, or dominant pole, approximation* and the *second order model*. In this section, the elementary dominant pole model is given only cursory attention because of its inadequacy with respect to frequency response predictions in state of the art monolithic circuit applications. On the other hand, the traditional second order model, which is embellished herewith through the pragmatic inclusion of a zero, receives the majority of present analytical attention.

4.3.1. Single Pole Open Loop Transfer Function

In the diagram of Fig. 4.1, assume a lowpass open loop whose transfer function reflects a single pole at $s = -p_{ol}$ and a zero frequency gain of $A_{ol}(0)$; that is,

$$A_{ol}(s) = \frac{A_{ol}(0)}{1 + \dfrac{s}{p_{ol}}}. \tag{4-8}$$

Assume further that the feedback factor, $f(s)$, is frequency invariant with a value of $f(0)$. If Eq. (4-8) is substituted into Eq. (4-5), the resultant closed loop transfer function, $A_{cl}(s)$, is found to preserve the first order nature of the open loop in that its analytical form is

$$A_{cl}(s) = \frac{A_{cl}(0)}{1 + \dfrac{s}{p_{cl}}}, \tag{4-9}$$

where the frequency, p_{cl}, of the closed loop transfer relationship is related to its open loop counterpart through

$$p_{cl} = [1 + f(0)A_{ol}(0)]p_{ol} = [1 + T(0)]p_{ol}, \tag{4-10}$$

and the zero frequency closed loop gain, $A_{cl}(0)$, is

$$A_{cl}(0) = \frac{A_{ol}(0)}{1 + T(0)}. \tag{4-11}$$

In Eqs. (4-10) and (4-11),

$$T(0) = f(0)A_{ol}(0) \tag{4-12}$$

defines the zero frequency value of the loop gain.

Because of the single pole nature of the open loop transfer function in Eq. (4-8), the pole frequency, p_{ol}, is identical to the open loop 3-dB bandwidth, say B_{ol}. It follows from Eq. (4-9) that the closed loop bandwidth, B_{cl}, is the frequency, p_{cl}, of the lone, or dominant, closed loop pole. Accordingly, Eq. (4-10) suggests that

$$\frac{B_{cl}}{B_{ol}} = 1 + T(0); \tag{4-13}$$

that is, if $T(0)$ is a positive number, a constraint shown ultimately to be a prerequisite for negative feedback and thus, network stability, negative feedback applied globally around a dominant pole open loop enhances the open loop bandwidth by one plus the zero frequency loop gain.

This bandwidth extension can be substantial since $T(0)$ is necessarily large if an adequate desensitization of the zero frequency closed loop gain is to be achieved with respect to potentially vagarious open loop parameters.

Before too much excitement is generated by the foregoing disclosure, the reader should be aware of the fact that single pole amplifiers can be realized only in the idealized confines of academic classrooms and can therefore only be approximated in the laboratory or on the production line. Even if the single pole structure can be synthesized, a price is paid for the foregoing bandwidth extension. In particular, the gain bandwidth product, say GBP_{cl}, of the closed loop, relates to the gain bandwidth product, GBP_{ol}, of the open loop as

$$GBP_{cl} = A_{cl}(0) p_{cl} = \left[\frac{A_{ol}(0)}{1+T(0)}\right][1+T(0)]p_{ol} = A_{ol}(0) p_{ol} \equiv GBP_{ol}. \tag{4-14}$$

In other words, the gain bandwidth product of both the closed and the open loops are identical when constant negative feedback is applied globally around a lowpass, dominant pole open loop. This observation means that the price paid for bandwidth enhancement is a reduction of the zero frequency gain by precisely the factor by which the bandwidth is extended. With reference to Example 4.1, Eq. (4-14) implies that if the op-amp whose open loop gain magnitude is 90,000 has an open loop 3-dB bandwidth of 100 Hz, a resultant closed loop circuit designed for a 3-dB bandwidth of 600 KHz delivers a gain magnitude of only 15.

A peripheral point of interest is the availability of adequate gain at very high signal frequencies. For radial signal frequencies satisfying $\omega \gg p_{cl}$, the closed loop gain in Eq. (4-9) becomes

$$A_{cl}(j\omega)|_{\omega \gg p_{cl}} \approx \frac{A_{cl}(0) p_{cl}}{j\omega} = \frac{A_{cl} B_{cl}}{j\omega} = \frac{GBP_{cl}}{j\omega} \equiv \frac{GBP_{ol}}{j\omega}, \tag{4-15}$$

which suggests that the magnitude of closed loop gain at very high frequencies degrades to unity at a signal frequency that is approximately equal to the gain bandwidth product of either the closed or open loops. To the extent that greater than unity gain is desired or mandated by a particular application, the gain bandwidth product of the open loop assumes the posture of the maximum practical frequency at which signals can be processed by the closed loop.

4.3.2. Second Order Open Loop Transfer Function

A more realistic mathematical model for the open loop transfer function in Fig. 4.1 is the second order relationship,

$$A_{ol}(s) = A_{ol}(0) \left[\frac{1 - \dfrac{s}{z_o}}{\left(1 + \dfrac{s}{p_{1o}}\right)\left(1 + \dfrac{s}{p_{2o}}\right)} \right], \qquad (4\text{-}16)$$

where two poles, whose respective frequencies are p_{1o} and p_{2o}, are incorporated, as is a zero at $s = +z_o$. If $z_o > 0$, the subject zero lies in the right half plane, while $z_o < 0$ implies a left half plane zero. The right half plane zero is included in the present analysis because such zeros invariably materialize in phase inverting amplifiers plagued by parasitic global capacitive feedback at high signal frequencies. Moreover, and is shortly demonstrated, the impact of right half plane zeros on stability margins and general closed loop response quality is potentially catastrophic. On the other hand, open loop stability requires that the two pole frequencies be positive numbers or, in the case of complex conjugate pole frequencies, stability constraints mandate positive real pole frequency parts. If the two pole frequencies are real numbers, $p_{1o} < p_{2o}$ is tacitly presumed.

An alternative to the transfer function form in Eq. (4-16) is

$$A_{ol}(s) = \frac{A_{ol}(0)\left(1 - \dfrac{s}{z_o}\right)}{1 + 2\zeta_{ol}\left(\dfrac{s}{\omega_{ol}}\right) + \left(\dfrac{s}{\omega_{ol}}\right)^2} = \frac{A_{ol}(0)\left(1 - \dfrac{s}{z_o}\right)}{1 + \left(\dfrac{s}{Q_{ol}\omega_{ol}}\right) + \left(\dfrac{s}{\omega_{ol}}\right)^2}, \qquad (4\text{-}17)$$

where

$$\zeta_{ol} = \frac{1}{2}\left(\sqrt{\frac{p_{2o}}{p_{1o}}} + \sqrt{\frac{p_{1o}}{p_{2o}}}\right), \qquad (4\text{-}18)$$

is the familiar damping factor of the second order characteristic polynomial for the open loop gain function,

$$Q_{ol} = 1/2\zeta_{ol}, \qquad (4\text{-}18)$$

is the open loop quality factor corresponding to ζ_{ol}, and

$$\omega_{ol} = \sqrt{p_{1o}p_{2o}} \qquad (4\text{-}19)$$

is the undamped natural frequency of the open loop characteristic polynomial. Recalling the relevant disclosures in Chapter 1, a damping factor, ζ_{ol}, considerably smaller than unity gives rise to excessive peaking in the frequency response and excessive time domain overshoot in the step and impulse responses. The excessive underdamping corresponding to $\zeta_{ol} \ll 1$ implies complex conjugate poles lying in the left half plane, but relatively close to the imaginary axis of the complex frequency plane. In turn, this pole distribution implies potential instability problems that are likely to be exacerbated by the presence of very high frequency poles and/or right half plane zeros that are not embodied in the second order model defined by Eq. (4-16). Recall further that to the extent that the magnitude of z_o is sufficiently large to allow tacit neglect of its impact on the open loop frequency response, ω_{ol} is a measure of the open loop 3-dB frequency. Indeed, if the magnitude of z_o approaches infinity and if ζ_{ol} is the inverse of root two, the open loop delivers a maximally flat magnitude frequency response, and ω_{ol} is precisely the 3-dB bandwidth of the open loop transfer function.

As in the preceding subsection of material, let the feedback factor, $f(s)$, be the frequency invariant constant, $f(0)$. Then, the loop gain, $T(s)$, of the closed loop depicted symbolically in Fig. 4.1 is simply

$$T(s) = \frac{f(0)A_{ol}(0)\left(1 - \frac{s}{z_o}\right)}{\left(1 + \frac{s}{p_{1o}}\right)\left(1 + \frac{s}{p_{2o}}\right)} = \frac{T(0)\left(1 - \frac{s}{z_o}\right)}{1 + 2\zeta_{ol}\left(\frac{s}{\omega_{ol}}\right) + \left(\frac{s}{\omega_{ol}}\right)^2}, \quad (4\text{-}20)$$

which is little more than a scaled version of the open loop gain function. This is to say that for frequency invariant feedback, the critical frequencies (poles and zero) of the loop gain are identical to those of the open loop gain. In turn, identical critical frequencies produce loop gain and open loop gain frequency responses, normalized to their respective zero frequency values, that are likewise identical. Using Eq. (4-5), the resultant closed loop transfer function assumes the generalized form,

$$A_{cl}(s) = \frac{A_{ol}(s)}{1 + T(s)} = \frac{A_{cl}(0)\left(1 - \frac{s}{z_o}\right)}{1 + 2\zeta_{cl}\left(\frac{s}{\omega_{cl}}\right) + \left(\frac{s}{\omega_{cl}}\right)^2}$$

$$= \frac{A_{cl}(0)\left(1 - \frac{s}{z_o}\right)}{1 + \left(\frac{s}{Q_{cl}\omega_{cl}}\right) + \left(\frac{s}{\omega_{cl}}\right)^2}, \quad (4\text{-}21)$$

where the zero frequency closed loop gain, $A_{cl}(0)$, remains given by Eq. (4-11). Obligatory algebraic manipulations give for the closed loop undamped natural frequency, ω_{cl}, and the closed loop damping factor, ζ_{cl},

$$\omega_{cl} = \omega_{ol}\sqrt{1 + T(0)}, \qquad (4\text{-}22)$$

and

$$\zeta_{cl} = \frac{1}{2Q_{cl}} = \frac{\zeta_{ol}}{\sqrt{1 + T(0)}} - \left(\frac{T(0)}{\sqrt{1 + T(0)}}\right)\left(\frac{\omega_{ol}}{2z_o}\right), \qquad (4\text{-}23)$$

respectively.

The algebra underpinning the delineation of Eqs. (4-22) and (4-23) may be annoying, but it is profitable in that it exposes the fundamental problems surrounding the incorporation of frequency invariant global feedback around a second order open loop. An insightful understanding of these revelations begins with a consideration of the special case of an open loop right half plane zero that lies at infinitely large frequency; that is, $z_{ol} = \infty$. For this special case, Eq. (4-23) shows that the damping factor of the closed loop second order polynomial is the corresponding open loop damping factor attenuated by a factor of the square root of one plus the zero frequency loop gain, which is often referenced as the *return difference*. Since the zero frequency loop gain, $T(0)$, is invariably large, a closed loop damping factor that is sufficiently large to ensure transient responses divorced of large time domain overshoots necessarily mandates a very large open loop damping factor and hence, a dominant pole open loop response. If, for example, $T(0) = 6,000$ (as is computed in Example 1), and a maximally flat magnitude closed loop frequency response is desired, which requires $\zeta_{cl} = 1/\sqrt{2}$, ζ_{ol} must be 54.78. From Eq. (4-18), the indicated open loop damping factor requires that the nondominant pole frequency, p_{2o} be about 12,000 times larger than the frequency, p_{1o}, of the open loop dominant pole. Thus, if the nondominant open loop pole frequency is 1 GHz, the dominant pole frequency can be no larger than approximately 83 KHz. The undertone of these elementary calculations is that acceptable closed loop damping mandates an open loop whose transfer function subscribes to the dominant pole approximation.

The engineering implications of the foregoing illustration are exacerbated by a finite frequency right half plane open loop zero. In particular, $z_o > 0$ forces the second term on the right hand side of Eq. (4-23) to be a positive number that subtracts from the first term, which already reflects a sharply attenuated effective damping factor. Indeed, the subject second term can be rendered larger than the first (damping factor) term, which results in a negative closed loop damping factor and hence, outright closed loop instability. This observation explains why many neophyte circuit designers become closet oscillator experts in the process of being asked to design suitable feedback amplifiers, not oscillators. On the other hand, $z_o < 0$, which implies a finite left half plane zero, provides a clue as to how feedback circuits might be stabilized prudently. Specifically, $z_o < 0$ renders the second term on the right hand side of Eq. (4-23) a negative number, thereby boosting the closed loop damping factor that otherwise accrues in the absence of a suitable open loop zero.

Finally, note in Eq. (4-22) that the closed loop undamped frequency parameter, ω_{cl}, which is a first order measure of closed loop bandwidth and the attainable time domain response speed, relates to its counterpart open loop frequency parameter as a multiplicative factor of the square root of one plus the zero frequency loop gain (return difference). This result contrasts with the single pole approximation, which yields a closed to open loop bandwidth ratio of simply the zero frequency return difference. In other words, appropriate negative feedback applied around a second order open loop increases the open loop bandwidth, as expected. However, the increase is not as dramatic as that predicted by the simple single pole open loop model.

4.3.3. Stability Issues

For steady state sinusoidal inputs, the second order open loop transfer function model in Eq. (4-16) becomes

$$A_{ol}(j\omega) = A_{ol}(0) \left[\frac{1 - \dfrac{j\omega}{z_o}}{\left(1 + \dfrac{j\omega}{p_{1o}}\right)\left(1 + \dfrac{j\omega}{p_{2o}}\right)} \right], \quad (4\text{-}24)$$

while, for a frequency invariant feedback factor, the corresponding loop gain in Eq. (4-20) is

$$T(j\omega) = \frac{f(0) A_{ol}(0) \left(1 - \dfrac{j\omega}{z_o}\right)}{\left(1 + \dfrac{j\omega}{p_{1o}}\right)\left(1 + \dfrac{j\omega}{p_{2o}}\right)} = \frac{T(0)\left(1 - \dfrac{j\omega}{z_o}\right)}{\left(1 + \dfrac{j\omega}{p_{1o}}\right)\left(1 + \dfrac{j\omega}{p_{2o}}\right)}, \quad (4\text{-}25)$$

where as usual, $T(0)$ represents the zero frequency value of the circuit loop gain. The resultant closed loop gain is, from Eq. (4-21),

$$A_{cl}(j\omega) = \frac{A_{ol}(j\omega)}{1 + T(j\omega)}. \quad (4\text{-}26)$$

The sheer simplicity of this last result virtually belies the discovery of a possible operating condition that is counterproductive to the realization of stable, linear closed loop circuits. In particular, if $T(j\omega) \equiv -1$ at a radial frequency, say ω_o, the closed loop gain, $A_{cl}(j\omega_o)$, is boundless; that is, $A_{cl}(j\omega_o) = \infty$. A plausible interpretation of this disclosure is that an output signal is produced despite the lack of input signal excitation, which is certainly indicative of circuit instability. At risk of oversimplification, it might therefore be proffered that the fundamental objective implicit to the design of stable analog signal processors is preclusion of the condition, $T(j\omega) \equiv -1$, by an engineering safety margin that is as large as circuit operational restrictions and performance specifications permit.

Before proceeding with a definitive investigation of the aforementioned design objective, it is worthwhile interjecting that $T(j\omega_o) \equiv -1$ at precisely one frequency, ω_o, is not necessarily an undesirable operating state when the resultant zero input signal value of the output response remains bounded for all time. In particular, $T(j\omega_o) \equiv -1$ at precisely one frequency, ω_o, quantifies the classic *Barkhausen criterion*, which is the necessary condition underlying the generation of a self-sustaining sinusoidal oscillation at radial frequency ω_o.[1] Sinusoidal oscillators are, of course, ubiquitous in communication system applications.

Two designable performance indices have emerged as measures of the relative stability of linear electronic circuits. By "relative stability", is meant the proximity of loop gain value to the undesirable condition, $T(j\omega) \equiv -1$, with the understanding that the ideal design circumstance is one reflecting the impossibility of attaining negative unity loop gain at any frequency throughout the entire frequency range of interest. The subject two performance indices are the *phase margin* and the *gain margin*, both of which are particularly utilitarian when the magnitude response of the loop gain is, as is typical of most electronic networks, a monotone decreasing function of signal frequency.

4.3.3.1. *Phase Margin*

The loop gain is a complex function of signal frequency and therefore, it may be designated symbolically as

$$T(j\omega) = |T(j\omega)|e^{j\varphi(\omega)}, \tag{4-27}$$

where $\phi(\omega)$ represents the phase angle of the loop gain as a function of radial frequency, ω. Let ω_u be the frequency at which the loop gain magnitude degrades to unity; that is,

$$|T(j\omega_u)| = 1. \tag{4-28}$$

Then, $\phi(\omega_u)$ is the value of the loop gain phase angle at the unity loop gain frequency, ω_u. Since $\phi(\omega_u) = -180°$ is disastrous from a circuit stability perspective, the phase margin, say ϕ_{pm}, might logically be defined as

$$\phi_{pm} = \phi(\omega_u) - (-180°) = \varphi(\omega_u) + 180°. \quad (4\text{-}29)$$

To ensure circuit stability, ϕ_{pm} must be a positive phase angle, which means that in the neighborhood of the frequency where the loop gain magnitude is one, the loop gain phase angle has yet to cross the negative 180-degree phase angle line. Note that for a dominant pole loop gain function whose zeros lie at infinitely large frequencies, the smallest possible value of $\phi(\omega_u)$ is $-90°$, thereby yielding a phase margin, ϕ_{pm}, of at least $+90°$. On the other hand, a two pole loop gain function whose zeros lie at infinitely large frequencies can yield $\phi(\omega_u)$ approaching $-180°$ and hence, a phase margin nearing a troublesome zero degrees. The two pole loop gain exuding a right half plane, finite frequency zero warrants calling "911", since the phase angle at $\omega = \omega_u$ can be more negative than $-180°$, which in turn implies a negative phase margin. Since a loop gain phase angle of $-180°$ at the frequency where the loop gain magnitude is one spells disaster in the sense of closed loop instability, the phase margin defined by Eq. (4-29) defines a kind of stability "headroom" with respect to the loop gain phase response.

From Eq. (4-25), the unity loop gain frequency, ω_u, satisfies the expression

$$|T(j\omega_u)| = \frac{T(0)\left|1 - \dfrac{j\omega_u}{z_o}\right|}{\left|\left(1 + \dfrac{j\omega_u}{p_{1o}}\right)\left(1 + \dfrac{j\omega_u}{p_{2o}}\right)\right|} = 1, \quad (4\text{-}30)$$

while the corresponding phase angle at this frequency is

$$\varphi(\omega_u) = -\tan^{-1}\left(\frac{\omega_u}{z_o}\right) - \tan^{-1}\left(\frac{\omega_u}{p_{1o}}\right) - \tan^{-1}\left(\frac{\omega_u}{p_{2o}}\right). \quad (4\text{-}31)$$

The task of solving Eq. (4-30) for frequency ω_u is algebraically daunting and of interest only to academic purists. An approximate solution boasting value to circuit designers exploits the previously established fact that closed loop stability, in the sense of adequately large closed loop damping factor, commands a reasonable facsimile of open loop pole dominance. Accordingly, if the frequencies, z_o and p_{2o}, of the open loop right half plane zero and the open loop nondominant pole, respectively, are each larger than the frequency, ω_u, of immediate interest,

$$|T(j\omega_u)| \approx \frac{T(0)}{\left|\left(1 + \frac{j\omega_u}{p_{1o}}\right)\right|} = 1, \qquad (4\text{-}32)$$

whence for large $T(0)$, which can be recalled as a desirable design objective,

$$\omega_u \approx T(0)p_{1o} = f(0)A_{ol}(0)p_{1o} \approx f(0)GBP_{ol}, \qquad (4\text{-}33)$$

where Eqs. (4-12) and (4-14) have been applied. Using this result and introducing the normalized frequency parameters,

$$k_p \triangleq \frac{p_{2o}}{\omega_u}, \qquad (4\text{-}34)$$

$$k_o \triangleq \frac{z_o}{\omega_u} \qquad (4\text{-}35)$$

(which is positive for a right half plane zero and negative for a left plane zero), and

$$k \triangleq \frac{k_p k_o - 1}{k_p + k_o}, \qquad (4\text{-}36)$$

Eq. (4-31) is expressible as

$$\varphi(\omega_u) = -\tan^{-1}\left(\frac{1}{k_o}\right) - \tan^{-1}[T(0)] - \tan^{-1}\left(\frac{1}{k_p}\right). \qquad (4\text{-}37)$$

The substitution of Eq. (4-37) into Eq. (4-29), followed by trigonometric gymnastics, results in a phase margin, ϕ_{pm}, given by

$$\phi_{pm} = \tan^{-1}\left(\frac{kT(0) + 1}{T(0) - k}\right). \qquad (4\text{-}38)$$

Since $T(0)$ is invariably large,

$$\varphi_{pm} \approx \tan^{-1}(k) = \tan^{-1}\left(\frac{k_p k_o - 1}{k_p + k_o}\right), \qquad (4\text{-}39)$$

which suggests that the phase margin of a closed loop network governed by frequency invariant feedback is essentially determined by the frequencies of the zero and the nondominant pole of the open loop transfer function.

Figure 4.3 plots the phase margin in Eq. (4-38) as a function of the critical frequency parameter, k, for various values of the zero frequency loop gain, $T(0)$. As anticipated, the phase margin, and thus the degree of network stability, diminishes for progressively larger loop gains. For any particular loop gain, note a phase margin that increases with the critical frequency parameter, k.

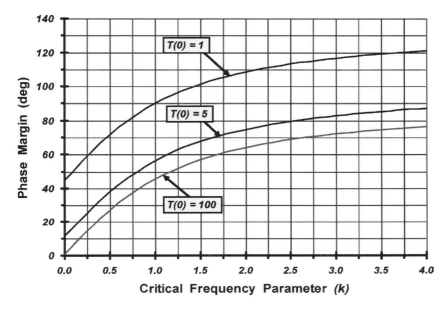

Figure 4.3. The phase margin of a second order network as a function of the critical frequency parameter, k, as defined by Eq. (4-36), for various values of the zero frequency loop gain, $T(0)$. Parameter k absorbs the effects of both the nondominant pole and any finite frequency zero that prevails in the open loop response.

4.3.3.2. Gain Margin

As noted previously, the gain margin is an alternative to the phase margin quantification of relative network stability. Its basic premise is that the magnitude of the loop gain, say $|T(j\omega_v)|$, at the frequency, ω_v, where the loop gain phase angle, $\phi(\omega_v)$, is $-180°$, must be less than one. The engineering logic underlying this design stipulation derives from Eq. (4-27), where with $\phi(\omega_v) = -180°$, observe that $T(j\omega_v) = -|T(j\omega_v)|$. If $|T(j\omega_v)|$ exceeds unity, $1 + T(j\omega)$ has a negative real part in the vicinity of frequency ω_v, thereby implying closed loop instability through the existence of at least one right half plane pole. Accordingly, if $|T(j\omega_v)| < 1$, there clearly exists a larger than unity number, say k_{gm}, such that

$$k_{gm}|T(j\omega_v)| = 1. \qquad (4\text{-}40)$$

The number, k_{gm}, is the gain margin of present interest. This metric effectively establishes the "headroom" between the actual loop gain at frequency ω_v and the unity loop gain that is commensurate with closed loop instability. Parameter k_{gm} is generally stipulated in units of decibels and from Eq. (4-40),

$$k_{gm} \text{ (in db)} = -20 \log_{10} |T(j\omega_v)|, \qquad (4\text{-}41)$$

which is positive in a stable closed loop.

From Eq. (4-25), the phase angle of the loop gain is

$$\varphi(\omega) = -\tan^{-1}\left(\frac{\omega}{z_o}\right) - \tan^{-1}\left(\frac{\omega}{p_{1o}}\right) - \tan^{-1}\left(\frac{\omega}{p_{2o}}\right). \qquad (4\text{-}42)$$

At $\omega = \omega_v$, $\phi(\omega_v) = -180°$, and from Eqs. (4-33) through (4-35), this result produces

$$\varphi(\omega_v) = -180° = -\tan^{-1}\left(\frac{\omega_v/\omega_u}{k_o}\right) - \tan^{-1}\left[T(0)\frac{\omega_v}{\omega_u}\right]$$

$$- \tan^{-1}\left(\frac{\omega_v/\omega_u}{k_p}\right). \qquad (4\text{-}43)$$

For the traditional design objective of large zero frequency loop gain, the second inverse tangent term on the right hand side of Eq. (4-43) approaches 90°, which constrains the algebraic sum of the first and third inverse tangent terms to $-90°$. It follows that

$$90° \approx \tan^{-1}\left(\frac{\omega_v/\omega_u}{k_o}\right) + \tan^{-1}\left(\frac{\omega_v/\omega_u}{k_p}\right) = \tan^{-1}\left[\frac{\dfrac{\omega_v/\omega_u}{k_o} + \dfrac{\omega_v/\omega_u}{k_p}}{1 - \dfrac{(\omega_v/\omega_u)^2}{k_o k_p}}\right].$$

$$(4\text{-}44)$$

Equation (4-44) is satisfied if

$$\omega_v = \omega_u \sqrt{k_o k_p}, \qquad (4\text{-}45)$$

which indicates that the frequency at which the loop gain phase angle degrades to $-180°$ is approximately the geometric mean of the frequencies associated with the right half plane zero and nondominant pole of the loop gain function. If this frequency is inserted into Eq. (4-25), the loop gain at $\omega = \omega_v$ is seen to be

$$T(j\omega_v) \approx \frac{T(0)\left(1 - j\sqrt{\frac{k_p}{k_o}}\right)}{\left[1 + jT(0)\sqrt{k_o k_p}\right]\left(1 + j\sqrt{\frac{k_o}{k_p}}\right)}, \quad (4\text{-}46)$$

whose magnitude for large $T(0)$ is

$$|T(j\omega_v)| \approx \frac{1}{k_o}\sqrt{\frac{k_p}{k_o}}. \quad (4\text{-}47)$$

From Eq. (4-40), the gain margin, k_{gm}, follows as

$$k_{gm} \approx k_o\sqrt{\frac{k_o}{k_p}}. \quad (4\text{-}48)$$

Not surprisingly, the gain margin, like the phase margin, is determined largely by the normalized frequencies of the loop gain right half plane zero and nondominant pole.

4.3.3.3. Alternative Damping and Undamped Frequency Expressions

The preceding two subsections succeed in relating the closed loop gain margin and phase margin, both of which are explicit measures of relative closed loop stability, to the zero frequency loop gain, $T(0)$, the nondominant normalized pole frequency, k_p, and the normalized frequency, k_o, of the loop gain right half plane zero. Since these stability metrics are implicitly related to closed loop damping factor and undamped natural frequency, ζ_{cl} and ω_{cl}, respectively, it ought to be possible to forge expressions for both damping factor and self-resonant frequency in terms of the aforementioned system parameters. To wit, Eqs. (4-18), (4-33), and (4-35) combine to deliver an open loop damping factor of

$$\zeta_{ol} = \frac{1 + k_p T(0)}{2\sqrt{k_p T(0)}}, \quad (4\text{-}49)$$

which, in the limit of large $T(0)$, collapses to the simple expression

$$\zeta_{ol} = \frac{\sqrt{k_p T(0)}}{2}. \quad (4\text{-}50)$$

A dominant pole open loop transfer function, which complements closed loop stability, corresponds to an open loop damping factor that is significantly larger than one. Accordingly, Eq. (4-50) reaffirms the desirability of large k_p (a high nondominant pole frequency in comparison to the unity gain frequency of the loop gain function) and/or a large zero frequency value of the loop gain.

Recalling Eq. (4-19), the open loop self-resonant frequency, ω_{ol}, is expressible as

$$\omega_{ol} = \sqrt{p_{1o}p_{2o}} = \omega_u \sqrt{\frac{k_p}{T(0)}}, \qquad (4\text{-}51)$$

while its closed loop counterpart, ω_{cl}, in Eq. (4-22) is

$$\omega_{cl} = \omega_u \sqrt{k_p \left[1 + \frac{1}{T(0)}\right]} \approx \omega_u \sqrt{k_p}. \qquad (4\text{-}52)$$

Using Eqs. (4-49) and (4-51) the damping factor, ζ_{cl}, associated with the closed loop characteristic polynomial is, by Eq. (4-23),

$$\zeta_{cl} = \frac{1}{2\sqrt{k_p T(0)[1 + T(0)]}} \left[1 + k_p T(0)\left(1 - \frac{1}{k_o}\right)\right] \approx \frac{\sqrt{k_p}}{2}\left(1 - \frac{1}{k_o}\right), \qquad (4\text{-}53)$$

where as usual, the indicated approximation is the result of tacitly presuming large $T(0)$. It is important to observe that the open and closed loop damping factors and self-resonant frequencies are approximately independent of the dominant pole open loop frequency. Instead, these metrics are determined essentially by the normalized frequencies of the open loop right half plane zero and nondominant pole.

Example 4.2. Frequency invariant feedback is applied globally around a second order open loop amplifier to provide a zero frequency loop gain, $T(0)$, of 25, or $T(0) = 28$ dB. The unity gain frequency of the loop gain is $\omega_u = 2\pi(1 \text{ GHz})$. The open loop amplifier is found to have a right half plane zero at $z_o = 2\pi(5 \text{ GHz})$. Determine the frequency of the nondominant open loop pole such that the resultant closed loop establishes a nominally maximally flat magnitude frequency response. For this response, calculate the phase and gain margins. Plot the magnitude and phase responses of the loop gain to confirm the propriety of the gain and phase margin calculations.

Feedback Circuit and System Theory

Solution 4.2.

(1) It must be understood at the outset that with $z_o = 2\pi(5\,\text{GHz})$ and $\omega_u = 2\pi(1\,\text{GHz})$, the normalized frequency of the open loop amplifier right half plane zero is $k_o z_o/\omega_u = 5$. Furthermore, since the subject right half plane zero is substantially larger than the unity gain frequency of the loop gain, a maximally flat magnitude closed loop response materializes if $\zeta_{cl} \approx 1/\sqrt{2}$.

(2) For $\zeta_{cl} \approx 1/\sqrt{2}$, Eq. (4-53) gives $k_p = 3.15$. Thus, the frequency of the nondominant open loop pole is

$$p_{2o} = k_p \omega_u = 2\pi(3.15\,\text{GHz}).$$

(3) With $k_p = 3.15$, Eq. (4-36) gives $k = 1.81$. Using Eq. (4-38), the phase margin follows as

$$\varphi_{pm} = 63.4°.$$

This result implies that the phase angle of the loop gain at the unity gain frequency is $-116.6°$.

(4) From Eq. (4-48), the gain margin is

$$k_{gm} = 6.3 = 16\,\text{dB}.$$

Accordingly, the loop gain magnitude at the signal frequency where the loop gain phase angle is $-180°$ is $1/6.3 = -16\,\text{dB}$. In the interest of completeness, the frequency at which the loop gain phase angle is $-180°$ is $\omega_v \approx 2\pi(3.97\,\text{GHz})$.

(5) Since $T(0) = 25$ and $\omega_u = 2\pi(1\,\text{GHz})$, the frequency of the open loop dominant pole is $p_{1o} = \omega_u/T(0) = 2\pi(40\,\text{MHz})$. It follows from Eq. (4-25) that the loop gain is

$$T(jf) = \frac{25\left(1 - \dfrac{jf}{5}\right)}{\left(1 + \dfrac{jf}{0.04}\right)\left(1 + \dfrac{jf}{3.15}\right)}, \quad \text{(E2-1)}$$

where it is understood that frequency f is in units of gigahertz. Moreover, the loop gain phase angle is

$$\phi(f) = -\tan^{-1}\left(\frac{f}{5}\right) - \tan^{-1}\left(\frac{f}{0.04}\right) - \tan^{-1}\left(\frac{f}{3.15}\right). \quad \text{(E2-2)}$$

Figure 4.4 depicts the frequency response implied by Eq. (E2-1) and the phase response corresponding to Eq. (E2-2).

Comments. A detailed examination of the calculations leading to the plots in Fig. 4.4 reveals that the true unity gain frequency is 972 MHz, which is 2.8% lower than the presumed 1 GHz value. Using this true unity gain frequency, the phase

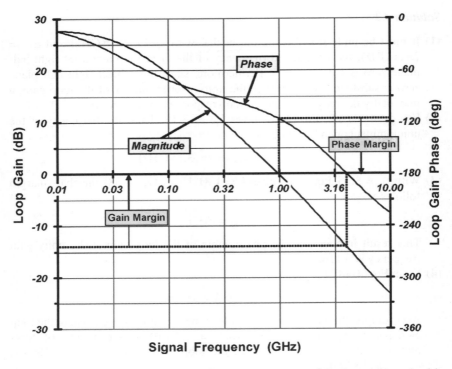

Figure 4.4. The magnitude and phase frequency responses of the loop gain evolved in Example 4.2.

margin is revealed as 64.2°, which is only 1.3% higher than the predicted phase margin of 63.4°. Moreover, the actual frequency at which the loop gain is −180° is 4.01 GHz, or about 1% higher than the calculated value of 3.97 GHz. Finally the true gain margin is 14.1 dB, which is 11.9% lower than the predicted gain margin of 16 dB. These errors are within acceptable bounds for first order design estimates.

It is also worth interjecting that while phase margins in the range of 40° to 50° are generally acceptable for representative industrial control systems that operate at very low frequencies, phase margins in the range of 50° to 70° are generally required to ensure stable broadband electronic networks. Similarly, gain margins in the range of 10 dB to 20 dB are also mandated of such networks.

4.3.4. Compensation for Closed Loop Stability

Circuits and systems for which the phase or gain margins are inadequate must be compensated to ensure closed loop stability and acceptable closed loop frequency and time domain responses. A definitive consideration of the compensation problem is best addressed in the context of specific classes of active networks and is therefore reserved for a forthcoming chapter. However, the problem can be broached herewith in an elementary fashion by observing that all global feedback and stability issues presented to this juncture are premised on a constant feedback factor, $f(s)$, which for analytical convenience is assigned its zero frequency value, $f(0)$. This observation suggests that plausible compensation schema incorporate suitable poles and/or zeros in the feedback factor so that $f(s)$ assumes the form,

$$f(s) = f(0)\frac{N(s)}{D(s)}, \tag{4-54}$$

where, assuming a lowpass feedback configuration, $N(s)$ and $D(s)$ are *monic polynomials*; that is, $N(0) = D(0) = 1$. Although the orders of the polynomials, $N(s)$ and $D(s)$, are arbitrary, pragmatic realization issues normally dictate that said orders be at most two.

A first order illustration of a viable, but not necessarily optimal, compensation strategy is best accomplished through a semi-numeric consideration of an extreme example. To this end, return to Eq. (4-25) to consider the special case in which the two pole frequencies, p_{1o} and p_{2o}, are identical and equal to the value, p_o. This multi-order pole circumstance is often an aftermath of an indiscriminate broadbanding exercise in which an active network is modified by increasing the frequency of the open loop dominant pole. Since a dominant pole limits the attainable 3-dB bandwidth, this broadbanding strategy is understandable. However, it is a reasonable design approach only to the extent that the frequency of the nondominant pole is correspondingly increased to ensure the open loop pole dominance that is commensurate with acceptable closed loop damping. Without increasing the nondominant pole frequency, the frequency of the broadbanded dominant pole ultimately approaches that of the nondominant pole; that is, $p_{1o} \approx p_{2o} \triangleq p_o$. As is to be demonstrated forthwith, a second order pole in the loop gain function can incur the curse of an unstable active network in the sense of establishing a negative phase margin and thus, right half plane closed loop poles.

With $p_{1o} \approx p_{2o} \triangleq p_o$, the compensated loop gain, $T_c(j\omega)$, is expressible as

$$T_c(j\omega) = T(j\omega)\frac{N(j\omega)}{D(j\omega)}, \tag{4-55}$$

where $f(0)$ in Eq. (4-25) is replaced by $f(s)$, in accordance with Eq. (4-54). Moreover, if the frequency, z_o, of the right half plane zero in Eq. (4-25) is viewed as a factor of the pole frequency, p_o, the uncompensated loop gain,

$T(j\omega)$, can be written as

$$T(j\omega) = \frac{T(0)\left(1 - \dfrac{j\omega}{k_z p_o}\right)}{\left(1 + \dfrac{j\omega}{p_o}\right)^2}, \qquad (4\text{-}56)$$

where k_z is a constant that is presumably larger than one. In terms of the normalized frequency,

$$y = \frac{\omega}{p_o}, \qquad (4\text{-}57)$$

the loop gain in Eq. (4-56) assumes the more convenient form,

$$T(jy) = \frac{T(0)\left(1 - \dfrac{jy}{k_z}\right)}{(1 + jy)^2}. \qquad (4\text{-}58)$$

The magnitude and phase responses implied by the loop gain function in Eq. (4-58) are depicted in Fig. 4.5 for the case of $T(0) = 15$ (or 23.5 dB)

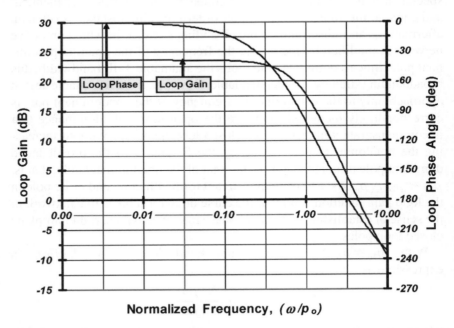

Figure 4.5. The magnitude and phase responses for the loop gain function in Eq. (4-56) for the case in which the zero frequency loop gain is 15 (23.5 dB) and the frequency (z_o) of the right half plane zero is five times the frequency (p_o) of the second order pole ($k_z = 5$).

and $k_z = 5$. The 3-dB bandwidth is found to be $y_b = \omega_b/p_o = 0.65$. If the right half plane zero in Eq. (4-58) is tacitly ignored, the approximate bandwidth can be shown to be

$$y_b = \frac{\omega_b}{p_o} = \sqrt{\sqrt{2}-1} = 0.644, \quad (4\text{-}59)$$

which is only 1.44% smaller than the true 3-dB bandwidth. On the other hand, the normalized unity gain frequency, say y_u, is 4.34, which, because of the second order and hence, nondominant pole, is considerably smaller than the observed normalized gain-bandwidth product of $T(0)y_b = (15)(0.65) = 9.75$. At the true unity gain frequency, the phase angle in Fig. 4.5 is $-195.1°$, which corresponds to a phase margin of $-15.1°$. Since closed loop stability requires a positive phase margin, the closed loop system is clearly unstable. This stability contention is reaffirmed by a gain margin evaluation. Specifically, the normalized frequency at which the loop phase angle is $-180°$ is $y_v = 3.32$, where the loop gain is 3.5 dB. From Eq. (4-41), the gain margin is resultantly negative.

Since the system at hand is unstable primarily because its open loop does not project a dominant pole response, a viable compensation strategy entails embedding a pole in the feedback factor to incur effective pole dominance in the loop gain. To this end, let $N(j\omega) = 1$ in Eq. (4-55), and

$$D(j\omega) = 1 + \frac{j\omega}{p_c} = 1 + \frac{jy}{k_c}, \quad (4\text{-}60)$$

where the normalized signal frequency, y, remains given by Eq. (4-57), and the frequency, p_c, of the introduced compensation pole is

$$p_c = k_c p_o, \quad (4\text{-}61)$$

with the understanding that k_c is a positive, less than unity constant. The compensated loop gain in Eq. (4-55) resultantly becomes

$$T_c(jy) = \frac{T(0)\left(1 - \dfrac{jy}{k_z}\right)}{(1+jy)^2 \left(1 + \dfrac{jy}{k_c}\right)}. \quad (4\text{-}62)$$

If the introduced pole is dominant, the normalized compensated unity gain frequency, say y_{uc}, is given approximately by

$$y_{uc} \approx k_c T(0). \quad (4\text{-}63)$$

A necessary condition underpinning acceptable closed loop responses is that this unity gain metric be smaller than the frequencies of all nondominant poles. The frequency of the nondominant pole for the system whose transfer characteristic is the function in Eq. (4-62) is the original, or uncompensated pole frequency, p_o. But since the original system pole is of order two, it is prudent to require that y_{uc} be smaller than the normalized bandwidth, y_b, which would be the original dominant pole (a kind of "virtual" dominant pole) frequency if the subject original pole were of order one. Thus, the fundamental design requirement is $y_{uc} < y_b$, or

$$k_c T(0) < 0.644. \tag{4-64}$$

The question now is the extent to which the normalized unity gain frequency, $k_c T(0)$, is smaller than the normalized virtual pole frequency, 0.644. Equivalently, the requisite value of constant k_c must be determined.

In order to ascertain the appropriate value of k_c, either a phase margin or a gain margin must be stipulated. The preceding example indicates that phase margins in the neighborhood of 60° appear to be acceptable. From Eq. (4-39), $\phi_{pm} = 60°$ corresponds to

$$\frac{k_p k_o - 1}{k_p + k_o} = \sqrt{3}. \tag{4-65}$$

Care must be exercised in the evaluation of constants k_p and k_o in this expression because Eq. (4-39), from which Eq. (4-65) derives, is premised on a dominant pole open loop. In the present exercise, open loop pole dominance is only emulated through the incorporation of the compensation pole at frequency p_c. Accordingly, recall that k_p is the frequency of the nondominant pole, normalized to the open loop unity gain frequency. Since the virtual nondominant pole in the present exercise is the 3-dB bandwidth, ω_b, and since the compensated open loop unity gain frequency is $T(0)p_c$,

$$k_p = \frac{\omega_b}{T(0)p_c} = \frac{0.644 p_o}{T(0)p_c} = \frac{0.644}{k_c T(0)}, \tag{4-66}$$

where Eqs. (4-59) and (4-61) are exploited. Recalling that k_o is the ratio of the right half plane zero frequency to the open loop unity gain frequency,

$$k_o = \frac{z_o}{T(0)p_c} = \frac{z_o/p_o}{T(0)p_c/p_o} = \frac{k_z}{k_c T(0)}. \tag{4-67}$$

The substitution of Eqs. (4-66) and (4-67) into Eq. (4-65) produces

$$\frac{0.644 k_z - [k_c T(0)]^2}{[k_c T(0)](k_z + 0.644)} = \sqrt{3}, \tag{4-68}$$

The solution for k_c in this relationship derives from

$$[k_c T(0)] = \frac{\sqrt{3}}{2}(k_z + 0.644)\left\{\left[1 + \frac{4(0.644)k_z}{3(k_z + 0.644)^2}\right]^{1/2} - 1\right\}, \quad (4\text{-}69)$$

whence for $k_z = 5$ and $T(0) = 15$, $k_c = 21.27(10^{-3})$. This result implies that the frequency of the introduced compensation pole must be about 47-times smaller than the frequency, p_o, of the second order open loop pole and slightly more than 30-times smaller than the uncompensated open loop 3-dB bandwidth. Note that $k_c T(0) = 0.319$, which satisfies the basic requirement highlighted by Eq. (4-64).

Figure 4.6 submits plots of the magnitude and phase responses of the compensated open loop gain function. The normalized value of the compensated unity gain frequency is 0.294, at which a phase angle corresponding to a phase margin of 58.1° results. This phase margin is within 2° of the 60° phase margin objective. The loop phase angle diminishes to −180° at a normalized frequency of 0.864, where the loop gain is −13.4 dB. It follows that a gain margin of +13.4 dB is evidenced.

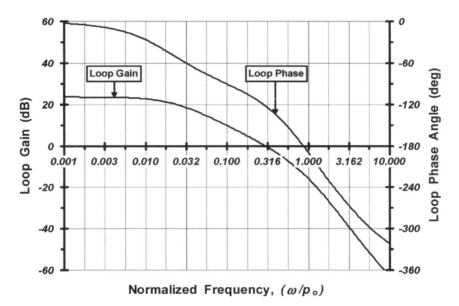

Figure 4.6. The magnitude and phase responses for the compensated loop gain function in Eq. (4-62) for the case in which the zero frequency loop gain is 23.5 dB and the frequency (z_o) of the right half plane zero is five times the frequency (p_o) of the second order pole. The compensation entails introduction of a feedback factor pole whose frequency is approximately 47-times smaller than that of the original second order open loop pole.

Although the foregoing compensation procedure leads to acceptable closed loop stability margins, it is difficult to dismiss the fact that the bandwidth of the compensated open loop is very near the frequency, p_c, of the compensation pole and is thus roughly 30-times smaller than the original 3-dB bandwidth. In effect, therefore, stability is accomplished herewith at the price of significant narrow banding. This bounty may be acceptable in many electronic system applications, including the ubiquitous general purpose operational amplifier. But it is assuredly an untenable price in broadband networking, high-speed data processing, and similar other broadband networks. For these latter applications, other compensation strategies, many of which are discussed in subsequent chapters, can be invoked.[2]

4.4.0. Time Domain Response

The time domain responses of overdamped, critically damped, and underdamped second order systems receive analytical scrutiny in Section 4.4 of Chapter 1. However, these analyses do not incorporate the effects of the right half plane zero embedded in the closed loop transfer function of Eq. (4-21). It is therefore both useful and instructive to adapt the relevant analyses in Chapter 1 to the specific closed loop circumstance defined by Eq. (4-21). To this end, only the case of an underdamped characteristic polynomial is addressed herewith, principally because the large loop gains mandated by parametric desensitization requirements invariably degrade open loop damping factors to smaller than unity closed loop damping coefficients.

4.4.1. Unit Step Response

Recalling the diagram of Fig. 4.1, the system level diagram for discerning the unit step response of the second order closed loop network is the abstraction posed in Fig. 4.7. From Eq. (4-21), the Laplace transform, $Y(s)$, of the time domain output response, $y(t)$, to a unit step input is

$$Y(s) = \frac{A_{cl}(s)}{s} = \frac{A_{ol}(s)}{s[1 + T(s)]} = \frac{A_{cl}(0)\left(1 - \dfrac{s}{z_o}\right)}{s\left[1 + 2\zeta_{cl}\left(\dfrac{s}{\omega_{cl}}\right) + \left(\dfrac{s}{\omega_{cl}}\right)^2\right]},$$

(4-70)

where the closed loop damping factor, ζ_{cl}, is presumed to be smaller than one. The desired time domain response can now be determined through

Feedback Circuit and System Theory

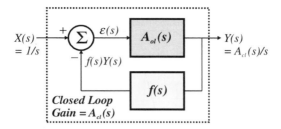

Figure 4.7. Block diagram abstraction of a linear feedback network to which a unit step input is applied.

an inverse Laplace transformation of the right hand side of Eq. (4-70). To expedite this computation, observe in Eq. (4-70) that the steady state response value is $A_{cl}(0)$, which conveniently allows for the determination of the normalized step response,

$$y_n(t) = \frac{y(t)}{A_{cl}(0)}. \tag{4-71}$$

Moreover, introduce variable x as a normalized time,

$$x = \omega_{cl} t, \tag{4-72}$$

where ω_{cl}, the self-resonant frequency of the closed loop characteristic polynomial, is given by Eq. (4-52). Finally, let

$$M \triangleq \frac{z_o}{\omega_{cl}} = \frac{k_o}{\sqrt{k_p \left[1 + \frac{1}{T(0)}\right]}} \approx \frac{k_o}{\sqrt{k_p}}, \tag{4-73}$$

where in addition to Eq. (4-52), Eq. (4-35) is exploited. With

$$\theta = \tan^{-1}\left(\frac{M\sqrt{1 - \zeta_{cl}^2}}{1 + M\zeta_{cl}}\right), \tag{4-74}$$

it can be shown that

$$y_n(x) = 1 - \left(\frac{e^{-\zeta_{cl} x}}{\sin \theta}\right) \sin\left(x\sqrt{1 - \zeta_{cl}^2} + \theta\right). \tag{4-75}$$

Figures 4.8 and 4.9 depict Eq. (4-75) in the normalized time domain for various values of the damping factor, ζ_{cl}, and the frequency parameter, M. As expected, Fig. 4.8 shows that diminished damping factor incurs potentially large step response overshoots and undershoots, which increase the

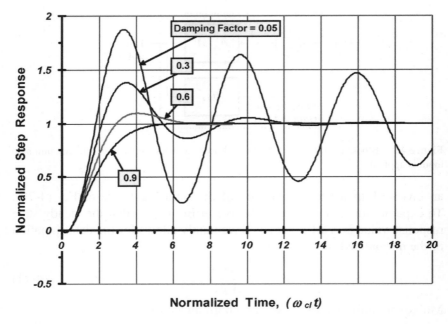

Figure 4.8. The unit step response of a second order closed loop network whose transformed output is given by Eq. (4-70). The right half plane zero is selected so that $M = z_o/\omega_{cl} = 5$. Four values of the characteristic polynomial damping factor are considered.

time required for the output response to achieve steady state operating conditions. In Fig. 4.9 the primary effect of progressively smaller frequencies for the right half plane zero (smaller M) is to incur observable response undershoot in the immediate neighborhood of the time, $t = 0$, at which the step excitation is applied to the input port of the feedback network. Parameter M is also observed to perturb, albeit moderately, the peak overshoot of the step response.

4.4.2. Settling Time

As explained in the first chapter, the settling time of an electronic network is an important performance metric because it stipulates the length of time required by a network to achieve steady state operation after the application of input excitation. In an ideal world, linear electrical and electronic networks are capable of instantaneously providing an output voltage or current that is simply proportional to the applied input signal. It is hardly shocking

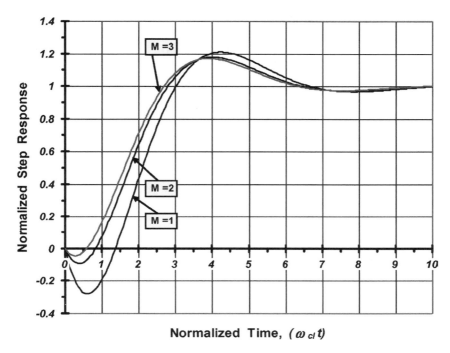

Figure 4.9. The unit step response of a second order closed loop network whose transformed output is given by Eq. (4-70). The characteristic polynomial damping factor is $\zeta_{cl} = 0.6$. Three values of the right half plane zero parameter, $M = z_o/\omega_{cl}$, are considered.

that no network is capable of such instantaneous responses. But networks can be designed to ensure that they respond sufficiently fast to preclude any substantive loss of input signal information. To these ends, the settling time, say t_s, is routinely specified as a desired maximum time required by the network to achieve an output response that achieves and remains within a designable, suitably small percentage of the steady state output response. Traditionally, this settling time is measured and/or computed for a step input excitation because step inputs force a network to respond abruptly at the time of signal application. By virtue of the classic final value theorem, step inputs conveniently allow a mathematical delineation of observable steady state outputs.[3]

In order to determine the settling time for the second order system whose time domain response is defined by Eq. (4-75), let the time function, $z(x)$, represent the difference between the steady state output (which is unity) and the actual response; namely,

$$z(x) = 1 - y_n(x) = \left(\frac{e^{-\zeta_{cl} x}}{\sin \theta}\right) \sin\left(x\sqrt{1 - \zeta_{cl}^2} + \theta\right). \quad (4\text{-}76)$$

The maxima associated with $y_n(x)$, and hence $z(x)$, in this relationship can be found by equating the derivative, $dz(x)/dx$, to zero. As is rendered transparent by Figs. (4.8) and (4.9), these maxima are periodic. In order to ensure that the output response is forever maintained within a specified percentage of its steady state value for all times $t \geq t_s$, it is essential to find the specific normalized time, say $x_s = \omega_{cl} t_s$, at which $dz(x)/dx = 0$ at the maximum magnitude of $z(x)$. This is to say that $|z(x_s)| \leq |z(x)|$ for all x, where it is understood that x_s is a value of the normalized time at which $dz(x)/dx \equiv 0$.

The determination of x_s in accordance with the preceding stipulations is a mathematically cumbersome, but certainly not impossible, task. The result can be shown to be

$$x_s = \frac{1}{\sqrt{1 - \zeta_{cl}^2}} \left[\pi + \tan^{-1}\left(\frac{\sqrt{1 - \zeta_{cl}^2}}{M + \zeta_{cl}}\right)\right]. \quad (4\text{-}77)$$

The substitution of this result into Eq. (4-76) yields

$$z(x_s) = \left(\frac{\sqrt{1 + 2M\zeta_{cl} + M^2}}{M}\right) e^{-\zeta_{cl} x_s}. \quad (4\text{-}78)$$

The error at hand is always positive in that it corresponds to the first positive maximum in the time domain plot of Fig. 4.8. For a right half plane zero lying at a frequency much larger then the frequency, ω_{cl}, M is large, and Eqs. (4-77) and (4-78) collapse respectively to

$$x_s \approx \frac{\pi}{\sqrt{1 - \zeta_{cl}^2}} + \frac{1}{M + \zeta_{cl}}, \quad (4\text{-}79)$$

and

$$z(x_s) \approx e^{-\zeta_{cl} x_s}. \quad (4\text{-}80)$$

Example 4.3. A second order feedback amplifier has a zero frequency loop gain of 15 and a right half plane zero whose frequency is roughly four times the undamped natural frequency of the closed loop characteristic polynomial. The subject amplifier is to be designed so that the output response to a unit step input settles to within 1% of steady state value within a time span no larger than 500 pSEC. Determine the requisite unity gain frequency of the open loop and the phase margin of the closed loop. Also, compute the frequencies of the open loop right half plane zero and the open loop nondominant pole.

Solution 4.3.

(1) The cornerstone of the design realization is the determination of the damping factor, ζ_{cl}, for the closed loop characteristic polynomial. This determination commands a solution of Eqs. (4-77) and (4-78) for $z(x_s) < 0.01$ (one percent settling error). Unfortunately, the requisite solution cannot be delineated in closed form when parameter M, which is 4 in this exercise, is finite and non-negligible. An iterative solution (which can be effected through MATLAB, MATHCAD, or comparable other software) shows that $\zeta_{cl} = 0.8275$ results in $z(x_s)$ slightly less than 0.01.

(2) With $\zeta_{cl} = 0.8275$ and $M = 4$, Eq. (4-77) gives a normalized settling time of $x_s = 5.802$. Since $x_s = \omega_{cl} t_s$ and $t_s \leq 500$ pSEC, ω_{cl} must be at least as large as $2\pi(1.847\,\text{GHz})$.

(3) Equation (4-52) can be solved for the requisite unity gain frequency, ω_u, of the open loop, provided the parameter, k_p, which represents the frequency of the nondominant open loop pole normalized to ω_u, is known. With ζ_{cl} found to be 0.8275, this parameter derives from Eq. (4-53), where by Eq. (4-73),

$$k_o = M\sqrt{k_p\left[1 + \frac{1}{T(0)}\right]}. \tag{E3-1}$$

The substitution of Eq. (E3-1) into Eq. (4-53) produces a relationship for closed loop damping factor in terms of parameter k_p that cannot by solved conveniently in closed form. However, the trusty computer readily provides $k_p = 3.673$ for $M = 4$ and $T(0) = 15$.

(4) With $k_p = 3.673$, $T(0) = 15$, and $\omega_{cl} = 2\pi(1.847\,\text{GHz})$, Eq. (4-52) delivers a required open loop unity gain frequency of

$$\omega_u = 2\pi(933.1\,\text{MHz}).$$

Using Eq. (4-73), $M = 4$ and $\omega_{cl} = 2\pi(1.847\,\text{GHz})$ yields for the frequency of the right half plane zero,

$$z_o = 2\pi(7.39\,\text{GHz}).$$

From Eq. (4-34), $k_p = 3.673$ and $\omega_u = 2\pi(933.1\,\text{MHz})$ delivers

$$p_{2o} = 2\pi(3.43\,\text{GHz})$$

for the frequency of the open loop nondominant pole. Finally, Eqs. (4-38) and (4-36) combine to give a phase margin of

$$\phi_{pm} = 72.7°.$$

Comments. This example illustrates that while the design-oriented computations for settling time and the critical frequencies of second order closed loop feedback networks are numerically intense, they are rendered straightforward by the analyses documented in preceding subsections. The example also shows that demanding settling time constraints place a substantial burden on the circuit designer. In particular, a right half plane zero at a frequency of almost 7.4 GHz is likely to prove challenging in deep submicron CMOS technology for which the forward device transconductance at minimal drain currents is only of the order of several millimhos. Moreover, a phase margin of almost 73° is a challenge in this monolithic technology, which is routinely comfortable with phase margins in the range of 50°-to-60°. This and the small transconductance challenges can, however, be addressed proactively through such design heroics as suitably compensated common gate cascode stages incorporated into traditional common source phase inverting amplifier cells.

References

1. B. Razavi, *Design of Analog CMOS Integrated Circuits* (McGraw-Hill Companies, Inc, New York, 2001), pp. 345–346.
2. J. J. D'Azzo and C. H. Houpis, *Feedback Control System Analysis and Synthesis* (McGraw-Hill Book Company, Inc., New York, 1960), chap. 14.
3. D. R. Cunningham and J. A. Stuller, *Circuit Analysis* (Houghton Miflin Company, Boston, 1995), pp. 502–503.

Exercises

Problem 4.1
The biasing of operational amplifiers, such as the unit used in Example 4.1, results in the flow into the inverting and noninverting input amplifier ports of small static currents, I_a and I_b. These so-called offset input currents are delineated in the feedback topology appearing in Fig. P4.1.

(a) Assuming that the operational amplifier sustains operation in its linear regime, derive an expression for the static output voltage, V_{odc}, in terms of the input offset currents, I_a and I_b. For this derivation, assume that the quiescent value of the signal voltage, V_s, is zero.

(b) Most commercially available operational amplifiers have offset currents that are reasonably well matched. If the currents, I_a and I_b are indeed equal, give a recommendation as to how the resistance, R, might be chosen to minimize the impact of offset currents on the static output port voltage.

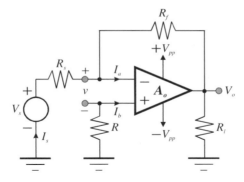

Figure P4.1.

Problem 4.2

The amplifier in Fig. 4.2(c) is modified by the addition of capacitance, C_f, in shunt with resistance R_f, as depicted in Fig. P4.2. Additionally, the open loop gain is not a simple constant but instead, it displays a dominant pole at a frequency, $p_{1o} = 2\pi\,(500\,\text{Hz})$. Accordingly, the constant, A_o, in Example 4.1 is replaced by the frequency variant function,

$$\frac{A_o}{1+\dfrac{s}{p_{1o}}}.$$

(a) Give general analytical expressions for the loop gain, feedback factor, and closed loop gain.

(b) Choose capacitor C_f so that a phase margin of approximately 70° is achieved.

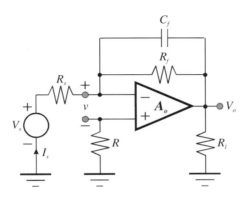

Figure P4.2.

(c) For the component selection discerned in Part (b), what are the closed loop damping factor and closed loop undamped natural frequency of oscillation?
(d) For the choice of capacitance C_f, in Part (b), at what time does the closed loop unit step response settle to within 5% of steady state value?

Problem 4.3

The operational amplifier in Fig. 4.2(c) has a nonzero output resistance, R_o, so that the resultant small signal equivalent circuit is the network displayed in Fig. P4.3.

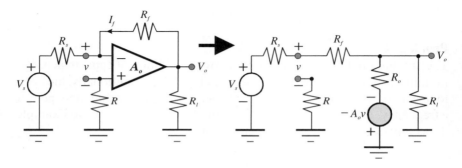

Figure P4.3.

(a) Derive expressions for the open loop gain, the loop gain, the feedback factor, and the closed loop gain.
(b) Simplify the expressions deduced in Part (a) for the case of a very large gain constant, A_o.
(c) Using the parameter values quoted in Example 4.1, and taking $R_o = 50\,\Omega$, numerically evaluate the metrics deduced in Part (a).

Problem 4.4

The op-amp in the amplifier of Fig. P4.4 has a large, but nonetheless finite, open loop gain, A_o, of 90,000, infinitely large input impedance at its inverting (−) and noninverting (+) input ports, and zero output port impedance. To first order, the static input/output (I/O) characteristic curve of the subject op-amp is as depicted in Fig. 4.2(b), where the applied biasing is $V_{pp} = 12$ volts. The circuit resistances, R_1 and R_2, are respectively, 1 KΩ and 9 KΩ. The Thévenin signal source resistance, R_s, is 75 Ω, and the resistance, R_l, terminating the amplifier output port is 500 Ω. Analyze the

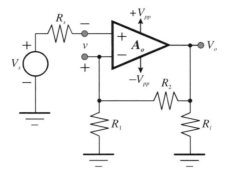

Figure P4.4.

feedback amplifier at hand to determine general expressions and numerical values for its loop gain, feedback factor, closed loop gain, and the maximum input signal amplitude commensurate with nominal I/O linearity of the op-amp for low signal frequencies. Use the basic op-amp model introduced in Example 4.1.

Problem 4.5

The feedback amplifier in Fig. P4.4 has a nonzero output resistance, R_o.

(a) Repeat Problem 4.4 by re-deriving expressions for the open loop gain, the loop gain, the feedback factor, and the closed loop gain.
(b) Simplify the expressions deduced in Part (a) for the case of a very large gain constant, A_o.
(c) Using the parameter values quoted in Problem 4.4, and taking $R_o = 50\,\Omega$, numerically evaluate the metrics deduced in Part (a).

Problem 4.6

An nth-order feedback amplifier delivers a loop gain, $T(s)$, given by

$$T(s) = \frac{T(0)}{\prod_{i=1}^{n}\left(1 + \dfrac{s}{p_i}\right)},$$

where the real parts of all pole frequencies, p_i, are positive numbers. Show that if the magnitude of the zero frequency loop gain, $T(0)$ exceeds one, but $T(0) < 0$, the subject feedback configuration is unstable, regardless of the values of all pole frequencies.

Problem 4.7

It can be shown that the loop gain of the Wien bridge oscillator depicted schematically in Fig. P4.7 is

$$T(s) = \left(\frac{sC_1}{1 + sR_1C_1}\right)\left[\frac{(1 - A_o)R_2}{1 + sR_2C_2}\right],$$

where A_o symbolizes the gain of the utilized amplifier (shown without requisite biasing), whose input impedance is infinitely large and whose output impedance is zero. The circuit is capable of sustaining a sinusoidal oscillation at radial frequency, ω_o. Determine the expression for ω_o and the amplifier gain condition commensurate with sustaining such oscillation.

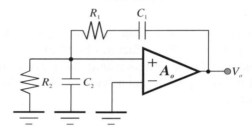

Figure P4.7.

Problem 4.8

For the Colpitts oscillator depicted in Fig. P4.8, the loop gain, $T(s)$, can be shown to be

$$T(s) = \frac{A_o + 1 + sR_o(C_1 + C_2)}{s^2 LC_1(1 + sR_oC_2)},$$

where A_o is the gain of the utilized amplifier (shown without requisite biasing), whose input impedance is infinitely large and whose output impedance

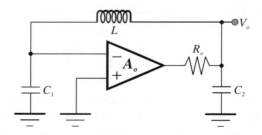

Figure P4.8.

is zero. The circuit is capable of sustaining a sinusoidal oscillation at radial frequency, ω_o. Determine the expression for ω_o and the condition commensurate with sustaining such oscillation.

Problem 4.9
For the Hartley oscillator depicted in Fig. P4.9, the loop gain, $T(s)$, can be shown to be

$$T(s) = \frac{s^2 R_o C (L_1 + L_2) + s^3 (A_o + 1) L_1 L_2 C}{R_o + s L_1},$$

where A_o is the gain of the utilized amplifier (shown without requisite biasing), whose input impedance is infinitely large and whose output impedance is zero. The circuit is capable of sustaining a sinusoidal oscillation at radial frequency, ω_o. Determine the expression for ω_o and the condition commensurate with sustaining such oscillation.

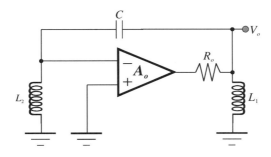

Figure P4.9.

Problem 4.10
Reconsider the feedback amplifier in Fig. P4.4.

(a) Derive a general expression for the indicated input port voltage, v, and show that this "error" voltage becomes progressively smaller for increasing loop gain.
(b) What closed loop gain results when the presumption, $v = 0$ is invoked *a priori*?

Problem 4.11
Frequency invariant feedback is applied globally around a second order open loop amplifier whose low frequency gain is 100 (40 dB). The closed loop is designed to establish unity gain at zero signal frequency. The unity

gain frequency of the open loop gain function is $2\pi(20\,\text{GHz})$, and the open loop is determined to have a right half plane zero at $z_o = 2\pi(60\,\text{GHz})$.

(a) Determine the frequency of the nondominant open loop pole such that the resultant closed loop establishes a nominally maximally flat magnitude frequency response.
(b) For the response discerned in Part (a), calculate the phase and gain margins.
(c) Plot the magnitude and phase responses of the loop gain to confirm the propriety of the gain and phase margin calculations.
(d) Plot the magnitude and phase responses of the loop gain to confirm the propriety of the gain and phase margin calculations.
(e) Compare the 3-dB frequency response bandwidths of the open and closed loops.

Problem 4.12
Repeat Problem 4.12 for the case in which the zero frequency closed loop gain is designed to be 5 (14 dB).

Problem 4.13
For the amplifier addressed in Problem 4.11, compute and plot the settling time as a function of settling errors over the range of 0.1%-to-10%.

Problem 4.14
An open loop amplifier is found to have a voltage transfer function, $A_v(s)$, of

$$A_v(s) = \frac{100}{\left(1 + \dfrac{s}{200}\right)\left(1 + \dfrac{s}{400}\right)\left(1 + \dfrac{s}{1000}\right)},$$

where the Laplace operator, "s", is understood to be in units of MHz. Feedback is applied globally around this open loop network to deliver a closed loop low frequency gain of one.

(a) Compute the gain and phase margins of the subject feedback configuration. Comment as to network quality from the perspective of closed loop stability.
(b) What are the open loop and resultant closed loop 3-dB bandwidths?

Problem 4.15
For the amplifier addressed in Problem 4.14, introduce a left half plane zero into the feedback factor while maintaining the desired unity closed loop gain

at low signal frequencies. What is the minimum frequency associated with the introduced left half plane zero that yields a 70° phase margin?

Problem 4.16
The loop gain of a second order network has a very large value at zero signal frequency, as well as a right half plane zero whose frequency is very large in comparison to the frequency at which the loop gain magnitude degrades to unity. Let parameter k_p be the frequency of the nondominant loop gain pole, normalized to the unity gain frequency. Derive in terms of k_p, a simplified expression for the settling time corresponding to a specified small unit step response error, say κ.

Chapter 5

Signal Flow Methods of Feedback Network Analysis

5.1.0. Introduction

In addition to delineating the classic two-port models that interrelate the input/output (I/O) volt-ampere characteristics of linear electrical and electronic circuits, Chapter 2 establishes a systematic procedure for studying feedback networks in terms of two-port model parameters. Although this straightforward procedure establishes a recipe for feedback circuit analysis, it suffers from at least three shortfalls. First, traditional two-port methods apply only to global feedback configurations; that is, circuit topologies whose feedback signal flow path initiates at the network output port and terminates at the network input port. This constraint is not conveniently adapted to routinely encountered electronic circuits for which intentional or parasitic feedback is applied locally between internal network ports or branches. Second, the aforementioned analytical recipe relies on a user ability to recognize the type of analog signal processing accomplished by the network undergoing study. This is to say that the type of two-port parameters (h-, g-, y-, or z-parameters) most appropriate to the determination of open loop gain, closed loop gain, loop gain, and I/O impedances or admittances depends on whether the network at hand functions as a voltage amplifier, current amplifier, transimpedance cell, or transadmittance cell. Such recognition is often obscured by topological complexities. It is also pragmatically unnecessary because an amplifying network optimally suited for a particular type of analog signal processing may actually be exploited for other dynamic purposes. For example, even though an operational amplifier to which an impedance element is connected between its output and inverting input ports is best suited for transimpedance amplification, this configuration is ubiquitously exploited for voltage amplification purposes. Finally, classic two-port methods of feedback network analysis are not convenient for a study of networks and systems in which multiple feedback loops are embedded.

This chapter mitigates the foregoing shortcomings by establishing a general analytical technique and circuit assessment procedure predicated on Kron's network partitioning theorem.[1,2] The relevant fruits of this powerful theorem, which were subsequently clarified and extended by Branin [3,4] and Rohrer,[5] unambiguously identify the open loop gain, loop gain, feedback factor, and feedforward factor for any linear feedback configuration, independent of its optimal or intended signal processing application. Even more notable is the fact that the partitioning-based approach to feedback circuit analysis, which has since become known as *signal flow theory*,[6,7] is germane to the study of multiloop feedback circuits and systems. The quantitative and definitive assessment of closed loop networks that incorporate multiple feedback paths is often perceived as daunting mathematical exercises principally because these relatively complex topologies cannot be conveniently studied by conventional two-port parametric methods.

5.2.0. Feedback Network Analysis Fundamentals

The development of a computationally efficient method for analyzing general feedback circuits commences with a consideration of the abstraction depicted in Fig. 5.1(a). In this diagram, a signal comprised of Thévenin voltage V_s connected in series with Thévenin impedance Z_s is applied to the network input port. In addition to this input signal voltage, a second signal, modeled as the current, I_c, is incident with an arbitrary terminal pair, which is referred to herewith as the cth port, of the network. The voltage and current inputs, V_s and I_c, manifest a voltage response, V_o, at the network output port, which is terminated in load impedance Z_l. The indicated current energy can be cast as a voltage source but in the interest of a semblance of generality, current excitation is chosen. At yet a fourth terminal pair, henceforth referenced as the kth intrinsic port, a voltage, V_k, is established in response to the two inputs, V_s and I_c. This kth port could, but need not, be coincident with either the network input port or output port.

Superposition concepts mandate

$$V_o = A_{os} V_s + Z_{oc} I_c \qquad (5\text{-}1)$$

and

$$V_k = A_{ks} V_s + Z_{kc} I_c, \qquad (5\text{-}2)$$

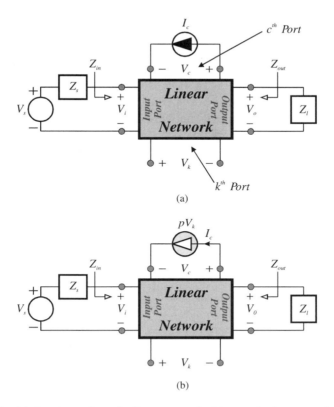

Figure 5.1. (a) An abstraction of a linear network driven at its input port by an independent voltage source and excited at an internal port by an independent current source. (b) The network in (a) with the independent current source, I_c, supplanted by a voltage controlled current source, pV_k. The controlling voltage, V_k, for this controlled current source is established at an arbitrary network port, which could be either the input or the output port.

where the proportionality factors, A_{os}, Z_{oc}, A_{ks}, and Z_{kc}, are constants that are independent of the excitation variables, V_s and I_c. Parameter A_{os} symbolizes a voltage gain from the source-to-the output port, and Z_{oc} is the transimpedance between the cth port and the output port. Moreover, A_{ks} represents a voltage gain from the signal source-to-the port at which voltage response V_k is extracted, while Z_{kc} is effectively a transimpedance between the two identified internal network ports.

Consider now the case in which current I_c is not an independent variable but instead, is a variable related to internal port voltage V_k in accordance with the simple linear relationship,

$$I_c = pV_k. \tag{5-3}$$

The immediate impact of the presumption reflected by this equation is that the current source, I_c, in Fig. 5.1(a) becomes the voltage controlled current source, pV_k, delineated in Fig. 5.1(b). A more subtle ramification of Eq. (5-3) is that the two internal network ports in the diagram of Fig. 5.1(a) are now coupled explicitly in that voltage V_k precipitates a current proportional to kth port voltage V_k that flows into the cth port where the original independent current, I_c, is incident. In effect, feedback, or possibly feedforward, is emulated from the kth port-to-the cth port. If feedback is indeed the applicable perspective, parameter p might logically be termed the *feedback factor* relevant to internal signal coupling from kth-to-cth ports. In the more general case in which feedforward from kth to cth ports is a possibility, p is more suitably referenced as a *critical parameter*. The combination of Eqs. (5-3) and (5-2) produces

$$I_c = \frac{pA_{ks}V_s}{1 - pZ_{kc}}. \tag{5-4}$$

Assuming that the gain parameter, A_{os}, is nonzero and finite, the insertion of Eq. (5-4) into Eq. (5-1) leads to the system voltage transfer function,

$$A_v(p, Z_s, Z_l) \triangleq \frac{V_o}{V_s} = A_{os} \left[\frac{1 - p\left(Z_{kc} - \dfrac{A_{ks}Z_{oc}}{A_{os}}\right)}{1 - pZ_{kc}} \right], \tag{5-5}$$

which is an explicit function of critical parameter p and an implicit function of both the source impedance, Z_s, and the load impedance, Z_l. Although this transfer function is developed for the voltage gain of a network for which local voltage-to-current feedback or feedforward prevails, the relationship is conceptually applicable to current gain, transconductance, or transimpedance gain, regardless of the nature of internal port coupling. Specifically, the feedback or feedforward from kth-to-cth ports can be, in addition to the considered voltage-to-current dynamic, current-to-voltage, voltage-to-voltage, or current-to-current.[8] In recognition of these contentions, Eq. (5-5) is more suitably expressed in the general form,

$$G(p, Z_s, Z_l) \triangleq \frac{\text{Output Response}}{\text{Signal Input}} \triangleq \frac{Y_o(s)}{X_s(s)} = G_{os} \left[\frac{1 + pQ_r(Z_s, Z_l)}{1 + pQ_s(Z_s, Z_l)} \right], \tag{5-6}$$

where $Y_o(s)$ designates the output response (voltage or current) to independent input signal excitation (voltage or current) $X_s(s)$. It is to be understood that the functions, $pQ_r(Z_s, Z_l)$ and $pQ_s(Z_s, Z_l)$, are dimensionless. These functions respectively denote the *null return ratio with respect to parameter p* and the *return ratio with respect to p*. In particular, the null return ratio with respect to p, $T_r(p, Z_s, Z_l)$, is

$$T_r(p, Z_s, Z_l) \triangleq pQ_r(Z_s, Z_l), \tag{5-7}$$

while

$$T_s(p, Z_s, Z_l) \triangleq pQ_s(Z_s, Z_l) \tag{5-8}$$

represents the return ratio with respect to parameter p. As is highlighted shortly, the return ratio with respect to a feedback parameter, p, is the effective loop gain of the closed loop network. On the other hand, the null return ratio with respect to feedback parameter p is a measure of I/O feedforward effects through the feedback signal flow path. A comparison of Eq. (5-6) with the voltage amplifier transfer function in Eq. (5-5) suggests that the so-called *normalized return ratio*, $Q_s(Z_s, Z_l)$, with respect to parameter p is

$$Q_s(Z_s, Z_l) = -Z_{kc}, \tag{5-9}$$

while the *normalized null return ratio*, $Q_r(Z_s, Z_l)$, with respect to p is

$$Q_r(Z_s, Z_l) = -\left(Z_{kc} - \frac{A_{ks}Z_{oc}}{A_{os}}\right) = Q_s(Z_s, Z_l) + \frac{A_{ks}Z_{oc}}{A_{os}}. \tag{5-10}$$

Equations (5-7) and (5-8) permit writing Eq. (5-6) as

$$G(p, Z_s, Z_l) \triangleq \frac{Y_o(s)}{X_s(s)} = G_{os}\left[\frac{1 + T_r(p, Z_s, Z_l)}{1 + T_s(p, Z_s, Z_l)}\right]. \tag{5-11}$$

This result gives rise to the network block diagram offered in Fig. 5.2. Note from this diagram and also from Eqs. (5-7), (5-8), and (5-11) that

$$G_{os} = G(0, Z_s, Z_l); \tag{5-12}$$

that is, the parameter, G_{os}, is the value of the input-to-output transfer function under the condition of a critical parameter null ($p = 0$), which reflects zero feedback or feedforward from the kth-to-the cth network ports. For this reason, G_{os} is often called the *null parameter gain* or the *zero feedback parameter gain* of the considered network. Moreover, observe that the feedback function applied around the block designated as having gain G_{os} is proportional to the return ratio with respect to parameter p and that

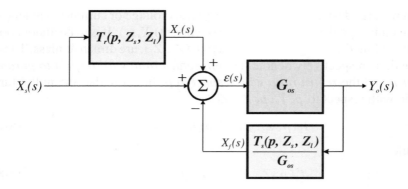

Figure 5.2. The signal flow block diagram of the feedback configuration given in Fig. 5.1.

the loop gain associated with this feedback path is identically said return ratio, $T_s(p, Z_s, Z_l)$.

Finally, the diagram at hand reveals that the error signal, $\epsilon(s)$, activating the input port of the null gain block is the algebraic sum of three signals; namely,

$$\epsilon(s) = X_s(s) + X_f(s) + X_r(s). \tag{5-13}$$

as delineated in Fig. 5.2. The first of these three signals is the actual source energy, and the second component is the feedback signal,

$$X_f(s) = \left[\frac{T_s(p, Z_s, Z_l)}{G_{os}}\right] Y_o(s), \tag{5-14}$$

which is proportional to the output response, $Y_o(s)$. In addition to these two typically encountered signals, a feedforward signal component,

$$X_r(s) = [T_r(p, Z_s, Z_l)] X_s(s), \tag{5-15}$$

is fomented as a linearly scaled signal source transform. If $T_r(p, Z_s, Z_l) = 0$, no feedforward phenomena prevail, and the block diagram in Fig. 5.2 collapses to the traditional forms set forth in Chapter 2.

5.2.1. Calculation of Feedback Network Parameters

It is important to underscore the fact that Eq. (5-11) is a transfer relationship that applies to any general linear feedback circuit or system. An inspection of the subject expression reveals that four signal flow parameters are required for its utilization. The first of these is the critical parameter, p, which the circuit analyst selects as the linear metric to which observable feedback or feedforward is clearly proportional. The selection process for p is clarified by several examples that follow. For the moment, suffice it to say that the identification process embodies a two-step procedure. First, the observed feedback or feedforward (usually feedback in commonly encountered electronics) subcircuit is represented by a two-port parameter model that emulates the electrical properties of the signal flow path linking kth and cth internal network ports. This two-port equivalent circuit is invariably a simple mathematical structure because, as justified in a previous chapter, the applied feedback is ordinarily a passive network. Second, the feedback revealed by the two-port model takes the form of a controlled source for which p becomes the associated voltage gain, current gain, transconductance, or transimpedance multiplier.

The remaining three parameters indigenous to Eq. (5-11) are the $p = 0$ value of gain (G_{os}), the normalized return ratio, $Q_s(p, Z_s, Z_l)$, and the normalized null return ratio, $Q_r(p, Z_s, Z_l)$. A computationally efficient methodology for enumerating these three metrics can be forged by using the original voltage amplifier in Fig. 5.1 as an analytical crutch. Note in this diagram that parameter p is a transconductance parameter identified as the gain associated with a voltage controlled current source that emulates feedback from the kth-to-cth network ports.

5.2.1.1. Null Parameter Gain

Figure 5.3(a) displays the voltage amplifier shown in Fig. 5.1(b) for which transconductance feedback in the form of the voltage controlled current source, pV_k, is applied from the kth to the cth internal network ports. Recalling Eq. (5-5), the null parameter gain, A_{os}, is identical to the input-to-output gain when the feedback parameter, p, is forced to zero. Accordingly, forcing $p = 0$ is tantamount to constraining $pV_k = 0$, with the result that the signal source-to-output port voltage gain is the desired metric, A_{os}, as is suggested in Fig. 5.3(b). Specifically,

$$A_{os} = A_v(0, Z_s, Z_l) = \left.\frac{V_o}{V_s}\right|_{pV_k=0} = G_{os}. \tag{5-16}$$

More generally, the null signal flow metric evaluated herewith is the gain, G_{os}, in accordance with Eq. (5-12). It should be understood that while the original network in Fig. 5.3(a) is characterized by a driving point input impedance, Z_{in}, and a driving point output impedance of Z_{out}, a feedback

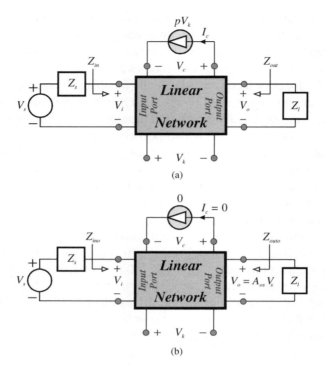

Figure 5.3. (a) The voltage amplifier of Fig. 5.1(b). Observe the presence of local feedback from kth-to-cth internal network ports. (b) The network in (a) configured to evaluate the null parameter voltage gain, A_{os}. Note that the feedback metric, parameter p, is set to zero, which forces the current, I_c, flowing into the cth network port to zero.

parameter null can be expected to alter these impedance levels to Z_{ino} and Z_{outo}, respectively, as is suggested in Fig. 5.3(b). These impedances are fundamental to the determination of the true impedance levels, Z_{in}, and Z_{out}, as is addressed shortly.

5.2.1.2. Normalized Return Ratio

From Eq. (5-9), the normalized return ratio with respect to parameter p for the voltage amplifier at hand is the negative of the transimpedance parameter, Z_{kc}. By Eqs. (5-2) and (5-9), $-Z_{kc}$, which is the normalized return ratio, $Q_s(Z_s, Z_l)$, for the subject voltage amplifier, derives as

$$-Z_{kc} = -\left.\frac{V_k}{I_c}\right|_{V_s=0} = \left.\frac{V_{kr}}{I_c}\right|_{V_s=0} = Q_s(Z_s, Z_l), \qquad (5\text{-}17)$$

where V_{kr} is inserted as the negative of V_k in an attempt to circumvent routinely encountered character flaws that drop minus signs inadvertently. This transaction suggests a four-step computational procedure. First, the signal source voltage, V_s, applied to the original voltage amplifier, which is redrawn for convenience in Fig. 5.4(a), is set to zero. Second, the controlled current source, pV_k, used to model internal network feedback, is supplanted by an independent current source, I_c, that flows in the same direction as does pV_k. Third, the stipulated polarity of the original kth port voltage, V_k, is reversed to account for the negative sign in Eq. (5-17). The resultant reversed polarity port voltage is symbolized by V_{kr}. Finally, the applicable network model is drawn as diagrammed in Fig. 5.4(b), and circuit analyses

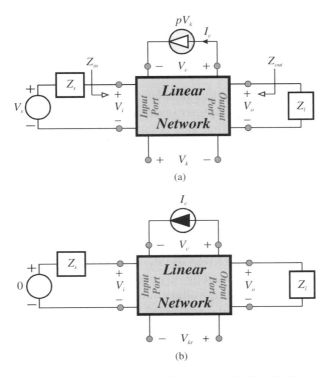

Figure 5.4. (a) The original feedback amplifier of Fig. 5.1(b). (b) The network in (a) configured to evaluate the normalized return ratio. The source signal, V_s, is reduced to zero, the controlled source, pV_k, is replaced by an independent source of current, I_c, and the original controlling voltage, V_k, is supplanted by its phase-reversed counterpart, V_{kc}. The desired normalized return ratio is the voltage-to-current ratio, V_{kc}/I_c.

are executed to evaluate the required normalized return ratio with respect to parameter p.

5.2.1.3. Normalized Null Return Ratio

In Eq. (5-1), constraining the output port voltage to zero requires a source voltage, V_s, of

$$V_s = -\frac{Z_{oc} I_c}{A_{os}}. \tag{5-18}$$

The insertion of this result into Eq. (5-2) leads to

$$V_k = -A_{ks}\left(\frac{Z_{oc}}{A_{os}}\right) I_c + Z_{kc} I_c, \tag{5-19}$$

whereupon,

$$-\frac{V_k}{I_c} = -Z_{kc} + \left(\frac{A_{ks} Z_{oc}}{A_{os}}\right), \tag{5-20}$$

which Eq. (5-10) confirms as identical to the desired normalized null return ratio, $Q_r(Z_s, Z_l)$, of the amplifier undergoing study. As in the case of the normalized return ratio, a four-step computational procedure is suggested. First, the dependent current source, pV_k, in the original amplifier, which is redrawn once again in Fig. 5.5(a), is replaced by an independent current source, I_c, whose directional flow mirrors that of pV_k. Second, the output response, which is presently the voltage, V_o, is held fast at zero while maintaining a nonzero signal source voltage, V_s. It must be understood that the output port is not short circuited to achieve $V_o = 0$, for such an action effectively adds a branch to the network undergoing study, thereby changing the network topology. Moreover, the signal source voltage, V_s, must not be reduced to zero for its value implicitly assumes that set forth by Eq. (5-18). Third, the original controlling voltage variable, V_k, is replaced by a voltage, V_{kr}, which is simply the phase-reversed value of V_k. Finally, the resultant computational model, which is depicted in Fig. 5.5(b), is analyzed within the constraint imposed by $V_o = 0$ to discern the ratio, V_{kr}/I_c. This voltage-to-current ratio is, in fact, the normalized null return ratio, $Q_r(Z_s, Z_l)$, with respect to feedback parameter p for the original voltage amplifier.

Example 5.1. Figure 5.6(a) is the low frequency, linear equivalent circuit of a transimpedance feedback amplifier realized in metal-oxide-semiconductor (MOS) device technology. The structure is actually the approximate, low frequency, small

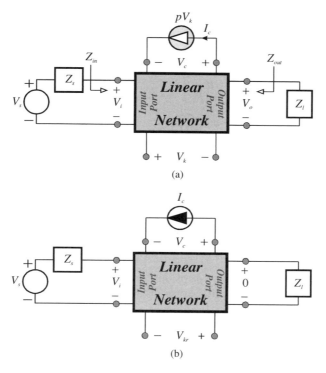

Figure 5.5. (a) The original feedback amplifier of Fig. 5.1(b). (b) The network in (a) configured to evaluate the normalized null return ratio. The response voltage, V_o, is clamped to zero, while maintaining a nonzero signal source voltage [which doubtlessly differs from its original value used in (a)]. The controlled source, pV_k, is replaced by an independent current source, I_c, and the original controlling voltage, V_k, is supplanted by its phase-reversed value, V_{kc}. The desired normalized null return ratio is the ratio, V_{kc}/I_c.

signal model of the amplifier depicted in Fig. 5.6(b). In the model, $R_s(600\,\Omega)$ is the Thévenin equivalent resistance of the signal source voltage, V_s, $R_{l1}(1\,\text{K}\Omega)$ and $R_{l2}(1\,\text{K}\Omega)$ are split drain resistances that collectively terminate the output port, and $C_l(500\,\text{fF})$ is a capacitive load that terminates the output port to ground. The resistance, $R_i(10\,\text{K}\Omega)$ is required for transistor biasing, and resistance $R_f(6.8\,\text{K}\Omega)$ is appended between the gate and the junction of the two load resistors to implement voltage feedback. Observe that the implemented feedback is not global since the feedback resistance, R_f, is not incident with the output port where voltage response V_o is extracted. Finally, parameter $g_m(50\,\text{mmho})$ is the forward transconductance of the MOS device. Subsequent to identifying the critical feedback parameter for the circuit, develop expressions for and compute the low frequency normalized return ratio, return ratio, normalized null return ratio, and null return ratio with respect

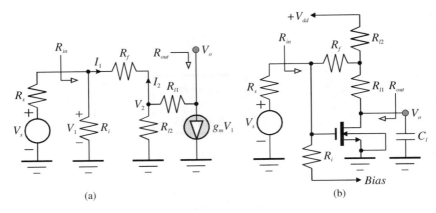

Figure 5.6. (a) The approximate low frequency, small signal equivalent circuit of a feedback amplifier realized in MOS device technology. (b) The schematic diagram of the feedback amplifier corresponding to the model in (a).

to this parameter. Also, derive expressions for and compute the low frequency null parameter I/O voltage gain, and the overall low frequency closed loop voltage gain V_o/V_s.

Solution 5.1.

(1) The first step to the solution process is, as silly as it might appear at first blush, to model the feedback element, R_f, with an appropriate two-port parameter equivalent circuit. Any two-port parameter model can be invoked. Since the subject resistance is in shunt with the input port of the amplifier and is also in shunt with the node at which voltage V_2 is developed, as shown in Fig. 5.6(a), short circuit admittance (y-) parameters are most suitable. To this end, resistance R_f is extricated from the diagram in Fig. 5.6(a) and shown in Fig. 5.7(a), together with its terminal voltages, V_1 and V_2, and input and output currents, I_1 and I_2. By inspection of the diagram in Fig. 5.6(a),

$$I_1 = -I_2 = \frac{V_1 - V_2}{R_f}, \tag{E1-1}$$

which leads immediately to the y-parameter equivalent circuit diagrammed in Fig. 5.7(b). This equivalent circuit may now be used to represent the resistive element in the model of Fig. 5.6(a), thereby resulting in the alternative amplifier model shown in Fig. 5.7(c). Aside from confirming student suspicions that academicians relish complicating simple phenomena, the procedure just invoked boasts an enormous advantage. In particular, it serves to reveal the feedback parameter, p, as the $1/R_f$ conductance multiplier of the voltage controlled current source, V_2/R_f, as is highlighted in the figure at hand. It must

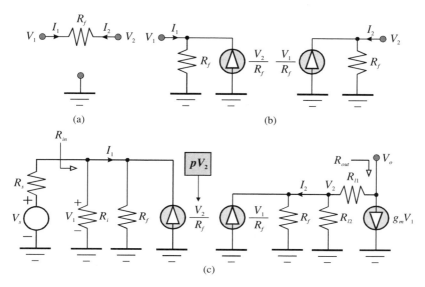

Figure 5.7. (a) The feedback resistor in the amplifier model of Fig. 5.6(a). (b) Short circuit admittance parameter model of the feedback resistor in (a). (c) The amplifier model in Fig. 5.6(a) modified through replacement of the resistive feedback element by its y-parameter model in (b).

be understood that this generator is indeed the feedback vehicle in the amplifier because it generates an input port current component, V_2/R_f, that is controlled by a voltage, V_2, which is proportional to the output port response, V_o, of the unit.

The alternative representation of the feedback resistance has two other attributes. One of these confirms that in addition to feedback, feedforward is evidenced because of the V_1/R_f controlled source established at the output port. The feedforward factor, $1/R_f$, is identical to the feedback factor, which is to be expected in light of the inherently bilateral nature of the feedback resistance, R_f. To be sure, the engineering significance of feedforward through R_f is debatable because in effect, the feedforward generator acts in tandem with the controlled source, $g_m V_1$, promoted by the transistor. Since the transistor is exploited expressly to realize forward gain in the overall amplifier, it is only natural to expect that the impact of $g_m V_1$, which emulates feedforward through the transistor, is far greater than that associated with the feedforward, V_1/R_f, through the feedback resistance.

A second additional bonus of the two-port approach is the clarion indication that the simple feedback element incurs additional loading at both the input and output amplifier ports. To this end, note the shunting resistance, R_f, at these ports.

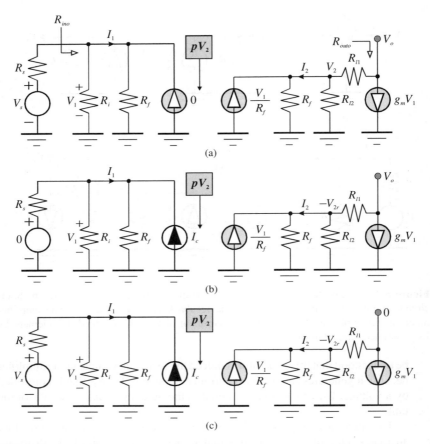

Figure 5.8. (a) The model in Fig. 5.7(c) configured for the evaluation of the low frequency null parameter voltage gain, $A_{os} = V_o/V_s$ of the amplifier in Fig. 5.6(b). Note that the feedback generator, V_2/R_f is reduced to zero. (b) The model in Fig. 5.7(c) configured for the computation of the low frequency normalized return ratio, $Q_s(R_s, \infty) = V_{2c}/I_c$. The signal source voltage is set to zero, the polarity of the controlling voltage, V_2, is reversed, and the feedback controlled source is replaced by an independent current generator. (c) The model in Fig. 5.7(c) configured for the computation of the low frequency normalized null return ratio, $Q_r(R_s, \infty) = V_{2c}/I_c$. In this analysis, the output voltage response is held to zero.

(2) The low frequency null parameter voltage gain, $A_{os} = V_o/V_s$, can now be evaluated with the help of Fig. 5.8(a), which exploits the fact that capacitance C_l emulates an open circuit at low signal frequencies. This figure is the diagram in Fig. 5.7(c), but with the V_2/R_f controlled generator set to zero; in other words, $p = 0$. It is crucial to note that A_{os} is *not* computed with the feedback resistance removed, since the feedforward generator, V_1/R_f, and the shunt R_f

resistive elements at both the input and output ports remain in tow. Instead, A_{os} derives from the removal of only the actual feedback component partitioned from the feedback resistance. From the diagram in Fig. 5.8(a),

$$V_1 = \left(\frac{R_i \| R_f}{R_i \| R_f + R_s}\right) V_s, \quad (E1\text{-}2)$$

and

$$V_o = -g_m R_{l1} V_1 + (R_{l2} \| R_f)\left(\frac{V_1}{R_f} - g_m V_1\right). \quad (E1\text{-}3)$$

The combination of the last two Kirchhoff relationships leads forthwith to

$$A_{os} = \frac{V_o}{V_s} = -\left[g_m R_{l1} + (g_m R_f - 1)\left(\frac{R_{l2}}{R_{l2} + R_f}\right)\right]\left(\frac{R_i \| R_f}{R_i \| R_f + R_s}\right). \quad (E1\text{-}4)$$

Numerically, this metric is $A_{os} = -81.4$, or $A_{os} = 38.2$ dB. Since feedforward through the transistor is expected to be much larger than feedforward through the resistive element, $R_f, g_m R_f \gg 1$, which is satisfied in the present example. Moreover, if the biasing resistance, R_i, is large enough to enable the assumption, $R_i \gg R_f$, which is not a valid simplification in this particular example,

$$A_{os} \approx -g_m[R_{l1} + (R_{l2} \| R_f)]\left(\frac{R_f}{R_f + R_s}\right) = 86.0, \quad (E1\text{-}5)$$

which is only about 5.6% larger than the true value deriving from Eq. (E1-4).
(3) The expression for the low frequency normalized return ratio derives from an analysis of the model given in Fig. 5.8(b). This metric is, in this case, an explicit function of only the Thévenin source resistance, R_s, since a capacitance terminates the output port, and only low signal frequencies are of immediate interest. Hence, the normalized return ratio is herewith symbolized as $Q_s(R_s, \infty)$, as opposed to the traditional notation, $Q_s(R_s, Z_l)$, since Z_l (the capacitive impedance connected directly across the output port) is effectively infinitely large. In this structure, the original V_2/R_f feedback generator is replaced by a constant current, I_c, flowing in the same direction as does the original feedback source. Additionally, the signal source voltage, V_s, is constrained to zero, and the polarity of the controlling voltage, V_2, for the feedback current source is reversed. In the subject diagram, this reversed polarity voltage is delineated as the node voltage, $-V_{2r}$. From the diagram in Fig. 5.8(b),

$$V_1 = (R_s \| R_i \| R_f) I_c, \quad (E1\text{-}6)$$

and

$$-V_{2r} = (R_{l2} \| R_f) \left(\frac{V_1}{R_f} - g_m V_1 \right), \tag{E1-7}$$

whence

$$Q_s(R_s, \infty) = \left(\frac{R_{l2}}{R_{l2} + R_f} \right) (R_s \| R_i \| R_f)(g_m R_f - 1) = 22.71 \text{ K}\Omega. \tag{E1-8}$$

It follows that the return ratio, $T_s(G_f, R_s, \infty)$ (with G_f representing the critical parameter conductance associated with R_f) is

$$T_s(G_f, R_s, \infty) = \left(\frac{R_{l2}}{R_{l2} + R_f} \right) \left(\frac{R_s \| R_i}{R_s \| R_i + R_f} \right) (g_m R_f - 1) = 3.34.$$

$$\tag{E1-9}$$

(4) Figure 5.8(c) pertains to the computation of the low frequency normalized null return ratio, $Q_r(R_s, \infty)$. As in the diagram of Fig. 5.8(b), the V_2/R_f feedback generator is supplanted by an independent current source, I_c. But unlike the aforementioned diagram, the signal source voltage is not set to zero. Instead, the output voltage response, V_o, is held null. The circuit model in Fig. 5.8(c) confirms

$$-V_{2r} = g_m R_{l1} V_1 \tag{E1-10}$$

and

$$-\frac{V_{2r}}{R_{l2} \| R_f} - \frac{V_1}{R_f} + g_m V_1 = 0. \tag{E1-11}$$

The substitution of Eq. (E1-10) into Eq. (E1-11) leads to the conclusion that $V_1 \equiv 0$. It follows from Eq. (E1-10) that $V_{2r} = 0$, whence

$$\left. \begin{array}{c} Q_r(R_s, \infty) = 0 \\ T_r(G_f, R_s, \infty) = 0 \end{array} \right\}. \tag{E1-12}$$

(5) The closed loop voltage gain of the amplifier undergoing investigation now follows directly from Eq. (5-11). In particular,

$$A_v(G_f, R_s, \infty) = \frac{V_o}{V_s} = A_{os} \left[\frac{1 + T_r(G_f, R_s, \infty)}{1 + T_s(G_f, R_s, \infty)} \right] = -18.76. \tag{E1-13}$$

or 25.5 dB.

Comments. The most challenging part of the problem solution methodology is the identification of the critical feedback parameter, which requires recognition of the feedback element or subcircuit, two-port parameter modeling of this feedback structure, and the incorporation of the two-port model into the equivalent circuit of the configuration undergoing study. Following this step, the analysis is effectively partitioned into three distinct circuit analyses. First, the gain is evaluated with the identified feedback parameter set to zero. Second, the normalized return ratio with respect to the feedback parameter is computed. Finally, the normalized null return ratio with respect to the critical parameter is evaluated. Accordingly, the computational procedure supplants the need to analyze a relatively cumbersome circuit topology with analyses executed on three simpler structures. Aside from this computational expediency, the signal flow method also forges design insights. For example, the foregoing solution reveals a feedback amplifier of only marginal quality in that the loop gain, $T_s(G_f, R_s, \infty)$, is relatively small. Accordingly, the implemented feedback cannot markedly desensitize the closed loop gain with respect to parameters embedded in the open loop gain expression, A_{os}.

5.2.2. Input and Output Impedances

The computational efficiency afforded by the circuit partitioning strategy outlined and exemplified in the preceding subsection is reinforced by the fact that the feedback metrics introduced at this juncture expedite the determination of closed loop I/O impedance levels, as well as closed loop gain. As is demonstrated herewith, the fruits of the analyses executed to evaluate the forward gain of a linear network can be exploited with but trivial modifications to the problem of determining the driving point input port and output port impedances. Aside from streamlining requisite analyses, the partitioning approach serves to reveal unambiguously the system level impact of various types of feedback on impedance levels, thereby conveying enhanced insights for circuit design purposes.

5.2.2.1 *Driving Point Input Impedance*

For the feedback amplifier abstracted in Fig. 5.1(b), the driving point input impedance, Z_{in}, is the voltage to current ratio, V_x/I_x, in the model offered in Fig. 5.9(a). Since the desired impedance is the effective impedance driven by the applied signal source in Fig. 5.1(b), observe in Fig. 5.9(a) that the entire Thévenin equivalent circuit of the signal source is supplanted by the current generator, I_x. The ratio of the indicated voltage, V_x, across this current source-to-I_x exploits Ohm's law to define quantitatively the input impedance.

The driving point input impedance expression mirrors a conventional transfer function for which the input excitation variable is current I_x and the output response is the voltage, V_x, which is the voltage resultantly developed across, and in disassociated polarity with, I_x. Accordingly, the mathematical

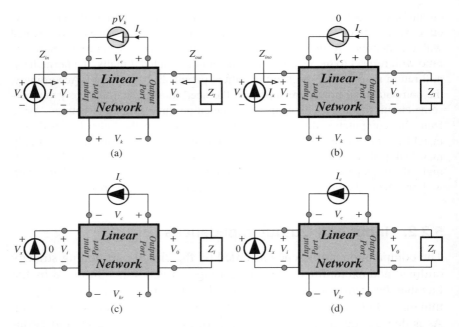

Figure 5.9. (a) The amplifier model used to determine the driving point input impedance, Z_{in}, as the transfer function, V_x/I_x. (b) The model appropriate to the determination of the null input impedance, Z_{ino}, where critical parameter p in the diagram of (a) is set to zero. (c) The equivalent circuit used to evaluate the normalized return ratio, $Q_{si}(Z_l)$, as the transimpedance, V_{kr}/I_c. Observe that the signal source, I_x, used to activate the input port for purposes of determining the driving point input impedance, is set to zero. (d) The equivalent circuit used to evaluate the normalized null return ratio, $Q_{ri}(Z_l)$, as the transfer function, V_{kr}/I_c. The voltage response, V_x, to the aforementioned signal source, I_x, is held fast at zero.

concepts implicit to the generalized gain relationship in Eq. (5-6) can be exploited to write the subject impedance relationship in the form,

$$Z_{\text{in}}(p, Z_l) \triangleq Z_{\text{in}} = Z_{ino}\left[\frac{1 + pQ_{ri}(Z_l)}{1 + pQ_{si}(Z_l)}\right], \tag{5-21}$$

where Z_{ino} is the $p = 0$ value of Z_{in} and is therefore the null input impedance. With reference to the critical parameter, p, $Q_{ri}(Z_l)$ is the normalized null return ratio pertinent to the input impedance calculation, and $Q_{si}(Z_l)$ is the normalized return ratio, also pertinent to the calculation of Z_{in}. It is worthwhile noting that both the normalized null return ratio and the normalized return ratio are independent of the Thévenin source impedance, Z_s, because the driving point input impedance is the effective load "seen" by the source and is therefore necessarily independent of Z_s.

The logical first step in the computational process for evaluating the driving point input impedance is the determination of the null input impedance, Z_{ino}. To this end, the diagram in Fig. 5.9(b) is the pertinent model, where the critical transconductance parameter, p is set to zero, thereby constraining the feedback controlled source, pV_k, to zero. A conventional circuit analysis, presumably simplified by virtue of the fact that $pV_k = 0$, is conducted to find Z_{ino} as

$$Z_{ino} = Z_{in}(0, Z_l) = \left.\frac{V_x}{I_x}\right|_{p=0}. \tag{5-22}$$

In the process of evaluating the null forward gain, A_{os}, in accordance with the network model of Fig. 5.3(b), parameter p is likewise zeroed, with the result that the input impedance indicated in Fig. 5.3(b) is properly delineated as Z_{ino}.

The normalized return ratio, $Q_{si}(Z_l)$, derives from replacing the feedback voltage controlled current source, pV_k, by the independent current, I_c, as shown in Fig. 5.9(c). Additionally, the polarity of the original controlling voltage, V_k, is reversed, with the result that this controlling variable becomes the indicated voltage, V_{kr}. Finally, the independent signal source that energizes the network undergoing investigation is set to zero. In the present case, this signal source is the applied current, I_x, whose null value is tantamount to an open-circuited input port. The pertinent model is depicted in Fig. 5.9(c), which projects an overt topological similarity with the circuit structure in Fig. 5.4(b). Recall that Fig. 5.4(b) is invoked to calculate the normalized return ratio, $Q_s(Z_s, Z_l)$, pertinent to the forward gain of the considered amplifier. Interestingly, the model in Fig. 5.4(b) collapses to that of Fig. 5.9(c) if the source impedance, Z_s, in the former structure is made infinitely large, thereby reflecting the input port open circuit implied by the constraint, $I_x = 0$. It follows that

$$Q_{si}(Z_l) \equiv Q_s(Z_s, Z_l)|_{Z_s=\infty} = Q_s(\infty, Z_l). \tag{5-23}$$

The fortuitous implication of Eq. (5-23) is that no additional circuit analysis is required to determine the normalized null return ratio pertinent to an input impedance evaluation. Instead, the normalized return ratio presumably found in the process of calculating the forward network gain need only be modified by allowing the source impedance in its mathematical relationship to assume infinitely large value.

The normalized null return ratio, $Q_{ri}(Z_l)$, likewise requires the replacement of the feedback voltage controlled current source, pV_k, by the independent current, I_c, as shown in Fig. 5.9(d). As in the case of $Q_{si}(Z_l)$, the controlling voltage, V_k, is reversed, thereby establishing the indicated voltage, V_{kr}. The normalized null return ratio mandates that the response to the applied source of energy be constrained to zero. Presently, this response is the voltage, V_x, whose null value is tantamount to a short-circuited input port. The pertinent model is offered in Fig. 5.9(d), which is topologically similar to the circuit structure in Fig. 5.4(b). The model in question reduces to that of Fig. 5.9(d) if the source impedance, Z_s, in the Fig. 5.4(b) is set to zero, thereby mirroring the input port short circuit implied by the constraint, $V_x = 0$. Accordingly,

$$Q_{ri}(Z_l) \equiv Q_s(Z_s, Z_l)|_{Z_s=0} = Q_s(0, Z_l). \tag{5-24}$$

It is worthwhile underscoring that $Q_{ri}(Z_l)$ bears no direct relationship to the normalized null return ratio, $Q_r(Z_s, Z_l)$, since this metric mandates the invocation of a null output response for the overall original network. In Fig. 5.9(b), it should be noted that the output response variable, V_o, is left untouched.

The upshot of the foregoing matters is that in general, the driving point input impedance of a linear network is, using Eqs. (5-21), (5-23), and (5-24),

$$Z_{in}(p, Z_l) \triangleq Z_{in} = Z_{ino}\left[\frac{1+pQ_s(0, Z_l)}{1+pQ_s(\infty, Z_l)}\right] = Z_{ino}\left[\frac{1+T_s(p, 0, Z_l)}{1+T_s(p, \infty, Z_l)}\right]. \tag{5-25}$$

The implication of this result is that on the assumption that the forward gain and all of its associated feedback metrics have already been evaluated, only one circuit analysis is required to enumerate the driving point impedance. In particular, the null impedance, Z_{ino}, whose calculation is invariably simplified by the requirement that parameter p be set to zero, is the only "new" network parameter.

5.2.2.2 Driving Point Output Impedance

The methodology for determining the driving point output impedance, Z_{out}, facing the load termination in Fig. 5.1(b) is analogous to the input impedance evaluation documented in the preceding subsection. The outline for this computational procedure, which derives conceptually from the general relationship,

$$Z_{out}(p, Z_s) \triangleq Z_{out} = Z_{outo}\left[\frac{1+pQ_{ro}(Z_s)}{1+pQ_{so}(Z_s)}\right], \quad (5\text{-}26)$$

is portrayed symbolically in Fig. 5.10. It must be understood that in the course of evaluating this driving point output impedance, the independent signal source, V_s, is reduced to zero, thereby leaving the input network port terminated in only the Thévenin source impedance, Z_s, as depicted in Fig. 5.10(a). Moreover, the subject impedance level is independent of the load impedance, Z_l, which means that the only extrinsic impedance to which

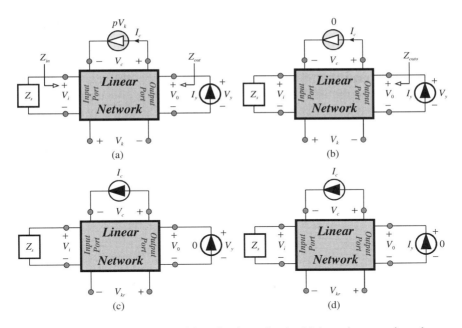

Figure 5.10. (a) The amplifier model used to determine the driving point output impedance, Z_{out}, as the transfer function, V_y/I_y. (b) The model appropriate to the determination of the null output impedance, Z_{outo}, where critical parameter p in the diagram of (a) is set to zero. (c) The equivalent circuit used to evaluate the normalized return ratio, $Q_{so}(Z_s)$, as the transimpedance, V_{kr}/I_c. Observe that the signal source, I_y, used to activate the output port for purposes of determining the driving point output impedance, is set to zero. (d) The equivalent circuit used to evaluate the normalized null return ratio, $Q_{ro}(Z_s)$, as the transfer function, V_{kr}/I_c. The voltage response, V_y, to the aforementioned signal source, I_y, is clamped to zero.

the null output impedance, Z_{outo}, the normalized null return ratio, $Q_{ro}(Z_s)$, and the normalized return ratio, $Q_{so}(Z_s)$, are functionally dependent is Z_s.

As implied by the diagram in Fig. 5.10(b), the computational procedure initiates with the evaluation of the null output impedance, Z_{outo}. This calculation entails a circuit analysis conducted to determine the voltage-to-current ratio, V_y/I_y, under the constraint of parameter p set to zero. In Fig. 5.10(c), the dependent current source, pV_k, is replaced by the independent current source, I_c, and the controlling voltage, V_k, is supplanted by its negative counterpart, V_{kr}. In concert with these manipulations, the applied independent current generator, I_y, is set to zero, an action that results in an effective open circuit at the output port. The desired normalized return ratio for output impedance purposes then derives as the ratio, V_{kr}/I_c. Since this analytical scheme mirrors that adopted in Fig. 5.4(b) for the special case of infinitely large load impedance, it follows that

$$Q_{so}(Z_s) \equiv Q_s(Z_s, Z_l)|_{Z_l=\infty} = Q_s(Z_s, \infty). \tag{5-27}$$

On the other hand, the evaluation of $Q_{ro}(Z_s)$ in Fig. 5.10(d) requires that the response, V_x, to the applied current excitation, I_x, be constrained to zero. This action results in a short-circuited output port, whence by comparison of the figure at hand with Fig. 5.4(b),

$$Q_{ro}(Z_s) \equiv Q_s(Z_s, Z_l)|_{Z_l=0} = Q_s(Z_s, 0). \tag{5-28}$$

It follows that the driving point output impedance of any linear network is given by

$$Z_{\text{out}}(p, Z_s) \triangleq Z_{\text{out}} = Z_{outo}\left[\frac{1 + pQ_s(Z_s, 0)}{1 + pQ_s(Z_s, \infty)}\right]$$

$$= Z_{outo}\left[\frac{1 + T_s(p, Z_s, 0)}{1 + T_s(p, Z_s, \infty)}\right]. \tag{5-29}$$

Example 5.2. Example 5.1 focuses on the problem of determining general expressions and numerical results for the forward gain characteristics of the amplifier offered in Fig. 5.6. Reconsider this same amplifier but now, find the low frequency driving point input impedance, Z_{in}, the low frequency driving point output impedance, Z_{out}, and the 3-dB bandwidth of the amplifier in question. Use the same elemental parameter values invoked in the preceding example.

Solution 5.2.

(1) The applicable low frequency equivalent circuit for the amplifier at hand appears in Fig. 5.7(c). With critical parameter p set to zero, an action that is tantamount to equating V_2/R_f in the model to zero. The null driving point input resistance, R_{ino}, is seen to be

$$R_{ino} = R_i \| R_f = 4{,}048\,\Omega. \tag{E2-1}$$

With $p = 0$ and additionally, V_s (independent signal source) set to zero, a tacit inspection of the model in Fig. 5.7(c) reveals $V_1 = 0$. It follows that the driving point null output resistance, R_{outo}, is

$$R_{outo} = R_{l1} + R_{l2} \| R_f = 1{,}872\,\Omega. \tag{E2-2}$$

(2) For analytical convenience, the return ratio with respect to the critical feedback parameter determined in Example 5.1 is repeated herewith:

$$T_s(G_f, R_s, \infty) = \left(\frac{R_{l2}}{R_{l2} + R_f}\right)\left(\frac{R_s \| R_i}{R_s \| R_i + R_f}\right)(g_m R_f - 1). \tag{E2-3}$$

It is immediately evident that

$$\left. \begin{aligned} T_s(G_f, 0, \infty) &= 0 \\ T_s(G_f, \infty, \infty) &= \left(\frac{R_{l2}}{R_{l2} + R_f}\right)\left(\frac{R_i}{R_i + R_f}\right)(g_m R_f - 1) = 25.87 \end{aligned} \right.$$

$$\tag{E2-4}$$

Using Eq. (5-25), the driving point input resistance is found to be

$$R_{in} = R_{ino}\left[\frac{1 + T_s(G_f, 0, \infty)}{1 + T_s(G_f, \infty, \infty)}\right] = 150.7\,\Omega. \tag{E2-5}$$

(3) For the output impedance determination, $T_s(G_f, R_s, \infty)$ derives directly from Eq. (E2-3). In particular, $T_s(G_f, R_s, \infty) = 3.34$, as computed in the preceding example. In order to obtain $T_s(G_f, R_s, 0)$, an intermediate analysis must be conducted since the load impedance, Z_l, does not appear explicitly in Eq. (E2-3). To this end, consider the circuit shown in Fig. 5.11(a), which is the model of Fig. 5.7(c), configured expressly for computing the normalized

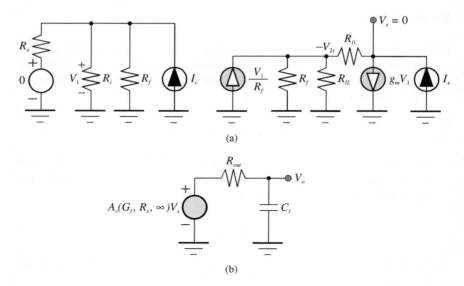

Figure 5.11. (a) The equivalent circuit used to determine the normalized null return ratio pertinent to the driving point output resistance of the amplifier in Fig. 5.6(b). (b) The Thévenin equivalent circuit that drives the capacitive load in the MOS technology amplifier of Fig. 5.6(b).

null return ratio pertinent to the driving point output impedance relationship. A conventional circuit analysis yields

$$\left. \begin{array}{c} V_1 = I_c\left(R_s \| R_i \| R_f\right) \\ \dfrac{V_1}{R_f} = -\dfrac{V_{2r}}{R_{l2} \| R_f} - \dfrac{V_{2r}}{R_{l1}} \end{array} \right\}. \tag{E2-6}$$

A combination of these two equilibrium relationships results in

$$Q_{ro}(R_s) = \frac{V_{2r}}{I_c} = -\left(R_{l1} \| R_{l2} \| R_f\right)\left(\frac{R_s \| R_i}{R_s \| R_i + R_f}\right), \tag{E2-7}$$

whence a null return ratio for output impedance purposes of

$$T_{ro}(G_f, R_s) = \frac{Q_{ro}(R_s)}{R_f}$$

$$= -\left(\frac{R_{l1} \| R_{l2}}{R_{l1} \| R_{l2} + R_f}\right)\left(\frac{R_s \| R_i}{R_s \| R_i + R_f}\right) = -5.263(10^{-3}).$$

$$\tag{E2-8}$$

The resultant driving point output resistance is

$$R_{out} = R_{outo}\left[\frac{1 + T_{ro}(G_f, R_s)}{1 + T_s(G_f, R_s, \infty)}\right] = 429.1\,\Omega. \quad \text{(E2-9)}$$

(4) The voltage gain, $A_v(G_f, R_s, \infty)$, determined as Eq. (E1-13), is a low frequency transfer function. Since the load imposed on the amplifier undergoing study is the only energy storage element indigenous to the amplifier and is strictly a capacitance, C_l, this gain, when multiplied by the signal source voltage, V_s, is effectively the Thévenin signal voltage driving the capacitive load. In view of the fact that R_{out} in Eq. (E2-9) is the driving point output resistance established at the terminals to which C_l is incident, the Thévenin equivalent model at the output port becomes the structure provided in Fig. 5.11(b). An analysis of this circuit produces an *I/O* transfer function of

$$\frac{V_o}{V_s} = \frac{A_v(G_f, R_s, \infty)}{1 + sR_{out}C_l}. \quad \text{(E2-10)}$$

An inspection of this relationship reveals a 3-dB bandwidth, say B, of

$$B = \frac{1}{R_{out}C_l} = 2\pi(741.8\,\text{MHz}). \quad \text{(E2-11)}$$

Comments. The normalized return and null return ratios are pivotal to determining, in addition to the forward gain of a linear network, the driving point input and output impedances. As this example illustrates, a complication arises with respect to the determination of the output impedance when no finite load is incident with the output port. But even this complication is straightforwardly mitigated simply through applying the basic analytical techniques that underpin the calculation of the return ratio metrics.

The bandwidth exercise in this example is a trivial endeavor because the amplifier undergoing investigation has only a single energy storage element and is therefore a first order network. In particular, the lone energy storage element is a capacitance incident with the amplifier output port, thereby rendering the low frequency gain determined in the preceding example independent of signal frequency. Since the subject capacitance establishes no zeros at finite frequencies, the bandwidth solution reduces to the simple problem of identifying the time constant associated with the pole frequency established by the load capacitance.

5.2.3 Output Port-to-Local Port Feedback

A particularly useful special case of the local feedback scenario investigated in Section 5.2.1 is the system level abstraction of Fig. 5.12(a), which depicts voltage controlled current feedback applied from the output port-to-a local

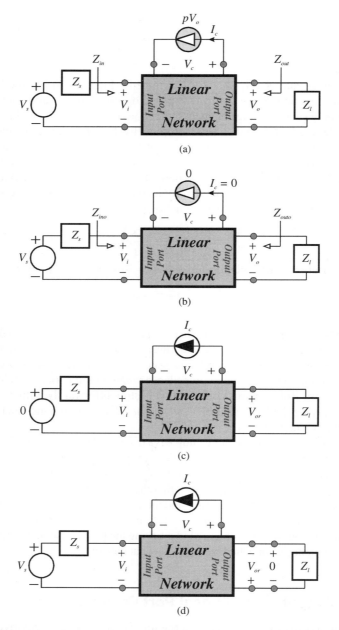

Figure 5.12. (a) Network with voltage controlled current feedback applied from the output port-to-an internal network port. (b) Network model for calculating the I/O gain with the feedback parameter set to zero. (c) Network model for calculating the normalized return ratio with respect to feedback parameter p. (d) Network model for calculating the normalized null return ratio with respect to parameter p.

cth network port. The voltage transfer function, V_o/V_s, derives from Eq. (5-6) and is given explicitly by

$$A_v(p, Z_s, Z_l) = \frac{V_o}{V_s} = A_{os}\left[\frac{1 + pQ_r(Z_s, Z_l)}{1 + pQ_s(Z_s, Z_l)}\right]. \tag{5-30}$$

In this expression, A_{os}, is the voltage gain with feedback parameter p equated to zero; that is, $A_{os} = A_v(0, Z_s, Z_l)$. In accordance with earlier disclosures, its calculation derives from a circuit analysis of the model in Fig. 5.12(b), where parameter p is null, thereby rendering zero applied current, I_c, to the terminals comprising the cth network port. Observe that the original driving point input and output impedances, Z_{in} and Z_{out} in Fig. 5.12(a), become their respective null parameter values, Z_{ino} and Z_{outo} in Fig. 5.12(b). The normalized return ratio, $Q_s(Z_s, Z_l)$, is the voltage-to-current transfer function, V_{or}/I_c, in Fig. 5.12(c), in which the original pV_o generator is supplanted by the independent current source, I_c. In Fig. 5.12(c), the signal source voltage, V_s, is set to zero, and since V_o is the original controlling voltage variable for the feedback current source, V_{or} is simply the negative of V_o.

On the other hand, the normalized null return ratio, $Q_r(Z_s, Z_l)$, with respect to parameter p is the impedance ratio, V_{or}/I_c, in Fig. 5.12(d). In this diagram, the signal source voltage is remanded to a value that results in a null output voltage response. But since the controlling feedback variable is the output voltage, V_{or}, which is the negative of V_o, V_{or} is identically zero. It follows that $Q_r(Z_s, Z_l) = 0$, which is a general result for any linear feedback network wherein the controlling electrical variable to which the feedback source is proportional is identical to the network output response. The immediate result of these arguments is that Eq. (5-30) reduces to the simpler expression,

$$A_v(p, Z_s, Z_l) = \frac{A_{os}}{1 + pQ_s(Z_s, Z_l)} = \frac{A_{os}}{1 + T_s(p, Z_s, Z_l)}, \tag{5-31}$$

where $T_s(p, Z_s, Z_l)$ is, of course, the return ratio with respect to parameter p or more simply, the loop gain of the considered network. It is worthwhile interjecting that Eqs. (5-25) and (5-29), which are not functionally dependent on the normalized null return ratio, remain applicable for the driving point input and output impedances, Z_{in} and Z_{out}, respectively.

5.3.0 Special Case Feedback Network Examples

Armed with the analytical methodology that underpins a computationally efficient investigation of linear feedback networks, it is expedient to examine the transfer and impedance dynamics of several routinely encountered feedback architectures. These special cases can be codified as global feedback and feedback that is not necessarily global in nature. In global feedback systems, a portion of the output voltage or current variable is fed back to the input port as either a controlled voltage or controlled current. In other feedback networks, the feedback is implemented often locally as a simple branch admittance or impedance, which can be resistive, capacitive, inductive, or even a short circuit.

5.3.1. Global Feedback

Four types of practical global feedback amplifiers prevail. These are the transimpedance amplifier, the transadmittance amplifier, the voltage amplifier, and the current amplifier. All of these amplifiers can be realized in MOS or bipolar device technologies. The transimpedance and transconductance units require phase inversion between input and output network ports, while voltage and current feedback amplifiers do not require such phase inversion.

5.3.1.1. *Transimpedance Feedback Amplifier*

The *transimpedance feedback amplifier* is characterized by small driving point input impedance, Z_{in}, small driving point output impedance, Z_{out}, and an input-to-output transimpedance, say $Z_m(Y, Z_s, Z_l)$, that is nominally independent of source and load terminations, Z_s and Z_l, respectively. Its system level schematic diagram appears in Fig. 5.13(a), wherein admittance parameter Y is cast as the critical feedback variable. Because of the presumably small input impedance, the input port excitation derives from a relatively high impedance signal source whose Norton equivalent current is I_s. In turn, the small output impedance gives rise to a voltage response, V_o, which is developed across the load impedance that terminates the output port of the network. Feedback appears as a voltage controlled current source, YV_o, in shunt and anti-phase with the applied current signal, I_s.

As usual, the first step in the problem of ascertaining the forward gain of the amplifier at hand is the determination of the null transimpedance, $Z_m(0, Z_s, Z_l)$. The pertinent equivalent circuit is given in Fig. 5.13(b),

Signal Flow Methods of Feedback Network Analysis 349

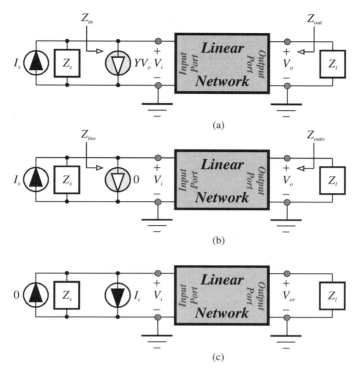

Figure 5.13. (a) System level schematic diagram of a transimpedance feedback amplifier. The feedback factor is the admittance, Y, the input signal is current I_s, and the resultant output response is voltage V_o. (b) Circuit used in the calculation of the forward transimpedance under the condition of null feedback parameter Y. (c) Circuit used to compute the return ratio, V_{or}/I_c, with respect to feedback variable Y.

where feedback parameter Y is set to zero. The resultant forward transimpedance is $Z_m(0, Z_s, Z_l) = V_o/I_s$, and the corresponding input and output impedance levels are Z_{ino} and Z_{outo}, as indicated in the diagram.

The return ratio with respect to the feedback parameter derives from an analysis of the model in Fig. 5.13(c). In this structure, the independent Norton signal current, I_s, is set to zero, the original YV_o controlled source is replaced by an independent current generator, I_c, and the original controlling voltage, V_o, becomes V_{or}, which is simply the negative of voltage V_o. The subject normalized return ratio, $Q_s(Z_s, Z_l)$ follows as the impedance function, V_{or}/I_c. But since current I_c is incident with the terminal pair to which the original signal current, I_s, is connected, I_c is anti-phase with I_s, and V_{or} is the negative of V_o, the normalized return ratio with respect to parameter Y is identical to the foregoing null transimpedance. In short,

$$Q_s(Z_s, Z_l) = \frac{V_{or}}{I_c} \equiv Z_m(0, Z_s, Z_l). \tag{5-32}$$

By virtue of the discourse in the preceding subsection of material, the normalized null return ratio, $Q_r(Z_s, Z_l)$ is itself null since the feedback current is controlled by the network output voltage response.

It follows that the overall gain, or closed loop transimpedance, of the subject linear feedback network is

$$Z_m(Y, Z_s, Z_l) = \frac{V_o}{I_s} = \frac{Z_m(0, Z_s, Z_l)}{1 + YZ_m(0, Z_s, Z_l)} = \frac{Z_m(0, Z_s, Z_l)}{1 + T_s(Y, Z_s, Z_l)}, \tag{5-33}$$

where the return ratio with respect to feedback variable Y, or network loop gain, is

$$T_s(Y, Z_s, Z_l) = YZ_m(0, Z_s, Z_l). \tag{5-34}$$

It is interesting to note that the closed loop transimpedance in Eq. (5-33) is a function of only two metrics; namely, the feedback parameter Y (which is selected by the circuit analyst or designer), and the null transimpedance, $Z_m(0, Z_s, Z_l)$, which is generally a straightforward upshot of analysis conducted on the model in Fig. 5.13(b). Note further that this null transimpedance is effectively the open loop network transimpedance, since both closed and open loop transimpedances are identical under the condition of $Y = 0$. If the loop gain magnitude, $|T_s(Y, Z_s, Z_l)|$, is very large over the signal processing frequency range of interest, the closed loop transimpedance reduces simply to $Z_m(Y, Z_s, Z_l) \approx 1/Y$. This reduction is significant in that the resultant closed loop transimpedance is rendered independent of source and load impedances and the vagaries pervasive to the active device model parameters indigenous to the null transimpedance function.

Recalling Eq. (5-25), the general expression for the driving point input impedance, Z_{in}, as a function of feedback parameter Y and load impedance Z_l, is

$$Z_{in}(Y, Z_l) = Z_{ino}\left[\frac{1 + YZ_m(0, 0, Z_l)}{1 + YZ_m(0, \infty, Z_l)}\right]. \tag{5-35}$$

Since $Z_m(0, Z_s, Z_l)$ is the open loop, or null feedback parameter value of, the transimpedance of the network in Fig. 5.13(a), the zero source impedance value, $Z_m(0, 0, Z_l)$, of the open loop transimpedance is zero in that $Z_s = 0$ effects a short-circuited network input port. Accordingly, Eq. (5-35) collapses to the simpler relationship,

$$Z_{in}(Y, Z_l) = \frac{Z_{ino}}{1 + YZ_m(0, \infty, Z_l)} = \frac{Z_{ino}}{1 + T_s(Y, \infty, Z_l)}. \quad (5\text{-}36)$$

The normalized return ratio, $Z_m(0, \infty, Z_l)$, is the open loop transimpedance for infinitely large signal source impedance ($Z_s = \infty$). An inspection of the diagram in Fig. 5.13(a) confirms that finite source impedance divides the Norton signal current between the source impedance branch and the network input port. It follows that $|Z_m(0, \infty, Z_l)| > |Z_m(0, Z_s, Z_l)|$; that is, the infinitely large source impedance value of the open loop transimpedance is necessarily larger than its finite source impedance counterpart. Thus,

$$Z_{in}(Y, Z_l) = \frac{Z_{ino}}{1 + YZ_m(0, \infty, Z_l)} < \frac{Z_{ino}}{1 + T_s(Y, Z_s, Z_l)} \quad (5\text{-}37)$$

is a valid conclusion. Moreover, very small driving point input impedance is inferred by the foregoing inequality because a large loop gain magnitude is mandated for an acceptable desensitization of the closed loop transimpedance with respect to source impedance, load impedance, and the parameters implicit to the open loop transimpedance function.

The driving point output impedance, Z_{out}, as a function of feedback parameter Y and source impedance Z_s is given by

$$Z_{out}(Y, Z_s) = Z_{outo}\left[\frac{1 + YZ_m(0, Z_s, 0)}{1 + YZ_m(0, Z_s, \infty)}\right]. \quad (5\text{-}38)$$

The zero load impedance value, $Z_m(0, Z_s, 0)$, of the null parameter open loop transimpedance is zero since $Z_l = 0$ is tantamount to short circuiting the network output port where response voltage V_o is developed. On the other hand, since the voltage developed across an open-circuited load that terminates a stable network is larger than that established across a termination of finite load impedance, the $Z_l = \infty$ value of the open loop transimpedance is larger than the open loop transimpedance established with finite load impedance. Resultantly,

$$Z_{out}(Y, Z_s) = \frac{Z_{outo}}{1 + YZ_m(0, Z_s, \infty)} < \frac{Z_{outo}}{1 + T_s(Y, Z_s, Z_l)}. \quad (5\text{-}39)$$

As in the case of the driving point input impedance, the driving point output impedance is likely to be very small, owing to the fact that a prudent feedback network design invariably entails a large loop gain magnitude.

5.3.1.2. *Transadmittance Feedback Amplifier*

The *transadmittance feedback amplifier* is the dual of the transimpedance signal processor in that it offers a large driving point input impedance, Z_{in}, a large driving point output impedance, Z_{out}, and an input-to-output transadmittance, say $Y_m(Z, Z_s, Z_l)$, that is nominally independent of source and load terminations, Z_s and Z_l, respectively. The pertinent schematic abstraction is offered in Fig. 5.14(a). The large network input impedance renders practical an input port excitation that derives from a low impedance signal source whose Thévenin equivalent voltage is V_s. On the other hand, large output port impedance is conducive to the current response, I_o, which is conducted by the terminating load impedance. Feedback appears as a current controlled voltage source, ZI_o, applied at the input port in anti-phase sense with the applied voltage signal, V_s. The feedback factor Z is measured in units of impedance.

The null transadmittance, $Y_m(0, Z_s, Z_l)$, is evaluated through an analytical consideration of the model in Fig. 5.14(b). In this equivalent circuit, the feedback parameter, Z, is set to zero and the resultant current-to-voltage ratio, I_o/V_s, is the desired null forward gain. With $Z = 0$, the driving point input and output impedances are Z_{ino} and Z_{outo}, respectively.

Figure 5.14(c) pertains to the computation of the normalized return ratio, $Q_s(Z_s, Z_l)$. In this diagram, the signal voltage source, V_s, is set to zero. Moreover, the original controlled feedback voltage, ZI_o, is replaced by the independent voltage generator, V_c, while the original controlling current, I_o, is reversed and identified as I_{or}. The resultant normalized return ratio with respect to feedback variable Z is $Q_s(Z_s, Z_l) = I_{or}/V_c$. Observe that voltage V_c effectively appears in series with the network input port as the negative of the currently zeroed signal source V_s. Additionally, I_{or} is clearly the negative of the original controlling feedback variable, I_o. It follows that $Q_s(Z_s, Z_l)$ is identical to the null transadmittance, $Y_m(0, Z_s, Z_l)$. Since the controlled feedback voltage source is directly proportional to the network output current, I_o, the normalized null return ratio is zero. The upshot of the foregoing matters is that the closed loop transadmittance, $Y_m(Z, Z_s, Z_l)$, is

$$Y_m(Y, Z_s, Z_l) = \frac{I_o}{V_s} = \frac{Y_m(0, Z_s, Z_l)}{1 + ZY_m(0, Z_s, Z_l)} = \frac{Y_m(0, Z_s, Z_l)}{1 + T_s(Y, Z_s, Z_l)}, \quad (5\text{-}40)$$

where $Y_m(0, Z_s, Z_l)$ is recognized as the open loop transadmittance of the subject feedback system, and $T_s(Y, Z_s, Z_l) = ZY_m(0, Z_s, Z_l)$ is the loop gain.

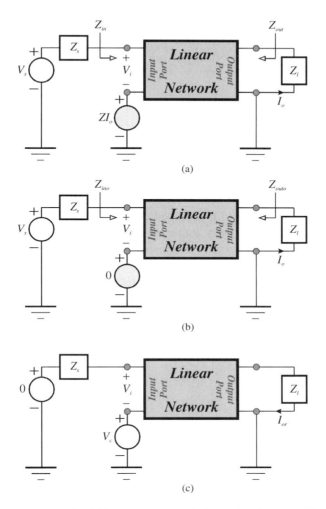

Figure 5.14. (a) System level diagram of a transadmittance feedback amplifier. The feedback factor is the impedance, Z, the input signal is voltage V_s, and the resultant output response is current I_o. (b) Circuit used in the calculation of the forward transadmittance under the condition of null feedback parameter Z. (c) Circuit used to compute the return ratio, I_{or}/V_c, with respect to feedback parameter Z.

Because $Y_m(0, Z_s, Z_l)$ represents a forward transadmittance from the Thévenin input signal voltage-to-the current conducted by the output port load termination, $Y_m(0, \infty, Z_l) = 0$. This rationale reflects the circumstance that infinitely large source impedance isolates the signal source voltage from the network input port. In contrast, $Y_m(0, 0, Z_l)$ yields maximal forward transadmittance since $Z_s = 0$ effects direct coupling of the signal source to the network input port. It follows that $|Y_m(0, 0, Z_l)| > |Y_m(0, Z_s, Z_l)|$, and

$$Z_{in}(Y, Z_l) = Z_{ino}\left[\frac{1 + ZY_m(0, 0, Z_l)}{1 + ZY_m(0, \infty, Z_l)}\right] = Z_{ino}[1 + ZY_m(0, 0, Z_l)]$$
$$> Z_{ino}[1 + T_s(Z, Z_s, Z_l)].$$

(5-41)

In short, the driving point input impedance of a transadmittance feedback amplifier is likely to be very large, owing to the mandate of large loop gain for adequate forward gain desensitization with respect to parametric vagaries.

In similar fashion, it can be demonstrated that the driving point output impedance of the amplifier at hand is also quite large. In particular,

$$Z_{out}(Y, Z_s) = Z_{outo}\left[\frac{1 + ZY_m(0, Z_s, 0)}{1 + ZY_m(0, Z_s, \infty)}\right] = Z_{outo}[1 + ZY_m(0, Z_s, 0)]$$
$$> Z_{outo}[1 + T_s(Z, Z_s, Z_l)].$$

(5-42)

This relationship derives from two observations. First, an open-circuited load impedance delivers a null forward transadmittance, $Y_m(0, Z_s, \infty)$, of zero. Second, a short circuit load termination delivers maximal transadmittance, $Y_m(0, Z_s, 0)$, so that $|Y_m(0, Z_s, 0)| > |Y_m(0, Z_s, Z_l)|$.

5.3.1.3. *Voltage Feedback Amplifier*

The *feedback voltage amplifier* abstracted in Fig. 5.15(a) approximates the I/O characteristics of an ideal voltage controlled voltage source in that it offers high input impedance, low output impedance, and a voltage gain that is relatively insensitive to source and load terminations. Like the transadmittance system, the voltage amplifier applies feedback to the amplifier input port in the form of a controlled voltage. But unlike the transadmittance amplifier, this controlled feedback voltage is dependent on

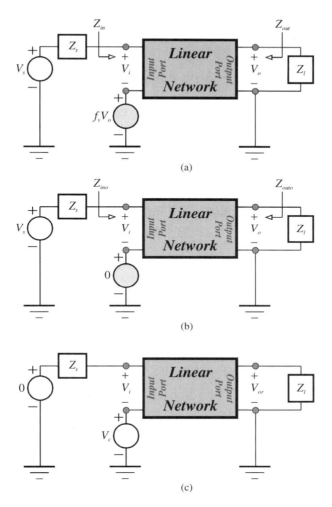

Figure 5.15. (a) System level diagram of a feedback voltage amplifier. The feedback factor is the dimensionless factor, f_v, the input is voltage V_s, and the resultant output response is voltage V_o. (b) Circuit used in the calculation of the forward voltage transfer function under the condition of null feedback parameter f_v. (c) Circuit used to compute the return ratio, V_{or}/V_c, with respect to feedback parameter f_v.

the output voltage response, V_o, as opposed to a dependence on the current conducted by the load impedance, Z_l. The high input impedance facilitates a voltage input, whence the signal source is modeled as a Thévenin equivalent circuit comprised of the voltage, V_s, and the source impedance, Z_s.

In general, the voltage gain of the amplifier in Fig. 5.15(a), as a function of feedback factor f_v, source impedance Z_s, and load impedance Z_l is

$$A_v(f_v, Z_s, Z_l) = \frac{V_o}{V_s} = A_v(0, Z_s, Z_l) \left[\frac{1 + f_v Q_r(Z_s, Z_l)}{1 + f_v Q_s(Z_s, Z_l)}\right], \quad (5\text{-}43)$$

where the null feedback parameter voltage gain, $A_v(0, Z_s, Z_l)$, is the voltage ratio, V_o/V_s, in Fig. 5.15(b), in which the controlled feedback generator, $f_v V_o$, is set to zero. The normalized null return ratio, $Q_r(Z_s, Z_l)$, is zero by virtue of the fact that the controlling variable of the voltage controlled feedback generator is the network output voltage, V_o. On the other hand, the normalized return ratio, $Q_s(Z_s, Z_l)$, is nonzero and given by the transfer function, V_{or}/V_c, in Fig. 5.15(c). In this diagram, the independent signal source voltage, V_s, is set to zero, and the original dependent feedback voltage, $f_v V_o$, is replaced by the independent source, V_c. Finally, the original controlling feedback variable, V_o, is phase reversed and resultantly identified as V_{or} in the subject figure. Since $V_{or} = -V_o$, $V_s = 0$, and V_c is anti-phase with the original V_s, it is clear that $Q_s(Z_s, Z_l)$ is identical to the null parameter voltage gain, $A_v(0, Z_s, Z_l)$. Accordingly, Eq. (5-43) becomes

$$A_v(f_v, Z_s, Z_l) = \frac{V_o}{V_s} = \frac{A_v(0, Z_s, Z_l)}{1 + f_v A_v(0, Z_s, Z_l)} = \frac{A_v(0, Z_s, Z_l)}{1 + T_s(f_v, Z_s, Z_l)}, \quad (5\text{-}44)$$

where $A_v(0, Z_s, Z_l)$ is, in addition to the normalized return ratio with respect to feedback factor f_v, the open loop gain of the network. Moreover,

$$T_s(f_v, Z_s, Z_l) = f_v A_v(0, Z_s, Z_l) \quad (5\text{-}45)$$

is the return ratio with respect to f_v and indeed, the loop gain of the feedback voltage amplifier.

In Eq. (5-45), it should be understood that

$$T_s(f_v, \infty, Z_l) = T_s(f_v, Z_s, 0) \equiv 0, \quad (5\text{-}46)$$

since infinitely large source impedance decouples the signal source from the network input port, thereby giving zero open loop gain. On the other hand, zero load impedance short circuits the output port to preclude nonzero open loop voltage gain. It follows that the driving point input impedance of the voltage amplifier in Fig. 5.15(a) is

$$Z_{in}(f_v, Z_l) = Z_{ino}\left[\frac{1 + T_s(f_v, 0, Z_l)}{1 + T_s(f_v, \infty, Z_l)}\right] = Z_{ino}[1 + T_s(f_v, 0, Z_l)], \quad (5\text{-}47)$$

while the driving point output impedance is

$$Z_{out}(f_v, Z_s) = Z_{outo}\left[\frac{1 + T_s(f_v, Z_s, 0)}{1 + T_s(f_v, Z_s, \infty)}\right] = \frac{Z_{outo}}{1 + T_s(f_v, Z_s, \infty)}. \quad (5\text{-}48)$$

Note that the input impedance in Eq. (5-47) is large because the zero source impedance value of open loop gain is larger than the nonzero, but finite, source impedance value of this metric. In contrast, the driving point output impedance is low since $|T_s(f_v, Z_s, \infty)| > |T_s(f_v, Z_s, Z_l)|$, which is traditionally large to ensure an acceptable desensitization of closed loop dynamics with respect to parametric uncertainties.

5.3.1.4. *Current Feedback Amplifier*

In contrast to the feedback voltage amplifier, the *feedback current amplifier*, whose system level schematic diagram appears in Fig. 5.16(a), approximates the I/O characteristics of an ideal current controlled current source in that it provides low input impedance, high output impedance, and a current gain that is relatively unaffected by source and load terminations. In the current amplifier, feedback is applied to the amplifier input port in the form of a controlled current. This controlled feedback current is proportional to the output current response, I_o, which is the current conducted by the load impedance, Z_l. The low input impedance encourages a signal current input, whence the signal source is modeled as a Norton equivalent circuit comprised of the current, I_s, and the source impedance, Z_s.

The current gain of the amplifier in Fig. 5.16(a), as a function of feedback factor f_i, source impedance Z_s, and load impedance Z_l is

$$A_i(f_i, Z_s, Z_l) = \frac{I_o}{I_s} = A_i(0, Z_s, Z_l)\left[\frac{1 + f_i Q_r(Z_s, Z_l)}{1 + f_i Q_s(Z_s, Z_l)}\right], \quad (5\text{-}49)$$

where the zero feedback parameter value of the current gain, $A_i(0, Z_s, Z_l)$, is the current ratio, I_o/I_s, in Fig. 5.16(b). Note therein that the controlled feedback generator, $f_i I_o$, is set to zero. Because the implemented feedback

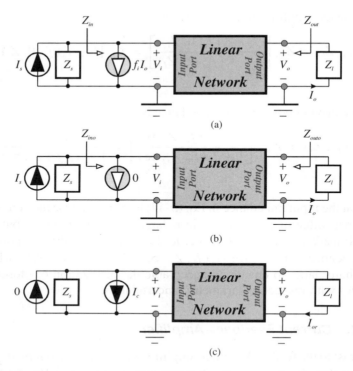

Figure 5.16. (a) System level diagram of a feedback current amplifier. The feedback factor is the dimensionless factor, f_i, the input signal is current I_s, and the resultant output response is current I_o. (b) Circuit used in the calculation of the forward current transfer function under the condition of null feedback parameter f_i. (c) Circuit used to compute the return ratio, I_{or}/I_c, with respect to feedback parameter f_i.

is directly proportional to the network output response, the normalized null return ratio, $Q_r(Z_s, Z_l)$, is zero. On the other hand, the normalized return ratio, $Q_s(Z_s, Z_l)$, is nonzero and given by the current transfer function, I_{or}/I_c, in Fig. 5.16(c). In this diagram, the independent signal source voltage, I_s, is set to zero, and the original dependent feedback voltage, $f_i I_o$, is replaced by the independent current, I_c. Finally, the original controlling feedback variable, I_o, is phase reversed and resultantly identified as I_{or} in the figure. Since $I_{or} = -I_o$, $I_s = 0$, and I_c is anti-phase with the original I_s, $Q_s(Z_s, Z_l)$ is seen as identical to the null parameter current gain, $A_i(0, Z_s, Z_l)$. Accordingly, Eq. (5-49) becomes

$$A_i(f_i, Z_s, Z_l) = \frac{I_o}{I_s} = \frac{A_i(0, Z_s, Z_l)}{1 + f_i A_i(0, Z_s, Z_l)} = \frac{A_i(0, Z_s, Z_l)}{1 + T_s(f_i, Z_s, Z_l)}, \quad (5\text{-}50)$$

where $A_i(0, Z_s, Z_l)$ is, in addition to the normalized return ratio with respect to feedback factor f_i, the open loop gain of the network. Furthermore,

$$T_s(f_v, Z_s, Z_l) = f_i A_i(0, Z_s, Z_l) \tag{5-51}$$

is the loop gain of the feedback current amplifier.

In Eq. (5-51), it is apparent that

$$T_s(f_i, 0, Z_l) = T_s(f_i, Z_s, \infty) \equiv 0, \tag{5-52}$$

since zero source impedance precludes the applied signal current from flowing into the amplifier input port, thereby resulting in zero current gain. Moreover, infinitely large load impedance effects an open-circuited load termination, which nulls the output current and again results in zero current gain. It follows that the driving point input impedance of the current amplifier in Fig. 5.16(a) is

$$Z_{\text{in}}(f_i, Z_l) = Z_{ino}\left[\frac{1 + T_s(f_i, 0, Z_l)}{1 + T_s(f_i, \infty, Z_l)}\right] = \frac{Z_{ino}}{1 + T_s(f_i, \infty, Z_l)}, \tag{5-53}$$

and the driving point output impedance is

$$Z_{\text{out}}(f_i, Z_s) = Z_{outo}\left[\frac{1 + T_s(f_i, Z_s, 0)}{1 + T_s(f_i, Z_s, \infty)}\right] = Z_{outo}[1 + T_s(f_i, Z_s, 0)]. \tag{5-54}$$

The current gain, and hence the loop gain, is larger for infinitely large source impedance than it is for finite source impedance. Similarly, both the current and loop gains are larger for a short-circuited load impedance than they are for nonzero load impedance. Accordingly, the driving point input impedance satisfies

$$Z_{\text{in}}(f_i, Z_l) = \frac{Z_{ino}}{1 + T_s(f_i, \infty, Z_l)} < \frac{Z_{ino}}{1 + T_s(f_i, Z_s, Z_l)}, \tag{5-55}$$

which is small because $T_s(f_i, Z_s, Z_l)$ is necessarily large for acceptable closed loop desensitization with respect to open loop parametric vagaries. Analogously, the output impedance is large because

$$Z_{\text{out}}(f_i, Z_s) = Z_{outo}[1 + T_s(f_i, Z_s, 0)] > Z_{outo}[1 + T_s(f_i, Z_s, Z_l)]. \tag{5-56}$$

Example 5.3. The circuit given in Fig. 5.17(a) is the small signal, low frequency model of the bipolar junction transistor amplifier whose schematic diagram appears

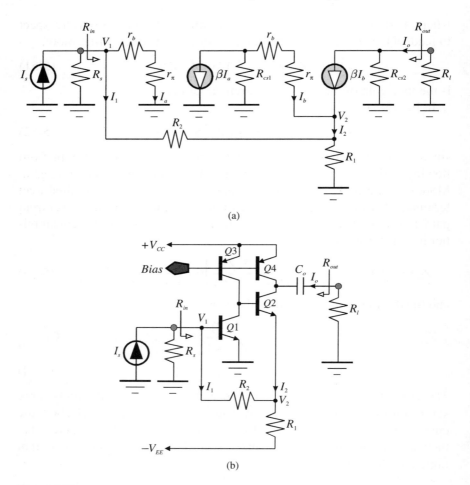

Figure 5.17. (a) Small signal, low frequency equivalent circuit of a current feedback amplifier realized in bipolar junction transistor technology. (b) The schematic diagram of the amplifier whose small signal model appears in (a).

in Fig. 5.17(b). The amplifier in question exploits shunt-series feedback through the divider network consisting of resistances R_1 and R_2 to achieve a current gain, $A_i = I_o/I_s$, that is nominally independent of the source resistance, R_s, and the load termination, R_l. Transistors $Q3$ and $Q4$ are biased to function as current sources and accordingly, their effective, small signal terminal resistances, R_{cs1} and R_{cs2}, in the model of Fig. 5.17(a) are large. For the purpose of this exercise, $R_{cs1} = R_{cs2} = 60\,\text{K}\Omega$. Transistors $Q1$ and $Q2$ are presumed to be identical transistors that are biased at the same current and voltage levels, thereby giving rise to identical small signal parameters, r_b, r_π, and β. In this problem, $r_b = 250\,\Omega$, $r_\pi = 1.5\,\text{K}\Omega$, and $\beta = 100$. The output coupling capacitor, C_o, is sufficiently large to emulate a short-circuited branch element over the signal frequency range of interest. The source resistance, R_s, is $10\,\text{K}\Omega$, the load resistance, R_l, is $300\,\Omega$, and the two resistances in the feedback subcircuit are such that $R_2 = 9R_1 = 4.5\,\text{K}\Omega$. Derive expressions for, and numerically evaluate, the open loop current gain, A_{io}, the loop gain, T_s, the closed loop current gain, A_i, and the driving point input resistance, R_{in}. Finally, determine the driving point output resistance, R_{out}, for the simplifying and reasonable case of $R_{cs2} \gg R_l$.

Solution 5.3.

(1) The solution initiates with a representation of the $R_1 - R_2$ feedback subcircuit by an appropriate two-port parameter set. Since this subcircuit is connected in shunt with the input port and in series with the output port of the amplifier, g-parameter modeling, with its inherent shunt-series architecture, is the most appropriate. To this end, the subcircuit at hand is redrawn in Fig. 5.18(a), which postures

$$\left. \begin{aligned} g_{11} &= \left.\frac{I_1}{V_1}\right|_{I_2=0} = \frac{1}{R_1 + R_2} \\ g_{21} &= \left.\frac{V_2}{V_1}\right|_{I_2=0} = \frac{R_1}{R_1 + R_2} \\ g_{12} &= \left.\frac{I_1}{I_2}\right|_{V_1=0} = -\frac{R_1}{R_1 + R_2} \\ g_{22} &= \left.\frac{V_2}{I_2}\right|_{V_1=0} = R_1 \| R_2 \end{aligned} \right\}. \tag{E3-1}$$

The g-parameter modeling equations follow as

$$\left. \begin{aligned} I_1 &= g_{11}V_1 + g_{12}I_2 \\ V_2 &= g_{21}V_1 + g_{22}I_2 \end{aligned} \right\}, \tag{E3-2}$$

where the feedback network port variables, I_1, V_2, V_1, and I_2 are delineated in Figs. 5.17 and 5.18(a). From Fig. 5.17(a),

$$I_2 = (\beta + 1)I_b, \tag{E3-3}$$

and

$$I_o = \left(\frac{R_{cs2}}{R_{cs2} + R_l}\right)\beta I_b. \tag{E3-4}$$

Letting

$$k_2 \triangleq \frac{R_{cs2}}{R_{cs2} + R_l}, \tag{E3-5}$$

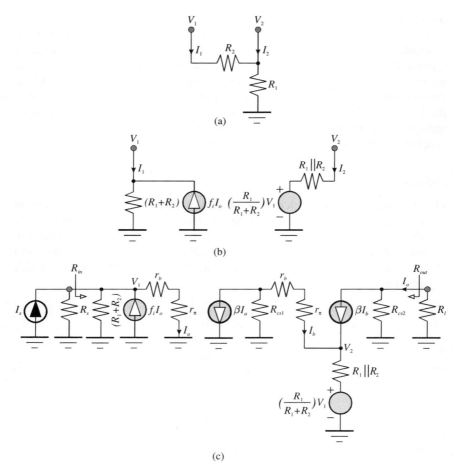

Figure 5.18. (a) The feedback subcircuit embedded in the model of Fig. 5.17(a). (b) An equivalent representation of the feedback subcircuit in (a). (c) The alternative small signal model of the current feedback amplifier in Fig. 5.17(a).

which is almost one by virtue of the fact that $R_{cs2} \gg R_l$, and

$$\alpha = \frac{\beta}{\beta+1}, \tag{E3-6}$$

which in bipolar technology jargon is referred to as the short circuit, common base current gain, and

$$f_i \triangleq \left(\frac{1}{\alpha k_2}\right)\left(\frac{R_1}{R_1+R_2}\right), \tag{E3-7}$$

Eq. (E3-2), with the help of Eq. (E3-1), can be rewritten in the form

$$I_1 = \frac{V_1}{R_1 + R_2} - f_i I_o$$

$$V_2 = \left(\frac{R_1}{R_1 + R_2}\right) V_1 + (R_1 \| R_2) I_2$$

(E3-8)

The last result allows the feedback subcircuit in Fig. 5.18(a) to be modeled by the structure offered in Fig. 5.18(b). In turn, this feedback subcircuit model gives rise to the alternative amplifier model provided in Fig. 5.18(c), which clearly reveals two network metrics. First, the controlled feedback source is $f_i I_o$. Since this controlled source appears directly in shunt with the Norton equivalent signal source current, I_s and its value is proportional to the amplifier output response, I_o, the implemented feedback is global, and the feedback factor is f_i, as defined by Eq. (E3-7). Second, feedforward through the feedback subcircuit materializes, as is measured by the controlled generator,

$$\left(\frac{R_1}{R_1 + R_2}\right) V_1 = \left(\frac{R_1}{R_1 + R_2}\right)(r_b + r_\pi) I_a.$$

(E3-9)

(2) Subsequent to the modeling phase of the solution procedure, circuit analyses are in order. To this end, focus on the equivalent circuit in Fig. 5.18(c) for the open loop case of $f_i I_o = 0$, which is drawn in the interest of analytical convenience as Fig. 5.19(a). Then,

$$\frac{I_a}{I_s} = \frac{R_s \| (R_1 + R_2)}{R_s \| (R_1 + R_2) + r_b + r_\pi},$$

(E3-10)

$$\frac{I_o}{I_b} = \beta \left(\frac{R_{cs2}}{R_{cs2} + R_l}\right),$$

(E3-11)

and

$$0 = R_{cs1}(\beta I_a + I_b) + [r_b + r_\pi + (\beta + 1)(R_1 \| R_2)] I_b$$

$$+ \left(\frac{R_1}{R_1 + R_2}\right)(r_b + r_\pi) I_a.$$

(E3-12)

The last Kirchhoff relationship implies

$$\frac{I_b}{I_a} = -\frac{\beta R_{cs1} + \left(\frac{R_1}{R_1 + R_2}\right)(r_b + r_\pi)}{R_{cs1} + r_b + r_\pi + (\beta + 1)(R_1 \| R_2)}.$$

(E3-13)

The combination of Eqs. (E3-10), (E3-11), and (E3-13) leads forthwith to an expression for the open loop current gain, A_{io}. In particular,

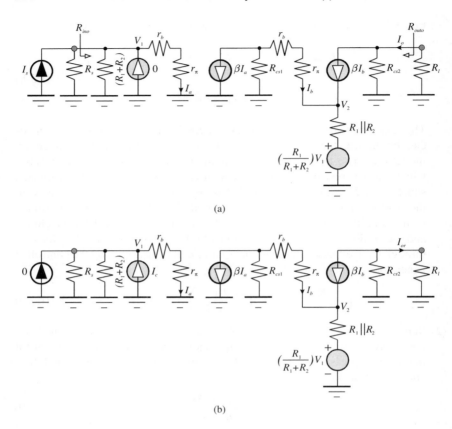

Figure 5.19. (a) Model of the current feedback amplifier in Fig. 5.17(a) configured for an evaluation of the open loop current gain. In this model, observe that the driving point input and output resistances are their respective null feedback parameter values, R_{ino} and R_{outo}. (b) Model of the subject current feedback amplifier configured for the computation of the normalized return ratio with respect to feedback parameter, f_i.

$$A_{io} = \left.\frac{I_o}{I_s}\right|_{f_i I_o = 0} = \frac{I_o}{I_b} \times \frac{I_b}{I_a} \times \frac{I_a}{I_s}$$

$$= -\beta^2 \left(\frac{R_{cs2}}{R_{cs2} + R_l}\right)$$

$$\times \left[\frac{R_s \| (R_1 + R_2)}{R_s \| (R_1 + R_2) + r_b + r_\pi}\right] \left[\frac{R_{cs1} + \left(\frac{R_1}{R_1 + R_2}\right)\left(\frac{r_b + r_\pi}{\beta}\right)}{R_{cs1} + r_b + r_\pi + (\beta + 1)(R_1 \| R_2)}\right]$$

(E3-14)

Numerically, $A_{io} = -3{,}652 = 71.25$ dB. To the extent that $R_{cs2} \gg R_l$, which is clear in this particular exercise, the open loop current gain is independent of the load resistance.

(3) Figure 5.19(b) displays the model appropriate to the calculation of the normalized return ratio, $Q_s(R_s, R_l)$, with respect to the feedback parameter, f_i. Note that the signal source current, I_s, is set herewith to zero, and the controlled feedback source, $f_i I_o$, is replaced by the independent current source, I_c. Moreover, the original controlling current, I_o, of this feedback source is reversed and signified by the current, I_{or}. An inspection of the model at hand reveals

$$Q_s(R_s, R_l) = \left.\frac{I_{or}}{I_c}\right|_{I_s=0} \equiv -A_{io}. \tag{E3-15}$$

From Eqs. (E3-5) to (E3-7), $k_2 = 0.995$, $\alpha = 0.9901$, and $f_i = 0.1015$. The loop gain follows as

$$T_s = f_i Q_s(R_s, R_l) = 370.7 = 51.4 \,\text{db}. \tag{E3-16}$$

Since the value of the feedback current source in the model of Fig. 5.18(b) is directly proportional to the amplifier current response, I_o, the null normalized return ratio is zero. It follows that the closed loop current gain, A_i, is

$$A_i = \frac{A_{io}}{1 + f_i Q_s(R_s, R_l)} = \frac{A_{io}}{1 - f_i A_{io}}. \tag{E3-17}$$

The corresponding closed loop current gain is $A_i = -9.825 = 19.85$ dB.

(4) An inspection of the circuit diagram in Fig. 5.19(b) shows that the null feedback parameter value of the driving point input resistance, R_{ino}, is

$$R_{ino} = (R_1 + R_2) \| (r_b + r_\pi) = 1{,}296 \,\Omega. \tag{E3-18}$$

With $R_s = 0$, $Q_s(R_s, R_l) = -A_{io} = 0$, while from Eqs. (E3-14) and (E3-15)

$$Q_s(\infty, R_l) = \beta^2 \left(\frac{R_{cs2}}{R_{cs2} + R_l}\right) \left[\frac{R_1 + R_2}{R_1 + R_2 + r_b + r_\pi}\right]$$

$$\times \left[\frac{R_{cs1} + \left(\frac{R_1}{R_1 + R_2}\right)\left(\frac{r_b + r_\pi}{\beta}\right)}{R_{cs1} + r_b + r_\pi + (\beta + 1)(R_1 \| R_2)}\right], \tag{E3-19}$$

whereupon, $Q(\infty, R_l) = 4{,}125$. Recalling Eq. (5-21), the driving point input resistance, R_{in}, is found to be

$$R_{in} = R_{ino} \left[\frac{1 + f_i Q_s(0, R_l)}{1 + f_i Q_s(\infty, R_l)}\right] = 3.09 \,\Omega. \tag{E3-20}$$

(5) A second inspection of the circuit in Fig. 5.19(b) confirms a null feedback parameter value of the output resistance of $R_{outo} = R_{cs2}$. To the extent that $R_{cs2} \gg R_l$ is assumed *a priori*, the normalized return ratio with respect to

parameter f_i is independent of R_l. This presumption means that $Q_s(R_s, 0) = Q_s(R_s, \infty)$, whence the driving point output resistance is

$$R_{\text{out}} = R_{outo} = R_{cs2} = 60 \, \text{K}\Omega. \tag{E3-21}$$

Comments. This example comprises a further demonstration of the computational effectiveness of signal flow theory applied to the analysis of feedback amplifiers. Although several foundational calculations are a prerequisite to forging analytical results, these entail only straightforward circuit considerations and associated algebraic manipulations.

In the case of the feedback current amplifier, two disclosures are particularly interesting. First, the subject amplifier achieves an enormously large impedance transformation ratio between the amplifier input and output ports. Second, the algebraic form of the closed loop gain relationship in Eq. (E3-17) is foreboding because of the appearance of the negative algebraic sign in the denominator of the term on the far right of this expression. Fortunately, a call to "911" is unnecessary because the open loop gain is a negative number. To the extent that the feedback factor is couched as a positive number in the course of feedback network modeling, the appearance of the aforementioned negative sign is inevitable whenever the amplifier undergoing investigation is a phase inverting structure.

5.3.2. Other Feedback Architectures

In addition to a study of the four basic forms of global feedback architectures, several other special case feedback examples are commonly encountered, thereby warranting their individual analytical scrutiny. These examples, which may or may not comprise global feedback, fundamentally address the case in which feedback is achieved through a single branch admittance or branch impedance.

5.3.2.1. Feedback Branch Admittance

Figure 5.20(a) displays a linear network to which feedback is implemented via a branch admittance, say Y_c, connected across the cth network port. As delineated in the figure, this admittance conducts a current, I_c, and supports a port voltage, V_c, which is shown in associated reference polarity with current I_c. Since I_c is obviously equal to $Y_c V_c$, the branch admittance in question can be supplanted by a controlled source of value, $Y_c V_c$, with the understanding that voltage V_c remains the voltage developed across the original branch admittance. Resultantly, the network in Fig. 5.20(a) is electrically identical to that of Fig. 5.20(b), where admittance Y_c is unambiguously

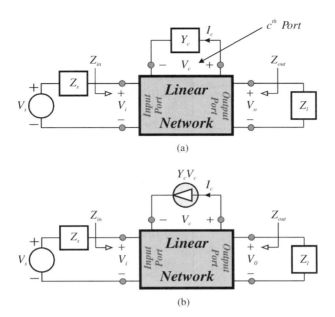

Figure 5.20. (a) A linear network in which feedback is implemented as a two-terminal branch admittance, Y_c, which is incident with the cth network port. (b) A network identical to that of (a), where the aforementioned branch admittance is replaced by the voltage controlled current source, $Y_c V_c$.

postured as the critical feedback parameter identified in earlier sections as the generic parameter, p.

The analysis of the network in Fig. 5.20(b) mirrors the three-step process developed in previous sections of material. In particular, the feedback controlled source, $Y_c V_c$, is set to zero to determine the null parameter gain and the null parameter values of the driving point input and output impedances, Z_{ino} and Z_{outo}, respectively. Setting $Y_c V_c$ to zero is tantamount to open circuiting the original branch admittance; that is, $Y_c = 0$. The situation at hand is illustrated in Fig. 5.21(a), where assuming that the input signal is the voltage, V_s, and the output response is the indicated voltage, V_o, the null gain, A_{vo}, is

$$A_{vo} = \left. \frac{V_o}{V_s} \right|_{Y_c V_c = 0}. \tag{5-57}$$

The next computation focuses on determining the normalized return ratio, $Q_s(Z_s, Z_l)$, with respect to the branch admittance, Y_c. To this end, Fig. 5.21(b) is applicable, where the signal source, V_s, is set to zero, the $Y_c V_c$

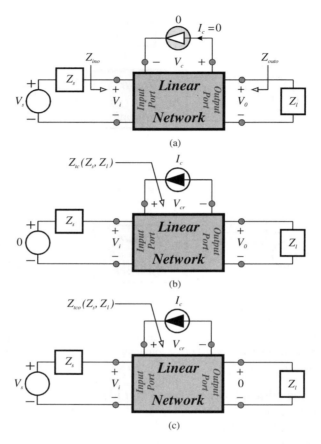

Figure 5.21. (a) The network of Fig. 5.20(a) configured for the evaluation of the voltage gain with $Y_c = 0$. (b) The network configuration for the determination of the normalized return ratio with respect to parameter Y_c. (c) Network configuration for the determination of the normalized null return ratio with respect to parameter Y_c.

controlled source is replaced by an independent current source, I_c (flowing in the same direction as does $Y_c V_c$), and the original controlling voltage, V_c, is reversed in polarity and identified as V_{cr}. The requisite normalized return ratio follows as V_{cr}/I_c. But in this case, the normalized return ratio of interest is little more than the Thévenin impedance, say $Z_{tc}(Z_s, Z_l)$, that electrically faces the original branch admittance. Accordingly,

$$Q_s(Z_s, Z_l) = \left.\frac{V_{cr}}{I_c}\right|_{V_s=0} \equiv Z_{tc}(Z_s, Z_l), \quad (5\text{-}58)$$

as suggested in Fig. 5.21(b).

Signal Flow Methods of Feedback Network Analysis 369

Finally, the normalized null return ratio, $Q_r(Z_s, Z_l)$, is computed in concert with the guidelines framed by Fig. 5.21(c). In this diagram, the signal source, V_s, is left untouched, the output response, V_o, is clamped to zero, the original controlled source, $Y_c V_c$, is supplanted by the independent source, I_c, and the phase-reversed form of V_c is delineated as the voltage, V_{cr}. The desired normalized null return ratio follows as the impedance function, V_{cr}/I_c, which can be interpreted as the *null Thévenin impedance*, say $Z_{tco}(Z_s, Z_l)$, seen by the original branch admittance, Y_c. The normalized null return ratio is termed a "null Thévenin impedance" in that unlike the conventional tack of setting the independent signal source (or sources) to zero while computing a Thévenin impedance, said source is sustained at a level that nulls the network output response. The interim result at hand is

$$Q_r(Z_s, Z_l) = \left.\frac{V_{cr}}{I_c}\right|_{V_o=0} \equiv Z_{tco}(Z_s, Z_l). \tag{5-59}$$

The closed loop (feedback branch admittance Y_c connected to cth port) voltage gain, A_v, driving point input impedance, Z_{in}, and driving point output impedance, Z_{out}, resultantly become

$$A_v = \frac{V_o}{V_s} = A_{vo}\left[\frac{1 + Y_c Z_{tco}(Z_s, Z_l)}{1 + Y_c Z_{tc}(Z_s, Z_l)}\right], \tag{5-60}$$

$$Z_{in} = Z_{ino}\left[\frac{1 + Y_c Z_{tc}(0, Z_l)}{1 + Y_c Z_{tc}(\infty, Z_l)}\right], \tag{5-61}$$

and

$$Z_{out} = Z_{outo}\left[\frac{1 + Y_c Z_{tco}(Z_s, 0)}{1 + Y_c Z_{tc}(Z_s, \infty)}\right]. \tag{5-62}$$

The special case of a feedback branch admittance itself spawns at least two peripheral cases worthy of attention. The first of these is the short-circuited branch depicted in the abstraction of Fig. 5.22. Since a branch short circuit implies $Y_c = \infty$, Eqs. (5-60) through (5-62) deliver

$$A_v = \frac{V_o}{V_s} = A_{vo}\left[\frac{Z_{tco}(Z_s, Z_l)}{Z_{tc}(Z_s, Z_l)}\right], \tag{5-63}$$

$$Z_{in} = Z_{ino}\left[\frac{Z_{tc}(0, Z_l)}{Z_{tc}(\infty, Z_l)}\right], \tag{5-64}$$

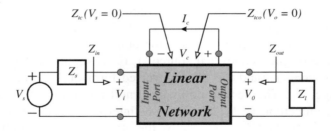

Figure 5.22. A feedback network for which feedback is implemented as a short circuit across the cth internal port of the circuit. The impedance, Z_{tc}, is the impedance "seen" by the branch short circuit under the condition of zero signal source voltage, V_s. In contrast, Z_{tco} is the impedance facing the short circuit when $V_s \neq 0$ and $V_o = 0$.

and

$$Z_{\text{out}} = Z_{outo}\left[\frac{Z_{tco}(Z_s, 0)}{Z_{tc}(Z_s, \infty)}\right]. \tag{5-65}$$

The evaluation of the impedances, $Z_{tc}(\cdot)$ and $Z_{tco}(\cdot)$, proceeds in a fashion reflective of the case of a finite branch admittance, Y_c. In the process of determining these impedances, the short-circuited branch is effectively broken, an independent current source is inserted into the break, and an Ohm's law analysis is conducted to calculate the subject impedances.

The second corollary to the feedback branch admittance case is the circumstance in which the admittance in question is that of a capacitor; that is, $Y_c = sC_c$, as shown in Fig. 5.23. For this network, the closed loop gain equation in Eq. (5-60) becomes

$$A_v = \frac{V_o}{V_s} = A_{vo}\left[\frac{1 + sC_c Z_{tco}(Z_s, Z_l)}{1 + sC_c Z_{tc}(Z_s, Z_l)}\right]. \tag{5-66}$$

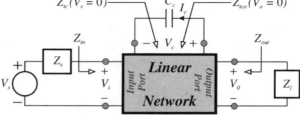

Figure 5.23. A linear feedback network for which feedback is implemented as a branch capacitance across the cth internal port of the circuit.

If the network in Fig. 5.23 is memoryless and if the source and load impedances are purely resistive, $Z_{tc}(\cdot)$ and $Z_{tco}(\cdot)$ in Eq. (5-66) are Thévenin resistances, say $R_{tc}(\cdot)$ and $R_{tco}(\cdot)$, that remain functionally dependent on the source and load resistances, R_s and R_l, respectively. In this event, Eq. (5-66) suggests that the time constant associated with the pole precipitated by capacitance C_c is $R_{tc}(R_s, R_l)C_c$, while the time constant associated with the zero generated by C_c is $R_{tco}(R_s, R_l)C_c$. These disclosures are handy when an engineering study compels an understanding of the frequency domain impact exerted on network performance by parasitic or otherwise undesirable capacitance. They are also useful when circuit bandwidth is to be approximated as a sum of open-circuited capacitive time constants[9] or when the zero frequency value of the network envelope delay is a required performance metric. It is worthwhile interjecting that $R_{tco}(R_s, R_l) = 0$ is a possibility, which implies that the zero produced by capacitance C_c lies at infinitely large frequency. Additionally, negative $R_{tco}(R_s, R_l)$ is also possible. In the latter case, the appended capacitance gives rise to a zero that lies in the right half complex frequency plane.

Example 5.4. The model depicted in Fig. 5.24(a) is the approximate, low frequency, small signal model of the CMOS feedback amplifier shown in Fig. 5.24(b). In this model, g_m represents the transconductance of the n-channel transistor, *MN*, and resistance r_o is the parallel interconnection of the channel resistances, r_{on} and r_{op}, of the n-channel and p-channel transistors, *MN* and *MP*, respectively. The signal source consists of the series interconnection of resistance R_s and Thévenin signal voltage V_s, to which the output response is the indicated voltage, V_o. The circuit resistance, R_g, is typically large and exploited almost exclusively for biasing purposes. The shunt interconnection of resistance R_f and capacitance C_f establishes a feedback path from the drain terminal to the gate terminal of transistor *MN*. Develop general expressions for the low frequency values of the closed loop voltage gain, the driving point input impedance, and the driving point output impedance. Exploit these disclosures to find the frequencies of the pole and zero established by capacitance C_f and the approximate 3-dB bandwidth of the amplifier. Approximate the gain and *I/O* impedance results for the typically encountered (but not necessarily always guaranteed) operating conditions of $R_g \gg R_s$, $g_m R_f \gg 1$, and $g_m(r_o \| R_l) \gg 1$. Investigate the possibility of achieving match terminated *I/O* port operation at low signal frequencies, in the sense of $R_{in} = R_{out} = R_l = R_s \triangleq R$. If match terminated operation is indeed possible, determine the requisite value of the feedback resistance, R_f, and the corresponding low frequency closed loop voltage gain.

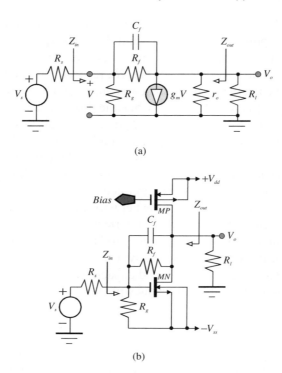

Figure 5.24. (a) Approximate, low frequency equivalent circuit of a CMOS amplifier for which feedback is established via the shunt $R_f - C_f$ subcircuit. (b) The CMOS amplifier modeled in (a).

Solution 5.4.

(1) The appropriate feedback parameter in this example is the admittance function, $Y_f = G_f + sC_f$, where $G_f = 1/R_f$, the conductance corresponding to the feedback resistance, R_f, is the effective feedback admittance evidenced at low signal frequencies. The admittance of this shunt $R_f - C_f$ interconnection is a more suitable signal flow feedback parameter than is the impedance of the interconnection for two practical reasons. In particular, setting Y_f to zero decouples the amplifier model input and output model sections, thereby simplifying the determination of the null parameter gain. If the impedance were to be selected, $Z_f = 0$ short circuits the input section to the output section and accordingly complicates the null parameter gain analysis. Moreover, $Z_f = 0$ does not remove the feedback branch as does $Y_f = 0$ and indeed, it effects short circuit feedback and feedforward between input and output amplifier ports. An inspection of Fig. 5.25(a), which depicts the model in Fig. 5.24(a) with $Y_f = 0$, reveals a null parameter voltage gain, A_{vo}, of

$$A_{vo} = \left. \frac{V_o}{V_s} \right|_{Y_f=0} = -g_m(r_o \| R_l)\left(\frac{R_g}{R_g + R_s}\right). \tag{E4-1}$$

While focused on this equivalent circuit, observe null input and output resistances, R_{ino} and R_{out}, respectively, of

$$R_{ino} = R_g, \tag{E4-2}$$

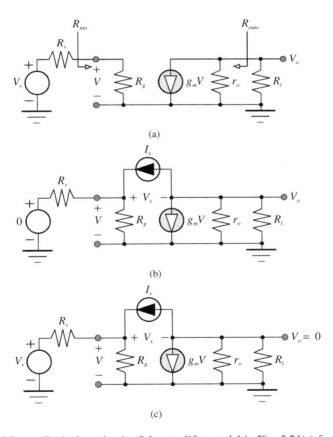

Figure 5.25. (a) Equivalent circuit of the amplifier model in Fig. 5.24(a) for computing the null parameter voltage gain, A_{vo}. (b) Circuit model used to determine the Thévenin impedance seen by the feedback admittance, $Y_f = G_f + sC_f$ in Fig. 5.24(a). This Thévenin impedance is the voltage-to-current ratio, V_x/I_x, under the condition of a signal source voltage, V_s, reduced to zero. (c) Circuit model used to determine the null Thévenin impedance seen by the feedback admittance in Fig. 5.24(a).

and
$$R_{outo} = r_o. \qquad (E4\text{-}3)$$

(2) The closed loop voltage gain, A_v, subscribes to the general relationship,

$$A_v = \frac{V_o}{V_s} = A_{vo} \left[\frac{1 + (G_f + sC_f) R_{tfo}(R_s, R_l)}{1 + (G_f + sC_f) R_{tf}(R_s, R_l)} \right], \qquad (E4\text{-}4)$$

where $R_{tf}(R_s, R_l)$ is the Thévenin resistance facing the feedback branch admittance, $Y_f = G_f + sC_f$, and $R_{tfo}(R_s, R_l)$ is the null Thévenin resistance seen by the subject admittance parameter. At low frequencies, the gain expression in Eq. (E4-4) collapses to its low frequency counterpart, $A_v(0)$,

$$A_v(0) = A_{vo} \left[\frac{1 + \dfrac{R_{tfo}(R_s, R_l)}{R_f}}{1 + \dfrac{R_{tf}(R_s, R_l)}{R_f}} \right], \qquad (E4\text{-}5)$$

Figure 5.25(b) is the model appropriate to finding the Thévenin resistance, $R_{tf}(R_s, R_l)$, facing the feedback branch admittance. In this diagram, observe a signal source voltage set to zero. Moreover, the feedback admittance is supplanted by an independent current source, I_x, across which a voltage, V_x, is established in disassociated polarity with I_x. A straightforward circuit analysis of the model at hand confirms

$$R_{tf}(R_s, R_l) = \left. \frac{V_x}{I_x} \right|_{V_s=0} = (r_o \| R_l) + [1 + g_m(r_o \| R_l)] \left(R_g \| R_s \right). \qquad (E4\text{-}6)$$

The null Thévenin resistance, $R_{tfo}(R_s, R_l)$, derives from an analysis of the equivalent circuit shown in Fig. 5.25(c). In this structure, the zero output response, $V_o = 0$, forces voltage V to be equal to the current source branch voltage, V_x. Moreover, $V_o = 0$ precludes current conduction through the net load resistance comprised of the shunt interconnection of resistances r_o and R_l, thereby forcing $I_x + g_m V = 0$. As a result,

$$R_{tfo}(R_s, R_l) = \left. \frac{V_x}{I_x} \right|_{V_o=0} = -\frac{1}{g_m}. \qquad (E4\text{-}7)$$

(3) Equations (E4-5) through (E4-7) combine with Eq. (E4-1) to produce a closed loop, low frequency voltage gain of

$$A_v(0) = -g_m(r_o \| R_l) \left(\frac{R_g}{R_g + R_s} \right)$$

$$\times \left[\frac{1 - \dfrac{1}{g_m R_f}}{1 + \dfrac{(r_o \| R_l) + [1 + g_m(r_o \| R_l)] \left(R_g \| R_s \right)}{R_f}} \right]. \qquad (E4\text{-}8)$$

For large $g_m(r_o \| R_l)$ and R_f selected to ensure $g_m R_f \gg 1$, this expression reduces to the recognizable simple gain result,

$$A_v(0) \approx -\frac{R_f}{R_s}. \tag{E4-9}$$

(4) The general expression for the low frequency driving point input resistance, R_{in}, is

$$R_{in} = R_{ino} \left[\frac{1 + \dfrac{R_{tf}(0, R_l)}{R_f}}{1 + \dfrac{R_{tf}(\infty, R_l)}{R_f}} \right], \tag{E4-10}$$

and by Eqs. (E4-2) and (E4-6),

$$R_{in} = R_g \left[\frac{1 + \dfrac{r_o \| R_l}{R_f}}{1 + \dfrac{(r_o \| R_l) + [1 + g_m(r_o \| R_l)]R_g}{R_f}} \right]. \tag{E4-11}$$

For large R_g and large $g_m(r_o \| R_l)$,

$$R_{in} \approx \frac{R_f + (r_o \| R_l)}{1 + g_m(r_o \| R_l)} \approx \frac{1}{g_m}\left(1 + \frac{R_f}{r_o \| R_l}\right). \tag{E4-12}$$

(5) In general, the driving point output resistance at low signal frequencies is

$$R_{out} = R_{outo} \left[\frac{1 + \dfrac{R_{tf}(R_s, 0)}{R_f}}{1 + \dfrac{R_{tf}(R_s, \infty)}{R_f}} \right], \tag{E4-13}$$

whence,

$$R_{out} = r_o \left[\frac{1 + \dfrac{R_g \| R_s}{R_f}}{1 + \dfrac{(R_g \| R_s) + [1 + g_m(R_g \| R_s)]r_o}{R_f}} \right]. \tag{E4-14}$$

For large r_o and large $g_m(R_g \| R_s)$,

$$R_{out} \approx \frac{R_f + (R_g \| R_s)}{1 + g_m(R_g \| R_s)} \approx \frac{1}{g_m}\left(1 + \frac{R_f}{R_g \| R_s}\right). \tag{E4-15}$$

It is notable that the expressions for the input and output impedances are similar and in either case, the impedance is only slightly larger than the inverse of the n-channel transistor transconductance. To crude first order, this transconductance increases as the square root of the biasing current conducted by the *n*-channel drain. Accordingly, low input and output impedances require suitably large drain biasing currents.

(6) An interesting and important sidebar derives from the observation in Eqs. (E4-11) and (E4-14) that R_{in} and R_{out} are identical if $R_s = R_l$ and $r_o = R_g$. If R_s and R_l are chosen to be a resistance of value R, and if in addition, a design scenario is adopted to ensure $R_{in} = R_{out} = R$, match terminated I/O ports are realized. This is to say that $R_{in} = R_{out} = R_l = R_s \triangleq R$ ensures the transfer of maximum signal power from the signal source to the amplifier input port, as well as from the amplifier output port to the load termination. Assuming $R_g \gg R_s = R$ in Eq. (E4-15), the value of feedback resistance, R_f, commensurate with this match terminated I/O port condition is

$$R_f = g_m R^2. \tag{E4-16}$$

For $r_o \gg R_l = R$ and $R_g \gg R_s = R$, the insertion of Eq. (E4-16) into Eq. (E4-8) produces a low frequency voltage gain of

$$A_v(0)|_{\substack{\text{Match} \\ \text{Terminated}}} \approx -\left(\frac{g_m R - 1}{2}\right). \tag{E4-17}$$

(7) The Thévenin resistance seen by capacitance C_f in the model of Fig. 5.24(a) is

$$R_f \| R_{tf}(R_s, R_l) = R_f \| \{(r_o \| R_l) + [1 + g_m(r_o \| R_l)](R_g \| R_s)\}$$
$$\approx R_f \| [R_l + (1 + g_m R_l) R_s]. \tag{E4-18}$$

It follows that the time constant, say τ_p, associated with pole established by C_f is

$$\tau_p \approx \{R_f \| [R_l + (1 + g_m R_l) R_s]\} C_f. \tag{E4-19}$$

For the match terminated condition implied by Eq. (E4-16), Eq. (E4-19) becomes

$$\tau_p|_{\substack{\text{Match} \\ \text{Terminated}}} \approx \left(\frac{2 + g_m R}{1 + g_m R}\right)\left(\frac{g_m R^2 C_f}{2}\right). \tag{E4-20}$$

On the other hand, the null Thévenin resistances faced by C_f is negative, provided $g_m R_f > 1$, and is given by

$$R_f \| R_{tfo}(R_s, R_l) = R_f \| \left(-\frac{1}{g_m}\right) = -\frac{R_f}{g_m R_f - 1}. \tag{E4-21}$$

The time constant, τ_z, precipitated by capacitance C_f follows as

$$\tau_z = -\frac{R_f C_f}{g_m R_f - 1}, \tag{E4-22}$$

which for match terminated conditions is

$$\tau_z|_{\substack{\text{Match} \\ \text{Terminated}}} = -\frac{g_m R^2 C_f}{(g_m R)^2 - 1}. \tag{E4-23}$$

The significance of the foregoing time constants is that the voltage transfer function of the feedback amplifier can be expressed in the form,

$$A_v = A_v(0) \left(\frac{1 + s\tau_z}{1 + s\tau_p} \right), \qquad \text{(E4-24)}$$

for which the corresponding 3-dB bandwidth is, by virtue of the finite frequency zero (albeit a right half plane zero) greater than $1/\tau_p$.

Comments. Amplifiers whose input and output ports are match terminated are pertinent to a diverse variety of broadband communication system applications. Despite the fact that signal flow analyses confirm theoretically that match terminated operation of the amplifier in Fig. 5.24(b) is possible, the actual closed loop gain results indicate at least one major shortcoming of the considered configuration. Specifically, Eq. (E4-17) verifies that a match terminated, low frequency magnitude of gain exceeding unity requires $g_m R > 3$. For the commonly desired match terminated resistance of $R = 50\,\Omega$, a forward device transconductance, g_m, of at least 60 mmho is required. Unfortunately, high frequency NMOS transistors featuring channel lengths that are no larger than 130 nanometers and biased to ensure minimal power dissipation are incapable of delivering transconductances larger than a few millimhos. This state of affairs implies that a single stage broadband realization of a CMOS *I/O* match terminated amplifier may comprise a daunting undertaking. However, match terminated operation can be achieved as a multistage cascade, in which the feedback admittance is connected between the *I/O* ports of an odd number of common source amplifiers, the architectures of which mirror that of the single stage cell in Fig. 5.24(b). The shortfall underpinning the multistage approach is that in the absence of design heroics, open loop pole dominance and hence closed loop stability, is not assured.

5.3.2.2. *Feedback Branch Impedance*

Feedback in a linear network is realized by means of a branch impedance, say Z_c, when open loop operating conditions are observed to prevail if said branch impedance is reduced to zero ohms. Fig. 5.26(a) abstracts the subject feedback branch impedance as an element incident with the *c*th network port. Observe that care is exercised to indicate the branch current, I_c, and the resultant *c*th port voltage, V_c, as circuit variables subscribing to the associated reference polarity convention. In view of this observation, Fig. 5.26(b) is electrically identical to the structure in Fig. 5.26(a) in that the

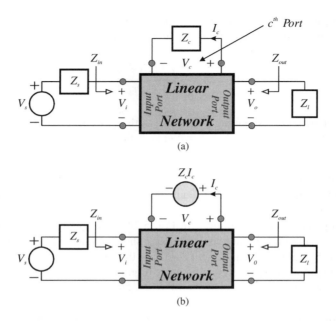

Figure 5.26. (a) A linear network in which feedback is implemented as a two terminal branch impedance, Z_c, which is incident with the cth network port. (b) A network identical to that of (a), in that the aforementioned branch impedance is replaced by the current controlled voltage source, $Z_c I_c$.

original branch impedance is supplanted by its branch voltage equivalent, which is modeled by the current controlled current generator, $Z_c I_c$.

The first analytical step entails the determination of the gain, presumed here to be the voltage gain, $A_{vo} = V_o/V_s$, under the condition of $Z_c = 0$, as illustrated in Fig. 5.27(a). Note that the immediate effect of $Z_c = 0$, or equivalently, $Z_c I_c = 0$, is a short-circuited cth network port. With Z_c sustained at zero ohms, the driving point input and output impedances can be evaluated as Z_{ino} and Z_{outo}, respectively.

The next analytical step embraces the computation of the normalized return ratio, $Q_s(Z_s, Z_l)$, with respect to the branch impedance, Z_c. Figure 5.27(b) is the pertinent electrical model, where the signal source, V_s, is set to zero, and the $Z_c I_c$ current controlled voltage source is replaced by an independent voltage, V_c, whose polarity mirrors that of the original controlled source. Moreover, the original controlling current, I_c, is reversed in polarity and identified as the current, I_{cr}. The desired normalized return ratio follows as I_{cr}/V_c. In this case, the subject normalized return ratio is

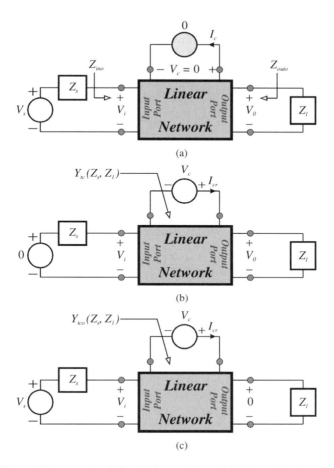

Figure 5.27. (a) The network of Fig. 5.26(a) configured for the evaluation of the voltage gain with $Z_c = 0$. (b) The network model for determining the normalized return ratio with respect to impedance parameter Z_c. (c) Network configuration for the determination of the normalized null return ratio with respect to parameter Z_c.

little more than the Thévenin admittance, say $Y_{tc}(Z_s, Z_l)$, that electrically faces the original branch admittance. Accordingly,

$$Q_s(Z_s, Z_l) = \left. \frac{I_{cr}}{V_c} \right|_{V_s = 0} \equiv Y_{tc}(Z_s, Z_l). \tag{5-67}$$

The normalized null return ratio, $Q_r(Z_s, Z_l)$, is evaluated in accordance with the diagram offered in Fig. 5.27(c). In particular, the signal source, V_s, is left unaltered, the output response, V_o, is clamped to zero, the original controlled source, $Z_c I_c$, is supplanted by the independent source, V_c, and the phase-reversed form of branch current I_c is delineated as I_{cr}. The required normalized null return ratio follows as the admittance function, I_{cr}/V_c, which can be interpreted as the *null Thévenin admittance*, say $Y_{co}(Z_s, Z_l)$, seen by the original branch impedance, Z_c. The interim result at hand is

$$Q_r(Z_s, Z_l) = \left.\frac{I_{cr}}{V_c}\right|_{V_o=0} \equiv Y_{tco}(Z_s, Z_l). \tag{5-68}$$

The closed loop (feedback branch impedance Z_c across the cth port not short circuited) voltage gain, A_v, driving point input impedance, Z_{in}, and driving point output impedance, Z_{out}, derive as

$$A_v = \frac{V_o}{V_s} = A_{vo}\left[\frac{1 + Z_c Y_{tco}(Z_s, Z_l)}{1 + Z_c Y_{tc}(Z_s, Z_l)}\right], \tag{5-69}$$

$$Z_{\text{in}} = Z_{ino}\left[\frac{1 + Z_c Y_{tc}(0, Z_l)}{1 + Z_c Y_{tc}(\infty, Z_l)}\right], \tag{5-70}$$

and

$$Z_{\text{out}} = Z_{outo}\left[\frac{1 + Z_c Y_{tco}(Z_s, 0)}{1 + Z_c Y_{tc}(Z_s, \infty)}\right]. \tag{5-71}$$

In Eq. (5-69), consider the special case in which Z_c is the inductive branch impedance, $Z_c = sL_c$. Then, if the linear network to which Z_c is connected to its cth port is memoryless, and if the source and load impedance terminations are purely resistive in the amount of R_s and R_l, respectively, $L_c Y_{tc}(R_s, R_l)$ is the time constant attributed to the network pole established by inductance L_c. Moreover, $L_c Y_{tco}(R_s, R_l)$ represents the time constant associated with the zero precipitated by L_c.

Example 5.5. The bipolar junction transistor transconductor amplifier in Fig. 5.28(a) utilizes an *emitter degeneration resistance*, R_e, to realize a series-series feedback architecture. In response to an applied signal that is modeled by a Thévenin voltage, V_s, placed in series with the Thévenin source resistance, R_s, the amplifier establishes an output current, I_o, conducted by the collector circuit load resistance, R_l. A proportion of this output current flows through the emitter resistance, R_e, to establish a signal voltage drop across R_e. This resistive drop subtracts from the signal source voltage (it therefore "degenerates" said signal voltage) to establish an "error" signal, v_{be}, across the base-emitter terminals of the transistors. Accordingly, $R_e = 0$ results in no such degenerative dynamics; that is, the implemented feedback is nullified when $R_e = 0$. Using the approximate, low frequency, small signal model provided in Fig. 5.28(b), determine the parameters, g_{mn} and

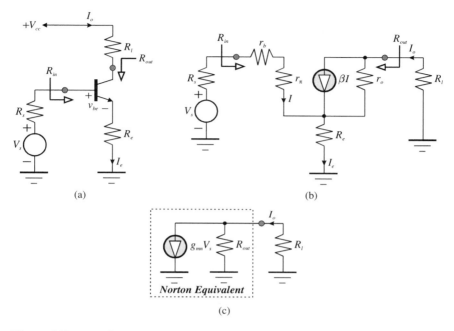

Figure 5.28. (a) Bipolar transistor feedback amplifier configured for deployment as a transconductor. (b) Approximate, low frequency, small signal equivalent circuit of the amplifier in (a). (c) The reduction of the model in (b) to a Norton equivalent circuit facing the load resistance, R_l. In this topology, g_{me} is the Norton transconductance between signal source and current output, while R_{out} is the driving point output resistance facing the load resistance, R_l.

R_{out}, of the Norton equivalent circuit that effectively drives the load resistance. For clarity, this Norton model is given in Fig. 5.28(c). Simplify the expressions for g_{mn} and R_{out} for the routinely encountered cases of substantively large transistor current gain (β) and substantial collector-emitter resistance (r_o). The latter transistor parameter is traditionally referred to as the *Early resistance* of the device.

Solution 5.5.

(1) Figure 5.29(a) is the pertinent model for calculating the null ($R_e = 0$) transconductance, say g_{me}, of the amplifier at hand. The load resistance is maintained in this model although ultimately, its value is to be set to zero to evaluate the Norton, or short circuit, forward transconductance, g_{mn}. The determination of g_{me} is tantamount to calculating the transconductance ratio, I_o/V_s, where I_o is the current conducted by the load. By inspection of the diagram in Fig. 5.29(a),

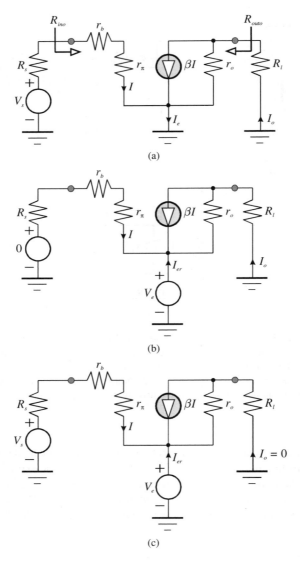

Figure 5.29. (a) Equivalent circuit of the amplifier in Fig. 5.28(a) for the calculation of the null parameter ($R_e = 0$) value of the transconductance, I_o/V_s. (b) Equivalent circuit pertinent to the computation of the normalized return ratio, I_{er}/V_e, with respect to resistance R_e. (c) Equivalent circuit for evaluating the normalized null return ratio, I_{er}/V_e, with respect to R_e.

this load current is given by

$$I_o = \beta I \left(\frac{r_o}{r_o + R_l}\right) = \left(\frac{\beta}{R_s + r_b + r_\pi}\right)\left(\frac{r_o}{r_o + R_l}\right) V_s, \quad \text{(E5-1)}$$

whence, the null transconductance, as a function of load and source resistances is

$$g_{meo}(R_s, R_l) = \left.\frac{I_o}{V_s}\right|_{R_e=0} = \left(\frac{\beta}{R_s + r_b + r_\pi}\right)\left(\frac{r_o}{r_o + R_l}\right). \quad \text{(E5-2)}$$

Also by inspection, the driving point null input and output resistances are

$$R_{ino} = r_b + r_\pi, \quad \text{(E5-3)}$$

and

$$R_{outo} = r_o, \quad \text{(E5-4)}$$

respectively.

(2) The closed loop (meaning resistance R_e is not short circuited) forward transconductance is given by the general signal flow relationship,

$$g_{me}(R_e, R_s, R_l) = \frac{I_o}{V_s} = g_{meo}(R_s, R_l)\left[\frac{1 + R_e G_{teo}(R_s, R_l)}{1 + R_e G_{te}(R_s, R_l)}\right]. \quad \text{(E5-5)}$$

In this expression, $G_{te}(R_s, R_l)$ is the normalized return ratio of the transconductance amplifier with respect to resistive element R_e. It is effectively the Thévenin conductance "seen" by resistance R_e under the condition of a null input signal ($V_s = 0$). With reference to the diagram of Fig. 5.29(b), this metric is the current to voltage ratio, I_{er}/V_e. Upon application of the Kirchhoff laws and after a bit of annoying algebra, it can be shown that

$$G_{te}(R_s, R_l) = \left(\frac{\beta + 1}{R_s + r_b + r_\pi}\right)\left(\frac{r_o + \dfrac{R_l + R_s + r_b + r_\pi}{\beta + 1}}{r_o + R_l}\right). \quad \text{(E5-6)}$$

Analogously, the diagram in Fig. 5.29(c) produces the normalized null return ratio, $G_{teo}(R_s, R_l)$ as the current to voltage ratio, I_{er}/V_e, under the constraint, $I_o = 0$. The result is

$$G_{teo}(R_s, R_l) = -\frac{1}{\beta r_o}. \quad \text{(E5-7)}$$

(3) Equations (E5-2), (E5-6), and (E5-7) can now be inserted into Eq. (E5-5) to arrive at a general expression, explicitly in terms of source resistance R_s, load resistance R_l, and emitter degeneration resistance R_e, for the forward transconductance of the amplifier in Fig. 5.28(a). In this particular example, however, only the Norton output conductance, g_{mn}, is required. Since the Norton current

derives from a short-circuited load ($R_l = 0$), it should be clear that

$$g_{mn} = g_{me}(R_e, R_s, 0) = g_{meo}(R_s, 0)\left[\frac{1 + R_e G_{teo}(R_s, 0)}{1 + R_e G_{te}(R_s, 0)}\right]. \tag{E5-8}$$

This observation and the foregoing results lead to

$$g_{mn} = \left(\frac{\beta}{R_s + r_b + r_\pi}\right)\left[\frac{1 - \dfrac{R_e}{\beta r_o}}{1 + \dfrac{R_e}{r_o \| \left(\dfrac{R_s + r_b + r_\pi}{\beta + 1}\right)}}\right]. \tag{E5-9}$$

For sufficiently large current gain β and sufficiently large Early resistance r_o, this result reduces to the satisfyingly simpler form,

$$g_{mn} \approx \frac{\beta}{R_s + r_b + r_\pi + (\beta + 1)R_e} \approx \frac{1}{R_e}. \tag{E5-10}$$

which is nominally independent of the signal source resistance.

(4) The general expression for the output resistance, R_{out}, in the Norton representation of the amplifier output port in Fig. 5.28(c) is

$$R_{\text{out}} = R_{outo}\left[\frac{1 + R_e G_{te}(R_s, 0)}{1 + R_e G_{te}(R_s, \infty)}\right]. \tag{E5-11}$$

It is important to interject that in the course of determining the metrics, $G_{te}(R_s, 0)$ and $G_{te}(R_s, \infty)$, no *a priori* liberties must be taken as regards the presumptions of large β and large r_o. In other words, it is necessary to apply the "exact" relationship for $G_{te}(R_s, R_l)$ while evaluating $G_{te}(R_s, 0)$ and $G_{te}(R_s, \infty)$. Simplifying assumptions regarding β, r_o, or any other transistor or circuit parameters can be invoked only after the functional forms of said return ratio metrics and the desired output resistance are meticulously delineated. Thus, from Eq. (E5-6),

$$G_{te}(R_s, 0) = \frac{1}{r_o \| \left(\dfrac{R_s + r_b + r_\pi}{\beta + 1}\right)}, \tag{E5-12}$$

while

$$G_{te}(R_s, 0) = \frac{1}{R_s + r_b + r_\pi}. \tag{E5-13}$$

It follows that the driving point output resistance is

$$R_{\text{out}} = r_o\left[\frac{1 + \dfrac{R_e}{r_o \| \left(\dfrac{R_s + r_b + r_\pi}{\beta + 1}\right)}}{1 + \dfrac{R_e}{R_s + r_b + r_\pi}}\right], \tag{E5-14}$$

which typically reduces to the approximate form,

$$R_{\text{out}} \approx r_o \left(1 + \frac{\beta R_e}{R_s + r_b + r_\pi + R_e}\right). \tag{E5-15}$$

Comments. Despite the topological simplicity of the transconductor in Fig. 5.28(a), its assessment entails surprisingly cumbersome algebra, thereby lending credence to the systematic analysis promoted by signal flow, or circuit partitioning, theory. In the present case, signal flow methods capture the dramatic influence exerted by the series-series feedback established by the emitter degeneration resistance, R_e. For example, large R_e (which actually may be detrimental from the perspectives of circuit power dissipation, electrical noise, and small forward transconductance), gives rise to a Norton transconductance in Eqs. (E5-9) and (E5-10) that approaches the inverse of R_e and is therefore independent of source resistance and the vagaries of transistor model parameters. Moreover, Eq. (E5-15) demonstrates that large R_e engenders a driving point output resistance, R_{out}, that can be substantively larger than the already reasonably large Early resistance, r_o.

5.3.3. Dual Loop Feedback

The feedback systems studied thus far are single loop architectures in that only one critical feedback parameter contributes to the closed loop response. For stipulated source and load terminations, the lone feedback parameter in these networks can be chosen to ensure a closed loop I/O gain that is adequately desensitized to potentially troublesome open loop parameters. Unfortunately, once the feedback parameter is prescribed in accordance with network gain objectives, no degrees of freedom are afforded the circuit designer interested in additionally satisfying particular input and output impedance specifications. Recall that the I/O impedances of all single loop systems studied thus far are either very large or very small, depending on the nature and interconnection of the feedback subcircuit. Moreover, these impedances are invariably dependent on open loop amplifier gain metrics whose precise values are rarely known. In numerous applications, the inability to realize accurately an input and/or output impedance can succumb to mere design assurances of sufficiently large or small impedance levels. In other applications, such as electronic circuits earmarked for broadband communication systems, match terminated input and output ports and hence, predictable driving point input and output impedances, are an absolute necessity.

In contrast to single loop systems, extra design degrees of freedom are provided by multiple loop feedback circuits. Although any number of feedback loops is theoretically permissible, two distinct loops appear to be the most practical from the perspectives of reliable and reproducible circuit design. To the latter end, Fig. 5.30 abstracts a dual loop linear feedback circuit for which two critical feedback parameters, p_1 and p_2, are evident. Although parameter p_1 is a transadmittance that emulates voltage controlled current feedback from the cth to the kth port and p_2 is a transimpedance reflecting current controlled voltage feedback from the mth to the nth port, minor and obvious interpretive changes to the analyses that follow allow for the consideration of any topological forms of feedback pairs.

The closed loop voltage gain, V_o/V_s, of the system in Fig. 5.30 is a function of the source impedance, Z_s, the load impedance, Z_l, and the two feedback parameters, p_1 and p_2. Using Eqs. (5-7), (5-8), and (5-11), this gain can be expressed as

$$A_v(p_1, p_2, Z_s, Z_l) = \frac{V_o}{V_s} = A_v(0, p_2, Z_s, Z_l) \left[\frac{1 + p_1 Q_{r1}(p_2, Z_s, Z_l)}{1 + p_1 Q_{s1}(p_2, Z_s, Z_l)} \right], \quad (5\text{-}72)$$

where the feedback loop embracing parameter p_1 is arbitrarily selected as the first focus of network analysis. In this expression, $A_v(0, p_2, Z_s, Z_l)$ is the null parameter gain with respect to parameter p_1. This null gain is necessarily dependent on p_2 because the second loop is left unaltered in the course of forging the expression in Eq. (5-72). Similarly, the normalized loop gain, $Q_{s1}(p_2, Z_s, Z_l)$, with respect to parameter p_1 is a function of p_2, as is the normalized null return ratio, $Q_{r1}(p_2, Z_s, Z_l)$, with respect to p_1. Since the null gain with respect to p_1 is also the $p_1 = 0$ value of the

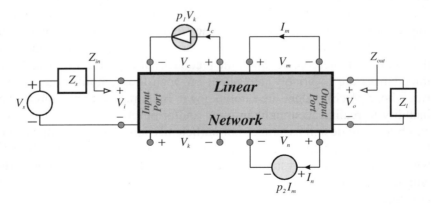

Figure 5.30. An abstraction of a linear network into which two feedback loops have been incorporated.

closed loop gain with reference to the second feedback loop established by parameter p_2, Eqs. (5-7), (5-8), and (5-11) can be reapplied to write

$$A_v(0, p_2, Z_s, Z_l) = A_v(0, 0, Z_s, Z_l)\left[\frac{1 + p_2 Q_{r2}(Z_s, Z_l)}{1 + p_2 Q_{s2}(Z_s, Z_l)}\right], \quad (5\text{-}73)$$

where $A_v(0, 0, Z_s, Z_l)$ is the voltage gain with both feedback parameters set to zero, $Q_{s2}(Z_s, Z_l)$ is the $p_1 = 0$ value of the normalized return ratio with respect to parameter p_2, and $Q_{r2}(Z_s, Z_l)$ is analogously the $p_1 = 0$ value of the normalized null return ratio with respect to p_2. Upon insertion of Eq. (5-73) into Eq. (5-72), the overall closed loop gain becomes

$$A_v(p_1, p_2, Z_s, Z_l) = A_v(0, 0, Z_s, Z_l)\left[\frac{1 + p_2 Q_{r2}(Z_s, Z_l)}{1 + p_2 Q_{s2}(Z_s, Z_l)}\right]$$

$$\times \left[\frac{1 + p_1 Q_{r1}(p_2, Z_s, Z_l)}{1 + p_1 Q_{s1}(p_2, Z_s, Z_l)}\right]. \quad (5\text{-}74)$$

If parameter p_1 is chosen judiciously in the sense that a null value of p_1 simplifies the analysis of the feedback network, $A_v(0, 0, Z_s, Z_l)$, $Q_{s2}(Z_s, Z_l)$, and $Q_{r2}(Z_s, Z_l)$ are likely to derive straightforwardly. In concert with the I/O impedance expressions deduced below, these feedback network metrics may even allow for an expedient and a relatively simple macromodel of the network pertinent to the condition, $p_1 = 0$. The resultant simplified model can then be exploited to deduce the requisite analytical expressions for $Q_{s1}(p_2, Z_s, Z_l)$ and $Q_{r1}(p_2, Z_s, Z_l)$. For the specific example delineated in Fig. 5.30, observe that $Q_{s1}(p_2, Z_s, Z_l)$ and $Q_{r1}(p_2, Z_s, Z_l)$ have units of impedance, while the dimensions of $Q_{s2}(Z_s, Z_l)$ and $Q_{r2}(Z_s, Z_l)$ reflect admittance.

An expression for the driving point input impedance of the dual loop feedback circuit in Fig. 5.30 can be developed in a fashion that mirrors the development of the closed loop gain relationship in Eq. (5-74). From Eq. (5-21), the input impedance, as a function of feedback parameters p_1 and p_2 and load impedance Z_l is

$$Z_{\text{in}}(p_1, p_2, Z_l) = Z_{\text{in}}(0, p_2, Z_l)\left[\frac{1 + p_1 Q_{s1}(p_2, 0, Z_l)}{1 + p_1 Q_{s1}(p_2, \infty, Z_l)}\right], \quad (5\text{-}75)$$

and in turn,

$$Z_{\text{in}}(0, p_2, Z_l) = Z_{\text{in}}(0, 0, Z_l)\left[\frac{1 + p_2 Q_{s2}(0, Z_l)}{1 + p_2 Q_{s2}(\infty, Z_l)}\right]. \quad (5\text{-}76)$$

It should be understood that $Q_{s2}(0, Z_l)$ and $Q_{s2}(\infty, Z_l)$ in Eq. (5-76) are evaluated with parameter p_1 set to zero; that is, the first feedback loop is deactivated in the process of evaluating these feedback network metrics. The combination of Eqs. (5-76) and (5-75) produces a final form closed loop input impedance of

$$Z_{\text{in}}(p_1, p_2, Z_l) = Z_{\text{in}}(0, 0, Z_l) \left[\frac{1 + p_2 Q_{s2}(0, Z_l)}{1 + p_2 Q_{s2}(\infty, Z_l)} \right]$$

$$\times \left[\frac{1 + p_1 Q_{s1}(p_2, 0, Z_l)}{1 + p_1 Q_{s1}(p_2, \infty, Z_l)} \right]. \qquad (5\text{-}77)$$

Similarly, it can be shown that for the driving point output impedance,

$$Z_{\text{out}}(p_1, p_2, Z_s) = Z_{\text{out}}(0, 0, Z_s) \left[\frac{1 + p_2 Q_{s2}(Z_s, 0)}{1 + p_2 Q_{s2}(Z_s, \infty)} \right]$$

$$\times \left[\frac{1 + p_1 Q_{s1}(p_2, Z_s, 0)}{1 + p_1 Q_{s1}(p_2, Z_s, \infty)} \right]. \qquad (5\text{-}78)$$

There are only two meaningful dual loop feedback architectures in the electronic circuits arena; namely the *series-series/shunt-shunt feedback* network and the *shunt-series/series-shunt* configuration.[10] Both of these feedback units require a multistage active network cascade in the open loop component of the feedback structure. The resultantly high open loop gain, coupled with the production of a multiplicity of open loop poles having potentially proximate frequencies, render these feedback amplifiers vulnerable to poor settling time characteristics, unacceptably large step response overshoot, and even outright dynamic instability. Fortunately, dominant pole and other forms of frequency compensation measures (to be addressed in a subsequent chapter) can mitigate the anemic phase and gain margins that are virtually inevitable with closed dual loop systems. Upon the implementation of satisfactory stability compensation elements or subcircuits, both the series-series/shunt-shunt and the shunt-series/series-shunt feedback amplifiers deliver a closed loop gain, input impedance, and output impedance that can be predicted accurately because the incorporated feedback loops render each of these performance metrics nominally independent of open loop parametric uncertainties.

5.3.4. Series-Series/Shunt-Shunt Feedback

Figure 5.31(a) is a system level schematic diagram of a series-series/shunt-shunt feedback amplifier. The amplifier block is a phase inverting configuration that is typically comprised of a cascade of an odd number

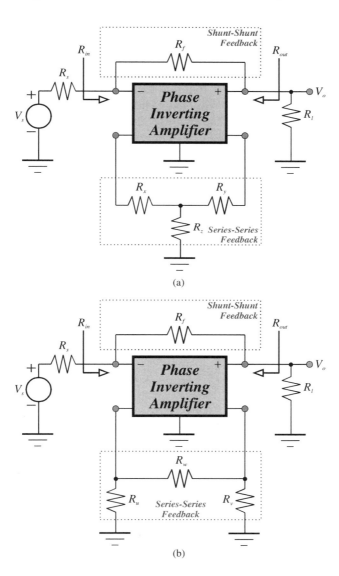

Figure 5.31. (a) A system level abstraction of a series-series/shunt-shunt dual loop feedback network. (b) An alternative form of the dual loop configuration in (a). If the resistances, R_u, R_v, and R_w are related to resistances R_x, R_y, and R_z in accordance with Eq. (5-79), the network is electrically identical to that of (a).

(generally three) of common source or common emitter amplifiers, depending on whether MOS technology or bipolar technology active devices are utilized. The signal excitation circuit is represented by the Thévenin equivalent circuit consisting of voltage V_s and source resistance R_s, while the load that terminates the amplifier output port to ground is taken as the resistance, R_l. It is assumed that the I/O response of interest is the voltage gain, V_o/V_s.

Shunt-shunt feedback in the diagram at hand derives from the global interconnection of the resistance, R_f, between the amplifier input and output ports. On the other hand, series-series feedback is established via the tee network interconnection of the three resistances, R_x, R_y, and R_z. An alternative series-series feedback subcircuit is the pi subcircuit comprised of the resistances, R_u, R_v, and R_w, shown in Fig. 5.31(b). A straightforward application of two-port parameter theory readily confirms that resistances R_u, R_v, and R_w can be chosen to ensure that the terminal volt ampere characteristics of the tee and pi structures are identical, whence the forward transfer and I/O impedance properties of the two networks in Fig. 5.31 are rendered respectively the same. To this end, R_u, R_v, and R_w are related to R_x, R_y, and R_z in accordance with the stipulations,

$$\left. \begin{array}{l} R_u = \left(1 + \dfrac{R_x}{R_y}\right)\left(R_z + R_x \| R_y\right) \\[1ex] R_v = \left(1 + \dfrac{R_y}{R_x}\right)\left(R_z + R_x \| R_y\right) \\[1ex] R_w = \left(1 + \dfrac{R_x \| R_y}{R_z}\right)\left(R_x + R_y\right) \end{array} \right\}. \qquad (5\text{-}79)$$

Either the tee or the pi embodiment of the series-series feedback subcircuit is acceptable for the dual loop feedback application considered herewith. The decision as to actual choice is dictated by such considerations as the nature of the requisite biasing networks used to linearize the active elements embedded within the amplifier and the ease of overall circuit integration. From a monolithic realization perspective, the pi network is usually more efficacious than is its tee counterpart since the resistances indigenous to the pi configuration tend to be slightly larger than the invariably small resistances of the tee network.

5.3.4.1. Analysis of the Series-Series/Shunt-Shunt Feedback Pair

The first step underlying the analysis of either of the two configurations shown in Fig. 5.31 consists of delineating the small signal model for the

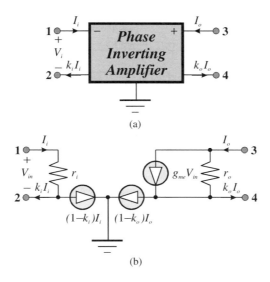

Figure 5.32. (a) The amplifier embedded within the feedback architectures shown in Fig. 5.31(a). (b) The low frequency macromodel of the amplifier in (a).

circuitry embedded within the amplifier, which is redrawn for convenience in Fig. 5.32(a). Assuming that only low signal frequencies are of immediate interest, the amplifier macromodel in Fig. 5.32(b) adequately services the present analytical discourse. Since the amplifier input port may not be a true series architecture, the current, I_i, flowing into input terminal 1 may differ from the current, designated as $k_i I_i$, flowing out of the input port return path at terminal 2. The difference, $(1 - k_i)I_i$, between these two input node currents is simply represented as an ideal current source directed from terminal 2 to ground. The two output port terminals are handled similarly. Finally, resistance r_i is the differential resistance of the amplifier input port, r_o represents the differential resistance associated with the amplifier output port, and g_{me} is the effective forward transconductance of the entire cascaded amplifier. Armed with the subject macromodel, the small signal model of the feedback network in Fig. 5.31(a) becomes the equivalent circuit offered in Fig. 5.33.

With reference to the discussion precipitating Eqs. (5-72) and (5-74), select feedback parameter p_1 as the conductance, G_f, associated with the shunt-shunt feedback resistance, R_f. It is to be noted that $G_f = 0$, which corresponds to $R_f = \infty$, results in the deactivation of the shunt-shunt

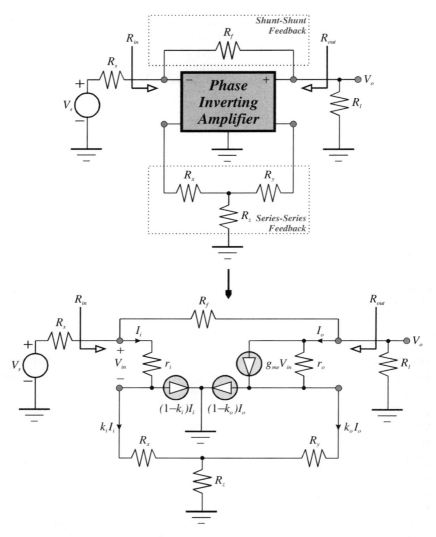

Figure 5.33. A viable low frequency macromodel of the dual loop feedback amplifier that appears in Fig. 5.31(a).

feedback loop. With $G_f = 0$, the equivalent circuit of Fig. 5.33 becomes the model in Fig. 5.34(a), whose only feedback entity is the series-series feedback loop established by the resistances, R_x, R_y, and R_z. With V_{f1} symbolizing the input voltage to this loop, and V_{f2} the output voltage, open

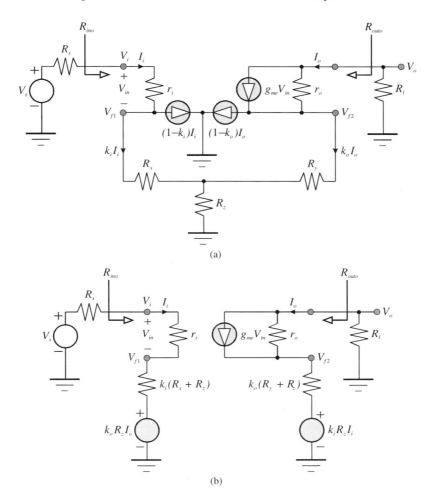

Figure 5.34. (a) The equivalent circuit of Fig. 5.33(b) with the shunt-shunt feedback loop established by the conductance, G_f, removed. (b) The circuit of (a) with the series-series feedback loop supplanted by its z-parameter equivalent circuit.

circuit impedance parameter (z-parameter) concepts give

$$\begin{bmatrix} V_{f1} \\ V_{f2} \end{bmatrix} = \begin{bmatrix} R_x + R_z & R_z \\ R_z & R_y + R_z \end{bmatrix} \begin{bmatrix} k_i I_i \\ k_o I_o \end{bmatrix}$$

$$= \begin{bmatrix} k_i(R_x + R_z) & k_o R_z \\ k_i R_z & k_o(R_y + R_z) \end{bmatrix} \begin{bmatrix} I_i \\ I_o \end{bmatrix}. \quad (5\text{-}80)$$

This expression allows the model of Fig. 5.34(a) to collapse to that of Fig. 5.34(b). Observe in the latter model that the second matrix term on the right hand side of Eq. (5-80) allows the currents, $(1-k_i)I_i$ and $(1-k_o)I_o$, to be removed, thereby allowing the elemental volt-ampere properties of the z-parameter representation of the series-series feedback loop to be referred to the amplifier input and output currents, I_i and I_o.

The feedback architecture depicted in Fig. 5.34(b) is clearly global in nature, since the feedback generator, $k_o R_z I_o$, with $k_o R_z$ serving as the effective feedback factor, is directly proportional to the output port current, I_o. Moreover, negative feedback is apparent, since current I_o is directly proportional to the amplifier input port voltage, or error signal, V_{in}, which is minimized by $k_o R_z I_o$ in that the polarity of this feedback generator opposes that of the applied signal source, V_s. It therefore follows that the effective forward transconductance, say $G_{fs}(k_o, R_s, R_l)$ of the model in Fig. 5.34(b) is of the form

$$G_{fs}(k_o, R_s, R_l) \triangleq \left.\frac{I_o}{V_s}\right|_{R_f=\infty} = \frac{G_{fo}(R_s, R_l)}{1 + k_o R_z G_{fo}(R_s, R_l)}, \tag{5-81}$$

where $G_{fo}(R_s, R_l)$ is the value, $G_{fs}(0, R_s, R_l)$, of $G_{fs}(k_o, R_s, R_l)$ with the feedback generator, $k_o R_z I_o$, set to zero. This transconductance metric is conveniently evaluated in conjunction with the diagram of Fig. 5.35, which

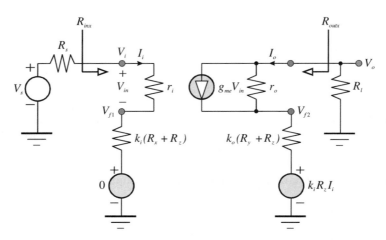

Figure 5.35. Equivalent circuit for the determination of the null feedback parameter values of the forward transconductance, I_o/V_s, input resistance, R_{inx}, and output resistance, R_{outx}, of the series-series feedback network in Fig. 5.34(b).

confirms

$$G_{fo}(R_s, R_l) \triangleq \left. \frac{I_o}{V_s} \right|_{\substack{R_f=\infty \\ k_o R_z I_o=0}}$$

$$= g_{me}\left(1 - \frac{k_i R_z}{g_{me} r_i r_o}\right)\left(\frac{r_i}{R_i + R_s}\right)\left(\frac{r_o}{R_o + R_l}\right), \quad (5\text{-}82)$$

where

$$\left. \begin{aligned} R_i &= r_i + k_i(R_x + R_z) \\ R_o &= r_o + k_o(R_y + R_z) \end{aligned} \right\}. \quad (5\text{-}83)$$

The diagram at hand also reveals that the null driving point input resistance is, in fact, $R_{inx} = R_i$, while the null driving point output resistance, R_{outx}, is identical to R_o. Recalling the generalizations of Eqs. (5-25) and (5-29), the resultant driving point input resistance, R_{ino}, to the series-series architecture, which is, in fact, the null input resistance with respect to shunt-shunt feedback parameter, G_f, is

$$R_{ino} = R_i\left[1 + k_o g_{me} R_z\left(1 - \frac{k_i R_z}{g_{me} r_i r_o}\right)\left(\frac{r_i}{R_i}\right)\left(\frac{r_o}{R_o + R_l}\right)\right]. \quad (5\text{-}84)$$

Similarly, the driving point output resistance, which is its null counterpart with respect to parameter G_f, is

$$R_{outo} = R_o\left[1 + k_o g_{me} R_z\left(1 - \frac{k_i R_z}{g_{me} r_i r_o}\right)\left(\frac{r_i}{R_i + R_s}\right)\left(\frac{r_o}{R_o}\right)\right]. \quad (5\text{-}85)$$

Equations (5-81), (5-84), and (5-85) allow for a utilitarian macromodel of the series-series subcircuit appearing in Fig. 5.35. This macromodel appears in Fig. 5.36(a), which clearly underscores a driving point input resistance of R_{ino} and a driving point output resistance of R_{outo}. Instead of displaying an output port voltage controlled current source that is dependent on the signal voltage, V_s, it is expedient to cast this controlled source as dependent on the input port voltage, V_i, in Fig. 5.35. To this end, observe that if $R_s = 0$, $V_s \equiv V_i$. Accordingly, the transconductance parameter, G_{so}, in Fig. 5.36(a) is, recalling Eq. (5-81),

$$G_{so} \triangleq G_{fs}(k_o, 0, R_l) \triangleq \left. \frac{I_o}{V_i} \right|_{R_f=\infty} = \frac{G_{fo}(0, R_l)}{1 + k_o R_z G_{fo}(0, R_l)}. \quad (5\text{-}86)$$

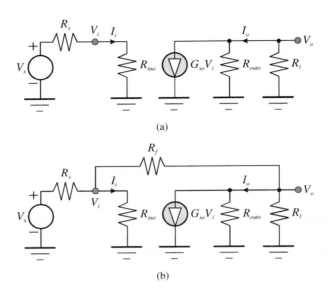

Figure 5.36. (a) Engineering macromodel for the series-series component of the dual loop feedback amplifier shown in Fig. 5.33. (b) The entire macromodel of the subject dual loop configuration.

In order to expedite forthcoming calculations, it is useful to stipulate the voltage gain, say A_{vo}, offered by this model as

$$A_{vo} \triangleq \left.\frac{V_o}{V_s}\right|_{R_f=\infty} = -G_{so}(R_{outo} \| R_l)\left(\frac{R_{ino}}{R_{ino}+R_s}\right). \tag{5-87}$$

With Fig. 5.36(a) serving as a valid engineering representation of the series-series component of the dual loop configuration undergoing investigation, the shunt-shunt subcircuit consisting of resistance R_f can now be strapped globally to the subject macromodel, with the result that Fig. 5.36(b) becomes the equivalent circuit of the entire dual loop feedback amplifier given in Fig. 5.31(a).

With G_f serving as the critical feedback parameter, the closed loop voltage gain of the amplifier modeled by the circuit in Fig. 5.36(b) is, recalling Eq. (5-60),

$$A_v = \frac{V_o}{V_s} = A_{vo}\left[\frac{1+G_f R_{tfo}(R_s, R_l)}{1+G_f R_{tf}(R_s, R_l)}\right], \tag{5-88}$$

where A_{vo} is given by Eq. (5-87), $R_{tf}(R_s, R_l)$ is the Thévenin resistance facing resistance R_f, and $R_{tfo}(R_s, R_l)$ is the null Thévenin resistance ($V_o = 0$) "seen" by R_f. With the aid of the model shown in Fig. 5.37(a), the Thévenin resistance facing R_f can be shown to be

$$R_{tf}(R_s, R_l) = \left.\frac{V_x}{I_x}\right|_{V_s=0} = (R_{outo} \| R_l) + [1 + G_{so}(R_{outo} \| R_l)](R_{ino} \| R_s).$$

(5-89)

On the other hand, a straightforward analysis of the companion equivalent circuit in Fig. 5.37(b) yields a null Thévenin resistance of

$$R_{tfo}(R_s, R_l) = \left.\frac{V_x}{I_x}\right|_{V_o=0} = -\frac{1}{G_{so}},$$

(5-90)

since $V_o = 0$ constrains V_i to V_x and I_x to $-G_{so}V_i$. The resultant closed loop voltage gain expression derives from the substitution of Eqs. (5-87), (5-89), and (5-90) into Eq. (5-88).

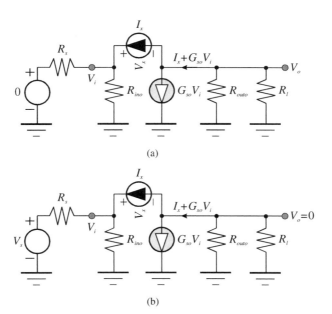

Figure 5.37. (a) Applicable circuit for the computation of the Thévenin resistance facing resistance R_f in the macromodel of Fig. 5.36(b). (b) Circuit for computing the null Thévenin resistance facing resistance R_f in the macromodel of Fig. 5.36(b).

The driving point input and output resistances now follow directly from Eqs. (5-61) and (5-62). In particular, the driving point input resistance is

$$R_{\text{in}} = R_{ino} \left\{ \frac{1 + \dfrac{(R_{outo} \| R_l)}{R_f}}{1 + \dfrac{(R_{outo} \| R_l) + [1 + G_{so}(R_{outo} \| R_l)]R_{ino}}{R_f}} \right\}, \quad (5\text{-}91)$$

which can be written more compactly as

$$R_{\text{in}} = R_{ino} \left\| \left\{ \frac{R_f + (R_{outo} \| R_l)}{[1 + G_{so}(R_{outo} \| R_l)]} \right\}. \quad (5\text{-}92)$$

For the output resistance,

$$R_{\text{out}} = R_{outo} \left\{ \frac{1 + \dfrac{(R_{ino} \| R_s)}{R_f}}{1 + \dfrac{(R_{ino} \| R_s) + [1 + G_{so}(R_{ino} \| R_s)]R_{outo}}{R_f}} \right\}, \quad (5\text{-}93)$$

or

$$R_{\text{out}} = R_{outo} \left\| \left\{ \frac{R_f + (R_{ino} \| R_s)}{[1 + G_{so}(R_{ino} \| R_s)]} \right\}. \quad (5\text{-}94)$$

5.3.4.2. Interpretation of Results and Design Considerations

The numerous analytical steps that underlie a definitive study of the series-series/shunt-shunt feedback pair can be daunting until the disclosed results are placed in proper engineering perspective. Before addressing these engineering perspectives, it may be prudent to review the entire analytical procedure adopted herewith.

Step 1: For the dual loop feedback amplifier given in Fig. 5.31(a), the *utilized phase inverting amplifier is modeled by a suitable small signal equivalent circuit*. If the topology embedded within the phase inverting structure is specified, this equivalent circuit derives directly from routine considerations of the device technologies exploited. In the present consideration, no such structure is delineated and thus, a reasonable macromodel is forged to emulate the requisite *I/O* electrical properties of the amplifier. This macromodel is the circuit provided in Fig. 5.32(b), where the amplifier is

Signal Flow Methods of Feedback Network Analysis

presumed to be a transconductor unit. Accordingly, the input port differential resistance, r_i, is large (of the order of a hundred thousand ohms or greater), as is the output port differential resistance, r_o (at least tens of thousands of ohms). Parameter g_{me} represents the effective forward transconductance of the phase inverting cell. Since the cell is likely a multistage cascade of high gain amplifiers g_{me} is of the order of hundreds to even thousands of millimhos. Finally, the constants, k_i and k_o, account for the fact that the input and output ports of the amplifier may not be true series architectures. For MOS or bipolar technologies, k_i can be very large, while k_o is generally close to unity. Once the amplifier model is established, it can be embedded into the given system level architecture. The result at hand is the diagram of Fig. 5.33.

Step 2: *One of the two feedback loops is deactivated.* There is no particular algorithm for selecting the loop to be removed other than such removal should lead to a simplified equivalent circuit. In the present case, the shunt-shunt loop comprised of the resistance, R_f, is removed, thereby leading to the series-series architecture depicted in Fig. 5.34(a).

Step 3: *The feedback subcircuit implicit to the resulting single loop configuration is modeled by conventional two-port network parameters.* Because the loop at hand is of a series-series nature, open circuit impedance parameters are chosen as the modeling vehicle. The advantage of two-port modeling is that the mathematical nature of the series-series feedback loop is clearly revealed, as is suggested by Fig. 5.34(b). The feedback observed presently is obviously global and negative, and the actual feedback parameter is $k_o R_z$, which is the factor multiplying the output current to establish a feedback voltage in series with the input port.

Step 4: *The forward gain and the I/O driving point resistances implied by the model of Fig. 5.34(b) are evaluated.* Since series-series feedback results in high input and high output resistance levels, the forward transconductance is selected as the gain of interest. This gain expression is the traditional global format of Eq. (5-81), where the null gain, $G_{fo}(R_s, R_l)$, requires an analysis of the circuit in Fig. 5.35 and is given by Eq. (5-82). The pertinent resistances are defined by Eqs. (5-83) through (5-85).

Step 5: The computations executed in the preceding step are intermediate to the determination of the final gain and *I/O* resistances of the overall dual loop system. Their principle impact is the *delineation of a model that incorporates the effects of the series-series subcircuit and allows for incorporation of the shunt-shunt feedback component*. This model is the structure of Fig. 5.36(a), the parameters of which are defined by Eqs. (5-84) through (5-86). Fig. 5.36(b) displays this model with the shunt-shunt feedback element, R_f, reconnected.

It is worthwhile interjecting that for general dual loop feedback architectures, it may not always be possible or convenient to construct a viable macromodel for the selected single loop subcircuit. For example, such modeling may render difficult the preservation of critical network terminals that must be delineated in the course of analyzing the electrical characteristics of the second feedback loop. But even if macromodeling proves to be impractical, the analytical fruits of the single loop analyses nevertheless expedite the delineation of ultimate closed loop amplifier response properties.

Step 6: The final step of the analytical procedure is the *determination of the requisite overall voltage gain and driving point input and output resistances*. The general gain expression is Eq. (5-88) and with the help of Fig. 5.37, the Thévenin parameters therein are straightforwardly evaluated. The pertinent resistance expressions are stipulated by Eqs. (5-91) through (5-94).

Although the foregoing analytical procedure is admittedly lengthy, each component of the procedure is computationally efficient and relatively straightforward. The alternative to the foregoing scenario is a direct circuit analysis of the structure in Fig. 5.33, which has two drawbacks, First, the algebra implicit to the direct method is sufficiently cumbersome to elicit four-letter descriptive words. Second, and most seriously, the direct method produces cumbersome performance expressions that cloud insightful comprehension of the operational dynamics of the subject feedback amplifier and therefore inhibits the formulation of meaningful design methodologies.

In an attempt to illustrate the design-oriented propriety of the foregoing signal flow investigation of the dual loop feedback amplifier, turn to Eq. (5-92) to examine the driving point input resistance. From Eq. (5-84), resistance R_{ino} is directly proportional to resistive parameter R_i, which Eq. (5-83) confirms is a linear function of the input port resistance, r_i, of the phase inverting amplifier. Since r_i is necessarily large in view of the transconductance nature of the phase inverting unit, it follows that R_{ino} is a large resistance. Similarly R_{outo} is a large resistance because this quantity is, by Eqs. (5-85) and (5-83), linearly related to the invariably large output port resistance, r_o, of the amplifier. It follows that Eq. (5-92) can be approximated as

$$R_{in} \approx \frac{R_f + R_l}{1 + G_{so}R_l}. \tag{5-95}$$

For reasons identical to those that lead to the foregoing expression, the driving point output resistance in Eq. (5-94) collapses to

$$R_{out} \approx \frac{R_f + R_s}{1 + G_{so}R_s}. \tag{5-96}$$

In both of the foregoing approximate relationships, recall that G_{so}, which is given by Eqs. (5-86) and (5-82), represents the zero source resistance value of the forward transconductance of the feedback amplifier with the shunt-shunt feedback resistance removed. It is interesting to observe that regardless of the numerical value of G_{so}, R_{in} and R_{out} are identical when the terminating source and load resistances are identical. This circumstance raises the specter of designing the feedback amplifier for maximum power transfer (at least at low signal frequencies) at both its input and output ports through the implementation of match terminated I/O constraints for which $R_s = R_{in} = R_{out} = R_l \triangleq R$. Either Eq. (5-95) or Eq. (5-96) leads to a shunt-shunt feedback resistance requirement for match terminated operation of

$$R_f = G_{so}R^2. \tag{5-97}$$

Note further that with or without a match terminated constraint and regardless of either the value or functional nature of parameter G_{so}, the driving point input and output resistances are not independent metrics since, by Eqs. (5-95) and (5-96),

$$\frac{R_{in}}{R_{out}} \approx \frac{R_f + R_l}{R_f + R_s}; \tag{5-98}$$

that is, for specified source and load terminations, the ratio of input to output resistances is determined by the shunt-shunt feedback element, R_f.

The rationale motivating the deployment of dual loop feedback is the desensitization of the input and output resistances. In Eqs. (5-95) and (5-96), the only dependence on relatively unreliable amplifier metrics materializes from the presence of parameter G_{so} in these functional forms. But from Eq. (5-86), G_{so} is determined dominantly by the passive resistance, R_z, and the current ratio parameter, k_o, which is invariably nearly unity for both MOS and bipolar technology devices. Specifically,

$$G_{so} \approx \frac{1}{k_o R_z}, \qquad (5\text{-}99)$$

provided $k_o R_z G_{fo}(0, R_l) \gg 1$. Clearly, the satisfaction of this inequality requires that $G_{fo}(0, R_l)$ be sufficiently large. Recalling Eq. (5-82),

$$G_{fo}(0, R_l) = g_{me}\left(1 - \frac{k_i R_z}{g_{me} r_i r_o}\right)\left(\frac{r_i}{R_i}\right)\left(\frac{r_o}{R_o + R_l}\right), \qquad (5\text{-}100)$$

which reduces to

$$G_{fo}(0, R_l) \approx g_{me}, \qquad (5\text{-}101)$$

if $g_{me} r_i \gg k_i R_z / r_o$, $r_i \gg k_i (R_x + R_z)$, and $r_o \gg [k_o(R_y + R_z) + R_l]$. The first of these inequalities is tantamount to tacit neglect of feedforward phenomena through the series-series feedback subcircuit, while the latter two constraints respectively require sufficiently large input port and output port resistances in the phase inverting amplifier. Accordingly, large $G_{fo}(0, R_l)$ requires a commensurately large forward transconductance in the phase inverting unit. Assuming that the foregoing inequalities are satisfied, Eq. (5-99) combines with Eqs. (5-95) and (5-96) to produce

$$R_{\text{in}} \approx \frac{R_f + R_l}{1 + \dfrac{R_l}{k_o R_z}}, \qquad (5\text{-}102)$$

and

$$R_{\text{out}} \approx \frac{R_f + R_s}{1 + \dfrac{R_s}{k_o R_z}}. \qquad (5\text{-}103)$$

These two analytical expressions portray the driving point input and output resistance functions as nominally independent of the parametric vagaries implicitly associated with the active amplifier block embedded in the dual loop feedback amplifier.

As might be expected, the overall feedback amplifier voltage gain defined by Eq. (5-88) is also relatively insensitive to active element uncertainties. In view of the various inequalities invoked at this juncture, the Thévenin resistance metrics, $R_{tf}(R_s, R_l)$, in Eq. (5-89) and $R_{tfo}(R_s, R_l)$ in Eq. (5-90), become, recalling Eq. (5-99),

$$R_{tf}(R_s, R_l) \approx R_l + \left[1 + \frac{R_l}{k_o R_z}\right] R_s, \qquad (5\text{-}104)$$

and

$$R_{tfo}(R_s, R_l) \approx -k_o R_z. \qquad (5\text{-}105)$$

Accordingly,

$$A_v = \frac{V_o}{V_s} \approx -\left[\frac{R_l}{R_l\left(1 + \frac{R_s}{k_o R_z}\right) + R_f + R_s}\right]\left(\frac{R_f}{k_o R_z} - 1\right). \qquad (5\text{-}106)$$

Using Eq. (5-103), this relationship can be cast in the form,

$$A_v \approx -\left(\frac{R_l}{R_l + R_{\text{out}}}\right)\left(\frac{\frac{R_f}{k_o R_z} - 1}{\frac{R_s}{k_o R_z} + 1}\right). \qquad (5\text{-}107)$$

For the I/O match terminated case in which R_f is chosen in accordance with Eq. (5-97) and $R_s = R_{\text{in}} = R_{\text{out}} = R_l \triangleq R$, Eq. (5-107) reduces to the simple result,

$$A_v \approx -\frac{1}{2}\left(\frac{R}{k_o R_z} - 1\right). \qquad (5\text{-}108)$$

Example 5.6. The series-series/shunt-shunt feedback amplifier of Fig. 5.31(a) is to be designed for a voltage gain of -10 (20 dB) when match terminated into 50 ohms. Assume that the current parameter constant, k_i, is 100, while $k_o = 1$. Assume further that noise constraints and the biasing of the phase-inverting amplifier require $R_x = R_y = 25\,\Omega$. Stipulate required constraints on the amplifier input resistance, r_i, the amplifier output resistance, r_o, and the amplifier forward transconductance, g_{me}. Realize the feedback amplifier in the topological form shown in Fig. 5.31(b).

Solution 5.6.

(1) With R = 50 Ω, $k_o = 1$, and $A_v = -10$, Eq. (5-108) yields $R_z = 2.38\,\Omega$.
(2) Recall that $g_{me} r_i \gg k_i R_z/r_o$, $r_i \gg k_i(R_x + R_z)$, and $r_o \gg [k_o(R_y + R_z) + R_l]$. Thus, with $R_z = 2.38\,\Omega$, $R_x = R_y = 25\,\Omega$, $R_l = 50\,\Omega$, $k_i = 100$, and $k_o = 1$, the amplifier input port resistance must satisfy $r_i \gg 2.74\,\text{K}\Omega$, while the amplifier output port resistance must abide by the constraint,

$r_o \gg 77.38\,\Omega$. Typical multistage transconductance amplifiers are readily designed with $r_i = 100\,\text{K}\Omega$ and $r_o = 10\,\text{K}\Omega$, both of which clearly satisfy the requisite inequalities. For $r_i = 100\,\text{K}\Omega$ and $r_o = 10\,\text{K}\Omega$, g_{me} must satisfy $g_{me} \gg 238.1$ nmho, which is hardly an issue. However, the discussion leading to Eqs. (5-100) and (5-101) reflects the presumption that $k_o R_z g_{me} \gg 1$, which implies $g_{me} \gg 420$ mmho. A safe value of the required transconductance might be 25 times this level, or $g_{me} \geq 10.5$ mho. Such a large transconductance mandates a multistage cascade realization of the phase inverting amplifier, regardless of whether MOS or bipolar technology devices are utilized.

(3) From Eqs. (5-97) and (5-99), the shunt-shunt feedback resistance, R_f, is obliged to be $R_f = R^2/k_o R_z = 1.05\,\text{K}\Omega$.

(4) With $R_x = R_y = 25\,\Omega$ and $R_z = 2.38\,\Omega$, Eq. (5-79) gives $R_u = R_v = 29.8\,\Omega$; moreover, $R_w = 312.5\,\Omega$. Figure 5.38 depicts the schematic diagram of the finalized design.

(5) Credence is lent to the foregoing design calculations by noting that if the 50 Ω load in Fig. 5.38 is removed, the resultant open circuit voltage gain is simply $-R_f/R_s = -1050/50 = -21$. In arriving at this result, use is made of the fact that the high input port impedance of the phase inverting amplifier inhibits the flow of appreciable input port current. But since the output resistance is matched to the terminating load resistance, the loaded voltage gain is this open

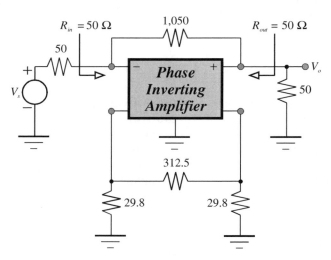

Figure 5.38. The amplifier designed in Example 5.6 for a voltage gain of 20 dB. All resistances in the diagram are in units of ohms. The target input port resistance of the gain block is 100 KΩ, the target output port resistance is 10 KΩ, and the recommended forward transconductance of the amplifier is 10.5 mhos. Moreover, the current gain parameters, k_i and k_o, are 100 and 1, respectively.

circuit voltage gain attenuated by a factor of two. Specifically, the terminated gain is $-21/2 = -10.5$, which is within 5% of the gain design objective of -10.

Comments. Despite the complexities surrounding the disclosures of analytical results for the series-series/shunt-shunt dual loop feedback amplifier, the laudable fruit of signal flow analysis is an elegantly straightforward design procedure. Moreover, the design tack nets reasonably accurate engineering results. To this end, a SPICE simulation of the circuit in Fig. 5.38 confirms a voltage gain magnitude of 9.788, a driving point input resistance of 51.996 Ω, and a driving point output resistance of 51.948 Ω, which differ from corresponding design objectives by only -2.12%, 3.93%, and 3.89%, respectively. The simulations also show that an outstanding desensitization of amplifier performance metrics with respect to the amplifier transconductance parameter, g_{me}, is achieved. In particular, a 20% increase in g_{me} to $g_{me} = 12.6$ mhos results in a gain magnitude increase of 0.358%, an input resistance decrease of 0.631%, and an output resistance decrease of 0.626%. On the other hand, a 20% decrease in g_{me} to $g_{me} = 8.4$ mhos yields a gain magnitude decrease of 0.531%, an increase in the input resistance of 0.945%, and an output resistance increase of only 0.938%.

5.3.5. Series-Shunt/Shunt-Series Feedback

The counterpart to the series-series/shunt-shunt dual loop feedback amplifier is the series-shunt/shunt-series configuration diagrammed in Fig. 5.39. The amplifier block is a non-inverting unit that is generally realized as a two- or four-stage common source or common emitter cascade. Series-shunt feedback is established by the $R_{e1} - R_{f1}$ resistive subcircuit, while shunt-series feedback is implemented as the $R_{e2} - R_{f2}$ network. As in the case of the series-series/shunt-shunt configuration, the source (R_s) and load (R_l) terminations are purely resistive in that only a low frequency analysis is undertaken herewith. The gain of interest is the voltage transfer function, $A_v = V_o/V_s$.

5.3.5.1. Analysis of the Series-Shunt/Shunt-Series Feedback Pair

The analysis of the feedback amplifier at hand initiates with a deactivation of either of the two feedback loops. If the shunt-series loop is selected for deactivation, either resistance R_{e1} can be supplanted by a short circuit or resistance R_{f1} can be open circuited. Choosing the latter approach, the reduced schematic diagram pertinent to the removal of the shunt-series

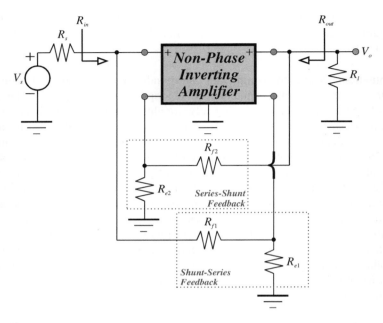

Figure 5.39. System level schematic diagram of a series-shunt/shunt-series dual loop feedback amplifier.

feedback subcircuit becomes that displayed in Fig. 5.40. In this diagram, the indicated driving point input resistance, R_{ino}, is understood to derive under the constraints that $R_{f1} = \infty$ and the shunt-series feedback loop is left untouched. The driving point output resistance, R_{outo}, and the voltage gain, A_{vo}, are computed for the same circuit conditions. It follows from Eq. (5-3) that the closed loop voltage gain, A_v, is

$$A_v = \frac{V_o}{V_s} = A_{vo} \left[\frac{1 + R_{to}(R_s, R_l)/R_{f1}}{1 + R_t(R_s, R_l)/R_{f1}} \right], \quad (5\text{-}109)$$

where $R_t(R_s, R_l)$ symbolizes the Thévenin resistance facing resistance R_{f1}, and $R_{to}(R_s, R_l)$ is the null ($V_o = 0$) Thévenin resistance seen by R_{f1}. It is to be understood that both of these Thévenin resistances are calculated with the series-shunt feedback loop connected. From Eqs. (5-70) and (5-71), the closed loop input and output resistances are

$$R_{in} = R_{ino} \left[\frac{1 + R_t(0, R_l)/R_{f1}}{1 + R_t(\infty, R_l)/R_{f1}} \right] \quad (5\text{-}110)$$

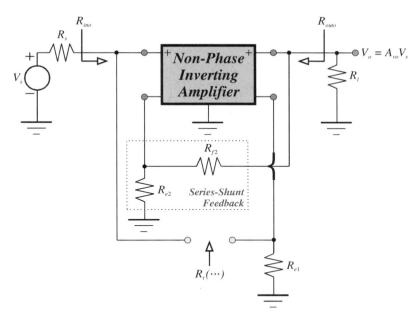

Figure 5.40. The feedback amplifier of Fig. 5.39 with the shunt-series feedback loop deactivated by replacing resistance R_{f1} by an open circuit.

and

$$R_{\text{out}} = R_{outo} \left[\frac{1 + R_t(R_s, 0)/R_{f1}}{1 + R_t(R_s, \infty)/R_{f1}} \right], \quad (5\text{-}111)$$

respectively.

To expedite the determination of the performance parameters, A_{vo}, R_{ino}, and R_{outo}, the circuit in Fig 5.40 is redrawn in Fig. 5.41(a) with the amplifier block conceptually represented by a macromodel whose form is similar to that appearing in Fig. 5.34(a). The principle difference between the presently utilized macromodel and its former version is that the polarity of the voltage controlled current source, $g_{me}V_{in}$, is reversed from the direction indicated in Fig. 5.34(a) to embrace the fact that the amplifier embedded in the system of Fig. 5.40 displays no I/O phase inversion. As in the previous macromodel, the current sources, $(1-k_i)I_i$ and $(1-k_o)I_o$, are incorporated to allow for the plausibility that the amplifier input and output ports may not comprise true series topologies.

408 Feedback Networks: Theory and Circuit Applications

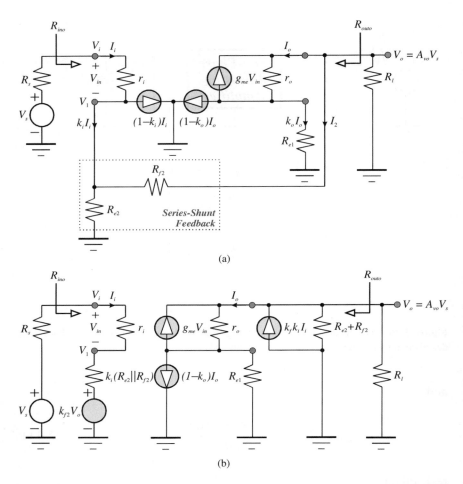

Figure 5.41. (a) The macromodel of the reduced feedback structure shown in Fig. 5.40. (b) The circuit of (a) with the series-shunt feedback subcircuit represented by its two-port h-parameter equivalent circuit.

The series-shunt nature of the feedback loop that prevails in Fig. 5.41(a) motivates exploiting hybrid h-parameters to emulate its terminal volt-ampere characteristics. To this end,

$$\begin{bmatrix} V_1 \\ I_2 \end{bmatrix} = \begin{bmatrix} h_{11} & h_{12} \\ h_{21} & h_{22} \end{bmatrix} \begin{bmatrix} k_i I_i \\ V_o \end{bmatrix} = \begin{bmatrix} k_i h_{11} & h_{12} \\ k_i h_{21} & h_{22} \end{bmatrix} \begin{bmatrix} I_i \\ V_o \end{bmatrix}, \qquad (5\text{-}112)$$

where

$$h_{11} = R_{e2} \| R_{f2}$$
$$h_{12} = -h_{21} = \frac{R_{e2}}{R_{e2} + R_{f2}} \triangleq k_{f2} \quad (5\text{-}113)$$
$$h_{22} = \frac{1}{R_{e2} + R_{f2}}$$

Equations (5-112) and (5-113) permit the replacement of the $R_{e2} - R_{f2}$ feedback subcircuit by its h-parameter model, thereby resulting in the equivalent circuit depicted in Fig. 5.41(b). It is worthwhile reinforcing the fact that this last model absorbs the series-shunt feedback loop embedded in the amplifier of Fig. 5.39, but it does not embody the electrical effects of the shunt-series feedback subcircuit. Accordingly, the input and output resistances, R_{ino} and R_{outo}, respectively, indicated in the subject diagram are effectively the open loop I/O resistances with respect to the shunt-series feedback network. Additionally, the voltage gain, A_{vo}, delineated implicitly in the diagram of Fig. 5.41(b) is the open loop gain; that is, it is the voltage gain of the dual loop amplifier under the condition that resistance R_{f1} is supplanted by an open circuit to deactivate the shunt-series feedback loop.

The diagram in Fig. 5.41(b) renders transparent the global negative feedback nature of the series-shunt loop. The critical feedback parameter is clearly the resistance ratio, k_{f2}, defined by the second of the expressions in Eq. (5-113). Thus, with $A_{vx}(R_s, R_l)$ designating the open loop gain with respect to parameter k_{f2} (and with resistance R_{f1} maintained at infinity), the pertinent closed loop gain, $A_{vo}(k_f, R_s, R_l)$, is

$$A_{vo}(k_f, R_s, R_l) = \frac{A_{vx}(R_s, R_l)}{1 + k_{f2} A_{vx}(R_s, R_l)}, \quad (5\text{-}114)$$

where, a conventional circuit analysis of the schematic diagram in Fig. 5.41(b) reveals

$$A_{vx}(R_s, R_l) = \left.\frac{V_o}{V_s}\right|_{k_{f2}V_o = 0}$$
$$= \left(\frac{g_{me} r_i}{1 + k_o R_{e1}/r_o} + k_{f2} k_i\right) \left[\frac{r_o \| (R_{e2} + R_{f2}) \| R_l}{R_s + r_i + k_i (R_{e2} \| R_{f2})}\right].$$
$$(5\text{-}115)$$

In Eq. (5-115), the shunt output resistance, r_o, of the noninverting transconductor amplifier is generally sufficiently large to permit the approximations, $r_o \gg k_o R_{e1}$ and $r_o \gg (R_{e2} + R_{f2}) \| R_l$. Moreover, feedforward through the series-shunt feedback loop is invariably small enough to validate the presumption, $g_{me} r_i \gg k_{f2} k_i$. As a result, Eq. (5-115) reduces to the simpler relationship,

$$A_{vx}(R_s, R_l) \approx g_{me} \left[(R_{e2} + R_{f2}) \| R_l \right] \left[\frac{r_i}{R_s + r_i + k_i (R_{e2} \| R_{f2})} \right]. \tag{5-116}$$

The substitution of Eq. (5-115) into Eq. (5-114) delivers an accurate expression for $A_{vo}(k_f, R_s, R_l)$, which is recalled to be the open loop voltage gain with respect to the shunt-series feedback loop that has been deactivated by setting $R_{f1} = \infty$. To the extent that a circuit design achieves $k_{f2} A_{vx}(R_s, R_l) \gg 1$, this open loop gain reduces to

$$A_{vo}(k_f, R_s, R_l) \approx \frac{1}{k_{f2}} = 1 + \frac{R_{f2}}{R_{e2}}. \tag{5-117}$$

The input resistance, R_{ino}, derives from

$$R_{ino} = R_{inx} \left[\frac{1 + k_{f2} A_{vx}(0, R_l)}{1 + k_{f2} A_{vx}(\infty, R_l)} \right], \tag{5-118}$$

where R_{inx}, the value of R_{ino} with $k_{f2} V_o = 0$, is seen in Fig. 5.41(b) to be

$$R_{inx} = r_i + k_i (R_{e2} \| R_{f2}). \tag{5-119}$$

Since Eq. (5-115) confirms $A_{vx}(\infty, R_l) = 0$, Eqs. (5-116) through (5-119) deliver

$$R_{ino} \approx \left[r_i + k_i (R_{e2} \| R_{f2}) \right] \left\{ 1 + k_{f2} g_{me} \left[(R_{e2} + R_{f2}) \| R_l \right] \right.$$
$$\left. \times \left[\frac{r_i}{r_i + k_i (R_{e2} \| R_{f2})} \right] \right\}$$
$$\approx (g_{me} r_i) R_{e2} \left(\frac{R_l}{R_l + R_{e2} + R_{f2}} \right), \tag{5-120}$$

where $r_i \gg k_i (R_{e2} \| R_{f2})$ and $k_{f2} g_{me} [(R_{e2} + R_{f2}) \| R_l] \gg 1$ are tacitly presumed. As is expected of the series-shunt feedback component of the dual loop amplifier at hand, the stipulated input resistance is large by virtue of its almost direct dependence on the amplifier forward gain metric, $(g_{me}r_i)$.

For the driving point output resistance, R_{outo},

$$R_{outo} = R_{outx} \left[\frac{1 + k_{f2}A_{vx}(R_s, 0)}{1 + k_{f2}A_{vx}(R_s, \infty)} \right]. \tag{5-121}$$

In order to determine R_{outx}, $k_{f2}V_o$ is set to zero in the model of Fig. 5.41(b). Since independent signal generators are always set to zero in the course of an output resistance determination, signal voltage V_s is also replaced by a null source. With $k_{f2}V_o$ and V_s both set to zero, current I_i vanishes, as does the input amplifier port voltage, V_{in}. Resultantly, an inspection of the diagram in Fig 5.41(b) shows that

$$R_{outx} = (r_o + k_o R_{e1}) \parallel (R_{e2} + R_{f2}) \approx R_{e2} + R_{f2}, \tag{5-122}$$

where the indicated approximation is the fruit of the valid assumption that r_o is sufficiently large to render $(r_o + k_o R_{e1}) \gg (R_{e2} + R_{f2})$. Equations (5-113), (5-116), (5-121), and (5-123) now combine to yield a driving point output resistance, R_{outo}, of

$$R_{outo} = \frac{R_{e2} + R_{f2}}{1 + k_{f2}g_{me}(R_{e2} + R_{f2}) \left[\frac{r_i}{r_i + R_s + k_i(R_{e2} \parallel R_{e1})} \right]}, \tag{5-123}$$

which reduces to

$$R_{outo} \approx \frac{1}{g_{me}} \left(1 + \frac{R_{f2}}{R_{e2}} \right) \tag{5-124}$$

for $r_i \gg [R_s + k_i(R_{e1} \parallel R_{e2})]$ and $g_{me}R_{e2} \gg 1$. Like the driving point input resistance, R_{ino}, defined by Eq. (5-120), the driving point output resistance, R_{outo}, given by Eq. (5-123) or Eq. (5-124) is the open loop output resistance with respect to the heretofore deactivated shunt-series feedback subcircuit highlighted by Fig. 5.40. The indicated inverse dependence of this resistance quantity on amplifier transconductance g_{me} renders said resistance relatively small, which is to be expected of a series-shunt feedback network.

The analysis of the series-shunt/shunt-series dual loop feedback amplifier concludes with a determination of the Thévenin resistance, $R_t(R_s, R_l)$, and the null Thévenin resistance, $R_{to}(R_s, R_l)$, facing the resistive feedback element, R_{f1}, in Fig. 5.39. Once these Thévenin resistance metrics are determined, Eqs. (5-109) through (5-111) can be exploited to evaluate the overall closed loop gain, A_v, driving point input resistance, R_{in}, and driving point output resistance, R_{out}. To this end, the applicable diagram is Fig. 5.42(a),

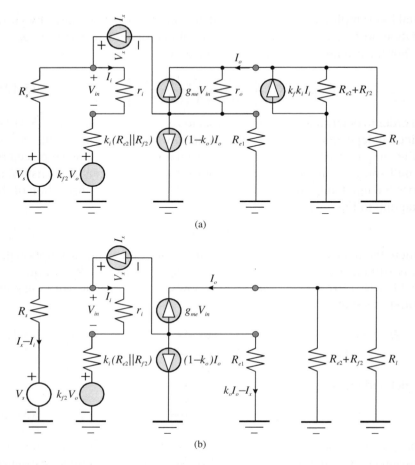

Figure 5.42. (a) Equivalent circuit pertinent to the computation of the Thévenin and the null Thévenin resistances facing the shunt-series feedback element, R_{f1}, in the dual loop feedback amplifier of Fig. 5.39. (b) Simplified version of the equivalent circuit in (a).

in which the series-shunt feedback subcircuit remains represented by its previously developed h-parameter equivalent circuit. The aforementioned Thévenin resistances derive as

$$R_t(R_s, R_l) = \left.\frac{V_x}{I_x}\right|_{V_s=0}$$
$$R_{to}(R_s, R_l) = \left.\frac{V_x}{I_x}\right|_{V_o=0} \quad (5\text{-}125)$$

Although the performance metrics in Eq. (5-125) can be evaluated without further modification of the architecture offered in Fig. 5.42(a), the preceding disclosures encourage at least two analytical liberties. The first of these liberties is the tacit neglect of the current conducted by the controlled source, $k_{f2}k_i I_i$, which accounts for feedforward through the series-shunt feedback component. The second is the presumption that the amplifier output port resistance, r_o, is sufficiently large to warrant its neglect. The temptation to presume $r_i = \infty$ is avoided because this resistance is connected in series with the source resistance, R_s, which is set to infinity to compute the driving point input resistance in Eq. (5-110). Accordingly, an anomalous input resistance result is potentially gleaned from an *a priori* presumption of infinitely large r_i. Such an anomaly does not prevail with respect to r_o because the load resistance, which is set to infinity in the process of determining the output resistance, is in shunt with the parallel combination of $(R_{e2} + R_{f2})$. The resultant simplified diagram is provided in Fig. 5.42(b).

An analysis of the circuit in Fig. 5.42(b) produces a cumbersome, but nonetheless mathematically tractable, expression for the Thévenin resistance, $R_t(R_s, R_l)$. In particular,

$$R_t(R_s, R_l) = \left\{ 1 + k_o g_{me} r_i \left[\frac{R_s}{R_s + k_i (R_{e2} \| R_{f2}) + (1 + k_{f2} g_{me} R_{lp}) r_i} \right] \right\} R_{e1}$$
$$+ R_s \| \{ k_i (R_{e2} \| R_{f2}) + (1 + k_{f2} g_{me} R_{lp}) r_i \}, \tag{5-126}$$

where R_{lp} denotes the parallel interconnection of resistances $(R_{e2} + R_{f2})$ and R_l; that is,

$$R_{lp} = \frac{(R_{e2} + R_{f2}) R_l}{R_{e2} + R_{f2} + R_l}. \tag{5-127}$$

For future reference, Eq. (5-126) verifies that

$$\left. \begin{array}{l} R_t(0, R_l) = R_{e1} \\ R_t(\infty, R_l) = (1 + k_o g_{me} r_i) R_{e1} + k_i (R_{e2} \| R_{f2}) \\ \qquad\qquad + (1 + k_{f2} g_{me} R_{lp}) r_i \end{array} \right\}, \tag{5-128}$$

and

$$\begin{aligned}R_t(R_s, 0) &= \left\{1 + k_o g_{me} r_i \left[\frac{R_s}{R_s + k_i (R_{e2} \| R_{f2}) + r_i}\right]\right\} R_{e1} \\ &\quad + R_s \| [k_i (R_{e2} \| R_{f2}) + r_i], \\ R_t(R_s, \infty) &= \left\{1 + k_o g_{me} r_i \left[\frac{R_s}{R_s + k_i (R_{e2} \| R_{f2}) + (1 + g_{me} R_{e2}) r_i}\right]\right\} R_{e1} \\ &\quad + R_s \| [k_i (R_{e2} \| R_{f2}) + (1 + g_{me} R_{e2}) r_i],\end{aligned} \quad (5\text{-}129)$$

From the diagram in Fig. 5.42(b), an output null ($V_o = 0$) forces $I_o = 0$, which renders $V_{\text{in}} = 0$ and $I_i = 0$. Consequently, $R_{to}(R_s, R_l)$ in Eq. (5-125) is the mercifully simple relationship,

$$R_{to}(R_s, R_l) = R_{e1}. \quad (5\text{-}130)$$

Recalling Eqs. (5-110), (5-120), and (5-128), the driving point input resistance of the subject dual loop feedback amplifier is, for sufficiently large amplifier transconductance g_{me} and amplifier differential input port resistance r_i,

$$R_{\text{in}} \approx \frac{\left(1 + \dfrac{R_{f1}}{R_{e1}}\right) R_l}{1 + \dfrac{R_{f2}}{R_{e2}} + \dfrac{R_l}{(k_o R_{e1}) \| R_{e2}}}. \quad (5\text{-}131)$$

If k_o is taken to be unity, which is rarely a contentious presumption for either a MOS technology or a bipolar technology realization of the series-shunt/shunt-series dual loop feedback amplifier,

$$R_{\text{in}} \approx \frac{\left(1 + \dfrac{R_{f1}}{R_{e1}}\right) R_l}{1 + \dfrac{R_{f2}}{R_{e2}} + \dfrac{R_l}{R_{e1} \| R_{e2}}} \quad (5\text{-}132)$$

is nominally independent of parameters that electrically define the required non-phase inverting amplifier. Similarly, large g_{me}, large r_i, and $k_o \approx 1$ deliver, from Eqs. (5-111), (5-123), and (5-129), an approximate driving point output resistance of

$$R_{\text{out}} \approx \frac{\left(1 + \dfrac{R_{f2}}{R_{e2}}\right) R_s}{1 + \dfrac{R_{f1}}{R_{e1}} + \dfrac{R_s}{R_{e1} \| R_{e2}}}. \quad (5\text{-}133)$$

Signal Flow Methods of Feedback Network Analysis   415

The closed loop gain follows from Eqs. (5-109), (5-117), (5-126), and (5-130). In particular,

$$A_v = \frac{V_o}{V_s} \approx \frac{\left(1 + \frac{R_{f2}}{R_{e2}}\right)\left(1 + \frac{R_{f1}}{R_{e1}}\right)}{1 + \frac{R_{f1}}{R_{e1}} + \frac{R_s}{R_{e1} \| R_{e2}} + \left(1 + \frac{R_{f2}}{R_{e2}}\right)\frac{R_s}{R_l}}, \qquad (5\text{-}134)$$

which, by Eq. (5-133), can be recast as

$$A_v \approx \left(1 + \frac{R_{f1}}{R_{e1}}\right)\left(\frac{R_{\text{out}} \| R_l}{R_s}\right). \qquad (5\text{-}135)$$

An alternative gain expression results from combining Eqs. (5-132) and (5-134) to obtain

$$A_v \approx \left(1 + \frac{R_{f2}}{R_{e2}}\right)\left(\frac{R_{\text{in}} \| R_s}{R_s}\right). \qquad (5\text{-}136)$$

It is interesting to note that for prescribed source and load terminations and desired or required driving point input and output resistances, the resistance ratios, R_{f1}/R_{e1} and R_{f2}/R_{e2}, are dependent designable parameters. Specifically, Eqs. (5-135) and (5-136) confirm

$$\frac{1 + R_{f2}/R_{e2}}{1 + R_{f1}/R_{e1}} = \frac{R_{\text{out}} \| R_l}{R_{\text{in}} \| R_s}. \qquad (5\text{-}137)$$

In addition to the interest held by the foregoing disclosure, it is also reassuring to observe that Eqs. (5-135) and (5-136) subscribe to previously established design concepts for relatively simple single loop feedback amplifiers. For example, the factor, $(1 + R_{f2}/R_{e2})$, on the right hand side of Eq. (5-136) represents the voltage gain, say A_{vsp}, of an ideal series-shunt feedback amplifier, which boasts infinitely large driving point input resistance. But the finite input resistance forged by the superimposed shunt-series feedback loop establishes an input port voltage divider between the source resistance, R_s, and the input resistance, R_{in}. Armed with this insight, it is hardly surprising that Eq. (5-136) can be written as

$$A_v \approx A_{vsp}\left(\frac{R_{\text{in}}}{R_{\text{in}} + R_s}\right). \qquad (5\text{-}138)$$

On the other hand, the multiplier, $(1 + R_{f1}/R_{e1})$ on the right hand side of Eq. (5-135) is the current gain, say A_{ips}, of an ideal shunt-series feedback amplifier, which is characterized by infinitely large driving point output resistance. Since the series-shunt feedback loop gives rise to a small, and assuredly finite, output resistance, this ideal current gain must be tempered by the output port current divider established between the output resistance, R_{out}, and the load resistance, R_l. Finally, if the signal current flowing in the terminating load resistance is taken in associated reference polarity with the signal load voltage, the observable voltage gain is simply the realized current gain multiplied by the load resistance to source resistance ratio. These contentions corroborate with Eq. (5-135), since the subject relationship is expressible as

$$A_v \approx \left(\frac{R_l}{R_s}\right) A_{ips} \left(\frac{R_{out}}{R_{out} + R_l}\right). \tag{5-139}$$

5.3.5.2. Design Restrictions

As forecast earlier, the series-shunt/shunt-series dual loop feedback amplifier, like its series-series/shunt-shunt counterpart, delivers a closed loop voltage gain, driving point input resistance, and driving point output resistance that are each nominally independent of vagarious parameters embedded in the requisite active gain block. Of these two fundamental dual loop feedback architectures, the series-shunt/shunt-series feedback amplifier is less popular, primarily because this configuration disallows concurrent match terminations at both input and output ports, thereby all but precluding its utilization in radio frequency and broadband electronics. A verification of this contention commences with an attempt to achieve a realization that boasts $R_s = R_{in} = R_{out} = R_l \triangleq R$. From Eq. (5-137) this constraint mandates $R_{f1}/R_{e1} \equiv R_{f2}/R_{e2}$, which means that the input resistance in Eq. (5-132) becomes

$$R_{in} \approx \left(\frac{1 + \dfrac{R_{f1}}{R_{e1}}}{1 + \dfrac{R_{f1}}{R_{e1}} + \dfrac{R}{R_{e1} \| R_{e2}}}\right) R. \tag{5-140}$$

Clearly, $R_{in} < R$, as opposed to the desired $R_{in} \equiv R$, and roughly approximates R only if $R \ll (R_{e1} \| R_{e2})$.

It is also impossible to realize a series-shunt/shunt-series feedback amplifier that features a matched input termination ($R_s = R_{in}$) that is not matched to a matched output termination ($R_l = R_{out}$). To wit, Eq. (5.137) demonstrates the necessity of

$$1 + \frac{R_{f2}}{R_{e2}} = \left(1 + \frac{R_{f1}}{R_{e1}}\right) \frac{R_l}{R_s}, \tag{5-141}$$

for which Eq. (5-132) stipulates

$$R_{in} \approx \frac{\left(1 + \frac{R_{f1}}{R_{e1}}\right) R_l}{\left(1 + \frac{R_{f1}}{R_{e1}}\right) \frac{R_l}{R_s} + \frac{R_l}{R_{e1} \| R_{e2}}} = \left(\frac{1 + \frac{R_{f1}}{R_{e1}}}{1 + \frac{R_{f1}}{R_{e1}} + \frac{R_s}{R_{e1} \| R_{e2}}}\right) R_s. \tag{5-142}$$

Obviously, $R_{in} < R_s$ and can be made to approximate R_s only if the value of the parallel interconnection of resistances R_{e1} and R_{e2} is significantly larger than R_s. Similarly, Eq. (5-133) verifies that

$$R_{out} \approx \frac{\left(1 + \frac{R_{f2}}{R_{e2}}\right) R_s}{\left(1 + \frac{R_{f2}}{R_{e2}}\right) \frac{R_s}{R_l} + \frac{R_s}{R_{e1} \| R_{e2}}} = \left(\frac{1 + \frac{R_{f2}}{R_{e2}}}{1 + \frac{R_{f2}}{R_{e2}} + \frac{R_l}{R_{e1} \| R_{e2}}}\right) R_l < R_l. \tag{5-143}$$

The last two disclosures confirm that individual impedance matches at the input and the output ports are not available options with the series-shunt/shunt-series dual loop feedback amplifier. However, and as the next example illustrates, it is possible to attain an impedance match at either, but not simultaneously both, of the two ports of the subject feedback amplifier.

Example 5.7. The series-shunt/shunt-series feedback amplifier of Fig. 5.39 is to be designed for a voltage gain of $+10$ (20 dB) when its input port is match terminated to a Thévenin source resistance of 50 ohms. The amplifier is to drive a 300 ohm resistive load incident at the output port, whose driving point resistance is only 20 ohms. Assume that the current parameter constant, k_i, is 100, while $k_o = 1$. Assume further that power dissipation and biasing constraints in the required non-phase inverting amplifier compel $R_{e1} \leq R_{e2}/4$. Stipulate required constraints with respect to the amplifier input resistance, r_i, the amplifier output resistance, r_o, and the amplifier forward transconductance, g_{me}. Test the propriety of the design through appropriate SPICE simulations.

Solution 5.7.

(1) With $R_{out} = 20\,\Omega$, $R_l = 300\,\Omega$, and $R_s = R_{in} = 50\,\Omega$, Eq. (5-137) requires

$$\frac{R_{f1}}{R_{e1}} = \frac{1}{3} + \frac{4}{3}\left(\frac{R_{f2}}{R_{e2}}\right). \tag{E7-1}$$

Recalling Eq. (5-136) $A_v = 10$ requires $R_{f2}/R_{e2} = 19$, whence by Eq. (E7-1), $R_{f1}/R_{e1} = 77/3$.

(2) Using Eq. (5-133) the preceding disclosures result in $(R_{e1} \| R_{e2}) = 2.143\,\Omega$. Since $R_{e1} \approx R_{e2}/4$, $R_{e2} = 10.72\,\Omega$ and $R_{e1} = 2.679\,\Omega$. It follows from the preceding computational step that $R_{f1} = 68.76\,\Omega$ and $R_{f2} = 203.7\,\Omega$.

(3) A plethora of approximations pervades the analytical disclosures upon which the foregoing design calculations are premised. The satisfaction of many of these simplifying approximations effectively defines the requisite electrical parameters of the non-phase inverting amplifier used in the realization of the dual loop feedback amplifier at hand.

  (a) Two previously disclosed approximations are $r_o \gg k_o R_{e1}$ and $r_o \gg (R_{e2} + R_{f2}) \| R_l$. The first of these approximations is not a problem since r_o is routinely of the order of tens of thousands of ohms, $k_o = 1$, and $R_{e1} = 2.68\,\Omega$. In light of the design calculations, $(R_{e2} + R_{f2}) \| R_l = 125.0\,\Omega$ and thus, $r_o \gg 125\,\Omega$ *is one of the design requirements*.

  (b) Recalling Eq. (5-113), $k_{f2} = 0.05$. A third approximation is $g_{me} r_i \gg k_{f2} k_i$, and with $k_i = 100$, $g_{me} r_i \gg 5$. The amplifier input port resistance, r_i, is typically in the range of the high tens to the low hundreds of kilo-ohms (especially for amplifiers realized in MOS technologies) and the amplifier forward transconductance, g_{me}, is at least several hundred millimhos for a multistage cascade realization of the non-phase inverting gain block. It is therefore difficult to dissatisfy the approximation, $g_{me} r_i \gg 5$.

  (c) A fourth approximation invoked in the preceding analyses is $r_i \gg R_s + k_i(R_{e1} \| R_{e2})$, or $r_i \gg 264.3\,\Omega$. If this approximation is to be satisfied to within an error implication of only 1%, $r_i \geq 26.43\,K\Omega$. Traditionally, and particularly with MOS technology devices, $r_i = 50\,K\Omega$ is readily achieved.

  (d) From Eq. (5-116), $A_{vx}(R_s, R_l) = 124.4\, g_{me}$ for $r_i = 50\,K\Omega$. Since $k_{f2} A_{vx}(R_s, R_l) \gg 1$ is a design requirement, $k_{f2} = 0.05$ yields $6.219\, g_{me} \gg 1$, or $g_{me} \gg 160.8$ mmho. Invoking the 1% error criterion once again, this stipulation translates to $g_{me} \geq 16.08$ mhos. An adequate transconductance objective is therefore $g_{me} = 20$ mho. For this transconductance value, note that $g_{me} R_{e2} = 214.4 \gg 1$, which is yet another design necessity, is easily satisfied.

  (e) The final design requirement is $r_o + k_o R_{e1} \gg R_{e2} + R_{f2}$, which implies $r_o \geq 21.44\,K\Omega$. Let $r_o = 25\,K\Omega$ be a design objective.

(4) Figure 5.43 displays the system level schematic diagram of the finalized design. In this diagram, the nominal values of the parameters indigenous to the non-phase inverting amplifier are a forward transconductance (g_{me}) of 20 mhos, an input port resistance (r_i) of 50 KΩ, and an output port resistance (r_o) of 25 KΩ.

Comments. SPICE simulations for the amplifier in Fig. 5.43, executed for $g_{me} = 20$ mhos $\pm$ 20%, yield the following tabulated results. Observe that for the nominal value of amplifier forward transconductance ($g_{me} = 20$ mhos), the simulated voltage gain (A_v) differs from the gain target by only -0.5%, and

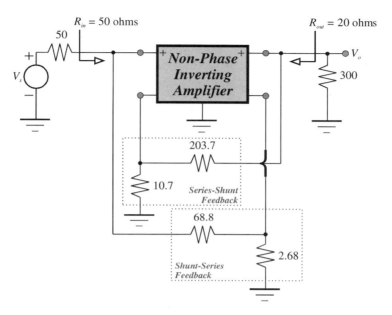

Figure 5.43. The amplifier designed in Example 5.7 for a voltage gain of 20 dB. All resistances in the diagram are in units of ohms. The input port resistance of the gain block is 50 KΩ, the output port resistance is 25 KΩ, and the nominal forward transconductance of the amplifier is 20 Mhos. Moreover, the amplifier current gain parameters, k_i and k_o, are 100 and 1, respectively.

g_{me} [mhos]	A_v [volts/volt]	R_{in} [Ω]	R_{out} [Ω]
16	9.93	50.16	21.08
20	9.95	50.13	20.87
24	9.96	50.11	20.73

the simulated driving point input resistance (R_{in}) is higher than its design target by only 0.26%. On the other hand, the simulated driving point output resistance (R_{out}) exceeds the pertinent design objective by 4.35%. The latter result is somewhat surprising in lieu of the extreme conservatism practiced with respect to the selection of appropriate amplifier transconductance, differential input port resistance, and differential output port resistance. This liberal design conservatism may prove impractical for broadband, low noise, and other high performance electronic system applications. In effect, the subject observation suggests relative difficulty in achieving a satisfactory design realization of the series-shunt/shunt-series dual loop feedback amplifier. On a positive note, a 50% variation in transconductance g_{me} from a low of 16 mhos to a high of 24 mhos impacts the voltage gain, input resistance, and output resistance by only 0.30%, −0.1%, and −1.66%, respectively.

References

1. G. Kron, *Tensor Analysis of Networks* (John Wiley and Sons, New York, 1939).
2. G. Kron, A set of principals to interconnect the solutions of physical systems, *J. Appl. Physics* **24** (1953), 965–980.
3. F. H. Branin, Jr., The relation between Kron's method and the classical methods of circuit analysis, *IEEE Conv. Rec.*, part 2 (1959), 3–28.
4. F. H. Branin, Jr., A sparse matrix modification of Kron's method of piecewise analysis, *Int'l. Symp. on Circuits and Systems* (1975), 383–386.
5. R. A. Rohrer, Circuit partitioning simplified, *IEEE Trans. on Circuits and Systems* **35** (1988), 2–5.
6. S. J. Mason, Feedback theory — some properties of signal flow graphs, *Proc. IRE* **41** (1953), 1144–1156.
7. S. J. Mason, Feedback theory — further properties of signal flow graphs, *Proc. IRE* **44** (1956), 920–962.
8. J. Choma, Jr., Advanced Network Analysis Concepts, *The Circuits and Filters Handbook* ed. W.-K. Chen. (CRC Press, 2003), Boca Raton, Florida, pp. 634–662.
9. J. Choma, Jr., *Electrical Networks*: *Theory and Analysis* (Wiley-Interscience, New York, 1985), Chapter 8.
10. G. Palumbo and J. Choma, Jr., An overview of analog feedback part II: Amplifier configurations in generic device technologies, *Analog Integrated Circuits and Signal Processing J.* **17** (1998), 195–219.

Exercises

Problem 5.1

Figure P5.1(a) depicts a shunt-series feedback amplifier realized in bipolar junction transistor (BJT) technology. Figure P5.1(b) provides the small signal, low frequency equivalent circuit for each transistor. In this small signal representation, r_b denotes the intrinsic base resistance, r_π symbolizes the small signal resistance associated with the base-emitter junction, r_o is the small signal collector-emitter resistance (often referred to as the "Early" resistance), and β is the common emitter short circuit current gain of the transistor. The model is capable of interrelating only signal induced, small perturbations of voltages and currents about respective quiescent operating, or biasing, levels. Accordingly, any static voltages and currents applied to, or generated within, the overall amplifier configuration are necessarily reduced to zero in the small signal amplifier equivalent circuit so that the only voltages and currents that remain in the model are those that materialize from applied signals. To this end, it should be understood in Fig. P5.1(a) that the currents, I_{iQ} and I_{oQ}, are biasing currents, V_{CC} and V_{EE} are biasing voltage sources, I_s is input signal current, and I_{os} is the signal component of net output current.

(a) What subcircuit in the amplifier comprises the feedback network?
(b) If the amplifier in Fig. P5.1(a) is to be modeled by a set of conventional two-port parameters, what two-port parameters are most appropriate for utilization?
(c) Use the model in Fig. P5.1(b) for both transistors (presumed to be identical devices that are biased identically) and signal flow partitioning theory to analyze the amplifier. In particular, select resistance R_1 as the critical feedback parameter to deduce analytical expressions for the following performance metrics:

  (i) the open loop current gain, I_{os}/I_s;
  (ii) the return ratio, T_s, with respect to resistance R_1;
  (iii) the null return ratio, T_r, with respect to resistance R_1;

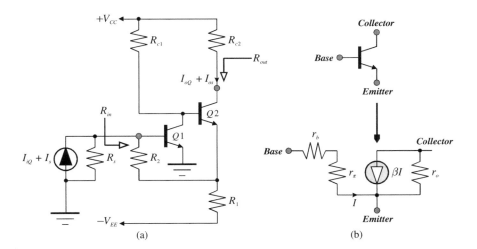

Figure P5.1.

(iv) the null driving point input resistance, say R_{ino};
(v) the null driving point output resistance, say R_{outo}.

(d) Assume tacitly that the transistors have large β and large r_o. Additionally, assume a large signal source resistance, R_s. Using the results generated in Part (c), give approximate expressions for the following circuit performance indices:

(i) the closed open loop current gain, $A_i = I_{os}/I_s$;
(ii) the closed loop driving point input resistance R_{in};
(iii) the closed loop driving point output resistance R_{out}.

(e) Let the transistors in the subject amplifier be characterized by $r_b = 150\,\Omega$, $r_\pi = 1.2\,\text{K}\Omega$, $r_o = 40\,\text{K}\Omega$, and $\beta = 120$ amps/amp. Additionally, let the circuit have $R_s = 25\,\text{K}\Omega$, $R_1 = 100\,\Omega$, $R_2 = 900\,\Omega$, and $R_{c1} = R_{c2} = 1.5\,\text{K}\Omega$. Compute the "exact" and the approximate values of the closed loop current gain, A_i, the driving point input resistance, R_{in}, and the driving point output resistance, R_{out}.

(f) Exploit SPICE to compare the simulated gain and resistance level results to both their "exact" and approximated values. Explain any observed discrepancies.

Problem 5.2

Reconsider the amplifier of Fig. P5.1(a) by allowing G_2, the conductance associated with resistance R_2, to be the critical feedback parameter. Repeat Parts (c) through (f) of Problem 5.1.

Problem 5.3

Figure P5.3(a) depicts a series-shunt feedback amplifier realized in metal-oxide-semiconductor field-effect transistor (MOSFET) technology. Figure P5.3(b) is the small signal, low frequency equivalent circuit for each MOS transistor. In this small signal representation, r_o denotes the drain-source channel resistance, and g_m symbolizes the forward transconductance of the transistor. The model is capable of interrelating only signal induced, small perturbations of voltages and currents about respective quiescent operating, or biasing, levels. Accordingly, any static voltages and currents applied to, or generated within, the overall amplifier configuration are necessarily reduced to zero in the small signal amplifier equivalent circuit so that the only voltages and currents that remain in the model are those that materialize from applied signals. To this end, it should be understood in Fig. P5.3(a) that the voltage, V_{oQ}, is a bias level, while V_{DD} and V_{SS} are biasing voltage sources.

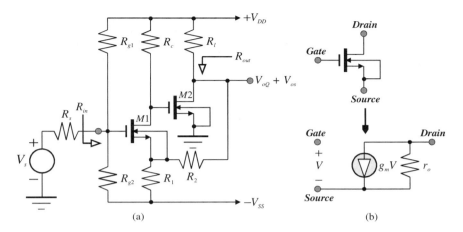

Figure P5.3.

(a) What subcircuit in the amplifier comprises the feedback network?
(b) If the amplifier in Fig. P5.3(a) is to be modeled by a set of conventional two-port parameters, what two port parameters are most appropriate for utilization?
(c) Use the model in Fig. P5.3(b) for both transistors (presumed to be identical devices that are biased identically) and signal flow partitioning theory to analyze the amplifier. In particular, select resistance R_1 as the critical feedback parameter to deduce analytical expressions for the following performance metrics:

(i) the open loop voltage gain, V_{os}/V_s;
(ii) the return ratio, T_s, with respect to resistance R_1;
(iii) the null return ratio, T_r, with respect to resistance R_1;
(iv) the null driving point input resistance, say R_{ino};
(iv) the null driving point output resistance, say R_{outo}.

(d) Assume tacitly that the transistors have large g_m and large r_o. Additionally, assume a signal source resistance, R_s, that is small. Using the results deduced in Part (c), give approximate expressions for the following circuit performance indices:

(i) the closed open loop voltage gain, $A_v = V_{os}/V_s$;
(ii) the closed loop driving point input resistance R_{in};
(iii) the closed loop driving point output resistance R_{out}.

(e) Let the transistors in the subject amplifier be characterized by $r_o = 30\,\text{K}\Omega$, and $g_m = 30\,\text{mmhos}$. Additionally, let the circuit have $R_s = 50\,\Omega$, $R_{g1} = R_{g2} = 1\,\text{Meg}\,\Omega$, $R_1 = 500\,\Omega$, $R_2 = 4.5\,\text{K}\Omega$, $R_c = 1\,\text{K}\Omega$, and $R_l = 2.5\,\text{K}\Omega$. Compute the "exact" and the approximate values of the closed loop voltage gain, A_v, the driving point input resistance, R_{in}, and the driving point output resistance, R_{out}.

(f) Exploit SPICE to compare the simulated gain and resistance level results to both their "exact" and approximated values. Explain any observed discrepancies.

Problem 5.4
Reconsider the amplifier of Fig. P5.3(a) by allowing G_2, the conductance associated with resistance R_2, to be the critical feedback parameter. Repeat Parts (c) through (f) of Problem 5.3.

Problem 5.5
The amplifier depicted in Fig. P5.5 has infinitely large shunt input resistance, zero Thévenin output port resistance, and a finite open loop voltage gain, A_o. The capacitance, C_i, represents the effective shunt input port capacitance and since no other amplifier capacitances are delineated, this capacitance is presumably the dominant energy storage element in the overall circuit. The amplifier is configured to function as an inverting buffer and accordingly, R_f is selected to first order as equal to the source resistance, R_s.

(a) Derive an expression for the closed loop voltage gain, $A_v(s) = V_o(s)/V_s(s)$. Execute this derivation be means of signal flow partitioning theory, using G_f, the conductance associated with resistance R_f, as the critical feedback parameter.
(b) Derive an expression for the 3-dB bandwidth, say B, of the circuit. Once again, apply signal flow theory to the solution methodology.

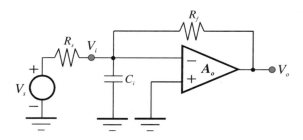

Figure P5.5.

(c) Using signal flow theory, derive an expression for the low frequency voltage gain, V_i/V_s, where voltage V_i is the signal voltage developed across the amplifier input port. Demonstrate that this voltage gain tends toward zero as the amplifier gain parameter, A_o, tends toward infinity.

(d) Since infinitely large open loop amplifier gains are observed only in academic environments, it is of engineering interest to investigate the response error precipitated by finite gain. To this end, define the error, ϵ, to be the difference between the magnitude of the input source signal voltage and the magnitude of the resultant response, V_o, under the simplifying condition of $V_s = 1$ volt. At low signal frequencies, what general condition must be satisfied by the gain parameter, A_o, if the design requirement is $\epsilon \leq 2\%$?

Problem 5.6

Numerous signal processing applications, such as transconductance amplifiers, phase detectors, and oscillators, demand current sources and sinks characterized by extremely high resistances at their current output ports. This design requirement is nontrivial when frequency response objectives mandate the use of deep submicron MOSFET technology transistors, which have relatively small drain-source channel resistances. The circuit in Fig. P5.6 begins responding to the foregoing requirement by incorporating a feedback voltage amplifier into a traditional cascode current sink. In this exercise, assume that the amplifier exuding voltage gain A_o is ideal in the senses of infinitely large input resistance, zero output resistance, and frequency-invariant open loop voltage gain. The indicated voltage, V_{bias}, is presumably constant in that it derives from some form of temperature

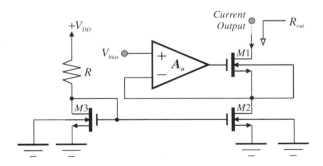

Figure P5.6.

stable bias supply. The small signal model for each transistor is the network displayed in Fig. P5.3(b). However, do not assume that the respective parameters of this model are the same for each transistor; that is, transistor $M1$ has $g_m = g_{m1}$ and $r_o = r_{o1}$, transistor $M2$ has $g_m = g_{m2}$ and $r_o = r_{o2}$, and so forth.

(a) Use the model of Fig. P5.3(b) to demonstrate that transistor behaves as a two-terminal resistor whose small signal resistance value, r_d, is

$$r_d = \frac{r_{o3}}{1 + g_{m3}r_{o3}}.$$

(b) Use the result of Part (a) and signal flow theory with amplifier gain A_o selected as the critical feedback parameter to derive an expression for the output resistance, R_{out}, indicated in Fig. P5.6.

(c) Give an engineering assessment of the impact exerted by parameter A_o on the subject output resistance.

Problem 5.7

A simple variation to the basic current sinking theme in Problem 5.6 is the Säckinger circuit displayed in Fig. P5.7. The biasing voltage, V_K, derives from an ideal, constant voltage source. All transistors abide by the model in Fig. P5.3(b), but they do not necessarily have the same respective small signal parameter values.

(a) Derive an expression for the indicated output resistance, R_{out}. Use signal flow theory with the transconductance, g_{m3}, of transistor $M3$ chosen as the critical feedback parameter.

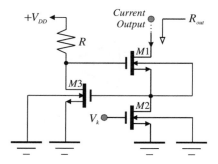

Figure P5.7.

(b) Give an engineering assessment of the impact exerted by transconductance g_{m3} on the output resistance of present interest.

Problem 5.8

The operational amplifier (op-amp) circuit shown in Fig. P5.8(a) exploits the resistance, R_f, to implement shunt-shunt global feedback. The signal source is the voltage, V_s, whose Thévenin resistance is R_s. The load termination is a resistance of value R_l. The simplified small signal dominant pole equivalent circuit of the op-amp is given in Fig. P5.8(b). In this structure, A_o symbolizes the positive and frequency invariant open loop gain, r_i is the effective input resistance, C_i is the effective input capacitance, and r_o is the Thévenin equivalent output resistance of the op-amp. Let the critical feedback parameter be the conductance, G_f, corresponding to resistance R_f. For the purpose of this problem, take $R_s = R_l = 500\,\Omega$, $R_f = 15\,\text{K}\Omega$, $r_i = 50\,\text{K}\Omega$, $r_o = 20\,\Omega$, $A_o = 3{,}500\,(70.9\,\text{dB})$, and $C_i = 1{,}000\,\text{pF}$.

(a) Derive an expression for the return ratio, $T_s(s, R_s, R_l)$, with respect to conductance G_f.
(b) Derive an expression for the null return ratio, $T_r(s, R_s, R_l)$, with respect to G_f.
(c) Derive an expression for the null parameter voltage gain, $A_{os}(s, R_s, R_l) = V_o/V_s(G_f = 0)$.
(d) Use the preceding results to deduce analytical relationships for the following circuit performance metrics:

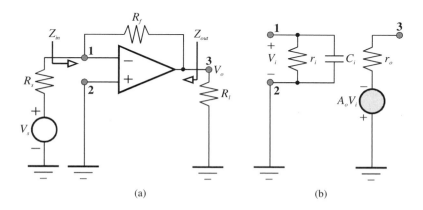

Figure P5.8.

(i) the open loop voltage gain, $A_{ol}(s)$, of the amplifier;
(ii) the loop gain, $T(s)$, of the amplifier;
(iii) the closed loop voltage gain, $A_{cl}(s)$, of the amplifier.

(e) Derive an expression for the open loop 3-dB bandwidth, B_{ol}, of the circuit.
(f) Derive an expression for the closed loop 3-dB bandwidth, B_{cl}.
(g) Discuss the relationship between the open loop and closed loop 3-dB bandwidths.
(h) Numerically evaluate the following metrics:

(i) the zero frequency return ratio;
(ii) the zero frequency null return ratio;
(iii) the zero frequency null voltage gain;
(iv) the zero frequency values of open loop gain, loop gain, and closed loop gain;
(v) the 3-dB open loop and closed loop bandwidths.

(i) Derive an approximate expression for, and numerically evaluate, the zero frequency closed loop driving point input impedance, $Z_{in}(0)$.
(j) Derive an approximate expression for, and numerically evaluate, the zero frequency closed loop driving point output impedance, $Z_{out}(0)$.

Problem 5.9
Repeat Problem 5.8 but now, choose as the critical feedback parameter the short circuit that interconnects resistance R_f to Terminal 1 of the operational amplifier.

Problem 5.10
Figure P5.10 is the simplified schematic diagram of a common gate amplifier that has been modified by incorporating transistor $M2$ to provide feedback between the source and gate terminals of transistor $M1$. The source current, I_s, is a low frequency signal, while I_Q is a constant current biasing element. All transistors can be modeled by the small signal equivalent circuit offered in Fig. P5.3(b). For analytical simplicity, however, assume all transistor channel resistances are infinitely large. In the inquiries that follow, select the transconductance, g_{m2}, of transistor $M2$ as the critical feedback parameter.

(a) Derive expressions for the following performance indices, assuming $g_{m2}R \gg 1$:

(i) the small signal null current gain, $A_{io} = I_o/I_s$;
(ii) the return ratio with respect to g_{m2};

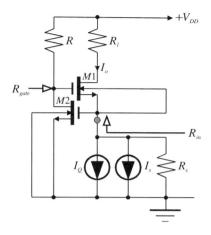

Figure P5.10.

- (iii) the null return ratio with respect to g_{m2};
- (iv) the open loop current gain;
- (v) the closed loop current gain.

(b) Exploit relevant preceding results to arrive at an expression for the driving point input resistance, R_{in}.

(c) If transistor $M2$ is supplanted by a constant current sink, $M1$ functions as a conventional common gate amplifier. In light of this observation and the analyses conducted herewith, discuss the attributes (if any) and shortfalls (if any) of utilizing $M2$ as indicated in the schematic diagram.

Problem 5.11

In the circuit of Fig. P5.10, use the model in Fig. P5.3(b) and signal flow partitioning theory to deduce an analytical expression for the resistance, R_{gate}, seen between the gate terminal of transistor $M1$ and circuit ground. Is the node at which resistance R_{gate} is evidenced susceptible to capacitive parasitics in the sense of significantly deterring the achievable 3-dB bandwidth of the amplifier?

Problem 5.12

In the feedback amplifier depicted in Fig. P5.12, the current sources, I_{iQ} and I_{oQ} are ideal biasing currents in the sense that their terminal resistances are infinitely large. In the interest of analytical simplification, the channel resistance, r_o, of the transistor can also be viewed as infinitely large in the small signal equivalent circuit given in Fig. P5.3(b). In the following inquiries, use signal flow partitioning theory with G_f, the conductance associated with the feedback resistance, R_f, as the critical feedback parameter.

(a) Show that the indicated driving point input resistance, R_{in}, is given by
$$R_{in} = \frac{R_f + R_l}{1 + g_{me} R_{ss}},$$
where g_{me} is the effective forward transconductance,
$$g_{me} = \frac{g_m}{1 + g_{me} R_{ss}}.$$

(b) Show that the indicated driving point output resistance, R_{out}, is given by
$$R_{out} = \frac{R_f + R_s}{1 + g_{me} R_{ss}}.$$

(c) Let the amplifier be designed for match terminated operation into a resistance R at both its input and output ports. In terms of resistance R and effective forward transconductance g_{me}, how must the feedback resistance, R_f, be selected to realize match-terminated performance at the input and output ports of the amplifier?

(d) In terms of resistance R and effective forward transconductance g_{me}, derive an expression for the voltage gain, $A_v = V_o/V_s$, for the match terminated condition specified in Part (c).

(e) What fundamental engineering limitations pervade the problem of realizing match terminated I/O performance with a voltage gain exceeding 0 dB in this amplifier architecture?

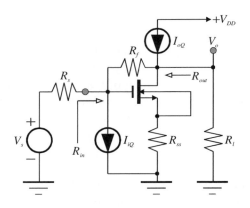

Figure P5.12.

Problem 5.13
Repeat Parts (a) through (d) of Problem 5.12 but now, choose resistance R_{ss} as the critical feedback parameter.

Problem 5.14
In the two-stage feedback amplifier depicted in Fig. P5.14, the response to the input signal represented by the Norton equivalent circuit comprised of current I_s and source resistance R_s is the small signal component, say I_{os}, of the indicated current, I_o, conducted by transistor, $M3$. The three transistors are identical, and all transistors, which abide by the small signal model in Fig. P5.3(b), have very large drain-source channel resistances. The coupling capacitor, C_c, is sufficiently large to enable its representation as a short circuit for the signal frequencies of interest. The source, I_Q, supplies constant current for biasing purposes only. Noting that no global feedback arises when the resistance, R_f, is infinitely large, choose $G_f = 1/R_f$ as the critical parameter in the course of applying signal flow partitioning theory to respond to the following issues.

(a) Determine the current gain, I_{os}/I_s, for the case of $G_f = 0$.
(b) Determine the return ratio with respect to conductance parameter G_f, and simplify the result for the special case of a very large resistance, R_q.

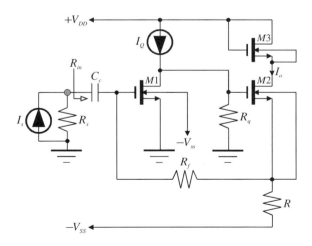

Figure P5.14.

(c) Determine the null return ratio with respect to conductance parameter G_f, and simplify the result for the special case of a very large resistance, R_q.
(d) What is the open loop current gain for large R_q?
(e) Use pertinent foregoing results to derive an expression for the closed loop input resistance, R_{in}. Once again, assume that R_q is a very large resistance.
(f) What small signal output resistance is "seen" by transistor $M3$?
(g) If the capacitor, C_c, is to emulate a short circuit for radial frequencies larger than ω_l, use pertinent preceding results to stipulate a design guideline for the nominal selection of this energy storage element.

Problem 5.15

A serious problem with radio frequency (RF) integrated circuits is the lack of high quality factor (high Q) inductors. Low Q is caused by resistive losses in the metallization spirals that comprise monolithic inductors, as well as by capacitive coupling of the spirals to the integrated circuit substrate. Since active feedback circuits can be designed to serve as an alternative to passive on chip inductance realization, the literature is understandably rife with proposals for active inductor synthesis. Unfortunately, most of these proposals suffer from several engineering problems, the most serious of which is limited dynamic range. This is to say that a design strategy succeeding in merely synthesizing a viable small signal inductance is not sufficient. The strategy must achieve an active inductance that operates over a wide range of signals whose boundaries are governed by the noise floor and the linearity ceiling indigenous to the proposed circuit architecture. This problem introduces the student to a particular type of active inductor module. This circuit, shown without the complexities of requisite biasing, appears in Fig. P5.15. The small signal model in Fig. P5.3(b) continues to apply.

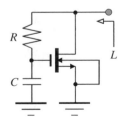

Figure P5.15.

(a) Derive an expression for the indicated effective inductance, L, established from transistor drain to signal ground. Assume that the resistance, R, the conductance of which can be selected to function as the critical feedback parameter, is much larger than the reactance of the generated inductor over all signal frequencies of interest.

(b) Derive an expression for the quality factor, Q, of the realized inductor.

Problem 5.16

Figure P5.16 depicts a series-series feedback amplifier realized in BJT technology. The applicable small signal model is given in Fig. P5.1(b). All transistors can be presumed to have respectively identical small signal parameters.

(a) What subcircuit in the amplifier comprises the feedback network?
(b) If the amplifier in Fig. P5.16 is to be modeled by a set of conventional two-port parameters, what two-port parameters are most appropriate for utilization?
(c) In the inquiries that follow, select resistance R_{ee} as the critical feedback parameter to deduce analytical expressions for the following performance metrics:

  (i) the open loop transconductance, I_{os}/V_s, where I_{os} designates the signal component of the net current (biasing and signal) conducted by the terminating load resistance, R_l;
  (ii) the return ratio, T_s, with respect to resistance R_{ee};
  (iii) the null return ratio, T_r, with respect to resistance R_{ee};

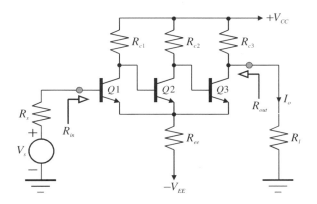

Figure P5.16.

(iv) the null driving point input resistance, say R_{ino};
(v) the null driving point output resistance, say R_{outo}.

(d) Assume tacitly that the transistors have large β and large r_o. Additionally, assume a small signal source resistance, R_s. Using the results deduced in Part (c), give approximate expressions for the following circuit performance indices:

(i) the closed open loop transconductance, $G_{me} = I_{os}/V_s$;
(ii) the closed loop driving point input resistance R_{in};
(iii) the closed loop driving point output resistance R_{out}.

(e) Let the transistors in the subject amplifier be characterized by $r_b = 100\,\Omega$, $r_\pi = 1.5\,\text{K}\Omega$, $r_o = 30\,\text{K}\Omega$, and $\beta = 140\,\text{amps/amp}$. Additionally, let the circuit have $R_s = 50\,\Omega$, $R_{ee} = 50\,\Omega$, $R_l = 300\,\Omega$, and $R_{c1} = R_{c2} = R_{c3} = 1.5\,\text{K}\Omega$. Compute the "exact" and the approximate values of the closed loop transconductance, G_{me}, the driving point input resistance, R_{in}, and the driving point output resistance, R_{out}.

(f) Exploit SPICE to compare the simulated gain and resistance level results to both their "exact" and approximated values. Explain any observed discrepancies.

Problem 5.17

Figure P5.17(a) is the symbolic diagram and corresponding simplified equivalent circuit of an operational transconductor amplifier (OTA). The OTA described by this figure is used in the simple first order filter shown in Fig. P5.17(b). Choosing the admittance, sC, of capacitor C in the schematic diagram of Fig. P5.17(b) as the critical feedback parameter for signal flow analysis purposes, show that the voltage transfer function of the filter, $H(s) = V_o/V_i$, is

$$H(s) = \frac{V_o}{V_i} = g_m R \left(\frac{1 + sC/g_m}{1 + sRC} \right).$$

Problem 5.18

Using the OTA model provided in Fig. P5.17(a) and signal flow methods with the admittance, sC_2, of capacitance C_2 selected as the critical feedback parameter, show that the voltage transfer function, $H(s) = V_o/V_i$, of the first order filter in Fig. P5.18 is

$$H(s) = \frac{V_o}{V_i} = \left(\frac{1 + sC_2/g_m}{1 + s(C_1 + C_2)/g_m} \right).$$

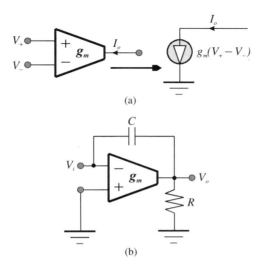

(a)

(b)

Figure P5.17.

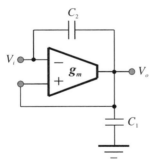

Figure P5.18.

Why is the filter a first order structure despite the presence of two capacitors in the circuit diagram?

Problem 5.19

Using the OTA model provided in Fig. P5.17(a) and signal flow methods with the admittance, sC_2, of capacitance C_2 selected as the critical feedback parameter, show that the voltage transfer function, $H(s) = V_o/V_i$, of the

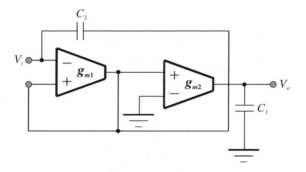

Figure P5.19.

first order filter in Fig. P5.19 is

$$H(s) = \frac{V_o}{V_i} = \left(\frac{g_{m1}}{g_{m1}+g_{m2}}\right)\left(\frac{1+\dfrac{sC_2}{g_{m1}}}{1+\dfrac{s(C_1+C_2)}{g_{m1}+g_{m2}}}\right).$$

Why is the filter a first order structure despite the presence of two capacitors in the circuit diagram?

Problem 5.20
Use the results documented in conjunction with Example 5.5 to derive an expression for the closed loop driving point input resistance, R_{in}, of the circuit in Fig. 5.28(a).

Problem 5.21
Return to Example 5.5 to derive an expression for the forward Norton transconductance in terms of a critical feedback parameter chosen to be the conductance associated with the Early resistance, r_o.

Problem 5.22
Derive the equations given in Eq. (5-79).

Problem 5.23
Repeat the design exercise of Example 5.6 for the case of matched I/O terminations of 75 ohms. Include SPICE simulations to verify the propriety of the design, inclusive of ascertaining the shifts incurred in gain and I/O resistance levels by ±20% perturbations in amplifier forward transconductance.

Problem 5.24
Repeat the design exercise of Example 5.6 for the case of matched I/O terminations of 300 ohms. Include SPICE simulations to verify the propriety of the design, inclusive of ascertaining the shifts incurred in gain and I/O resistance levels by $\pm 20\%$ perturbations in amplifier forward transconductance.

Problem 5.25
Repeat the design exercise of Example 5.7 for the case of an input port match termination of 300 ohms and a load resistance of 50 ohms. Include SPICE simulations to verify the propriety of the design, inclusive of ascertaining the shifts incurred in gain and I/O resistance levels by $\pm 20\%$ perturbations in amplifier forward transconductance.

Problem 5.26
Use SPICE to assess the frequency response of the amplifier designed in Example 5.7 for the case in which a 300 fF capacitance shunts both the input and the output port.

(a) Calculate the time constant associated with the input port capacitance under the condition of an output port capacitance that is removed from the circuit.
(b) Calculate the time constant associated with the output port capacitance under the condition of an input port capacitance that is removed from the circuit.
(c) In relationship to the 3-dB bandwidth projected by the SPICE simulation, what is the significance of the sum of the two time constants deduced in Parts (a) and (b)?

Problem 5.27
An active network produces sinusoidal oscillations in the steady state at radial frequency ω_o if its return ratio with respect to a feedback element is precisely negative one at $\omega = \omega_o$. Determine the oscillation criteria for the circuit model offered in Fig. P5.27.

Problem 5.28
Repeat Problem 5.27 for the oscillator model depicted in Fig. P5.28.

Problem 5.29
Figure P5.29 is an approximate high frequency model of a voltage buffer realized in BJT technology. Select the admittance of capacitance C as the

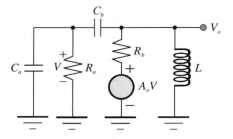

Figure P5.27.

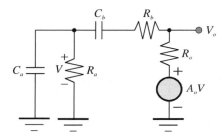

Figure P5.28.

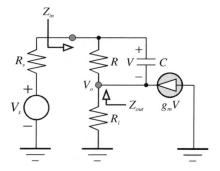

Figure P5.29.

critical feedback parameter while pursuing signal flow strategies to respond to the following inquiries.

(a) Derive an expression for the driving point input impedance, Z_{in}.
(b) Derive an expression for the driving point output impedance, Z_{out}.

(c) Derive an expression for the closed loop voltage transfer function, $A_v = V_o/V_s$.
(d) Comment as to the projected frequency response capabilities of the buffer at hand.

Problem 5.30

The Wilson current amplifier depicted schematically in Fig. P5.30 utilizes active feedback via transistor $M3$ to achieve a low driving point input resistance, R_{in}, and a high driving point output resistance, R_{out}. The MOSFET model in Fig. P5.3(b) remains applicable, subject to two provisos. First, the channel resistances of transistors $M2$ and $M3$ can, in the interest of analytical expediency, be tacitly ignored. However, the channel resistance of $M1$ cannot be ignored. Second, transistors $M1$ and $M2$ are identical and are biased identically so that $g_{m1} = g_{m2}$. On the other hand, transistor $M3$ has a transconductance that satisfies the simple linear relationship, $g_{m3} = g_{m2}/k$, where k is a constant that, as it materializes, is determined solely by the relative geometries of transistors $M2$ and $M3$.

(a) Derive an expression for the value, say R_d, of the two terminal resistance established by transistor $M2$.
(b) Derive expressions for the short circuit admittance parameters that interrelate the small signal values of the voltage and current variables, I_1, V_1, I_2, and V_2.

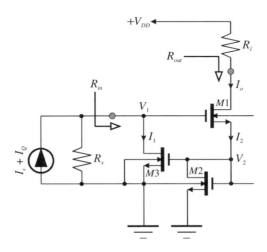

Figure P5.30.

(c) Examine the y-parameter model derived in Part (b) to deduce an appropriate critical feedback parameter.
(d) For the critical feedback parameter gleaned in Part (c), use signal flow partitioning theory to derive expressions for the return ratio and null return ratio.
(e) Use the results generated in Part (d) to derive general expressions for R_{in}, R_{out}, and the closed loop current gain, $A_i = I_{os}/I_s$, where I_{os} represents the signal component of the net current, I_o, conducted by load resistance R_l.
(f) Why is it imprudent to assume that the channel resistance of $M1$ is infinitely large? On the other hand, why is it prudent to make this presumption for both transistors $M2$ and $M3$?
(g) What reasonable assumptions render a closed loop current gain that is effectively determined by the transconductance ratio, $k = g_{m2}/g_{m3}$?
(h) For the assumptions adopted in Part (f), give the resultant approximate expressions for R_{in} and R_{out}.

Chapter 6

Multiple Loop Feedback Amplifiers

6.1.0. Introduction

Chapters 4 and 5, and particularly Chapter 5, exploit signal flow analytical methods to circumvent the limitations inherent to the ideal feedback model in the course of studying the properties of feedback amplifiers. The idealized feedback model is useful only if the fundamental constituents of a feedback structure can be separated into the basic amplifier, $\mu(s)$, and the feedback network, $\beta(s)$. The procedure is difficult and sometimes virtually impossible, because the forward path may not be strictly unilateral, the feedback path is usually bilateral, and the input and output coupling networks are often complicated. Thus, the ideal feedback model may not be adequate to represent a practical amplifier. In this chapter, Bode's classic feedback theory is developed and assessed and in the process, a firm theoretical foundation is imparted to the disclosures in Chapter 5. Since Bode's technique is applicable to general feedback network configurations, it avoids the explicit necessity of identifying the forward and feedback transfer functions, $\mu(s)$ and $\beta(s)$, respectively.

Bode's feedback theory is based on the concept of return difference, which is defined in this chapter in terms of network determinants. The return difference, which is a generalization of the feedback factor concept implicit to the ideal feedback model, is a physical performance metric in that it can be measured directly. The null return difference and its physical significance follow straightforwardly from the return difference. Since Bode's feedback theory is formulated in terms of the first and second order cofactors of the elements of the indefinite admittance matrix of a feedback circuit, it is appropriate to review the mathematical concepts that underpin the indefinite admittance matrix for a linear network.

6.2.0. Indefinite Admittance Matrix

Figure 6.1 is an n-terminal network, N, composed of an arbitrary number of active and passive network elements connected in any electrically meaningful manner. Let $V_1, V_2, \ldots, V_n$ be the Laplace-transformed potentials measured between terminals $1, 2, \ldots, n$ and some arbitrary, but unspecified, reference point, and furthermore, let $I_1, I_2, \ldots, I_n$ be the Laplace-transformed currents entering the terminals $1, 2, \ldots, n$ from outside the network. Since network N and its load are linear, the terminal currents and voltages are related by the equation

$$\begin{bmatrix} I_1 \\ I_2 \\ \vdots \\ I_n \end{bmatrix} = \begin{bmatrix} y_{11} & y_{12} & \cdots & y_{1n} \\ y_{21} & y_{22} & \cdots & y_{2n} \\ \vdots & \vdots & \vdots & \vdots \\ y_{n1} & y_{n2} & \cdots & y_{nn} \end{bmatrix} \begin{bmatrix} V_1 \\ V_2 \\ \vdots \\ V_n \end{bmatrix} + \begin{bmatrix} J_1 \\ J_2 \\ \vdots \\ J_n \end{bmatrix} \quad (6\text{-}1)$$

or more compactly,

$$\boldsymbol{I}(s) = \boldsymbol{Y}(s)\boldsymbol{V}(s) + \boldsymbol{J}(s) \quad (6\text{-}2)$$

where $\boldsymbol{Y}(s)$, is called the *indefinite admittance matrix* because the reference point for all network potentials is an arbitrary node extrinsic to the network, The current, J_k, $(k = 1, 2, \ldots, n)$ denotes the current flowing into the kth terminal under the special condition of all terminals of N grounded to the reference point. This short circuit current can be construed as independent sources deriving from initial energy conditions established in the network interior. When no initial conditions prevail, $\boldsymbol{J}(s)$ is necessarily a null vector,

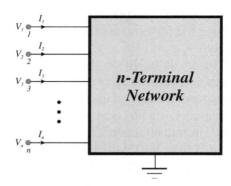

Figure 6.1. The general symbolic representation of an n-terminal network. All of the indicated terminal voltages are referenced to the system ground.

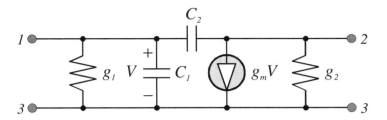

Figure 6.2. A small signal equivalent circuit of a transistor. The model is generally applicable to both bipolar and MOS technology devices.

whence

$$I(s) = Y(s)V(s), \qquad (6\text{-}3)$$

where the elements, y_{ij}, of $Y(s)$ can be obtained as

$$y_{ij} = \left.\frac{I_i}{V_j}\right|_{V_x=0, x \neq j} \qquad (6\text{-}4)$$

To illustrate, consider the small signal equivalent model of a transistor shown in Fig. 6.2. Its indefinite admittance matrix is

$$Y(s) = \begin{bmatrix} g_1 + sC_1 + sC_2 & -sC_2 & -g_1 - sC_1 \\ g_m - sC_2 & g_2 + sC_2 & -g_2 - g_m \\ -g_1 - sC_1 - g_m & -g_2 & g_1 + g_2 + g_m + sC_1 \end{bmatrix} \qquad (6\text{-}5)$$

Observe, as has been demonstrated in general in Chapter 2, that the sum of elements of each row and column is equal to zero.

If Y_{uv} denotes the submatrix obtained from an indefinite admittance matrix, $Y(s)$, by deleting the uth row and vth column, the *first order cofactor*, denoted by the symbol Y_{uv}, of the element y_{uv} in $Y(s)$, is

$$Y_{uv} = (-1)^{u+v} \det Y_{uv}. \qquad (6\text{-}6)$$

Because of the aforementioned zero row sum and zero column sum properties, all cofactors of the elements of the indefinite admittance matrix are equal. Such a matrix is referred to as an *equicofactor matrix*. It follows that if Y_{uv} and Y_{ij} are any two cofactors of an equicofactor matrix, $Y(s)$,

$$Y_{uv} = Y_{ij} \qquad (6\text{-}7)$$

for all $u, v, i,$ and j. For the indefinite admittance matrix, $Y(s)$, of Eq. (6-5), it is easily verified that all of its nine cofactors are equal to

$$Y_{uv} = s^2 C_1 C_2 + s(C_1 g_2 + C_2 g_1 + C_2 g_2 + g_m C_2) + g_1 g_2 \qquad (6\text{-}8)$$

for $u, v = 1, 2, 3$.

Denote by $Y_{rp,sq}$ the submatrix obtained from $Y(s)$ by striking out rows r and s and columns p and q. Then the *second order cofactor*, denoted by the symbol $Y_{rp,sq}$ of the elements y_{rp} and y_{sq} of $Y(s)$ is the scalar quantity defined by the expression,

$$Y_{rp,sq} = \text{sgn}(r-s)\text{sgn}(p-q)(-1)^{r+p+s+q} \det Y_{rp,sq} \qquad (6\text{-}9)$$

where $r \neq s$ and $p \neq q$, and

$$\left.\begin{array}{ll} \text{sgn } u = +1 & \text{if } u > 0 \\ \text{sgn } u = -1 & \text{if } u < 0 \\ \text{sgn } u \stackrel{\Delta}{=} 0 & \text{if } u = 0 \end{array}\right\}. \qquad (6\text{-}10)$$

In attempt to forestall confusion, the reader should understand that in Eq. (6-9) and related other equations, boldface type refers to a matrix, where conventional type denotes a scalar quantity.

As a further example, consider the hybrid-pi equivalent network of a bipolar junction transistor shown in Fig. 6.3. Assume that each node is an accessible terminal of a four-terminal network. Its indefinite admittance matrix is

$$Y(s) = \begin{bmatrix} 0.02 & 0 & -0.02 & 0 \\ 0 & 5 \times 10^{-12}s & 0.2 - 5 \times 10^{-12}s & -0.2 \\ -0.02 & -5 \times 10^{-12}s & 0.024 + 105 \times 10^{-12}s & -0.004 - 10^{-10}s \\ 0 & 0 & -0.204 - 10^{-10}s & 0.204 + 10^{-10}s \end{bmatrix}. \qquad (6\text{-}11)$$

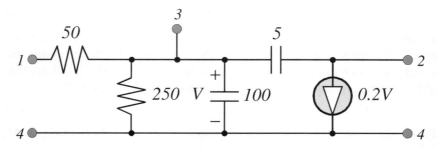

Figure 6.3. The hybrid-pi equivalent circuit of a bipolar junction transistor. All resistances in the model are in units of ohms, all capacitances are in picofarads, and the transconductance multiplier in the voltage controlled current source is in units of mhos, or siemens.

The second-order cofactor $Y_{31,42}$ and $Y_{11,34}$ of the elements of $Y(s)$ of Eq. (6-11) are computed as follows:

$$Y_{31,42} = \text{sgn}(3-4)\text{sgn}(1-2)(-1)^{3+1+4+2}$$

$$\times \det \begin{bmatrix} -0.02 & 0 \\ 0.2 - 5 \times 10^{-12}s & -0.2 \end{bmatrix} = 0.004. \quad (6\text{-}12)$$

$$Y_{11,34} = \text{sgn}(1-3)\text{sgn}(1-4)(-1)^{1+1+3+4}$$

$$\times \det \begin{bmatrix} 5 \times 10^{-12}s & 0.2 - 5 \times 10^{-12}s \\ 0 & -0.204 - 10^{-10}s \end{bmatrix}$$

$$= 5 \times 10^{-12}s \left(0.204 + 10^{-10}s\right). \quad (6\text{-}13)$$

The engineering utility of the indefinite admittance matrix lies in the fact that it facilitates the computation of the driving point impedance presented by any pair of nodes or the transfer function from any nodal pair to any other pair. This contention is demonstrated by the material that follows.

Consider Fig. 6.4, which abstracts a linear network for which a current source is connected between any two nodes r and s so that a current I_{sr} is injected into the rth node and at the same time is extracted from the sth node. An ideal voltmeter is connected from node p to node q to monitor the potential rise from q to p. Then the *transfer impedance*, or *transimpedance*, $z_{rp,sq}$, between the node pairs rs and pq of the subject network is simply

$$z_{rp,sq} = \frac{V_{pq}}{I_{sr}}, \quad (6\text{-}14)$$

where it is understood that all initial conditions and associated independent energy sources inside N are set to zero; specifically, $J(s) = 0$ in Eq. (6-2). The representation is, of course, quite general. When $r = p$ and $s = q$,

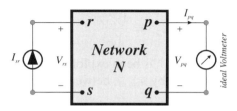

Figure 6.4. System abstraction for the measurement of a network transfer impedance, or transimpedance.

the transfer impedance $z_{rp,sq}$ becomes the *driving point impedance* $z_{rr,ss}$ between the terminal pair rs.

In Fig. 6.4, choose terminal q to be the zero voltage reference node for all other terminals. In terms of the equations of Eq. (6-1), this operation, coupled with the zero initial condition constraint, is tantamount to setting $J(s) = 0$, $V_q = 0$, $I_x = 0$ for $x \neq r, s$ and $I_r = -I_s = I_{sr}$. Since $Y(s)$ is an equicofactor matrix, the equations of Eq. (6-1) are not linearly independent; that is, one of these equations is superfluous. To this end, suppress the sth equation in Eq. (6-1), which resultantly reduces to

$$I_{-s} = Y_{sq} V_{-q}, \qquad (6\text{-}15)$$

where I_{-s} and V_{-q} denote the subvectors obtained from I and V of Eq. (6-2) by deleting the sth and qth rows. Applying Cramer's rule, the voltage V_p, referenced to terminal q, is

$$V_p = \frac{\det \tilde{Y}_{sq}}{\det Y_{sq}}, \qquad (6\text{-}16)$$

where $\tilde{Y}_{sq}$ is the matrix derived from Y_{sq} by replacing the column corresponding to V_p by I_{-s}. It should be noted that I_{-s} is in the pth column if $p < q$ but in the $(p-1)$th column if $p > q$. Furthermore, the row in which I_{sr} appears is the rth row if $r < s$ but in the $(r-1)$th row if $r > s$. Thus,

$$(-1)^{s+q} \det \tilde{Y}_{sq} = I_{sr} Y_{rp,sq}. \qquad (6\text{-}17)$$

In addition,

$$\det Y_{sq} = (-1)^{s+q} Y_{sq}. \qquad (6\text{-}18)$$

Substituting these relationships into Eq. (6-16),

$$z_{rp,sq} = \frac{Y_{rp,sq}}{Y_{uv}} \qquad (6\text{-}19)$$

$$z_{rr,ss} = \frac{Y_{rr,ss}}{Y_{uv}} \qquad (6\text{-}20)$$

where the fact that $Y_{sq} = Y_{uv}$ has been exploited.

The *voltage gain*, denoted by $g_{rp,sq}$, between node pairs rs and pq of the network of Fig. 6.4 is

$$g_{rp,sq} = \frac{V_{pq}}{V_{rs}}, \qquad (6\text{-}21)$$

provided null initial conditions prevail in network N. It follows from Eqs. (6-19) and (6-20) that

$$g_{rp,sq} = \frac{z_{rp,sq}}{z_{rr,ss}} = \frac{Y_{rp,sq}}{Y_{rr,ss}}. \quad (6\text{-}22)$$

In an attempt to avoid confusion, it is worthwhile to review the mathematical logistics underpinning the subscripts invoked in Eqs. (6-19) through (6-22). In the numerators of these relationships, r is the current injecting node, p symbolizes the voltage measurement node, s denotes the current extracting node, and q represents the voltage reference node. Moreover, nodes r and p correspond to input and output transfer measurement.

To highlight the engineering utility of the foregoing theoretical disclosures, consider the hybrid-pi transistor equivalent network offered in Fig. 6.5. A 100-Ω load resistor is incident at nodes 2 and 4, and a voltage source, V_{14}, excites the amplifier input port. In the interest of analytical simplicity, let p denote the normalized frequency, $p = 10^{-9}s$. The indefinite admittance matrix of the amplifier is found to be

$$Y(s) = \begin{bmatrix} 0.02 & 0 & -0.02 & 0 \\ 0 & 0.01 + 0.005p & 0.2 - 0.005p & -0.21 \\ -0.02 & -0.005p & 0.024 + 0.105p & -0.004 - 0.1p \\ 0 & -0.01 & -0.204 - 0.1p & 0.214 + 0.1p \end{bmatrix}. \quad (6\text{-}23)$$

To compute the voltage gain $g_{12,44}$, Eq. (6-22) can be exploited to obtain

$$g_{12,44} = \frac{V_{24}}{V_{14}} = \frac{Y_{12,44}}{Y_{11,44}} = \frac{p - 40}{5p^2 + 21.7p + 2.4}. \quad (6\text{-}24)$$

The input impedance facing the voltage source V_{14} is determined as

$$z_{11,44} = \frac{V_{14}}{I_{41}} = \frac{Y_{11,44}}{Y_{uv}} = \frac{Y_{11,44}}{Y_{44}} = \frac{50p^2 + 217p + 24}{p^2 + 4.14p + 0.08}. \quad (6\text{-}25)$$

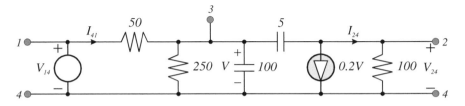

Figure 6.5. Model of a transistor amplifier used to illustrate the computation of the voltage gain, $g_{rp,sq}$. All resistances in the model are in units of ohms, all capacitances are in picofarads, and the transconductance multiplier in the voltage controlled current source is in units of siemens.

On the other hand, the current gain, defined as the ratio of current I_{24} in the 100-Ω resistor to the input current, I_{41}, Eq. (6-19) can be applied to arrive at

$$\frac{I_{24}}{I_{41}} = 0.01\frac{V_{24}}{I_{41}} = 0.01 z_{12,44} = 0.01\frac{Y_{12,44}}{Y_{44}} = \frac{0.1p - 4}{p^2 + 4.14p + 0.08}. \quad (6\text{-}26)$$

Finally, to compute the transfer admittance, defined as the ratio of load current I_{24} to input voltage V_{14}, we appeal to Eq. (6-22), which delivers

$$\frac{I_{24}}{V_{14}} = 0.01\frac{V_{24}}{V_{14}} = 0.01 g_{12,44} = 0.01\frac{Y_{12,44}}{Y_{11,44}} = \frac{p - 40}{500p^2 + 2170p + 240}. \quad (6\text{-}27)$$

It is notable that the delineation of the voltage gain, current gain, input impedance, and I/O transfer admittance metrics for a network entails a mere straightforward application of the indefinite admittance matrix of the network undergoing examination.

6.2.1. Return Difference

The study of feedback amplifier responses generally entails how a particular element or parameter of the subject amplifier affects that response. The selected element or parameter is crucial either in terms of its effect on the entire system or possibly, it is of engineering concern because of routinely encountered monolithic processing vagaries or manufacturing uncertainties. Ordinarily, the transfer function of an active device, the gain of an amplifier, or the immittance of a one-port network is of particular interest with respect to an assessment of the response sensitivity of a feedback network. For the present, assume that the selected crucial, or critical, parameter is x, the controlling parameter of a voltage controlled current source; that is,

$$I = xV. \quad (6\text{-}28)$$

To focus attention on parameter x, Fig. 6.6 is the general configuration of a feedback amplifier in which the controlled source is highlighted as a two-port network connected to a general four-port network, along with the input source combination of I_s and admittance Y_1, and load admittance Y_2. The two-port representation of a controlled source in Eq. (6-28) is quite general. It includes the special situation where a one-port element is characterized

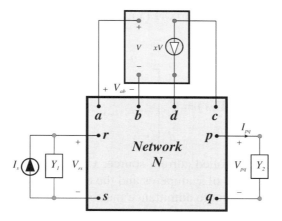

Figure 6.6. A generalized abstraction of a feedback circuit. The abstraction delineates local feedback imposed between two arbitrary internal ports of Network N.

by its immittance. In this case, the controlling voltage, V, is the terminal voltage of the controlled current source, I, whence x becomes the one-port admittance.

The *return difference*, $F(x)$, of a feedback amplifier with respect to a parameter x is defined as the ratio of the two functional values assumed by the first order cofactor of an element of its indefinite admittance matrix, under the condition that parameter x assumes its nominal value and the condition that parameter x assumes a null value. To emphasize the importance of the feedback critical parameter, x, express the indefinite admittance matrix Y of the amplifier as a function of x, even though it is also a function of the complex frequency variable s, and write $Y = Y(x)$. Then

$$F(x) \equiv \frac{Y_{uv}(x)}{Y_{uv}(0)}, \tag{6-29}$$

where

$$Y_{uv}(0) = Y_{uv}(x)|_{x=0}. \tag{6-30}$$

The physical significance of the return difference can now be addressed. In the network of Fig. 6.6, the input, the output, the controlling branch, and the controlled source are labeled as indicated. Then, parameter x enters the indefinite admittance matrix, $Y(x)$, in a rectangular pattern as shown below:

$$Y(x) = \begin{array}{c} \\ a \\ b \\ c \\ d \end{array} \begin{array}{c} a \quad b \quad c \quad d \\ \left[\begin{array}{cccc} & & & \\ & & & \\ & & x & -x \\ & & -x & x \end{array} \right] \end{array}. \tag{6-31}$$

If in Fig. 6.6 the controlled current source, xV, is supplanted by an independent current source of x amperes and the excitation current source, I_s, is set to zero, the indefinite admittance matrix of the resulting network is simply $Y(0)$. By appealing to Eq. (6-24), the new voltage V'_{ab} appearing at terminals a and b of the controlling branch is found to be

$$V'_{ab} = x \frac{Y_{da,cb}(0)}{Y_{uv}(0)} = -x \frac{Y_{ca,db}(0)}{Y_{uv}(0)}. \tag{6-32}$$

Notice that the current injecting point is terminal d, not terminal c.

The foregoing operation is reflected electrically by the schematic diagram in Fig. 6.7. Observe that the controlling branch for the voltage controlled current source is severed and unity voltage is applied to the right of the break. This 1-volt sinusoidal voltage of a fixed angular frequency produces a current of x amperes at the controlled current source. The voltage established at the left of the circuit break by this 1-volt excitation is V'_{ab}, as

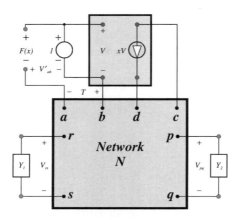

Figure 6.7. The physical interpretation of the return difference, $F(x)$, with respect to the controlling parameter, x, of a voltage controlled current source, xV.

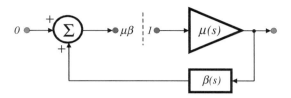

Figure 6.8. System level interpretation of the loop transmission metric.

indicated. This returned voltage, V'_{ab}, has the same physical significance as the loop transmission $\mu\beta$ defined for the ideal feedback model. A confirmation of this contention derives from setting the input excitation to the ideal feedback model to zero, breaking the forward path, and applying a unit input to the right of the break, as suggested by Fig. 6.8. The signal appearing at the left of the break is precisely the loop transmission.

The foregoing analytical disclosures all promote the introduction of the concept of the *return ratio*, T, which is defined as the negative of the voltage appearing at the controlling branch when the controlled current source is replaced by an independent current source of x amperes and the input excitation is set to zero. Thus, return ratio T is simply the negative of the returned voltage, V'_{ab} or $T = -V'_{ab}$. In view of this observation, the difference between the 1-volt excitation and the returned voltage, V'_{ab}, can be evaluated as

$$1 - V'_{ab} = 1 + x \frac{Y_{ca,db}}{Y_{uv}(0)} = \frac{Y_{uv}(0) + xY_{ca,db}}{Y_{uv}(0)} = \frac{Y_{db}(0) + xY_{ca,db}}{Y_{db}(0)}$$

$$= \frac{Y_{db}(x)}{Y_{db}(0)} = \frac{Y_{uv}(x)}{Y_{uv}(0)} = F(x), \qquad (6\text{-}33)$$

in which the identities, $Y_{uv} = Y_{ij}$, and

$$Y_{db}(x) = Y_{db}(0) + xY_{ca,db} \qquad (6\text{-}34)$$

have been invoked.

Of particular pertinence is the ability to write $Y_{ca,db}(x)$ as $Y_{ca,db}$ because $Y_{ca,db}(x)$ is independent of x. In other words, the return difference, $F(x)$, is simply the difference of the 1-volt excitation and the returned voltage, V'_{ab}, as illustrated in Fig. 6.7. Since

$$F(x) = 1 + T = 1 - \mu\beta, \qquad (6\text{-}35)$$

the return difference can be seen to be identical to the feedback factor of the ideal feedback model. The significance of the above analytical disclosures is

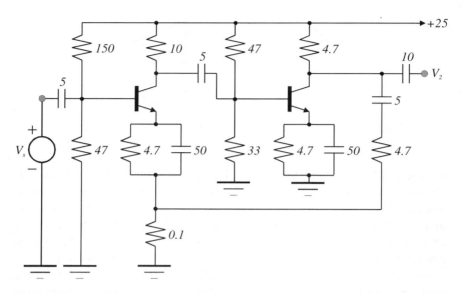

Figure 6.9. A series-shunt feedback amplifier shown with its requisite biasing and signal coupling circuitry. All resistances are in units of kilo-ohms, all capacitances are in units of microfarads, and the biasing supply is in units of volts.

the suggestion that the return ratio, T, or $-\mu\beta$, is deterministic via direct measurement. Once the return ratio is measured, other critical feedback network metrics, such as return difference and loop transmission, follow forthwith.

As an example of the engineering utility of the preceding arguments, consider the series-shunt feedback amplifier of Fig. 6.9. Assume that the two transistors are identical and are characterized by the following small signal hybrid h-parameters:

$$h_{ie} = 1.1 \text{ K}\Omega, \quad h_{fe} = 50, \quad h_{re} = h_{oe} = 0. \quad (6\text{-}36)$$

After the biasing and coupling circuitry have been removed, the equivalent network is represented by the model in Fig. 6.10. The effective load of the first transistor is composed of the parallel combination of the 10-KΩ, 33-KΩ, 47-KΩ and 1.1-KΩ resistors. The effects of the 150-KΩ and 47-KΩ resistors can be ignored; they are included in the equivalent network only to show their insignificance in the computation.

To simplify notation, let

$$\tilde{\alpha}_k = \alpha_k \times 10^{-4} = \frac{h_{fe}}{h_{ie}} = 455 \times 10^{-4}, \quad k = 1, 2. \quad (6\text{-}37)$$

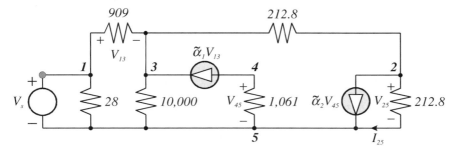

Figure 6.10. A low frequency equivalent circuit of the feedback amplifier in Figure 6.9. The resistive elements are represented as conductances with values in units of microsiemens.

The subscript k is used to distinguish the transconductances of the first and the second transistors. The indefinite admittance matrix of the feedback amplifier of Fig. 6.9 is found to be

$$Y = 10^{-4} \begin{bmatrix} 9.37 & 0 & -9.09 & 0 & -0.28 \\ 0 & 4.256 & -2.128 & \alpha_2 & -2.128 - \alpha_2 \\ -9.09 - \alpha_1 & -2.128 & 111.218 + \alpha_1 & 0 & -100 \\ \alpha_1 & 0 & -\alpha_1 & 10.61 & -10.61 \\ -0.28 & -2.128 & -100 & -10.61 - \alpha_2 & 113.018 + \alpha_2 \end{bmatrix}.$$

(6-38)

By applying Eq. (6-22), the voltage gain of the amplifier is found to be

$$g_{12,55} = \frac{V_{25}}{V_s} = \frac{Y_{12,55}}{Y_{11,55}} = \frac{211.54 \times 10^{-7}}{4.66 \times 10^{-7}} = 45.39. \quad (6\text{-}39)$$

To calculate the return differences with respect to the transconductances, $\tilde{\alpha}_k$, of the transistor, the input signal source, V_s, is short circuited. The resulting indefinite admittance matrix is obtained from Eq. (6-38) by adding the first row to the fifth row and the first column to the fifth column and then deleting the first row and column. Its first order cofactor is simply $Y_{11,55}$. Thus, the return differences with respect to $\tilde{\alpha}_k$ are

$$\left. \begin{array}{l} F(\tilde{\alpha}_1) = \dfrac{Y_{11,55}(\tilde{\alpha}_1)}{Y_{11,55}(0)} = \dfrac{466.1 \times 10^{-9}}{4.97 \times 10^{-9}} = 93.70 \\[2mm] F(\tilde{\alpha}_2) = \dfrac{Y_{11,55}(\tilde{\alpha}_2)}{Y_{11,55}(0)} = \dfrac{466.1 \times 10^{-9}}{25.52 \times 10^{-9}} = 18.26 \end{array} \right\}. \quad (6\text{-}40)$$

6.2.2. Null Return Difference

The *null return difference*, $\hat{F}(x)$, of a feedback amplifier, with respect to a parameter x is an important feedback network metric from at least the perspective of network sensitivity with respect to a crucial parameter, x. It is defined to be the ratio of the two functional values assumed by the second order cofactor, $Y_{rp,sq}$, of the elements of the network indefinite admittance matrix, Y, under the condition that element x assumes its nominal value and the condition that element x assumes a null value. It is to be understood that r and s are the input terminals, and p and q are the output terminals of the amplifier undergoing study. It follows that

$$\hat{F}(x) = \frac{Y_{rp,sq}(x)}{Y_{rp,sq}(0)}. \tag{6-41}$$

Similarly, the *null return ratio*, $\hat{T}$, with respect to a voltage controlled current source, $I = xV$, is the negative of the voltage appearing at the controlling branch when the controlled current source is replaced by an independent current source of x amperes and when the input excitation is adjusted so that the output of the amplifier is identically zero.

The null return difference is simply the return difference in the network under the constraint that the input excitation, I_s, has been adjusted so that the output is identically zero. In the network of Fig. 6.6, suppose that the controlled current source is supplanted by an independent current source of x amperes. Then by applying Eq. (6-19) and the superposition principle, the output current, I_{pq}, at the load is found to be

$$I_{pq} = Y_2 \left[I_s \frac{Y_{rp,sq}(0)}{Y_{uv}(0)} + x \frac{Y_{dp,cq}(0)}{Y_{uv}(0)} \right]. \tag{6-42}$$

Setting $I_{pq} = 0$ or $V_{pq} = 0$ yields

$$I_s \equiv I_0 = -x \frac{Y_{dp,cq}(0)}{Y_{rp,sq}(0)}, \tag{6-43}$$

in which $Y_{dp,cq}$ is independent of x. This adjustment is possible only if there is a direct transmission from the input to the output when x is set to zero. Thus, in the network of Fig. 6.7, if an independent current source of strength I_0 is connected at the network input port, the voltage, V'_{ab}, is the negative of the null return ratio $\hat{T}$. Using Eq. (6-19),

$$\hat{T} = -V'_{ab} = -x \frac{Y_{da,cb}(0)}{Y_{uv}(0)} - I_0 \frac{Y_{ra,sb}(0)}{Y_{uv}(0)}$$

$$= -\frac{x[Y_{da,cb}(0)Y_{rp,sq}(0) - Y_{ra,sb}(0)Y_{dp,cq}(0)]}{Y_{uv}(0)Y_{rp,sq}(0)}$$

$$= \frac{x\dot{Y}_{rp,sq}}{Y_{rp,sq}(0)} = \frac{Y_{rp,sq}(x)}{Y_{rp,sq}(0)} - 1, \qquad (6\text{-}44)$$

where

$$\dot{Y}_{rp,sq} \equiv \frac{dY_{rp,sq}(x)}{dx}. \qquad (6\text{-}45)$$

Consequently

$$\hat{F}(x) = 1 + \hat{T} = 1 - V'_{ab}, \qquad (6\text{-}46)$$

which shows that the null return difference, $\hat{F}(x)$, is simply the difference of the 1-volt excitation applied to the right of the severed controlling branch of the controlled source and the returned voltage, V'_{ab}, appearing at the left of the break under the situation that the input signal, I_s, is adjusted to null the output port response to zero.

Reconsider the series-shunt feedback amplifier of Fig. 6.9, for which the pertinent small signal model appears in Fig. 6.10. Using the indefinite admittance matrix of Eq. (6-38) in conjunction with Eq. (6-37), the null return differences with respect to $\tilde{\alpha}_k$ are

$$\left. \begin{array}{l} \hat{F}(\tilde{\alpha}_1) = \dfrac{Y_{12,55}(\tilde{\alpha}_1)}{Y_{12,55}(0)} = \dfrac{211.54 \times 10^{-7}}{205.24 \times 10^{-12}} = 103.07 \times 10^3 \\[2ex] \hat{F}(\tilde{\alpha}_2) = \dfrac{Y_{12,55}(\tilde{\alpha}_2)}{Y_{12,55}(0)} = \dfrac{211.54 \times 10^{-7}}{104.79 \times 10^{-10}} = 2018.70 \end{array} \right\}. \qquad (6\text{-}47)$$

Alternatively, $\hat{F}(\tilde{\alpha}_1)$ can be computed directly through use of its physical interpretation. To this end, replace the controlled source, $\tilde{\alpha}_1 V_{13}$, in Fig. 6.10 by an independent current source of $\tilde{\alpha}_1$ amperes. Adjust the voltage source, V_s, so that the output current, I_{25}, is identically zero. Let I_0 be the input current resulting from this adjusted signal source. The corresponding network

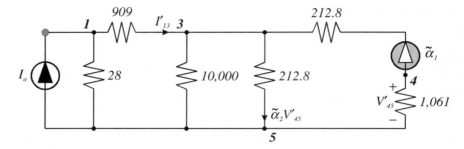

Figure 6.11. The network used to compute the null return difference with respect to parameter α_1. The resistive elements are represented as conductances with values in units of microsiemens.

is the structure in Fig. 6.11. From this network, an analysis reveals

$$\hat{F}(\tilde{\alpha}_1) = 1 + \hat{T} = 1 - V'_{13} = 1 - \frac{100 V'_{35} + \alpha_2 V'_{45} - \alpha_1}{9.09} = 103.07 \times 10^3.$$

(6-48)

An analogous procedure computes the return difference, $\hat{F}(\tilde{\alpha}_2)$.

6.3.0. Network Functions and Feedback

Refer to the generalized feedback configuration of Fig. 6.6. Let w be a transfer function. As before, to emphasize the importance of the feedback element x, write $w = w(x)$. To be definitive, let $w(x)$ be the current gain, which is intrinsically dependent on critical parameter x, from the input port to the output port. Then from Eq. (6-19),

$$w(x) = \frac{I_{pq}}{I_s} = \frac{Y_2 V_{pq}}{I_s} = \frac{Y_{rp,sq}(x)}{Y_{uv}(x)} Y_2,$$

(6-49)

which produces

$$\frac{w(x)}{w(0)} = \frac{Y_{rp,sq}(x)}{Y_{uv}(x)} \frac{Y_{uv}(0)}{Y_{rp,sq}(0)} = \frac{\hat{F}(x)}{F(x)},$$

(6-50)

provided that $w(0) \neq 0$. It follows that the current gain of interest is expressible as the simple relationship,

$$w(x) = w(0) \frac{\hat{F}(x)}{F(x)}.$$

(6-51)

Equation (6-51) remains valid if $w(x)$ denoted the transfer impedance $z_{rp,sq} = V_{pq}/I_s$, as opposed to the current gain.

6.3.1. Blackman's Formula

When $r = p$ and $s = q$, $w(x)$ represents the driving point impedance $z_{rr,ss}(x)$ established at the r-s terminal pair. In this case, $F(x)$ is the return difference with respect to parameter x under the condition, $I_s = 0$. Thus, $F(x)$ is the return difference for the case when the port at which the input impedance is defined is left open circuited; that is, the port is left without current source excitation. For this case, clarity compels writing $F(x) = F$(input open circuited). Likewise, from Fig. 6.6, $\hat{F}(x)$ is the return difference with respect to x for the input excitation I_s and output response V_{rs} under the condition that I_s is adjusted to clamp voltage response V_{rs} to zero. In other words, $\hat{F}(x)$ is the return difference for the situation when the port at which the input impedance is defined is short circuited. Accordingly, write $\hat{F}(x) = F$ (input short circuited). Resultantly, the input impedance, $Z(x)$, looking into a terminal pair can be conveniently expressed as

$$Z(x) = Z(0) \frac{F(\text{input short circuited})}{F(\text{input open circuited})}. \tag{6-52}$$

Equation (6-52) is the well-known *Blackman's formula* for computing an impedance presented at any terminal pair of any linear, active or passive, network. The formula is extremely useful because its right hand side can usually be determined easily. If x represents the controlling parameter of a controlled source in a single loop feedback amplifier, setting $x = 0$ opens the feedback loop and $Z(0)$ is simply the zero feedback value of the impedance of interest. The return difference for x when the input port is short circuited or open circuited is relatively simple to compute because shorting or opening a terminal pair frequently breaks the feedback loop. In addition, Blackman's formula can be used to determine the return difference by measurements. Because it involves two return differences, only one of them can be identified and the other must be known in advance. In the case of a single loop feedback amplifier, it is usually possible to choose a terminal pair so that either the numerator or the denominator on the right hand side of Eq. (6-52) is unity. If F(input short circuited) $= 1$, F(input open circuited) becomes the return difference under normal operating conditions, whence

$$F(x) = \frac{Z(0)}{Z(x)}. \tag{6-53}$$

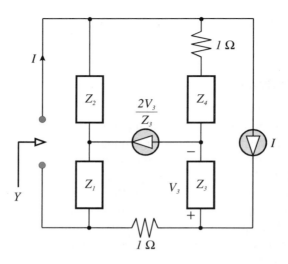

Figure 6.12. A general active RC one port realization of a rational function.

On the other hand, if F(input open circuited) $= 1$, F(input short circuited) becomes the return difference under normal operating conditions and thus,

$$F(x) = \frac{Z(x)}{Z(0)}. \tag{6-54}$$

Example 6.1. The network of Fig. 6.12 is a normalized active RC one port realization of a rational impedance. Use Blackman's formula to verify that the indicated input admittance of the subject network is given by

$$Y = 1 + \frac{Z_3 - Z_4}{Z_1 - Z_2}.$$

Solution 6.1.
Appealing to Eq. (6-52), the input admittance, $Y = Y(x)$, can be written as

$$Y(x) = Y(0) \frac{F(\text{input open circuited})}{F(\text{input short circuited})}, \tag{E6-1}$$

where $x = 2/Z_3$. By setting x to zero, the network used to compute $Y(0)$ becomes the topology offered in Fig. 6.13. Its input admittance is found to be

$$Y(0) = \frac{Z_1 + Z_2 + Z_3 + Z_4 + 2}{Z_1 + Z_2}. \tag{E6-2}$$

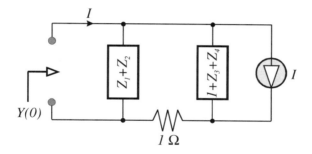

Figure 6.13. The network used to compute the admittance, $Y(0)$.

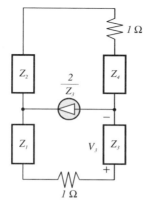

Figure 6.14. The network used to compute the function $F(\text{input open circuited})$.

When the input port is open circuited, the network of Fig. 6.1 degenerates to that of Fig. 6.14. The return difference with respect to x is found to be

$$F(\text{input open circuited}) = 1 - V_3' = \frac{Z_1 + Z_3 - Z_2 - Z_4}{2 + Z_1 + Z_2 + Z_3 + Z_4}, \quad \text{(E6-3)}$$

where the returned voltage, V_3', at the controlling branch is

$$V_3' = \frac{2(1 + Z_2 + Z_4)}{2 + Z_1 + Z_2 + Z_3 + Z_4}. \quad \text{(E6-4)}$$

To compute the return difference when the input port is short circuited, use the network in Fig. 6.15, to deduce

$$F(\text{input short circuited}) = 1 - V_3'' = \frac{Z_1 - Z_2}{Z_1 + Z_2}, \quad \text{(E6-5)}$$

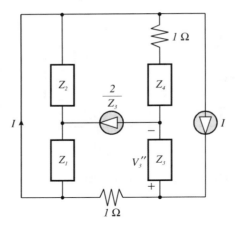

Figure 6.15. The network used to compute the function F (*input short circuited*).

where the returned voltage, V_3'', at the controlling branch is

$$V_3'' = \frac{2Z_2}{Z_1 + Z_2}. \tag{E6-6}$$

Substituting Eqs. (E6-2), (E6-3) and (E6-5) into Eq. (E6-1) yields the desired result,

$$Y = 1 + \frac{Z_3 - Z_4}{Z_1 - Z_2}. \tag{E6-7}$$

To determine the effect of feedback on the input and output impedances, consider the series-shunt feedback configuration given in Fig. 6.16. Since

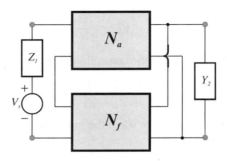

Figure 6.16. Generalized series-shunt feedback configuration.

shorting the terminals of Y_2 interrupts the feedback loop, Eq. (6-53) applies and the output impedance across load admittance Y_2 becomes

$$Z_{\text{out}}(x) = \frac{Z_{\text{out}}(0)}{F(x)}, \tag{6-55}$$

which verifies that the impedance measured across the path of the feedback is reduced by the factor that is the normal value of the return difference with respect to the element x. For the input impedance of the amplifier seen by the voltage source, V_s, in Fig. 6.16, by open-circuiting or removing the voltage source V_s, the feedback loop is broken. Thus, Eq. (6-6) applies, and the input impedance becomes

$$Z_{\text{in}}(x) = F(x) Z_{\text{in}}(0), \tag{6-56}$$

which suggests that the impedance measured in series lines is increased by the return difference, $F(x)$. Similar conclusions can be reached for other types of configurations.

Refer once again to the general feedback configuration of Fig. 6.6. If $w(x)$ represents either the voltage gain, V_{pq}/V_{rs}, or the transfer admittance, I_{pq}/V_{rs}, Eq. (6-22) offers

$$\frac{w(x)}{w(0)} = \frac{Y_{rp,sq}(x)}{Y_{rp,sq}(0)} \frac{Y_{rr,ss}(0)}{Y_{rr,ss}(x)}. \tag{6-57}$$

The first term in the product on the right hand side of this expression is the null return difference, $\hat{F}(x)$, with respect to x for the input terminals r and s and output terminals p and q. The second term is the reciprocal of the null return difference with respect to x for the same input and output port at terminals r and s. This reciprocal can then be interpreted as the return difference with respect to x when the input port of the amplifier is short circuited. Thus, the voltage gain or the transfer admittance can be expressed as

$$w(x) = w(0) \frac{\hat{F}(x)}{F(\text{input short circuited})}. \tag{6-58}$$

If $w(x)$ denotes the short circuit current gain I_{pq}/I_s as Y_2 approaches infinity,

$$\frac{w(x)}{w(0)} = \frac{Y_{rp,sq}(x)}{Y_{rp,sq}(0)} \frac{Y_{pp,qq}(0)}{Y_{pp,qq}(x)}. \tag{6-59}$$

The second term in the product on the right hand side in Eq. (6-59) is the reciprocal of the return difference with respect to x when the output port of the amplifier is short circuited, thereby enabling an expression for the short circuit current gain in the form of

$$w(x) = w(0) \frac{\hat{F}(x)}{F(\text{output short circuited})}. \tag{6-60}$$

Consider once again the series-shunt feedback amplifier of Fig. 6.9, the equivalent network of which is depicted in Fig. 6.10. The return differences, $F(\tilde{\alpha}_k)$, the null return differences, $\hat{F}(\tilde{\alpha}_k)$, and the voltage gain, w, have been computed as Eqs. (6-40), (6-47) and (6-39), respectively, and are repeated herewith for reader convenience; namely,

$$\left. \begin{array}{l} F(\tilde{\alpha}_1) = 93.70, \quad F(\tilde{\alpha}_2) = 18.26 \\ \hat{F}(\tilde{\alpha}_1) = 103.07 \times 10^3, \quad \hat{F}(\tilde{\alpha}_2) = 2018.70 \\ w = \dfrac{V_{25}}{V_S} = w(\tilde{\alpha}_1) = w(\tilde{\alpha}_2) = 45.39 \end{array} \right\}. \tag{6-61}$$

Applying Eq. (6-21), the voltage gain, w, evolves as

$$w(\tilde{\alpha}_1) = w(0) \frac{\hat{F}(\tilde{\alpha}_1)}{F(\text{input short circuited})}$$

$$= 0.04126 \frac{103.07 \times 10^3}{93.699} = 45.39, \tag{6-62}$$

where

$$\left. \begin{array}{l} w(0) = \left. \dfrac{Y_{12,55}(\tilde{\alpha}_1)}{Y_{11,55}(\tilde{\alpha}_1)} \right|_{\tilde{\alpha}_1=0} = \dfrac{205.24 \times 10^{-12}}{497.41 \times 10^{-11}} = 0.04126 \\ F(\text{input short circuited}) = \dfrac{Y_{11,55}(\tilde{\alpha}_1)}{Y_{11,55}(0)} = \dfrac{466.07 \times 10^{-9}}{4.9741 \times 10^{-9}} = 93.699 \end{array} \right\}, \tag{6-63}$$

and

$$w(\tilde{\alpha}_2) = w(0) \frac{\hat{F}(\tilde{\alpha}_2)}{F(\text{input short circuited})} = 0.41058 \frac{2018.70}{18.26} = 45.39, \tag{6-64}$$

with the understanding that

$$w(0) = \left.\frac{Y_{12,55}(\tilde{\alpha}_2)}{Y_{11,55}(\tilde{\alpha}_2)}\right|_{\tilde{\alpha}_2=0} = \frac{104.79 \times 10^{-10}}{255.22 \times 10^{-10}} = 0.41058$$

$$F(\text{input short circuited}) = \frac{Y_{11,55}(\tilde{\alpha}_2)}{Y_{11,55}(0)} = \frac{466.07 \times 10^{-9}}{25.52 \times 10^{-9}} = 18.26$$

(6-65)

6.3.2. Sensitivity Function

One of the most important effects of negative feedback is its ability to make an amplifier less sensitive to parametric variations caused by the ravages of aging, temperature variations, and other environmental changes. A useful quantitative measure for the degree of dependence of an amplifier on a particular parameter is known as the sensitivity. The *sensitivity function*, written as $S(x)$, for a given transfer function with respect to a parameter, x, is defined as the ratio of the fractional change in a transfer function to the fractional change in x for the circumstance in which all of interest are differentially small. Thus, if $w(x)$ is the transfer function, the sensitivity of this function with respect to small perturbations in parameter x is

$$S(x) = \lim_{\Delta x \to 0} \frac{\Delta w/w}{\Delta x/x} = \frac{x}{w}\frac{\partial w}{\partial x} = x\frac{\partial \ln w}{\partial x} \qquad (6\text{-}66)$$

Refer to the general feedback configuration of Fig. 6.6, and let $w(x)$ represent either the current gain, I_{pq}/I_s, or the transfer impedance, V_{pq}/I_s. Then from Eq. (6-19),

$$w(x) = Y_2 \frac{Y_{rp,sq}(x)}{Y_{uv}(x)}. \qquad (6\text{-}67)$$

Letting

$$\dot{Y}_{uv}(x) \stackrel{\Delta}{=} \frac{\partial Y_{uv}(x)}{\partial x}$$

$$\dot{Y}_{rp,sq}(x) \stackrel{\Delta}{=} \frac{\partial Y_{rp,sq}(x)}{\partial x}$$

(6-68)

which precipitates

$$Y_{uv}(x) = Y_{uv}(0) + x\dot{Y}_{uv}(x)$$
$$Y_{rp,sq}(x) = Y_{rp,sq}(0) + x\dot{Y}_{rp,sq}(x)$$

(6-69)

Substituting Eq. (6-22) in Eq. (6-21) in conjunction with Eq. (6-24) yields

$$S(x) = x\frac{\dot{Y}_{rp,sq}(x)}{Y_{rp,sq}(x)} - x\frac{\dot{Y}_{uv}(x)}{Y_{uv}(x)} = \frac{Y_{rp,sq}(x) - Y_{rp,sq}(0)}{Y_{rp,sq}(x)} - \frac{Y_{uv}(x) - Y_{uv}(0)}{Y_{uv}(x)}$$

$$= \frac{Y_{uv}(0)}{Y_{uv}(x)} - \frac{Y_{rp,sq}(0)}{Y_{rp,sq}(x)} = \frac{1}{F(x)} - \frac{1}{\hat{F}(x)}. \tag{6-70}$$

Recalling Eq. (6-3),

$$S(x) = \frac{1}{F(x)}\left[1 - \frac{w(0)}{w(x)}\right]. \tag{6-71}$$

Note that for $w(0) = 0$, Eq. (6-71) becomes

$$S(x) = \frac{1}{F(x)}, \tag{6-72}$$

which implies that the aforementioned sensitivity metric with respect to a parameter x is equal to the reciprocal of the return difference, also evaluated with respect to x. For the ideal feedback model, the feedback path is unilateral. Hence, $w(0) = 0$ and

$$S = \frac{1}{F} = \frac{1}{1+T} = \frac{1}{1-\mu\beta}. \tag{6-73}$$

In a practical amplifier, $w(0)$ is usually very much smaller than $w(x)$ in the passband, and $F \approx 1/S$ may be used as a reasonable estimate of the reciprocal of the sensitivity in the same frequency band. A single loop feedback amplifier comprised of a cascade of common emitter stages with a passive network providing the desired feedback fulfills this requirement. If in such a structure any one of the transistors fails, the forward transmission is nearly zero and $w(0)$ is practically zero. Accordingly, if the failure of any element interrupts signal transmission through the amplifier to nearly zero, the sensitivity is approximately equal to the reciprocal of the return difference with respect to that element. In the case of driving point impedance, $w(0)$ is generally not much smaller than $w(x)$, which renders the reciprocity approximation invalid.

Assume now that $w(x)$ represents the voltage gain. Substituting Eq. (6-22) in Eq. (6-21) results in

$$S(x) = x\frac{\dot{Y}_{rp,sq}(x)}{Y_{rp,sq}(x)} - x\frac{\dot{Y}_{rr,ss}(x)}{Y_{rr,ss}(x)} = \frac{Y_{rp,sq}(x) - Y_{rp,sq}(0)}{Y_{rp,sq}(x)}$$

$$-\frac{Y_{rr,ss}(x) - Y_{rr,ss}(0)}{Y_{rr,ss}(x)}$$

$$= \frac{Y_{rr,ss}(0)}{Y_{rr,ss}(x)} - \frac{Y_{rp,sq}(0)}{Y_{rp,sq}(x)} = \frac{1}{F(\text{input short circuited})} - \frac{1}{\hat{F}(x)}.$$

(6-74)

Combining this result with Eq. (6-71) gives

$$S(x) = \frac{1}{F(\text{input short circuited})}\left[1 - \frac{w(0)}{w(x)}\right]. \quad (6\text{-}75)$$

Finally, if $w(x)$ denotes the short circuit current gain, I_{pq}/I_s, as Y_2 approaches infinity, the sensitivity function can be written as

$$S(x) = \frac{Y_{pp,qq}(0)}{Y_{pp,qq}(x)} - \frac{Y_{rp,sq}(0)}{Y_{rp,sq}(x)} = \frac{1}{F(\text{output short circuited})} - \frac{1}{\hat{F}(x)},$$

(6-76)

which when combined with Eq. (6-71) yields

$$S(x) = \frac{1}{F(\text{output short circuited})}\left[1 - \frac{w(0)}{w(x)}\right]. \quad (6\text{-}77)$$

As an example, consider the network of Fig. 6.17, which schematically portrays a common emitter transistor amplifier. After removing the biasing circuit and using the common emitter hybrid model for the transistor at low frequencies, an appropriate equivalent circuit of the amplifier is the structure presented in Fig. 6.18 with

$$\left. \begin{array}{l} I'_s = \dfrac{V_s}{R_1 + r_x} \\[6pt] G'_1 = \dfrac{1}{R'_1} = \dfrac{1}{R_1 + r_x} + \dfrac{1}{r_\pi} \\[6pt] G'_2 = \dfrac{1}{R'_2} = \dfrac{1}{R_2} + \dfrac{1}{R_c} \end{array} \right\rangle. \quad (6\text{-}78)$$

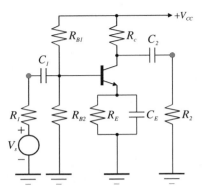

Figure 6.17. A common emitter transistor feedback amplifier.

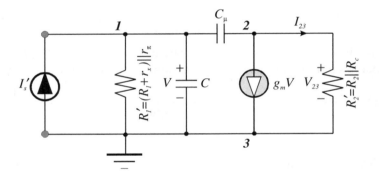

Figure 6.18. Equivalent circuit of the feedback amplifier in Figure 6.17.

The indefinite admittance matrix of the amplifier is seen to be

$$Y = \begin{bmatrix} G_1' + sC_\pi + sC_\mu & -sC_\mu & -G_1' - sC_\pi \\ g_m - sC_\mu & G_2' + sC_\mu & -G_2' - g_m \\ -G_1' - sC_\pi - g_m & -G_2' & G_1' + G_2' + sC_\pi + g_m \end{bmatrix} \quad (6\text{-}79)$$

Assume that the controlling transconductance, g_m, is the critical element of immediate interest. The return difference and the null return difference with respect to g_m in Fig. 6.18, with I_s' applied to the input port and R_2' terminating the output port, are found to be

$$F(g_m) = \frac{Y_{33}(g_m)}{Y_{33}(0)} = \frac{(G_1' + sC_\pi)(G_2' + sC_\mu) + sC_\mu(G_2' + g_m)}{(G_1' + sC_\pi)(G_2' + sC_\mu) + sC_\mu G_2'} \quad (6\text{-}80)$$

and

$$\hat{F}(g_m) = \frac{Y_{12,33}(g_m)}{Y_{12,33}(0)} = \frac{sC_\mu - g_m}{sC_\mu} = 1 - \frac{g_m}{sC_\mu}. \quad (6\text{-}81)$$

The current gain, I_{23}/I'_s, as defined in Fig. 6.18, is computed as

$$w(g_m) = \frac{Y_{12,33}(g_m)}{R'_2 Y_{33}(g_m)} = \frac{sC_\mu - g_m}{R'_2\left[(G'_1 + sC_\pi)(G'_2 + sC_\mu) + sC_\mu(G'_2 + g_m)\right]}. \quad (6\text{-}82)$$

Substituting these in Eq. (6-74) or Eq. (6-75) gives

$$\mathsf{S}(g_m) = -\frac{g_m(G'_1 + sC_\pi + sC_\mu)(G'_2 + sC_\mu)}{(sC_\mu - g_m)\left[(G'_1 + sC_\pi)(G'_2 + sC_\mu) + sC_\mu(G'_2 + g_m)\right]}. \quad (6\text{-}83)$$

Finally, compute the sensitivity for the driving point impedance facing the current source, I'_s. From Eq. (6-74),

$$\mathsf{S}(g_m) = \frac{1}{F(g_m)}\left[1 - \frac{Z(0)}{Z(g_m)}\right]$$

$$= -\frac{sC_\mu g_m}{(G'_1 + sC_\pi)(G'_2 + sC_\mu) + sC_\mu(G'_2 + g_m)}, \quad (6\text{-}84)$$

where

$$Z(g_m) = \frac{Y_{11,33}(g_m)}{Y_{33}(g_m)} = -\frac{G'_2 + sC_\mu}{(G'_1 + sC_\pi)(G'_2 + sC_\mu) + sC_\mu(G'_2 + g_m)}. \quad (6\text{-}85)$$

6.4.0. Measurement of Return Difference

The zeros of a network determinant are called the *natural frequencies* of the network. Their locations in the complex frequency plane are extremely important in that they determine the stability and both the frequency domain and time domain responses of the network. A network is said to be *stable* if all of its natural frequencies are restricted to the open left-half of the complex frequency plane (LHS). If a network determinant is known, its roots can readily be computed explicitly with the aid of a computer if necessary, and the stability problem can then be settled directly. However, for a physical network there remains the difficulty of getting an accurate formulation of the network determinant itself, because every equivalent network is, to a greater or lesser extent, an idealization of physical reality. As frequency is increased, parasitic effects of the physical elements must be taken into account. What is really needed is some kind of experimental verification that the network is stable and remains stable under prescribed operational and environmental conditions. The measurement of the return difference provides an elegant solution to this problem.

The return difference with respect to an element x in a feedback amplifier is given as

$$F(x) = \frac{Y_{uv}(x)}{Y_{uv}(0)}. \tag{6-86}$$

Since $Y_{uv}(x)$ denotes the nodal determinant and $Y_{uv}(0)$ is said determinant evaluated for $x = 0$, the zeros of the return difference are exactly the same as the zeros of the nodal determinant, provided that there is no cancellation of common factors between $Y_{uv}(x)$ and $Y_{uv}(0)$. Therefore, if $Y_{uv}(0)$ is known to have no zeros in the closed right-half of the complex frequency plane (RHS), which is usually the case in a single loop feedback amplifier, $F(x)$ gives precisely the same information about the stability of a feedback amplifier, as does the nodal determinant itself. The difficulty inherent in the

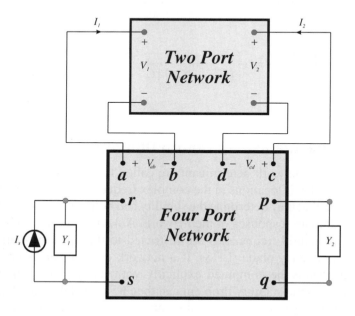

Figure 6.19. The general configuration of a feedback amplifier with a two port device.

measurement of the return difference with respect to the controlling parameter of a controlled source is that, in a physical system, the controlling branch and the controlled source invariably form part of a single device, such as a transistor, and cannot be physically separated. The measurement schema presented herewith does not require the physical decomposition of a device.

Let a device of interest be brought out as a two-port network connected to a general four-port network as shown in Fig. 6.19. Assume that this device can be characterized by its y-parameters, thereby enabling the electrical representation in Fig. 6.20, where parameter y_{21} controls signal transmission in the forward direction through the device while y_{12} gives the reverse transmission deriving from intrinsic device feedback. The fundamental objective here is the measurement of the return difference with respect to the forward short circuit transfer admittance y_{21}.

6.4.1. Blecher's Procedure

Let the two-port device in question be a transistor operated in the common emitter configuration with terminals a, b, c, and d respectively representing, the base, emitter, collector, and ground terminals. To simplify notation, let $a = 1, b = d = 3$ and $c = 2$, as exhibited in Fig. 6.21.

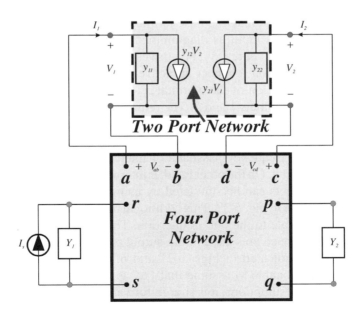

Figure 6.20. The system of Figure 6.19 with the two port network represented by its short circuit admittance (y-) parameters.

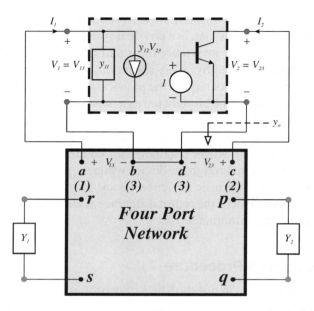

Figure 6.21. Physical interpretation of the return difference, $F(y_{11})$, for a transistor operated in common emitter mode and modeled by its short circuit admittance parameters.

To measure $F(y_{21})$, break the base terminal of the transistor and apply a 1-V excitation at its input as diagrammed in Fig. 6.21. To ensure that the controlled current source, $y_{21}V_{13}$, drives a replica of the load it sees during normal operation, connect an active one-port network composed of a parallel combination of the admittance, y_{11}, and a controlled current source, $y_{12}V_{23}$, at terminals 1 and 3. The returned voltage, V_{13}, is precisely the negative of the return ratio with respect to the element, y_{21}. If the externally applied feedback is large compared with the internal feedback of the transistor over the frequency passband of interest, the controlled source, $y_{12}V_{23}$, can be ignored. On the other hand, if internal device feedback cannot be ignored, its effects can be simulated by using an additional transistor, connected as shown in Fig. 6.22. This additional transistor must be matched as closely as possible to the one in question. The one-port admittance y_o denotes the admittance presented to the output port of the transistor under consideration as indicated in Figs. 6.21 and 6.22. For a common emitter state, it is reasonable to assume that $|y_o| \gg |y_{12}|$ and $|y_{11}| \gg |y_{12}|$. In view of these presumptions, it is straightforward to show that the Norton equivalent network looking into the two-port network at terminals 1 and

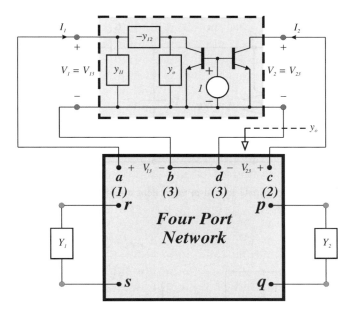

Figure 6.22. The measurement of the return difference, $F(y_{21})$, for a transistor operated as a common emitter amplifier and represented electrically by its short circuit admittance parameters.

3 of Fig. 6.22 can be approximated by the parallel combination of y_{11} and $y_{12}V_{23}$ as indicated in Fig. 6.21. In Fig. 6.22, voltage sources having very low internal impedances can be joined together at the two transistor base terminals, which can then be driven by a single voltage source of low internal impedance. But for the foregoing measurement procedure to be feasible, it must be demonstrated that the admittances, y_{11} and $-y_{12}$, can be realized as positive real admittances presented by suitable one-port passive networks.

Consider the small signal hybrid-pi model of a common emitter transistor shown in Fig. 6.23. The short circuit admittance matrix of the this circuit model is

$$Y_{sc} = \frac{1}{g_x + g_\pi + sC_\pi + sC_\mu}$$

$$\times \begin{bmatrix} g_x \left(g_\pi + sC_\pi + sC_\mu \right) & -g_x sC_\mu \\ g_x \left(g_n - sC_\mu \right) & sC_\mu \left(g_x + g_\pi + sC_\pi + g_m \right) \end{bmatrix}. \quad (6\text{-}87)$$

It is easy to confirm that the admittances, y_{11} and $-y_{12}$, can be realized by the one-port networks of Fig. 6.24.

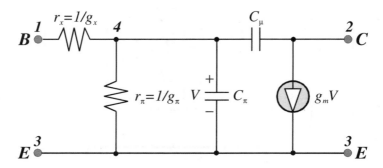

Figure 6.23. The approximate hybrid-pi equivalent circuit of a transistor operated as a common emitter amplifier.

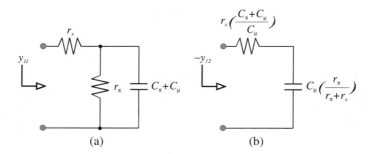

Figure 6.24. (a) The realization of the common emitter short circuit admittance parameter, y_{11}. (b) The realization of the short circuit admittance function, $-y_{12}$.

6.4.2. Impedance Measurements

Refer again to the general feedback configuration of Fig. 6.6. Suppose that the return difference with respect to the forward short circuit transfer admittance y_{21}, is to be evaluated. The controlling parameters, y_{12} and y_{21}, enter the indefinite admittance matrix, $\mathbf{Y}$, in the rectangular patterns abstracted below:

$$Y(x) = \begin{array}{c} \\ a \\ b \\ c \\ d \end{array} \begin{array}{c} a   b c d \end{array} \left[\begin{array}{cccc} & & y_{12} & -y_{12} \\ & & -y_{12} & y_{12} \\ y_{21} & -y_{21} & & \\ -y_{21} & y_{21} & & \end{array} \right]. \quad (6\text{-}88)$$

To emphasize the importance of y_{12} and y_{21}, write $Y_{uv}(x)$ as $Y_{uv}(y_{12}, y_{21})$ and $z_{aa,bb}(x)$ as $z_{aa,bb}(y_{12}, y_{21})$. By appealing to Eq. (6-20), the impedance looking into terminals a and b of Fig. 6.6 is

$$z_{aa,bb}(y_{12}, y_{21}) = \frac{Y_{aa,bb}(y_{12}, y_{21})}{Y_{dd}(y_{12}, y_{21})}. \tag{6-89}$$

The return difference with respect to parameter y_{21} follows as

$$F(y_{21}) = \frac{Y_{dd}(y_{12}, y_{21})}{Y_{dd}(y_{12}, 0)}. \tag{6-90}$$

Combining Eqs. (6-89) and (6-90),

$$F(y_{21})z_{aa,bb}(y_{12}, y_{21}) = \frac{Y_{aa,bb}(y_{12}, y_{21})}{Y_{dd}(y_{12}, 0)} = \frac{Y_{aa,bb}(0, 0)}{Y_{dd}(y_{12}, 0)}$$

$$= \frac{Y_{aa,bb}(0, 0)}{Y_{dd}(0, 0)} \frac{Y_{dd}(0, 0)}{Y_{aa,bb}(y_{12}, 0)} = \frac{z_{aa,bb}(0, 0)}{F(y_{12})|_{y_{21}=0}}, \tag{6-91}$$

which engenders

$$F(y_{12})|_{y_{21}=0} F(y_{21}) = \frac{z_{aa,bb}(0, 0)}{z_{aa,bb}(y_{12}, y_{21})} \tag{6-92}$$

as an interrelationship among the return differences and the driving point impedances. $F(y_{12})|_{y_{21}=0}$ is the return difference with respect to y_{12} when y_{21} is set to zero. This quantity can be measured by the arrangement of Fig. 6.25. In this diagram, $z_{aa,bb}(y_{12}, y_{21})$ is the driving point impedance looking into terminals a and b of the network of Fig. 6.6. Finally, $z_{aa,bb}(0, 0)$ is the impedance to which $z_{aa,bb}(y_{12}, y_{21})$ reduces when the controlling parameters, y_{12} and y_{21}, are both set to zero. This impedance can be measured by the configuration of Fig. 6.26. Note that in all three measurements, the independent current source, I_s, is removed.

Suppose now that the return difference, $F(y_{21})$, with respect to the forward transfer admittance, y_{21}, of the common emitter transistor shown in Fig. 6.20 is to be measured. For all practical purposes, the return difference, $F(y_{12})$, when y_{21} is set to zero, is indistinguishable from unity. Therefore, Eq. (6-92) reduces to the simpler form,

$$F(y_{21}) \approx \frac{z_{11,33}(0, 0)}{z_{11,33}(y_{12}, y_{21})}, \tag{6-93}$$

which shows that the return difference, $F(y_{21})$, is effectively the ratio of the two functional values assumed by the driving point impedance looking into terminals 1 and 3 of Fig. 6.6 under the conditions that the controlling parameters, y_{12} and y_{21}, are both set to zero and that these parameters assume their nominal values. These two impedances can be measured by the network arrangements of Figs. 6.27 and 6.28.

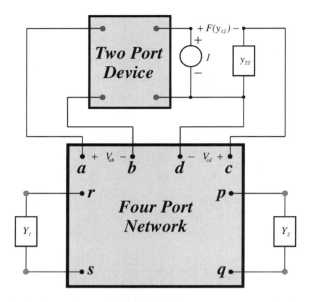

Figure 6.25. The measurement of the return difference, $F(y_{12})$, with short circuit admittance parameter y_{21} set to zero.

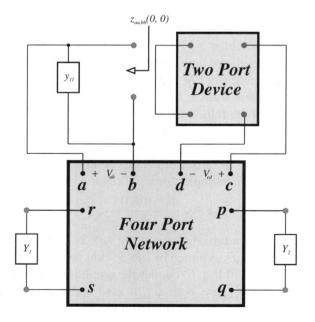

Figure 6.26. The measurement of the driving point impedance, $z_{aa,bb}(0, 0)$.

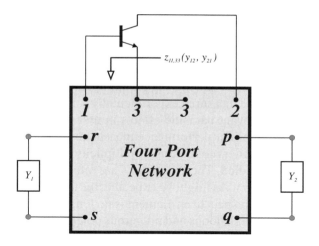

Figure 6.27. The measurement of the driving point impedance, $z_{11,33}(y_{12}, y_{21})$.

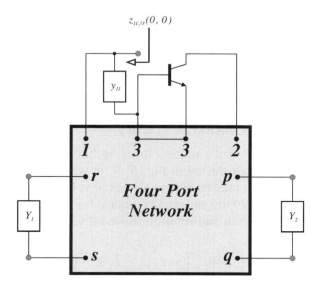

Figure 6.28. The measurement of the driving point impedance, $z_{11,33}(0, 0)$.

6.5.0. Multiloop Feedback

At this juncture, single loop feedback amplifiers have been studied in depth, wherein the return difference has been postured as a pivotally important characteristic. The return difference is the difference between a unit applied signal and the resultantly returned signal. The returned signal has the same physical meaning as the loop transmission in the ideal feedback model. It plays an important role in the study of amplifier stability, the sensitivity of amplifier responses to variations of parameters, and the determination of feedback amplifier transfer and driving point impedances. The fact that the return difference can be measured experimentally for most practical amplifiers suggests that relevant parasitic effects in assessing relative network stability and other amplifier performance measures.

In this section, amplifiers containing a multiplicity of inputs, outputs, and feedback loops are studied. These networks are referred to as the *multiple loop feedback amplifiers*. As might be expected, the traditional concept of a return difference with respect to an element is no longer applicable because of the presence of multiple loops and numerous potentially critical parameters. Accordingly, the return difference concept for a controlled source must be expanded to the notion of a return difference matrix for a multiplicity of controlled sources. For measurement situations, the null return difference matrix is introduced, and its engineering significance is discussed. It is shown herewith that the determinant of the overall transfer function matrix of a multiple loop feedback network can be expressed explicitly in terms of the determinants of the return difference and the null return difference matrices, thereby allowing a generalization of Blackman's impedance formula.

6.5.1. Multiloop Feedback Theory

The general configuration of a multiple input, multiple output, multiple loop feedback amplifier is abstracted in Fig. 6.29, in which the input, output, and feedback variables may be either currents or voltages. For the specific arrangement of Fig. 6.29, the input and output variables are represented by an n-dimensional vector $\boldsymbol{u}$ and an m-dimensional vector $\boldsymbol{y}$ as

$$\boldsymbol{u}(s) = \begin{bmatrix} u_1 \\ u_2 \\ \vdots \\ u_k \\ u_{k+1} \\ u_{k+2} \\ \vdots \\ u_n \end{bmatrix} = \begin{bmatrix} I_{s1} \\ I_{s2} \\ \vdots \\ I_{sk} \\ V_{s1} \\ V_{s2} \\ \vdots \\ V_{s(n-k)} \end{bmatrix}, \quad \boldsymbol{y}(s) = \begin{bmatrix} y_1 \\ y_2 \\ \vdots \\ y_r \\ y_{r+1} \\ y_{r+2} \\ \vdots \\ y_m \end{bmatrix} = \begin{bmatrix} I_1 \\ I_2 \\ \vdots \\ I_r \\ V_{r+1} \\ V_{r+2} \\ \vdots \\ V_m \end{bmatrix}, \quad (6\text{-}94)$$

Multiple Loop Feedback Amplifiers

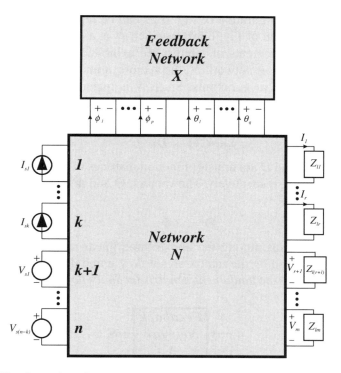

Figure 6.29. General configuration of a multiple input, multiple output, multiple loop feedback amplifier.

respectively. The elements of interest can be represented by a rectangular matrix, X, of order $q \times p$ relating the controlled and controlling variables by the matrix equation

$$\Theta = \begin{bmatrix} \theta_1 \\ \theta_2 \\ \vdots \\ \theta_q \end{bmatrix} = \begin{bmatrix} x_{11} & x_{12} & \cdots & x_{1p} \\ x_{21} & x_{22} & \cdots & x_{2p} \\ \vdots & \vdots & \vdots & \vdots \\ x_{q1} & x_{q2} & \cdots & x_{qp} \end{bmatrix} \begin{bmatrix} \phi_1 \\ \phi_2 \\ \vdots \\ \phi_p \end{bmatrix} = X\Phi, \qquad (6\text{-}95)$$

where the p-dimensional vector, Φ, is called the *controlling vector*, and the q-dimensional vector, Θ, the *controlled vector*. The controlled variables, θ_k, and the controlling variables, ϕ_k, can either be currents or voltages. The matrix X can represent either a transfer function matrix or a driving point function matrix. If X represents a driving point function matrix, the vectors, Θ and Φ, are of the same dimension ($q = p$), and their components are the currents and voltages of a p-port network.

The general configuration of Fig. 6.29 can be represented equivalently by the block digraph of Fig. 6.30, in which N is a $(p+q+m+n)$-port network and the elements of interest are exhibited by block X. For the $(p+q+m+n)$-port network N, the vectors, u and Θ, are its inputs, and the vectors, Φ and y, its outputs. Since N is linear, the input and output vectors are related by the matrix equation

$$\left.\begin{array}{l}\Phi = A\Theta + Bu \\ y = C\Theta + Du\end{array}\right\}, \qquad (6\text{-}96)$$

where A, B, C, and D are transfer function matrices of orders $p \times q$, $p \times n$, $m \times q$ and $m \times n$, respectively. The vectors, Θ and Φ, are not independent and are related by

$$\Theta = X\Phi. \qquad (6\text{-}97)$$

The relationships among the above three linear matrix equations can also be represented by the matrix signal flow graph depicted in Fig. 6.31, which is known as the *fundamental matrix feedback flow graph*. The overall

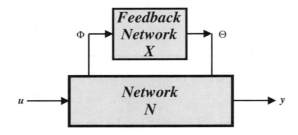

Figure 6.30. The block diagram representation of the general feedback configuration of Figure 6.29.

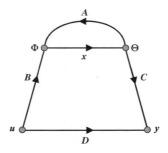

Figure 6.31. The fundamental matrix feedback signal flow graph.

closed loop *transfer function matrix*, $W(X)$, of the multiple loop feedback amplifier derives from the equation

$$y = W(X)u, \quad (6\text{-}98)$$

where $W(X)$ is of order $m \times n$. As before, to emphasize the importance of X, the matrix W is written as $W(X)$ for the present discussion, even though it is also a function of the complex frequency variable, s. Combining the above matrix equations, the transfer function matrix is found to be

$$\left. \begin{array}{l} W(X) = D + CX\left(1_p - AX\right)^{-1} B \\ W(X) = D + C\left(1_q - XA\right)^{-1} XB \end{array} \right\}, \quad (6\text{-}99)$$

where 1_p denotes the identity matrix of order p. Clearly,

$$W(0) = D. \quad (6\text{-}100)$$

In particular, when X is square and nonsingular, Eq. (6-99) can be written as

$$W(X) = D + C\left(X^{-1} - A\right)^{-1} B. \quad (6\text{-}101)$$

As an illustration, consider the feedback amplifier of Fig. 6.9. An equivalent network is shown in Fig. 6.32, in which the two transistors are presumed identical with $h_{ie} = 1.1\,\text{k}\Omega$, $h_{fe} = 50$, and $h_{re} = h_{oe} = 0$. Let the controlling parameters of the two controlled sources be the elements of

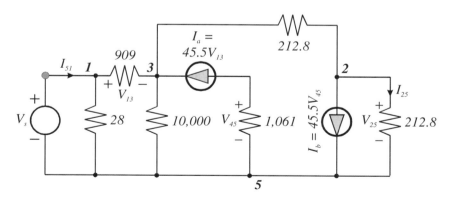

Figure 6.32. A low frequency equivalent circuit of the feedback amplifier offered in Figure 6.9. The resistive branch elements are represented as conductances with values in units of microsiemens, while the transconductance parameters associated with the voltage controlled current sources are in units of millisiemens.

interest. Then,

$$\Theta = \begin{bmatrix} I_a \\ I_b \end{bmatrix} = 10^{-4} \begin{bmatrix} 455 & 0 \\ 0 & 455 \end{bmatrix} \begin{bmatrix} V_{13} \\ V_{45} \end{bmatrix} = X\Phi. \quad (6\text{-}102)$$

Assume that the output voltage, V_{25}, and the input current, I_{51}, are the output variables. Then the seven-port network, N, defined by the variables V_{13}, V_{45}, V_{25}, I_{51}, I_a, I_b and V_s can be characterized by the matrices,

$$\left. \begin{aligned} \Phi &= \begin{bmatrix} V_{13} \\ V_{45} \end{bmatrix} = \begin{bmatrix} -90.782 & 45.391 \\ -942.507 & 0 \end{bmatrix} \begin{bmatrix} I_a \\ I_b \end{bmatrix} \\ &+ \begin{bmatrix} 0.91748 \\ 0 \end{bmatrix} [V_s] = A\Theta + Bu \\ y &= \begin{bmatrix} V_{25} \\ I_{51} \end{bmatrix} = \begin{bmatrix} 45.391 & -2372.32 \\ -0.08252 & 0.04126 \end{bmatrix} \begin{bmatrix} I_a \\ I_b \end{bmatrix} \\ &+ \begin{bmatrix} 0.041260 \\ 0.000862 \end{bmatrix} [V_s] = C\Theta + Du \end{aligned} \right\} . \quad (6\text{-}103)$$

According to Eq. (6-98), the transfer function matrix of the amplifier is defined by the matrix equation

$$y = \begin{bmatrix} V_{25} \\ I_{51} \end{bmatrix} = \begin{bmatrix} w_{11} \\ w_{21} \end{bmatrix} [V_s] = W(X)u. \quad (6\text{-}104)$$

Since X is square and nonsingular, Eq. (6-101) can be exploited to calculate $W(X)$:

$$W(X) = D + C\left(X^{-1} - A\right)^{-1} B = \begin{bmatrix} 45.387 \\ 0.369 \times 10^{-4} \end{bmatrix} = \begin{bmatrix} w_{11} \\ w_{21} \end{bmatrix}, \quad (6\text{-}105)$$

where

$$\left(X^{-1} - A\right)^{-1} = 10^{-4} \begin{bmatrix} 4.856 & 10.029 \\ -208.245 & 24.914 \end{bmatrix}. \quad (6\text{-}106)$$

These relationships can be used to obtain the closed loop voltage gain, w_{11}, and the input impedance, Z_{in}, facing the voltage source, V_s. In particular,

$$w_{11} = \frac{V_{25}}{V_s} = 45.387, \quad Z_{\text{in}} = \frac{V_s}{I_{51}} = \frac{1}{w_{21}} = 27.1 \text{ K}\Omega. \quad (6\text{-}107)$$

6.5.2. Return Difference Matrix

In the fundamental feedback flow graph of Fig. 6.31, break the input of the branch with transmittance X, set the input excitation vector, u, to zero, and apply a signal p-vector, g, to the right of the break, as depicted in Fig. 6.33.

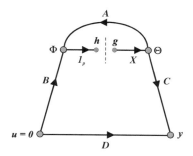

Figure 6.33. The engineering interpretation of the loop transmission matrix.

The resultant signal p-vector, h, returned to the left of the break is

$$h = AXg. \tag{6-108}$$

The square matrix, AX, is called the *loop transmission matrix*, and its negative, referred to as the *return ratio matrix*, is denoted by

$$T(X) = -AX. \tag{6-109}$$

The difference between the applied signal vector, g, and the returned signal vector, h, is

$$g - h = (1_p - AX)g. \tag{6-110}$$

The square matrix, $[1_p - AX]$, relating the applied signal vector, g, to the difference of applied signal vector g and the returned signal vector, h, is called the *return difference matrix* with respect to X and is symbolized by

$$F(X) = 1_p - AX. \tag{6-111}$$

Combining this expression with Eq. (6-109) gives

$$F(X) = 1_p + T(X). \tag{6-112}$$

For the series-shunt feedback amplifier of Fig. 6.32, let the controlling parameters of the two controlled current sources be the elements of interest. The corresponding return ratio matrix is found from Eqs. (6-102) and (6-103) to be

$$T(X) = -AX = -\begin{bmatrix} -90.782 & 45.391 \\ -942.507 & 0 \end{bmatrix} \begin{bmatrix} 455 \times 10^{-4} & 0 \\ 0 & 455 \times 10^{-4} \end{bmatrix}$$

$$= \begin{bmatrix} 4.131 & -2.065 \\ 42.884 & 0 \end{bmatrix}, \tag{6-113}$$

which yields

$$F(X) = 1_2 + T(X) = \begin{bmatrix} 5.131 & -2.065 \\ 42.884 & 1 \end{bmatrix}. \quad (6\text{-}114)$$

6.5.3. Null Return Difference Matrix

A direct extension of the null return difference for the single loop feedback amplifier is the null return difference matrix for the multiple loop feedback networks. Refer again to the fundamental matrix feedback flow graph of Fig. 6.31. Once again, break the branch with transmittance X and apply a signal p-vector, g, to the right of the break, as illustrated in Fig. 6.34. Next, adjust the input excitation n-vector, u, so that the total output m-vector, y, resulting from the inputs g and u is null. From Fig. 6.34, the required input excitation, u, is

$$Du + CXg = 0, \quad (6\text{-}115)$$

whence

$$u = -D^{-1}CXg, \quad (6\text{-}116)$$

provided that matrix D is square and nonsingular. An underlying prerequisite to this constraint is that output y be of the same dimension as input u; namely, $m = n$. Physically, this requirement is reasonable because the effects at the output caused by g can be neutralized by a unique input excitation u only when u and y are of the same dimension. Given these inputs, u and g, the returned signal, h, to the left of the break in Fig. 6.34 computes as

$$h = Bu + AXg = \left(-BD^{-1}CX + AX\right)g, \quad (6\text{-}117)$$

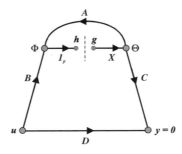

Figure 6.34. The engineering interpretation of the null return difference matrix.

whereby
$$g - h = \left(1_p - AX + BD^{-1}CX\right)g. \tag{6-118}$$

The square matrix,
$$\hat{F}(X) = 1_p + \hat{T}(X) = 1_p - AX + BD^{-1}CX = 1_p - \hat{A}X, \tag{6-119}$$

relating input signal vector g to the difference of input signal vector g and returned signal vector h is commonly referred to as the *null return difference matrix* with respect to X, where

$$\left.\begin{array}{l} \hat{T}(X) = -AX + BD^{-1}CX = -\hat{A}X \\ \hat{A} = A - BD^{-1}C \end{array}\right\}. \tag{6-120}$$

The square matrix, $\hat{T}(X)$, is the *null return ratio matrix*.

As yet another example, consider the voltage-series feedback amplifier of Fig. 6.9, for which the equivalent network is the structure in Fig. 6.32. Assume that the voltage, V_{25}, is the output variable. From Eq. (6-103),

$$\left.\begin{array}{l}
\Phi = \begin{bmatrix} V_{13} \\ V_{45} \end{bmatrix} \\
\quad = \begin{bmatrix} -90.782 & 45.391 \\ -942.507 & 0 \end{bmatrix} \begin{bmatrix} I_a \\ I_b \end{bmatrix} + \begin{bmatrix} 0.91748 \\ 0 \end{bmatrix} [V_s] = A\Theta + Bu \\
y = [V_{25}] \\
\quad = \begin{bmatrix} 45.391 & -2372.32 \end{bmatrix} \begin{bmatrix} I_a \\ I_b \end{bmatrix} + [0.04126][V_s] = C\Theta + Du
\end{array}\right\} \tag{6-121}$$

Using the coefficient matrices in Eq. (6-120),
$$\hat{A} = A - BD^{-1}C = \begin{bmatrix} -1100.12 & 52{,}797.6 \\ -942.507 & 0 \end{bmatrix}, \tag{6-122}$$

which gives the null return difference matrix with respect to X as
$$\hat{F}(X) = 1_2 - \hat{A}X = \begin{bmatrix} 51.055 & -2402.29 \\ 42.884 & 1 \end{bmatrix}. \tag{6-123}$$

Suppose that the input current, I_{51}, is chosen as the output variable. Then from Eq. (6-103),

$$y = [I_{51}] = \begin{bmatrix} -0.08252 & 0.04126 \end{bmatrix} \begin{bmatrix} I_a \\ I_b \end{bmatrix} + [0.000862][V_s] = C\Theta + Du. \tag{6-124}$$

The corresponding null return difference matrix is

$$\hat{F}(X) = 1_2 - \hat{A}X = \begin{bmatrix} 1.13426 & -0.06713 \\ 42.8841 & 1 \end{bmatrix}, \qquad (6\text{-}125)$$

where

$$\hat{A} = \begin{bmatrix} -2.95085 & 1.47543 \\ -942.507 & 0 \end{bmatrix}. \qquad (6\text{-}126)$$

6.5.4. Transfer Function Matrix

This section demonstrates the effect of feedback on the transfer function matrix, $W(X)$. Specifically, express the determinant, $\det[W(X)]$, in terms of the determinant, $\det[X(0)]$, and the determinants of the return difference and null return difference matrices, thereby expanding Blackman's impedance formula for a single input to that of a formulation applicable to a multiplicity of inputs.

Before proceeding, it is necessary to state the following determinant identity for two arbitrary matrices, M and N, of orders $m \times n$ and $n \times m$, respectively:

$$\det(1_m + MN) = \det(1_n + NM). \qquad (6\text{-}127)$$

A proof of this identity relationship can be found in Chen [1991]. Armed with this stipulation, the following generalization of Blackman's formula for input impedance can be proffered.

Theorem 6.1. In a multiple loop feedback amplifier, if $W(0) = D$ is nonsingular, the determinant of the transfer function matrix, $W(X)$, is related to the determinants of the return difference matrix, $F(X)$, and the null return difference matrix, $\hat{F}(X)$, by

$$\det W(X) = \det W(0) \frac{\det \hat{F}(X)}{\det F(X)}. \qquad (6\text{-}128)$$

Proof. From Eq. (6-99),

$$W(X) = D \left[1_n + D^{-1} C X \left(1_p - AX \right)^{-1} B \right], \qquad (6\text{-}129)$$

which yields

$$\det W(X) = [\det W(0)] \det \left[1_n + D^{-1}CX \left(1_p - AX\right)^{-1} B \right]$$

$$= [\det W(0)] \det \left[1_p + BD^{-1}CX \left(1_p - AX\right)^{-1} \right]$$

$$= \frac{\det W(0) \det \hat{F}(X)}{\det F(X)}. \tag{6-130}$$

The second line follows directly from Eq. (6-127).

As indicated in Eq. (6-52), the input impedance, $Z(x)$, looking into a terminal pair can be conveniently expressed as

$$Z(x) = Z(0) \frac{F(\text{input short circuited})}{F(\text{input open circuited})} \tag{6-131}$$

A similar expression can be derived from Eq. (6-128) if $W(X)$ denotes the impedance matrix of the n-port network in Fig. 6.29. In this case, $F(X)$ is the return difference matrix with respect to X for the situation when the n ports at which the impedance matrix is defined are left open circuited without any applied sources. Hence, $F(X) = F(\text{input open circuited})$. Likewise, $\hat{F}(X)$ is the return difference matrix with respect to X for the input port current vector I_s and the output port voltage vector V under the condition that I_s is adjusted so that the port voltage vector V is identically zero. In other words, $\hat{F}(X)$ is the return difference matrix for the circumstance in which the n ports where the impedance matrix is defined are short circuited. Consequently, $\hat{F}(X) = F(\text{input short circuited})$. Resultantly, the determinant of the impedance matrix, $Z(X)$, of an n-port network can be expressed as, recalling Eq. (6-128),

$$\det [Z(X)] = \det [Z(0)] \frac{\det \left[F(\text{input short circuited}) \right]}{\det \left[F(\text{input open circuited}) \right]} \tag{6-132}$$

As an illustration, refer again to the feedback amplifier of Fig. 6.9 and its linearized model of Fig. 6.32. As computed in Eq. (6-114), the return difference matrix with respect to the two controlling parameters is given by

$$F(X) = 1_2 + T(X) = \begin{bmatrix} 5.131 & -2.065 \\ 42.884 & 1 \end{bmatrix}, \tag{6-133}$$

the determinant of which is found to be

$$\det[F(X)] = 93.68646. \tag{6-134}$$

If V_{25} of Fig. 6.32 is chosen as the output and V_s as the input, the null return difference matrix is, from Eq. (6-123),

$$\hat{F}(X) = 1_2 - \hat{A}X = \begin{bmatrix} 51.055 & -2402.29 \\ 42.884 & 1 \end{bmatrix} \tag{6-135}$$

the determinant of which is found to be

$$\det\left[\hat{F}(X)\right] = 103{,}071. \tag{6-136}$$

From Eq. (6-128), the feedback amplifier voltage gain, V_{25}/V_s, is seen to be

$$w(X) = \frac{V_{25}}{V_s} = w(0)\frac{\det\left[\hat{F}(X)\right]}{\det[F(X)]} = 0.04126\frac{103{,}071}{93.68646} = 45.39, \tag{6-137}$$

where $w(0) = 0.04126$, as given in Eq. (6-121).

Suppose, instead, that the input current, I_{51}, is chosen as the output and V_s as the input. Using Eq. (6-125), the null return difference matrix becomes

$$\hat{F}(X) = 1_2 - \hat{A}X = \begin{bmatrix} 1.13426 & -0.06713 \\ 42.8841 & 1 \end{bmatrix}, \tag{6-138}$$

for which the determinant is

$$\det\left[\hat{F}(X)\right] = 4.01307. \tag{6-139}$$

By applying Eq. (6-128), the amplifier input admittance is

$$w(X) = \frac{I_{51}}{V_s} = w(0)\frac{\det[\hat{F}(X)]}{\det[F(X)]} = 8.62 \times 10^{-4}\frac{4.01307}{93.68646} = 36.92\,\mu\text{mho} \tag{6-140}$$

or 27.2 KΩ, confirming Eq. (6-107), where $w(0) = 862\,\mu$mho is found from Eq. (6-124).

Another application of the generalized Blackman's formula is its use as a basis of a procedure for the indirect measurement of return difference. Refer to the general feedback network of Fig. 6.20. Suppose the return difference, $F(y_{21})$, with respect to the forward short circuit transfer admittance, y_{21}, of a two-port device is to be measured. Choose the two controlling parameters, y_{21} and y_{12}, to be the elements of interest. From Fig. 6.21,

$$\Theta = \begin{bmatrix} I_a \\ I_b \end{bmatrix} = \begin{bmatrix} y_{21} & 0 \\ 0 & y_{12} \end{bmatrix} \begin{bmatrix} V_1 \\ V_2 \end{bmatrix} = X\Phi, \qquad (6\text{-}141)$$

where I_a and I_b are the currents of the voltage controlled current sources. By appealing to Eq. (6-132), the impedance looking into terminals a and b of Fig. 6.20 can be written as

$$z_{aa,bb}(y_{12}, y_{21}) = z_{aa,bb}(0,0) \frac{\det\left[F(\text{input short circuited})\right]}{\det\left[F(\text{input open circuited})\right]}. \qquad (6\text{-}142)$$

When input terminals a and b are open circuited, the resulting return difference matrix is exactly the same as that found under normal operating conditions, and accordingly,

$$F(\text{input open circuited}) = F(X) = \begin{bmatrix} F_{11} & F_{12} \\ F_{21} & F_{22} \end{bmatrix}. \qquad (6\text{-}143)$$

Since

$$F(X) = 1_2 - AX, \qquad (6\text{-}144)$$

the elements, F_{11} and F_{21}, are calculated with $y_{12} = 0$, whereas F_{12} and F_{22} are evaluated with $y_{21} = 0$. When input terminals a and b are short circuited, the feedback loop is interrupted and only the second row and first column element of matrix A is nonzero, thereby resulting in

$$\det\left[F(\text{input short circuited})\right] = 1. \qquad (6\text{-}145)$$

Since X is diagonal, the return difference function, $F(y_{21})$, can be expressed in terms of $\det[F(X)]$ and the cofactor of the first row and first column element of $F(X)$ as

$$F(y_{21}) = \frac{\det[F(X)]}{F_{22}}. \qquad (6\text{-}146)$$

Substituting these results into Eq. (6-142) generates the result,

$$F(y_{12})|_{y_{21}=0} F(y_{21}) = \frac{z_{aa,bb}(0,0)}{z_{aa,bb}(y_{12}, y_{21})}, \qquad (6\text{-}147)$$

where

$$F_{22} = 1 - a_{22}y_{12}|_{y_{21}=0} = F(y_{12})|_{y_{21}=0}, \qquad (6\text{-}148)$$

and a_{22} is the second row and second column element of A.

6.5.5. Sensitivity Matrix

The sensitivity with respect to a perturbation in a particular element of a transfer function has been studied for a single loop feedback network. In multiple loop feedback networks, interest generally focuses on the sensitivity of a transfer function with respect to the variation of a set of network elements. This set may include either elements that are inherently sensitive to variation or elements whose effect on the overall amplifier performance is of paramount concern to circuit designers. To this end, a sensitivity matrix is introduced, and formulas for computing multiparameter sensitivity functions for a multiple loop feedback amplifier are developed.

Figure 6.35 is the block diagram of a multivariable open loop control network having n inputs and m outputs, whereas Fig. 6.36 shows the corresponding general feedback structure. If all feedback signals derive from the output variable and if the controllers are linear, there is no loss of generality by assuming the controller to be of the form shown in Fig. 6.37.

Denote the set of Laplace transformed input signals by the n-vector $\boldsymbol{u}$, the set of inputs to the network X in the open loop configuration of Fig. 6.35 by the p-vector $\boldsymbol{\Phi}_o$, and the set of outputs of network X in Fig. 6.35 by the m-vector $\boldsymbol{y}_o$. Let the corresponding signals for the closed loop configuration of Fig. 6.37 be denoted by the n-vector, $\boldsymbol{u}$, the p-vector, $\boldsymbol{\Phi}_c$, and the

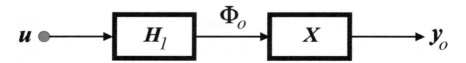

Figure 6.35. The block diagram of a multivariable open loop system.

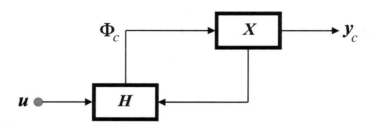

Figure 6.36. General feedback structure.

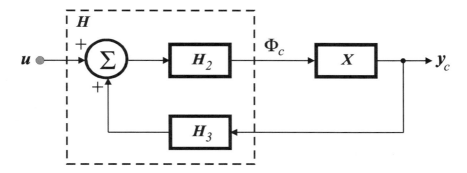

Figure 6.37. General feedback configuration.

m-vector, y_c. Figures 6.35 and 6.37 produce the linear relationships,

$$\begin{aligned} y_o &= X\Phi_o \\ \Phi_o &= H_1 u \\ y_c &= X\Phi_c \\ \Phi_c &= H_2(u + H_3 y_c) \end{aligned} \Bigg\}, \quad (6\text{-}149)$$

where the transfer function matrices, $X, H_1, H_2,$ and H_3 are of orders $m \times p$, $p \times n$, $p \times n$ and $n \times m$, respectively. Combining Eqs. (6-149) and (6-110) results in

$$(1_m - X H_2 H_3) y_c = X H_2 u \quad (6\text{-}150)$$

from which

$$y_c = (1_m - X H_2 H_3)^{-1} X H_2 u. \quad (6\text{-}151)$$

Since the closed loop transfer function matrix, $W(X)$, relating input vector u to output vector y_c is implicitly defined by

$$y_c = W(X) u, \quad (6\text{-}152)$$

$$W(X) = (1_m - X H_2 H_3)^{-1} X H_2. \quad (6\text{-}153)$$

If parameter matrix X is perturbed from X to $X + \delta X$, the outputs of the open loop and closed loop systems of Figs. 6.35 and 6.37 no longer assume their original nominal values. Distinguishing the new from the old variables by the superscript, $+$,

$$\begin{aligned} y_o^+ &= X^+ \Phi_o \\ y_c^+ &= X^+ \Phi_c^+ \\ \Phi_c^+ &= H_2(u + H_3 y_c^+) \end{aligned} \Bigg\}, \quad (6\text{-}154)$$

where Φ_o retains its original value.

It is now necessary to compare the relative effects of the variations of X on the performance of the open loop and the closed loop systems. For a meaningful comparison, assume that H_1, H_2, and H_3 are such that when there is no variation of X, $y_o = y_c$. Define the error vectors resulting from the perturbation in X as

$$\left. \begin{array}{l} E_o = y_o - y_o^+ \\ E_c = y_c - y_c^+ \end{array} \right\}. \qquad (6\text{-}155)$$

The square matrix relating vector E_o to vector E_c is called the *sensitivity matrix*, $S(X)$, for the transfer function matrix, $W(X)$, with respect to the variations of X; that is,

$$E_c = S(X) E_o. \qquad (6\text{-}156)$$

The subject sensitivity matrix, $S(X)$, is expressible in terms of the system matrices, X, H_2, and H_3.

The input and output relationship for the perturbed system can be written as

$$y_c^+ = \left(1_m - X^+ H_2 H_3\right)^{-1} X^+ H_2 u. \qquad (6\text{-}157)$$

Substituting Eq. (6-151) into Eq. (6-155) gives

$$E_c = y_c - y_c^+ = \left[(1_m - X H_2 H_3)^{-1} X H_2 \right.$$
$$\left. - \left(1_m - X^+ H_2 H_3\right)^{-1} X^+ H_2\right] u$$
$$= \left(1_m - X^+ H_2 H_3\right)^{-1}$$
$$\times \left\{[1_m - (X + \delta X) H_2 H_3] (1_m - X H_2 H_3)^{-1} X H_2 \right.$$
$$\left. - (X + \delta X) H_2\right\} u$$
$$= \left(1_m - X^+ H_2 H_3\right)^{-1}$$
$$\times \left[X H_2 - \delta X H_2 H_3 (1_m - X H_2 H_3)^{-1} X H_2 - X H_2 - \delta X H_2\right] u$$
$$= -\left(1_m - X^+ H_2 H_3\right)^{-1} \delta X H_2 [1_n + H_3 W(X)] u. \qquad (6\text{-}158)$$

From Eqs. (6-149) and (6-152),

$$\Phi_c = H_2 [1_n + H_3 W(X)] u. \qquad (6\text{-}159)$$

Since $y_o = y_c$,

$$\Phi_o = \Phi_c = H_2 [1_n + H_3 W(X)] u, \qquad (6\text{-}160)$$

Multiple Loop Feedback Amplifiers

which yields

$$E_o = y_o - y_o^+ = (X - X^+)\Phi_o = -\delta X \Phi_o$$
$$= -\delta X H_2[1_n + H_3 W(X)]u. \quad (6\text{-}161)$$

Combining Eqs. (6-158) and (6-161) leads to an expression that interrelates the error vectors, E_c and E_o, of the closed loop and open loop systems in accordance with

$$E_c = (1_m - X^+ H_2 H_3)^{-1} E_o, \quad (6\text{-}162)$$

thereby giving rise to the sensitivity matrix,

$$S(X) = (1_m - X^+ H_2 H_3)^{-1}. \quad (6\text{-}163)$$

For a small variation in X, X^+ is approximately equal to X. It follows in Fig. 6.37 that if the matrix triple product, XH_2H_3, is regarded as the *loop transmission matrix* and $-XH_2H_3$ is interpreted as the *return ratio matrix*, the difference between the unit matrix and the loop transmission matrix,

$$1_m - X^+ H_2 H_3, \quad (6\text{-}164)$$

can be defined as the *return difference matrix*. Therefore, Eq. (6-163) is a direct extension of the sensitivity function defined for a single input, single output system to which focus is given to only a single parameter. This result is synergistic with Eq. (6-28), where it was demonstrated that, using the ideal feedback model, the sensitivity function of the closed loop transfer function with respect to the forward amplifier gain is equal to the reciprocal of its return difference with respect to the same parameter.

In particular, when $W(X)$, δX, and X are square and nonsingular, from Eqs. (6-149), (6-152), (6-155) combine to deliver

$$\left. \begin{array}{l} E_c = y_c - y_c^+ = [W(X) - W^+(X)]u = -\delta W(X)u \\ E_o = y_o - y_o^+ = (XH_1 - X^+H_1)u = -\delta X H_1 u \end{array} \right\}. \quad (6\text{-}165)$$

If H_1 is nonsingular, u in the expressions in Eq. (6-165) combine to give

$$E_c = \delta W(X) H_1^{-1} (\delta X)^{-1} E_o. \quad (6\text{-}166)$$

Since $y_o = y_c$,

$$XH_1 = W(X). \quad (6\text{-}167)$$

From Eq. (6-166),

$$E_c = \delta W(X) W^{-1}(X) X (\delta X)^{-1} E_o, \quad (6\text{-}168)$$

whence

$$S(X) = \delta W(X)W^{-1}(X)X(\delta X)^{-1}. \qquad (6\text{-}169)$$

This result compares favorably with the scalar sensitivity function in Eq. (6-21), which can be put in the form

$$S(x) = (\delta w)w^{-1}x(\delta x)^{-1}. \qquad (6\text{-}170)$$

6.5.6. Multi-Parameter Sensitivity

In this section, formulas are derived to relate the effect of a change in X on a scalar transfer function $w(X)$. Let x_k, $k = 1, 2, \ldots, pq$, be the elements of X. The multivariable Taylor series expansion of $w(X)$ with respect to x_k is

$$\delta w = \sum_{k=1}^{pq} \frac{\partial w}{\partial x_k}\delta x_k + \sum_{j=1}^{pq}\sum_{k=1}^{pq} \frac{\partial^2 w}{\partial x_j \partial x_k}\frac{\delta x_j \delta x_k}{2!} + \cdots. \qquad (6\text{-}171)$$

The first order perturbation can then be written as

$$\delta w \approx \sum_{k=1}^{pq} \frac{\partial w}{\partial x_k}\delta x_k. \qquad (6\text{-}172)$$

From Eq. (6-21),

$$\frac{\delta w}{w} \approx \sum_{k=1}^{pq} S(x_k)\frac{\delta x_k}{x_k}. \qquad (6\text{-}173)$$

This expression gives the fractional change of the transfer function, w, in terms of the scalar sensitivity functions $S(x_k)$.

Refer once again to the fundamental matrix feedback flow graph of Fig. 6.31. If the amplifier has a single input and a single output, Eq. (6-129) gives the overall transfer function, $w(X)$, of the multiple loop feedback as

$$w(X) = D + CX\left(\mathbf{1}_p - AX\right)^{-1} B. \qquad (6\text{-}174)$$

When X is perturbed to $X^+ = X + \delta X$, the corresponding expression to Eq. (6-174) is

$$w(X) + \delta w(X) = D + C(X + \delta X)\left(\mathbf{1}_p - AX - A\delta X\right)^{-1} B, \qquad (6\text{-}175)$$

or

$$\delta w(X) = C\lfloor(X + \delta X)(\mathbf{1}_p - AX - A\delta X)^{-1} - X(\mathbf{1}_p - AX)^{-1}\rfloor B.$$

$$(6\text{-}176)$$

Multiple Loop Feedback Amplifiers

As δX approaches zero,

$$\begin{aligned}\delta w(X) &= C\lfloor (X+\delta X) - X\left(\mathbf{1}_p - AX\right)^{-1}\left(\mathbf{1}_p - AX - A\delta X\right)\rfloor \\ &\quad \times \left(\mathbf{1}_p - AX - A\delta X\right)^{-1} B \\ &= C\lfloor \delta X + X\left(\mathbf{1}_p - AX\right)^{-1} A\delta X\rfloor \left(\mathbf{1}_p - AX - A\delta X\right)^{-1} B \\ &= C\left(\mathbf{1}_q - XA\right)^{-1}(\delta X)\left(\mathbf{1}_p - AX - A\delta X\right)^{-1} B \\ &\approx C\left(\mathbf{1}_q - XA\right)^{-1}(\delta X)\left(\mathbf{1}_p - AX\right)^{-1} B, \end{aligned} \quad (6\text{-}177)$$

where C is a row q-vector and B is a column p-vector. Write

$$\left.\begin{aligned} C &= [c_1 \quad c_2 \quad \cdots \quad c_q] \\ B' &= [b_1 \quad b_2 \quad \cdots \quad b_p] \\ \tilde{W} &= X\left(\mathbf{1}_p - AX\right)^{-1} = \left(\mathbf{1}_q - XA\right)^{-1} X = [\tilde{w}_{ij}] \end{aligned}\right\}. \quad (6\text{-}178)$$

The increment, $\delta w(X)$, can be expressed in terms of the elements of Eq. (6-178) and those of X. In the case where X is diagonal with

$$X = \text{diag}\lfloor x_1 \quad x_2 \quad \cdots \quad x_p \rfloor, \quad (6\text{-}179)$$

where $p = q$, the expression for $\delta w(X)$ can be compactly written as

$$\begin{aligned}\delta w(\mathbf{X}) &= \sum_{i=1}^{p}\sum_{k=1}^{p}\sum_{j=1}^{p} c_i \left(\frac{\tilde{w}_{ik}}{x_k}\right)(\delta x_k)\left(\frac{\tilde{w}_{kj}}{x_k}\right) b_j \\ &= \sum_{i=1}^{p}\sum_{k=1}^{p}\sum_{j=1}^{p} \frac{c_i \tilde{w}_{ik} \tilde{w}_{kj} b_j}{x_k}\frac{\delta x_k}{x_k}\end{aligned} \quad (6\text{-}180)$$

Comparing this equation with Eq. (6-173), obtain an explicit form for the single parameter sensitivity function becomes

$$\mathsf{S}(x_k) = \sum_{i=1}^{p}\sum_{j=1}^{p} \frac{c_i \tilde{w}_{ik} \tilde{w}_{kj} b_j}{x_k w(X)}. \quad (6\text{-}181)$$

Thus, knowing Eqs. (6-178) and (6-179), the multiparameter sensitivity function for the scalar transfer function, $w(X)$, follows immediately.

As an example, consider again the voltage-series feedback amplifier of Fig. 6.9 and its small signal model in Fig. 6.32. Assume that V_s is the input and V_{25} is the output response. The transfer function of interest is the amplifier voltage gain V_{25}/V_s. The elements of concern are the two controlling parameters of the controlled sources. Thus, let

$$X = \begin{bmatrix} \tilde{\alpha}_1 & 0 \\ 0 & \tilde{\alpha}_2 \end{bmatrix} = \begin{bmatrix} 0.0455 & 0 \\ 0 & 0.0455 \end{bmatrix}. \tag{6-182}$$

From Eq. (6-121),

$$\left. \begin{array}{l} A = \begin{bmatrix} -90.782 & 45.391 \\ -942.507 & 0 \end{bmatrix} \\ B' = \begin{bmatrix} 0.91748 & 0 \end{bmatrix} \\ C = \begin{bmatrix} 45.391 & -2372.32 \end{bmatrix} \end{array} \right\}, \tag{6-183}$$

which produces

$$\tilde{W} = X(1_2 - AX)^{-1} = 10^{-4} \begin{bmatrix} 4.85600 & 10.02904 \\ -208.245 & 24.91407 \end{bmatrix}. \tag{6-184}$$

Also, from Eq. (6-107),

$$w(X) = \frac{V_{25}}{V_s} = 45.387. \tag{6-185}$$

To compute the sensitivity functions with respect to $\tilde{\alpha}_1$ and $\tilde{\alpha}_2$, apply Eq. (6-181) to obtain

$$\left. \begin{array}{l} \mathsf{S}(\tilde{\alpha}_1) = \sum_{i=1}^{2} \sum_{j=1}^{2} \frac{c_i \tilde{w}_{i1} \tilde{w}_{1j} b_j}{\alpha_1 w(X)} \\ \quad = \frac{c_1 \tilde{w}_{11} \tilde{w}_{11} b_1 + c_1 \tilde{w}_{11} \tilde{w}_{12} b_2 + c_2 \tilde{w}_{21} \tilde{w}_{11} b_1 + c_2 \tilde{w}_{21} \tilde{w}_{12} b_2}{\tilde{\alpha}_1 w} \\ \quad = 0.01066 \\ \mathsf{S}(\tilde{\alpha}_2) = \frac{c_1 \tilde{w}_{12} \tilde{w}_{21} b_1 + c_1 \tilde{w}_{12} \tilde{w}_{22} b_2 + c_2 \tilde{w}_{22} \tilde{w}_{21} b_1 + c_2 \tilde{w}_{22} \tilde{w}_{22} b_2}{\tilde{\alpha}_2 w} \\ \quad = 0.05426 \end{array} \right\}. \tag{6-186}$$

As a check, Eq. (6-25) can be exploited to compute the foregoing sensitivities. From Eqs. (6-40) and (6-47),

$$\left.\begin{array}{l} F(\tilde{\alpha}_1) = 93.70 \\ F(\tilde{\alpha}_2) = 18.26 \\ \hat{F}(\tilde{\alpha}_1) = 103.07 \times 10^3 \\ \hat{F}(\tilde{\alpha}_2) = 2018.70 \end{array}\right\}. \tag{6-187}$$

Substituting these results into Eq. (6-25), the pertinent sensitivities are

$$\left.\begin{array}{l} \mathsf{S}(\tilde{\alpha}_1) = \dfrac{1}{F(\tilde{\alpha}_1)} - \dfrac{1}{\hat{F}(\tilde{\alpha}_1)} = 0.01066 \\ \mathsf{S}(\tilde{\alpha}_2) = \dfrac{1}{F(\tilde{\alpha}_2)} - \dfrac{1}{\hat{F}(\tilde{\alpha}_2)} = 0.05427 \end{array}\right\}, \tag{6-188}$$

which agree with Eq. (6-186).

Suppose that $\tilde{\alpha}_1$ is changed by 4% and $\tilde{\alpha}_2$ by 6%. The fractional change of the voltage gain, $w(X)$, is found from Eq. (6-173) to be

$$\frac{\delta w}{w} \approx \mathsf{S}(\tilde{\alpha}_1)\frac{\delta \tilde{\alpha}_1}{\tilde{\alpha}_1} + \mathsf{S}(\tilde{\alpha}_2)\frac{\delta \tilde{\alpha}_2}{\tilde{\alpha}_2} = 0.003683 \tag{6-189}$$

or 0.37%.

References

1. F. H. Blecher, Design principles for single loop transistor feedback amplifiers, *IRE Trans. Circuit Theory* **CT-4** (1957), 145–156.
2. H. W. Bode, *Network Analysis and Feedback Amplifier Design* (Van Nostrand, Princeton, New Jersey, 1945).
3. W. K. Chen, Indefinite admittance matrix formulation of feedback amplifier theory, *IEEE Trans. Circuits and Systems* **CAS-23** (1976), 498–505.
4. W. K. Chen, On second-order cofactors and null return difference in feedback amplifier theory, *Int. J. Circuit Theory and Applications* **6** (1978), 305–312.
5. W. K. Chen, *Active Network and Feedback Amplifier Theory* (McGraw-Hill, New York, 1980), Chapters 2, 4, 5 and 7.
6. W. K. Chen, *Active Network Analysis* (World Scientific, Singapore, 1991), Chapters 2, 4, 5 and 7.
7. J. B. Cruz, Jr. and W. R. Perkins, A new approach to the sensitivity problem in multivariable feedback system design, *IEEE Trans. Automatic Control* **AC-9** (1964), 216–223.

8. S. S. Haykin, *Active Network Theory* (Addison-Wesley, Reading, MA, 1970).
9. E. S. Kuh and R. A. Rohrer, *Theory of Linear Active Networks* (Holden-Day, San Francisco, CA, 1967).
10. I. W. Sandberg, On the theory of linear multi-loop feedback systems, *Bell Sys. Tech. J.* **42** (1963), 355–382.

Chapter 7

Analog MOS Technology Circuits

7.1.0. Introduction

Innovative electronic systems servicing diverse applications in industrial, home, and military communities are a mantra of the modern technological age. Electronic circuits necessarily underpin and define the properties and performance of these systems. All electronic circuits, and particularly those intended for broadband and very high frequency applications, are inherently feedback structures. Since the theories and analytical methods documented in the preceding chapters inspire insights into the electrical characteristics of feedback networks and systems, it is prudent to apply these methodologies to the engineering task of assessing the performance attributes and limitations of practical active circuits. This chapter is assuredly not a treatise on modern electronic systems. Instead, its focus is the formulation of a general analytical strategy for exploiting the signal flow concepts and theories that underpin feedback networks to define and assess the performance characteristics of commonly encountered linear analog active circuits.

A single chapter cannot embrace the salient features of all modern electronics. But the nature of analog electronic circuit technologies renders possible an analytical methodology that is broadly applicable to fundamentally important analog architectures fabricated in various types of monolithic technologies. Specifically, the vast majority of active analog circuits destined for linear signal processing applications are predicated, somewhat surprisingly, on only three basic architectures.[1] Equally significant is the fact that the majority of all other linear analog configurations derives from straightforward modifications to these canonical architectures. In metal-oxide-semiconductor field-effect transistor (MOSFET) technologies, these fundamental units, which might be termed the canonic cells of analog MOSFETs, are the common source amplifier, the common drain amplifier, and the common gate amplifier. In bipolar junction transistor (BJT) technologies, inclusive of silicon-germanium (SiGe) heterostructures, the respective canonic equivalents to their MOSFET counterparts are the common emitter, common collector, and common base amplifiers. Of particular interest is the fact that the linearized, or small signal, model of each of these canonic cells is topologically similar, independent of either the specific cell of interest or the utilized fabrication technology. The resultant implication is that an understanding of the electrical characteristics of one cell facilitates comprehending the performance properties indigenous to other canonic cells.

This chapter focuses on the analysis and performance assessment of linear amplifiers and other forms of linear active networks realized in MOSFET device technologies. The preferential treatment of MOSFET technologies, as opposed to BJT constructs, is premised on commercial trends. In particular, most modern electronic systems embody mixed signal circuit architectures; that is circuits utilizing both digital and analog subcircuits on a single monolithic chip. Digital subcircuits are fundamental because many modern systems require analog to digital and/or digital to analog signal conversion, as well as digital signal processing (DSP) within the input/output (I/O) transmission path. Other systems utilize programmable hardware or field programmable gate arrays (FPGAs) to control the observable performance of incorporated analog subcircuits in light of process vagaries and environmental uncertainties.[2] Since most commercial digital structures are realized in MOSFET or complementary metal-oxide-semiconductor (CMOS) device technologies, it is advantageous from a digital–analog interface perspective to forge the requisite analog units with MOS or CMOS transistors. But despite the prejudice shown herewith to MOS technologies, many of the forthcoming analytical disclosures apply equally well to bipolar and even compound semiconductor circuits.

As noted above, the three canonic cells of analog MOSFET technology circuits are the *common source amplifier*, the *common drain amplifier*, and the *common gate amplifier*. The common source amplifier is the foundation of almost all MOS technology amplifiers and systems in that it is capable of greater than unity voltage and current gain magnitudes, as well as substantial forward transadmittance and transimpedance. The input signal to this amplifier is applied between the gate terminal of a MOSFET and signal ground, while the output response is extracted as either a signal current conducted by the transistor drain terminal or a signal voltage developed between the drain terminal and signal ground. Of the three basic analog cells, the common source unit is the only topology that offers phase inversion between gate-ground and drain-ground signal voltages. In a common drain cell, which is often called a *source follower*, the traditional response to a voltage input signal applied between the gate terminal and signal ground is a signal voltage developed with respect to signal ground at the source terminal. As such, it operates as a voltage buffer in that its resultant voltage gain is slightly less than one, its low frequency input resistance is almost infinitely large, and its low frequency output resistance is low. The common gate amplifier is effectively the dual of its source follower counterpart in the sense that its current gain is slightly smaller than one, its low frequency input resistance is small, and its low frequency output resistance is large. Because the input signal to the common gate configuration is generally a current applied to its source terminal, it is rarely utilized as a self-contained active circuit. Instead, its most common application is that of an effective current buffer inserted between the output port of a common source amplifier and the load impedance that would otherwise be incident with the drain port of the common source cell.

Feedback prevails implicitly between the input and output ports of each of the three basic analog MOSFET cells. Accordingly, the didacticism of the preceding chapter, coupled with the theoretical disclosures of Chapter 4, can be gainfully exploited for purposes of characterizing the small signal performance and establishing both the performance attributes and shortfalls of each of these cells. Although a myriad of topological variants to each of the respective cells is possible, the following discourse renders apparent the fact that an analytical commonality prevails among all such variations.

7.2.0. MOS Transistor Models

Before attempting a meaningful investigation of the feedback dynamics implicit to MOS technology analog integrated circuits, an understanding of the fundamentals underlying both the static and high frequency models of MOS transistors is essential. A physically sound modeling treatment, complete with relevant derivations and discussions of pertinent charge versus voltage and current versus voltage relationships is beyond the scope of this text. Accordingly, the following discussion focuses almost exclusively on the circuit-level implications of the simplified volt-ampere relationships that define the first order terminal and transfer characteristics of MOS technology devices. The reader interested in more definitive transistor modeling information is referred to the extensive device modeling literature.[3]–[8]

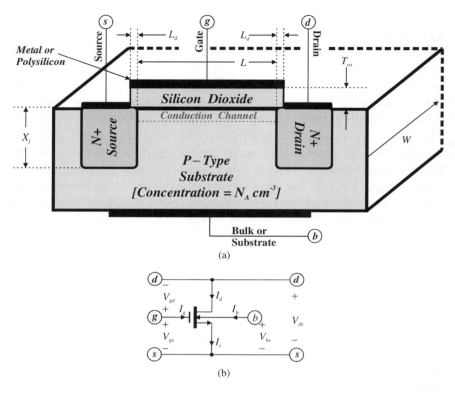

Figure 7.1. (a) The cross-section diagram of an N-channel enhancement mode MOSFET. The diagram is not drawn to scale. (b) The electrical schematic symbol of the N-channel MOSFET shown in (a).

7.2.1. *Transistor Cross-Section and Electrical Symbol*

MOS technology transistors used in present day analog, digital, and mixed signal circuits are invariably enhancement mode devices, which differ from older generation depletion mode transistors in that they require the application of a nonzero input voltage to activate device current flow. Two types of enhancement MOSFETs prevail in integrated circuits. The *N-channel* or *NMOS* transistor, whose cross-section diagram and electrical schematic symbol appear in Fig. 7.1, is the favored device, particularly when very high speed or very wide bandwidth signals must be processed. For comparable geometries, the *P-channel* or *PMOS* transistor, whose corresponding diagrams are offered in Fig. 7.2, is slower than is its N-channel brethren. It is resultantly relegated to bias subcircuits and to signal flow paths whose

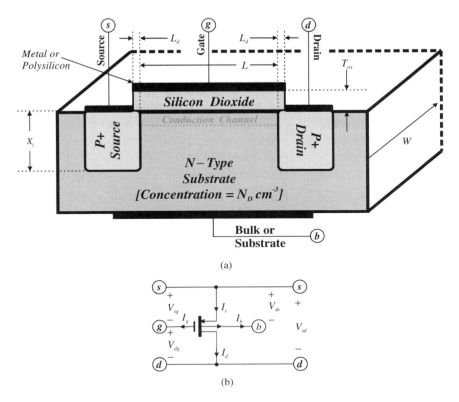

Figure 7.2. (a) Cross-section diagram of a P-channel enhancement mode MOSFET. The diagram is not drawn to scale. (b) The corresponding electrical schematic symbol of the P-channel MOSFET.

overall frequency or time responses are not dominated by the charge transport times of transistors.

Figures 7.1 and 7.2 depict the MOSFET as a device having four accessible terminals; namely, the drain (d), the source (s), the gate (g), and the bulk (b), which is often called the substrate. The four currents carried by the leads connected to these terminals are the drain current, I_d, the source current, I_s, the gate current, I_g, and the bulk current, I_b, for which positive reference polarity conventions are identified in Figs. 7.1(b) and 7.2(b). Note that the positive directions of all NMOS currents are the reverse of the respective currents in the PMOS unit. By Kirchhoff, both types of transistors abide by the charge conservation constraint,

$$I_s = I_d + I_g + I_b. \tag{7-1}$$

The schematic symbols in the subject two figures also delineate the positive senses of four transistor voltages. In the N-channel transistor, the voltages of interest are the drain-source voltage, V_{ds}, the gate-source voltage, V_{gs}, the gate-drain voltage, V_{gd}, and the bulk-source voltage, V_{bs}. For the P-channel device, the corresponding critical voltages are the source-drain voltage, V_{sd}, the source-gate voltage, V_{sg}, the drain-gate voltage, V_{dg}, and the source-bulk voltage, V_{sb}. Kirchhoff's voltage law requires

$$\left.\begin{array}{l}V_{ds} = V_{gs} - V_{gd}\\ V_{sd} = V_{sg} - V_{dg}\end{array}\right\}. \qquad (7\text{-}2)$$

The semiconductor foundation of the NMOS transistor abstracted in Fig. 7.1(a) is a P-type substrate whose impurity concentration, N_A, is of the order of 10^{14} to 10^{15} acceptor atoms per cubic centimeter. An electrical contact is made to the substrate to establish the bulk electrical terminal, (b). Two heavily doped N-type regions are diffused or implanted into the lightly doped substrate to form the source and drain volumes of the transistor cross-section. Metal or polysilicon contacts formed at the top surfaces of these regions provide for the source and drain electrical terminals of the transistor, (s) and (d), respectively. The two heavily doped regions are separated from one another by the indicated length, L, which is termed the channel length of the device. State of the art fabrication methods achieve $L = 0.065$ microns (65 nanometers), but many circuits continue to exploit 0.13 micron, 0.18 micron, 0.25 micron, or even larger geometries. It is illuminating to put these astonishingly small dimensions in perspective. To this end, the average thickness of a human hair (for those of you still blessed with hair) is about 70 microns, which implies that the 0.09 micron channel length of a state of the art MOSFET is more than 800-times smaller than human hair thickness! The simplicity of the figure belies the fact that the heavily doped N-type (denoted in the figure as "$N+$") regions are three-dimensional, extending into the page by nominally the net width, W, of the transistor. Width W can be any reasonable number larger than L and is a designable circuit parameter chosen in consideration of the amount of drain current that a transistor is compelled to conduct in a given application. The depth, X_j, of the N+ regions is such that the net perimeter, say P, (not explicitly shown in the figure) of each PN junction interface is approximately 10% to 20% larger than W.

Grown atop the semiconductor in the spacing between the N+ regions is silicon dioxide to which metal or polysilicon is attached to form the gate electrical terminal, (g). The thickness, T_{ox}, of the insulating oxide layer is of the order of tens of angstroms. Ideally, the oxide interface between the gate electrical contact and the semiconductor is precisely aligned with the length, L, of the spacing between source and drain regions. In practice, however, this oxide overlaps both of the aforementioned regions by the indicated amount, L_d. For deep submicron transistors, which are devices featuring channel lengths that are smaller than nominally 0.25 micron, L_d is in the neighborhood of 5% to 15% of X_j.

The foregoing commentary applies equally well to the P-channel device depicted in Fig. 7.2(a), subject to two provisos that a comparison of Figs. 7.1(a) and 7.2(a) reveals. The first of these provisos is that in the P-channel transistor, the substrate is N-type, as opposed to P-type. Like the acceptor concentration, N_A, in the substrate of NMOS, the donor concentration, N_D, in the P-channel substrate is also relatively low. The second proviso is that the heavily doped source and drain regions of PMOS devices are P-type, instead of N-type.

In normal operation, the bulk terminal of a MOSFET is connected to a circuit potential that ensures a reverse bias of both the substrate–source and substrate–drain PN junctions for all applied bias voltages and input signals. For NMOS devices, such an assurance derives from connecting the bulk terminal of the P-type substrate to the most negative potential postured by the circuit in which the considered transistor is embedded. In the P-channel case, substrate reverse biasing requires the bulk terminal of the N-type substrate to be incident with the most positive of available circuit potentials. At low frequencies where signal currents flowing through the substrate–source and substrate–drain junction depletion capacitances can be tacitly ignored, the bulk region reverse bias ensures $I_b \approx 0$. Implicit to this approximation is the presumption that the reverse leakage currents of the two reverse biased PN junctions are inconsequentially small.

The gate current, I_g, is also zero at low signal frequencies, since the silicon dioxide interface between the gate metal and the semiconductor channel region between N+ implants is an insulator. At high frequencies, the presumption of zero gate current is dubious because the gate contact, oxide interface, and underlying semiconductor form an effective parallel plate capacitance that is assuredly capable of high frequency current conduction. Other phenomena, such as hot (or highly energetic) current carriers[9] and transit time delays with respect to the transport of charge carriers from the source region to the drain region[10] also precipitate non-zero high frequency gate, as well as substrate, current. The latter two phenomena prevail especially in deep submicron devices. The immediate ramification of $I_g \approx 0$, as well as $I_b \approx 0$, is, recalling Eq. (7-1), $I_d \approx I_s$. This is to say that at low signal frequencies, all of the current conducted by the drain lead is mirrored by the current flowing in the source lead.

7.2.2. Static Volt-Ampere Relationships

An observable current in both the source and the drain leads requires that at least a thin conduction channel, such as is hypothesized in Figs. 7.1(a) and 7.2(a) and capable of transporting free charge carriers from the source region to the drain region, be induced. Without such a conduction channel, the current path from drain to source is comprised of a series interconnection of two PN junction diodes, one of which is always reversed biased, regardless of the polarity of voltage applied between drain and source. In NMOS, the conduction channel is forged by electrons attracted to the oxide-semiconductor interface by the action of a suitably large, positive gate to source voltage. In contrast, the conduction channel of PMOS materializes when holes are attracted to the semiconductor surface because of the application of a sufficiently large, positive source to gate voltage. Once launched by the applied gate-source or source-gate potential, the conduction channel establishes an effective closed current path between the drain and the source. For NMOS, the conduction channel is the vehicle by which electrons can be transported from the source to the drain in response to an applied positive drain-source voltage, V_{ds}. On the other hand, the conduction channel of PMOS is a source to drain conduit for holes when a suitably positive source to drain voltage, V_{sd}, is applied.

The steady state maximum possible response speed of a MOSFET is limited by the average time required by charge carriers to transit the conduction channel from the source to the drain. It can be shown that this average transit time is inversely proportional of the square of channel length L, which explains the omnipresent penchant toward a reduction of channel length. The transit time is also proportional to the mobility (meaning literally, the ability to move) of charge carriers. The fact that electron mobility μ_n is roughly 40% larger than hole mobility μ_p for a given background impurity concentration explains why N-channel devices are preferred over their P-channel counterparts in high speed circuit applications.

Carrier mobility is also inversely related to background concentration; that is, mobility progressively decreases with increasing background dopant levels. Without delving into the relevant physics of charge carrier mobility, this dependence on concentration can be explained via trivial analogy. To this end, consider the N+ source of an N-channel device as the western sidewalk of an avenue, and view the N+ drain as the eastern side of the street. The street itself is the P-type semiconductor between source and drain, while the impurity concentration, N_A, of the substrate is the steady state traffic flowing down the avenue. The free electron of mobility μ_n flowing from the source to the drain can be represented by an individual intent on crossing the street from its west side to its east side. Clearly, the mobility (ability to move from one side of the street to the other) of this individual (the free electron) is impaired when the avenue (the substrate) experiences high traffic volume (large N_A). The need for relatively small substrate dopant levels is rendered apparent by this superficial physical model.

To first order, the mathematical delineation of the static volt-ampere relationships of MOSFETs is facilitated by a consideration of three distinct circuit operating conditions. These conditions embrace device operation in (1) cutoff, (2) the ohmic regime, and (3) the saturation region.

7.2.2.1. Cutoff

Cutoff is the electrical condition wherein the voltage applied between the gate and source terminals of a MOSFET is insufficient to establish the source to drain conduction channel that is a prerequisite to free charge carrier transit and thus, current conduction. In cutoff, the drain current, I_d, is zero, which at low signal frequencies is tantamount to asserting $I_s = 0$. The cutoff condition is manifested in NMOS when the gate-source voltage, V_{gs}, is smaller than the transistor threshold voltage, say V_h, which is a performance barometer of the MOSFET. While $V_{gs} < V_h$ produces NMOS cutoff, the comparable condition for PMOS is $V_{sg} < V_h$, with the understanding that V_h remains a positive voltage.

In either transistor type, the threshold voltage, V_h, is the superposition of two voltage components, V_{ox} and V_F, such that

$$V_h = V_{ox} + 2V_F, \tag{7-3}$$

where V_{ox} is the voltage developed across the thin, insulating silicon dioxide layer, and V_F is termed the *Fermi* or *flatband potential*. In an N-channel transistor, the Fermi potential is given by

$$V_F = \frac{kT}{q} \ln\left(\frac{N_A}{N_i}\right), \tag{7-4}$$

where N_A is the average dopant concentration of the P-type substrate, N_i is the *intrinsic carrier concentration* of the utilized semiconductor, k is Boltzmann's constant [1.38(10^{-23}) joules/°K], q represents electronic charge magnitude [1.6(10^{-19}) coulombs], and T is the absolute temperature [in Kelvin degrees (°K)] of the conduction channel. The intrinsic concentration depends on both the material nature (silicon, germanium, etc.) and temperature of the semiconductor. For silicon, a routinely invoked empirical relationship is

$$N_i \approx \left[2^{(T-300.2)/11}\right]\left(10^{10}\right), \tag{7-5}$$

with the understanding that T expressed in units of Kelvin degrees produces N_i in units of atoms/cm^3. The relationship at hand reflects a doubling of the intrinsic carrier concentration for each 11°C rise in semiconductor body temperature. For a P-channel transistor, Eq. (7-4) changes to

$$V_F = \frac{kT}{q}\ln\left(\frac{N_D}{N_i}\right), \tag{7-6}$$

with N_D symbolizing the average dopant concentration of the N-type substrate.

The voltage, V_{ox}, in Eq. (7-3) is that amount of the applied gate-source potential that is "lost" as a voltage developed across the insulating oxide layer in either type of MOSFET. The remaining voltage, $2V_F$, which is literally twice the Fermi potential of the conducting channel region, is the potential commensurate with establishing the channel of free charges between the source and drain regions. In effect, twice the Fermi potential developed in the substrate at the oxide-semiconductor interface inverts the electrical character of the interfacial region. In other words, the preponderance of interfacial free carriers in NMOS, prior to the establishment of twice Fermi potential at this interface, are holes arising from the ionization of the acceptor atoms that dope the substrate. When twice the Fermi potential is imposed at the interface because of applied gate-source voltage, the interfacial region effectively inverts to N-type material in that the region in question now contains a preponderance of free electrons as charge carriers. Indeed, Boltzmann statistics confirm that when the interfacial potential in a P-type substrate is $2V_F$, the average concentration of interfacial free electrons is twice the average concentration of holes elsewhere in the substrate. Analogous contentions can be proffered for a conducting channel of holes in PMOS.

For a fixed oxide potential, which is determined largely by oxide thickness, and a given channel temperature, Eq. (7-3) implies a constant threshold potential. Unfortunately, the threshold voltage is not constant in that it is impacted by the voltage applied between bulk and source. To this end, it can be shown that to first order, an N-channel device has[11]

$$V_h = V_{ho} + 2\sqrt{V_\theta (V_F - V_T)} \left[\sqrt{1 - \frac{V_{bs}}{2(V_F - V_T)}} - 1 \right], \quad (7\text{-}7)$$

where V_{bs} is the applied bulk to source voltage, V_{ho} is the $V_{bs} = 0$ value of threshold potential, as stipulated by the right hand side of Eq. (7-3), V_F is the Fermi potential given by Eq. (7-4) and its Eq. (7-5) companion relationship, and

$$V_T = kT/q \quad (7\text{-}8)$$

is the Boltzmann voltage corresponding to an absolute conduction channel temperature of T. Because the substrate in NMOS is ordinarily connected to the most negative of available circuit potentials, it should be understood that $V_{bs} \leq 0$ in Eq. (7-7). It follows that the threshold voltage rises with increases in bulk-source reverse bias.

The voltage V_θ, in Eq. (7-7) is related to oxide thickness T_{ox} and average substrate dopant level N_A in accordance with

$$V_\theta = q N_A \varepsilon_s \left(\frac{T_{ox}}{\varepsilon_{ox}} \right)^2, \quad (7\text{-}9)$$

where ϵ_s is the dielectric constant of silicon [1.05 pF/cm], and ϵ_{ox} is the dielectric constant of silicon dioxide [345 fF/cm]. An expression identical in form to that of Eq. (7-7) prevails for P-channel transistors, wherein the generally nonpositive bulk to source voltage, V_{bs}, is supplanted by its nonpositive source to bulk counterpart, V_{sb}, and Fermi potential V_F derives from Eq. (7-6), as opposed to Eq. (7-4). Finally, N_A in Eq. (7-9) must be replaced by the P-channel substrate donor concentration, N_D.

It should be noted in Eq. (7-7) that if V_θ were zero, the threshold voltage is independent of the bulk-source voltage, V_{bs}. Constant threshold voltage is a laudable attribute in MOSFETs because signal voltages in certain circuit topologies can modulate transistor bulk-source voltage, which incurs changes in effective threshold voltage. In turn, such changes can produce signal distortion, as well as unanticipated uncertainties in the electrical conditions that ensure adequate transistor current conduction. In view of the idealized objective of constant threshold voltage and the dependence of V_θ on the square of oxide thickness, it is hardly surprising that integrated circuit foundries continue to develop manufacturing methods that ensure progressively thinner, but nonetheless uniformly thin, gate oxide layers.

Example 7.1. A silicon N-channel transistor is fabricated with an average substrate acceptor impurity concentration, N_A, of $(5)(10^{15})$ atoms/cm^3. The oxide layer, whose thickness is 25 angstroms, is observed to support a potential, V_{ox}, of 100 mV, for practical values of applied gate-source voltage and is independent of channel temperature. Plot the threshold voltage, V_h, of the transistor as a function of temperature over the range of 0°C to 100°C for $V_{bs} = 0$ volt, -1.5 volts, and -3 volts. Examine and discuss the generated curves and specifically, give the threshold potentials at 27°C and 75°C for each of the considered three bulk-source voltages.

Solution 7.1.

(1) With $T_{ox} = 25\text{Å} = (25)(10^{-8})$ cm and $N_A = (5)(10^{15})$ cm^{-3}, the voltage parameter, V_θ, in Eq. (7-9) is $V_\theta = 441.6\,\mu$V. At 27°C, which is typically taken as reflective of room temperature, even though 27°C corresponds to a very warm room whose temperature is 80.6°F, the intrinsic carrier concentration in Eq. (7-5) is 10^{10} cm^{-3}, while the Boltzmann voltage defined by Eq. (7-8) is $V_T = 25.89$ mV. At room temperature, Eq. (7-4) yields a Fermi potential of $V_F = 339.8$ mV, whence a $V_{bs} = 0$, room temperature value of threshold voltage from Eq. (7-3) of $V_{ho} = 779.5$ mV.
(2) Figure 7.3 displays the dependence of threshold voltage on the temperature of the induced conductive channel for two particular values of the bulk-source voltage. A careful examination of the simulated data reveals the following results.

$$\begin{aligned}&\text{For } V_{bs} = 0 \text{ volts:} && V_h = 779.5 \text{ mV } @ \ T = 27°C \\ & && V_h = 706.5 \text{ mV } @ \ T = 75°C \\ &\text{For } V_{bs} = -1.5 \text{ volts:} && V_h = 799.3 \text{ mV } @ \ T = 27°C \\ & && V_h = 727.1 \text{ mV } @ \ T = 75°C \\ &\text{For } V_{bs} = -3.0 \text{ volts:} && V_h = 812.6 \text{ mV } @ \ T = 27°C \\ & && V_h = 740.5 \text{ mV } @ \ T = 75°C\end{aligned}$$

Comments. As predicted, the threshold voltage increases with increasing reverse bias applied to the bulk-source terminals. Additionally, the threshold voltage is observed to decrease with increasing channel temperature for any bulk-source voltage. An investigation of this temperature-induced decrease suggests that the threshold voltage decreases at a rate of about 1.5 mV/°C, which is a handy rule of thumb designers can invoke in their first draft circuit designs.

Analog MOS Technology Circuits 509

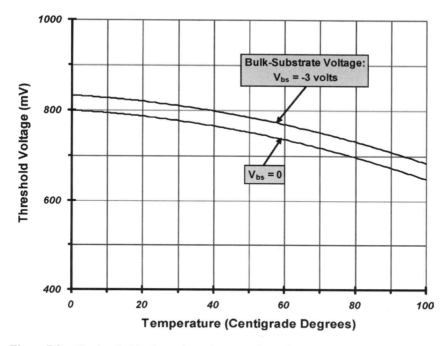

Figure 7.3. The threshold voltage dependence on channel temperature as a function of the applied bulk-source voltage for the transistor addressed in Example 7.1.

7.2.2.2. Ohmic Electrical Regime

When the gate-source voltage, V_{gs}, of NMOS or the source-gate voltage, V_{sg}, of its PMOS counterpart is at least equal to the threshold voltage, V_h, current conduction through either type of transistor is enabled. If such conduction in NMOS is incurred for drain-source voltages that are sufficiently small to satisfy the inequality, $V_{ds} < V_{gs} - V_h$, the subject transistor is said to operate in its *ohmic regime* where

$$I_d = K_n \left(\frac{W}{L}\right) V_{ds} \left(V_{gs} - V_h - \frac{V_{ds}}{2}\right)\left(1 + \frac{V_{ds}}{V_\lambda}\right). \qquad (7\text{-}10)$$

In this expression, dimensions W and L are delineated in Fig. 7.1(a). The ratio, W/L, of these geometric parameters is termed the *gate aspect ratio*. The constant, K_n, is curiously called the *transconductance coefficient* despite the fact that the units of K_n are mhos/volt. This constant is given by

$$K_n = \mu_n C_{ox}, \qquad (7\text{-}11)$$

where μ_n is the previously discussed mobility of electrons in the induced conduction channel. The parameter, C_{ox}, relates to the oxide dielectric constant, ϵ_{ox}, and thickness T_{ox} as

$$C_{ox} = \frac{\varepsilon_{ox}}{T_{ox}}, \qquad (7\text{-}12)$$

which symbolizes the density of net oxide capacitance; that is, C_{ox} is the net capacitance-per-unit of gate area, which is WL. Finally, V_λ is termed the *channel length modulation voltage*. This parameter is nominally proportional to the length, L, of the channel. In devices whose channel lengths are at least 0.5 micron, V_λ is likely to be in the range of 30 to 60 volts, while for deep submicron units, V_λ is generally below 20 volts and can be only a few volts.

Although MOSFETS used in linear signal processing applications are rarely biased for operation in their ohmic regimes, three interesting sidebars are implicit to Eq. (7-10). The first of these observations derives from the requisite constraints, $V_{gs} \geq V_h$ and $V_{ds} < (V_{gs} - V_h)$. The requirement, $V_{gs} \geq V_h$, ensures channel inversion in at least the neighborhood of the heavily doped source region. But recalling Eq. (7-2), $V_{ds} < (V_{gs} - V_h)$, coupled with $V_{gs} \geq V_h$, is tantamount to requiring $V_{gd} > V_h$, which ensures that channel inversion is also evidenced at the drain site. In short, the two constraints, $V_{gs} \geq V_h$ and $V_{ds} < (V_{gs} - V_h)$, ensure channel inversion throughout the entire length, L, of the spacing between the source and drain regions.

A second important issue is the fact that the mobility of electrons in the induced conductive channel is inversely related to channel temperature. Specifically, the mobility, $\mu_n(T)$, at any absolute temperature T relates empirically to the mobility, $\mu_n(T_o)$, at a reference temperature T_o as

$$\mu_n(T) = \mu_n(T_o) \left(\frac{T_o}{T}\right)^{3/2}, \qquad (7\text{-}13)$$

where unless noted otherwise, T_o is taken as the "room temperature," 300.2°K. Since the transconductance coefficient, K_n, is directly proportional to carrier mobility and in turn, the ohmic regime drain current is proportional to K_n, the drain current displays a negative temperature coefficient. A transistor current that is independent of operating temperature comprises a highly idealistic operating condition. But if said current must modulate with temperature, it is arguably better to incur decreasing current than increasing current with temperature rises. If indeed a device current increases with temperature, as it does in bipolar technology transistors, thermal runaway induced by unavoidable intrinsic self-heating is an unfortunate possibility.

The third and final issue explains the "ohmic" reference to the operating regime under current consideration. In Eq. (7-10), assume that $V_{ds} \ll V_\lambda$, as is invariably the case in the ohmic operating region where the drain-source voltage is necessarily small. Consider the case in which the gate-source voltage, V_{gs}, is held fast while allowing the drain current, I_d, to respond to small changes in the drain-source voltage, V_{ds}. The resultant drain current effect of the small change in drain-source voltage can be quantified as the small signal drain-source conductance, g_{ds}, which is

$$g_{ds} \triangleq \frac{\partial I_d}{\partial V_{ds}} \approx K_n \left(\frac{W}{L}\right)(V_{gs} - V_h - V_{ds}), \tag{7-14}$$

which portrays a small signal conductance whose value depends on the fixed gate-source bias, V_{gs}. More significantly, Eq. (7-10) combines with this result to deliver

$$g_{ds} \triangleq \frac{\partial I_d}{\partial V_{ds}} \approx \frac{I_d}{V_{ds}} \left(\frac{V_{gs} - V_h - V_{ds}}{V_{gs} - V_h - \frac{V_{ds}}{2}} \right). \tag{7-15}$$

If V_{ds} is very small so that it satisfies the inequality, $V_{ds} \ll (V_{gs} - V_h)$, Eq. (7-15) collapses to

$$g_{ds} \triangleq \frac{\partial I_d}{\partial V_{ds}} \approx \frac{I_d}{V_{ds}}, \tag{7-16}$$

which suggests that for very small drain-source voltages, the drain-source port of a MOSFET behaves as an approximately linear resistance controlled by gate-source voltage; that is, the MOSFET behaves as a voltage controlled linear resistance, which is tantamount to an electronic potentiometer. In addition to portending of a possible application in tunable electronic circuits, Eq. (7-16) ostensibly justifies the "ohmic" descriptive to the operating regime currently undergoing study.

The preceding disclosures can be straightforwardly modified to embrace P-channel transistors. In particular, Eq. (7-10) becomes

$$I_d = K_p \left(\frac{W}{L}\right) V_{sd} \left(V_{sg} - V_h - \frac{V_{sd}}{2}\right)\left(1 + \frac{V_{sd}}{V_\lambda}\right), \tag{7-17}$$

where K_n in Eq. (7-11) is modified to

$$K_p = \mu_p C_{ox}, \tag{7-18}$$

with the understanding that μ_p symbolizes the mobility of holes in the conductive channel of P-channel devices. Equation (7-13) applies equally well to hole mobility, and, subject to obvious symbolic alterations, Eqs. (7-14) through (7-16) are also applicable.

7.2.2.3. Saturation Regime

In contrast to the ohmic electrical regime, the saturation operating region of NMOS entails $V_{ds} \geq (V_{gs} - V_h)$, in addition to the usual conduction requirement, $V_{gs} \geq V_h$. For these constraints, Eq. (7-10) is supplanted by the nominally square law relationship,

$$I_d = \frac{K_n}{2}\left(\frac{W}{L}\right)(V_{gs} - V_h)^2\left(1 + \frac{V_{ds}}{V_\lambda}\right). \qquad (7\text{-}19)$$

While the channel length modulation term in V_{ds}/V_λ is invariably inconsequential in the ohmic regime where the drain-source voltage is small, it can exert a significant impact on the drain current in saturation, where the drain-source voltage is relatively large. The voltage difference, $(V_{gs} - V_h)$, is defined as the *drain saturation voltage*, V_{dss}; that is,

$$V_{dss} \triangleq V_{gs} - V_h. \qquad (7\text{-}20)$$

The drain current corresponding to $V_{ds} = V_{dss}$ is termed the *drain saturation current*, I_{ds}. Continuity at the crossover voltage between the ohmic and saturated operating regimes is assured since by either Eq. (7-10) or Eq. (7-19),

$$I_{ds} = \frac{K_n}{2}\left(\frac{W}{L}\right)V_{dss}^2\left(1 + \frac{V_{dss}}{V_\lambda}\right). \qquad (7\text{-}21)$$

It is interesting to note that for $V_{ds} \geq (V_{gs} - V_h)$, Eq. (7-2) implies $V_{gd} \leq V_h$. The relevant implication is that when the drain-source voltage, V_{ds}, assumes its saturated value, V_{dss}, V_{gd} is precisely the threshold voltage, V_h, which infers that channel inversion at the semiconductor-oxide interface is barely evident at the drain end of the channel. In effect, the conductive channel is "pinched" to zero thickness when $V_{ds} \equiv V_{dss}$, and the channel inversion layer is said to be "pinched off" at the drain. In recognition of the fact that V_{dss} is precisely the drain-source voltage commensurate with channel pinch off at the drain site, drain-source voltages larger than V_{dss} must incur pinch off at the interface somewhere between the drain and source implants. The excess voltage, $(V_{ds} - V_{dss})$ is accordingly dropped in the depletion region formed between the channel pinch off point and the drain site. The electric field established laterally in this depletion zone in response to drain-source voltage excesses sustains the drain current that would be otherwise predicted by an inverted conductive channel presumed to extend throughout the source to drain volume.

For PMOS, saturation ensues when the source to drain voltage satisfies the inequality, $V_{sg} \geq (V_{sg} - V_h)$, and the resultant drain current corresponding to Eq. (7-19) becomes

$$I_d = \frac{K_p}{2}\left(\frac{W}{L}\right)(V_{sg} - V_h)^2\left(1 + \frac{V_{sd}}{V_\lambda}\right). \tag{7-22}$$

The drain saturation voltage is now the source to drain saturation voltage,

$$V_{sds} \triangleq V_{sg} - V_h, \tag{7-23}$$

and the companion drain saturation current is

$$I_{ds} = \frac{K_p}{2}\left(\frac{W}{L}\right)V_{sds}^2\left(1 + \frac{V_{sds}}{V_\lambda}\right). \tag{7-24}$$

Example 7.2. Consider a silicon N-channel MOSFET having a substrate doping concentration of $N_A = (5)(10^{14})$ atoms/cm^3, a gate oxide thickness of $T_{ox} = 30$Å, a gate aspect ratio of $W/L = 10$, a channel length modulation voltage of $V_\lambda = 20$ volts, and a gate oxide potential of $V_{ox} = 50$ mV. Exploit Eqs. (7-10), (7-13), (7-19), and (7-21) to generate the static drain characteristic curves for this transistor. The reference temperature is taken as $T_o = 300.2°$K, the electron mobility, $\mu_n(T_o)$, at the reference temperature can be estimated to be $\mu_n(T_o) = 550$ cm^2/volt-sec, and a bulk-source voltage of $V_{bs} = -3$ volts is presumed.

Solution 7.2.

(1) With $T_{ox} = 30$Å $= (30)(10^{-8})$ cm and $N_A = (5)(10^{14})$ cm^{-3}, V_θ, in Eq. (7-9) is $V_\theta = 63.5 \,\mu$V. At 27°C, the intrinsic carrier concentration in Eq. (7-5) is 10^{10} cm^{-3}, while the Boltzmann voltage defined by Eq. (7-8) is $V_T = 25.89$ mV. At room temperature, Eq. (7-4) yields a Fermi potential of $V_F = 280.1$ mV, whence a $V_{bs} = -3$ volts, room temperature value of threshold voltage from Eq. (7-3) of $V_h = 610.4$ mV.

(2) Figure 7.4 displays the requisite set of characteristic curves, which derive from Eqs. (7-10) and (7-19). Superimposed on the static characteristics is a dashed curve deriving from Eq. (7-21) that represents the electrical boundary between ohmic and saturation regimes. Any volt-ampere point to the right of the boundary curve corresponds to transistor operation in its saturated domain. To the left of the boundary curve, the transistor is immersed in its ohmic regime.

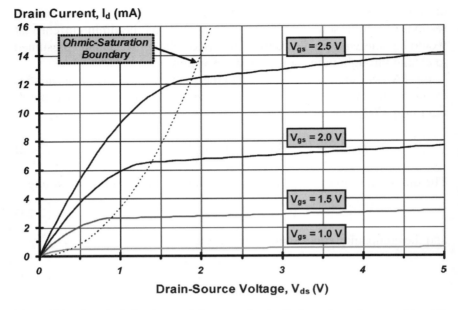

Figure 7.4. Sample drain characteristic curves for the N-channel transistor considered in Example 7.2.

Comments. In Fig. 7.4, consider the separation between V_{gs}-specific curves at any particular value of the drain-source voltage, V_{ds}. The consideration of a particular value of V_{ds} is equivalent to holding V_{ds} fixed at an arbitrary operating point, thereby disallowing any changes in V_{ds} that might be incurred as a result of a gate-source voltage perturbed by applied input signals. Since the individual curves in the considered static characteristics correspond to particular values of gate-source voltage, the separation between curves at fixed V_{ds} is equivalent to examining the low frequency, forward, short circuit transconductance with respect to V_{gs}. Accordingly, the separation between individual V_{gs} curves is at least a crude measure of the forward gain afforded by the transistor undergoing characterization. An inspection of the plot at hand clearly shows that such separation is more pronounced in the saturation domain than it is in the ohmic region. This observation is one of several reasons that MOSFETs utilized in amplifiers and other linear signal processing applications are almost always biased to constrain their operation in saturated regimes.

7.2.3. Small Signal Models

As noted in the commentary of the preceding example, MOSFETs intended for linear circuit applications are invariably biased in saturation. Accordingly, Eq. (7-19) is the applicable static volt-ampere characteristic for an N-channel transistor, while Eq. (7-22) applies to its PMOS counterpart. In Eq. (7-19), the drain current is functionally dependent on three transistor voltages; namely, the gate-source voltage, V_{gs}, the drain-source voltage, V_{ds}, and, because of the dependence of threshold voltage on bulk-source voltage, V_{bs}. Under zero signal, or *quiescent operating conditions*, let the subject transistor be biased at $V_{gs} = V_{gsQ}$, $V_{ds} = V_{dsQ}$, and $V_{bs} = V_{bsQ}$, such that the resultant quiescent drain current, $I_d = I_{dQ}$, lies in the region where $V_{gsQ} > V_h$ and $V_{dsQ} > (V_{gsQ} - V_h)$. Because of Eq. (7-19), a small signal superimposed with the gate-source bias voltage perturbs the drain current and conceivably, the remaining two device voltages. Assume that the signal applied to the gate with respect to the source is sufficiently small to ensure correspondingly small changes in the two other transistor voltages. It follows that the resultant perturbation, $(I_d - I_{dQ})$, in drain current can be approximated by the linear terms of the Taylor series expansion of the drain current about its quiescent operating point, or more simply, about its *Q–point*. Specifically,

$$I_d - I_{dQ} \approx \left.\frac{\partial I_d}{\partial V_{gs}}\right|_Q (V_{gs} - V_{gsQ}) + \left.\frac{\partial I_d}{\partial V_{ds}}\right|_Q (V_{ds} - V_{dsQ})$$

$$+ \left.\frac{\partial I_d}{\partial V_{bs}}\right|_Q (V_{bs} - V_{bsQ}). \tag{7-25}$$

It is vital to understand that by ignoring the infinity of high order terms that prevail in a Taylor series expansion of the drain current, at least one of two operating conditions must prevail. First, the biasing voltages must establish a Q-point that, in addition to lying in the saturated domain of transistor operation, lies in a nominally linear portion of the saturated regime. To the right of the ohmic-saturation boundary, the curves in Fig. 7.4 depict apparent reasonable linearity of the drain current with respect to the drain-source voltage. Unfortunately, such linearity is not evidenced insofar as the drain current dependence on gate-source voltage is concerned. But if the signal-induced change in gate-source voltage is sufficiently small, the inherently square law nature of the I_d versus V_{gs} dependence can be represented adequately by a linear function. If, in fact, the drain current were precisely linearly related to all three transistor voltages, all higher order Taylor series terms, which are respectively proportional to high order current derivatives with respect to individual voltages, would vanish, thereby rendering Eq. (7-25) an exact relationship.

In Eq. (7-25), the signal induced perturbation, say I, in drain current about its saturation regime Q-point value is

$$I = I_d - I_{dQ}, \qquad (7\text{-}26)$$

while the corresponding signal changes in gate-source, drain-source, and bulk-source voltages are given by

$$\left. \begin{array}{l} V_g = V_{gs} - V_{gsQ} \\ V_d = V_{ds} - V_{dsQ} \\ V_b = V_{bs} - V_{bsQ} \end{array} \right\}. \qquad (7\text{-}27)$$

Accordingly, Eq. (7-25) is expressible as

$$I \approx \left. \frac{\partial I_d}{\partial V_{gs}} \right|_Q V_g + \left. \frac{\partial I_d}{\partial V_{ds}} \right|_Q V_d + \left. \frac{\partial I_d}{\partial V_{bs}} \right|_Q V_b, \qquad (7\text{-}28)$$

which is divorced of quiescent currents and voltages and serves to interrelate only the signal components of transistor electrical variables.

The coefficient of the gate-source signal voltage, V_g, on the right hand side of Eq. (7-28) is necessarily in units of conductance. More significantly, this coefficient is the forward transconductance, say g_{mf}, that relates the signal drain current, I, to the signal gate-source voltage. A similar statement applies to the coefficient of the signal bulk-source voltage, V_b, with the proviso that this multiplier represents the transconductance, say, g_{mb}, that couples the signal drain current to the signal bulk-source voltage. The factor multiplying the signal drain-source voltage, V_d, in the second term on the right hand side of Eq. (7-28) couples signal drain current directly to signal drain-source voltage. This factor is the conductance of the small signal resistance, say r_o, of the drain-source port. Using Eqs. (7-19) and (7-7), it is a straightforward, if not somewhat laborious, matter to demonstrate that

$$g_{mf} \stackrel{\Delta}{=} \left. \frac{\partial I_d}{\partial V_{gs}} \right|_Q = \sqrt{2K_n \left(\frac{W}{L}\right)\left(1 + \frac{V_{dsQ}}{V_\lambda}\right) I_{dQ}}, \qquad (7\text{-}29)$$

$$\frac{1}{r_o} \stackrel{\Delta}{=} \left. \frac{\partial I_d}{\partial V_{ds}} \right|_Q = \frac{V_\lambda + V_{dsQ}}{I_{dQ}}, \qquad (7\text{-}30)$$

and

$$g_{mb} \stackrel{\Delta}{=} \left. \frac{\partial I_d}{\partial V_{bs}} \right|_Q = \lambda_b g_{mf}, \qquad (7\text{-}31)$$

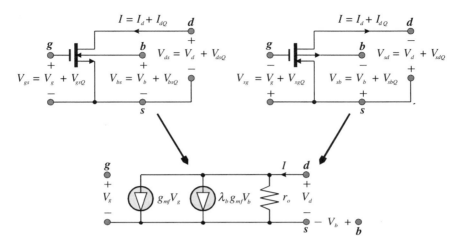

Figure 7.5. Small signal, low frequency equivalent circuit of NMOS or PMOS transistor. Assuming saturation region biasing, the small signal model parameters are given by Eqs. (7-29) through (7-32).

where

$$\lambda_b = \sqrt{\frac{V_\theta/2}{2(V_F - V_T) - V_{bsQ}}}. \tag{7-32}$$

Equation (7-28) now becomes

$$I \approx g_{mf}V_g + \frac{V_d}{r_o} + \lambda_b g_{mf} V_b, \tag{7-33}$$

which gives rise to the low frequency, small signal MOSFET model drawn in Fig. 7.5. The model at hand clearly portrays g_{mf} as a measure of the small signal gate to drain gain, or forward transconductance of a MOSFET, $\lambda_b g_{mf}$ as a secondary transconductance effected from the bulk to the drain, and r_o as a shunting resistance that renders the two current sources, $g_{mf} V_g$ and $\lambda_b g_{mf} V_b$, that shunt the drain-source circuit nonideal. The gate terminal is left open circuited because of the insulating gate oxide layer interposed between gate metal (or polysilicon) and the semiconductor channel region. The bulk terminal is also represented as an open circuit because, as is routinely the case, the bulk is reverse biased with respect to both the source and the drain implants. As inferred by the figure, the model is applicable to both NMOS and PMOS because its capability is limited exclusively to examining changes of transistor electrical variables about their respective quiescent values.

7.2.3.1. Small Signal Model at High Frequencies

When the signals applied to the gate or any other terminal of a MOSFET contain high frequency components, capacitances must be appended to the low frequency model principally between gate-source, gate-drain, bulk-drain, and bulk-source terminals. The resultant equivalent circuit is the structure provided in Fig. 7.6. To crude first order, the gate-source capacitance, C_{gs}, in this model relates to the MOSFET cross section of Fig. 7.1(a) and physical parameters introduced in preceding sections of material as

$$C_{gs} \approx \frac{2}{3}WLC_{ox} + WL_dC_{ox} = \frac{2}{3}WLC_{ox}\left(1 + \frac{3L_d}{2L}\right), \qquad (7\text{-}34)$$

where L_d is the length of the gate oxide overlap at the source site. The indicated two-thirds factor is an estimate of the source region distribution of the net oxide capacitance, WLC_{ox}. This factor empirically and experimentally accounts for a channel inversion layer that does not extend completely from source to drain when the transistor operates in saturation. The gate-drain capacitance, C_{gd}, is almost exclusively determined by the oxide overlap over the drain region, whence

$$C_{gd} \approx WL_dC_{ox}. \qquad (7\text{-}35)$$

The two remaining capacitances in the model of Fig. 7.6 are voltage dependent. The capacitance, C_{bd}, is attributed to the depletion zone surrounding the back biased bulk-drain junction in the cross-section diagram of Fig. 7.1(a). It can be expressed as

$$C_{bd} \approx \frac{P_d C_{jod}}{\left(1 - \dfrac{V_{bdQ}}{V_{jd}}\right)^{m_d}}, \qquad (7\text{-}36)$$

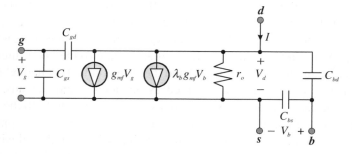

Figure 7.6. Small signal, high frequency equivalent circuit of NMOS or PMOS transistor.

where C_{jod} is the zero bulk-drain junction bias ($V_{bdQ} = 0$) value of the transition capacitance per unit length of the net perimeter, P_d, indigenous to the bulk-drain junction, V_{jd} is the built-in potential of this junction, and m_d is the junction grading coefficient. Recall that P_d is 10% to 20% longer than the gate width, W, while V_{jd} is in the range of 700 mV to 900 mV, and m_d lies between 0.25 and 0.33. Analogously, the bulk-source capacitance, C_{bs}, is

$$C_{bs} \approx \frac{P_s C_{jos}}{\left(1 - \frac{V_{bsQ}}{V_{js}}\right)^{m_s}}. \tag{7-37}$$

It should be underscored that because the bulk is generally incident with a node supporting the most negative potential available in the circuit in which the considered MOSFET is embedded, $V_{bdQ} \leq 0$ and $V_{bsQ} \leq 0$ in Eqs. (7-36) and (7-37), respectively.

For the commonly encountered case in which the bulk and source terminals are incident with one another for small signal operating conditions, the model of Fig. 7.6 simplifies to that offered in Fig. 7.7. Comparing this latter topology with its predecessor, the transition capacitance, C_{bs}, as well as the controlled current source, $\lambda_b g_{mf} V_b$, disappear owing to a bulk to source signal voltage, V_b, that is rendered null by the interconnection of source and bulk terminals.

The model in Fig. 7.7 is reminiscent of the generalized y-parameter pi topology presented in Fig. 2.15, which is redrawn in Fig. 7.8(a) in a form appropriate to the small signal operation of a MOSFET having $V_b = 0$. Each of the admittance parameters, y_i, y_r, y_f, and y_o, in this equivalent circuit

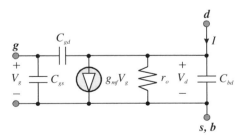

Figure 7.7. Simplification of the small signal model in Fig. 7.6 for the special case in which the bulk and source terminals support identical signal voltages.

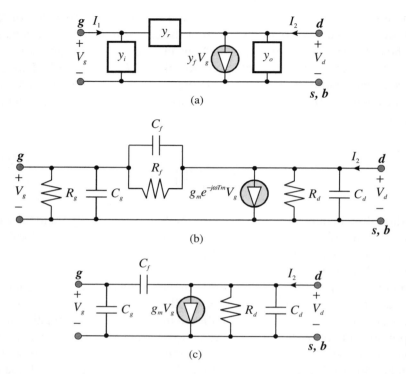

Figure 7.8. (a) Generalized pi topology architecture of the y-parameter model for a linear circuit. (b) High frequency, small signal MOSFET model applicable to the special case when zero signal voltage is developed between bulk and source terminals. The parameters of this model derive from measured or simulated scattering parameters evaluated at a specific quiescent operating point and over a suitable range of signal frequencies. (c) Simplified version of the model in (b).

relate to the short circuit admittance parameters, y_{11}, y_{12}, y_{21}, and y_{22}, of the subject MOSFET as per Eq. (2-29). In turn, the short circuit admittance parameters can be deduced from the MOSFET scattering parameters, S_{ij}, which can be either measured or simulated over a signal frequency range of interest for a specific quiescent operating point. If the scattering parameters derive from computer-based simulations of transistor characteristics, care must be exercised to ensure accurate device modeling, such as that afforded by the LEVEL 49 model in HSPICE. The upshot of this exercise is that the model in Fig. 7.8(a) can be recast in the form shown in Fig. 7.8(b), which effectively decomposes the pi network parameters into their respective real and imaginary components. Specifically, the resistive and transconductance parts of the model can be evaluated in accordance with

$$\left. \begin{array}{l} \dfrac{1}{R_g} = Re\,(y_i) = Re\,(y_{11}) + Re\,(y_{12}) \\ \dfrac{1}{R_d} = Re\,(y_o) = Re\,(y_{22}) + Re\,(y_{12}) \\ \dfrac{1}{R_f} = Re\,(y_r) = -Re\,(y_{12}) \\ g_m = Re\,(y_f) = Re\,(y_{21}) - Re\,(y_{12}) \end{array} \right\}. \qquad (7\text{-}38)$$

On the other hand, the capacitive elements are calculated as

$$\left. \begin{array}{l} C_g = \dfrac{Im\,(y_i)}{\omega} = \dfrac{Im\,(y_{11}) + Im\,(y_{12})}{\omega} \\ C_d = \dfrac{Im\,(y_o)}{\omega} = \dfrac{Im\,(y_{22}) + Im\,(y_{12})}{\omega} \\ C_f = \dfrac{Im\,(y_r)}{\omega} = -\dfrac{Im\,(y_{12})}{\omega} \end{array} \right\}. \qquad (7\text{-}39)$$

Finally, the delay, T_m, appended to the transconductance, g_m, in Fig. 7.8(b) exploits the reasonable presumption that over the frequency range of interest, $Im(y_f) \ll Re(y_f)$. This delay metric is resultantly given by

$$T_m \approx \dfrac{Im\,(y_f)}{\omega Re\,(y_f)} = \dfrac{Im\,(y_{21}) - Im\,(y_{12})}{\omega[Re\,(y_{21}) - Re\,(y_{12})]}. \qquad (7\text{-}40)$$

All of the parameters defined by the preceding three expressions can be expected to vary somewhat with signal frequency and most assuredly with biasing level. The biasing dependence is best handled by ensuring that the original S-parameters on which the model elements are computationally premised are discerned for the quiescent operating point of immediate interest. The frequency dependencies of resistance R_o, all three capacitances, and delay T_m are rarely pronounced. Accordingly, respective averaged parametric values over the frequency range undergoing scrutiny can be computed without significantly impairing the accuracy of circuit analyses premised on the model in Fig. 7.8(b). In contrast, the frequency dependencies of resistances R_i and R_f are dramatic but fortunately, their respective values are so large as to justify their tacit neglect in most analytical situations. One potentially notable exception to this directive entails the analysis of electrical noise phenomena, which is not a focus of this chapter. Finally, the delay term, T_m, is invariably small enough to warrant its neglect in analytical endeavors that do not embrace device operation at frequencies approaching the upper limits of practical transistor utilization. The resultant simplified small signal equivalent circuit is the structure of Fig. 7.8(c). The elements of this model differ from those of Fig. 7.7 only insofar as the elements of the latter configuration represent effective values based on measured or simulated data.

7.2.3.2. Unity Gain Frequency

A popular barometer of MOSFET transistor quality is its unity gain frequency, $\omega_T = 2\pi f_T$, which is defined as the signal frequency at which the magnitude of the short circuit current gain, I_{out}/I_{in}, in the grounded source configuration of Fig. 7.9(a) is unity. In this idealized test fixture, the transistor is biased at a constant drain current, I_{dQ}, which is selected to ensure saturation region operation. Because the load imposed between drain and ground is a capacitance, C_{big}, no drain biasing current is diverted from the transistor. Moreover, the subject capacitance is chosen sufficiently large to emulate a short circuit over the signal frequency range of interest and as a result, the drain is effectively shorted to ground for signal conditions. The output current, I_{out}, flows through this large capacitance in response to the current signal, I_{in}, applied from ground to gate terminal of the transistor.

Since the bulk terminal is short circuited to the MOSFET source terminal, either the equivalent circuit in Fig. 7.7 or that of Fig. 7.8(c) is appropriate for the small signal analysis of the circuit at hand. Choosing the former structure, the applicable model is the topology given in Fig. 7.9(b). It must be recalled that a small signal equivalent circuit allows for an analytical consideration of only the signal components of respective circuit voltages and currents. Accordingly, the network in Fig. 7.9(b) derives from that of Fig. 7.9(a) by (1) replacing the MOSFET by its small signal model,

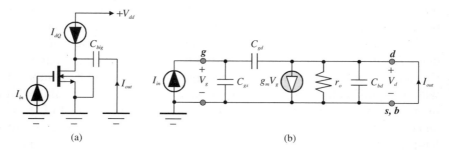

Figure 7.9. (a) Idealized circuit test configuration appropriate for the measurement of the grounded source, short circuit current gain, I_{out}/I_{in}. (b) High frequency, small signal MOSFET equivalent circuit for the test structure in (a).

(2) setting to zero (short circuiting) the presumably ideal power supply voltage, V_{dd}, and (3) setting to zero (open circuiting) the presumably ideal constant current source, I_{dQ}.

An analysis of the model in Fig. 7.9(b) reveals

$$\left. \begin{array}{l} I_{\text{in}} = j\omega \left(C_{gs} + C_{gd} \right) V_g \\ I_{\text{out}} = \left(g_m - j\omega C_{gd} \right) V_g \end{array} \right\}. \qquad (7\text{-}41)$$

whence a current gain of

$$\frac{I_{\text{out}}}{I_{\text{in}}} = \frac{g_m - j\omega C_{gd}}{j\omega \left(C_{gs} + C_{gd} \right)}. \qquad (7\text{-}42)$$

On the tacit assumption that $g_m \gg \omega C_{gd}$ for $\omega \leq \omega_T$, Eq. (7-42) reduces to the simple expression,

$$\frac{I_{\text{out}}}{I_{\text{in}}} \approx -j\frac{\omega_T}{\omega}, \qquad (7\text{-}43)$$

where

$$\omega_T = \frac{g_m}{C_{gs} + C_{gd}}. \qquad (7\text{-}44)$$

Clearly, ω_T is the objective unity gain frequency since $\omega = \omega_T$ in Eq. (7-43) renders $|I_{\text{out}}/I_{\text{in}}| \approx 1$. Recalling Eqs. (7-11), (7-19), (7-29), (7-34), and (7-35), Eq. (7-44) can be cast into the alternative form,

$$\omega_T = 2\pi f_T = \frac{3\mu_n \left(V_{gsQ} - V_h \right) \sqrt{1 + \dfrac{V_{dsQ}}{V_\lambda}}}{2L^2 \left(1 + \dfrac{3L_d}{L} \right)}, \qquad (7\text{-}45)$$

which depicts the transistor unity gain frequency as nominally proportional to channel carrier mobility and inversely proportional to the square of the channel length between the source and drain regions.

The unity gain frequency, f_T, provides an expedient quantitative index for comparing the high frequency performance capabilities of candidate MOSFET devices for circuit applications. In other words, MOSFETs boasting large f_T-values are generally construed as "better" than those projecting smaller f_T-values. Obviously, "better" in this context is limited to only small signal, high frequency performance attributes, for f_T, as defined by Eq. (7-45), offers no information regarding requisite power dissipation levels, available forward transconductance, electrical noise characteristics, nonlinearity shortfalls, and other performance barometers.

While the transistor f_T arguably comprises a useful comparison index, albeit in a limited sense, for individual transistors, it is a grossly optimistic figure of merit for assessing high frequency performance attributes at the circuit level. Most of the reasons in support of this declaration derive from a scrutiny of pertinent circuit time constants. For example, the signal short circuit imposed between drain and source and between bulk and source remove from analytical consideration the potentially significant impact exerted on frequency response by the time constants associated with both the bulk-drain and the bulk-source transition region capacitances. The subject signal short circuit also effectively places the gate-drain capacitance, C_{gd}, in shunt with the gate-source capacitance, C_{gs}. This shunting connection renders incomplete a consideration of the charging time of gate-drain capacitance. Specifically, the gate-drain capacitance, which is connected between a phase inverting node pair, is capable of large time constants because of Miller multiplication (more is said about this phenomenon shortly), which is precluded by the signal short circuit at the drain output port. In effect, the f_T figure of merit explicitly factors into the frequency response problem only the effects of the charging time of the net input port capacitance and the channel transit time, as measured by carrier mobility μ_n (or μ_p in PMOS). Time constant considerations notwithstanding, identifying an application that is more meaningless than a grounded source amplifier in which the drain output port is shorted to ground and the gate input port is driven by an ideal current source is a daunting task.

7.3.0. Common Source Amplifier

Figure 7.10(a) schematically depicts a basic common source amplifier. The divider formed of the resistances, R_1 and R_2, establishes a quiescent gate-source voltage commensurate with a Q-point drain current and a Q-point drain-source voltage that ensure transistor operation in its saturation regime. The applied input signal, which is comprised of the series interconnection of the Thévenin signal source voltage, V_s, and the Thévenin source resistance, R_s, is coupled to the transistor gate terminal through a capacitor, C_s, whose value is sufficiently large to ensure that its terminal impedance approximates a short circuit over all signal frequencies of interest. The output response to signal excitation V_s is extracted as a voltage, V_o, at the transistor drain. It is understood that only the small signal component, V_{os}, of this output voltage is linearly proportional to the applied input signal, V_s. Resistance R_l is construed as the load imposed on the amplifier and while its value affects the available voltage gain of the network, it also influences the Q-point

Analog MOS Technology Circuits 525

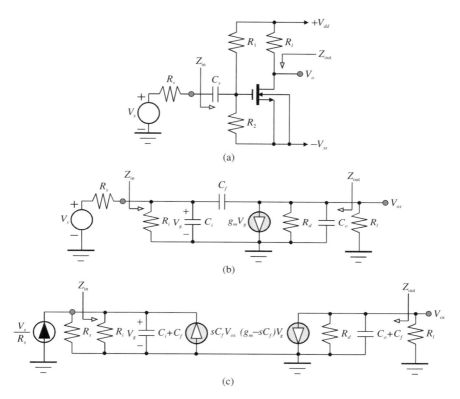

Figure 7.10. (a) Circuit schematic diagram of a common source amplifier. (b) High frequency, small signal equivalent circuit for the amplifier in (a). Capacitances C_i and C_o include effective transistor input port and output port capacitances, respectively, as well as parasitic I/O port capacitances attributed to the circuit topology and its layout. Voltage V_{os} represents the small signal component of the output voltage, V_o, delineated in (a). (c) Alternative circuit model for which the feedback capacitance, C_f, in (b) has been supplanted by its y-parameter equivalent circuit.

value of drain-source voltage. The two utilized power supplies, V_{dd} and V_{ss}, are presumed to be ideal voltage sources. Practical power supply lines, which are commonly referred to as *power busses*, are plagued by parasitic series resistance and inductance, which can affect the nominal performance, as well as the stability, of the amplifier. For this reason, relatively large *decoupling capacitances*, which emulate signal short circuits throughout the signal passband, are often appended between ground and the circuit nodes to which the supply lines are incident.

Recalling the approximate transistor model of Fig. 7.8(c), the equivalent circuit of the amplifier at hand is the structure offered in Fig. 7.10(b). In this circuit, capacitance C_i embraces both net effective gate-source transistor capacitance and any parasitic capacitance evidenced at the amplifier input port. Similarly, capacitance C_o accounts for both transistor bulk-drain and, if necessary, parasitic output port capacitances. The power supply lines are represented by short circuits to ground, while capacitance C_s is supplanted by the short circuit it emulates over the range of frequencies to which analytical attention is focused. Finally, the resistance, R_i, is the parallel combination of the biasing resistances, R_1 and R_2.

It is clear that the capacitance, C_f, in Fig. 7.10(b) is a shunt-shunt feedback element, thereby suggesting that the amplifier undergoing scrutiny can be viewed analytically as a transimpedance network. Accordingly, the subject model is recast into the topological form shown in Fig. 7.10(c), where the Thévenin signal source is replaced by its Norton equivalent. Moreover, the branch impedance associated with the feedback element, C_f, is modeled by short circuit admittance parameters. This representation of the feedback element is tantamount to its replacement by four network elements. In particular, capacitance C_f is placed in shunt with the input port, while capacitance C_f shunts the output port as well. Additionally, a feedforward current source, $sC_f V_g$, is placed in anti-phase shunt with the controlled source, $g_m V_g$, at the drain, or amplifier output port. Finally, a global feedback current source, $sC_f V_{os}$, is appended in shunt and in phase with the Norton current generator, V_s/R_s.

7.3.1. Voltage Transfer Function

The global feedback nature of the phase inverting circuit model diagrammed in Fig. 7.10(c) allows the forward transimpedance function, $Z_m(s)$, to be written in the form,

$$Z_m(s) = \frac{V_{os}}{V_s/R_s} = \frac{Z_{mo}(s, R_s, R_l)}{1 - sC_f Z_{mo}(s, R_s, R_l)}. \qquad (7\text{-}46)$$

In this expression, $Z_{mo}(s, R_s, R_l)$, the open loop value (meaning $sC_f V_{os} = 0$) of the forward transimpedance, is given by

$$Z_{mo}(s, R_s, R_l) = \left. \frac{V_{os}}{V_s/R_s} \right|_{sC_f V_{os}=0}$$

$$= -\left[\frac{R_{ss}}{1 + sR_{ss}(C_i + C_f)} \right] \left[\frac{(g_m - sC_f) R_{ll}}{1 + sR_{ll}(C_o + C_f)} \right], \qquad (7\text{-}47)$$

where
$$R_{ss} = R_s \| R_i = R_s \| R_1 \| R_2 \\ R_{ll} = R_l \| R_d \Bigg\}. \quad (7\text{-}48)$$

Typically, R_1 and R_2 are very large biasing resistances so that $R_{ss} \approx R_s$. Moreover, R_d is often (but not always) significantly larger than the drain circuit load resistance, R_l. The loop gain, $T(s, R_s, R_l)$, of the amplifier at hand follows as

$$T(s, R_s, R_l) = -sC_f Z_{mo}(s, R_s, R_l) = \left[\frac{sR_{ss}C_f}{1+sR_{ss}(C_i+C_f)}\right]$$

$$\times \left[\frac{(g_m - sC_f)R_{ll}}{1+sR_{ll}(C_o+C_f)}\right], \quad (7\text{-}49)$$

with the result that the overall voltage gain, $A_v(s)$, can be evaluated from

$$A_v(s) = \frac{V_{os}}{V_s} = \frac{Z_{mo}(s, R_s, R_l)/R_s}{1+T(s, R_s, R_l)} = \frac{A_{vo}(s, R_s, R_l)}{1+T(s, R_s, R_l)}. \quad (7\text{-}50)$$

In the interest of clarity, the open loop component, $A_{vo}(s, R_s, R_l)$, of the closed loop voltage gain for the common source amplifier is

$$A_{vo}(s) = \left.\frac{V_{os}}{V_s}\right|_{sC_f V_{os}=0}$$

$$= -\frac{g_m R_{ll}\left(1 - \dfrac{sC_f}{g_m}\right)\left(\dfrac{R_i}{R_i+R_s}\right)}{[1+sR_{ss}(C_i+C_f)][1+sR_{ll}(C_o+C_f)]}. \quad (7\text{-}51)$$

If Eqs. (7-49) and (7-51) are substituted into Eq. (7-50), the transfer function defining the common source amplifier voltage gain is seen to exude two left half plane poles and a single right half plane zero, not unlike the network generalities propounded in Chapter 4. In particular,

$$A_v(s) = \frac{V_{os}}{V_s} = \frac{A_v(0)\left(1 - \dfrac{s}{z_f}\right)}{\left(1+\dfrac{s}{p_1}\right)\left(1+\dfrac{s}{p_2}\right)} = \frac{A_v(0)\left(1 - \dfrac{s}{z_f}\right)}{1+\left(\dfrac{1}{p_1}+\dfrac{1}{p_2}\right)s + \dfrac{s^2}{p_1 p_2}}, \quad (7\text{-}52)$$

where
$$A_v(0) = -g_m R_{ll}\left(\frac{R_i}{R_i+R_s}\right) \quad (7\text{-}53)$$

is the voltage gain at zero signal frequency, and

$$z_f = g_m/C_f \tag{7-54}$$

is the frequency of the right half plane zero. The zero frequency gain in Eq. (7-53) can be discerned through mere inspection of the model in Fig. 7.10(b). It is to be understood that $A_v(0)$ is the low frequency voltage gain only to the extent that "low" refers to the lowest signal frequency at which C_s functions viably as a signal short circuit. Algebraic manipulations confirm that the two pole frequencies, p_1, and p_2, satisfy the dual constraints,

$$\frac{1}{p_1} + \frac{1}{p_2} = R_{ss}\big[C_i + (1 + g_m R_{ll})C_f\big] + R_{ll}(C_o + C_f), \tag{7-55}$$

and

$$\frac{1}{p_1 p_2} = (R_{ss} C_i)(R_{ll} C_o)\left(1 + \frac{C_f}{C_i} + \frac{C_f}{C_o}\right). \tag{7-56}$$

7.3.1.1. Poles and Time Constants

Equations (7-55) and (7-56) can be solved simultaneously for the pole frequencies, p_1 and p_2. An awareness of these characteristic frequencies, as well as the frequency defined by the right half plane zero in Eq. (7-54), is vital for quantifying numerous circuit performance metrics, such as the 3-dB bandwidth, unity gain frequency, and phase margin. Before attempting to undertake the depressing algebra that underpins this solution, a definitive examination of the preceding two pole frequency expressions proves enlightening and productive.

The examination commences by resorting to the commonplace, albeit potentially invalid or overtly inappropriate, reduction of a second order transfer function to a first order, or single pole, approximation. The essence of this reduction is that for the signal frequency range of immediate interest, the second order term, $s^2/p_1 p_2$, in the denominator on the right hand side of Eq. (7-52) is presumed to be mathematically inconsequential. This presumption is tantamount to stipulating a highest frequency of interest that is substantially smaller than the circuit self-resonant frequency, which is recalled to be the geometric mean of the two pole frequencies. In view of a tacit dominant pole presumption, Eq. (7-52) collapses to

$$A_v(s) \approx \frac{A_v(0)\left(1 - \dfrac{s}{z_f}\right)}{1 + \left(\dfrac{1}{p_1} + \dfrac{1}{p_2}\right)s}, \tag{7-57}$$

for which the approximate radial 3-dB bandwidth, B, follows directly as

$$B \approx \frac{1}{\dfrac{1}{p_1}+\dfrac{1}{p_2}} = \frac{1}{R_{ss}\left[C_i + (1+g_m R_{ll})C_f\right] + R_{ll}(C_o + C_f)}, \quad (7\text{-}58)$$

provided that $B \ll z_f$; that is, the bandwidth is much smaller than the frequency of the right half plane zero.

The compelling attraction of the first order approximation is the simplicity of the resultant bandwidth expression. Complementing the inclination to ignore the second order term in the network characteristic polynomial is the fact that this bandwidth approximation can be computed directly from circuit considerations, without resorting to an actual transfer function enumeration. In particular, the right hand side of Eq. (7-58) is clearly a sum of capacitive time constants. Specifically, it is the superposition of time constants attributed to each circuit capacitance, with all other capacitances open circuited and the signal source reduced to zero. To demonstrate this contention, return to Fig. 7.10(b), reduce the signal voltage, V_s, to zero, and determine the time constant, say τ_i, attributed to capacitance C_i, under the condition that capacitances C_f and C_o are remanded to open circuits. The applicable diagram is the circuit in Fig. 7.11(a), for which the pertinent "open circuit" time constant is the product of C_i and the resistance, V_x/I_x, it faces when the signal source is reduced to zero and all other network capacitances are replaced by open circuits. An inspection of the subject diagram reveals

$$\tau_i = C_i \left.\frac{V_x}{I_x}\right|_{V_s=0} = R_{ss} C_i. \quad (7\text{-}59)$$

With capacitances C_i and C_f supplanted by open circuits, Fig. 7.11(b) confirms an open circuit time constant, τ_o, due to capacitance C_o of

$$\tau_o = C_o \left.\frac{V_x}{I_x}\right|_{V_s=0} = R_{ll} C_o. \quad (7\text{-}60)$$

Finally, open circuiting C_i and C_o in Fig. 7.10(b) gives rise to the model in Fig. 7.11(c), for which the time constant, τ_f, precipitated by capacitance C_f can be shown to be

$$\tau_f = C_f \left.\frac{V_x}{I_x}\right|_{V_s=0} = [R_{ll} + (1+g_m R_{ll})R_{ss}]C_f. \quad (7\text{-}61)$$

It should be noted that the time constant impact of capacitance C_f, which is connected in Fig. 7.10(b) between phase inverting nodes that respectively support signal voltages V_g and V_{os}, is considerably different than the

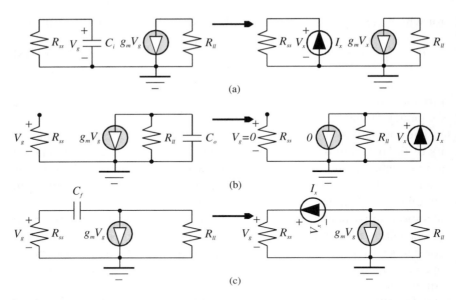

Figure 7.11. (a) Computation of the zero input, open circuit time constant attributed to capacitance C_i in the equivalent circuit of Fig. 7.10(b). (b) Computation of the zero input, open circuit time constant due to capacitance C_o in the equivalent circuit of Fig. 7.10(b). (c) Computation of the zero input, open circuit time constant attributed to capacitance C_f in the equivalent circuit of Fig. 7.10(b).

time constant implications of the other two network capacitances. Equation (7-61) suggests that from a time constant perspective, C_f behaves as two capacitances. Specifically, time constant τ_f is equivalent to a capacitance, C_f, in shunt with the load resistance, R_{ll}, and an "amplified" capacitance, $(1 + g_m R_{ll})C_f$, in parallel with the effective input port resistance, R_{ss}. The multiplied component can engender significant input port capacitance because the product, $g_m R_{ll}$, is related intimately, by Eq. (7-53), to the network voltage gain. Indeed, said product is precisely the gain magnitude, $|V_{os}/V_g|$, effected between the two phase inverting nodes to which capacitance C_f is incident. The literature commonly references the feedback capacitance multiplier, $(1 + g_m R_{ll})C_f$, as the *Miller effect*. One might proffer that it is "Miller time" when the bandwidth of a phase inverting amplifier is dominantly determined by the time constant precipitated by the Miller effect.

The sum of τ_i, τ_o, and τ_f is precisely the right hand side Eq. (7-58), which in turn is the s-term coefficient in a monic format of the network characteristic polynomial. Although this exercise is predicated on a relatively simple three capacitor equivalent circuit, the concluding result is general and applicable to any N-capacitor model. In particular, the coefficient of the s-term in any monic representation of the characteristic polynomial for a lumped network, and hence the inverse of the first order approximation to network 3-dB bandwidth, is exactly the sum of the open circuit time constants attributed to each of the N-capacitors.[12]

The foregoing open circuit time constant methodology can also be exploited to determine the coefficient of the s-term in a monic form of the numerator polynomial in the network transfer function.[13] For this determination, the output response, which is voltage V_{os} in the present case, is clamped to zero, as opposed to setting the input signal source to zero. The applicable circuit models are delineated in Fig. 7.12. In Fig. 7.12(a), the zero output value of the time constant, τ_{io}, associated with capacitance C_i is zero. This result follows from the fact that $V_{os} = 0$ constrains $g_m V_x$, and

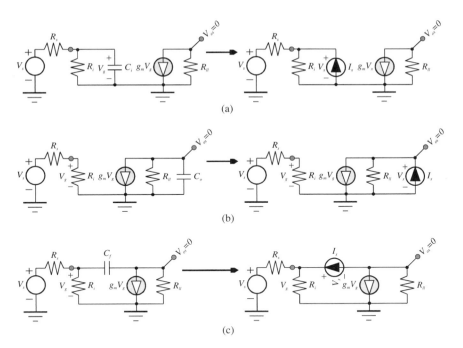

Figure 7.12. (a) Computation of the zero output, open circuit time constant attributed to capacitance C_i in the equivalent circuit of Fig. 7.10(b). (b) Computation of the zero output, open circuit time constant due to capacitance C_o in the equivalent circuit of Fig. 7.10(b). (c) Computation of the zero output, open circuit time constant attributed to capacitance C_f in the equivalent circuit of Fig. 7.10(b).

hence V_x, to zero, whence the effective resistance, V_x/I_x, facing C_i is zero. Similarly, the zero output time constant, τ_{oo}, due to capacitance C_o is zero because $V_{os} = 0$ in Fig. 7.12(b) obviously forces $V_x = 0$, whereby the resistance, V_x/I_x, implicit to this capacitive time constant is zero. In Fig. 7.12(c), $V_{os} = 0$ clamps V_x to V_g and additionally, $V_{os} = 0$ precludes any current flow in the net load resistance, R_{ll}. Hence, $I_x + g_m V_g = I_x + g_m V_x = 0$, which results in a null Thévenin resistance, V_x/I_x, seen by capacitance C_f of $-1/g_m$. Accordingly, the zero output time constant, τ_{fo}, arising from C_f is $-C_f/g_m$. The resultant sum of zero output capacitive time constants is simply $\tau_{fo} = -C_f/g_m$, which synergizes with the parenthesized numerator factor on the right hand side of Eq. (7-52).

Although using the sum of the zero input open circuit time constants as a vehicle for estimating the network 3-dB bandwidth is fraught with engineering peril, the expression in question is often useful in identifying the principle source of bandwidth degradation. The algebraic nature of Eq. (7-58) suggests that a large 3-dB bandwidth requires small open circuit time constants, thereby implying that prudent broadbanding necessarily addresses design methods that reduce the largest of the computed time constants. But as is the case with virtually any circuit compensation strategy, time constant reduction scenarios can have a pejorative impact on frequency response optimization. The most glaring problem is that the resultantly compensated amplifier may not exude a dominant pole frequency response, which is mandated if the amplifier is to be utilized in conjunction with either purposefully incorporated or parasitically encountered feedback. To this end, a necessary, but not sufficient, condition for pole dominance is that only one of the open circuit time constants be significantly larger than all other computed time constants. The challenge, therefore, is to ensure that one (and only one) of the open circuit time constants contributes dominantly to the sum of all open circuit time constants, but that this dominant time constant is not so large as to overly compress the attainable 3-dB bandwidth.

7.3.1.2. Miller-Limited Frequency Response

In an attempt to demonstrate the tradeoffs implicit between broadband and inherently stable network responses, return to Eq. (7-55) and assume that capacitance C_f establishes the dominant open circuit time constant. The Miller multiplication of this capacitive element renders plausible its time constant dominance. In addition, since a small value of C_f establishes a time constant that is significantly larger than might be routinely anticipated because of its Miller multiplication, appending an extrinsic gate to drain capacitance, which amounts to shunting C_f, with a compensation capacitance, is actually an efficient embodiment of stability compensation. While such a strategy arguably results in a big compensation bang for the femtofarad spent, it is hardly indicative of foolproof compensation. To wit, Eq. (7-54) verifies that progressively larger effective values of C_f reduce the frequency of the right half plane zero, which in turn degrades phase margin.

Because a dominant pole response is highly desirable, if not essential, Eq. (7-55) becomes, with C_f establishing the dominant open circuit time constant,

$$\frac{1}{p_1} + \frac{1}{p_2} \approx \frac{1}{p_1} = [R_{ll} + (1 + g_m R_{ll}) R_{ss}] C_f, \qquad (7\text{-}62)$$

where the lower of the two pole frequencies, p_1 and p_2, is arbitrarily designated as p_1. Recalling Eq. (7-56), it follows that the frequency of the nondominant pole is

$$p_2 \approx \frac{[R_{ll} + (1 + g_m R_{ll}) R_{ss}] C_f}{(R_{ss} C_i)(R_{ll} C_o)\left(1 + \dfrac{C_f}{C_i} + \dfrac{C_f}{C_o}\right)}. \qquad (7\text{-}63)$$

In a voltage amplifier, the Thévenin source resistance, R_s, and hence, the resistance, R_{ss}, in Eq. (7-48) is generally a small resistance to ensure minimal voltage signal loss between the signal source and the gate input terminal of the common source amplifier. Moreover, the load resistance, R_l, and hence the resistance, R_{ll}, must be sufficiently large if acceptably large values of low frequency gain are to be achieved. It is therefore realistic to presume the adoption of design procedures resulting in $R_{ll} \gg R_{ss}$. For the ubiquity of practical cases in which this inequality is readily satisfied, Eqs. (7-62) and (7-63) respectively reduce to

$$p_1 \approx \frac{1}{R_{ll}(1 + g_m R_{ss}) C_f}, \qquad (7\text{-}64)$$

and

$$p_2 \approx \frac{(1 + g_m R_{ss})\left(\dfrac{C_f}{C_o}\right)}{R_{ss} C_i \left(1 + \dfrac{C_f}{C_i} + \dfrac{C_f}{C_o}\right)}. \qquad (7\text{-}65)$$

To the extent that p_1 is indeed the frequency of the dominant circuit pole, p_1 closely approximates the 3-dB bandwidth. Moreover, a dominant pole response implies that the unity gain frequency, say ω_u, is simply the product of this observed bandwidth and the magnitude of the zero frequency gain. Using Eqs. (7-53) and (7-64),

$$\omega_u \approx |A_v(0)| p_1 \approx \left(\frac{g_m R_{ss}}{1 + g_m R_{ss}} \right) \frac{1}{R_s C_f}. \tag{7-66}$$

Network pole dominance is emulated, and thus the validity of the unity gain frequency expression is reaffirmed, if and only if the frequency, p_2, of the higher frequency pole exceeds the unity gain frequency of the amplifier. Letting k_p symbolize the ratio, p_2/ω_u, it follows that

$$k_p \triangleq \frac{p_2}{\omega_u} \approx \left(\frac{1 + g_m R_{ss}}{g_m R_{ss}} \right)^2 \left[\frac{g_m R_s \left(\dfrac{C_f}{C_o} \right)}{\left(1 + \dfrac{C_i}{C_o} + \dfrac{C_i}{C_f} \right)} \right] > 1 \tag{7-67}$$

is a design restriction that underpins network pole dominance. Note that satisfying Eq. (7-67) without incurring the bandwidth impairment caused by large C_f requires sufficiently large $g_m R_s$.

The preceding two relationships disclose several interesting, and perhaps unanticipated, observations. The first of these annotations is that to the extent that the dominant network pole is largely established by the feedback capacitance, C_f, the unity gain frequency of the common source amplifier is limited by the time constant comprised of the product of Thévenin source resistance and net feedback capacitance. A second observation derives from the fact that since the unity gain frequency in Eq. (7-66) is neither directly proportional to the transistor forward transconductance nor inversely dependent on effective gate-source capacitance, the circuit unity gain frequency, ω_u, is divorced of any direct relationship to the transistor unity gain frequency, ω_T. This revelation reinforces earlier arguments to the extent that transistor ω_T is virtually a useless figure of merit from the perspective of analog circuits. Third, Eq. (7-66) points out the inevitable quandary that is pervasive of broadbanding exercises that are necessarily constrained by stability considerations. In particular, the dominant pole response that is a prerequisite to closed loop amplifier stability mandates a sufficiently large value of the feedback capacitance, C_f, which in turn potentially limits the achievable unity gain frequency of the circuit. The only remedy to this shortfall is a sufficiently small source resistance, which justifies the common use of small, generally 50-ohm, source terminations. Finally, it is critical to understand that k_p in Eq. (7-67) measures the propriety of the dominant pole approximation. This is to say that the greater the factor by which k_p exceeds unity, the more pole-dominant is the realized frequency response. Large k_p is seen to require commensurately large g_m; that is, the utilized transistor must have appropriate gain potential, as is measured by its effective forward transconductance. It is also true that large k_p accrues from small C_i and/or small C_o, but these capacitances are tacitly assumed to be acceptably small if the net feedback capacitance, C_f, establishes the dominant open circuit time constant of the considered network.

Example 7.3. The common source amplifier shown in Fig. 7.10(a) is to be driven from a 50 Ω signal source and is to be designed for a low frequency voltage gain magnitude of 15 (23.5 dB) over a passband extending through a 3-dB bandwidth of 250 MHz. Assume that the power supply voltages, V_{dd} and V_{ss}, can be selected to ensure acceptable quiescent operation of the transistor in its saturation domain. When properly biased and with reference to the small signal model appearing in Fig. 7.10(b), the transistor delivers a gate-source capacitance (C_i) of 30.3 fF, a gate-drain capacitance (C_f) of 10.1 fF, a bulk-drain capacitance (C_o) of 7.6 fF, a forward transconductance (g_m) of 15.2 mmhos, and finally, a drain-source channel resistance (R_d) of 6.4 KΩ. The parallel combination, R_i, of the two biasing resistances, R_1 and R_2, is 86 KΩ. For the purpose of this example, the input port coupling capacitance, C_s, need not be calculated, but assume that C_s functions as a signal short circuit at the lowest frequency of concern. Calculate the required value of the drain load resistance, R_l, and the amount of capacitance that must be appended between gate and drain terminals to ensure a dominant pole response. Simulate the small signal model to confirm design adequacy. Additionally, calculate the phase margin of the design realization for the worst case stability situation of unity closed loop gain magnitude. To the latter end, the disclosures of Chapter 4 are likely to prove helpful.

Solution 7.3.

(1) Assuming a dominant pole response, the unity gain frequency, ω_u, of the desired amplifier is approximately the gain-bandwidth product, which is

$$\omega_u \approx |A_v(0)|B = (15)[2\pi(250\text{ MHz})] = 2\pi(3.75\text{ GHz}). \quad \text{(E3-1)}$$

It is interesting to point out that the unity gain frequency of the transistor itself is

$$\omega_T = \frac{g_m}{C_i + C_f} = 2\pi(59.88\text{ GHz}).$$

(2) In Eq. (7-66), R_{ss} is the parallel combination of source resistance R_s and shunt biasing resistance R_i, ergo $R_{ss} = 49.97\Omega$. Using Eq. (7-66), the requisite updated value, say C_{fu}, of the net feedback capacitance is found to be

$$C_{fu} = \left(\frac{g_m R_{ss}}{1 + g_m R_{ss}}\right)\left(\frac{1}{\omega_u R_s}\right) = 366.4\text{ fF}. \quad \text{(E3-2)}$$

This computation means that a compensation capacitance, C_{fc}, must be appended between the gate and drain terminals of the transistor embedded in the amplifier. The value of this compensating capacitance is

$$C_{fc} = C_{fu} - C_f = 356.3 \, \text{fF}. \tag{E3-3}$$

Owing to the numerous approximations invoked in the course of arriving to Eq. (E3-3), it is wise to back off 5% or so from this calculated compensation capacitance. Accordingly, the rounded compensation capacitance actually used in this design is

$$C_{fc} = 338 \, \text{fF}, \tag{E3-4}$$

which is tantamount to a net updated feedback capacitance of

$$C_{fu} = 348.1 \, \text{fF}. \tag{E3-5}$$

(3) Recalling Eq. (7-53), the net output resistance, R_{ll}, in shunt with the amplifier output port must be

$$R_{ll} = \frac{|A_v(0)|}{g_m \left(\frac{R_i}{R_i + R_s} \right)} = 987.4 \, \Omega. \tag{E3-6}$$

Since R_{ll} is the shunt interconnection of drain load resistance R_l and transistor channel resistance R_d,

$$R_l = \frac{1}{\frac{1}{R_{ll}} - \frac{1}{R_d}} = 1,168 \, \Omega. \tag{E3-7}$$

(4) It is worthwhile confirming that the desired 3-dB bandwidth of 250 MHz subscribes to Eq. (7-58). To this end, the time constant, τ_i, in Eq. (7-59) is

$$\tau_i = R_{ss} C_i = 1.514 \, \text{pSEC}, \tag{E3-8}$$

while from Eq. (7-60),

$$\tau_o = R_{ll} C_o = 7.504 \, \text{pSEC}. \tag{E3-9}$$

Using Eq. (7-61), the time constant attributed to the updated net feedback capacitance, C_{fu}, is found to be

$$\tau_f = [R_{ll} + (1 + g_m R_{ll}) R_{ss}] C_{fu} = 622.2 \, \text{pSEC}, \tag{E3-10}$$

which clearly overshadows the two previously computed time constants. The 3-dB bandwidth now follows as

$$B = \frac{1}{\tau_i + \tau_o + \tau_f} = 2\pi (252.1 \, \text{MHz}), \tag{E3-11}$$

which is close enough for government work to the desired bandwidth of 250 MHz.

(5) From Eq. (7-54), the frequency, z_f, of the zero established by the updated value of net feedback capacitance is

$$z_f = \frac{g_m}{C_{fu}} = 2\pi(6.95 \text{ GHz}), \qquad \text{(E3-12)}$$

which is a factor,

$$k_o = \frac{z_f}{\omega_u} = 1.85, \qquad \text{(E3-13)}$$

larger than the amplifier unity gain frequency. Recall that closed loop stability mandates that the frequency of any established right half plane zero be no smaller than the network unity gain frequency.

(6) Using Eq. (7-67), with capacitance C_f supplanted by capacitance C_{fu}, the ratio of the frequency of the nondominant circuit pole, p_2, to the amplifier unity gain frequency is seen to be

$$k_p \triangleq \frac{p_2}{\omega_u} \approx \left(\frac{1 + g_m R_{ss}}{g_m R_{ss}}\right)^2 \left[\frac{g_m R_s \left(\frac{C_{fu}}{C_o}\right)}{\left(1 + \frac{C_i}{C_o} + \frac{C_i}{C_{fu}}\right)}\right] = 37.05, \qquad \text{(E3-14)}$$

which surely reflects a dominant pole response.

(7) A review of Chapter 4 and in particular, Section 4.3.3, reveals that with parameter k defined as

$$k \triangleq \frac{k_p k_o - 1}{k_p + k_o} = 1.736, \qquad \text{(E3-15)}$$

the phase margin is, from Eq. (4-38)

$$\varphi_{pm} = \tan^{-1}\left(\frac{kT(0) + 1}{T(0) - k}\right) = 63.9°, \qquad \text{(E3-16)}$$

where $T(0) \equiv A_v(0)$ for the special case of unity magnitude closed loop gain; that is, a feedback factor in the present case of negative one. The calculated phase margin is within a generally acceptable stability range of operation.

(8) Figure 7.13(a) depicts the schematic diagram of the completed design, while Fig. 7.13(b) offers its small signal equivalent circuit. An HSPICE small signal simulation of the latter structure divulges the frequency and phase responses depicted in Fig. 7.14. A low frequency gain of 23.51 dB, which reflects the design objective, is postured by the frequency response graph. An assiduous consideration of the data underpinning the plots in Fig. 7.14 reveals a 3-dB bandwidth of 253 MHz, which is only 1.2% larger than the 250 MHz target. Evidently, the arbitrary 5% reduction in computed compensation capacitance is fractionally overzealous. In the immediate neighborhood of the circuit bandwidth, the rate of frequency response degradation is almost −20 dB/decade and

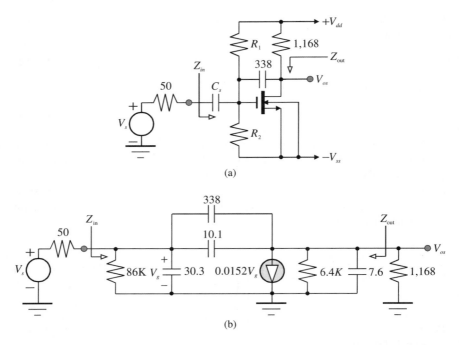

Figure 7.13. (a) Common source amplifier designed in Example 7.3. (b) Small signal equivalent circuit of the amplifier in (a). In both diagrams, resistances are in units of ohms, and capacitances are in femtofarads.

at this bandwidth, the phase angle is $-227.2°$. This phase angle differs from its low frequency value of $-180°$ by $-47.2°$, which is only $2.7°$ different than the phase differential advanced by an ideal single pole frequency response.

(9) In an attempt to substantiate further the well-behaved nature of the amplifier designed above, Fig. 7.15 submits an HSPICE simulation of the transient pulse response of the network. In this simulation, V_s in Fig. 7.13(b) is taken as a periodic pulse train of amplitude $-1/15$ volts, which produces a steady state output signal voltage of 1 volt. The applied pulse has a width of 10 nSEC, a period of 20 nSEC, and rise and fall times of 2 pSEC. The good news here is that the output response displays no drama insofar as overshoot and undershoot is concerned, a state of affairs that is indicative of a dominant pole network response. To be sure, the ramification of input signal plunges to zero is almost indiscernible response overshoots. This overshoot is precipitated by the right half plane zero. It is, in fact, inconsequential because the frequency of the right half plane zero is significantly larger than the amplifier 3-dB bandwidth.

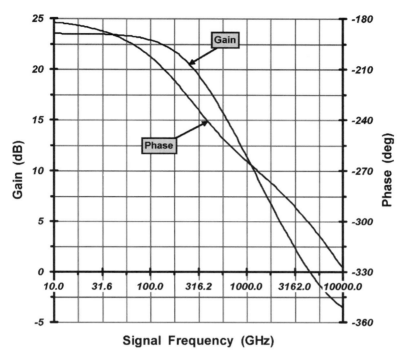

Figure 7.14. The simulated frequency and phase responses of the amplifier considered in Example 7.3.

Comments. The HSPICE simulations confirm the propriety of all calculations leading to gain and bandwidth predictions, and they verify a nominally dominant pole response through and appreciably beyond the circuit bandwidth. However, the simulations show a unity gain frequency of 4.51 GHz, in contrast to the predicted unity gain metric of 3.75 GHz. The pessimistic prediction of unity gain frequency arises from tacitly neglecting the effects on magnitude response of the right half plane zero. The impact of this zero is significant since its frequency is far closer to the unity gain frequency of the circuit than it is to the frequency of the nondominant pole. Interestingly, however, the phase margin prediction in Eq. (E3-16) is reasonably accurate. Since the amplifier at hand is phase inverting, a stable unity closed loop gain magnitude requires a feedback factor of −1, which means that the applicable loop gain is the negative of the amplifier gain function. In turn, this observation implies that the loop gain phase response is the phase response shown in Fig. 7.14, provided 180° is added to each phase angle data point. Thus, for example, the zero frequency value of the loop gain phase is $-2.4°$, as opposed to the $-182.4°$ depicted by the phase plot. At the unity gain frequency of 4.51 GHz, the loop gain phase is $-121.5°$, not the depicted $-301.5°$. The resultant simulated

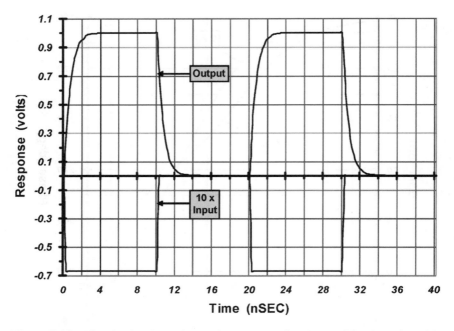

Figure 7.15. The simulated transient pulse response for the amplifier investigated in Example 7.3. In the interest of clarity, the input voltage is plotted as a 10-times signal level.

value of phase margin follows as $-121.5° - (-180°)$, or $+58.5°$, which differs from the computed phase margin by only about 8.5%.

7.3.1.3. *Output Port Time Constant Dominance*

At least one popular application of the common source amplifier and one transistor process technology compels a voltage transfer function investigation for the special case in which the dominant time constant is established at the network output port. The circuit application entails using a common source amplifier as a transconductor that is loaded in substantial output port capacitance. The resultant configuration is known as an *operational transconductance amplifier-capacitor (OTA-C)* integrator, which finds widespread use in the realization of high performance, active biquadratic filters.[14] To be sure, a large capacitance placed in parallel with the common source output port, which is characterized by relatively large resistance, destroys any hopes of broadbanding the amplifier. But an OTA-C is not designed for wideband voltage amplification. Instead the OTA-C is designed to function as an integrator, which is to say that its idealized I/O transfer function is of the form, $A_v(s) = K/s$; that is, the dominant pole lies at zero frequency. The device issue involves self-aligning gate technology, wherein the ends of the gate oxide layer align with the channel length between source and drain regions to ensure minimal, if not zero, value for the dimension, L_d, in the device cross-section of Fig. 7.1. Recalling Eq. (7-35), minimal L_d reduces the gate-drain feedback capacitance, C_{gd}, for which the value extracted from scattering parameter measurements is C_f in Fig. 7.10(b). The substantive reduction of C_f all but precludes its time constant dominance in the network characteristic polynomial.

With C_o serving as the dominant capacitance of a common source unit, Eq. (7-55) implies

$$\frac{1}{p_1} + \frac{1}{p_2} \approx \frac{1}{p_1} = R_{ll} C_o, \qquad (7\text{-}68)$$

where C_o is sufficiently large to satisfy the approximation, $R_{ll} C_o \gg R_{ss}[C_i + (1 + g_m R_{ll})C_f] + R_{ll} C_f$. This approximation is viable if no external capacitance is appended to the gate-drain terminals, a self-aligned gate reduces capacitance C_f to almost zero, and/or the Thévenin source resistance is small. It follows from Eq. (7-56) that the resultant frequency of the nondominant amplifier pole is

$$p_2 \approx \frac{1}{R_{ss} C_i \left(1 + \dfrac{C_f}{C_i} + \dfrac{C_f}{C_o}\right)}, \qquad (7\text{-}69)$$

which collapses to the inverse time constant at the amplifier input port if $C_f \ll C_i C_o/(C_i + C_o)$. Using Eq. (7-66), the unity gain frequency becomes

$$\omega_u \approx |A_v(0)| p_1 \approx \frac{g_m \left(\dfrac{R_i}{R_i + R_s}\right)}{C_o} \approx \frac{g_m}{C_o}, \qquad (7\text{-}70)$$

since R_i is invariably much larger than the signal source resistance, R_s. The preceding two expressions give rise to

$$k_p \triangleq \frac{p_2}{\omega_u} \approx \left(\frac{R_i + R_s}{R_i}\right)^2 \left[\frac{\dfrac{C_o}{C_i}}{g_m R_s \left(1 + \dfrac{C_f}{C_i} + \dfrac{C_f}{C_o}\right)}\right] \approx \frac{C_o}{g_m R_s C_i}, \qquad (7\text{-}71)$$

which must exceed unity if a dominant pole response is to be assured. A comparison of Eq. (7-71) with Eq. (7-67) reveals the interesting fact that while large $g_m R_s$ is required for common source pole dominance when capacitance C_f establishes the dominant time constant, small $g_m R_s$ is mandated to ensure pole dominance in the face of a dominant time constant forged at the amplifier output port.

7.3.2. Input and Output Impedances

Although the determination of the high frequency input and output impedances, Z_{in} and Z_{out}, respectively, for the common source amplifier modeled in Fig. 7.10(b) can be algebraically taxing, the methodology for such determination is straightforward in light of the disclosures in the preceding chapter. Specifically, the driving point input impedance, $Z_{in}(s)$, as a function of complex frequency, is expressible as

$$Z_{in}(s) = Z_{ino}(s)\left[\frac{1+T(s,0,R_l)}{1+T(s,\infty,R_l)}\right], \tag{7-72}$$

where

$$Z_{ino}(s) = \frac{R_i}{1+sR_i(C_i+C_f)} \tag{7-73}$$

is the zero feedback ($sC_f V_{os} = 0$) value of the desired input impedance. From Eqs. (7-49) and (7-48), $T(s, 0, R_l) = 0$, while

$$T(s,\infty,R_l) = \left[\frac{sC_f R_i}{1+sR_i(C_i+C_f)}\right]\left[\frac{(g_m - sC_f)R_{ll}}{1+sR_{ll}(C_o+C_f)}\right]. \tag{7-74}$$

It follows that $Z_{in}(s)$ in Eq. (7-72) can be written in the form,

$$Z_{in}(s) = R_i\left[\frac{1+as}{1+(a+b)s+cs^2}\right], \tag{7-75}$$

where

$$\left.\begin{array}{l}a = R_{ll}(C_o + C_f)\\ b = R_i[C_i + (1+g_m R_{ll})C_f]\\ c = (R_i C_i)(R_{ll} C_o)\left(1 + \dfrac{C_f}{C_o} + \dfrac{C_f}{C_i}\right)\end{array}\right\} \tag{7-76}$$

Equation (7-75) can be written as,

$$Z_{in}(s) = \frac{1}{\dfrac{1}{R_i} + \left(\dfrac{b}{R_i}\right)s + \dfrac{(c-ab)s^2}{R_i(1+as)}} \approx \frac{1}{\dfrac{1}{R_i} + \left(\dfrac{b}{R_i}\right)s}, \tag{7-77}$$

where the indicated approximation limits the frequency range of applicability of the input impedance expression to signal frequencies through and slightly beyond the frequency, $(1/b)$, established by the time constant associated with the amplifier input port. The subject approximate result confirms a capacitive input impedance over the passband of immediate interest. It also infers the common source amplifier input port model depicted in Fig. 7.16(a), which clearly shows the dramatic impact of Miller multiplication of the feedback capacitance, C_f.

Repeating the foregoing analytical methodology, the driving point output impedance, $Z_{\text{out}}(s)$, for the common source amplifier derives from

$$Z_{\text{out}}(s) = Z_{outo}(s) \left[\frac{1 + T(s, R_s, 0)}{1 + T(s, R_s, \infty)} \right], \qquad (7\text{-}78)$$

where an inspection of the model in Fig. 7.10(c) reveals

$$Z_{outo}(s) = \frac{R_d}{1 + s R_d (C_o + C_f)}. \qquad (7\text{-}79)$$

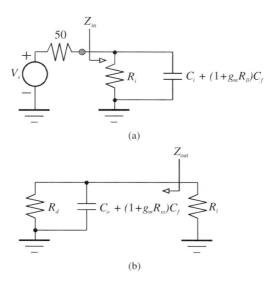

Figure 7.16. (a) The approximate input port equivalent circuit for the computation of the driving point input impedance of the common source amplifier in Fig. 7.10(a). (b) The approximate output port equivalent circuit for the computation of the driving point output impedance of the amplifier in Fig. 7.10(a).

From Eqs. (7-49) and (7-48), $T(s, R_s, 0) = 0$, and

$$T(s, R_s, \infty) = \left[\frac{sC_f R_{ss}}{1 + sR_{ss}(C_i + C_f)}\right]\left[\frac{(g_m - sC_f)R_d}{1 + sR_d(C_o + C_f)}\right]. \quad (7\text{-}80)$$

Accordingly, Eq. (7.78) delivers the capacitive impedance expression,

$$Z_{\text{out}}(s) = R_d\left[\frac{1 + ds}{1 + (d+e)s + fs^2}\right], \quad (7\text{-}81)$$

where

$$\left.\begin{array}{l} d = R_{ss}(C_i + C_f) \\ e = R_d[C_o + (1 + g_m R_{ss})C_f] \\ f = (R_{ss}C_i)(R_d C_o)\left(1 + \dfrac{C_f}{C_o} + \dfrac{C_f}{C_i}\right) \end{array}\right\}. \quad (7\text{-}82)$$

Following the algebraic exercise leading to Eq. (7-77), it is a simple matter to show that Eq. (7-81) implies the approximate common source amplifier output port impedance model offered in Fig. 7.16(b).

7.3.3. Variants of the Common Source Topology

There are two commonly exploited variations to the common source amplifier theme projected by the schematic diagram in Fig. 7.10(a). The first of these variants supplants the passive load resistance, R_l, by an N-channel transistor whose gate and drain terminals are returned to the positive power bus. The second replaces R_l by a P-channel device biased for operation in its saturation regime.

7.3.3.1. NMOS Load

Figure 7.17(a) is the circuit schematic diagram of a common source amplifier in which the drain circuit load imposed on the driver transistor, $M1$, is the NMOS transistor, $M2$. Superimposed on this diagram in the interest of ultimate model clarity are the pertinent device capacitances, which are delineated as dashed circuit branches. Absent from this capacitance map are the bulk-source capacitance of $M1$ and the bulk-drain and gate-drain capacitances of $M2$ since the source and bulk terminals of $M1$, as well as the drain, gate, and bulk terminals of $M2$, are incident with signal ground (assuming ideal power busses). An interesting aspect of the load transistor is that both its gate and drain terminals are returned to the positive power supply line. This connection forces an identity between the gate-source

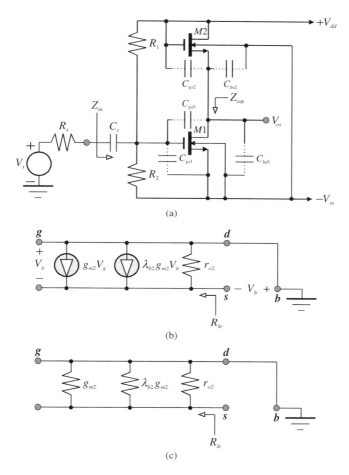

Figure 7.17. (a) Common source amplifier with NMOS load. (b) The low frequency equivalent circuit of transistor $M2$ In the schematic diagram of (a). (c) The effective resistance, R_{le}, presented to the $M1$ driver by the NMOS load transistor, $M2$.

and drain-source voltages, V_{gs2} and V_{ds2}, respectively, of transistor $M2$. Recalling that V_{ds} must exceed the gate-source voltage, V_{gs}, less a threshold potential, to effect saturated transistor operation, transistor $M2$ operates in its saturated domain, regardless of the drain current conducted by either transistor. Assuming that the driver transistor is biased in saturation, it follows that the model shown in Fig. 7.6 is the applicable equivalent circuit for both the driver and the load transistors.

Because the common connection of the gate and drain terminals of transistor $M2$ forces this device to operate effectively as a two terminal element in the subject circuit diagram, its replacement by the low frequency version of the model in Fig. 7.6 remands its small signal circuit level implication to that associated with a two terminal effective resistance, say R_{le}. The low frequency, small signal load model of record appears in Fig. 7.17(b), where it can be observed that the controlling signal voltages, V_g and V_b, are identical. Moreover, the fact that voltage V_g appears directly across, and in associated reference polarity with, the voltage controlled current source, $g_{m2}V_g$, means that this controlled current branch is equivalent to a two terminal linear conductance of value g_{m2}. Similarly, the presence of the controlling voltage, V_b, directly across the controlled current source, $\lambda_{b2}g_{m2}V_b$, as shown in Fig. 7.17(b), implies that this second controlled current branch behaves electrically as a two terminal linear conductance of value $\lambda_{b2}g_{m2}$. The upshot of this due diligence is that the model of Fig. 7.17(b) collapses to the simple structure shown in Fig. 7.17(c). A mere inspection of the last structure projects an effective load resistance, R_{le}, presented by the load transistor to the driver cell of

$$R_{le} = \frac{r_{o2}}{1 + (1 + \lambda_{b2})g_{m2}r_{o2}} \approx \frac{1}{g_{m2}}, \qquad (7\text{-}83)$$

where the indicated approximation exploits the fact that the bulk threshold modulation factor, λ_{b2}, is invariably much smaller than one and $g_{m2}r_{o2} \gg 1$.

The upshot of the analytical matter at hand is that the topological form of the small signal equivalent circuit of the amplifier in Fig. 7.17(a) is identical to that of the basic common source model portrayed in Fig. 7.10(c). Topological similarities notwithstanding, a few of the branch element parameters in Fig. 7.10(c) must be altered to ensure model tracking with the amplifier undergoing study. These changes are summarized by the following stipulations.

(a) The branch capacitance, $(C_i + C_f)$, must be changed to $(C_{gs1} + C_{gd1})$, with the proviso that any parasitic capacitance established by a signal input pad connection at the input port must be added to the sum of the gate-source and gate-drain capacitances of transistor $M1$.
(b) The feedback controlled source, $sC_f V_{os}$, becomes $sC_{gd1} V_{os}$. Additionally, the controlled source, $(g_m - sC_f)V_g$, changes to $(g_{m1} - sC_{gd1})V_g$.
(c) The effective channel resistance, R_d, in the structure of Fig. 7.10(b) is now the channel resistance, r_{o1}, of transistor $M1$ in Fig. 7.17(a).
(d) Load resistance R_l becomes the effective load resistance, R_{le}, defined by Eq. (7-83).

(e) The output port capacitance sum, $(C_o + C_f)$, becomes $(C_{bd1} + C_{bs2} + C_{gs2} + C_{gd1})$, with the understanding that any parasitic capacitance established by a signal output pad must be added to this sum of four capacitances.

In addition to rendering the model in Fig. 7.10(c) applicable to the small signal analysis of the amplifier at hand, the foregoing parametric adjustments enable a straightforward utilization of all previously derived relationships for gain, I/O impedance, and other common source performance metrics. Such utilization spawns at least two interesting observations. First, the zero frequency voltage gain becomes, from Eq. (7-53),

$$A_v(0) = -g_{m1} R_{ll} \left(\frac{R_i}{R_i + R_s} \right) = -g_{m1} R_{le} \left(\frac{R_i}{R_i + R_s} \right) \approx -\frac{g_{m1}}{g_{m2}}, \quad (7\text{-}84)$$

where the approximation in Eq. (7-83) is invoked, and the source resistance, R_s, is presumed much smaller than the parallel combination, R_i, of the input port biasing resistors. But if transistors $M1$ and $M2$ in Fig. 7.17(a) are identical, save possibly for differing gate aspect ratios, the conduction of the same quiescent drain current through both driver and load transistors allows Eq. (7-84) to be expressed as

$$A_v(0) \approx -\sqrt{\frac{W_1/L_1}{W_2/L_2}}, \quad (7\text{-}85)$$

where Eq. (7-29) is applied to each of the transconductance parameters in Eq. (7-84), and the channel length modulation voltage, V_λ, is presumed much larger than the respective drain-source Q-point voltages of the transistors. In short, the zero frequency voltage gain is determined almost exclusively by transistor geometries and in particular, by an accurately predictable ratio of gate aspect ratios.

The second observation pertains to the establishment of the amplifier dominant pole. With an effective load resistance of $R_{le} \approx 1/g_{m2}$, it is unlikely that the frequency response is limited by the time constant associated with the Miller-multiplied feedback capacitance. In fact, it is probable that for a small source resistance, no significant external load capacitance, and a small zero frequency voltage gain, none of the three time constants given respectively by Eqs. (7-59) through (7-61) is dominant. To mitigate phase and/or gain margin shortfalls, it may therefore be necessary to append additional circuit capacitance across an amplifier terminal pair that features a sufficiently large open circuit resistance.

7.3.3.2. PMOS Load

The second of the previously inferred two basic variants to the common source amplifier is the complementary MOS, or CMOS, configuration shown in Fig. 7.18(a). The load presented to the common source driver transistor, $M1$, in this amplifier is the active impedance established by the PMOS transistor, $M2$. The constant Thévenin equivalent voltage sources, V_{g1} and V_{g2}, are selected to ensure that $M1$ and $M2$ are biased in saturation. To the extent that channel length modulation phenomena in both transistors can be tacitly ignored, this biasing constrains the operation of transistor $M1$ as a nominally constant current sink, while $M2$ emulates a constant current source. Since the series interconnection of virtually constant current branch elements renders problematic the accurate determination of the static output voltage, V_{oQ}, established at the junction of the two transistor

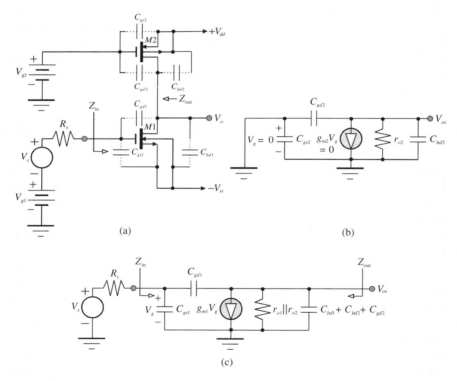

Figure 7.18. (a) Common source amplifier with PMOS load. (b) The small signal equivalent circuit of transistor $M2$ in the schematic diagram of (a). (c) The small signal model of the entire amplifier in (a).

drain terminals, additional biasing compensation, which is not shown in the figure at hand, is required to stabilize the quiescent node voltages and branch currents of the amplifier.

Because the two transistors in the CMOS amplifier operate in saturation with their bulk terminals grounded for small signal conditions, the applicable small signal model for both transistors is the network depicted in Fig. 7.10(b). The model for transistor $M2$, however, is rendered elementary by the grounding of its gate and source terminals for signal conditions. The resultant $M2$ equivalent circuit is the structure given in Fig. 7.18(b), which clearly underscores the facts that the controlled source, $g_{m2}V_g$, is null, and the gate-source capacitance, C_{gs2}, is inconsequential. Indeed, the subject model confirms that for small signal operation, $M2$ behaves as an RC impedance comprised of the shunt interconnection of the transistor channel resistance, r_{o2}, bulk-drain capacitance, C_{bd2}, and gate-drain capacitance, C_{gd2}. It follows that the small signal equivalent circuit of the entire CMOS amplifier is the network provided in Fig. 7.18(c). The latter model is topologically identical to that of Fig. 7.10(b), which means that the feedback equivalent circuit of Fig. 7.10(c), as well as all subsequently derived gain, impedance, and critical frequency expressions, can be invoked for the amplifier under present consideration, subject to the parametric interpretations itemized herewith.

(a) The input port capacitance, $(C_i + C_f)$, must be changed to $(C_{gs1} + C_{gd1})$, with the proviso that any parasitic capacitance established by a signal input pad connection at the input port must be added to the sum of the gate-source and gate-drain capacitances of transistor $M1$.

(b) The feedback controlled source, $sC_f V_{os}$, becomes $sC_{gd1} V_{os}$. Additionally, the controlled source, $(g_m - sC_f)V_g$, changes to $(g_{m1} - sC_{gd1})V_g$.

(c) The output port resistance, R_d, in the structure of Fig. 7.10(b) is now the parallel combination, $r_{o1}||r_{o2}$, of respective transistor channel resistances.

(d) The load resistance, R_l, is now infinitely large because no external load termination is explicitly considered in the amplifier of Fig. 7.18(a). Accordingly, the effective load resistance, R_{ll}, in Eq. (7-48) is simply the aforementioned shunt combination of channel resistances.

(e) The output port capacitance sum, $(C_o + C_f)$, becomes $(C_{bd1} + C_{bd2} + C_{gd2} + C_{gd1})$, with the understanding that any parasitic capacitance established by a signal output pad must be added to this sum of four capacitances.

550　Feedback Networks: Theory and Circuit Applications

Example 7.4. The N-channel transistor in the CMOS amplifier of Fig. 7.18(a) has a forward transconductance, g_{m1}, of 18 mmhos, a drain-source channel resistance, r_{o1}, of 8.8 KΩ, and gate-source, gate-drain, and bulk-drain capacitances of $C_{gs1} = 35$ fF, $C_{gd1} = 12$ fF, and $C_{bd1} = 20$ fF. The resistance, R_s, of the signal source that drives the cell is 50 Ω, and the output port is terminated in a capacitance (not shown in the diagram) of $C_l = 60$ fF. The P-channel device has $r_{o2} = 18$ KΩ, $C_{gd2} = 19$ fF, and $C_{bd2} = 38$ fF. Assume that the power supply voltages, V_{dd}, V_{ss}, V_{g1}, and V_{g2}, are ideal and that they collectively ensure acceptable quiescent operation of both transistors in their saturated regimes. Deduce the transfer function, V_{os}/V_s, of the small signal voltage gain and estimate the resultant 3-dB bandwidth of the amplifier.

Solution 7.4.

(1) With appropriate parametric interpretation, the model in Figs. 7.10(b) and 7.10(c) apply, which means that the pertinent voltage transfer function is the expression in Eq. (7-52). Since no input port biasing resistances are used in the schematic diagram of Fig. 7.18(a), Eq. (7-48) confirms that the effective source resistance, R_{ss}, is simply R_s. Similarly, the absence of an external load resistance, R_l, constrains the effective drain load resistance to resistance R_d. In accordance with the discussion preceding this problem, R_d, in turn, is the parallel combination of the two transistor channel resistances. In short,

$$\left. \begin{array}{l} R_{ss} = R_s = 50\,\Omega \\ R_{ll} = R_d = r_{o1}\|r_{o2} = 5.91\,\text{K}\Omega \end{array} \right\} . \quad \text{(E4-1)}$$

The preceding discussion also confirms model capacitances of values

$$\left. \begin{array}{l} C_i = C_{gs1} = 35\,\text{fF} \\ C_f = C_{gd1} = 12\,\text{fF} \\ C_o = C_{bd1} + C_{bd2} + C_{gd2} + C_l = 137\,\text{fF} \end{array} \right\} . \quad \text{(E4-2)}$$

(2) Using Eq. (7-53), the zero frequency gain of the amplifier is

$$A_v(0) = -g_{m1} R_{ll} \left(\frac{R_i}{R_i + R_s} \right) = -g_{m1}(r_{o1}\|r_{o2}) = -106.4 \text{ V/V}. \quad \text{(E4-3)}$$

(3) By Eq. (7-54), the frequency of the right half plane zero of the amplifier is

$$z_f = g_{m1}/C_f = g_{m1}/C_{gd1} = 1{,}500 \text{ GRPS}, \quad \text{(E4-4)}$$

or 238.7 GHz, which appears large enough to void engineering significance.

(4) The calculation of the open circuit time constants is the next order of business. From Eq. (7-59), the open circuit input port time constant is

$$\tau_i = R_{ss} C_i = R_s C_{gs1} = 1.75 \text{ pSEC}, \quad \text{(E4-5)}$$

while the open circuit output port time constant derives from Eq. (7-60) as

$$\tau_o = R_{ll}C_o = (r_{o1} \| r_{o2})(C_{bd1} + C_{bd2} + C_{gd2} + C_l)$$
$$= 809.7 \text{ pSEC}. \tag{E4-6}$$

The time constant precipitated by the feedback capacitance, which is the gate-drain capacitance of transistor $M1$, is

$$\tau_f = [R_{ll} + (1 + g_{m1}R_{ll})R_{ss}]C_f$$
$$= \{(r_{o1}\|r_{o2}) + [1 + g_{m1}(r_{o1}\|r_{o2})]R_s\}C_{gd1} = 135.4 \text{ pSEC}. \tag{E4-7}$$

Despite the relatively large Miller multiplication factor of $[1 + g_{m1}(r_{o1}\|r_{o2})] = 107.4$, the output port time constant, τ_o, appears to be dominant in that it is larger than the feedback capacitance time constant, τ_f, by a factor of almost six. The net open circuit time constant attributed to the poles of the amplifier, which is the sum of the inverse frequencies of the two amplifier poles, is

$$\tau_i + \tau_o + \tau_f = \frac{1}{p_1} + \frac{1}{p_2} = 946.9 \text{ pSEC} = 1/(1.056 \text{ GRPS}). \tag{E4-8}$$

(5) From Eq. (7-56), the inverse product of the two amplifier pole frequencies is

$$\frac{1}{p_1 p_2} = (R_{ss}C_i)(R_{ll}C_o)\left(1 + \frac{C_f}{C_i} + \frac{C_f}{C_o}\right) = 946.9 \text{ pSEC}.$$
$$= 2.027 \left(10^{-21}\right) \text{SEC}^{-2} = 1/\left[493.3 \left(10^{18}\right) RPS^2\right]. \tag{E4-9}$$

(6) The preceding numerical results combine with Eqs. (7-52) and (7-53) to deliver an input to output voltage transfer function, $A_v(s)$, of

$$A_v(s) = -\frac{106.4\left(1 - \dfrac{s}{1500}\right)}{1 + \dfrac{s}{1.056} + \left(\dfrac{s}{22.21}\right)^2}, \tag{E4-10}$$

where the complex frequency variable, s, is understood to be in units of giga-radians. Because pole dominance evidently prevails and because the frequency of the zero is extremely large in comparison to the open circuit time constant associated with the amplifier poles, Eq. (7-58) applies to the problem of bandwidth enumeration. In particular,

$$B \approx \frac{1}{p_1} + \frac{1}{p_2} = 1.056 \text{ GRPS} = 2\pi(168.1 \text{ MHz}). \tag{E4-11}$$

is the approximate 3-dB bandwidth of the subject amplifier.

Comments. This example reaffirms the straightforward undertone to deducing the second order transfer function relationship of a common source amplifier. It is left as an exercise for the reader to confirm, through circuit simulation, the dominant pole nature of the amplifier frequency response. To this end, HSPICE verifies that the computed bandwidth is within 2% of its simulated value. Moreover, HSPICE shows that the simulated unity gain frequency is within 2% of the computed gain-bandwidth product and that the difference between the phase angles at zero signal frequency and the simulated 3-dB bandwidth is very nearly 45°. Both of the latter two observations support the theoretic contention of a dominant pole frequency response.

7.4.0. Common Drain Amplifier

Figure 7.19(a) portrays the basic schematic diagram of a *common drain amplifier*, which is often referred to as a *source follower*. While the common source amplifier is the fundamental gain stage in MOS and CMOS technologies, the source follower functions as a voltage buffer. Specifically, it emulates an ideal voltage buffer at low signal frequencies by providing very large input impedance, small output impedance, and a positive voltage gain that is slightly smaller than one. As is the case with the previously studied common source configurations, the amplifier is driven at the transistor gate

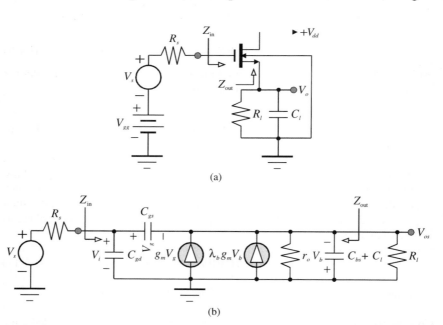

Figure 7.19. (a) Basic schematic diagram of a common drain amplifier. The biasing voltages, V_{dd} and V_{gg}, establish transistor quiescent operation in saturation. (b) The small signal model of the amplifier depicted in (a).

terminal by a signal source represented as the series interconnection of a Thévenin voltage, V_s, and a Thévenin resistance, R_s. The output port is the transistor source terminal, which is terminated to ground by a shunt interconnection of load resistance R_l and load capacitance C_l. The biasing voltages, V_{dd} and V_{gg}, are selected to ensure that the transistor in the follower configuration operates in its saturated regime. The resultant small signal equivalent circuit of the source follower, which exploits the basic model of Fig. 7.6, is depicted in Fig. 7.19(b). Note that the bulk-source capacitance has been absorbed into the load termination and as a result, the driving point output impedance, Z_{out}, is the effective impedance seen by the shunt interconnection of load resistance R_l and capacitances C_{bs} and C_l.

Prior to formulating the equations that underpin the transfer characteristic of the source follower, several topological simplifications can be effected. For example, a voltage, V_b, appears directly across, and in associated reference polarity with, the current conducted by the controlled source, $\lambda_b g_m V_b$, as highlighted in Fig. 7.20(a). Accordingly, this controlled current branch is electrically equivalent to a conductance of value $\lambda_b g_m$. Moreover, the voltage, V_g, upon which the controlled current source, $g_m V_g$, is dependent, is related to the input port voltage, V_i, and the output small signal voltage response, V_{os}, by

$$V_g = V_i - V_{os}. \tag{7-86}$$

It follows that the voltage controlled current source, $g_m V_g$, is equivalent to the anti-phase shunt interconnection of two controlled sources, $g_m V_i$ and $g_m V_{os}$, as shown in Fig. 7.20(b). But since the controlling voltage, V_{os}, appears across, and in associated reference polarity with, the current source, $g_m V_{os}$, said current source is equivalent to a simple conductance of value g_m, as is also depicted in Fig. 7.20(b). The upshot of the foregoing matters is that the equivalent circuit of Fig. 7.19(b) collapses to the electrically equivalent structure offered in Fig. 7.20(c). Aside from an inherent model simplification, the structure at hand is topologically identical to that of Fig. 7.10(b). This observation renders most of the transfer function and I/O impedance results forged in conjunction with the common source amplifier applicable to the present amplifier undergoing study, provided the following parametric modifications are made.

(a) The shunt input capacitance, C_i, in Fig. 7.10(b) is now the gate-drain capacitance, C_{gd}, of the transistor in the source follower. Additionally, the shunt input port resistance, R_i, can now be taken as an infinitely

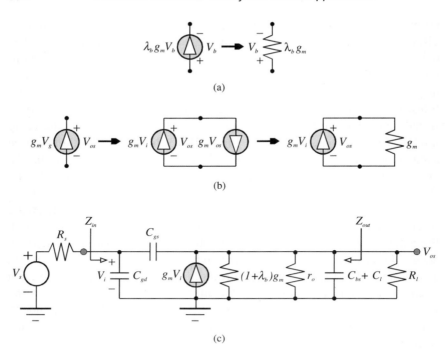

Figure 7.20. (a) Transformation of the bulk transconductance controlled source to a simple branch conductance. (b) Transformation of the forward transconductance controlled source to the shunt interconnection of a controlled source and a branch conductance. (c) The modified high frequency equivalent circuit of the source follower in Fig. 7.19(a).

large resistance since no biasing resistances are explicitly depicted in the schematic diagram of Fig. 7.19(a). Note from Eq. (7-48), $R_i = \infty$ renders R_{ss} identical to the Thévenin source resistance, R_s.

(b) The feedback capacitance, C_f, in Fig. 7.10(b) becomes the transistor gate-source capacitance, C_{gs}.

(c) It should be noted that the input port voltage, V_g, in the original model is now designated V_i in the source follower equivalent circuit. Accordingly, the controlled source, $g_m V_g$, in Fig. 7.10(b) corresponds to $-g_m V_i$ in the source follower model. Specifically, g_m in relevant previous analyses is now supplanted by $-g_m$.

(d) The output port capacitance, C_o, becomes the sum, $(C_{bs} + C_l)$, of bulk-source and load capacitances.

(e) The shunt output resistance, R_d, in Fig. 7.10(b) is now the resistance established by the parallel combination of transistor channel resistance, r_o, and the effective transistor conductance, $(1 + \lambda_b)g_m$; namely,

$$R_d = \frac{r_o}{1 + (1 + \lambda_b) g_m r_o}, \quad (7\text{-}87)$$

which reduces to $1/g_m$ for large r_o and small λ_b.

7.4.1. Source Follower Transfer Function

Equation (7-52) remains applicable for the voltage transfer function of the amplifier under present consideration. Armed with the foregoing parametric modifications, the zero frequency voltage gain, $A_v(0)$, is, from Eqs. (7-53) and (7-48),

$$A_v(0) = \frac{g_m (r_o \| R_l)}{1 + (1 + \lambda_b) g_m (r_o \| R_l)}, \quad (7\text{-}88)$$

which exudes no I/O phase inversion and is clearly less than unity. In the limit of large channel resistance and a reasonably large extrinsic load resistance that combine to allow for the satisfaction of the inequality, $(1 + \lambda_b) g_m (r_o \| R_l) \gg 1$,

$$A_v(0) = \frac{g_m (r_o \| R_l)}{1 + (1 + \lambda_b) g_m (r_o \| R_l)} \approx \frac{1}{1 + \lambda_b}. \quad (7\text{-}89)$$

When the source follower in Fig. 7.19(a) drives a purely capacitive load, provisions must be made to allow for a path to ground of the biasing current conducted by the source follower transistor. The most common embodiment of this requirement is the current sinking circuit depicted in Fig. 7.21(a). In addition to allowing for the proper biasing of transistor $M1$, the current sink subcircuit formed by transistor $M2$ in this diagram establishes an effective load resistance value that nominally equates to the large drain-source channel resistance of transistor $M2$. Accordingly, the current sink supports the inequality that underpins Eq. (7-89), thereby producing a zero frequency gain approaching unity. However, care must be exercised from a frequency response perspective since the incorporation of the current sinking subcircuit enhances the net output port capacitance by an amount equal to the sum of the drain-bulk and gate-drain capacitances of transistor $M2$.

A further enhancement to the achievable zero frequency voltage gain is theoretically promoted by returning the bulk terminal of transistor $M1$ to the $M1$ source terminal, as is suggested in Fig. 7.21(b). Although this tack effectively renders parameter λ_b null for transistor $M1$ and additionally

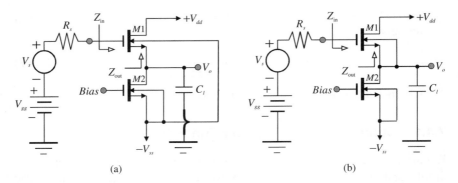

Figure 7.21. (a) Source follower terminated in a capacitive load. The biasing of transistor $M1$ is established by a constant current sink formed of transistor $M2$. (b) The circuit of (a) with the bulk terminal of transistor $M1$ connected to the $M1$ source terminal to achieve effectively zero bulk transconductance factor, λ_b.

circumvents any frequency response deterioration incurred by the bulk-source capacitance, C_{bs}, of transistor $M1$, care must be exercised, through either circuit or process means to ensure that the bulk-drain junction is never forward biased for all practical signal operating environments.

In the transfer relationship of Eq. (7-52), the indicated zero now lies in the left half plane since with transconductance g_m replaced by its negative value and recalling that $C_f = C_{gs}$, Eq. (7-54) gives

$$z_f = -(g_m/C_{gs}). \tag{7-90}$$

Using Eq. (7-59), the open circuit input port time constant, τ_i, associated with the circuit poles is

$$\tau_i = R_s C_{gd}, \tag{7-91}$$

which is rarely a dominant contributor to the net sum of open circuit time constants, especially if the transistor is fabricated in a self-aligned gate technology. On the other hand, the output port open circuit time constant, τ_o, is, by Eq. (7-60),

$$\tau_o = \frac{(r_o \| R_l)(C_{bs} + C_l)}{1 + (1 + \lambda_b) g_m (r_o \| R_l)} \approx \frac{C_{bs} + C_l}{(1 + \lambda_b) g_m}, \tag{7-92}$$

which is potentially dominant in view of the fact that the source follower is often called upon to drive strongly capacitive load terminations. Output port time constant dominance is especially likely when the transistor utilized in the follower is biased at low drain currents, which produces relatively small

Analog MOS Technology Circuits

device forward transconductance. From Eq. (7-61), the open circuit time constant arising from the feedback capacitance, C_{gs}, is

$$\tau_f = \left\{ \frac{r_o \| R_l}{1 + (1+\lambda_b) g_m (r_o \| R_l)} + \left[\frac{1 + \lambda_b g_m (r_o \| R_l)}{1 + (1+\lambda_b) g_m (r_o \| R_l)} \right] R_s \right\} C_{gs}$$

$$\approx \left(\lambda_b R_s + \frac{1}{g_m} \right) \left(\frac{C_{gs}}{1+\lambda_b} \right). \tag{7-93}$$

Finally, the product of the buffer pole frequencies is, from Eq. (7-56),

$$p_1 p_2 = \frac{1}{\tau_i \tau_o \left(1 + \dfrac{C_{gs}}{C_{gd}} + \dfrac{C_{gs}}{C_{bs} + C_l} \right)}. \tag{7-94}$$

where it should be remembered that

$$\frac{1}{p_1} + \frac{1}{p_2} \equiv \tau_i + \tau_o + \tau_f. \tag{7-95}$$

If Eqs. (7-92), (7-93), and (7-94) are substituted into Eq. (7-95), the sum of the amplifier open circuit time constants attributed to network poles can be approximated as

$$\frac{1}{p_1} + \frac{1}{p_2} \approx \frac{C_{bs} + C_{gs} + C_l}{(1+\lambda_b) g_m} + R_s \left[C_{gd} + \left(\frac{\lambda_b}{1+\lambda_b} \right) C_{gs} \right], \tag{7-96}$$

where the analytical liberties taken in conjunction with Eqs. (7-92) through (7-94) are exploited herewith. If the transistor utilized in the follower is fabricated in a self-aligning gate technology, the effective gate-drain capacitance, C_{gd}, is small. A small bulk-induced transconductance factor, λ_b, can be additionally presumed. Then, if pole dominance prevails, which is likely if time constant τ_o is dominant, as inferred earlier,

$$\frac{1}{p_1} + \frac{1}{p_2} \approx \frac{1}{p_1} \approx \frac{C_{bs} + C_{gs} + C_l}{g_m}. \tag{7-97}$$

This result and Eq. (7-94) combine to deliver

$$\frac{1}{p_2} \approx \left(\frac{C_{bs} + C_l}{C_{bs} + C_{gs} + C_l} \right) \left(\frac{g_m R_s}{\omega_T} \right), \tag{7-98}$$

where by Eq. (7-44) ω_T symbolizes the unity gain frequency of the utilized transistor. Recalling Eq. (7-52), the resultant I/O voltage transfer function,

$A_v(s)$, is

$$A_v(s) = \frac{A_v(0)\left(1 - \dfrac{s}{z_f}\right)}{\left(1 + \dfrac{s}{p_1}\right)\left(1 + \dfrac{s}{p_2}\right)}$$

$$\approx \frac{\left[\dfrac{g_m(r_o \| R_l)}{1 + (1 + \lambda_b)g_m(r_o \| R_l)}\right]\left(1 + \dfrac{sC_{gs}}{g_m}\right)}{\left[1 + \dfrac{s(C_{bs} + C_{gs} + C_l)}{g_m}\right]\left[1 + g_m R_s\left(\dfrac{C_{bs} + C_l}{C_{bs} + C_{gs} + C_l}\right)\left(\dfrac{s}{\omega_T}\right)\right]}.$$

(7-99)

The engineering significance of the foregoing transfer function relationship is best digested with the aid of its asymptotic frequency response, which is offered in Fig. 7.22. As expected, the depicted low frequency gain is $A_v(0)$, as defined by Eq. (7-88). But since the frequency of the first pole is smaller than the frequency of the left half plane zero, a range of frequencies given by $p_1 < \omega < |z_f|$ delivers a follower voltage gain that is nominally frequency invariant but is potentially significantly smaller than one. This degraded gain is

$$A_v(0)\frac{p_1}{|z_f|} = A_v(0)\left(\frac{C_{gs}}{C_{gs} + C_{bs} + C_l}\right), \qquad (7\text{-}100)$$

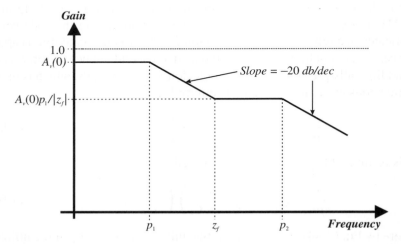

Figure 7.22. The asymptotic frequency response of the source follower in Fig. 7.19(a). The transfer function corresponding to this response is given as Eq. (7-99).

which is indeed considerably smaller than the zero frequency gain when the load capacitance driven by the source follower is large. Equation (7-100) explains the bad gain rap endured by MOS technology followers that are compelled to operate in mid-passband frequency ranges.

The preceding equation also suggests that if source followers are to sustain near unity voltage gain for signal frequencies extending from zero to approximately the frequency, p_2, of the nondominant pole, the frequency, p_1, of the dominant pole must nominally equate to the frequency of the left half plane zero. Unfortunately, $p_1 \approx |z_f|$ if and only if $C_{gs} \gg (C_{bs} + C_l)$, which is hardly likely in practical electronic systems. But the validity of the subject inequality can be forced if a compensating capacitance, say C_c, is appended across the gate-source terminals of the transistor in question, as is exemplified in the schematic diagram shown in Fig. 7.23. In this case, the transistor gate-source capacitance, C_{gs}, is effectively increased to a value of $(C_{gs} + C_c)$ so that Eq. (7-100) becomes

$$A_v(0)\frac{p_1}{|z_f|} = A_v(0)\left(\frac{C_{gs} + C_c}{C_{gs} + C_c + C_{bs} + C_l}\right), \qquad (7\text{-}101)$$

with the understanding that C_c is selected to ensure $(C_{gs} + C_c) \gg (C_{bs} + C_l)$. The upshot of this design methodology is that the resultant right half plane zero cancels the dominant pole in the transfer function of Eq. (7-99). In turn, the source follower voltage gain is nominally constant at near unity value from zero signal frequency to a 3-dB bandwidth, say B_{sf}, given by

$$B_{sf} \approx p_2 \approx \frac{1}{R_s\left(C_{gs} + C_c\right)}\left(1 + \frac{C_{gs} + C_c}{C_{bs} + C_l}\right) \approx \frac{1}{R_s\left(C_{bs} + C_l\right)}. \qquad (7\text{-}102)$$

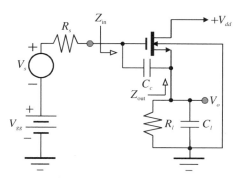

Figure 7.23. Source follower whose frequency response is compensated through the addition of capacitance C_c across the gate-source terminals of the transistor.

In arriving at Eq. (7-102), Eqs. (7-98) and (7-99) are used, subject to the proviso that the radial unity gain frequency, ω_T, of the transistor, which is roughly g_m/C_{gs} (ignoring gate-drain capacitance C_{gd}), is now $g_m/(C_{gs} + C_c)$.

7.4.2. Source Follower I/O Impedances

Subject to the previously discussed parametric redefinitions that allow exploiting the small signal model of the common source amplifier in the analysis of the source follower, Eqs. (7-75) and (7-76) apply to the problem of determining the driving point input impedance of the follower in Fig. 7.19(a). The lone caveat to this declaration is that in the follower model of Fig. 7.19(b), a resistance, R_i, must be imagined in shunt with the input port capacitance, C_{gd}, to mirror the input port topology of the common source equivalent circuit. In the source follower of Fig. 7.19(a), input port biasing is accomplished by a simple battery placed in series with the signal source, thereby obviating the necessity of including a shunt input port resistance in its model. Accordingly, Eqs. (7-75) and (7-76) can be applied straightforwardly to produce valid results, provided that the constraint, $R_i = \infty$, is invoked.

The foregoing strategy and the presumption of large $g_m(r_o \| R_l)$ deliver a driving point input impedance, $Z_{in}(s)$, of

$$Z_{in}(s) \approx \frac{1 + s\tau_{in}}{sC_{ix}(1 + sk_{in}\tau_{in})}, \tag{7-103}$$

where

$$C_{ix} = C_{gd} + (1 - g_m R_{ll})C_{gs} \approx C_{gd} + \left(\frac{\lambda_b}{1 + \lambda_b}\right)C_{gs} \tag{7-104}$$

represents the effective input capacitance at low to moderately large signal frequencies,

$$\tau_{in} = R_{ll}(C_{bs} + C_{gs} + C_l), \tag{7-105}$$

$$k_{in} = \frac{C_{gd} + \dfrac{C_{gs}(C_{bs} + C_l)}{C_{bs} + C_{gs} + C_l}}{C_{gd} + \left(\dfrac{\lambda_b}{1 + \lambda_b}\right)C_{gs}} = \frac{C_{gd} + \dfrac{C_{gs}(C_{bs} + C_l)}{C_{bs} + C_{gs} + C_l}}{C_{ix}}, \tag{7-106}$$

Analog MOS Technology Circuits 561

which is invariably larger than one, and from Eqs. (7-48) and (7-87),

$$R_{ll} = \frac{r_o \| R_l}{1 + (1 + \lambda_b) g_m (r_o \| R_l)} \approx \frac{1}{(1 + \lambda_b) g_m}. \quad (7\text{-}107)$$

Note that the Miller effect ramification, $(1 - g_m R_{ll}) C_{gs}$, of feedback capacitance C_{gs} is minimal in the source follower because the output port response is in phase with the input port excitation at low frequencies, and the I/O port gain is almost unity. Indeed, this Miller-multiplied feedback capacitance vanishes if the bulk transconductance factor, λ_b, of the utilized transistor is negligible.

Equation (7-103) projects capacitive driving point input impedance. As is depicted in Fig. 7.24(a), the effective input capacitance at low frequencies is C_{ix}, which Eq. (7-104) confirms is slightly larger than the transistor gate-drain capacitance, C_{gd}. It is precisely equal to C_{gd} if λ_b is zero. On the other hand, a high frequency input port capacitance of

$$k_{in} C_{ix} = C_{gd} + \frac{C_{gs}(C_{bs} + C_l)}{C_{bs} + C_{gs} + C_l} \quad (7\text{-}108)$$

prevails, which, as is inferred by the model in Fig. 7.24(b), is the net capacitance resulting from C_{gd} placed in shunt with the series interconnection of gate-source capacitance, C_{gs}, and the net output port capacitance, $(C_{bs} + C_l)$. The latter input port model derives intuitively from the equivalent circuit in Fig. 7.20(c). In particular, at very high signal frequencies, the net shunt output capacitance in Fig. 7.20(c) ultimately dominates the interconnection of the two indicated shunt output port resistances. Moreover, voltage V_i, which controls the current source, $g_m V_i$, collapses because the impedance of the capacitance, C_{gd}, across which V_i is

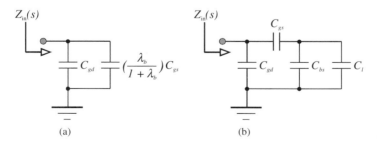

Figure 7.24. (a) Low frequency model of the driving point input impedance of the source follower in Fig. 7.19(a). (b) High frequency model of the driving point input impedance of the source follower in Fig. 7.19(a).

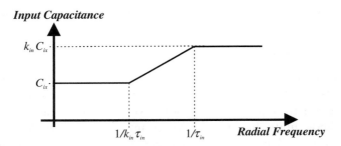

Figure 7.25. The asymptotic frequency domain behavior of the effective input port capacitance of the source follower in Fig. 7.19(a).

developed, approaches zero at high frequencies. The salient aspects of the overall frequency domain behavior of the net input port capacitance are underscored by the asymptotic plot shown in Fig. 7.25.

A distressing observation is that the subject input impedance is not a positive real function of signal frequency, thereby portending of potential circuit instability. In particular, the input admittance, $Y_{in}(j\omega)$, corresponding to $Z_{in}(s)$ in Eq. (7-103) is

$$Y_{in}(j\omega) = G_{in}(\omega) + jB_{in}(\omega), \qquad (7\text{-}109)$$

where

$$\left. \begin{array}{l} G_{in}(\omega) = -\omega^2 \tau_{in} C_{ix} \left[\dfrac{k_{in} - 1}{1 + (\omega \tau_{in})^2} \right] \\[2mm] B_{in}(\omega) = \omega C_{ix} \left[\dfrac{1 + k_{in}(\omega \tau_{in})^2}{1 + (\omega \tau_{in})^2} \right] \end{array} \right\rangle . \qquad (7\text{-}110)$$

The positive susceptance, $B_{in}(\omega)$, in Eq. (7-110) validates the capacitive arguments proffered above. Note, however, that since the constant, k_{in}, in Eq. (7-106) is likely to exceed unity, the conductive component, $G_{in}(\omega)$, of the driving point input admittance for the source follower is negative for all frequencies. Concern over this observation is exacerbated by the fact that a source follower is traditionally utilized to mitigate the undesirable effects of an unacceptably large Thévenin signal source resistance, R_s. It follows that the net gate to ground conductance, $[G_s + G_{in}(\omega)]$, can be negative at high signal frequencies, thereby precipitating a right half plane pole attributed to the input port time constant of the amplifier. In such an event, it is necessary either to adjust device geometries to ensure $k_{in} \leq 1$ or to modify the simple biasing structure depicted in Fig. 7.19(a) with the

Analog MOS Technology Circuits

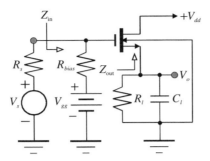

Figure 7.26. Modification of the input port biasing of the follower in Fig. 7.19(a) for the purpose of ensuring I/O network stability.

biasing subcircuit submitted in Fig. 7.26. It is understood that in the latter diagram, the conductance, G_{bias}, of biasing resistance R_{bias} is sufficiently large to ensure $[G_s + G_{bias} + G_{in}(\omega)] > 0$ for all frequencies within the passband of the source follower.

Equations (7-81) and (7-82) apply directly to the problem of ascertaining the driving point output impedance, $Z_{out}(s)$, of the source follower, subject to the previously stipulated adjustments in model parameters. In particular, if R_d in Eq. (7-87) is approximated as $1/g_m$,

$$Z_{out}(s) \approx \left(\frac{1}{g_m}\right)\left[\frac{1+ds}{1+(d-e)s+fs^2}\right], \quad (7\text{-}111)$$

where

$$\left.\begin{array}{l} d = R_s\left(C_{gd}+C_{gs}\right) = g_m R_s/\omega_T \\ e = R_s C_{gs} - \dfrac{C_{bs}+C_{gs}+C_l}{g_m} \\ f = \dfrac{k_{in} R_s C_{ix}\left(C_{bs}+C_{gs}+C_l\right)}{g_m} \end{array}\right\} \quad (7\text{-}112)$$

At signal frequencies for which the term, in fs^2, can be tacitly ignored, Eq. (7-111) reduces to

$$Z_{out}(s) \approx \left(\frac{1}{g_m}\right)\left[\frac{1+ds}{1+(d-e)s}\right]. \quad (7\text{-}113)$$

This impedance relationship is interesting in that the open circuit time constant, $(d-e)$, attributed to the poles of the output impedance function is smaller than the open circuit time constant, d, that derives from the impedance function zero, provided $e > 0$. In turn, $e > 0$ is manifested by the constraint,

$$g_m R_s > 1 + \frac{C_{bs} + C_l}{C_{gs}}, \qquad (7\text{-}114)$$

which is invariably satisfied in view of the fact that a source follower is typically biased for substantive forward transconductance and is used, as noted earlier, to process input signals deriving from a high resistance signal source. The engineering implication of Eq. (7-114) is that for signal frequencies that allow the term, fs^2, to be neglected, the output impedance is inductive. The resultant effective output inductance, L_{eff}, presented to an extrinsic load by the source follower derives from the low frequency approximation of $Z_{out}(s)$ in Eq. (7-113) as

$$Z_{out}(s) \approx \left(\frac{1}{g_m}\right)(1 + ds)[1 - (d - e)s]$$

$$= \left(\frac{1}{g_m}\right)\left[1 + es - d(d - e)s^2\right]. \qquad (7\text{-}115)$$

The first two terms on the far right hand side of Eq. (7-115) suggest a resistance, $(1/g_m)$, placed in series with the effective inductance. It follows that

$$L_{eff} \approx \frac{e}{g_m} = \frac{R_s C_{gs}}{g_m}\left[1 - \frac{1 + (C_{bs} + C_l)/C_{gs}}{g_m R_s}\right]. \qquad (7\text{-}116)$$

In contrast, the subject output impedance at very high signal frequencies converges toward a pure capacitance, say C_{eff}, which by Eqs. (7-111) and (7-112) is

$$C_{eff} = \frac{f g_m}{d} = C_{bs} + C_l + \frac{C_{gs} C_{gd}}{C_{gs} + C_{gd}}. \qquad (7\text{-}117)$$

Like the effective input capacitance, $k_{in} C_{ix}$, evident at very high frequencies, the topological nature of the effective high frequency output capacitance, C_{eff}, is obvious by inspection of the follower model in Fig. 7.20(c). Figure 7.27 schematically summarizes the results leading to Eqs. (7-116) and (7-117).

Example 7.5. In the source follower model of Fig. 7.20(a), the forward transconductance, g_m, of the utilized N-channel transistor is 18 mmhos, the drain-source channel resistance, r_o, is 18.8 KΩ, and the gate-source, gate-drain, and bulk-source

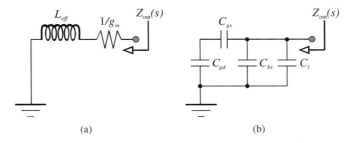

Figure 7.27. (a) Low to moderately high frequency model of the driving point output impedance of the source follower in Fig. 7.19(a). The effective inductance, L_{eff}, is given analytically by Eq. (7-116). (b) Very high frequency model of the driving point output impedance of the source follower in Fig. 7.19(a).

capacitances are $C_{gs} = 35\,\text{fF}$, $C_{gd} = 12\,\text{fF}$, and $C_{bs} = 20\,\text{fF}$, respectively. The transistor has a bulk transconductance factor, λ_b, of 0.05. The follower is driven by an input signal voltage whose internal resistance, R_s, is 500 Ω, and the amplifier drives a load comprised of a shunt interconnection of a 1 KΩ resistance (R_l) and a 60 fF capacitance (C_l). Assume that the transistor is biased for operation in its saturation regime. Use HSPICE to simulate the voltage transfer function, input impedance, and output impedance established by the model in question. Confirm the accuracy of foregoing approximate results for low frequency gain and both low and high frequency I/O impedances.

Solution 7.5.

(1) Without invoking any simplifying approximations, the voltage gain expression in Eq. (7-99) is

$$A_v(s) = \frac{0.902\left[1 + \dfrac{s}{2\pi\,(81.85)}\right]}{1 + \dfrac{s}{2\pi\,(11.81)} + \left[\dfrac{s}{2\pi\,(14.77)}\right]^2}, \qquad \text{(E5-1)}$$

where complex frequency s is in units of giga-radians per second. The damping factor associated with the characteristic polynomial of this transfer function is 0.658, which indicates that complex conjugate poles prevail. Note an undamped natural frequency of oscillation of 15.55 GHz and a zero whose frequency is better than five times larger than this self-resonant frequency. Because the frequency of the left half plane zero is large in comparison to the self-resonant frequency, the frequency response is largely determined by the complex pole pair. Figure 7.28 displays the simulated voltage gain response for the considered follower model.

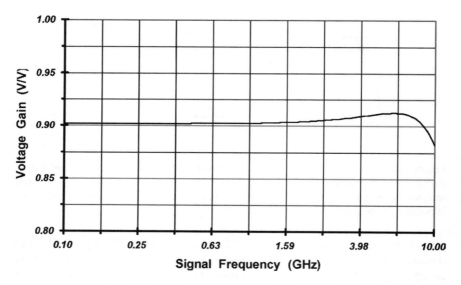

Figure 7.28. The voltage gain frequency response of the source follower considered in Example 7.5. The indicated response peaking is due largely to the fact that the amplifier model produces complex conjugate poles for the given parameters.

(2) The input impedance expression in Eq. (7-103) provides a driving point input admittance, $Y_{in}(s)$, of

$$Y_{in}(s) = \frac{13.67s\left[1 + \dfrac{s}{2\pi(10.39)}\right]}{1 + \dfrac{s}{2\pi(27.61)}}, \quad \text{(E5-2)}$$

whose nature is predominantly a capacitance rising monotonically with signal frequency. It is understood that s in giga-radians per second produces $Y_{in}(s)$ in units of micromhos. Figure 7.29 submits the simulated frequency domain plots of the effective shunt input port resistance and the effective shunt input port capacitance.

(3) Without approximations, Eqs. (7-87) and (7-111) yield a driving point output impedance of

$$Z_{out}(s) = \frac{52.76\left[1 + \dfrac{s}{2\pi(6.77)}\right]}{1 + \dfrac{s}{2\pi(12.85)} + \left[\dfrac{s}{2\pi(15.16)}\right]^2}, \quad \text{(E5-3)}$$

where s in giga-radians per second results in $Z_{out}(s)$ in units of ohms. Figure 7.30 submits the simulated frequency domain plots of the effective series output port resistance and inductance.

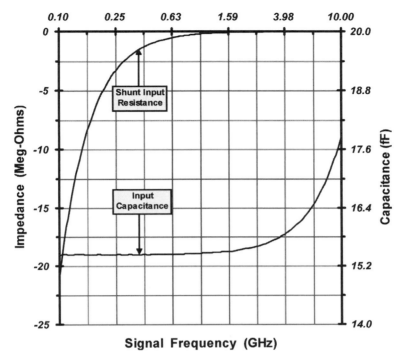

Figure 7.29. Frequency domain plots of the effective shunt input port resistance and effective input port capacitance of the source follower addressed in Example 7.5.

Comments. The simulated low frequency gain is 0.902, which agrees precisely with Eq. (E5-1), but is about 5.3% smaller than the approximate gain value given in Eq. (7-89). The 3-dB bandwidth is larger than 10 GHz, which is to be expected in view of the left half plane zero evidenced in Eq. (E5-1) and the fact that the frequency associated with the open circuit time constant due to network poles is 11.81 GHz.

As predicted, the shunt input port resistance is negative throughout the entire range of considered frequencies. The approximate low frequency shunt input capacitance set forth by Eq. (E5-2) is lower than the simulated value of 15.4 fF by about 11%.

Not surprisingly, the simulated low frequency output resistance agrees precisely with that predicted by Eq. (E5-3). Moreover, the simulated value of this resistance differs from the approximate $1/g_m = 55.56\,\Omega$ value by only 5%. The simulated output port inductance is 0.557 nH. On the other hand, Eq. (7-116) predicts $L_{\mathit{eff}} = 0.617$ nH, which is in error by under 11%.

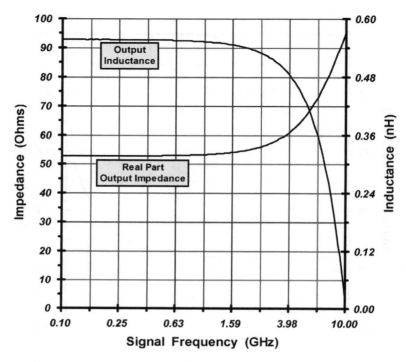

Figure 7.30. Frequency domain plots of the effective series output port resistance and inductance of the source follower addressed in Example 7.5.

7.5.0. Common Gate Amplifier

The common gate amplifier is the dual of the common drain circuit analyzed in the preceding section of material. Whereas the common drain unit projects very high input impedance, low output impedance, and slightly less than unity voltage gain at low signal frequencies, its common gate counterpart delivers low input impedance, high output impedance, and a slightly less than unity current gain. In contrast to the source follower, which functions as a voltage buffer, the common gate amplifier is a current buffer. In particular, its low input and high output impedances enable coupling a high impedance signal source to a moderately large load termination without incurring a substantial loss in signal source to load current gain.

The basic schematic diagram of an NMOS common gate configuration appears in Fig. 7.31, where R_s represents the presumably large signal source resistance and R_l is the load resistance that terminates the transistor drain terminal to signal ground. In a traditional embodiment of the common

Analog MOS Technology Circuits 569

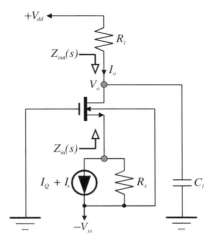

Figure 7.31. Basic schematic diagram of a common gate amplifier realized in NMOS technology.

gate stage, the current, I_Q, is a biasing current supplied by an N-channel common source amplifier (not shown in the subject diagram) whose drain terminal is directly coupled to the source terminal of the common gate cell. This architecture is known as a *common source-common gate cascode amplifier*, whose characteristics and purposes are discussed subsequently. At this juncture, it suffices to view I_Q as a simple biasing source that, in conjunction with the supply voltages, V_{dd} and V_{ss}, and the biasing voltage, V_{bias}, establishes saturation regime operation of the NMOS transistor in the considered circuit diagram. The response to the signal input current, I_s, is taken to be the small signal component, I_{os}, of the indicated load resistance current, I_o. A parasitic load capacitance, C_l, is incorporated between ground and the amplifier output port.

The small signal equivalent circuit of the amplifier in Fig. 7.31 appears in Fig. 7.32(a), where V_{is} is the signal voltage developed at the amplifier input port, and V_{os} is the signal component of the voltage established across load resistance R_l; that is,

$$V_{os} = -I_{os} R_l. \tag{7-118}$$

In this model, which reflects that of Fig. 7.6, g_o is the conductance associated with the transistor channel resistance, r_o,

$$C_o = C_{bd} + C_{gd} + C_l \tag{7-119}$$

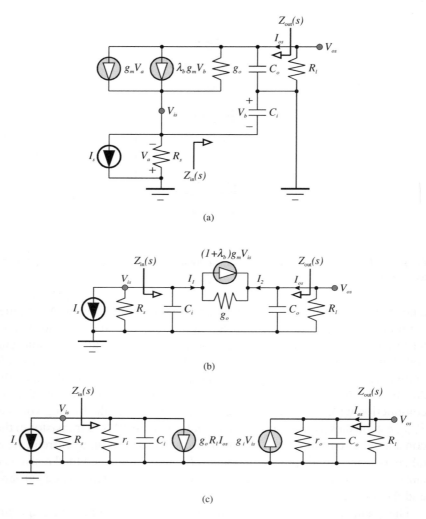

Figure 7.32. (a) Small signal, high frequency equivalent circuit of the common gate amplifier in Fig. 7.31. (b) An equivalent representation of the model in (a). (c) Global feedback form of the common gate equivalent circuit.

is the net output port capacitance comprised of transistor bulk-drain (C_{bd}), transistor gate-drain (C_{gd}), and actual load capacitances. Note specifically that the parasitic load capacitance is absorbed into the net capacitance presented to the amplifier output port by the transistor. Furthermore,

$$C_i = C_{gs} + C_{bs} \qquad (7\text{-}120)$$

is the net input port capacitance, whose constituents are transistor gate-source (C_{gs}) and bulk-source (C_{bs}) capacitances. Since the control voltages, V_a and V_b are identical and equal to the negative of voltage V_{is}, the model in Fig. 7.32(a) can be couched as the architecture depicted in Fig. 7.32(b). In the latter equivalent circuit, the indicated current, I_1, is expressible as

$$I_1 = (1 + \lambda_b) g_m V_{is} + g_o (V_{is} - V_{os}), \qquad (7\text{-}121)$$

while the current, I_2, in the circuit is clearly the negative of I_1. Letting

$$g_i = (1 + \lambda_b) g_m + g_o \triangleq \frac{1}{r_i}, \qquad (7\text{-}122)$$

Eq. (7-122) and the observation, $I_1 = -I_2$, produce

$$I_1 = -I_2 = g_i V_{is} - g_o V_{os}. \qquad (7\text{-}123)$$

In turn, this result and Eq. (7-118), coupled with the model in Fig. 7.32(b), precipitate Fig. 7.32(c) as an alternative model, which clearly postulates a closed loop feedback configuration whose global feedback parameter is the product of channel conductance and load resistance. In this final modeling disclosure, the shunt input port resistance, r_i, of the common gate amplifier is seen, by way of Eq. (7-123), to be equivalent to the "exact" low frequency output port resistance of the common drain cell. The earth hardly shatters at this revelation since the input port to a common gate amplifier is the transistor source terminal, which also comprises the output port of a common drain amplifier. Moreover, the model in Fig. 7.32(c) shows that the common gate amplifier collapses to an open loop topology when the channel conductance, g_o, is negligibly small. This observation is synergistic with the schematic diagram in Fig. 7.31, which confirms that the only feedback path from the drain output port to the source input port is the channel resistance, r_o, that implicitly couples transistor drain and source terminals.

7.5.1. Common Gate I/O Characteristics

The topological form of the model shown in Fig. 7.32(c) immediately produces a closed loop current gain, $A_{ic}(s)$, of the form

$$A_{ic}(s) = \frac{I_{os}}{I_s} = \frac{A_{io}(s)}{1 + T(s, R_s, R_l)}. \qquad (7\text{-}124)$$

In this relationship, $A_{io}(s)$ symbolizes the open loop current gain, which is the ratio, I_{os}/I_s, under the zero feedback constraint of $g_o R_l I_{os} = 0$. By inspection of the model at hand,

$$A_{io}(s) = \left.\frac{I_{os}}{I_s}\right|_{g_o R_l I_{os}=0} = \frac{\left(\dfrac{R_s}{R_s+r_i}\right)\left(\dfrac{r_o}{r_o+R_l}\right)}{[1+s(R_s\|r_i)C_i][1+s(R_l\|r_o)C_o]}, \quad (7\text{-}125)$$

while the loop gain, $T(s, R_s, R_l)$, is

$$T(s, R_s, R_l) = g_o R_l[-A_{io}(s)] = -\frac{\left(\dfrac{R_s}{R_s+r_i}\right)\left(\dfrac{R_l}{R_l+r_o}\right)}{[1+s(R_s\|r_i)C_i][1+s(R_l\|r_o)C_o]}. \quad (7\text{-}126)$$

The substitution of Eqs. (7-126) and (7-125) into Eq. (7-124) establishes a closed loop gain expressible as

$$A_{ic}(s) = \frac{A_{ic}(0)}{1+as+bs^2}, \quad (7\text{-}127)$$

where

$$A_{ic}(0) = \left(\frac{R_s}{R_s+r_i}\right)\left[\frac{r_o(1+R_s/r_i)}{r_o(1+R_s/r_i)+R_l}\right] \quad (7\text{-}128)$$

is the closed loop gain at zero signal frequency. This gain is obviously less than one, although it tends toward unity in the limits of large r_o and large R_s. The expression suggests the low frequency closed loop equivalent circuit shown in Fig. 7.33 for which an inspection of Eq. (7-128) reveals a low

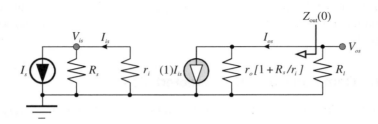

Figure 7.33. The low frequency equivalent circuit of the common gate amplifier in Fig. 7.31. The circuit is applicable to determining the closed loop, low frequency values of the forward current gain and the driving point output impedance.

frequency driving point output impedance, $Z_{\text{out}}(0)$, of

$$Z_{\text{out}}(0) = r_o \left(1 + \frac{R_s}{r_i}\right) = R_s + [1 + (1 + \lambda_b) g_m R_s] r_o. \qquad (7\text{-}129)$$

The current amplification propriety of the common gate cell is affirmed by this impedance expression and the low frequency model offered in Fig. 7.33. In particular, a current signal source is characterized by a large source resistance, R_s. For reasonable values of transistor forward transconductance (g_m) and invariably large device channel resistance (r_o), Eq. (7-129) predicts a potentially huge driving point output resistance, thereby reflecting an output port behaving as an ideal, controlled source of signal current.

The time constant, a, and the inverse squared frequency parameter, b, in Eq. (7-127) can be shown to be

$$a = \frac{R_s C_i}{1 + \dfrac{R_s r_o}{r_i(R_l + r_o)}} + \frac{R_l C_o}{1 + \dfrac{R_l r_i}{r_o(R_s + r_i)}}, \qquad (7\text{-}130)$$

and

$$b = \frac{(R_s \| r_i) R_l C_i C_o}{1 + \dfrac{R_l r_i}{r_o(R_s + r_i)}}. \qquad (7\text{-}131)$$

For the commonly encountered case of large channel resistance, the zero frequency closed loop current gain reduces to

$$A_{ic}(0) \approx \frac{R_s}{R_s + r_i}, \qquad (7\text{-}132)$$

which is simply the current divider established between the signal source termination and the amplifier input port. Moreover,

$$a \approx (R_s \| r_i) C_i + R_l C_o, \qquad (7\text{-}133)$$

which, to the extent that $r_o \gg R_l$, is the sum of open circuit pole time constants indigenous to the open loop amplifier response. Finally,

$$b \approx (R_s \| r_i) R_l C_i C_o. \qquad (7\text{-}134)$$

When considered in light of Eq. (7-127), the last two approximations imply that the closed loop common gate amplifier response projects two real poles. In particular,

$$A_{ic}(s) \approx \frac{A_{ic}(0)}{[1 + s(R_s \| r_i) C_i](1 + s R_l C_o)}. \qquad (7\text{-}135)$$

Ordinarily, the time constant established by capacitance C_o in Eq. (7-135) is dominant, but either of the two delineated time constants can be dominant. When neither time constant dominates, the inverse of the parameter, a, in Eq. (7-133) is a suitable approximation of the amplifier 3-dB bandwidth.

Classic signal flow theory allows the driving point input impedance, $Z_{\text{in}}(s)$, to be expressed as

$$Z_{\text{in}}(s) = Z_{ino}(s) \left[\frac{1 + T(s, 0, R_l)}{1 + T(s, \infty, R_l)} \right], \qquad (7\text{-}136)$$

where, by an inspection of the model in Fig. 7.32, the open loop input impedance, $Z_{ino}(s)$, is

$$Z_{ino}(s) = \frac{r_i}{1 + sr_i C_i}. \qquad (7\text{-}137)$$

Recalling Eq. (7-126), $T(s, 0, R_l) = 0$, while

$$T(s, \infty, R_l) = -\frac{\dfrac{R_l}{R_l + r_o}}{(1 + sr_i C_i)\left[1 + s\left(R_l \| r_o\right) C_o\right]}. \qquad (7\text{-}138)$$

It follows that Eq. (7-136) becomes

$$Z_{\text{in}}(s) = \frac{r_i \left(1 + \dfrac{R_l}{r_o}\right)\left[1 + s\left(R_l \| r_o\right) C_o\right]}{1 + s\left[r_i \left(1 + \dfrac{R_l}{r_o}\right) C_i + R_l C_o\right] + s^2 r_i R_l C_i C_o}. \qquad (7\text{-}139)$$

For commonly encountered applications in which $r_o \gg R_l$, the driving point input impedance exudes two real poles, one of which cancels with the left half plane zero that is established by the circuit model. Specifically, $Z_{\text{in}}(s)$ collapses to the single pole function,

$$Z_{\text{in}}(s) \approx \frac{r_i}{1 + sr_i C_i}, \qquad (7\text{-}140)$$

which is identical to the open loop value of the input impedance. Although this input impedance, and indeed, the expression in Eq. (7-139), suggest a capacitive input port, the presence of any parasitic impedance in series with the gate lead in the schematic portrayal of Fig. 7.31 can render the input impedance inductive over wide frequency passbands. The condition underlying this inductance generation is the subject of a problem exercise at the conclusion of this chapter.

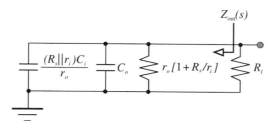

Figure 7.34. The approximate output port model of the common gate amplifier in Fig. 7.31. The model is a valid output port representation at low to moderately high signal frequencies.

An analogous analytical tack results in a common gate driving point output impedance, $Z_{out}(s)$, of

$$Z_{out}(s) = \frac{r_o\left(1 + \frac{R_s}{r_i}\right)[1 + s(R_s \| r_i)C_i]}{1 + s\left[r_o\left(1 + \frac{R_s}{r_i}\right)C_o + R_s C_i\right] + s^2 R_s r_o C_i C_o}, \qquad (7\text{-}141)$$

where it should be noted that the resultant zero frequency value, $Z_{out}(0)$, of this impedance function agrees precisely with the formulation in Eq. (7-129). The signal source resistance, R_s, is generally much larger than the low frequency, open loop common gate impedance, r_i. This fact renders impossible the reduction of the driving point output impedance to a single pole function, as is the case for the driving point input impedance. At low-to- moderately high frequencies where the term in s^2 can be ignored and the contribution of the left half plane zero is likewise negligible,

$$Z_{out}(s) \approx \frac{r_o\left(1 + \frac{R_s}{r_i}\right)}{1 + s r_o \left(1 + \frac{R_s}{r_i}\right)\left[C_o + \frac{(R_s \| r_i)C_i}{r_o}\right]}, \qquad (7\text{-}142)$$

which suggests the approximate output port model offered in Fig. 7.34.

7.5.2. Common Source-Common Gate Cascode

As suggested earlier, the common gate stage is rarely exploited as a standalone analog signal processor. In its most common application, the common source-common gate cascode configuration, the cell functions as a current buffer between the output port of a common source amplifier and a load termination. Under certain conditions, the impact of this buffering can be a potentially significant increase in the gain-bandwidth product evidenced by the common source cell alone.

The process of examining current buffering strategy and pinpointing the precise operating conditions that may herald a significant enhancement in circuit gain-bandwidth product, begins with a consideration of Fig. 7.35. Figure 7.35(a) is simply a replica of the basic common source stage depicted in Fig. 7.10(a), while Fig. 7.35(b) depicts the cascode modification of the basic cell. In the latter topology, transistor $M1$ is identical to the MOSFET utilized in the former structure. To the extent that capacitor C_b is selected to ensure that it emulates a signal short circuit over the passband of interest, transistor $M2$ operates as a common gate stage. The resistors, R_1, R_2, and R_3, form a voltage divider off of the two supply voltages, $+V_{dd}$ and $-V_{ss}$, to bias $M1$ and $M2$ in their saturated domains.

From Eqs. (7-48), (7-52), and (7-53), the low frequency (but not so low that capacitors C_s and C_b can no longer be approximated as signal short circuits) the voltage gain, $A_{v1}(0)$, of the stage in Fig. 7.35(a) is

$$A_{v1}(0) = -g_{m1}\left(\frac{R_1\|R_2}{R_1\|R_2 + R_s}\right)(r_o\|R_l) \approx -g_{m1}R_l, \qquad (7\text{-}143)$$

where the approximation reflects the presumptions that $(R_1\|R_2) \gg R_s$ and $r_o \gg R_l$. Formal gain relationships aside, Eq. (7-143) can be deduced by mere inspection of the diagram in Fig. 7.35(a). This intuitive process

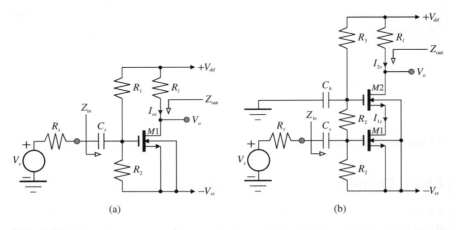

Figure 7.35. (a) The common source amplifier of Fig. 7.10(a). (b) Common source-common gate cascode modification of the amplifier cell in (a).

is premised on recalling from Fig. 7.10(c) that at low frequencies, a large transistor channel resistance and a negligible or null bulk transconductance factor combine to deliver a drain to source signal current that is simply the transistor transconductance multiplied by the small signal value of the gate to source voltage. Thus, from Fig. 7.35(a), the gate to source signal voltage is seen to be the input signal voltage, V_s, provided that capacitance C_s behaves as a short circuit for all frequencies of interest and $(R_1||R_2) \gg R_s$. It follows that the indicated drain to source signal current, I_{os}, is approximately $g_{m1}V_s$, which implies an output signal voltage, V_{os}, of $-g_{m1}R_l V_s$, whence a voltage gain identical to the approximate result set forth by Eq. (7-143).

In the cascode configuration of Fig. 7.35(b), the gate to source signal voltage applied to transistor $M1$ is the signal voltage, V_s, assuming that both of the circuit capacitors behave as short circuits throughout the frequency range of interest and $(R_1||R_2) \gg R_s$. It follows that the drain to source signal current conducted by $M1$ is approximately $g_{m1}V_s$. But since zero gate current is conducted at low frequencies by transistor $M2$, the resultant drain to source signal current flowing in $M2$ is also $g_{m1}V_s$. This current gives rise to an output signal voltage of $V_{os} = -g_{m1}R_l V_s$. Accordingly, the low frequency voltage gain of the subject circuit is $A_{v2}(0) \approx -g_{m1}R_l$, which is identical to the approximate gain provided by the simple common source structure in Fig. 7.35(a). The actual voltage gain magnitude of the second circuit is somewhat less than $g_{m1}R_l$ because by Eq. (7-128), the current gain of the common gate unit formed of transistor $M2$ is less than one. In Eq. (7-128), it should be understood that R_s is the output or drain-source channel resistance, r_{o1}, seen looking into the drain of transistor $M1$, r_o is the channel resistance, r_{o2}, indigenous to $M2$, and r_i is the effective resistance seen looking into the source terminal of transistor $M2$. In particular, Eq. (7-122) yields

$$r_i = \frac{r_{o2}}{1 + (1 + \lambda_{b2})g_{m2}r_{o2}} \approx \frac{1}{g_{m2}}. \qquad (7\text{-}144)$$

What could possibly be the advantage of using a two-transistor circuit to deliver a gain magnitude that is actually slightly smaller than that afforded by the simpler one-transistor cell?

The answer to the foregoing question is rooted in considerations related to the achievable 3-dB bandwidth. In particular, return to Eq. (7-58) and assume that the load resistance, R_l, in the drain circuit is sufficiently large so that the Miller multiplication of the feedback capacitance, $C_f = C_{gd1}$, establishes the dominant open circuit time constant in the common source amplifier of Fig. 7.35(a). The 3-dB bandwidth, B_1, of this circuit is resultantly

$$B_1 \approx \frac{1}{(1 + g_m R_{ll}) R_{ss} C_f} \approx \frac{1}{(1 + g_{m1} R_l) R_s C_{gd1}}, \qquad (7\text{-}145)$$

where the typical approximations, $r_{o1} \gg R_l$ and $(R_1 \| R_2) \gg R_s$ are invoked. The corresponding gain-bandwidth product, GBP_1, is

$$GBP_1 = |A_{v1}(0) B_1| \approx \frac{g_{m1} R_l}{(1 + g_{m1} R_l) R_s C_{gd1}} \approx \frac{1}{R_s C_{gd1}}. \qquad (7\text{-}146)$$

where the zero frequency gain, $|A_{v1}(0)| = g_{m1} R_l$, is presumed to be much larger than unity. The actual gain-bandwidth product is smaller than the approximate result in Eq. (7-146) because the time constants precipitated by the transistor gate-source capacitance, $C_i = C_{gs1}$, and the net output port capacitance, C_o, arising from the superposition of bulk-drain (C_{bd1}) and parasitic load (C_l) capacitances, while not dominant, contribute to the net open circuit time constant, whose inverse approximates the bandwidth.

Several time constants contribute to the frequency response properties of the cascode cell in Fig. 7.35(b). Equation (7-55) facilitates understanding the impact on frequency response of the time constants generated by transistor $M1$, while Eq. (7-133) is useful for digesting the significance of the time constants attributed to the $M2$ subcircuit.

(a) The gate-source capacitance, C_{gs1}, of transistor $M1$ forges a time constant of approximately $R_s C_{gs1}$. In view of the fact that the signal source resistance, R_s, is invariably 50 ohms or smaller in high frequency circuit applications, it is unlikely that this time constant appreciably degrades the attainable 3-dB bandwidth of the circuit.

(b) The effective capacitive load imposed on the drain of transistor $M1$ is the sum of the $M1$ bulk-drain capacitance, C_{bd1}, the bulk-source capacitance, C_{bs2}, of $M2$, and the gate-source capacitance, C_{gs2}, of $M2$. Since the resistance that prevails between the drain node at which these capacitances are incident and signal ground is nominally r_i, as defined by Eq. (7-144), the open circuit time constant associated with this net capacitance is $(C_{bd1} + C_{bs2} + C_{gs2})/g_{m2}$. The drain-bulk and bulk-source capacitances in this time constant expression can be reasonably large, particularly when transistor gate widths are large. But the subject time constant is invariably non-dominant if the $M2$ transistor geometry and biasing are chosen to ensure large transconductance.

(c) Without the inserted common gate current buffer, the time constant attributed to the feedback capacitance, $C_f = C_{gd1}$ is approximately

$$[(1 + g_{m1}R_l)R_s + R_l]C_{gd1},$$

where the usual approximations are invoked. With the common gate stage incorporated as shown in Fig. 7.35(b), the effective load resistance driven by the gate of transistor $M1$ is r_i, which by Eq. (7-144) is roughly the inverse forward transconductance of transistor $M2$. It follows that to first order, the open circuit time constant due to C_{gd1} becomes

$$\left[\left(1 + \frac{g_{m1}}{g_{m2}}\right)R_s + \frac{1}{g_{m2}}\right]C_{gd1}. \tag{7-147}$$

The Miller multiplication factor, $(1 + g_{m1}R_l)$, which can be substantial if the common source stage is designed for large gain magnitude, supports the plausibility that capacitance C_{gd1} produces the dominant, and hence bandwidth-determining, time constant in the basic common source circuit. But in the cascode version of the common source amplifier, to which Eq. (7-147) applies, the dominance of the time constant arising from C_{gd1} is dubious if the geometry of transistor $M2$ is selected to ensure $g_{m2}R_l \gg 1$. The subject inequality implies that the effective low frequency load resistance imposed on the drain of transistor $M1$ in the cascode architecture is much smaller than the resistance load seen by $M1$ in the basic common source circuit.

(d) The capacitances evidenced at the drain of transistor $M2$ in Fig. 7.35(b) are the $M2$ bulk-drain capacitance, C_{bd2}, the $M2$ gate-drain capacitance, C_{gd2}, and any load capacitance, C_l, (not explicitly shown in the diagram) encountered at the amplifier output port. By virtue of Eq. (7-141), the low frequency value, $Z_{out}(0)$, of the driving point output impedance is extremely large in that

$$Z_{out}(0) = r_o\left(1 + \frac{R_s}{r_i}\right) = r_{o2}\left(1 + \frac{r_{o1}}{r_i}\right) \approx (1 + g_{m1}r_{o1})r_{o2}. \tag{7-148}$$

It follows that the time constant incurred by the net output port capacitance is

$$R_l(C_{bd2} + C_{gd2} + C_l). \tag{7-149}$$

Because the load resistance, R_l, is necessarily large in high gain applications, the bulk-drain capacitance, C_{bd2}, is invariably large in transistors implemented on chip with large gate widths. Moreover, MOS technology amplifiers routinely drive strongly capacitive load terminations (C_l), whence the time constant defined by Eq. (7-149) is most probably the dominant time constant in the cascode configuration. It follows that the gain-bandwidth product, GBP_2, of the circuit in Fig. 7.35(b) is

$$GBP_2 \approx \frac{g_{m1} R_l}{R_l \left(C_{bd2} + C_{gd2} + C_l\right)} = \frac{g_{m1}}{C_{bd2} + C_{gd2} + C_l}. \qquad (7\text{-}150)$$

A comparison of the last result with Eq. (7-146) confirms that the common source-common gate cascode amplifier is superior from a gain-bandwidth product perspective to its basic common source counterpart if

$$g_{m1} R_s > \frac{C_{bd2} + C_{gd2} + C_l}{C_{gd1}}. \qquad (7\text{-}151)$$

Although the satisfaction of the foregoing inequality is probable whenever the frequency response of a high gain common source stage is dominated by Miller multiplication phenomena, the exploitation of common gate buffering in common source amplifiers by no means guarantees an enhancement of the observable gain-bandwidth product. Fundamentally, the common gate stage serves to attenuate the time constant associated with the gate-drain capacitance of the utilized transistor. In the process of such attenuation, however, the common gate buffer introduces additional time constants attributed to the capacitances implicit to the inserted additional transistor. Answers to questions about the ability of a common gate stage to aid and abet the never-ending quest for additional bandwidth therefore reduce to resolving one fundamental issue. In particular, is the reduction in gate-drain capacitance time constant incurred by the inserted common gate buffer larger than the sum of the additional time constants incurred by the capacitances of the inserted common gate transistor? If the answer is "yes", the common source-common gate cascode amplifier features a gain-bandwidth product that is indeed larger than that evidenced in the simple common source structure.

7.5.3. Enhanced Common Gate Cell

Equation (7-148) bears out the contention that the low frequency driving point output resistance of a common gate stage placed in cascode with a companion common source amplifier is large. The implicitly large transconductance-channel resistance product, $g_{m1}r_{o1}$, of the common source driver underpins this large impedance level at low signal frequencies. But the subject impedance is directly proportional to the channel resistance, r_{o2}, of the common gate subcircuit, which, if realized with minimal geometry devices featuring channel lengths of 130 nanometers or less, is of order of only a few kilo-ohms for routine biasing. In numerous applications, such as transconductor-based integrators used in continuous time biquadratic filters,[14] high gain current amplifiers, and high impedance current sinks and sources, this channel resistance quandary is unacceptable. For these and related other circuit functions, the enhanced common gate amplifier depicted schematically in Fig. 7.36 delivers requisite mitigation of the performance shortfalls incurred by anemic channel resistance.[15] In this diagram, note that active local feedback in the form of the common source amplifier formed of transistor $M3$ and drain circuit resistance R is the

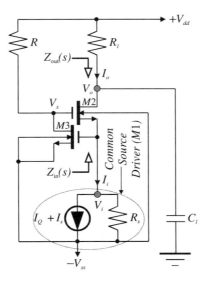

Figure 7.36. Schematic diagram of the enhanced common gate cell. The cell is typically driven by a common source amplifier, which is represented within the dashed oval as a low frequency Norton output port equivalent circuit.

582 Feedback Networks: Theory and Circuit Applications

topological feature that distinguishes the enhanced common gate amplifier from the basic common gate configuration of Fig. 7.31.

The engineering impact of the enhancement scheme postured in Fig. 7.36 is most easily understood by investigating its small signal performance at low signal frequencies. To this end, the pertinent equivalent circuit of the entire enhancement cell is given in Fig. 7.37(a), where use is made of the fact that the both the bulk and the source terminals of transistor $M3$ lie at signal ground, thereby nullifying any bulk transconductance effects in this transistor. Observe that the zero frequency voltage gain, denoted herewith as $-A_{v3}(0)$, of the $M3$ subcircuit is

$$-A_{v3}(0) = \frac{V_{xs}}{V_{is}} = -g_{m3}(r_{o3}\|R). \tag{7-152}$$

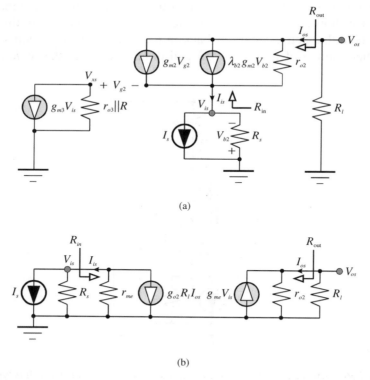

(a)

(b)

Figure 7.37. (a) The low frequency, small signal equivalent circuit of the enhanced common gate amplifier shown in Fig. 7.36. (b) Reduced version of the model in (a).

Analog MOS Technology Circuits

The gate-source signal voltage, V_{g2}, applied to transistor $M2$ is seen to be

$$V_{g2} = V_{xs} - V_{is} = -[1 + A_{v3}(0)]V_{is}, \qquad (7\text{-}153)$$

while the bulk-source signal voltage applied to the same transistor is $V_{b2} = -V_{is}$. Recall in the circuit of Fig. 7.32 that the gate-source signal voltage developed on the common gate transistor is simply the negative of the input port voltage, V_{is}. In contrast, the gate-source voltage established on the common gate device in the enhanced cell is larger by roughly a factor of the gain magnitude of the $M3$ local feedback amplifier. Recalling Eqs. (7-10) and (7-29), this amplified gate-source signal is tantamount to amplifying the transconductance of the common gate transistor by a factor nominally equal to the aforementioned gain magnitude.

Continuing with the analysis of the model in Fig. 7.37(a), the output and input signal currents, I_{os} and I_{is}, respectively, are identical and given by

$$I_{os} \equiv I_{is} = g_{m2}V_{g2} + \lambda_{b2}g_{m2}V_{b2} + g_{o2}(V_{os} - V_{is}),$$

which the foregoing disclosures reduce to

$$I_{os} \equiv I_{is} = -g_{me}V_{is} + \frac{V_{os}}{r_{o2}} = -g_{me}V_{is} - g_{o2}R_l I_{os}, \qquad (7\text{-}154)$$

where

$$g_{me} = g_{o2} + [1 + \lambda_{b2} + A_{v3}(0)]g_{m2}. \qquad (7\text{-}155)$$

Since the conductance, g_{o2}, associated with the channel resistance, r_{o2}, is small, as is the bulk transconductance parameter, λ_{b2}, Eq. (7-155) implies an approximate effective transconductance in the common gate transistor of

$$g_{me} \approx A_{v3}(0)g_{m2} \approx (g_{m3}R)\,g_{m2}, \qquad (7\text{-}156)$$

where the liberty of assuming $r_{o3} \gg R$ has been taken. Equation (7-154) leads immediately to the reduced small signal model of Fig. 7.37(b). In the latter structure, the indicated shunt input resistance, r_{me}, is the inverse of the effective transconductance, g_{me}; that is,

$$r_{me} = \frac{1}{g_{me}} = \frac{r_{o2}}{1 + [1 + \lambda_{b2} + A_{v3}(0)]g_{m2}r_{o2}} \approx \frac{1}{(g_{m3}R)\,g_{m2}}. \qquad (7\text{-}157)$$

7.5.3.1. Low Frequency Circuit Properties

It is a simple matter to show that the global feedback nature of the model in Fig. 7.37(b) yields a low frequency open loop current gain, $A_{io}(0)$, of

$$A_{io}(0) = \left(\frac{r_{o2}}{r_{o2} + R_l}\right)\left(\frac{R_s}{R_s + r_{me}}\right), \quad (7\text{-}158)$$

while the associated loop gain at low frequencies is

$$T_s(0, R_s, R_l) = -g_{o2} R_l A_{io}(0) = -\left(\frac{R_l}{R_l + r_{o2}}\right)\left(\frac{R_s}{R_s + r_{me}}\right). \quad (7\text{-}159)$$

It follows that the resultant low frequency closed loop current gain, $A_{ic}(0)$, is

$$A_{ic}(0) = \frac{A_{io}(0)}{1 + T_s(0, R_s, R_l)} = \left(\frac{R_s}{R_s + r_{me}}\right)\left[\frac{r_{o2}(1 + R_s/r_{me})}{r_{o2}(1 + R_s/r_{me}) + R_l}\right]. \quad (7\text{-}160)$$

This result, which portends a current gain very near unity for reasonable values of source and load resistances, renders transparent a low frequency closed loop output resistance, R_{out}, of

$$R_{\text{out}} = r_{o2}\left(1 + \frac{R_s}{r_{me}}\right) \approx (g_{m2} r_{o2})(g_{m3} R) R_s. \quad (7\text{-}161)$$

In view of the fact that R_s is a device channel resistance when the enhanced common gate cell is driven by a common source driver, it is easy to appreciate the distinct possibility that R_{out} can be enormous. A comparison with Eq. (7-148) specifically reveals that the subject driving point output resistance is larger than that delivered by the traditional common source-common gate cascode amplifier by a factor of the voltage gain magnitude indigenous to the $M3$ subcircuit; that is, by an amount roughly equal to $g_{m3} R$.

The application of signal flow concepts to the model of Fig. 7.37(b) leads directly to an expression for the driving point input resistance, R_{in}. In particular,

$$R_{\text{in}} = \frac{r_{me}}{1 + T_s(0, \infty, R_l)} = r_{me}\left(1 + \frac{R_l}{r_{o2}}\right) \approx \frac{1 + R_l/r_{o2}}{(g_{m3} R) g_{m2}}. \quad (7\text{-}162)$$

Obviously, this input resistance is very small for suitably large values of the subcircuit gain magnitude, $g_{m3}R$. When coalesced with Eq. (7-161) and the fact that the current gain of the circuit at hand is very nearly unity, Eq. (7-162) confirms the propriety of the enhanced common gate cell for current buffering applications.

7.5.3.2. High Frequency Circuit Properties

The foregoing disclosures render transparent the enhanced common gate amplifier as a more effective current buffer than is the traditional, single transistor common gate topology. However, questions naturally arise as to the possible deleterious impact exerted by the additional transistor and resistor (transistor $M3$ and resistor R) on the frequency response properties of the resultantly compensated current buffer.

To the foregoing end, it is unlikely that the net effective capacitance at the source terminal of transistor $M2$ in Fig. 7.36 forms a dominant time constant because of the very small input resistance established at this terminal. On the other hand, time constant dominance invariably prevails at the output port because of the large closed loop driving point output resistance projected by Eq. (7-161). Accordingly, a high frequency investigation of circuit characteristics mandates that the load resistance, R_l, in preceding analytical disclosures be supplanted by a load impedance, $Z_l(s)$, where

$$Z_l(s) = \frac{R_l}{1 + sR_lC_o}, \qquad (7\text{-}163)$$

and capacitance C_o represents the net output port capacitance. It is crucial to underscore the fact that this output capacitance includes actual load capacitance C_l and pertinent transistor capacitances.

The only remaining node of interest is the $M3$ transistor drain, which exudes the bulk-drain capacitance of $M3$ and the Miller multiplied gate-drain capacitance of transistor $M2$. Because the source terminal of $M2$, at which the gate of $M3$ is incident, is a low resistance node, an additional capacitance nominally equal to the sum of the gate-source capacitance of $M2$ and the gate-drain capacitance of $M3$ is manifested at the $M3$ drain. Owing to the low resistance nature of the node to which the $M3$ gate is incident, the net sum of the aforementioned three capacitances at the $M3$ drain site can be expected to establish time constant dominance within the $M3$–R subcircuit. It is therefore reasonable to speculate that the low frequency gain given by Eq. (7-152) can be modified to embrace the salient aspects of high frequency phenomena by writing

$$-A_{v3}(s) = \frac{V_{xs}}{V_{is}} \approx -\frac{g_{m3}(r_{o3}\|R)}{1+s/B_3} = -\frac{A_{v3}(0)}{1+s/B_3} \approx -\frac{g_{m3}R}{1+s/B_3}, \quad (7\text{-}164)$$

where resistance R is tacitly presumed to be significantly smaller than the channel resistance of $M3$, and B_3 is the radial 3-dB bandwidth observed in the $M3$–R amplifier.

Since an insufficient bandwidth, B_3, can degrade the performance enhancement afforded by the $M3$–R compensation circuit, an examination of its dependence on designable circuit parameters and device parameters is a worthwhile undertaking. This exploration commences logically with a determination of the effective small signal resistance, say R_g, established at the drain terminal of transistor $M3$. A problem exercise at the conclusion of this chapter asks the reader to verify that for a circuit resistance, R, that is much larger than the channel resistance, r_{o3}, of $M3$ this resistance is

$$R_g \approx \frac{R}{1+g_{m3}R} \approx \frac{1}{g_{m3}}\left[\frac{A_{v3}(0)}{1+A_{v3}(0)}\right], \quad (7\text{-}165)$$

which is certainly smaller than $(1/g_{m3})$. The effective capacitance, say C_g, with which this resistance interacts to forge a dominant circuit time constant is, making use of the discussion in the preceding paragraph,

$$C_g \approx C_{bd3} + \left(1 - \frac{V_{os}}{V_{xs}}\right)C_{gd2} + C_{gs2} + C_{gd3}, \quad (7\text{-}166)$$

where the voltage ratio, V_{os}/V_{xs}, represents the small signal, low frequency voltage gain from the gate of transistor $M2$ to the drain of the same transistor. For realistic values of $M2$ transconductance and channel resistance, the magnitude of this gain is large, thereby implying that the Miller-multiplied gate-drain capacitance, which is the second term on the right hand side of the foregoing relationship, can be substantial and even dominant. Yet another exercise remanded to the reader at the end of this chapter is the demonstration that

$$\frac{V_{os}}{V_x} \approx -\left(\frac{R_l}{R_l+r_{o2}}\right)\left\{\frac{1+[1+A_{v3}(0)]g_{m2}r_{o2}}{A_{v3}(0)}\right\}$$

$$\approx -\left\{\frac{[1+A_{v3}(0)]g_{m2}(R_l\|r_{o2})}{A_{v3}(0)}\right\}. \quad (7\text{-}167)$$

Assuming that the magnitude of this gain is indeed large, the 3-dB bandwidth, B_3, of the $M3$–R subcircuit derives from

$$\frac{1}{B_3} \approx \left[\frac{A_{v3}(0)}{1+A_{v3}(0)}\right]\left(\frac{C_{bd3}+C_{gd3}+C_{gs2}}{g_{m3}}\right)$$

$$+ \left(\frac{g_{m2}}{g_{m3}}\right)(R_l\|r_{o2})\,C_{gd2}. \quad (7\text{-}168)$$

Whenever the effective value of the load resistance, R_l, is large, the last term in this expression is the most significant. In this event, a large 3-dB bandwidth suggests the necessity of a large $M3$ transconductance, g_{m3}. However, and as is discussed shortly, increasing the transistor transconductance begets increases in device capacitances, which can compromise any response advantages ostensibly accrued by increased transconductance.

An immediate ramification of Eq. (7-164) is that the effective transconductance, g_{me}, in Eq. (7-156) can be replaced by the frequency dependent transconductance, $g_{me}(s)$, such that

$$g_{me}(s) \approx A_{v3}(s)g_{m2} = \frac{g_{meo}}{1+s/B_3}, \quad (7\text{-}169)$$

where

$$g_{meo} \approx A_{v3}(0)g_{m2} \approx (g_{m3}R)\,g_{m2} \quad (7\text{-}170)$$

is the originally deduced approximate transconductance at low signal frequencies. A second upshot is that the first parenthesized factor on the right hand side of the current gain expression of Eq. (7-160) becomes

$$\frac{R_s}{R_s+r_{me}(s)} = \frac{g_{me}(s)R_s}{1+g_{me}(s)R_s} = \frac{g_{meo}R_s}{(1+g_{meo}R_s)(1+s/B_3)}. \quad (7\text{-}171)$$

As expected, this factor degrades from its low frequency value by the effects of the dominant left half plane pole whose frequency is the 3-dB bandwidth of the enhancement subcircuit formed of transistor $M3$ and resistor R. A third ramification is the incurred frequency dependence of the driving point output impedance, $Z_{out}(s)$. Using Eq. (7-161) as an analytical foundation, it can be demonstrated that

$$Z_{out}(s) = R_{out}\left(\frac{1+s/\omega_{z3}}{1+s/B_3}\right)$$

$$\approx (g_{m3}R)(g_{m2}r_{o2})\,R_s\left(\frac{1+s/\omega_{z3}}{1+s/B_3}\right), \quad (7\text{-}172)$$

where the low frequency output impedance value, R_{out}, is given by Eq. (7-161) and

$$\omega_{z3} = (1 + g_{meo}R_s)B_3. \tag{7-173}$$

It is interesting to record that the frequency, ω_{z3}, of the left half plane zero in the foregoing impedance expression is larger, and quite possibly significantly larger, than the unity gain frequency of the enhancement amplifier. Recalling Eq. (7-164), the subject unity gain frequency, say ω_{u3}, is nominally $(g_{m3}R)B_3$. But from Eq. (7-170), Eq. (7-173) is seen to equate to

$$\omega_{z3} = B_3 + (g_{m2}R_s)\omega_{u3}. \tag{7-174}$$

Because ω_{z3} is assuredly larger than B_3, the driving point output impedance at hand is capacitive. Indeed, it is trivial to show that for $g_{meo}R_s \gg 1$, whose validity is the proverbial slam-dunk when the effective source resistance, R_s, derives as the output resistance of a common source driver, Eq. (7-172) can be expanded as

$$Z_{\text{out}}(s) \approx \cfrac{1}{\cfrac{1}{(g_{meo}R_s)r_{o2}} + \cfrac{1}{\cfrac{(g_{m2}R_s)r_{o2}\omega_{u3}}{s} + r_{o2}}}. \tag{7-175}$$

In turn, this result gives rise to the output port impedance model given in Fig. 7.38(a).

If attention now focuses on deducing a Norton, or short circuit, output port model of the enhanced common gate cell, the recollection of Eq. (7-160) renders Eq. (7-171) as the apparent short circuit current gain, $A_{\text{in}}(s)$, of the network. This is to say that load resistance R_l in Eq. (7-160), which must now be replaced by $Z_l(s)$, is set to zero. Thus,

$$A_{\text{in}}(s) = \left.\frac{I_{os}}{I_s}\right|_{Z_l(s)=0} = \frac{g_{meo}R_s}{(1+g_{meo}R_s)(1+s/B_3)} \approx \frac{1}{1+s/B_3}. \tag{7-176}$$

This result, Eq. (7-175), and the impedance model of Fig. 7.38(a) combine to produce the equivalent output port model offered in Fig. 7.38(b).

The transimpedance of the cell in question is, using Fig. 7.38(b), Eqs. (7-163), (7-172), and (7-176), of the form,

$$Z_f(s) = \frac{V_{os}}{I_s} \approx -\left(\frac{1}{1+s/B_3}\right)[Z_l(s)\|Z_{\text{out}}(s)]$$

$$= -\frac{(R_l\|R_{\text{out}})(1+s/\omega_{z3})}{(1+s/B_3)(1+as+bs^2)}, \tag{7-177}$$

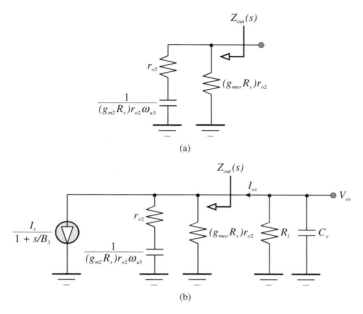

Figure 7.38. (a) The approximate high frequency representation of the driving point output impedance for the enhanced common gate cell of Fig. 7.36. The primary approximation is $g_{meo}R_s \gg 1$. (b) The approximate Norton equivalent circuit of the output port for the enhanced cell.

where

$$a = (R_l \| R_{out})C_o + \frac{1}{B_3}\left(\frac{R_l}{R_l + R_{out}}\right) + \frac{1}{\omega_{z3}}\left(\frac{R_{out}}{R_l + R_{out}}\right), \quad (7\text{-}178)$$

$$b = \frac{(R_l \| R_{out})C_o}{\omega_{z3}}$$

If the time constant established at the amplifier output port is truly dominant, as has been inferred,

$$C_o \gg \frac{1}{R_{out}B_3} + \frac{1}{R_l\omega_{z3}}, \quad (7\text{-}179)$$

whence, Eq. (7-177) collapses to

$$Z_f(s) = \frac{V_{os}}{I_s} \approx -\frac{(R_l \| R_{out})}{(1 + s/B_3)[1 + s(R_l \| R_{out})C_o]}. \quad (7\text{-}180)$$

To the extent that the net output port capacitance, C_o, satisfies the inequality set forth in Eq. (7-179), the time constant, $(R_l \| R_{out})C_o$, establishes a dominant closed loop pole in the enhanced common gate configuration.

This means that the pole evidenced on the right hand side of Eq. (7-180) at $s = -B_3$ is inconsequential to pertinent frequency domain dynamics. In turn, the bandwidth, B_3, of the incorporated common source amplifier formed of transistor $M3$ and resistance R is seen as minimally influencing the closed loop transimpedance response of the enhanced current buffer.

Finally, the high frequency driving point input impedance, $Z_{\text{in}}(s)$, to the enhancement cell is interesting in that it is dominantly capacitive at relatively low signal frequencies, but inductive at high signal frequencies. Upon replacing resistance R_l by load impedance $Z_l(s)$ in Eq. (7-162), and recalling Eqs. (7-169) and (7-170), Eq. (7-162) leads to

$$Z_{\text{in}}(s) = R_{\text{in}} \left\{ \frac{(1 + s/B_3)\,[1 + s\,(R_l \| r_{o2})\,C_o]}{1 + sR_l C_o} \right\}, \qquad (7\text{-}181)$$

where R_{in} is the low frequency input resistance defined by Eq. (7-162). Figure 7.39 depicts the asymptotic frequency domain behavior of this impedance function. Observe that for frequencies below the bandwidth metric, B_3, the impedance mirrors the driving point properties of a low-pass resistance-capacitance network. Above B_3, the monotonic rise of the input impedance magnitude reflects an inductive nature to this impedance function. Indeed, for radial frequencies larger than B_3,

$$Z_{\text{in}}(s) \approx \frac{1}{g_{meo}} + \frac{s}{g_{meo} B_3}, \qquad (7\text{-}182)$$

which suggests an effective input inductance, L_{in}, at high signal frequencies of

$$L_{\text{in}} \approx \frac{1}{g_{meo} B_3}. \qquad (7\text{-}183)$$

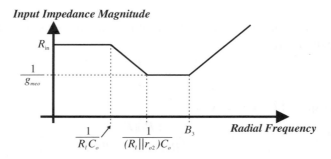

Figure 7.39. The asymptotic frequency domain plot of the driving point input impedance, $Z_{\text{in}}(j\omega)$, for the enhanced common gate circuit in Fig. 7.36.

It is enlightening to note that an effective inductance prevails at high signal frequencies only because the gate of transistor $M2$ is terminated in an effective nonzero resistance that is required implicitly for launching a dominant time constant at the node to which both the gate of $M2$ and the drain of transistor $M3$ are incident.

7.5.3.3. Integrator Application

An implicit signature of the enhanced common gate cell is its potentially huge driving point output impedance. This high impedance level renders its output port naturally vulnerable to even small amounts of capacitive loading, thereby bringing into question the propriety of the enhancement cell for broadband circuit applications. The cell assuredly is not suitable for wideband signal processing. Instead, its high output impedance encourages its utilization as a current output stage which, when loaded in a suitable capacitance, delivers a time domain output response emulating mathematical integration of the applied input signal. As noted earlier, integrators are the electronic core of high performance biquadratic filters synthesized with operational transconductor amplifiers. Other applications of the integrator include frequency to voltage conversion in instrumentation systems and in a ubiquity of network signal processors.

In an attempt to illustrate the engineering foundation that underpins the realization of an electronic integrator, consider Fig. 7.40(a), which abstracts an idealized operational transconductor amplifier (OTA) terminated in a capacitive load. The OTA is ideal in the senses of delivering infinitely large driving point output impedance, infinitely large driving point input impedance, and a short circuit output current that is linearly proportional to the applied input signal, $v_s(t)$. Accordingly, its output port electrical model is the network shown in Fig. 7.40(b), where g_m, which is the frequency domain ratio of the short circuit output current to the transformed input signal voltage, $V_s(s)$, is the transconductance indigenous to the OTA. The model in Fig. 7.40(b) offers

$$V_o(s) = -\frac{g_m V_s(s)}{sC} = -\frac{\omega_m V_s(s)}{s}, \qquad (7\text{-}184)$$

where

$$\omega_m = g_m/C \qquad (7\text{-}185)$$

can be viewed rationally as the unity gain frequency (in units of radians per second) of the circuit at hand. Observe that in the real frequency domain,

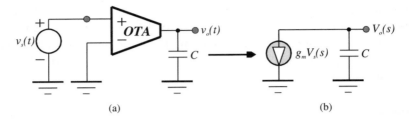

Figure 7.40. (a) An ideal operational transconductance amplifier terminated at its output port in a capacitance. (b) The electrical equivalent circuit of the idealized system in (a).

the magnitude of the voltage transfer function, $V_o(j\omega)/V_s(j\omega)$, diminishes with signal frequency at a rate of 20 decibels/decade. In the time domain, Eq. (7-184) implies

$$v_o(t) = -\omega_m \int_0^t v_s(\tau)\, d\tau, \qquad (7\text{-}186)$$

which codifies the time domain response as a voltage that is proportional to the integral of the input signal; hence, the OTA with capacitive load, or the OTA-C network, functions as an "integrator".

The enhanced common gate amplifier in Fig. 7.36, whose pertinent transimpedance function is the approximate relationship of Eq. (7-180), provides a sturdy foundation upon which a viable integrator can be built. To begin this construction, the subcircuit encircled by the ellipse in the subject figure is supplanted by a simple common source amplifier, thereby resulting in the configuration shown in Fig. 7.41. In this embellished configuration, the nominal value of the signal component, I_{is}, of current I_i approximates the signal source current, I_s, which is $g_{m1}V_s$ in the present case. The rationale underpinning this leap of engineering faith is as follows.

(a) The channel resistance, r_{o1}, of transistor $M1$ in the network undergoing present consideration is significantly larger than the zero frequency value, $Z_{in}(0)$, of the indicated impedance, $Z_{in}(s)$. Equations (7-167) and (7-162) confirm the propriety of this assumption at low to moderate signal frequencies.

(b) It is unlikely that the bulk-drain capacitance of transistor $M1$, as well as any other capacitances effectively appearing at the drain node of $M1$, exert a substantive impact on circuit transfer function and resultant frequency response because the aforementioned zero frequency impedance, $Z_{in}(0)$, is small. The same reasoning ensures that the Miller-multiplied gate-drain capacitance of $M1$ is insignificant.

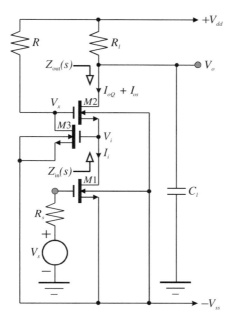

Figure 7.41. The enhanced common gate cell of Fig. 7.36 configured to operate as an integrator. The Norton equivalent circuit of the signal source in the former diagram is realized herewith by transistor $M1$, which functions as a common source amplifier.

(c) The gate-source capacitance of transistor $M1$, as well as any other capacitances effectively appearing at the gate node of $M1$, are unlikely to affect the circuit transfer function and corresponding frequency response if the source resistance, R_s, is suitably small. In effect, the time constant formed by the product of signal source resistance and the net capacitance appearing at the gate node must be much smaller than the dominant time constant established at the amplifier output port.

The foregoing observations allow Eq. (7-180) to be written as

$$\frac{V_{os}}{V_s} \approx -\frac{g_{m1}(R_l \| R_{out})}{(1 + s/B_3)[1 + s(R_l \| R_{out})C_o]}. \quad (7\text{-}187)$$

Let the time constant, $(R_l \| R_{out})C_o$, be selected so that at the lowest radial signal frequency, say ω_L, of interest,

$$\omega_L(R_l \| R_{out})C_o \gg 1. \quad (7\text{-}188)$$

Then, Eq. (7-187) reduces to

$$\frac{V_{os}}{V_s} \approx -\frac{g_{m1}}{sC_o\left(1+s/B_3\right)}. \qquad (7\text{-}189)$$

If in addition to satisfying Eq. (7-187), the bandwidth, B_3, of the M3–R amplifier is large enough to exceed the effective unity gain frequency, $g_{m1}/C_o \triangleq \omega_m$, the integrator design objective is emulated, albeit in a somewhat restricted range of frequencies, in that

$$\left.\frac{V_{os}}{V_s}\right|_{B_3 > g_{m1}/C_o} \approx -\frac{g_{m1}}{sC_o} = -\frac{\omega_m}{s} \qquad (7\text{-}190)$$

for signal frequencies in the range, $\omega_L < \omega < \omega_m$.

The design requirements reflected by Eqs. (7-188) and (7-190) engender two other engineering issues that warrant design attention. The first of these issues is the lowest frequency of integrator operation. In particular, Eq. (7-190) shows that the effective output port capacitance, C_o, must be kept small to achieve an acceptably large unity gain frequency. For small ω_L, this constraint mandates, from Eq. (7-188), a correspondingly large parallel resistance, $(R_l \| R_{\text{out}})$. The driving point output resistance, R_{out}, seen looking into the drain of transistor $M2$ in the diagram of Fig. 7.41 is assured to be large because of the incorporated $M3$–R enhancement subcircuit. Clearly, the indicated load resistance, R_l, must be commensurately large. But since a large passive load resistance is impractical from a circuit biasing perspective, large R_l commands its realization as an active subcircuit functioning as a current source. A single transistor current source delivers a nominal output port resistance of the order of only the drain-source channel resistance of the utilized transistor. This resistance contrasts sharply with resistance R_{out} in Eq. (7-161), which, when applied to the architecture of Fig. 7.141, is the large value gleaned from the product, $(g_{m2}r_{o2})(g_{m3}R)r_{o1}$. Engineering prudence therefore dictates the active load delineated in Fig. 7.42(a), which is the PMOS analog to the previously addressed NMOS enhanced common gate cell. It is not necessary that the equivalent load resistance established by this current source be equal to or even greater than R_{out}; it is necessary only that the effective R_l be comparable to R_{out} so as to abrogate difficulties with respect to satisfying the constraint imposed by Eq. (7-188).

A combination of the circuit diagrams of Figs. 7.42(a) and 7.41 leads to the compensated integrator diagrammed in Fig. 7.42(b). Virtually any engineering modification aimed toward neutralizing a design dilemma

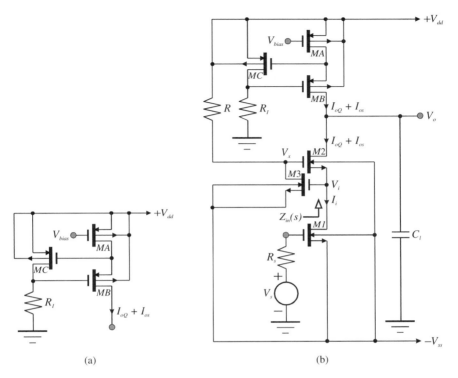

Figure 7.42. (a) An enhanced common gate circuit cell configured with PMOS transistors to operate as a current source. The cell replaces the load resistance, R_l, in the integrator schematic diagram of Fig.7.41. (b) The resultant integrator network, wherein the resistance, R_l, in Fig.7.41 is supplanted by the current source of (a).

precipitates possible other problems. In the present case, the replacement of the load resistance, R_l, in the original version of the integrator diagram shown in Fig. 7.41 by the *MA–MB–MC* current source adds capacitance to the output port, primarily because of the bulk-drain capacitance at the drain node of transistor *MB*. This additional capacitance requires a commensurate reduction in the utilized load capacitance, C_l, to ensure that the net effective output port capacitance, C_o, is maintained at a sufficiently small value appropriate to the unity gain frequency objective of the design realization. Another shortfall of the proposed compensation scheme is the necessity of elevated power supply voltages since static voltage drops must now be impressed across the source to drain terminals of transistors *MA* and *MB*, in addition to the drain to source voltage requirements of *M*1 and *M*2.

A second issue derives from the stipulation that the bandwidth, B_3, of the $M3-R$ amplifier must be larger than the approximate unity gain frequency, g_{m1}/C_o, of the integrator. To the extent that B_3 derives from a dominant pole established at the gate node of transistor $M2$ (or, equivalently, the drain node of $M3$) in Fig. 7.42(b), Eq. (7-168) confirms that a large B_3 translates into a requirement for sufficiently large transconductance, g_{m3}, in transistor $M3$. For a given biasing current conducted by the drain of $M3$, one might be moved to increase g_{m3} through an upward scaling of the gate width, W, of transistor $M3$. Recall, however, from Eq. (7-29) that the forward transconductance is approximately dependent on the square root of device gate dimension W. Thus, increasing g_{m3} by a factor of five, for example, demands a 25-fold increase in gate width W. Now, the unity gain frequency, ω_{T3}, of $M3$ is defined by the transconductance to capacitance ratio, $g_{m3}/(C_{gs3} + C_{gd3})$. Since Eq. (7-45) conveys a nominal independence of device unity gain frequency on gate width W, an increase in g_{m3} by a particular factor is necessarily matched by increases in gate-source and gate-drain capacitances by the same factor. In the present application, capacitive enhancements may prove acceptable because the gate-source and Miller multiplied gate-drain capacitances of $M3$ interact with the implicitly small resistance at the source node of $M2$ to establish a presumably nondominant time constant at the gate input port of transistor $M3$. Care must nonetheless be exercised when adopting the foregoing design tack because the bulk-drain capacitance of $M3$ is proportional to gate width. This embellished capacitance is an implicit component of the presumably dominant time constant evidenced at the $M3$ drain node. It can accordingly degrade bandwidth B_3 to an unacceptable value.

The gate width game played in conjunction with transistor $M3$ can also be applied to the common source driver transistor, $M1$. The apparent ramifications are an increase in transconductance g_{m1}, whence by Eq. (7-190), an increase in the overall unity gain frequency, ω_m, of the integrator ensues. Despite the tantamount increases in relevant gate-source, gate-drain, and bulk-drain capacitances, gate width increases in transistor $M1$ may prove to be a laudable design option in light of the inherently small resistance forged at the gate node of transistor $M3$ and as long as the Thévenin source resistance, R_s, seen by the $M1$ gate is suitably small.

Example 7.6. The integrator shown in Fig. 7.42(b) is to be designed so that is functional in the frequency range, $10\,\text{KHz} < f < 1\,\text{GHz}$. The propriety of the design realization is to be confirmed via small signal HSPICE simulations. The signal source resistance (R_s) is 50 Ω. For the purpose of this exercise, assume the availability of suitably biased, minimal geometry NMOS transistors that deliver the following small signal parameters.

Forward Transconductance	g_m	18 mmhos
Drain-Source Channel Resistance	r_o	22 KΩ
Bulk Transconductance Factor	λ_b	0.04
Gate-Drain Capacitance	C_{gd}	4.9 fF
Gate-Source Capacitance	C_{gs}	66.7 fF
Bulk-Drain Capacitance	C_{bd}	21.7 fF
Bulk-Source Capacitance	C_{bs}	21.7 fF

Assume that the $MA-MB-MC$ current source in Fig. 7.42(b) is designed so that the small signal resistance seen looking into the drain terminal of transistor MB is matched to the driving point output resistance observed in the enhanced common gate cell. In other words, the effective load resistance, R_l, in relevant preceding analytical disclosures is identical to the subject output resistance, R_{out}. Moreover, assume that the capacitances of each PMOS device are nominally 50% larger than the values of respective NMOS device capacitances.

Solution 7.6.

(1) Since the gates of transistors MB and $M2$ comprise low impedance nodes, the nominal net output port capacitance, C_o, is

$$C_o \approx C_l + C_{bd2} + C_{bdB} + C_{gd2} + C_{gdB} = C_l + 66.5 \text{ fF}. \quad \text{(E6-1)}$$

In order to ensure that the unity gain frequency, $\omega_m = g_{m1}/C_o$, of the integrator is relatively independent of the uncertainties that implicitly pervade transistor capacitances, C_l must be suitably larger than the computed parasitic capacitance value of 66.5 fF. Taking 2% as an acceptable parametric uncertainty, let $C_l \approx (50)(66.5 \text{ fF}) = 3.325$ pF. Choosing $C_l = 3.3$ pF, $C_o = 3.367$ pF, which implies that the resultant unity gain frequency of $\omega_m = 2\pi(851.0 \text{ MHz})$ is deficient by a factor of (1 GHz)/(851 MHz) = 1.175. It is therefore necessary to increase the transconductance of transistor $M1$ by at least 17.5% through an increase in its gate width, W_1. To allow for parametric uncertainties and analytical approximations, increase g_{m1} by 25%, which requires that the gate width of $M1$ be elongated by 56.3%. Thus,

$$\left. \begin{array}{l} g_{m1} = (1.25)(18 \text{ mmhos}) = 22.5 \text{ mmhos} \\ C_{gs1} = (1.25)(66.7 \text{ fF}) = 83.38 \text{ fF} \\ C_{gd1} = (1.25)(4.9 \text{ fF}) = 6.125 \text{ fF} \\ C_{bd1} = (1.25)^2(21.7 \text{ fF}) = 33.91 \text{ fF} \end{array} \right\}. \quad \text{(E6-2)}$$

With

$$C_l = 3.3\,\text{pF}, \qquad (\text{E6-3})$$

whence $C_o = 3.367\,\text{pF}$, $\omega_m = g_{m1}/C_o = 2\pi(1.064\,\text{GHz})$, which satisfies the unity gain frequency design objective.

(2) For $R_l = R_{\text{out}}$, $C_o = 3.367\,\text{pF}$, and $\omega_L \leq 2\pi(10\,\text{KHz})$, Eq. (7-188) delivers $R_{\text{out}} \gg 9.454\,\text{MEG}\Omega$, thereby requiring, albeit conservatively, $R_{\text{out}} \geq 94.54\,\text{MEG}\Omega$. Since R_s in Eq. (7-161) is now $r_{o1} = 22\,\text{K}\Omega$, $g_{m3}R \geq 10.85$. Assuming transistor $M3$ is a minimal geometry device, R must be at least $602.8\,\Omega$. Allowing for a nominal 10% analytical uncertainty, choose $R = 670\,\Omega$ as a first iteration of the drain circuit load on transistor $M3$. This choice gives an $M3$–R subcircuit voltage gain magnitude at low signal frequencies of $A_{v3}(0) = g_{m3}(R\|r_{o3}) = 11.7$, or 21.37 dB.

(3) The 3-dB bandwidth, B_3, of the $M3$–R enhancement subcircuit must be larger than the unity gain frequency, ω_m, of the integrator under present consideration. Indeed, B_3 should be at least 2-times to even 3-times larger than ω_m if it is to impact minimally both the magnitude and phase responses of the integrator. In other words, B_3 should approach $2\pi(3 \times 1.064\,\text{GHz}) = 2\pi(3.192\,\text{GHz})$. Equation (7-168) provides the vehicle for testing analytically the propriety of the realized value of B_3. The first term on the right hand side of this time constant relationship computes as 4.775 pSEC, while the second term evaluates as 107.8 pSEC, which confirms Miller multiplication of the gate-drain capacitance in transistor $M2$ as the dominant source of B_3 degradation. Accordingly, $B_3 = 2\pi(1.414\,\text{GHz})$. This bandwidth is indeed larger than the unity gain frequency of the integrator. However at $\omega = \omega_m$, the left half plane, presumably dominant, pole at frequency B_3 contributes $\tan^{-1}(1.065/1.414)$, or almost a $37°$ perturbation to the $90°$ phase shift that characterizes the transfer characteristic of an ideal integrator.

To reduce this phase perturbation, attempts can be made to increase B_3 by increasing g_{m3}, while adjusting resistance R so that the previously determined gain, $A_{v3}(0)$, which sets the driving point output resistance at low frequencies, is maintained. To this end, it is prudent to increase g_{m3} by 3-fold through a 9-fold increase in the $M3$ gate width. As a result,

$$\left.\begin{aligned}
g_{m3} &= (3)(18\,\text{mmhos}) = 54\,\text{mmhos} \\
C_{gs3} &= (3)(66.7\,\text{fF}) = 200.1\,\text{fF} \\
C_{gd3} &= (3)(4.9\,\text{fF}) = 14.7\,\text{fF} \\
C_{bd3} &= (3)^2(21.7\,\text{fF}) = 195.3\,\text{fF}
\end{aligned}\right\}. \qquad (\text{E6-4})$$

With $g_{m3} = 54\,\text{mmhos}$, $g_{m3}(R\|r_{o3}) = 11.7$ requires

$$R = 219\,\Omega. \qquad (\text{E6-5})$$

The first term on the right hand side of Eq. (7-168) now becomes 4.721 pSEC, which is the anticipated modest reduction from the formerly computed first term value. On the other hand, the second term computes to be 35.93 pSEC, which is 3-times smaller than the value calculated earlier. Accordingly, the revised value of B_3 is $B_3 = 2\pi(3.92\,\text{GHz})$, which is larger than the original bandwidth by a factor of almost 2.8.

(4) It should be noted that transistor $M2$ remains a minimal geometry transistor in this design so that its relevant small signal model parameter are

$$\left.\begin{aligned} g_{m2} &= 18\,\text{mmhos} \\ \lambda_{b2} &= 0.04 \\ C_{gs2} &= 66.7\,\text{fF} \\ C_{gd2} &= 4.9\,\text{fF} \\ C_{bd2} &= C_{bs2} = 21.7\,\text{fF} \end{aligned}\right\}. \tag{E6-6}$$

Figure 7.43 displays the salient features of the small signal model pertinent to the completed integrator design. The fruits of HSPICE simulations of this model can confirm the propriety of the design tack exercised herewith. Additionally, these simulations can identify operational shortfalls that warrant more definitive and/or more pedantic design attention.

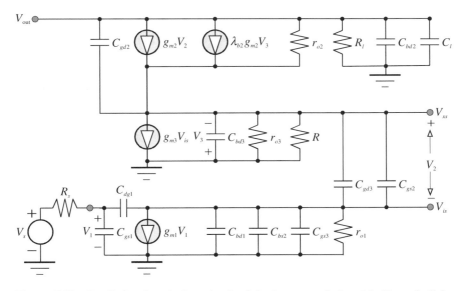

Figure 7.43. Small signal equivalent circuit of the integrator designed in Example 7.6. The Thévenin signal source resistance, R_s, is 50 ohms, while the load resistance, R_l, which is realized by a PMOS enhanced common gate current source, is 105.1 meg-ohms. Capacitance C_l, which is 3.3 pF, establishes a dominant pole at the output port of the network. The transconductances, channel resistances, and internal capacitances for all three utilized transistors are defined in Example 7.6 By Eqs. (E6-2), (E6-4), and (E6-6), the $M2$ bulk transconductance coefficient, λ_{b2}, is 0.04.

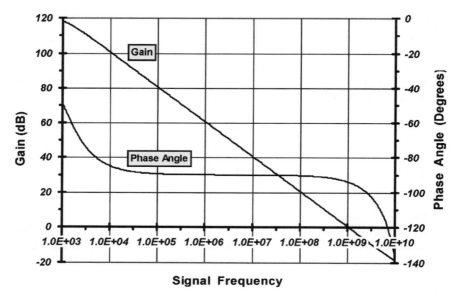

Figure 7.44. Simulated magnitude and phase responses of the integrator designed in Example 7.6. The phase plot is referenced to the zero frequency phase angle value ($-180°$) of the I/O function. The unity gain frequency achieved in this design is 1.08 GHz; at this signal frequency the phase angle is $-94.2°$.

(5) Figure 7.44 displays the HSPICE simulations of both the magnitude and phase responses of the integrator. Over the frequency range, 10 KHz $< f <$ 10 GHz, the magnitude response diminishes with signal frequency at the expected nominal rate of 20 dB/decade. A careful examination of the simulated data infers an achieved unity gain (0 dB) frequency of 1.082 GHz, which differs from the predicted value of 1.064 GHz by only 1.7%.

The phase response delineated in Fig. 7.44 plots I/O phase angle referred to the $-180°$ phase angle evidenced at zero signal frequency. Ideally, this plot should exude a constant $-90°$ phase shift between input signal and output response over the passband of interest. At the unity gain frequency, the observed phase shift is about $-94°$, or $4°$ less than the ideal result. This 4.4% deviation is probably acceptable for many system applications of the integrator. Unfortunately, a 4.4% error is evidenced at the low end of the frequency spectrum at slightly more than 13 KHz, which is better than 3 KHz above the lowest 10 KHz frequency of interest. The "error" at the low edge of the circuit passband can be corrected through an increase in the utilized load capacitance, C_l, or an increase in the effective load resistance, R_l. The latter is the more preferable optimization tack, since an increase in C_l mandates a commensurate increase in the transconductance, g_{m1} of transistor $M1$ if the desired unity gain frequency of the integrator is to be maintained. An increase in either R_l or R_{out} can also be effected by increasing the gain of the $M3-R$ subcircuit, as well as the gain of the corresponding enhancement subcircuit in the PMOS current source. However, it is worth interjecting that an integrator spanning a five-decade frequency range of operation is an ambitious undertaking for most CMOS processes.

(6) The frequency response in Fig. 7.45 is submitted to test the adequacy of the $M3-R$ subcircuit realization. The observed low frequency gain of 21.4 dB precisely matches the design value. However, the 3-dB bandwidth evaluates to only 2.59 GHz. Although this bandwidth satisfies the basic operational constraint of exceeding the unity gain frequency of the integrator, it is 34% smaller than its corresponding design value. Recall that a dominant pole approximation has been invoked to ascertain analytically the bandwidth value. A clue as to the possible impropriety of this presumption is offered by an assiduous inspection of the high frequency rate of magnitude response roll-off in the figure at hand. In particular, the magnitude attenuation rate with high signal frequencies is slightly less than the 6 dB/octave rate that is indicative of a true dominant pole frequency response. Accordingly, the presence of a significant

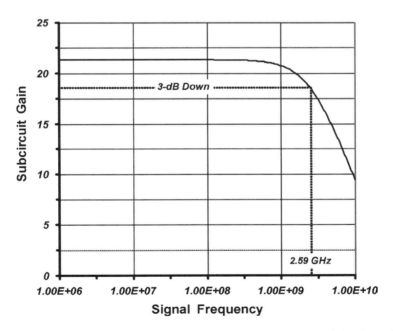

Figure 7.45. Simulated small signal magnitude response of the $M3-R$ subcircuit amplifier in the integrator designed in Example 7.6 and depicted schematically in Fig. 7.42(b). The 3-dB bandwidth achieved in this compensating subcircuit is 2.59 GHz.

transfer function zero and/or nondominant critical circuit frequencies in the $M3-R$ subcircuit can be suspected.

The hypothesis of a nondominant pole frequency response can be scrutinized by appropriate transient response simulations. To this end, let the input signal source, V_s, in Fig. 7.43 and Fig. 7.42(b) be a symmetrical, 100 mV square wave featuring rise and fall times of 2 pSEC, individual pulse widths of 1 nSEC (which synergizes with the 1 GHz high frequency objective), and 2 nSEC period.

The ±100 mV amplitude stipulation is a moot point in light of the linearity implicit to the small signal model of Fig. 7.43. In other words, any amplitude can be chosen without fear of encountering nonlinearity issues, but 100 mV is a practical signal level. In Fig. 7.46, this signal source excitation is superimposed

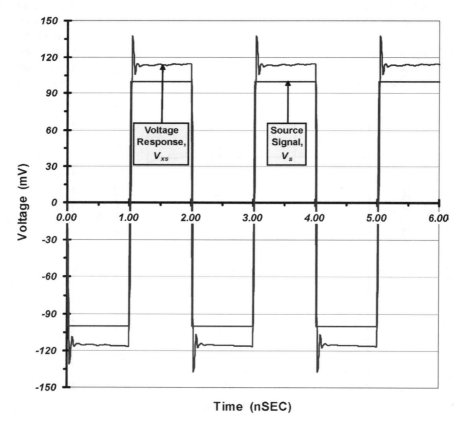

Figure 7.46. The transient square wave response of the $M3-R$ enhancement subcircuit in the integrator designed in Example 7.6. The applied input signal is a symmetrical 100 mV square wave featuring 2 pSEC rise and fall times, 500 pSEC pulse widths, and a 2 nSEC period.

with the resultant voltage response, V_{xs}, which is the small signal component of the voltage, V_x, appearing at the drain terminal of transistor $M3$ in Fig. 7.42(b). Observe the appreciable overshoot and slight undershoot in the V_{xs} waveform at the times where signal V_s undergoes a change of voltage level. Therefore, a second order response in the $M3$–R subcircuit apparently prevails, which renders dubious a bandwidth estimate predicated on pole dominance. Moreover, this viscerally deduced second order characteristic suggests the possibility of optimizing the $M3$–R subcircuit for maximally flat magnitude frequency response, thereby allowing for the large 3-dB bandwidth, B_3, that conduces minimal phase angle error in the neighborhood of the integrator unity gain frequency.

(7) The proverbial proof of the design pudding derives from transient simulations that highlight the output response to a prescribed input signal excitation. The input signal, V_s, can be taken to be the symmetrical square wave described in the preceding section of material. Figure 7.47 depicts the resultant response, V_{out}, shown in Fig. 7.43, where it is understood that V_{out} is the small signal

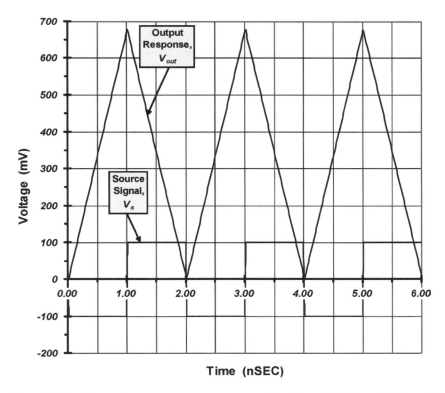

Figure 7.47. The transient square wave response of the integrator designed in Example 7.6. The applied input signal is a symmetrical 100 mV square wave featuring 2 pSEC rise and fall times, 1 nSEC pulse widths, and a 2 nSEC period.

component of the integrator output port voltage, V_o, flagged in Fig. 7.42(b). Since the time domain integration of a constant is a ramp, the responses offered in Fig. 7.47 are reasonable. Because of the phase inversion underscored by the time domain I/O relationship in Eq. (7-186), the output voltage response ramps upward over time when the input signal is negative and vice versa. The maximum value of the triangular wave response is seen to be 675.8 mV. With $\omega_m = 2\pi(1.082\,\text{GHz})$ and $V_s = -100\,\text{mV}$ in the nominal time interval of $0 < t < 1\,\text{nSEC}$, Eq. (7-186) gives a maximum output voltage of $V_{omax} = -2\pi(1.082\,\text{GHz})(-100\,\text{mV})(1\,\text{nSEC}) = 679.8\,\text{mV}$, which is only 0.6% larger than the simulated value. The bottom line is that the circuit appears to be functioning as a viable integrator.

Comments. This example serves to underscore the basic fact that design is not the analytical inverse of analysis; that is, design does not entail the unique solution of a set of N-equations in N-unknown circuit variables. Indeed, a design realization is rarely unique, for the number of equilibrium equations that can be written is invariably not the same as the number of available variables to which suitable numerical values can be assigned. Instead, the "art" of design implicitly embodies the insightful understanding of network dynamics that derives from creative analytical measures, such as those predicated on signal flow theory. In the example at hand, these analyses reveal the pivotally important variables that underpin desired network performance. The solution for these foundational designable elements leads to a first iteration design that can then be refined or optimized by computer aids. Note that the computer aids do not directly embody the undertaken design strategy; instead, they are exploited to refine design results gleaned from realistic first order analytical expositions. And because computer-aided analyses are executed subsequent to definitive circuit analysis, the specific engineering methods conducive to requisite refinements and modifications can often be deduced with relative ease.

References

1. J. Choma, Jr., *Electrical Networks*: *Theory and Analysis* (Wiley-Interscience, New York, 1985), Chapter 12.
2. K. Bondalapati and V. Prasanna, Reconfigurable computing systems, *Proc. IEEE* **90** (2002) 1201–1217.
3. D. P. Foty, *MOSFET Modeling With SPICE*: *Principles and Practice* (Prentice Hall PTR, Upper Saddle River, New Jersey, 1997).
4. W. Liu, *MOSFET Models for Spice Simulation, Including BSIM3v3 and BSIM4* (Wiley-Interscience, New York, 2001).
5. H. Shichman and D. Hodges, Modeling and simulation of insulated-gate field-effect transistor switching circuits, *IEEE J. Solid-State Circuits* **SC-3** (1968) 285–289.

6. J. Meyer, MOS models and circuit simulation, *RCA Review* **32** (1971) 42–63.
7. Y. Tsividis, *The MOS Transistor* (McGraw-Hill Book Co., New York, 1987).
8. D. Ward and R. Dutton, A charge-oriented model for MOS transistor capacitances, *IEEE J. Solid-State Circuits* **SC-13** (1978) 703–708.
9. T. H. Lee, *The Design of CMOS Radio-Frequency Integrated Circuits* (Cambridge University Press, United Kingdom 1998), Chapter 3.
10. A. van der Ziel, *Noise in Solid State Devices and Circuits* (John Wiley and Sons, New York, 1986).
11. J. Choma, Circuit level models of CMOS technology transistors, Univ. Southern California, EE 533ab, Lecture Notes, Fall Semester 2002.
12. R. D. Thornton, C. L. Searle, D. O. Pederson, R. B. Adler and E. J. Angelo, Jr., *Multistage Transistor Circuits*. Semiconductor Electronics Education Committee (SEEC), Vol. 5. (John Wiley & Sons, Inc., New York, 1965), pp. 13–25.
13. J. Choma, Canonic cells of analog MOS/CMOS technology, Univ. Southern California, EE 533ab, Lecture Notes, Fall Semester 2002.
14. R. L. Geiger and E. Sánchez-Sinencio, Active filter design using operational transconductance amplifiers: A tutorial, *IEEE Circuits and Devices Mag.* (1985) 20–32.
15. D. Johns and K. Martin, *Analog Integrated Circuit Design* (John Wiley & Sons, Inc., New York, 1997), pp. 260–264.

Exercises

Problem 7.1
A silicon N-channel transistor is fabricated with an average acceptor impurity concentration in the substrate of 10^{15} atoms/cm^3. The oxide layer has a thickness of 50Å and is observed to support a potential, V_{ox}, of 175 mV, for most practical values of applied gate-source voltage. This oxide layer voltage drop can be presumed to be independent of channel temperature.

(a) Plot the threshold voltage, V_h, of the transistor as a function of temperature over the range of 0°C to 125°C for $V_{bs} = 0$ volt, -2 volts, and -4 volts.
(b) Examine and discuss the generated curves and specifically give the threshold potentials at 27°C and 80°C for each of the three bulk-source voltages of interest.

Problem 7.2
At a reference temperature of 27°C, a silicon N-channel MOSFET is characterized by the following parameters.

Average Substrate Doping Concentration	$(5.5)(10^{14})$ atoms/cm^3
Average Drain/Source Implant Concentration	$(5.0)(10^{21})$ atoms/cm^3
Intrinsic Carrier Concentration	$(1.0)(10^{10})$ atoms/cm^3
Silicon Dielectric Constant	1.05 pF/cm
Silicon Dioxide Dielectric Constant	345 fF/cm
Electron Mobility	480 cm^2/volt-sec
Gate Oxide Thickness	18 Å
Channel Length	180 nM
Gate Width	1800 nM
Zero Bias Threshold Voltage	600 mV
Channel Length Modulation Voltage	16 V

(a) For a null bulk-source bias voltage, calculate and plot the common source, static, reference temperature, volt-ampere characteristics of the subject transistor for gate-source biases of 1 volt, 2 volts, and 3 volts. Allow the plots to span a drain-source voltage range extending from 0 volt to 4 volts.
(b) Repeat Part (a) for the case of an operating temperature of 75°C.
(c) Repeat Part (a) for the case of a bulk-source bias of −1.5 volts.
(d) Repeat Part (a) for the case of a bulk-source bias of −1.5 volts and an operating temperature of 75°C.

Problem 7.3
Reconsider the transistor whose room temperature parameters are delineated in Problem 7.2.

(a) For a null bulk-source bias voltage, calculate and plot the drain current versus gate-source voltage reference temperature characteristics of the subject transistor for drain-source biases of 1 volt, 2 volts, and 3 volts. Allow the plots to span a gate-source voltage range extending from threshold level to 3 volts.
(b) Repeat Part (a) for the case of an operating temperature of 75°C.
(c) Repeat Part (a) for the case of a bulk-source bias of −1.5 volts.
(d) Repeat Part (a) for the case of a bulk-source bias of −1.5 volts and an operating temperature of 75°C.

Problem 7.4
Once again, consider the transistor whose room temperature parameters are delineated in Problem 7.2.

(a) For a null bulk-source bias voltage, calculate the reference temperature values of the small signal forward transconductance, g_{mf}, the small signal bulk transconductance, g_{mb}, and the small signal channel resistance, r_o, for a drain-source bias of 1.8 volts and a drain current of 1.1 mA. Assume saturation regime biasing for the transistor.

(b) Repeat Part (a) for the case of an operating temperature of 75°C.

(c) Repeat Part (a) for the case of a bulk-source bias of −1.5 volts.

(d) Repeat Part (a) for the case of a bulk-source bias of −1.5 volts and an operating temperature of 75°C.

Problem 7.5

A certain NMOS transistor has the following HSPICE Level 3 model parameters, where *MNT* identifies the transistor to the simulator. At this juncture, one need not be concerned about the engineering definitions of the indicated parameters. Indeed, since the Level 3 model is semi-empirical, several of these parameters cannot be interpreted or calculated in terms of physical considerations of either the fabrication process or the layout geometry of the transistor. The transistor is biased with its bulk and source terminals connected together. In the simulation exercises that follow, assume room temperature conditions, and specify $L = 1\,\text{u}$, $W = 100\,\text{u}$, $PS = 103\,\text{u}$, $PD = 103\,\text{u}$, $AS = 150\,\text{p}$, and $AD = 150\,\text{p}$ for the utilized transistor. These latter stipulations are used to calculate intrinsic device capacitances and a few other characteristics. Once again, the indicated numerical values are not necessarily physically sound in the Level 3 model.

.MODEL MNT NMOS(LEVEL = 3 PHI = 0.67 TOX = 9.8 n XJ = 230 n TPG = 1 VTO = 0.63 DELTA = 1.01 + LD = 45 n KP = 168 u UO = 460 THETA = 260 m RD = 2 RS = 2.25 GAMMA = 210 m NSUB = 6.25E17 + NFS = 7.8E11 VMAX = 180 k ETA = 30.83 m KAPPA = 95 m CGDO = 341.5 p CGSO = 328.3 p CGBO = 479 p + CJ = 530.1 u CJSW = 330.1 p MJ = 0.58 MJSW = 0.32 PB = 1.11)

(a) Submit a simulated plot of the static common drain volt-ampere characteristics of the transistor; that is, submit a plot of drain current, I_d, versus drain-source voltage, V_{ds}, for various values of gate-source voltage, V_{gs}. The drain-source voltage range should span 0 through 4 volts, while the considered values of V_{gs} should embrace 1 volt, 2 volts, and 3 volts.

(b) Submit a simulated plot of the static characteristic, I_d versus V_{gs} for suitable values of V_{ds}.

(c) At a quiescent operating point of $V_{gs} = 930\,\text{mV}$, and $V_{ds} = 2.5$ volts, submit frequency response plots of the real and imaginary parts of each of the four common source, small signal, short circuit admittance (y-) parameters.

(d) Use the results gleaned from the simulation exercise of Part (c) to determine the numerical values at a signal frequency of 1 GHz of each of the eight parameters implicit to the generic small signal device model provided in Fig. 7.8(b).

(e) Use relevant y-parameter results disclosed in Part (c) to plot the frequency response of the transistor short circuit current gain.

(f) What transistor unity gain frequency is implied by the model deduced in Part (d)?

Problem 7.6

In addition to operating in cutoff, triode, and saturation regimes, MOSFETs can deliver nonzero drain currents, I_d, in the so-called *subthreshold regime* for which the gate source voltage, V_{gs}, is below the threshold level, V_h, and satisfies the constraint, $0 < V_{gs} < (V_h - 2nV_T)$, where V_T is the Boltzmann voltage and parameter n is numerically between one and two. Subthreshold operation of MOSFETs is useful in applications where low power dissipation is a pivotal consideration and the requisite frequency response is not demanding. An example of such an application is the audio amplifier used in hearing aids. For a drain-source voltage, V_{ds}, satisfying the inequality, $V_{ds} \geq 2nV_T$, the pertinent low frequency drain current expression is

$$I_d = \left(\frac{2K_n}{e^2}\right)\left(\frac{W}{L}\right)(nV_T)^2 e^{(V_{gs}-V_h)/nV_T},$$

where e is the base of natural logarithms, and the remaining parameters reflect those used in conjunction with earlier MOSFET modeling disclosures.

(a) Derive an expression for the subthreshold, small signal forward transconductance with respect to gate-source signal voltage.

(b) Derive an expression for the subthreshold, small signal bulk transconductance with respect to bulk-source signal voltage.

(c) What small signal drain-source resistance is implied by the foregoing volt-ampere expression?

Problem 7.7

In the two-stage feedback amplifier depicted in Fig. P7.7, the response to the input signal represented by the Norton equivalent circuit comprised of

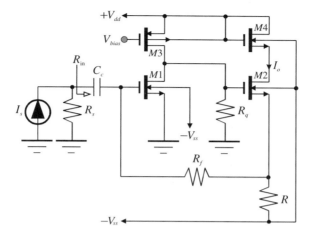

Figure P7.7.

current I_s and source resistance R_s is the small signal component, say I_{os}, of the indicated current, I_o, conducted by transistor, M4. The three N-channel transistors are identical, and all transistors, which are biased for operation in their respective saturation regimes, have negligible drain-source channel conductances, as well as negligibly small bulk-induced threshold voltage modulation. The coupling capacitor, C_c, is sufficiently large to enable its representation as a short circuit for the signal frequencies of interest. Noting that no global feedback arises when the resistance, R_f, is infinitely large, choose $G_f = 1/R_f$ as the "critical feedback parameter" in this problem. Assume that all N-channel transistors are biased identically.

(a) Derive an expression for the current gain, $A_o = I_{os}/I_s$, for the case of $G_f = 0$.
(b) Derive an expression for the return ratio, say T_f, with respect to conductance parameter G_f, and simplify the result for the special case of very large resistance R_q.
(c) Determine an analytical expression for the null return ratio, say T_{fo}, with respect to G_f and simplify the result for the special case of very large R_q.
(d) What is the open loop current gain, say A_{io}, of the circuit if R_q is a large resistance?
(e) Use pertinent foregoing results to derive an expression for the closed loop input resistance, R_{in}. Once again, assume that R_q is a very large resistance.

(f) If the coupling capacitance, C_c, is to emulate a short circuit for radial frequencies larger than ω_L, exploit appropriate preceding results to stipulate a design guideline for the selection of this capacitance.

(g) What small signal output resistance, R_{os}, is "seen" by transistor $M4$?

(h) What effective small signal resistance, R_{d1}, is established between signal ground and the drain terminal of transistor $M1$?

(i) What effective small signal resistance, R_{s1}, is established between signal ground and the source terminal of transistor $M2$?

(j) What effective small signal resistance, R_{d2}, is established between signal ground and the drain terminal of transistor $M2$?

Problem 7.8

Return to the circuit shown in Fig. P7.7 to examine its high frequency performance. Recall that at high signal frequencies, the response of each transistor is affected by numerous device capacitances; namely, gate-source, gate-drain, bulk-drain, and bulk-source capacitances.

(a) Identify the net capacitances that appear, with respect to ground, at the following device nodes:

   i. the drain of transistor $M1$;
   ii. the drain of transistor $M2$;
   iii. the source of transistor $M2$;
   iv. the gate of transistor $M2$.

(b) Using pertinent results gleaned in the preceding problem, stipulate the transistor node, along with the corresponding capacitance appearing at this node, that is likely to establish the dominant pole of the open loop response? Give an expression for the frequency of this pole.

(c) Under open loop operating circumstances, why is self-aligning gate technology far more critical in transistor $M1$ than it is in transistor $M2$?

(d) Assume that the amplifier loop gain delivers a phase angle of ϕ_u at a signal frequency where the magnitude of the loop gain degrades to unity. If this phase angle is inadequate for unity gain closed loop stability, a capacitor, say C_f, might be placed in shunt with resistance R_f in an attempt to circumvent the phase margin shortfall. By what approximate amount does C_f improve or degrade the phase margin that is evidenced before the incorporation of capacitance C_f?

(e) What node is likely to support the dominant pole of the closed loop configuration? Provide engineering rational in support of your predilection.

Problem 7.9

In the source-degenerated, common source amplifier depicted in Fig. P7.9, the feedback resistance, R_{ss}, is 50 Ω, the drain circuit load resistance, R_l, is 1.3 KΩ, and the Thévenin equivalent signal source resistance, R_s, is 75 Ω. The two resistances, R_1 and R_2, which are used to deliver the proper biasing voltage to the gate-source terminals of the transistor, are 1 MEGΩ and 500 KΩ, respectively. The indicated coupling capacitance, C_c, which serves to isolate the signal voltage, V_s, from circuit standby voltages, is chosen large enough to emulate a short circuit over the signal passband of interest. At the quiescent operating point forged by the power supply voltage, V_{dd}, and the two circuit resistances, R_1 and R_2, the transistor operates in saturation and delivers a forward transconductance, g_m, of 35 mmhos, a drain-source channel resistance, r_o, of 18 KΩ, and a bulk transconductance coefficient, λ_b, of 0.07. In the problems that follow, signal flow analysis techniques are to be invoked, with resistance R_{ss} taken as the critical feedback parameter, to evaluate the effective forward transconductance, $G_m = I_{os}/V_s$, where I_{os} is the small signal component of the indicated current, I_o. In this problem, only low signal frequencies are of interest so that all internal device capacitances can be tacitly ignored.

(a) Derive expressions for the return and null return ratios with respect to resistance R_{ss}.
(b) With R_{ss} set to zero, derive expressions for the small signal values of the effective forward transconductance, G_{mo}, driving point input resistance, R_{ino}, and driving point output resistance, R_{outo}.

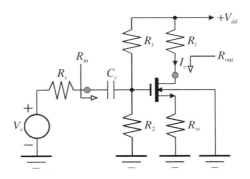

Figure P7.9.

(c) Use the foregoing results to derive closed loop expressions for the small signal values of the effective transconductance, G_m, driving point input resistance, R_{in}, and driving point output resistance, R_{out}.

(d) Numerically evaluate G_m, R_{in}, and R_{out} and from these disclosures, give appropriate analytical approximations of these three circuit performance indices.

(e) If the lowest signal frequency of interest is 10 MHz, what is a suitable value for capacitance C_c?

Problem 7.10

The performance of the amplifier considered in the previous problem is to be assessed at high signal frequencies. To this end, assume that for the quiescent operating point at which the transistor operates, the observed gate-source capacitance, C_{gs}, is 286 fF, the gate-drain capacitance, C_{gd}, is 23.8 fF, the bulk-drain capacitance, C_{bd}, is 89 fF, and the transistor bulk-source capacitance, C_{bs}, is 122 fF. Additionally, parasitic capacitances, C_s and C_l, respectively, are engaged at the amplifier input and output ports, as shown in the diagram of Fig. P7.10. Specifically, $C_s = 2$ pF, while $C_l = 70$ fF.

(a) Derive expressions for, and numerically evaluate, the $R_{ss} = 0$ values of the open circuit pole and zero time constants associated with each of the six (6) delineated capacitances; that is, the four (4) transistor capacitances and the two (2) capacitances that prevail at the I/O ports of the amplifier.

(b) Derive expressions for, and numerically evaluate, the closed loop values of the open circuit pole and zero time constants associated with each of the six (6) network capacitances.

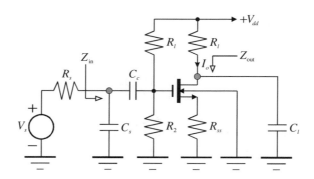

Figure P7.10.

(c) Execute a SPICE frequency response simulation of the entire network. Use the value of coupling capacitance C_c computed in the previous problem. Ensure that the simulation highlights the zero frequency value of the forward transconductance, the so-called *transverse cutoff frequency* at which this transconductance degrades by a factor of the inverse of root two, and the phase angle evidenced at the transverse frequency.

(d) Given the two most significant open circuit time constants associated with the network poles, deduce a second order transfer function appropriate to estimating the frequency response of the effective forward transconductance. Make sure that any zeros precipitated by these two time constants are incorporated in the transfer relationship. Use EXCEL, MATLAB, MATHCAD, or any other appropriate software to plot both the magnitude and phase responses associated with the approximated transfer relationship. Compare these analytical plots with their simulated counterparts and rationalize any significant differences observed over a passband extending to the simulated transverse cutoff frequency.

Problem 7.11

The amplifier in Fig. P7.9 is modified by appending shunt-shunt feedback in the form of resistance R_f between the gate and drain terminals of the transistor, as offered in Fig. P7.11. In view of this modification, the small signal output response is now V_{os}, with the understanding that V_{os} is the small signal component of the indicated output port voltage, V_o. For the numerical portions of the following inquiries, use the numbers provided

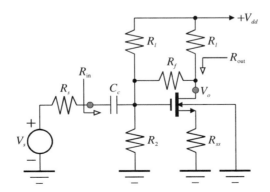

Figure P7.11.

in Problem 7.9, for which the analytical results deduced can be gainfully employed herewith. As usual, the transistor operates in saturation.

(a) Derive expressions for the return and null return ratios with respect to the conductance, G_f, associated with the shunt-shunt feedback resistance, R_f.
(b) With G_f set to zero, derive expressions for the small signal values of the voltage gain, A_{vo}, driving point input resistance, R_{ino}, and driving point output resistance, R_{outo}.
(c) Use the foregoing results to derive closed loop expressions for the small signal values of the voltage gain, A_v, driving point input resistance, R_{in}, and driving point output resistance, R_{out}.
(d) Determine the value of resistance R_f such that the match terminated operating condition of $R_s = R_{in} = R_{out} = R_l \triangleq R$ is realized.
(e) In terms of the match terminated resistance, R, introduced in the preceding part of this problem, give an expression for the match terminated closed loop voltage gain, say A_{vm}, of the feedback amplifier.
(f) Argue the engineering significance of match terminated amplifier operation.

Problem 7.12
The high frequency characteristics of the common source amplifier shown in Fig. 7.10 were analyzed in Section 7.3.0 from two perspectives. The first of these embraced a transfer function whose frequency response was limited by Miller multiplication of transistor gate-drain capacitance, while the second perspective adopted output port time constant dominance as the fundamental approximation. Repeat the analyses documented in Section 7.3.1.3 but now, assume that the input port is the source of time constant dominance. In other words, derive expressions for the frequency of the dominant pole, p_1, the frequency of the nondominant pole, p_2, and the unity gain frequency, ω_u, of the amplifier for the case in which the net effective capacitance across the amplifier input port produces the dominant time constant of the network.

Problem 7.13
Verify the propriety of the simplified input and output impedance models of the common source amplifier offered in Fig. 7.16 by comparing their SPICE I/O impedance simulations with those executed on the "true" small signal model depicted in Fig. 7.13(b). Use the numerical disclosures offered and calculated in Example 7.3. Be sure to compare both impedance magnitude

and phase angle results over a frequency range extending to the neighborhood of twice the 3-dB bandwidth of the considered amplifier. Comment on and rationalize any observed disparities.

Problem 7.14
Reconsider the common source amplifier configured in Fig. 7.17 with an active NMOS load. The transistors embedded in the given circuit architecture are identical, save for the fact that the gate aspect ratio of transistor $M1$ is k^2-times larger than that of transistor $M2$. For simplicity, assume that the biasing resistances, R_1 and R_2, are very large, all drain-source channel resistances (r_o) are sufficiently large to warrant their tacit neglect, and all bulk transconductance factors (λ_b) are negligibly small.

(a) Determine the open circuit time constants attributed to each transistor capacitance.
(b) Assuming a dominant pole response and a relatively large signal source resistance, R_s, give an approximate expression for the 3-dB bandwidth pertinent to the amplifier voltage gain frequency response.
(c) Use the results of Part (b) to derive an approximate expression for the unity gain frequency, say ω_u, of the amplifier.

Problem 7.15
With reference to the CMOS technology amplifier in Fig. 7.18, the P-channel transistor is characterized by a transconductance coefficient of K_p, a zero bias threshold potential of V_{hp}, a channel length modulation voltage of $V_{\lambda p}$, and a gate aspect ratio of $\eta_p = W_p/L_p$. The corresponding parameters of the NMOS device are $K_n, V_{hn}, V_{\lambda n}$, and $\eta_n = W_n/L_n$, respectively. The indicated biasing voltage, V_{g2}, and power supply voltage V_{dd} are selected to ensure that $V_{dd} - V_{g2} > V_{hp}$.

(a) Perform a zero signal ($V_s = 0$), static analysis of the circuit, culminating in a plot of static output voltage (V_{oQ}) versus static input voltage (V_{g1}). The range of considered V_{g1} should embrace $V_{g1} + V_{ss} < V_{hn}$ to $V_{g1} > V_{omax}$, where V_{omax} is the largest attainable static voltage at the circuit output port.
(b) If the circuit at hand is to be utilized as a nominally linear amplifier, what value of V_{g1} ensures maximum symmetrical signal swing about the quiescent output port voltage? Comment as to the possible engineering difficulties that may be encountered in both setting and reliably sustaining this Q-point input voltage.

(c) What is the quiescent output port voltage corresponding to the quiescent input port voltage determined in Part (b)?
(d) The considered circuit can also be used as a logic inverter. To this end, set $V_{ss} = 0$ and determine the logic "1", or "high", level of the output port voltage.
(e) If V_{g1} is set to the logic "1" voltage level found in Part (d), what is the corresponding logic "0", or logic "low", voltage response?

Problem 7.16

The CMOS amplifier in Fig. 7.18 is modified to the topological structure shown in Fig. P7.16, wherein the gate terminals of both the NMOS and the PMOS devices are simultaneously driven by the signal source comprised of voltage V_s and resistance R_s. The biasing supplied by the two voltage sources, V_{dd}, and V_{gg}, are selected to ensure saturation regime operation of both active devices.

(a) By using superposition theory, adapt the CMOS amplifier analysis documented in Section 7.3.3.2 to determine the zero frequency, small signal voltage gain, V_{os}/V_s, of the circuit, where as usual, V_{os} is the small signal component of the net output port voltage, V_o.

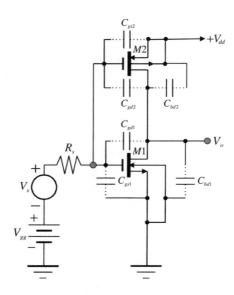

Figure P7.16.

(b) Compare the result obtained in Part (a) with the gain disclosure delineated in Section 7.3.3.2 to ascertain any advantage the modified CMOS amplifier enjoys over its traditional counterpart.

(c) Derive expressions for the open circuit pole and zero time constants attributed to each and every active device capacitance in the circuit of Fig. P7.16. Compare these time constant results with counterpart generalized disclosures provided in Example 7.4 to ascertain any disadvantage the modified CMOS amplifier suffers with respect to the traditional CMOS architecture.

(d) Assuming that the Miller–multiplied component of the net input port capacitance in Fig. P7.16 forges the dominant circuit time constant, what is the resultant 3-dB bandwidth, B, and unity gain frequency, ω_u, of the circuit?

Problem 7.17

Supplement the CMOS amplifier analysis provided in Example 7.4 by deriving approximate, closed form expressions for the driving point input and output impedances, $Z_{in}(s)$ and $Z_{out}(s)$, respectively. Use appropriate software to plot the real and imaginary components of these impedances and conclude whether said impedances are predominantly capacitive or inductive over signal frequency. The numerical circuit and device parameters supplied in Example 7.4 apply herewith.

Problem 7.18

Consider the compensated source follower shown in Fig. 7.23. Assume that the two transistors are identical and are biased in their saturation regions. In the following inquiries, use the numerical parameters provided in Example 7.5.

(a) Design the amplifier so that the zero attributed to the net capacitance observed between the gate and source terminals of transistor $M1$ approximately cancels the dominant circuit pole evidenced prior to the introduction of the compensating capacitance, C_c.

(b) Repeat the gain and impedance analyses implicit to Example 7.5 for the resultantly compensated follower.

(c) Compare the results obtained in Part (b) with SPICE simulations of the small signal model of the compensated source follower. Explain any significant differences between analytical and simulated results.

Problem 7.19

Fig. P7.19 is the basic schematic diagram of the *Wilson current amplifier*, which fundamentally functions as a common gate amplifier, but with potentially greater than unity current gain. The applied input excitation is a current source whose quiescent component is I_Q and whose small signal component is I_s. The Thévenin equivalent resistance of the input source is denoted as R_s. The response to small signal input current I_s is the small signal component, say I_{os}, of the indicated output port current I_o, which is seen as flowing through a load resistance, R_l. All three transistors are biased for operation in their saturated domains. These transistors are identical, save for differences in their gate aspect ratios. In particular, the gate aspect ratio, $\eta_2 = W_2/L_2$, of transistor $M2$ is identical to that of transistor $M1$, but η_2 is k-times larger than the gate aspect ratio, $\eta_3 = W_3/L_3$, of transistor $M3$. Since transistor $M3$ serves as a feedback element that couples the source terminal of M_1 to the M_1 gate, it is natural to choose the transconductance, g_{m3}, of transistor $M3$ as the critical feedback parameter. In the following queries, bulk-induced threshold modulation in all transistors can be ignored, and the drain-source channel resistance of transistor $M2$ can be taken as infinitely large. Do not ignore channel resistances in either $M1$ or $M3$.

(a) Why is it acceptable to ignore the channel resistance in transistor $M2$, but not in either transistors $M1$ or $M3$?
(b) Using signal flow analytical methods, derive expressions for both the open loop and the closed loop values of the low frequency, small signal

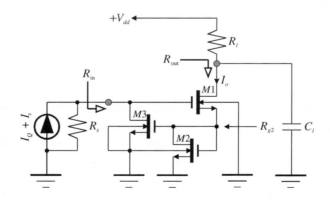

Figure P7.19.

current gain, I_o/I_s. Simplify these expressions for the case of large device channel resistances.

(c) Using signal flow analyses, derive expressions for both the open loop and the closed loop values of the low frequency, small signal driving point input and output resistances, R_{in} and R_{out}, respectively. Simplify these expressions for the case of large device channel resistances.

(d) Repeat Part (c) for the determination of the indicated resistance R_{g2}, established between signal ground and the gate terminal of transistor $M2$. Simplify these expressions for the case of large device channel resistances.

(e) Determine the net effective capacitance that appears between the following circuit nodes:

  i. the drain of transistor $M1$ and ground;
  ii. the source of transistor $M1$ and ground;
  iii. the gate-source terminals of transistor $M1$;
  iv. the gate of transistor $M1$.

Problem 7.20

Reconsider the Wilson amplifier of the preceding problem at high signal frequencies.

(a) Deduce expressions for the open circuit pole and zero time constants generated by each of the net effective capacitances determined in Part (e) of the foregoing exercise.

(b) In view of the fact that the Wilson amplifier enjoys utility in applications that feature relatively large source and load resistances, R_s and R_l, respectively, specify the frequency of the dominant amplifier pole.

(c) In light of the conclusion arrived at in Part (b) of this problem, give expressions for the 3-dB bandwidth, B, and the unity gain frequency, ω_u, of the amplifier.

Problem 7.21

The Wilson amplifier addressed in the preceding two problems is connected in cascode with a PMOS common source amplifier, as shown in Fig. P7.21. The signal component, V_s, of the net input excitation, $V_Q + V_s$, delivers a small signal voltage component, V_{os}, to the indicated net output port voltage, V_o. Given the guidelines for transistors $M1$, $M2$, and $M3$ documented in conjunction with Problem 7.19, and if transistor $M4$ operates in saturation, use the results of Problem 7.19 to give an expression for the low frequency voltage gain, V_{os}/V_s. Do not ignore the channel resistance of transistor $M4$.

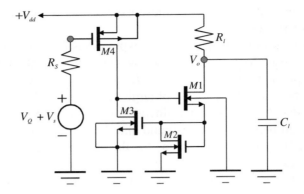

Figure P7.21.

Problem 7.22
In the source follower of Fig. 7.26, give the criterion that the biasing resistance, R_{bias}, must satisfy to ensure that the circuit remains stable even when the signal source, inclusive of its Thévenin resistance, R_s, is disconnected from the circuit input port.

Problem 7.23
In the common gate amplifier schematic diagram of Fig. 7.31, let the gate of the transistor be biased by a voltage source, V_{gg}, whose internal resistance

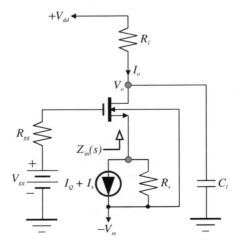

Figure P7.23.

is R_{gg}, as depicted in Fig. P7.23. Despite the additional biasing source, the transistor remains in its saturated domain.

(a) Re-derive the input impedance expression given as Eq. (7-139) for the circuit at hand.
(b) What is the resultant nature of the driving point input impedance, $Z_{in}(s)$, at both low and high signal frequencies?

Problem 7.24

The active voltage divider depicted in Fig. P7.24 is commonly used as a biasing vehicle for the gate terminal of a MOSFET used in an amplifier that is not shown in the figure. In particular, the circuit divides the supply line voltage, V_{dd}, to a voltage, V_{bias}, which is suitable for the gate input port of a subsequent stage. Unfortunately, and as is often the case in electronic circuits, an interference signal, V_{int}, couples to the supply line to perturb the biasing level, V_{bias}. The *decoupling capacitance*, C_d, is therefore appended to the output port of the divider to mitigate the effects of this interference signal. The utilized transistors are identical except for the fact that the gate aspect ratio, $\eta_1 = W_1/L_1$, of transistor $M1$ is not necessarily the same as the gate aspect ratio, $\eta_2 = W_2/L_2$, indigenous to transistor $M2$. Assume that the channel length modulation voltage of each device is much larger than the respective drain-source voltages developed across the individual devices. Assume throughout insignificant bulk-related threshold voltage modulation.

(a) Derive an expression for the indicated biasing voltage, V_{bias}.
(b) The lowest radial frequency implicit to V_{int} is ω_{int}, and the amplitude of the interfering signal at this frequency is 15% of V_{dd}. If the ramification of V_{int} at the output bias port of the divider is to be clamped to no

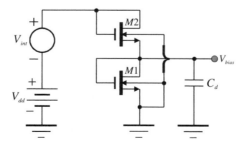

Figure P7.24.

more than 0.5% of V_{bias}, what design constraint must be satisfied by capacitance C_d? Assume that at the frequency, ω_{int}, the impact of all transistor capacitances is miniscule.

Problem 7.25

The circuit in Fig. P7.25 functions as a broadband MOS technology amplifier that is capable of delivering a nominally maximally flat frequency response. All transistors are biased to ensure their operation in saturation. The gate aspect ratios of transistors $M1$ and $M2$ are identical and are k-times larger than the gate aspect ratio of transistor $M3$; the three active devices are otherwise identical. For analytical simplicity, assume that all transistors have negligibly small bulk-induced threshold modulation factors (λ_b) and infinitely large small signal channel resistances (r_o). The resistance, R_2, is very large, while the capacitor, C_c, which couples the signal source comprised of voltage V_s and Thévenin resistance R_s to the gate of transistor $M1$, is chosen to be sufficiently large to ensure that it emulates a signal short circuit over the frequency range of interest. The capacitances delineated as dashed branch elements represent significant capacitances intrinsic to the relevant transistors.

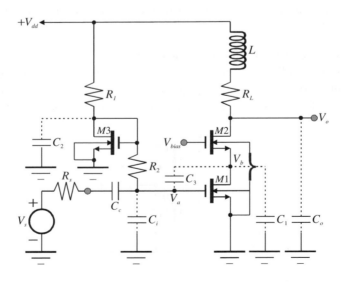

Figure P7.25.

(a) Identify each of the capacitances, C_1, C_2, C_3, C_i, and C_o, in terms of gate-drain, gate-source, bulk-drain, and bulk-source capacitances of respective transistors.

(b) Assuming that the only dominant capacitances are C_i and C_o, draw the small signal equivalent circuit of the entire amplifier. Remember that all transistors have $\lambda_b = 0$ and $r_o = \infty$, and remember further that capacitance C_c behaves as a short circuit for the signal frequencies of interest.

(c) Recalling that resistance R_2 is very large, use the model deduced in Part (b) to determine an expression for the small signal voltage gain, $A_v(s) = V_o(s)/V_s(s)$.

(d) Relate the time constant, L/R_L, to the time constant associated with the capacitance, C_i, so that the amplifier delivers a two pole frequency response having no finite frequency zeros.

(e) In view of the constraint imposed by the result deduced in Part (d), select resistance R_L to achieve a maximally flat magnitude frequency response; that is, a frequency response whose damping factor due to poles is the inverse of the square root of two.

(f) In light of the result gleaned in Part (e), give an expression for the resultant 3-dB bandwidth, B, of the amplifier.

Problem 7.26

Return to the circuit given in Fig. 7.25. Recall that the gate aspect ratio of transistors $M1$ and $M2$ is k-times larger than the gate aspect ratio of transistor $M3$, all transistors have negligibly small λ_b and infinitely large r_o, resistance R_2 is very large, and capacitor C_c emulates an AC short circuit over passband of interest.

(a) If resistance R_1 is chosen to allow transistor $M3$ to conduct a static drain current of I_{d3}, what static drain current is conducted by transistors $M1$ and $M2$?

(b) At low signal frequencies, what is the voltage gain, V_b/V_a?

(c) If the transistors conduct relatively large static drain currents, why is capacitance C_1 unlikely to degrade significantly the frequency response of the network?

(d) Recalling that resistance R_2 is large and assuming that resistance R_1 is relatively small, why is capacitance C_2 unlikely to degrade significantly the frequency response of the network?

(e) Give two reasons why capacitance C_3 is unlikely to degrade significantly the frequency response of the network?

(f) Because capacitance C_3 is connected between the phase-inverting terminals of a common source transistor, $M1$, it is vulnerable to the Miller effect. For forward gain calculations, this capacitance can resultantly be replaced by an effective capacitance placed in shunt with capacitance C_i. What is the value of this effective shunting capacitance?

(g) The deployment of the so-called *shunt peaking* inductance, L, reflects design practices dating back to ancient vacuum tube history. What fundamental purpose is served by this inductance and what possible drawback is indigenous to its use?

Problem 7.27

In the three-stage configuration shown in Fig. P7.27(b), each amplifier is identical and can be modeled in accordance with the small signal equivalent circuit provided in Fig. P7.27(a). Global feedback in the form of the network comprised of resistances R_1, R_2, and R_3 and capacitance C is imposed around the three-stage cascade. The amplifier drives a load consisting of the shunt interconnection of the indicated resistance, R_L, and the capacitance, C_L. Capacitance C is connected across the resistance, R_3, for the purpose of broadbanding the closed loop response. The response variable can be taken as the current, I_o, or the voltage, V_o.

(a) Is the architecture in Fig. P7.27(b) an example of series-series, series-shunt, shunt-shunt, or shunt-series feedback?

(b) Would you expect the input impedance of this architecture to be larger or smaller than the input impedance attained with all three "M" terminals grounded?

(c) Is the amplifier in Fig. P7.27(b) best suited for voltage gain, transimpedance gain, transadmittance gain, or current gain?

(d) Would you expect the input impedance of the subject architecture to be larger or smaller than the input impedance attained with all three "M" terminals grounded?

(e) Use signal flow analysis techniques to derive an expression for the open loop transadmittance of the amplifier.

(f) Determine the loop gain of the amplifier.

(g) Assuming a very large loop gain over the frequency range of interest, derive an expression for the closed loop voltage gain, $A_v(s) = V_o/V_s$.

(h) How might the time constant, $R_3 C$, be selected to achieve a first order broadbanding of the closed loop voltage gain?

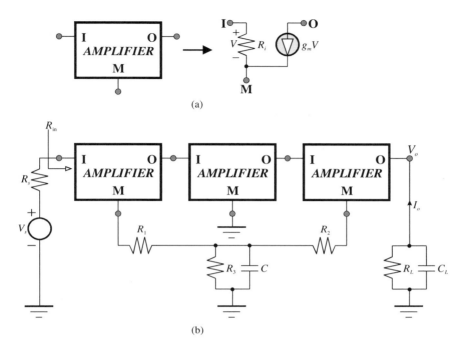

Figure P7.27.

(i) Assuming a very large loop gain, use signal flow theory and concepts to derive an approximate expression for the zero frequency input resistance, R_{in}, seen by the signal source.

Problem 7.28

In a simple common source amplifier utilizing an NMOS transistor biased in its saturation regime, the dominant pole is often established at the output port, principally because of load capacitance, bulk-drain transistor capacitance, and a load resistance that must be sufficiently large to yield a reasonably large voltage gain in the face of traditionally anemic forward device transconductance. One way of compensating such an amplifier for increased circuit bandwidth is to incorporate RC source degeneration, as is depicted by the schematic diagram in Fig. P7.28. In this circuit, C_o represents the net output capacitance, inclusive of all transistor energy storage parasitics, and R_1 and C_1 are the introduced source circuit compensating elements. For simplicity, the transistor channel resistance can be presumed infinitely large, the bulk transconductance factor can be taken to be zero, and the

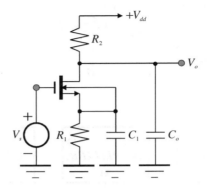

Figure P7.28.

signal source resistance (not shown in the diagram) can be presumed to be zero.

(a) Give a frequency domain expression for the small signal voltage gain, V_{os}/V_s.
(b) Let the zero established by the compensation elements, R_1 and C_1, be chosen to cancel the dominant pole evidenced prior to introduction of the shunt interconnection of R_1 and C_1. For large $g_m R_1$, what is the resultant 3-dB bandwidth of the circuit?
(c) The 3-dB bandwidth determined in Part (b) is the "new," or compensated, dominant pole of the amplifier. What does this pole frequency represent from a circuit-level perspective?
(d) What is the factor by which the original amplifier is broadbanded through introduction of the source degeneration impedance?
(e) Within the context of the stated approximations, why is Miller multiplication of the gate-drain transistor capacitance inconsequential?

Problem 7.29
A typical value for the inductance density of a straight piece of wire is 1 nH/mm. Consequently, it is difficult to construct any "real" circuits with parasitic inductances much smaller than a few nanohenries. This engineering fact of life poses potentially serious problems in common source and common gate cells. To investigate at least one facet of these problems, consider the simple source follower in Fig. P7.29, where parasitic inductance in the amount of L is introduced in series with the gate lead. Although the entire biasing scheme is not depicted, assume that the transistor operates

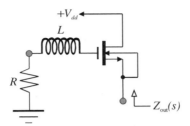

Figure P7.29.

in its saturated domain where the effects of channel resistance r_o can be tacitly ignored. Assume further that self-aligning gate technology reduces the net gate-drain capacitance, C_{gd}, to negligible proportions.

(a) Derive an expression for the indicated output impedance, $Z_{out}(s)$.
(b) Determine the real part of the input impedance function found in Part (a). Cast this result in terms of the unity gain frequency, ω_T, of the transistor.
(c) Above what critical frequency, say ω_{crit}, does a potential instability problem arise?
(d) What inequality must be satisfied by the dimensionless product, $g_m R$, to preclude potential instability?

Problem 7.30
For the enhancement cell depicted in Fig. 7.36, derive an expression for the net resistance established between signal ground and the drain terminal of transistor $M3$. In the process, verify the propriety of Eq. (7-165) when appropriate approximations are invoked.

Problem 7.31
For the integrator of Fig. 7.42(b), develop a second order model that adequately characterizes the small signal frequency response evidenced between the applied signal source, V_s, and the small signal voltage developed across resistance R. Use any approximations deemed appropriate, but rationalize them in view of the relevant integrator discussion propounded in Section 7.5.3.3. Test the validity of the developed circuit model by simulating it under transient square wave input conditions to see if the simulated results garnered herewith reflect those offered in Fig. 7.46.

628   Feedback Networks: Theory and Circuit Applications

Problem 7.32
Derive the gain expression in Eq. (7-167) for the circuit diagrammed as Fig. 7.42(b).

Problem 7.33
Re-simulate the transient square wave response of the integrator modeled in Fig. 7.43 for input signal periods of 500 pSEC and then, 100 pSEC. Compare the results obtained to the simulated plots in Fig. 7.46 and explain any observed differences in the general response characteristics.

Chapter 8

MOS Technology Operational Amplifiers

8.1.0. Introduction

Three reasons motivate an investigation of the monolithic operational amplifier, or op-amp. The first of these reasons is that prudent engineering strategies underlying the design of an operational amplifier can be cultivated with the feedback network and signal flow theories developed in earlier chapters. Second, since its introduction some four decades ago, fundamental op-amp architectures have been continually modified and optimized so that the state of the art, general purpose op-amp is currently among the most widely used of analog integrated circuit cells. Op-amps today find widespread utility in such applications as active filters, signal amplification at moderately high frequencies, low frequency precision amplification, impedance transformation and conversion, and signal processing for data conversion, multimedia, and other applications. The third reason warranting a study of the op-amp derives from the insights garnered about the characteristics and proper utilization of the canonic electronic structures studied in conjunction with MOS technology devices in the preceding chapter. In particular, the topologies of conventional op-amps are an interconnection of these canonic structures. Additionally, the engineering insights that necessarily underpin the analyses of these circuit interconnections, and hence the realization of reliable, high performance op-amps, force engineers to exploit creatively such basic circuit theoretic tools and concepts as Thévenin's and Norton's theorems, frequency and time domain responses, dominant time constants, gain margin, and phase margin.

This chapter begins with a review of the salient properties of prevailing op-amp system architectures. Although the roots of these architectures pertain to bipolar junction transistor technologies, most of them are applicable to op-amp realization in MOS technologies, which are the exclusive focus of this chapter. The circuits indigenous to MOS technology op-amps, many of which are traditional feedback topologies, are then studied. Although the similarity between the circuit schematic diagrams of at least the simpler bipolar op-amp topologies and counterpart MOS op-amps is transparent, their respective design philosophies differ. Bipolar op-amps were first conceived as, and are still largely designed for, standalone active networks that respond to diverse analog circuit and system requirements. As a result, these op-amps must be capable of driving a variety of resistive, capacitive, and even inductive load terminations. In contrast, MOS technology op-amps were originally developed because of burgeoning interests in implementing subcircuits on integrated circuit chips that perform both analog and digital signal processing functions. The fact that the digital subcircuits implicit to virtually all such mixed signal processors are fabricated in MOS technology motivated analog MOS technology as a means to circumvent potential interface dilemmas between digital and analog components. But because MOS op-amps utilized in mixed signal integrated circuits generally drive only the gate terminals of MOS logic inverters and other digital cells, most need be designed to function properly with only high impedance and otherwise dominantly capacitive loads.

8.2.0. Op-Amp System Architectures

The majority of operational amplifiers can be codified as one-stage, two-stage, or three-stage architectures. Two-stage structures predominate in the extant state of the art. In particular, application specific MOS technology op-amps used in switched capacitor and mixed signal large scale integration (LSI) or very large scale integration (VLSI) systems are two-stage topologies. Single stage op-amp configurations enjoy popularity when the primary design objective is a broadband closed loop response, as opposed to very large open loop gain. Of the three-fundamental architectures, three-stage op-amps, which are not addressed in this chapter, are the most difficult to realize as unconditionally stable entities.[1] Nonetheless, they are utilized in precision instrumentation systems and in other applications for which high open loop gain is a priority.

8.2.1. Single Stage Architecture

The system level diagram of a single stage op-amp is shown in Fig. 8.1(a). Gain in this system is provided exclusively by the input stage, which

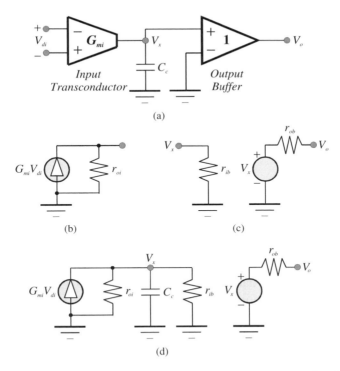

Figure 8.1. (a) The system schematic diagram of a one-stage operational amplifier. (b) Small signal, low frequency equivalent circuit of the output port for the input transconductor. (c) Small signal, low frequency equivalent circuit of the output stage voltage buffer. (d) Complete small signal, low frequency equivalent circuit of the entire single stage op-amp. The output port is held open circuited.

functions as a transconductor that converts the applied *differential input signal*, V_{di}, which is not referenced to signal ground, to a *single ended output voltage*, V_x, that is developed at the interstage node with respect to ground. Assuming that all capacitors intrinsic to the input transconductor can be ignored, its output port can be represented as the memoryless Norton equivalent circuit depicted in Fig. 8.1(b). In this model, G_{mi} is the short circuit transconductance of the input transconductor, and r_{oi} is its high output port resistance. The output stage, whose first order model is offered in Fig. 8.1(c), is a unity gain voltage buffer featuring a very high input port resistance, r_{ib}, and a very low output resistance, r_{ob}. The interstage shunt capacitance, C_c, whose value is generally large enough to mandate its implementation as an off chip element, assures closed loop stability by establishing a dominant open loop pole.

The resultant small signal model of the single stage op-amp architecture depicted in Fig. 8.1(d) suggests a dominant pole at the radial frequency, ω_c, such that

$$\omega_c = \frac{1}{(r_{oi} \| r_{ib})C_c}. \tag{8-1}$$

Moreover, the open circuit, open loop voltage gain, $A_{ol}(s)$, implied by the model at hand is

$$A_{ol}(s) = \frac{V_o(s)}{V_{di}(s)} = \frac{G_{mi}(r_{oi} \| r_{ib})}{1 + \dfrac{s}{\omega_c}}. \tag{8-2}$$

Three conditions must be met to render viable the single stage op-amp configuration for general purpose analog signal processing applications. First, the buffer output resistance, r_{ob}, must be significantly smaller than the magnitude of any impedance appended as an open loop load termination on the subject op-amp. Second, a large zero frequency open loop gain, $A_{ol}(0) = G_{mi}(r_{oi} \| r_{ib})$, must prevail to ensure that closed loop responses are nearly independent of the parametric vagaries implicit to the transconductance and driving point resistance parameters evidenced on the right hand side of Eq. (8-2). If capacitor C_c is chosen to ensure that the frequency of the resultant dominant pole of the network is indeed ω_c, the third mandate is that the unity gain frequency,

$$\omega_u \approx \frac{G_{mi}}{C_c}, \tag{8-3}$$

of the open loop response must be sufficiently smaller than the individual frequencies associated with corresponding nondominant poles and any right half plane zeros that may be manifested by the circuit realization. The last requirement supports the fundamental design objectives of adequate phase and gain margins for all practical closed loop responses.

8.2.2. Two-Stage Architecture

As in its single stage counterpart, a transconductor comprises the first stage of the two-stage op-amp abstracted in Fig. 8.2(a). The Norton output current, $G_{mi}V_{di}$, of this stage drives the input port of a phase inverting voltage amplifier whose open circuit voltage gain is $(-A_2)$, and whose input port and output port resistances are r_{i2} and r_{o2}, respectively, as modeled in

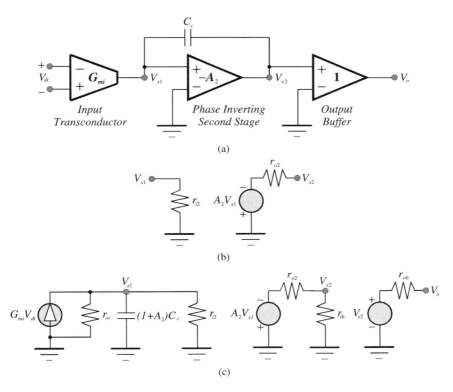

Figure 8.2. (a) The system schematic diagram of a two-stage operational amplifier. (b) Small signal, low frequency equivalent circuit of the phase inverting second stage amplifier. (c) Complete small signal, low frequency equivalent circuit of the entire two-stage op-amp. The output port is held open circuited.

Fig. 8.2(b). In addition to providing enhanced open loop gain, the second stage exploits Miller capacitance multiplication to realize dominant pole compensation without the need for off chip capacitance. In particular, the Miller effect of the compensation capacitance, C_c, results in a comparatively large effective capacitance, $(1 + A_2)C_c$, across the input port of the second stage. Assuming that the output resistance, r_{oi}, of the transconductor, the second stage input resistance, r_{i2}, and the second stage gain magnitude, A_2, are sufficiently large, a dominant open loop pole can be forged with a small, and therefore monolithically realizable, capacitance, C_c. The output response of the second stage amplifier is applied to the voltage buffer, which is incorporated to facilitate the general purpose nature of the op-amp architecture for a diversity of load terminations. The resultant open circuit, open loop, small signal model of the two-stage op-amp is the network depicted in Fig. 8.2(c), in which the feedback element, C_c, is presumed to be the only significant capacitance in the basic architecture of Fig. 8.2(a).

The simplified equivalent circuit in Fig. 8.2(c) confirms a dominant pole radial frequency, ω_c, of

$$\omega_c = \frac{1}{(r_{oi} \| r_{i2})(1 + A_2)C_c}. \tag{8-4}$$

If the second stage output port resistance, r_{o2}, is much smaller than the input port resistance, r_{ib}, of the buffer, the open circuit, open loop voltage gain is seen to be

$$A_{ol}(s) = \frac{V_o(s)}{V_{di}(s)} = -\frac{G_{mi} A_2 (r_{oi} \| r_{i2})}{1 + \dfrac{s}{\omega_c}}. \tag{8-5}$$

8.2.3. Input Stage Transconductor

The preceding two subsections of material make clear the fact that both the single stage and the two-stage architectures exploit transconductance signal processing of applied input port signals as the first stage of an operational amplifier. Apart from providing input voltage to output current conversion, the input transconductor stage also realizes *differential to single ended conversion*. Specifically, the transconductor converts a differentially applied input signal that is not referenced to ground to an output response that is referred to ground and is directly proportional to the differential input excitation. In MOS technology, the quest for topological simplicity has engendered only two forms of op-amp input stage transconductors. These two structures are the *P-channel driver transconductor* and the *N-channel driver transconductor*.

The system level schematic diagram of the *P*-channel driver transconductor appears in Fig. 8.3(a), where V_{s1} and V_{s2} denote independent signal voltages applied to the two input terminals of the identified differential amplifier. Any significant Thévenin resistances associated with these two signal sources are presumed absorbed into the two input ports of the amplifier. The difference between signal voltages V_{s1} and V_{s2}, which is cleverly termed the *differential input signal*, V_{di}, is

$$V_{di} = V_{s1} - V_{s2}. \tag{8-6}$$

It is expedient to introduce the concept of the *common mode input signal*, say V_{ci}, which is the high brow equivalent of the average of the two applied

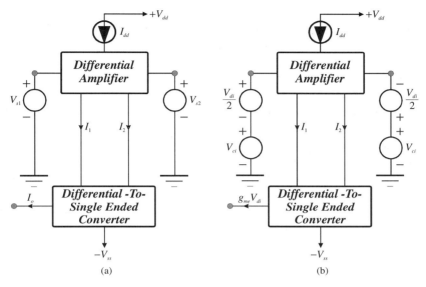

Figure 8.3. (a) System schematic abstraction of a P-channel transconductor serving as the input stage driver in an op-amp. The input signals are represented as the voltages, V_{s1} and V_{s2}. (b) The system diagram of (a), but with the applied input signals decomposed into their differential mode, V_{di}, and common mode, V_{ci}, components.

signal voltages, V_{s1} and V_{s2}; that is,

$$V_{ci} = \frac{1}{2}(V_{s1} + V_{s2}). \tag{8-7}$$

Equations (8-6) and (8-7) can be solved for V_{s1} and V_{s2} to write

$$\left. \begin{array}{l} V_{s1} = V_{ci} + \dfrac{V_{di}}{2} \\[6pt] V_{s2} = V_{ci} - \dfrac{V_{di}}{2} \end{array} \right\}, \tag{8-8}$$

which suggests the alternative depiction of the P-channel driver transconductor offered in Fig. 8.3(b).

At least two important engineering points underlie the representation in Fig. 8.3(b). First, it is clear that V_{di} is literally the *difference* of the two applied input signal voltages, V_{s1} and V_{s2}. On the other hand, V_{ci} represents a component of these two signals that is *commonly applied* to the two input ports of the differential amplifier. Note, for example, that if V_{di} were zero, which would imply $V_{s1} = V_{s2}$, each of the two input ports of the differential amplifier is excited by the same level of signal; namely, V_{ci}. Despite the fact that each differential amplifier input port is energized by a superposition of a common mode component and a component proportional to the differential mode input signal, the idealized response of the input stage transconductor is a signal current that ideally is dependent on only the differential mode input signal. This current response is denoted in the diagram as $g_{me}V_{di}$, where g_{me} symbolizes an effective forward transconductance linking the differential input signal voltage to the single ended, Norton output current. In effect, the input stage transconductor rejects the common mode input excitation in that it responds only to difference signals applied across the differential amplifier input ports.

Common mode signal rejection addresses the fact that undesirable input signals are ubiquitous in all electronic systems. These signals derive from spurious electromagnetic radiation, the radiation fields emanating from local fluorescent lights and other electronics or electrical hardware proximately located to the input transconductor, and numerous other sources. If the two differential amplifier input ports are laid out closely to one another, any parasitic signal energy fields in which the differential pair is immersed are likely to produce equivalent parasitic voltages at the two input ports. If spurious inputs are indeed observed as common mode inputs, the indicated response at the output port of the differential to single ended converter is ideally oblivious of such spurs. In effect, the properly designed differential amplifier is a "smart" circuit in that it selectively processes desirable (differential) inputs and rejects undesirable (common mode) energies.

A second noteworthy point is that a necessary condition for the realization of the foregoing rejection characteristics is that the differential amplifier be a *balanced circuit*.[2,3] Balance, as applied to differential circuits, enjoys a plethora of design and performance implications. For the present, it suffices to state that balance in a differential amplifier requires the appropriate interconnection of two identical amplifier cells. Moreover, balance implies short circuit, or Norton equivalent, values of the indicated net output currents, I_1 and I_2, of the differential amplifier that are expressible as

$$\left.\begin{aligned} I_1 &= I_Q + g_{mc}V_{ci} + g_{md}\left(\frac{V_{di}}{2}\right) \\ I_2 &= I_Q + g_{mc}V_{ci} - g_{md}\left(\frac{V_{di}}{2}\right) \end{aligned}\right\}, \qquad (8\text{-}9)$$

where I_Q is the quiescent current required to support nominally linear operation of the differential amplifier, g_{mc} is the *common mode transconductance* of the subject amplifier, and g_{md} is the *differential mode transconductance*. Observe that the quiescent current, I_Q, is effectively a component of the common mode output current response in that I_Q is common to both output currents. Accordingly, small perturbations in I_Q incurred by thermal sensitivities, power supply variations, and the like are rejected by the differential output current, $(I_1 - I_2)$, which is simply $g_{md}V_{di}$, independent of I_Q and the common mode input signal. The transconductance parameters, g_{md} and g_{mc}, as well as other circuit performance indices of balanced differential amplifiers, can be extracted efficiently from pertinent small signal equivalent circuits configured as half circuit model topologies. The reader lacking confidence in half circuit analysis methodologies is well-advised to review the basic electronics literature.[3,4]

If the common mode input signal is to exert minimal influence on the individual, single ended output currents, I_1 and I_2, it is necessary that the common mode transconductance, g_{mc}, be much smaller than its differential counterpart, g_{md}. To this end, a commonly invoked figure of merit is the *common mode rejection ratio*, ρ, where

$$\rho \triangleq \frac{g_{md}}{g_{mc}}. \tag{8-10}$$

Ideally, the magnitude of ρ tends toward infinity, thereby implying $g_{mc} = 0$, whence single ended Norton current responses divorced of common mode components are established. As is illustrated shortly, an acceptably large common mode rejection ratio hinges on the ability to realize the current source, I_{dd}, in Fig. 8.3 with sufficiently large Thévenin equivalent signal resistance.

The N-channel version of the input stage transconductor is shown in Fig. 8.4. The architectures of the two systems are virtually identical. They differ only in the fact that while a quiescent current of I_{dd} is sourced to the P-channel differential amplifier, a corresponding biasing current, I_{ss} is sunk from the N-channel differential network. The circuit performance disclosures proffered for the P-channel circuit apply equally well to the N-channel structure except that large common mode rejection ratio in the N-channel unit relies now on a sufficiently large Thévenin resistance associated with the current sink.

8.3.0. CMOS Input Stage Analysis

As already noted, the input stage of a two-stage op-amp is an interconnection of three subcircuits. The differential amplifier functions as a transconductor that converts an input differential signal voltage-to-balanced

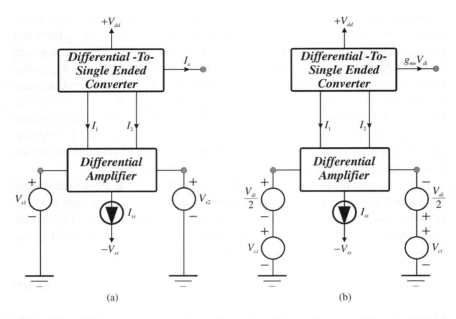

Figure 8.4. (a) System schematic abstraction of an N-channel transconductor serving as the input stage driver in an op-amp. The input signals are represented as the voltages, V_{s1} and V_{s2}. (b) The system diagram of (a), but with the applied input signals decomposed into their differential mode, V_{di}, and common mode, V_{ci}, components.

Norton, or short circuit, output currents. These output differential currents activate a differential to single ended converter whose response can be viewed as either a single ended Thévenin signal voltage proportional to the applied differential input excitation or a single ended Norton current that is likewise linearly related to the differential input. The third subcircuit is either a current source or a current sink whose small signal terminal resistance must be sufficiently large to achieve adequate common mode rejection in the amplifier-converter topology.

8.3.1. *P*-Channel Transconductor

The schematic diagram of a CMOS P-channel driver transconductor is given in Fig. 8.5, where it is understood that all transistors are biased to operate in their saturation regimes. For simplicity, the diagram does not depict the connection of the substrate bulk terminals of all P-channel transistors

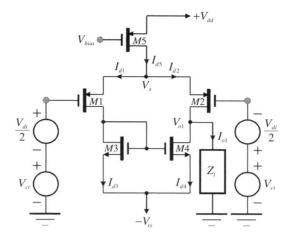

Figure 8.5. Schematic diagram of the input P-channel driver transconductor for a two-stage op-amp. All transistors operate in saturation and have their substrate bulk terminals connected to the appropriate power supply voltage lines.

to the positive, $+V_{dd}$, power rail, nor does the figure show the incidence of the bulk terminals of all N-channel devices with the negative, $-V_{ss}$, supply line. These traditional substrate connections are tacitly presumed in all forthcoming MOS technology circuit investigations in this chapter.

Transistors $M1$ and $M2$ comprise the active elements of the differential amplifier. These transistors are ideally an identical pair, inclusive of equal gate aspect ratios. Transistor $M5$ supplies nominally constant current to the source terminal interconnections of $M1$ and $M2$. If $M1$ and $M2$ operate as balanced devices in the sense of the current stipulations in Eq. (8-9), transistor $M5$ necessarily sources twice the quiescent drain current conducted by either $M1$ or $M2$. In order to avoid an excessive thermal gradient between the current source transistor, $M5$, and each active device in the differential transistor pair, $M1$ and $M2$, the gate aspect ratio of $M5$ must be double that of either $M1$ or $M2$. This necessity derives from the fact that the heat produced by the power dissipation incurred by current flow in narrow device channels is proportional to the density of channel current. Increasing the gate aspect ratio of the device forced to conduct large drain current by increasing its gate width proportionately increases the channel cross section area pierced by the drain-to-source channel current, thereby leading to an equalization of current density. This design tack also ensures that the requisite biasing of the source-gate terminals of transistor $M5$ is constrained to a level that approximates the source-gate bias voltages imposed on $M1$ and $M2$.

The current outputs of $M1$ and $M2$ drive the matched N-channel transistors, $M3$ and $M4$, which function as a current mirror to produce the output current, I_{o1}, flowing in the generalized load impedance, Z_l. While $M1$ and $M2$ are matched devices, as are $M3$ and $M4$, the identical gate aspect ratios of $M3$ and $M4$ are not necessarily the same as those of $M1$ and $M2$. This situation, apart from the fact that $M1$ and $M2$ are P-channel units while $M3$ and $M4$ are N-channel devices, means that the small signal parameters of $M3$ and $M4$ are likely to differ from the corresponding counterpart parameters of $M1$ and $M2$.

8.3.1.1. Transconductor Small Signal Analysis

The low frequency small signal analysis of the transconductor in Fig. 8.5 is best accomplished by first determining the small signal Norton equivalent circuits established at the drain terminals of transistors $M1$ and $M2$. To this end, the differential mode half circuit schematic diagram of the $M1-M2$ pair is given in Fig. 8.6(a), while the corresponding small signal equivalent circuit appears as Fig. 8.6(b). For common mode excitation, Fig. 8.6(c) depicts the applicable circuit schematic diagram, and its small signal model is the topology in Fig. 8.6(d). In the two small signal models at hand, g_{m1} is the forward transconductance of transistor $M1$, r_{o1} is the drain-source channel resistance of $M1$, λ_{b1} is the bulk transconductance factor of $M1$, and finally, r_{o5} symbolizes the drain-source channel resistance of transistor $M5$. With reference to Fig. 8.6(a), the current, $I_{d1s}/2$, is one-half of the differential mode small signal component of the net current, I_{d1}, delineated in Fig. 8.5. In concert with the small signal MOS transistor models developed in the preceding chapter, the direction of this current is taken as conduction from the drain of $M1$-to-the source of $M1$. On the other hand, the current, I_{d1n}, in Fig. 8.6(b) is the Norton, or short circuit value, of current I_{d1s}. Analogously, I_{c1s} in Fig. 8.6(c) is the common mode signal component of I_{d1}, while I_{c1n} in Fig. 8.6(d) is the Norton value of I_{c1s}.

An inspection of the model in Fig. 8.6(b) leads to

$$\frac{I_{d1n}}{2} = g_{m1} V_a = g_{m1} \left(\frac{V_{di}}{2} \right),$$

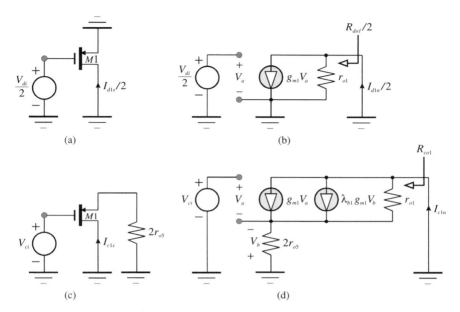

Figure 8.6. (a) Differential mode half circuit schematic diagram of the $M1-M2$ differential amplifier in Fig. 8.5. (b) Small signal model of the differential mode half circuit in (a). (c) Common mode half circuit schematic diagram of the $M1-M2$ differential amplifier in Fig. 8.5. (d) Small signal model of the common mode half circuit in (c).

whence a differential mode Norton transconductance, g_{md}, of

$$g_{md} = \frac{I_{d1n}/2}{V_{di}/2} = g_{m1}, \qquad (8\text{-}11)$$

which is simply the forward transconductance of either $M1$ or $M2$. Also by inspection, the shunt resistance presented to the output port of the model undergoing investigation is r_{o5}. This resistance is one-half of the small signal differential output resistance, say R_{do1}, presented between the drain terminals of the $M1-M2$ pair. Accordingly,

$$R_{do1} = 2r_{o1}. \qquad (8\text{-}12)$$

A straightforward analysis of the common mode model in Fig. 8.6(d) delivers a common mode Norton transconductance, g_{mc}, that is expressible as

$$g_{mc} = \frac{I_{c1n}}{V_{ci}} = \frac{g_{m1}}{\rho_1}, \qquad (8\text{-}13)$$

where ρ_1, the common mode rejection ratio of the differential amplifier formed of transistors $M1$, $M2$, and $M5$, is

$$\rho_1 = 1 + 2[1 + (1 + \lambda_{b1})g_{m1}r_{o1}]\left(\frac{r_{o5}}{r_{o1}}\right) \approx 1 + 2g_{m1}r_{o5}, \quad (8\text{-}14)$$

and the approximation reflects the reasonable assumptions, $\lambda_{b1} \ll 1$ and $g_{m1}r_{o1} \gg 1$. Clearly, large common mode rejection ratio relies on large channel resistance, r_{o5}, which may require that transistor $M5$ be synthesized as a relatively long channel device. A continuing analysis of the model at hand produces a small signal, common mode output resistance, R_{co1}, of

$$R_{co1} = \rho_1 r_{o1}, \quad (8\text{-}15)$$

which can be enormously large if the common mode rejection ratio, ρ_1, of the input differential circuit is large.

8.3.1.2. Transconductor Output Macromodel

In view of the tacit presumption of reader familiarity with differential electronics, it may be fruitful to place Eqs. (8-11) through (8-15) into engineering perspective. To this end, Fig. 8.7(a) captures the salient features of the small signal schematic version of the $M1-M2-M5$ subcircuit in Fig. 8.5. Figure 8.7(b) depicts the Norton macromodel effectively witnessed by the differential-to-single ended converter under small signal operating conditions. In view of the differential transconductance found in Eq. (8-11), a current source, $g_{m1}(V_{di}/2)$, directed from the drain to the source of $M1$ is shown on the left side of this macromodel. On the other hand, a dependent current, whose magnitude is precisely the same as that of the dependent source delineated on the right side, or $M2$ transistor drain site, of the macromodel appears with a polarity that is opposite to that of the first source. Although this latter current is still formally directed from drain-to-source terminals of $M2$, its indicated polarity reversal derives from the fact that for differential mode, the gate of $M2$ is driven by $-(V_{di}/2)$, while the gate of $M1$ is excited by $+(V_{di}/2)$. Additionally, and in concert with Eq. (8-13), current sources, each of value $g_{m1}V_{ci}/\rho_1$, appear in shunt with the respective former dependent sources to account for common mode responses. No polarity reversal of these common mode currents is required since each of the $M1$ and $M2$ gate terminals is driven by a signal of value $+V_{ci}$ for common mode excitation.

The common mode output resistance, $\rho_1 r_{o1}$, as defined by Eq. (8-15), is now placed in shunt with each dependent current source pair, as diagrammed

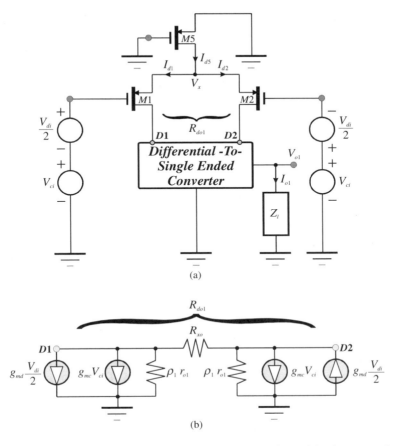

Figure 8.7. (a) The small signal schematic diagram for the differential pair transconductor that drives the differential to single ended converter in Fig. 8.5. (b) Small signal macromodel of the output port of the differential transconductor driver.

in Fig. 8.7(b). This configuration is reasonable in view of the fact that under common mode circumstances, for which $V_{di} = 0$, the common excitation of both the $M1$ and $M2$ gates results in identical signal responses at the respective drains of these transistors. Resultantly, no common mode current flows through the indicated resistance, R_{xo}, which electrically interconnects the two transistor drain terminals. In effect, the two transistor drains are isolated from one another for common mode signals and are each coupled only to signal ground through a resistance given by Eq. (8-15).

The resistance, R_{xo}, is introduced to ensure that the macromodel in Fig. 8.7(b) delivers the differential output port resistance predicted by Eq. (8-12). Since the differential output resistance, $R_{do1} = 2r_{o1}$, is literally the small signal resistance "seen" between the drain of $M1$ and the drain of $M2$, $2r_{o1}$ is necessarily the parallel combination of resistance R_{xo} and the series interconnection of the two common mode resistances, $\rho_1 r_{o1}$. It is a simple matter to show that this constraint mandates

$$R_{xo} = \frac{2r_{o1}}{1 - 1/\rho_1}. \qquad (8\text{-}16)$$

Note that for large common mode rejection ratio, which produces a very large common mode output resistance, R_{xo} collapses as expected to the differential output resistance, $(2r_{o1})$.

For purely common mode excitation, the differential component, V_{di}, of applied input signal is zero. The circuit model in Fig. 8.7(b) resultantly collapses to the structure given in Fig. 8.8(a). Since the two controlled generators, $g_{mc}V_{ci}$, perturb the nodes delineated in this figure as $D1$ and $D2$ in the same direction by precisely the same amount, the current conducted by the resistance, R_{xo}, is zero, which implies that R_{xo} can be supplanted by an open circuit. It follows that the applicable common mode equivalent circuit of the output port for the differential amplifier undergoing investigation is the second structure shown in Fig. 8.8(a). This topology confirms the utility of executing an analysis on only the common mode half circuit since the branch variable responses computed for the left half of the circuit are identical to their corresponding variable values on the right side of the representation.

In differential mode, the common mode input voltage, V_{ci}, is set to zero, and the model in Fig. 8.7(b) simplifies to that shown in Fig. 8.8(b). In the first of the three given diagrams, the controlled source on the left sinks current from node $D1$, while the controlled generator on the right, whose value is identical to the source at the left of the model, sources current to node $D2$. Consequently, node $D2$ is pulled up in voltage, while node $D1$ is pulled down by an equivalent amount. It follows that the middle of resistance R_{xo} lies at signal ground potential, the implication of which is the second circuit diagram in Fig. 8.8(b). But in the latter diagram, resistance $\rho_1 r_{o1}$ is in shunt with resistance $R_{xo}/2$ for both circuit halves. Using Eq. (8-16), it is a simple matter to confirm that the resultant resistance value of this parallel interconnection is simply r_{o1}, thereby producing the third representation in

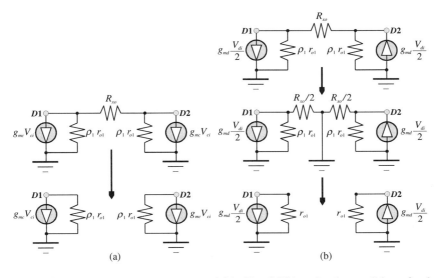

Figure 8.8. (a) The output port macromodel in Fig. 8.7(b) under the condition of only common mode excitation; that is, zero differential input signal. (b) The output port macromodel in Fig. 8.7(b) under the condition of only differential mode excitation; that is, zero common mode input signal.

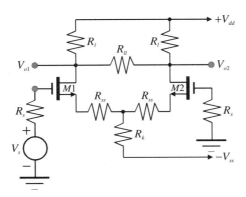

Figure 8.9. The differential amplifier studied in Example 8.1. The substrate terminals of both transistors are connected to the negative voltage supply bus.

Fig. 8.8(b). As in the case of common mode responses, half circuit analysis measures apply, subject to the proviso that while nodes $D1$ and $D2$ are perturbed by identical voltage magnitudes, these two voltage responses are phase inverted.

Example 8.1. The two transistors used in the differential amplifier of Fig. 8.9 are matched devices biased at identical quiescent operating points. For simplicity, the drain-source channel resistances of both devices can be presumed to be infinitely large, and their bulk transconductance factors can be taken to be zero. The signal source, which is represented as the series interconnection of the Thévenin voltage, V_s, and the Thévenin resistance, R_s, produces voltage responses, V_{o1s} and V_{o2s}, which are respectively the small signal components of the indicated $M1$ and $M2$ drain node voltages. Although the signal source resistance is inconsequential at low signal frequencies, where transistor and circuit capacitive impedances are very large, it may comprise a critical branch element at high frequencies. Accordingly, a physical resistance, matched to the Thévenin signal source resistance, is introduced in the gate circuit of transistor $M2$ to ensure balanced operation throughout the amplifier passband. Observe no incidence of external load impedances at either drain terminal. Determine the Thévenin equivalent small signal macromodel with respect to the drain of transistor $M2$. Numerically evaluate the parameters of this macromodel, as well as the circuit common mode rejection ratio, if the transistors have a forward transconductance (g_m) of 12 mmhos, and the indicated circuit elements are $R_s = 300\,\Omega$, $R_l = 2\,\text{K}\Omega$, $R_{ll} = 20\,\text{K}\Omega$, $R_{ss} = 25\,\Omega$, and $R_k = 25\,\text{K}\Omega$.

Solution 8.1.

(1) With only a single signal source applied to the differential amplifier, Eqs. (8-6) and (8-7) yield a differential input signal voltage of $V_{di} = V_s$ and a common mode input signal voltage of $V_{ci} = V_s/2$. The resultant small signal drain node responses are

$$\left. \begin{aligned} V_{o1s} &= V_{co} + \frac{V_{do}}{2} = A_c V_{ci} + A_d \frac{V_{di}}{2} = \left(\frac{A_c + A_d}{2}\right) V_s \\ V_{o2s} &= V_{co} - \frac{V_{do}}{2} = A_c V_{ci} - A_d \frac{V_{di}}{2} = \left(\frac{A_c - A_d}{2}\right) V_s \end{aligned} \right\}, \quad \text{(E1-1)}$$

where V_{co} and V_{do} respectively symbolize the common mode and differential mode small signal components of output voltage, and A_c and A_d are respectively the common mode voltage gain and the differential mode voltage gain. It follows that the voltage gain, A_{vt}, of present interest, which is the Thévenin gain in lieu of an external load, is

$$A_{vt} = \frac{V_{o2s}}{V_s} = \frac{A_c - A_d}{2}. \quad \text{(E1-2)}$$

With

$$\rho = \frac{A_d}{A_c} \quad \text{(E1-3)}$$

designating the common mode rejection ratio of the amplifier, Eq. (E1-2) is expressible as

$$A_{vt} = -\frac{A_d}{2}\left(1 - \frac{1}{\rho}\right). \tag{E1-4}$$

If the common mode rejection ratio is large in comparison to one, the Thévenin voltage gain is seen to be one-half of the differential gain of the amplifier. Specifically, the magnitude of the single ended gain is one-half that of the gain magnitude resulting from an output response extracted differentially between the two transistor drain nodes.

(2) In view of the stipulations in (1), Fig. 8.10(a) depicts the common mode small signal schematic diagram, while Fig. 8.10(b) is the corresponding common mode half circuit schematic diagram. In the latter diagram, resistance R_k is replaced by twice its value because of three phenomena. First, each of the common mode input voltages generate identical currents in the transistor source terminals. Second, the sum of these currents is necessarily returned to ground via R_k. Third, since the common mode half circuit embraces the source lead of only transistor $M1$, R_k must be doubled to establish a signal voltage drop across its terminals that incorporates the Kirchhoff voltage law ramifications of the foregoing current observations. Note further that resistance R_{ll} does not appear in the common mode half circuit of Fig. 8.10(b) because the presence of common mode output signal voltages at each drain node precludes any current from flowing through R_{ll}. The resistance established with respect to ground at either drain terminal in Fig. 8.10(a), as well as at the transistor drain node in the half circuit schematic of Fig. 8.10(b), is the common mode output resistance, R_{co}.

Figure 8.10(c) offers the differential mode counterpart to the common mode schematic diagram of Fig. 8.10(a), and Fig. 8.10(d) is the relevant half circuit schematic diagram. Since the gate of $M1$ is driven by a voltage, $V_{di}/2$, that is precisely the negative of the voltage exciting the gate of transistor $M2$, the node at which the two resistances, R_{ss}, and the resistance, R_k, are incident lies at signal ground. In particular, voltage $V_{di}/2$ at the $M1$ gate tries to pull the subject node upwards in voltage. Because the circuit is balanced, it follows that the voltage, $-V_{di}/2$, applied to the $M2$ gate tries to pull the node in question downward by a mirrored amount. The result is no movement of the subject node, which is tantamount to a signal ground therein, as suggested by Fig. 8.10(d). The same commentary applies to the resistance, R_{ll}. In particular, its midpoint lies at ground, which effectively places one-half of R_{ll} in shunt with the drain load resistance, R_l, as offered in Fig. 8.10(d). The effective resistance seen with respect to ground at either drain node in Fig. 8.10(c) and at the drain node in the corresponding half circuit topology is $R_{do}/2$, where R_{do} is understood to be the differential resistance established between the two drain nodes in Fig. 8.10(c).

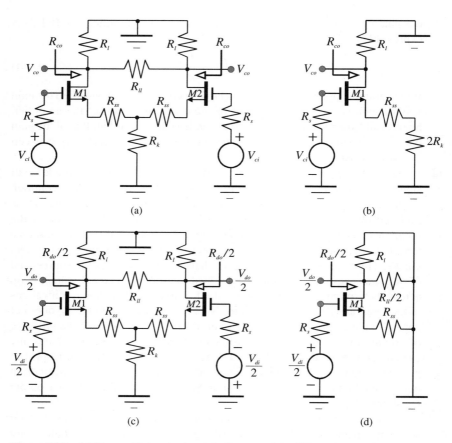

Figure 8.10. (a) The small signal schematic diagram of the differential amplifier in Fig. 8.9 under the condition of a common mode input. (b) The common mode half circuit diagram corresponding to (a). (c) The small signal schematic diagram of the differential amplifier in Fig. 8.9 under differential input signal conditions. (d) The differential mode half circuit diagram corresponding to (c).

(3) Figure 8.11(a) offers the small signal model of the differential mode half circuit. his model derives from pertinent considerations in the preceding chapter. In the interests of clarity and completeness, the network branches involving transistor model parameters that are tacitly ignored herewith (λ_b and r_o) are depicted as dashed elements. By inspection, the differential mode voltage gain, A_d, is

$$A_d = \frac{V_{do}/2}{V_{di}/2} = -\left(\frac{g_m}{1 + g_m R_{ss}}\right)\left(R_l \left\| \frac{R_{ll}}{2}\right.\right) = -15.38, \quad \text{(E1-5)}$$

MOS Technology Operational Amplifiers

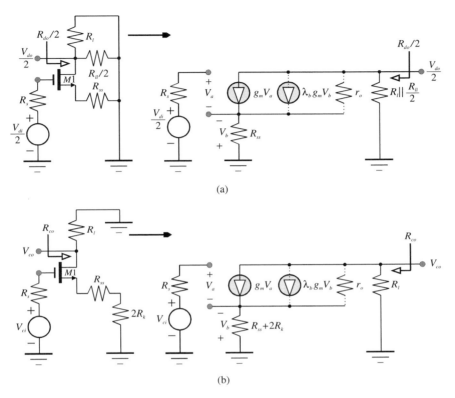

Figure 8.11. (a) Small signal model of the differential mode half circuit shown in Fig. 8.10(d). (b) Small signal model of the common mode half circuit shown in Fig. 8.10(b).

while the differential mode output resistance, R_{do}, is

$$R_{do} = 2\left(R_l \left\| \frac{R_{ll}}{2} \right.\right) = 3.33 \text{ K}\Omega. \tag{E1-6}$$

From Fig. 8.11(b), the common mode voltage gain, A_c, is seen to be

$$A_c = \frac{V_{co}}{V_{ci}} = -\frac{g_m R_l}{1 + g_m (R_{ss} + 2R_k)} = -39.91\left(10^{-3}\right). \tag{E1-7}$$

Clearly, the current sinking resistance, R_k, dramatically influences the achievable stage gain. Finally, the common mode output resistance, R_{co}, is

$$R_{co} = R_l = 2 \text{ K}\Omega. \tag{E1-8}$$

(4) The common mode rejection ratio now follows as

$$\rho = \frac{A_d}{A_c} = \left(\frac{R_{ll}}{R_{ll} + 2R_l}\right)\left(1 + \frac{2g_m R_k}{1 + g_m R_{ss}}\right) = 385.4 = 51.72\,\text{dB}, \quad \text{(E1-9)}$$

and the Thévenin voltage gain is

$$A_{vt} = \frac{V_{o2s}}{V_s} = \frac{A_c - A_d}{2} = +7.67 = 17.70\,\text{dB}. \quad \text{(E1-10)}$$

(5) Figure 8.12(a) submits the resistive macromodel for the two transistor drain ports of the differential amplifier under consideration. In concert with preceding disclosures, the resistance, R_{xo}, is chosen to ensure that the differential resistance, R_{do}, measured between terminals $D1$ and $D2$ matches the result promulgated by Eq. (E1-6). Accordingly,

$$R_{xo} = \frac{(2R_{co})R_{do}}{2R_{co} - R_{do}}. \quad \text{(E1-11)}$$

Using Eqs. (E1-6) and (E1-8),

$$R_{xo} = \frac{(2R_{co})R_{do}}{2R_{co} - R_{do}} = \frac{(2R_l)\left[2\left(R_l \,\Big\|\, \frac{R_{ll}}{2}\right)\right]}{2R_l - 2\left(R_l \,\Big\|\, \frac{R_{ll}}{2}\right)} \equiv R_{ll} = 20\,\text{K}\Omega, \quad \text{(E1-12)}$$

which is hardly apocryphal in view of the fact that with device channel resistances ignored, the circuit resistance, R_{ll}, is the only available conductive path between the two transistor drain nodes. The driving point output, or Thévenin output, resistance follows as

$$R_{\text{out}} = R_{co} \| (R_{xo} + R_{co}) = \frac{R_l(R_{ll} + R_l)}{2R_l + R_{ll}} = 1.83\,\text{K}\Omega. \quad \text{(E1-13)}$$

The resultant Thévenin equivalent circuit appears in Fig. 8.12(b).

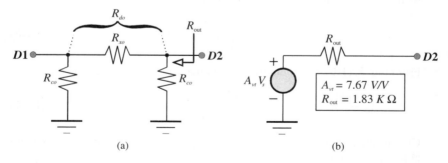

Figure 8.12. (a) Driving point resistance macromodel for the output ports of the differential amplifier addressed in Example 8.1. (b) Thévenin equivalent circuit of the single ended output port of the differential amplifier.

Comments. Half circuit analysis strategies underpin systematic and computationally efficient analyses of balanced differential amplifiers. This example demonstrates that for at least low signal frequencies, the realization of high common mode rejection ratio is not a particularly daunting endeavor. For the design considered herewith, the differential voltage gain exceeds the common mode gain by almost 52 dB! With reference to Eq. (E1-4), this rejection level implies that the common mode impact on the idealized differential gain of $-A_d/2$ is only about 0.26%.

Equation (E1-9) underscores the critical impact resistance R_k has on the common mode rejection ratio, ρ. In this example, $R_k = 25\,\text{K}\Omega$, which is doubtlessly best realized as a simple N-channel current sink. If this current sink were to be optimized, perhaps by exploiting the transconductance enhancement technology assessed in the preceding chapter, R_k values of 250 KΩ or larger are feasible. In the case of $R_k = 250\,\text{K}\Omega$, the common mode rejection ratio is about 10-times, or 20 dB larger than that realized with a 25 KΩ current sink.

8.3.1.3. First Stage Output Macromodel

The output voltage (V_{o1}) and current (I_{o1}) responses of the first stage of the dual stage op-amp are extracted at the drain terminal interconnection of transistors $M2$ and $M4$. The high impedance nature of transistor drain nodes encourages a Norton methodology for macromodeling the output port of the first op-amp stage. Unfortunately and unlike the balanced amplifier featured in the preceding example, half circuit analytical methods are not applicable herewith since the loads driven by the $M1-M2$ differential pair in Fig. 8.5 are unbalanced. In particular, the load imposed on the drain of transistor $M1$ is a diode-connected transistor, $M3$, while the load on the drain of $M2$ is a common source amplifier terminated in an impedance, Z_l. But since the Norton equivalent circuit at the output ports of the differential pair is a balanced configuration, the transconductor output port macromodel developed in the preceding section of material and depicted schematically in Fig. 8.7(b) is applicable to the present Norton analysis task. Moreover, a simplifying approximation can be invoked for at least relatively low frequencies. In particular and as is inferred by Example 8.1, the common mode rejection ratio indigenous to the Norton model of the $M1-M2$ pair can be expected to be very large because of the presumably large channel resistance postured by the current source transistor, $M5$. Accordingly, common mode responses in the input transductor can be ignored, which means that the applicable Norton output port model for the differential pair is the simple network shown in Fig. 8.8(b). It follows that the small signal schematic

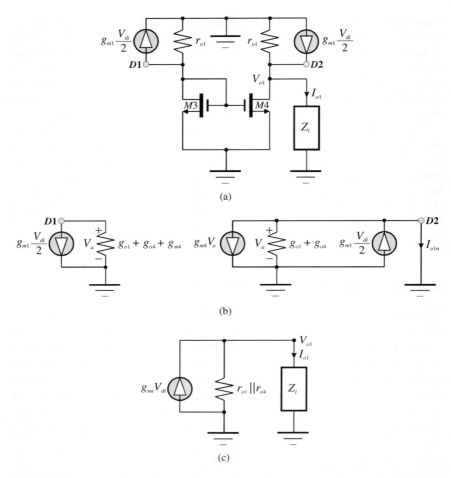

Figure 8.13. (a) Small signal schematic diagram of the signal source drive applied to the M3–M4 current mirror in the op-amp first stage shown in Fig. 8.5. (b) Low frequency, small signal equivalent circuit for the conceptual diagram in (a). The diagram is used to determine the Norton, or short circuit, signal current available at the amplifier output port. (c) The resultant small signal macromodel for the output port of the first stage amplifier.

diagram of the entire first stage (transconductor and differential-to-single ended converter) is the circuit given in Fig. 8.13(a), where use has been made of Eq. (8-11). Figure 8.13(b) is the corresponding small signal model for determining the Norton equivalent circuit at the stage output port, where in general, g_{oi} is used to symbolize the conductance pertinent to the drain-source channel resistance, r_{oi}, of the ith transistor. This model exploits the fact that transistors $M3$ and $M4$ are matched devices biased at nominally the same quiescent operating point. Moreover, the diode-connected transistor, $M3$, is supplanted by its equivalent drain-source resistance, whose conductance is $(g_{m3} + g_{o3}) \equiv (g_{m4} + g_{o4})$.

An inspection of the model in Fig. 8.13(b) readily leads to a short circuit output signal current, I_{on}, of

$$I_{on} = g_{m1}\left(1 + \frac{g_{m4}}{g_{m4} + g_{o4} + g_{o1}}\right)\frac{V_{di}}{2}, \qquad (8\text{-}17)$$

which can be cast as

$$I_{on} = g_{me} V_{di}. \qquad (8\text{-}18)$$

In Eq. (8-18), the effective Norton transconductance, g_{me}, is

$$g_{me} = \frac{g_{m1}}{2}\left(2 - \frac{1}{1 + g_{m4}(r_{o1}\|r_{o4})}\right), \qquad (8\text{-}19)$$

which reduces to g_{m1}, the transconductance of either transistor $M1$ or transistor $M2$, provided $g_{m4}(r_{o1} \| r_{o4}) \gg 1$. A further inspection of the model in Fig. 8.13(b) reveals an equivalent shunt output resistance, r_{oe}, of

$$r_{oe} = (r_{o1}\|r_{o4}). \qquad (8\text{-}20)$$

Figure 8.13(c) offers the final form macromodel of the output port undergoing assessment.

The propriety of the foregoing analysis can be supported by a direct, approximate analysis of the schematic diagram of the first stage amplifier in Fig. 8.5. If transistors $M1$ and $M2$ are matched devices that are biased at nominally identical quiescent operating points, the concepts underpinning Eq. (8-9) permit writing the currents, I_{d1} and I_{d2}, in the form

$$\left.\begin{aligned} I_{d1} &= I_{d1q} - g_{mc} V_{ci} - g_{m1}\left(\frac{V_{di}}{2}\right) \\ I_{d2} &= I_{d1q} - g_{mc} V_{ci} + g_{m1}\left(\frac{V_{di}}{2}\right) \end{aligned}\right\}. \qquad (8\text{-}21)$$

In Eq. (8-21), I_{d1q} is the Q-point drain current conducted by $M1$ and $M2$. This Q-point current is precisely one-half of the static current, I_{d5q}, supplied by the current source transistor, $M5$. The common mode current components are $g_{mc}V_{ci}$, while the differential currents are $\pm g_{m1}(V_{di}/2)$. The negative algebraic signs associated with the common mode and differential mode current components in the first of the relationships in Eq. (8-21) derive from the fact that small signal MOSFET currents are always directed from drain to source terminals. In the second of these relationships, the algebraic signs of the signal components derive from the fact that the gate of transistor $M2$ is excited by a net voltage of $(V_{ci} - V_{di}/2)$. Because the gates of $M3$ and $M4$ conduct no low frequency currents, the current, I_{d3}, flowing through $M3$ is the same as I_{d1}. Moreover, the fact that the gate source voltages of $M3$ and $M4$ are identical renders I_{d4} a mirror image of I_{d3}, if channel length modulation phenomena in $M3$ and $M4$ are negligible. It should be noted that the tacit neglect of channel length modulation phenomena is tantamount to the presumption of negligibly small drain-source channel conductances, g_{o3} and g_{o4}, in $M3$ and $M4$, respectively. At this juncture, therefore,

$$I_{d4} = I_{d3} = I_{d1} = I_{d1q} - g_{mc}V_{ci} - g_{m1}\left(\frac{V_{di}}{2}\right). \tag{8-22}$$

The load current, I_{o1}, in Fig. 8.5 is the difference between currents I_{d2} and I_{d4}, whence

$$I_{o1} = I_{d2} - I_{d4} = g_{m1}V_{di}, \tag{8-23}$$

which synergizes with Eq. (8-18) if the channel conductances are indeed negligibly small.

8.3.1.4. *First Stage Static Analysis*

A static analysis of the P-channel transconductor proceeds from a consideration of the network diagram in Fig. 8.14, which differs from that of Fig. 8.5 in that the applied differential and common mode input signal voltages are null to highlight the focus on static responses. Moreover, the impedance, Z_l, is presumed to comprise a capacitive load and accordingly, the current, I_{o1q}, it conducts is zero under quiescent operating circumstances. In the analyses that follow, let the static transconductance parameter and gate aspect ratio, K_{ni} and η_i, respectively be introduced for the ith transistor, such that

$$K_{ni} = \mu_i C_{oxi} \tag{8-24}$$

and

$$\eta_i = W_i/L_i. \tag{8-25}$$

In Eqs. (8-24) and (8-25), μ_i is the average mobility of majority carriers (holes in P-channel units and electrons in N-channel devices) in the source to drain channel of the ith transistor, which features a gate width of W_i, a

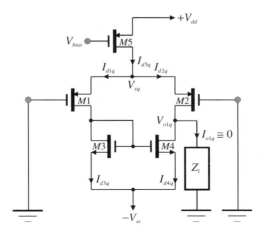

Figure 8.14. The driver transconductor of Fig. 8.5 under quiescent operating conditions. The load impedance, Z_l, that terminates the output port is presumed to be dominantly capacitive so that the quiescent current, I_{olq}, it conducts is essentially zero.

nominal channel length of L_i, a channel length modulation voltage of $V_{\lambda i}$, and a gate-source threshold voltage of V_{hi}. The parameter, C_{oxi}, symbolizes the capacitance density of the oxide-channel interface for the ith transistor. For analytical tractability, bulk induced modulation of threshold voltage is tacitly ignored in all transistors.

Transistor $M5$ is saturated if, with reference to Fig. 8.14, $V_{bias} \geq V_{xq} - V_{h5}$. For this biasing constraint, the static current, I_{d5q}, conducted by the drain of $M5$ is

$$I_{d5q} = \frac{K_{n5}}{2} \eta_5 (V_{dd} - V_{bias} - V_{h5})^2 \left(1 + \frac{V_{dd} - V_{xq}}{V_{\lambda 5}}\right), \qquad (8\text{-}26)$$

where V_{xq}, the static voltage developed with respect to ground at the $M5$ drain, is also the source-gate biasing voltages of both transistors $M1$ and $M2$. If $M1$ and $M2$ are indeed identical devices that support equivalent source to drain potentials, their drains each conduct $I_{d5q}/2$; that is, I_{d5q} splits evenly between the subject two devices. Since $M3$ and $M4$ are identical transistors, it follows that

$$\frac{I_{d5q}}{2} = \frac{K_{n4}}{2} \eta_4 (V_{gs4} - V_{h4})^2 \left(1 + \frac{V_{olq} + V_{ss}}{V_{\lambda 4}}\right) \qquad (8\text{-}27)$$

and

$$\frac{I_{d5q}}{2} = \frac{K_{n4}}{2} \eta_4 (V_{gs3} - V_{h4})^2 \left(1 + \frac{V_{gs3}}{V_{\lambda 4}}\right). \qquad (8\text{-}28)$$

The topological interconnection of $M3$ and $M4$ is such as to force $V_{gs3} \equiv V_{gs4}$. Accordingly, Eqs. (8-27) and (8-28) yield $V_{o1q} + V_{ss} = V_{gs4}$, which implies

$$V_{o1q} = V_{h4} - V_{ss} + \sqrt{\frac{I_{d5q}}{K_{n4}\eta_4}}. \tag{8-29}$$

Equations (8-29) and (8-26) reveal that the no load, quiescent output voltage, V_{o1q}, depends in a nominally linear fashion on the biasing control voltage V_{bias}. Specifically, increases in V_{bias} effect decreases in V_{o1q} and vice versa, since if $V_{dd} - V_{xq} \equiv V_{dd} - V_{sg1} \equiv V_{dd} - V_{sg2} \ll V_{\lambda 5}$,

$$V_{o1q} \approx V_{h4} - V_{ss} + (V_{dd} - V_{bias} - V_{h5})\sqrt{\left(\frac{K_{n5}}{2K_{n4}}\right)\left(\frac{\eta_5}{\eta_4}\right)}. \tag{8-30}$$

Observe that the sensitivity of V_{o1q} with respect to V_{bias} increases slowly with increases in the ratio, (η_5/η_4), of $M5$ and $M4$ gate aspect ratios. For a fixed V_{bias} setting, thermally induced fluctuations in threshold potentials, as well as in carrier mobilities, is not likely to pose significant engineering problems, principally because V_{o1q} in the last expression depends on a threshold voltage difference, $(V_{h4} - V_{h5})$, and a ratio, K_{n5}/K_{n4}, of mobility-dependent transconductance coefficients.

8.3.2. N-Channel Transconductor

Figure 8.15 is the electrical schematic diagram of a CMOS N-channel driver transconductance amplifier. A comparison of this topology to its PMOS form underscores clarion similarities. Transistor $M5$ functions as a current sink, as opposed to a current source, for the balanced differential amplifier comprised of identical NMOS transistors, $M1$ and $M2$. The current outputs of this pair drive P-channel loads synthesized with matched devices $M3$ and $M4$, which function as a current mirror to produce the single ended output current, I_{o1}, and corresponding output voltage, V_{o1}. The small signal results deduced for the P-channel unit apply without modification to the present NMOS topology.

The static analysis conducted in the preceding section of material can also be adapted straightforwardly to the N-channel circuit. In particular, if transistor $M5$ sinks a quiescent current, I_{d5q},

$$I_{d5q} = \frac{K_{n5}}{2}\eta_5(V_{bias} + V_{ss} - V_{h5})^2\left(1 + \frac{V_{xq} + V_{ss}}{V_{\lambda 5}}\right). \tag{8-31}$$

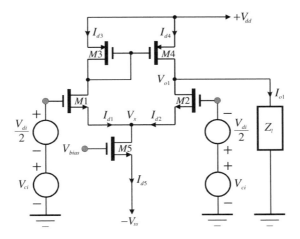

Figure 8.15. Schematic diagram of the input N-channel driver transconductor stage for a two-stage op-amp. All transistors operate in saturation and have their substrate bulk terminals connected to the appropriate power supply voltage lines.

For matched $M3$ and $M4$ and negligible load current conduction under static operating circumstances,

$$\frac{I_{d5q}}{2} = \frac{K_{n4}}{2}\eta_4\left(V_{sg4} - V_{h4}\right)^2\left(1 + \frac{V_{dd} - V_{o1q}}{V_{\lambda 4}}\right) \tag{8-32}$$

and

$$\frac{I_{d5q}}{2} = \frac{K_{n4}}{2}\eta_4\left(V_{sg4} - V_{h4}\right)^2\left(1 + \frac{V_{sg4}}{V_{\lambda 4}}\right). \tag{8-33}$$

It follows that $(V_{dd} - V_{o1q}) = V_{sg4}$, whence by Eq. (8-32),

$$V_{o1q} \approx V_{dd} - V_{h4} - \sqrt{\frac{I_{d5q}}{K_{n4}\eta_4}}. \tag{8-34}$$

8.4.0. Phase Inverting Second Stage

Because of the similarity between the P-channel and the N-channel transconductor drivers, a second op-amp stage appropriate for utilization with only the P-channel driver topology is studied herewith. The basic form of the schematic diagram for the requisite second stage is offered in Fig. 8.16. In this CMOS unit, gain is provided by an N-channel common source amplifier that is actively loaded in a P-channel current source. The

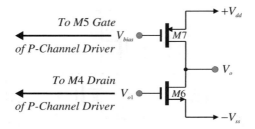

Figure 8.16. Basic schematic diagram of the phase inverting second stage of an op-amp that utilizes an input P-channel transconductor driver. Both transistors operate in saturation and have their substrate bulk terminals connected to the appropriate power supply voltage lines.

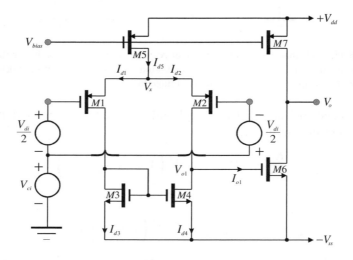

Figure 8.17. Basic schematic diagram of the two-stage operational amplifier.

interconnection of this second stage with the first stage transconductor is shown in Fig. 8.17.

8.4.1. Low Frequency Small Signal Analysis

The low frequency, small signal model of the phase inverting amplifier of Fig. 8.16 is depicted in Fig. 8.18(a), which postures an open circuit voltage gain, A_{v2}, of

$$A_{v2} = \frac{V_{os}}{V_{o1s}} = -\frac{g_{m6}}{g_{o6} + g_{o7}} = -g_{m6}(r_{o6} \| r_{o7}). \tag{8-35}$$

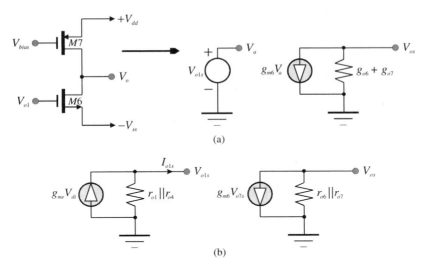

Figure 8.18. (a) The small signal equivalent circuit of the second stage inverting amplifier of the two-stage op-amp. (b) Low frequency, small signal equivalent circuit of the operational amplifier shown in Fig. 8.17. Common mode responses are presumed insignificant.

The model also conveys an output port resistance of $(r_{o6} \| r_{o7})$. Recalling Fig. 8.13(c), the resultant small signal equivalent circuit for the entire op-amp is the structure depicted in Fig. 8.18(b). Note that because the output port of the first stage amplifier drives the gate terminal of the common source amplifier utilized in the second stage, the signal current, I_{os}, supplied by the first stage to the second stage is effectively zero at low frequencies. The latter network clearly conveys an open circuit, open loop voltage gain, A_{ol}, of

$$A_{ol} = \frac{V_{os}}{V_{di}} = -[g_{me}(r_{o1}\|r_{o4})][g_{m6}(r_{o6}\|r_{o7})], \qquad (8\text{-}36)$$

which can be large for correspondingly large device channel resistances. Note that the large output resistance, $(r_{o6} \| r_{o7})$, of the two stage configuration all but precludes the ability of the op-amp to deliver large, low frequency voltage gain to all but high impedance or capacitive load terminations. A voltage buffer appended to this output port mitigates this performance shortfall whenever the performance objectives of the subject op-amp embody general purpose signal processing applications.

8.4.2. Op-Amp Static Analysis

Figure 8.19 submits the schematic diagram of the entire operational amplifier under static operating conditions. Recall Eqs. (8-29) and (8-30), which stipulate the no load quiescent voltage, V_{o1q}, established with respect to ground at the output terminal of the first stage of amplification. These expressions remain applicable, despite the interconnection of this output node to the gate terminal of transistor $M6$, because the $M6$ gate draws no low frequency current. Thus, assuming that transistor $M6$ is biased in its saturated domain, the indicated static current, I_{d6q}, is

$$I_{d6q} = \frac{K_{n6}}{2} \eta_6 \left(V_{o1q} + V_{ss} - V_{h6} \right)^2 \left(1 + \frac{V_{oq} + V_{ss}}{V_{\lambda 6}} \right), \quad (8\text{-}37)$$

or by Eq. (8-29),

$$I_{d6q} = \left(\frac{\eta_6}{2\eta_4} \right) I_{d5q} \left(1 + \frac{V_{oq} + V_{ss}}{V_{\lambda 6}} \right). \quad (8\text{-}38)$$

The last expression exploits the fact that, save for possible differences in gate aspect ratios, transistors $M4$ and $M6$ are identical N-channel devices having equal transconductance coefficients ($K_{n4} = K_{n6}$) and equal threshold potentials ($V_{h4} = V_{h6}$).

If no conductive load is imposed on the drain terminals of transistors $M6$ and $M7$ under quiescent circumstances, current I_{d6q} in Eq. (8-38) is

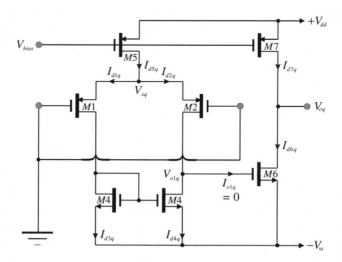

Figure 8.19. The op-amp of Fig. 8.17 under quiescent operating conditions.

conducted by transistor $M7$. Assuming saturation regime operation of $M7$, it follows that

$$I_{d6q} = \frac{K_{n7}}{2}\eta_7 (V_{dd} - V_{bias} - V_{h7})^2 \left(1 + \frac{V_{dd} - V_{oq}}{V_{\lambda 7}}\right). \tag{8-39}$$

Transistors $M7$ and $M5$ in Fig. 8.19 are matched P-channel units having $K_{n7} = K_{n5}$, $V_{h7} = V_{h5}$, and $V_{\lambda 7} = V_{\lambda 5}$, but possibly different gate aspect ratios, η_7 and η_5. Thus, Eqs. (8-39) and (8-26) combine to deliver

$$I_{d6q} = \left(\frac{\eta_7}{\eta_5}\right) I_{d5q} \left[\frac{1 + \dfrac{V_{dd} - V_{oq}}{V_{\lambda 5}}}{1 + \dfrac{V_{dd} - V_{xq}}{V_{\lambda 5}}}\right]. \tag{8-40}$$

If transistor $M5$ is chosen to ensure that its channel length modulation voltage, $V_{\lambda 5}$, is significantly larger than the value, V_{dd}, of the positive power bus voltage, Eqs. (8-40) and (8-38) impose a simple constraint on relevant gate aspect ratios. In particular,

$$\frac{\eta_6}{\eta_4} \approx 2\left(\frac{\eta_7}{\eta_5}\right). \tag{8-41}$$

Equation (8-41) ensures a proper biasing interface between the input port of the phase inverting second stage of the op-amp and the output port of the transconductor first stage.

8.5.0. Frequency Compensation

As illustrated in the system level diagram of Fig. 8.2, a two-stage CMOS op-amp is compensated to realize a dominant pole open loop response by incorporating shunt-shunt, capacitive global feedback around the phase inverting second stage. As such, this so called *pole splitting capacitance* is chosen to ensure that the resultant open loop bandwidth and unity gain frequency are limited by Miller effect phenomena.[5]–[8] To this end, the frequency compensated version of the op-amp diagrammed in Fig. 8.17 is offered as Fig. 8.20, wherein C_c is the incorporated compensation capacitance, and C_l represents the effective load capacitance terminating the output port of the op-amp. It is worthwhile interjecting that the consideration of a purely capacitive load is pragmatic from at least two perspectives. In particular, if the two-stage configuration is utilized in conjunction with an output buffer, the output port of the op-amp in Fig. 8.17 drives a transistor gate, which can be modeled adequately by a superposition of gate-drain and

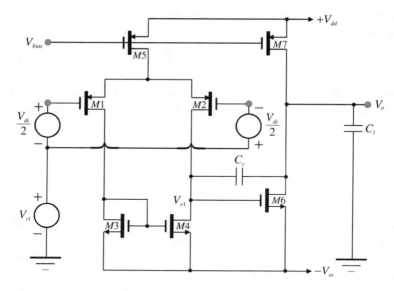

Figure 8.20. The op-amp of Fig. 8.17 with a pole splitting, or Miller effect, compensation capacitance, C_c, incorporated in shunt-shunt connection with the phase inverting second stage. The capacitance, C_l, represents the load impedance driven by the compensated op-amp.

Miller multiplied gate-source capacitances of the buffering transistor. An analogous contention applies if the op-amp is employed without an output buffer, as is the case in a switched capacitor filter or other types of mixed signal, low to moderate speed, signal processing applications.

8.5.1. Approximate High Frequency Analysis

The model appropriate to a design-oriented analysis of the op-amp given in Fig. 8.20 appears in Fig. 8.21 and derives from the following observations, considerations, and approximations.

(a) The common mode rejection ratio can be rendered very large through appropriate design of the current source forged by transistor $M5$. To this end, $M5$ must be biased in its saturation regime, and its channel length must be suitably long to ensure an adequately large channel length modulation voltage. If the common mode response is indeed inconsequential, the common mode signal voltage, V_{ci}, in Fig. 8.17 can be approximated as zero, which means that the Norton model in

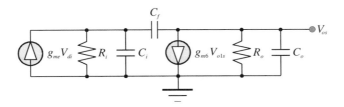

Figure 8.21. Approximate high frequency, small signal equivalent circuit of the compensated operational amplifier of Fig. 8.20.

Fig. 8.13(c) is a suitable representation of the low frequency, small signal characteristics of the first stage output port. In the interest of mathematical simplicity, let

$$R_i \stackrel{\Delta}{=} r_{o1} \| r_{o4}. \tag{8-42}$$

(b) Assume that the gates of the differential pair comprised of transistors $M1$ and $M2$ are terminated in relatively small signal source resistances. In light of these presumed modest resistances, it is unlikely that the gate-source and Miller multiplied gate-drain capacitances of $M1$ and $M2$ establish dominant time constants at high signal frequencies. Since the source terminal interconnection of $M1$ and $M2$ lies at a virtual ground potential for exclusively differential input excitation, the bulk-source capacitances of these two transistors are inconsequential.

(c) Because transistor $M3$ is configured as a diode exuding a relatively small terminal resistance of roughly $1/g_{m3}$, the bulk-drain capacitance of $M1$ is not likely to generate a dominant time constant. A similar contention can be proffered with respect to the gate-source capacitance of $M3$, the gate-source capacitance of $M4$, and the Miller multiplied gate-drain capacitance of $M4$.

(d) The drain interconnection of transistors $M2$ and $M4$ is a high resistance node. Accordingly, an account must be made of the high frequency effects of the bulk-drain capacitance, C_{bd2}, of transistor $M2$, the bulk-drain capacitance, C_{bd4}, of $M4$, the gate-source capacitance, C_{gs6}, of $M6$, and any parasitic stray capacitance, say C_{si}, encountered at the subject node. Moreover, Miller multiplication of the net $M6$ gate to drain capacitance, C_f, is significant, particularly because C_c, which is a part of this net capacitance, is purposefully incorporated to establish a dominant amplifier pole. In the model of Fig. 8.21, the foregoing capacitances are implicit to the shunt input capacitance,

$$C_i = C_{bd2} + C_{bd4} + C_{gs6} + C_{si}, \tag{8-43}$$

and the capacitive feedback element

$$C_f = C_{gd6} + C_c, \qquad (8\text{-}44)$$

where C_{gd6} symbolizes the net intrinsic gate-drain capacitance of $M6$, inclusive of overlap effects at the drain implant site.

(e) The shunt output port resistance in the model of Fig. 8.21 is comprised of the shunt interconnection of the drain-source channel resistances, r_{o6}, and r_{o7}, of transistors $M6$ and $M7$, respectively. Define this net resistance, R_o, as

$$R_o \triangleq r_{o6} \| r_{o7}. \qquad (8\text{-}45)$$

(f) The high resistance nature of the drain interconnection of transistors $M6$ and $M7$ fosters the potential vulnerability of amplifier high frequency performance to capacitance loading imposed on the output port. Apart from the obvious load capacitance, C_l, an engineering account must be made of the bulk-drain capacitance, C_{bd6}, of $M6$, the bulk-drain capacitance, C_{bd7}, of $M7$, and, to the extent that voltage V_{bias} derives from a reasonably well-regulated biasing source, the net gate-drain capacitance, C_{gd7}, of transistor $M7$. In the model of Fig. 8.21,

$$C_o = C_{bd6} + C_{bd7} + C_{gs7} + C_l, \qquad (8\text{-}46)$$

where it is tacitly presumed that any stray output port capacitance is absorbed into the numerical definition of load capacitance C_l.

8.5.2. Miller Compensation

As might have been expected, the small signal model in Fig. 8.21 is topologically identical to the generic common source model appearing in Fig. 6.10(b) of this text. As such, the voltage transfer function and related expressions documented in Section 6.3.1 are applicable with but minor changes made to the network parametric elements implicit to these expressions. It follows that the voltage transfer function, $A_{ol}(s)$ of the op-amp open loop is

$$A_{ol}(s) = \frac{V_{os}}{V_{di}} = -\frac{A_{ol}(0)\left(1 - \dfrac{s}{z_f}\right)}{\left(1 + \dfrac{s}{p_1}\right)\left(1 + \dfrac{s}{p_2}\right)} = -\frac{A_{ol}(0)\left(1 - \dfrac{s}{z_f}\right)}{1 + \left(\dfrac{1}{p_1} + \dfrac{1}{p_2}\right)s + \dfrac{s^2}{p_1 p_2}}, \qquad (8\text{-}47)$$

where

$$A_{ol}(0) = (g_{me}R_i)(g_{m6}R_o) \tag{8-48}$$

is the magnitude of the open circuit, zero frequency, open loop voltage gain, p_1 is the frequency of the dominant left half plane pole for the open loop, p_2 is the frequency of the nondominant pole, and z_f is the frequency of the right half plane zero incurred by the net feedback capacitance, C_f. The frequency of the subject zero is

$$z_f = \frac{g_{m6}}{C_f}, \tag{8-49}$$

while the pole frequencies, p_1 and p_2, satisfy the relationships,

$$\frac{1}{p_1} + \frac{1}{p_2} = R_i[C_i + (1 + g_{m6}R_o)C_f] + R_o(C_o + C_f) \tag{8-50}$$

and

$$\frac{1}{p_1 p_2} = (R_i C_i)(R_o C_o)\left(1 + \frac{C_f}{C_i} + \frac{C_f}{C_o}\right). \tag{8-51}$$

If the feedback capacitance, C_f, which is comprised of the effective shunt interconnection of $M6$ gate-drain capacitance, C_{gd6}, and the introduced compensating element, C_c, is to establish the dominant time constant of the open loop amplifier, C_c must be selected to satisfy the inequality,

$$R_i(1 + g_{m6}R_o)C_f \gg R_i C_i + R_o(C_o + C_f), \tag{8-52}$$

where $(1 + g_{m6}R_o)C_f$ is recognized as the Miller effect value of the net feedback capacitance, C_f. It follows that the resultant frequency, p_1, of the dominant open loop pole is given approximately by

$$p_1 \approx \frac{1}{R_i(1 + g_{m6}R_o)C_f} \approx \frac{1}{g_{m6}R_i R_o C_f}, \tag{8-53}$$

where $g_{m6}R_o \gg 1$ is tacitly presumed. A combination of this expression with Eq. (8-51) produces

$$p_2 \approx \frac{1 + g_{m6}R_o}{R_o\left(C_i + C_o + \frac{C_i C_o}{C_f}\right)} \approx \frac{g_{m6}}{C_i + C_o + \frac{C_i C_o}{C_f}}. \tag{8-54}$$

A comparison of Eq. (8-53) with Eq. (8-54) suggests that because of the utilized feedback capacitor, progressively larger values of the $M6$ transconductance, g_{m6}, result in correspondingly smaller values of p_1 and simultaneously, larger values of p_2. In other words, the frequencies of the two open loop poles split farther apart from one another as g_{m6} increases. In view of the fact that the frequency of the open loop zero is, like p_2, proportional to transconductance g_{m6}, large g_{m6} is therefore seen as a principle engineering vehicle for realizing a dominant pole open loop frequency response.

In light of the apparent dominant pole established by the utilized feedback capacitance, C_c, an assessment of the resultantly achievable phase and gain margins is appropriate. To this end, let

$$\omega_u = A_v(0)p_1 = \left(\frac{g_{m6}R_o}{1+g_{m6}R_o}\right)\frac{g_{me}}{C_f} \approx \frac{g_{me}}{C_f}, \qquad (8\text{-}55)$$

which is seen to approximate the open loop unity gain frequency, provided p_2 and z_f are large frequencies. Moreover, introduce the factor, k_p, such that

$$k_p \triangleq \frac{p_2}{\omega_u} = \left(\frac{1+g_{m6}R_o}{g_{m6}R_o}\right)^2 \left(\frac{g_{m6}}{g_{me}}\right)\left(\frac{C_f}{C_i+C_o+\dfrac{C_iC_o}{C_f}}\right)$$

$$\approx \left(\frac{g_{m6}}{g_{me}}\right)\left(\frac{C_f}{C_i+C_o+\dfrac{C_iC_o}{C_f}}\right), \qquad (8\text{-}56)$$

which represents the frequency of the nondominant pole normalized to the approximate unity gain frequency. Noting that z_f in Eq. (8-49) is expressible as

$$z_f = \frac{g_{m6}}{C_f} = \left(\frac{g_{m6}}{g_{me}}\right)\left(\frac{1+g_{m6}R_o}{g_{m6}R_o}\right)\omega_u \approx \left(\frac{g_{m6}}{g_{me}}\right)\omega_u, \qquad (8\text{-}57)$$

the open loop gain function of Eq. (8-47) becomes

$$A_{ol}(s) \approx -A_{ol}(0)\frac{\left(1-\dfrac{g_{me}s}{g_{m6}\omega_u}\right)}{\left[1+\dfrac{A_{ol}(0)s}{\omega_u}\right]\left[1+\dfrac{s}{k_p\omega_u}\right]}. \qquad (8\text{-}58)$$

It should be understood that Eq. (8-58) represents the negative of the loop gain, $T_u(s)$, indigenous to unity closed loop gain operation of the op-amp. In an attempt to clarify this contention, consider the system level abstraction of Fig. 8.22, which depicts the op-amp in question configured for unity closed loop gain operation. In this diagram, note that the differential input signal voltage, V_{di}, is polarized with respect to the noninverting op-amp input node (+) as a positive voltage at the inverting op-amp node (−). This

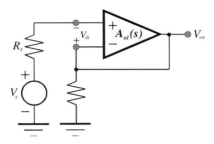

Figure 8.22. The op-amp of Fig. 8.20 configured for unity gain closed loop operation.

polarization mirrors the electrical implications of the op-amp schematic diagram in Fig. 8.20 and its companion small signal model in Fig. 8.21. Assuming no input current flows into either op-amp input terminals, it is clear that the signal voltage, V_s, the differential input voltage, V_{di}, and the output response voltage, V_{os}, interrelate as

$$V_s = -V_{di} + V_{os}. \tag{8-59}$$

Since $V_{os} = A_{ol}(s)V_{di}$, it is trivial matter to show that

$$\frac{V_{os}}{V_s} = -\frac{A_{ol}(s)}{1 - A_{ol}(s)}, \tag{8-60}$$

which understandably converges to unity for large open loop gain magnitude, $|A_{ol}(j\omega)|$. More importantly, Eqs. (8-60) and (8-58) confirm a pertinent loop gain, $T_u(s)$, of

$$T_u(s) = -A_{ol}(s) \approx A_{ol}(0)\frac{\left(1 - \dfrac{g_{me}s}{g_{m6}\omega_u}\right)}{\left[1 + \dfrac{A_{ol}(0)s}{\omega_u}\right]\left[1 + \dfrac{s}{k_p\omega_u}\right]}. \tag{8-61}$$

The stability analyses prescribed in Chapter 4 remind of an additional engineering implication of the loop gain function in Eq. (8-60). In particular, the utility of an electronic circuit is fundamentally determined by its ability to provide a closed loop gain magnitude that is no smaller than unity for all signal frequencies of interest. For this gain constraint, the worst case, from a closed loop stability perspective, is unity closed loop gain. Accordingly, adequate phase and gain margins with respect to the loop gain, $T_u(j\omega)$, comprise necessary and sufficient conditions for stable responses under all other practical closed loop gain conditions.

The relevant results in Section 4.3.3 can be adapted directly to the problem of discerning both the phase and gain margins implied by Eq. (8-60).

In particular, the phase margin, say ϕ_{pm}, can be shown to be

$$\varphi_{pm} = \tan^{-1}\left[\frac{\alpha_p\left(\dfrac{g_{m6}}{g_{me}}\right) - 1}{\left(\dfrac{g_{m6}}{g_{me}}\right) + \alpha_p}\right], \qquad (8\text{-}62)$$

where

$$\alpha_p = k_p\left[\frac{1 + \dfrac{1}{k_p A_{ol}(0)}}{1 - \dfrac{k_p}{A_{ol}(0)}}\right] \approx k_p. \qquad (8\text{-}63)$$

On the other hand, the gain margin, say k_{gm}, is, by Eq. (4-48),

$$k_{gm} \approx \frac{g_{m6}}{g_{me}}\sqrt{\frac{C_i + C_o + \dfrac{C_i C_o}{C_f}}{C_f}}, \qquad (8\text{-}64)$$

where the indicated approximation exploits large $A_{ol}(0)$.

Figure 8.23 plots the phase margin of the op-amp as a function of the metric, α_p, which Eq. (8-63) poses as a reasonable approximation of the ratio of the nondominant amplifier pole frequency, p_2, to the approximate unity gain amplifier frequency, ω_u. For any given ratio, g_{m6}/g_{me}, of transistor transconductances, the phase margin increases, and hence the stability of the amplifier improves, with increasing α_p. This graphical observation synergizes with engineering expectations, since progressively larger α_p implies a progressively more dominant pole open loop response. For stipulated α_p, stability is also seen to improve with increasing g_{m6}/g_{me}, which is also expected in view of the fact that larger g_{m6}/g_{me} effects a progressively more dramatic split between the dominant and the nondominant pole frequencies. Unfortunately, the subject plots also convey a modicum of design concern. In particular, Chapter 4 underscores the fact that a maximally flat magnitude closed loop response mandates a phase margin of 63.43°, which Fig. 8.23 shows is unattainable for any value of α_p unless g_{m6}/g_{me} is suitably larger than two. In general, peaking in the closed loop response is precluded if and only if the phase margin tangent exceeds two, which requires

$$\alpha_p \geq \frac{2\left(\dfrac{g_{m6}}{g_{me}}\right) + 1}{\left(\dfrac{g_{m6}}{g_{me}}\right) - 2}. \qquad (8\text{-}65)$$

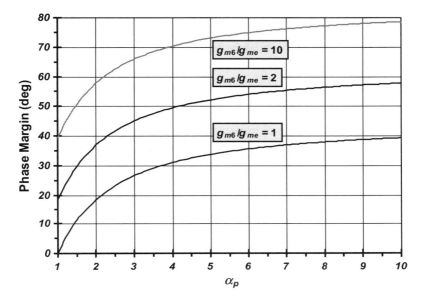

Figure 8.23. The phase margin of the two-stage op-amp as a function of the approximate ratio, α_p, of the nondominant pole frequency-to-the approximate unity gain frequency. Three values of the transconductance ratio, g_{m6}/g_{me}, are considered.

For $g_{m6}/g_{me} = 4$, which is an entirely reasonable transconductance ratio, $\alpha_p \geq 4.5$. The nondominant pole frequency implied by this constraint may comprise a daunting design challenge in light of the vagaries implicit to high frequency network responses.

8.5.3. Improved Frequency Compensation

The compensation shortfall to which the preceding section alludes derives fundamentally from the right half plane zero incurred by traditional Miller, or pole splitting, compensation. In particular, it is difficult to force the frequency, z_f, of this zero to be very large because by Eq. (8-57), z_f is proportional to transconductance g_{m6}. In turn, rendering g_{m6} large requires large $M6$ biasing current, which causes heartburn associated with undesirably large power dissipation. Alternatively, the gate aspect ratio of $M6$ can be increased to effect increases in its transconductance, but large gate aspect ratio incurs increased device capacitances, which leads to a potentially degraded unity gain frequency. At least two modifications to the basic Miller compensation strategy, each of which incurs neither significantly increased power dissipation nor frequency response penalties, are available to mitigate the problems surrounding the right half plane zero. These two schemes might be termed *buffered capacitive feedback* and *passive highpass feedback*.

8.5.3.1. Buffered Capacitive Feedback

Buffered capacitive feedback entails the use of a voltage buffer in conjunction with a capacitor to emulate feedback from the output port to the input port of a phase inverting amplifier, without establishing feedforward in the opposite signal flow direction. A conventional capacitor placed in a feedback loop is an obviously bilateral circuit element, whereas active feedback approximates unilateral feedback. Unilateral capacitive feedback obviates the pesky right half plane zero by nullifying any feedforward path through the capacitance from the input port to the output port of a compensated amplifier.

When applied to the phase inverting amplifier subcircuit of Fig. 8.20, buffered capacitive feedback assumes the topological form shown in Fig. 8.24(a). In this circuit diagram, transistor $M8$ functions as a voltage buffer

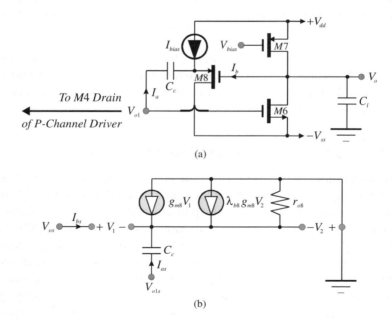

Figure 8.24. (a) Buffered capacitive feedback compensation of the phase inverting second stage of a two-stage op-amp. (b) Approximate small signal model of the $M8$-C_c feedback signal flow path.

that is biased in its saturation regime through a constant current source, I_{bias}. If the gate-drain capacitance, C_{gd8}, of $M8$ is superimposed with the load capacitance imposed on the op-amp output port, the small signal component, I_{bs}, of the net current, I_b, conducted by the gate of $M8$ is essentially zero. To be sure, I_{bs} is zero only to a first order approximation, since high frequency current flows through the gate-source capacitance, C_{gs8} of transistor $M8$. But if the geometry of $M8$ is chosen to ensure a voltage gain, say A_{ob}, from the gate of $M8$ to its source of nearly one, the current conducted by this gate-source capacitance is miniscule. The incorporation of the source follower buffer in the manner shown in the subject diagram precludes signal coupling through capacitance C_c from the gate of transistor $M6$-to-the $M6$ drain. But the approximately unity voltage gain of this buffer does allow for the connection of capacitance C_c from the gate node of $M6$ to a signal voltage that mirrors that established at the $M6$ drain, thereby effectively emulating Miller multiplication of C_c. In effect, the $M8$-C_c topology realizes a feedback capacitance that is capable of current conduction in only one direction; namely, from the $M6$ drain-to-the $M6$ gate. Given the approximation of $I_{bs} = 0$, the absorption of C_{gd8} into the load capacitance, and a presumed negligible time constant that is established by the $M8$ bulk-source capacitance at the relatively low impedance $M8$ source node, the resultant equivalent circuit of the feedback compensation path is the structure offered in Fig. 8.24(b).

Let

$$R_{ob} = \frac{r_{o8}}{1 + (1 + \lambda_{b8}) g_{m8} r_{o8}} \approx \frac{1}{g_{m8}} \qquad (8\text{-}66)$$

represent the driving point output resistance of the incorporated $M8$ buffer and

$$A_{ob} = \frac{g_{m8} r_{o8}}{1 + (1 + \lambda_{b8}) g_{m8} r_{o8}} = g_{m8} R_{ob} \qquad (8\text{-}67)$$

symbolize the gate to source voltage gain of the buffer. Then, a straightforward analysis of the model in Fig. 8.24(b) reveals the admittance parameter matric,

$$\begin{bmatrix} I_{as} \\ I_{bs} \end{bmatrix} = \begin{bmatrix} \dfrac{sC_c}{1 + sR_{ob}C_c} & -\dfrac{sA_{ob}C_c}{1 + sR_{ob}C_c} \\ 0 & 0 \end{bmatrix} \begin{bmatrix} V_{o1s} \\ V_{os} \end{bmatrix}, \qquad (8\text{-}68)$$

which suggests Fig. 8.25(a) as a model for the $M8$-C_c feedback subcircuit in Fig. 8.24(b). This model can be merged with that of the first and second stages to produce the resultant high frequency equivalent circuit, offered

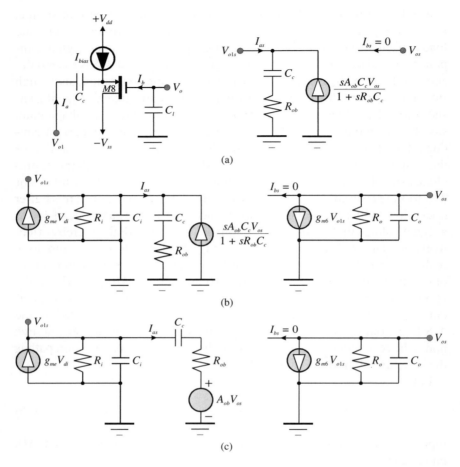

Figure 8.25. (a) The short circuit admittance parameter equivalent circuit of the $M8$-C_c compensation subcircuit. (b) Small signal, high frequency model of the op-amp with active capacitance compensation. The gate-drain capacitance of transistor $M8$ is factored into the op-amp load capacitance, and hence into the net output port capacitance, C_o. (c) An alternative form to the model in (b).

in Fig. 8.25(b), for the compensated op-amp. An alternative form of this latter model, derived through application of Thévenin's theorem to the parallel combination of the voltage controlled current feedback source and the R_{ob}-C_c branch, is the structure offered in Fig. 8.25(c). This topology shows that for $A_{ob} \approx 1$ and small R_{ob}, capacitance C_c effectively couples the gate of transistor $M6$ to its drain signal voltage, without the baggage of a

MOS Technology Operational Amplifiers

bilateral feedback element incident with the $M6$ gate and drain nodes. In other words, the feedback parameter is no longer the admittance of capacitance C_c. Rather, it is the almost unity gain, A_{ob}, associated with the buffer inserted into the feedback path.

The model depicted in Fig. 8.25(b) is an open loop equivalent circuit for the compensated operational amplifier. But despite its open loop nature, local shunt-shunt negative feedback is evidenced through the voltage controlled current source whose feedback transadmittance factor is $sA_{ob}C_c/(1+sR_{ob}C_c)$. Accordingly, the signal flow analysis methods developed in Chapter 5 produce an open loop voltage gain function that is expressible in the form,

$$A_{ol}(s) = \frac{V_{os}}{V_{di}} = \frac{g_{me} Z_{fo}(s)}{1 + T_{fo}(s)}, \qquad (8\text{-}69)$$

where the transimpedance function, $Z_{fo}(s)$, is

$$Z_{fo}(s) = \frac{V_{os}}{g_{me} V_{di}}\bigg|_{A_{ob}=0} = -\left[\frac{g_{m6} R_o R_i}{1 + sR_o C_o}\right]$$

$$\times \left\{ \frac{1 + sR_{ob}C_c}{1 + s\left[R_i C_i + (R_i + R_{ob})C_c\right] + s^2 R_i R_{ob} C_i C_c} \right\}$$

(8-70)

and the pertinent loop gain, $T_{fo}(s)$, is

$$T_{fo}(s) = -\left[\frac{sA_{ob}C_c}{1 + sR_{ob}C_c}\right] Z_{fo}(s). \qquad (8\text{-}71)$$

The substitution of Eqs. (8-71) and (8-70) into Eq. (8-69), which requires championship algebraic endurance, results in

$$A_{ol}(s) = \frac{V_{os}}{V_{di}} = -\frac{A_{ol}(0)(1 + sR_{ob}C_c)}{D_{ol}(s)}, \qquad (8\text{-}72)$$

where $A_{ol}(0)$ is given by Eq. (8-48), and the characteristic polynomial, $D_{ol}(s)$, is

$$D_{ol}(s) = 1 + s\left[R_i C_i + R_o C_o + R_{ob} C_c + (1 + g_{m6} A_{ob} R_o) R_i C_c\right]$$

$$+ s^2 \left[R_i R_o (C_i + C_c) C_o + R_{ob} (R_i C_i + R_o C_o) C_c\right]$$

$$+ s^3 R_{ob} R_i R_o C_c C_i C_o. \qquad (8\text{-}73)$$

As in the simple capacitive compensation scheme, the loop gain, $T_u(s)$, pertinent to active capacitive compensation and a closed loop gain of one is

$$T_u(s) = -A_{ol}(s) = \frac{A_{ol}(0)\,(1 + sR_{ob}C_c)}{D_{ol}(s)}. \tag{8-74}$$

Cumbersome algebra notwithstanding, the loop gain expression of Eq. (8-74) clearly confirms the presence of a left half plane zero, as opposed to the right half plane zero indigenous to the simple capacitive compensation scheme. Moreover, if resistance R_{ob} is small, which Eq. (8-66) suggests is plausible if the geometry and/or biasing of transistor $M8$ is implemented judiciously, Eq. (8-73) offers an approximate dominant pole frequency of

$$p_1 \approx \frac{1}{R_i\,(1 + g_{m6}A_{ob}R_o)\,C_c} \approx \frac{1}{g_{m6}A_{ob}R_iR_oC_c}, \tag{8-75}$$

which is similar in form to the dominant pole frequency defined by Eq. (8-53). The corresponding frequency of the nondominant pole is

$$p_2 \approx \frac{g_{m6}A_{ob}}{\left(1 + \dfrac{C_i}{C_c}\right)C_o}. \tag{8-76}$$

Assuming a sufficiently large product, $g_{m6}A_{ob}$, the resultant unity gain frequency, say ω_u, of the buffered compensation network is

$$\omega_u = A_v(0)p_1 \approx \frac{g_{me}}{A_{ob}C_c}. \tag{8-77}$$

Although the phase margin delivered by buffered compensation of the two-stage op-amp is potentially larger than that of the conventionally compensated op-amp because of the absence of a right half plane zero in the buffered approach, it is nonetheless instructive to examine the actual phase margin, albeit to a crude first order approximation. To this end, introduce the transconductance ratio,

$$k_o = g_{me}/g_{m8}, \tag{8-78}$$

so that by Eqs. (8-67) and (8-77), the time constant associated with the left half plane zero in the loop gain function of Eq. (8-72) is $R_{ob}C_c = k_o/\omega_u$. Note that for large g_{m8} in comparison to g_{me}, the frequency of the subject zero can be considerably larger than the approximate unity gain frequency of the loop gain metric. Moreover, let

$$k_p = \frac{p_2}{\omega_u} \approx A_{ob}^2 \left(\frac{g_{m6}}{g_{me}}\right) \left[\frac{C_c}{\left(1+\dfrac{C_i}{C_c}\right)C_o}\right], \qquad (8\text{-}79)$$

where Eqs. (8-76) and (8-77) are exploited. Finally, the inverse product of the three pole frequencies must equate to the coefficient of the s^3 term in the characteristic polynomial of Eq. (8-73); that is, $1/p_1 p_2 p_3 = R_{ob} R_i R_o C_c C_i C_o$. In view of Eqs. (8-75) and (8-76), this constraint requires

$$p_3 \approx \left(1+\frac{C_c}{C_i}\right)\left(\frac{\omega_u}{k_o}\right) = \frac{k_c \omega_u}{k_o}, \qquad (8\text{-}80)$$

which is larger than the frequency of the left half plane zero, since

$$k_c \triangleq 1 + \frac{C_c}{C_i} \qquad (8\text{-}81)$$

is obviously larger than unity.

The loop gain function in Eq. (8-74) can now be written as

$$T_u(s) \approx \frac{A_{ol}(0)\left(1+\dfrac{k_o s}{\omega_u}\right)}{\left[1+\dfrac{A_{ol}(0)s}{\omega_u}\right]\left(1+\dfrac{s}{k_p \omega_u}\right)\left(1+\dfrac{k_o s}{k_c \omega_u}\right)}, \qquad (8\text{-}82)$$

and its corresponding phase angle, $\phi_u(\omega_u)$, at $\omega = \omega_u$ is

$$\varphi_u(\omega_u) \approx \tan^{-1}(k_o) - \tan^{-1}[A_{ol}(0)] - \tan^{-1}\left(\frac{1}{k_p}\right) - \tan^{-1}\left(\frac{k_o}{k_c}\right). \qquad (8\text{-}83)$$

For very large open loop gain at low signal frequencies, the second term on the right hand side of the last relationship converges toward 90°. Thus, the phase margin, which is simply $180° + \phi_u(\omega_u)$, becomes

$$\varphi_{pm} \approx 90° + \tan^{-1}(k_o) - \tan^{-1}\left(\frac{1}{k_p}\right) - \tan^{-1}\left(\frac{k_o}{k_c}\right). \qquad (8\text{-}84)$$

Figure 8.26 plots the phase margin in Eq. (8-84) as a function of the nondominant pole frequency factor, k_p, for $k_o = g_{me}/g_{m8} = 1$ and three realistic values of the capacitance factor, k_c, defined by Eq. (8-81). Small k_c, which corresponds to minimal compensation capacitance, expectedly

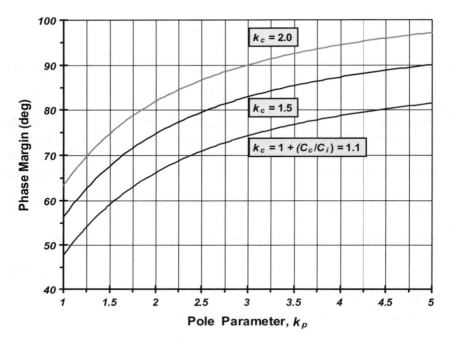

Figure 8.26. The phase margin of the two-stage op-amp with buffered capacitive compensation as a function of parameter k_p, which is the nondominant pole frequency normalized to the open loop unity gain frequency. The effective transconductance of the first stage and the transconductance of transistor $M8$ in the compensation subcircuit are presumed to be the same. The plots reflect the presumption of a very large zero frequency value of loop gain.

comprises the worst case for stability. But even for $k_c = 1.1$, a k_p value of only about 1.75 achieves the laudable objective of an approximate 63° phase margin, which is indicative of a maximally flat closed loop, unity gain frequency response. For progressively larger k_c, stability problems rapidly disintegrate. In particular, $k_c = 2$ results in a maximally flat closed loop magnitude response even for $k_p = 1$, which corresponds to a nondominant pole frequency that matches the open loop amplifier unity gain frequency.

In the interest of completeness, Fig. 8.27 schematically portrays the buffer compensated two-stage op-amp. In this diagram, the differential and common mode signal notation is supplanted by arbitrary input excitations, V_1 and V_2. The $(-)$ algebraic sign associated with input V_1 designates a phase inverting input node; that is, an increasing V_1 in the time or frequency domains yields a decreasing response, V_o. The $(+)$ sign appended to the

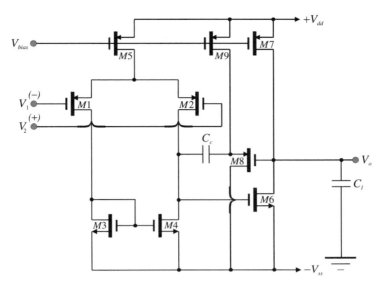

Figure 8.27. Schematic diagram of a two-stage operational amplifier employing active capacitance compensation to ensure a nominally dominant pole open loop response and sufficient phase margin.

V_2 input denotes a nonphase inverting input node. Transistor $M8$ serves as the requisite voltage buffer for the compensation scheme, capacitance C_c effectively sets the open loop dominant pole and unity gain frequencies, and transistor $M9$ supplies biasing current to the $M8$ buffer.

8.5.3.2. Passive Highpass Feedback

As is depicted in Fig. 8.28(a), passive highpass frequency compensation amounts to connecting a series combination of resistance R_c and capacitance C_c between the input and output ports of the phase inverting second stage of a two-stage op-amp. Fig. 8.28(b) offers the corresponding high frequency, small signal equivalent circuit. In the latter configuration, the small signal currents, I_{as} and I_{bs}, flowing through the series R_c-C_c branch are related to the indicated I/O signal voltages, V_{o1s} and V_{os}, via

$$I_{as} = -I_{bs} = \left(\frac{sC_c}{1 + sR_cC_c}\right)(V_{o1s} - V_{os}). \tag{8-85}$$

Accordingly, the R_c-C_c branch in question can be represented by the two-port equivalent circuit shown in Fig. 8.29(a). In turn, this model coalesces

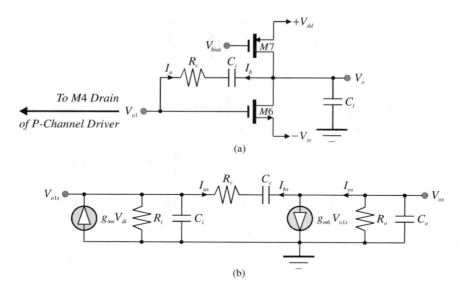

Figure 8.28. (a) Schematic diagram of the phase inverting second stage to which passive highpass frequency compensation is applied. (b) Small signal, high frequency model of the compensated network in (a).

with the op-amp equivalent circuit in Fig. 8.28(b) to produce the circuit in Fig. 8.29(b). In the latter model, the two parallel branches on the left hand side of the representation in Fig. 8.29(a) have been supplanted with a Thévenin equivalent circuit to underscore the Miller multiplication of capacitance C_c, particularly if resistance R_c is small.

The current source, $[sC_cV_{ols}/(1+sR_cC_c)]$, at the output port of the model in Fig. 8.29(b) is a feedforward signal flow path through the R_c-C_c feedback branch in the phase inverting amplifier. This feedforward current should pale in comparison to the voltage controlled current source, $g_{m6}V_{ols}$, which represents the desired feedforward through the $M6$ common source amplifier in the phase inverting amplifier. The two controlled currents at hand add algebraically to produce an effective forward transadmittance, $Y_f(s)$, of

$$Y_f(s) = \left(g_{m6} - \frac{sC_c}{1+sR_cC_c}\right) = \frac{g_{m6}\left(1+\dfrac{s}{z_f}\right)}{1+\dfrac{s}{p_f}}, \tag{8-86}$$

which posits a zero at a frequency,

$$z_f = \frac{g_{m6}}{(g_{m6}R_c - 1)C_c}, \tag{8-87}$$

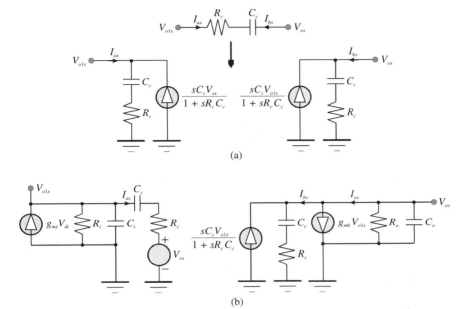

Figure 8.29. (a) Two-port parameter equivalent circuit of R_c-C_c feedback branch in the phase inverting amplifier of Fig. 8.28(a). (b) Small signal, high frequency model of the two-stage op-amp to which passive highpass frequency compensation is applied. The model is an alternative to that offered in Fig. 8.28(b).

and a pole at a frequency,

$$p_f = \frac{1}{R_c C_c}. \tag{8-88}$$

Since the voltage gain of the compensated inverter is directly proportional to $Y_f(s)$, it is obviously advantageous from a stability perspective to ensure that the zero indicated in Eq. (8-86) lies in the left half s-plane; that is, it is desirable that z_f in Eq. (8-87) be a positive frequency. To this end, the facts that $z_f > 0$ results when $g_{m6} R_c > 1$ and z_f is infinitely large if $g_{m6} R_c = 1$ are noteworthy.

The plausibility of varying the product, $g_{m6} R_c$, to enhance stability margins, coupled with the unavoidable uncertainty underlying the accurate prediction of transistor transconductance g_{m6}, motivates an electronic means of implementing an adjustable resistance, R_c. A suitable design strategy is suggested by Fig. 8.30, wherein a MOSFET operating in the ohmic regime of its static characteristic curves is utilized to realize R_c as a voltage

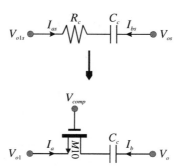

Figure 8.30. The realization of highpass stability compensation with an active voltage controlled resistance.

variable resistance. Since capacitance C_c precludes static current from flowing through the drain of transistor $M10$, its quiescent drain-source voltage is necessarily zero when its static gate-source voltage, $(V_{comp} - V_{o1q})$, exceeds the $M10$ threshold potential, V_{h10}. A review of the disclosures leading to Eqs. (6-14) through (6-16) leads to the conclusion that in Fig. 8.30,

$$R_c = \frac{1}{K_{n10}\left(\dfrac{W_{10}}{L_{10}}\right)\left(V_{comp} - V_{iq} - V_{h10}\right)}, \tag{8-89}$$

which portrays resistance R_c as a well-behaved inverse function of the compensation bias, V_{comp}. The exploitation of this compensation strategy results in the op-amp whose schematic diagram appears in Fig. 8.31.

Example 8.2. Figure 8.28(b) depicts the small signal equivalent circuit of a two-stage operational amplifier whose frequency response is compensated with a passive highpass network consisting of the series interconnection of resistance R_c and capacitance C_c. Investigate the loop gain of the network corresponding to unity closed loop gain operation of the op-amp. In the course of this investigation, assume that R_c is selected to ensure $\omega_u R_c C_c \ll 1$, where ω_u designates the compensated unity gain frequency of the aforementioned op-amp loop gain. Additionally, let the transconductance, g_{m6}, of transistor $M6$ in Fig. (8.27) or (8.31) be selected so that $g_{m6} R_c \equiv 1$. Derive approximate expressions for the frequency, p_1, of the amplifier dominant pole, the frequency, p_2, of the amplifier nondominant pole, the unity gain frequency, and the phase margin, ϕ_{pm}. What constraint must be satisfied by the transconductance, g_{m6}, of the phase inverting second stage amplifier and transconductance, g_{me}, of the differential first stage of amplification if resistance R_c is indeed an electrically negligible branch element and the open loop response

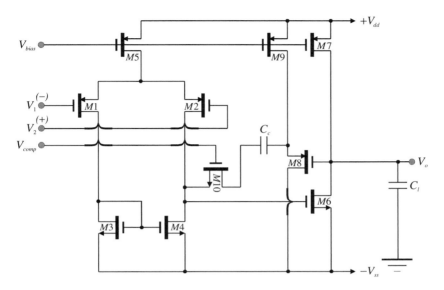

Figure 8.31. The op-amp of Fig. 8.27 implemented with highpass compensation that utilizes a MOSFET ($M10$) to realize a series resistance controlled by the biasing voltage, V_{comp}.

exudes a dominant pole response? Assuming that these constraints are satisfied, what frequency of the nondominant open loop amplifier pole, in relationship to the approximate unity gain frequency, is required for a 65° phase margin?

Solution 8.2.

(1) If $\omega_u R_c C_c \ll 1$, the radial frequency, $1/R_c C_c$, of the pole incurred by the compensation subcircuit lies outside the relevant frequency range of op-amp utility. It follows that the y-parameter equivalent circuit of the R_c-C_c feedback branch shown in Fig. 8.29(a) collapses to the more manageable structure offered in Fig. 8.32(a). With the subject pole frequency rendered insignificant, the constraint, $g_{m6} R_c = 1$, implies, recalling Eqs. (8-86)-through-(8-88), that the net forward transconductance of the second stage of the op-amp is little more than the transconductance, g_{m6}, of transistor $M6$. In effect, no appreciable feedforward through the R_c-C_c feedback branch is resultantly evidenced. Moreover, resistance R_c is significantly smaller than the capacitive impedance, $1/\omega C_c$, for all radial frequencies ω that are smaller than or equal to the op-amp unity gain frequency, ω_u. The upshot of these stipulations is that the small signal equivalent circuit of the two-stage op-amp depicted in Fig. 8.29(b) converges to the approximate model appearing in Fig. 8.32(b).

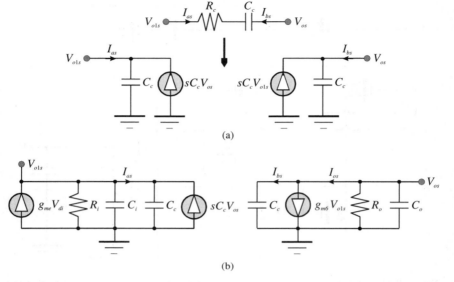

Figure 8.32. (a) The approximate y-parameter equivalent circuit of the R_c-C_c compensation branch for the case in which the pole frequency incurred by the compensation subcircuit is significantly larger than the unity gain frequency of the op-amp. (b) Approximate small signal, high frequency model of the compensated two-stage op-amp. The model reflects the presumption discussed in (a) and reflects the design constraint, $g_{m6} R_c = 1$.

(2) Signal flow theory readily enables an open loop gain relationship, $A_{ol}(s)$, that can be expressed in the form

$$A_{ol}(s) = \frac{V_{os}}{V_{di}} = \frac{g_{me} Z_{ol}(s)}{1 - sC_c Z_{ol}(s)}, \quad \text{(E2-1)}$$

where

$$Z_{ol}(s) = \left.\frac{V_{os}}{g_{me} V_{di}}\right|_{sC_c V_{os}=0} = -\left[\frac{g_{m6} R_o}{1 + s R_o (C_o + C_c)}\right]\left[\frac{R_i}{1 + s R_i (C_i + C_c)}\right] \quad \text{(E2-2)}$$

is the open loop transimpedance of the op-amp. It follows that the loop gain, $T_u(s)$, pertinent to unity gain closed loop operation of the op-amp is the second order relationship

$$T_u(s) = -A_{ol}(s) = \frac{A_{ol}(0)}{1 + as + bs^2}, \quad \text{(E2-3)}$$

where

$$A_{ol}(0) = (g_{me} R_i)(g_{m6} R_o) \quad \text{(E2-4)}$$

is the magnitude of the open loop voltage gain,

$$a = R_o(C_o + C_c) + R_i[C_i + (1 + g_{m6}R_o)C_c], \quad \text{(E2-5)}$$

and

$$b = R_i R_o (C_i + C_c)(C_o + C_c). \quad \text{(E2-6)}$$

It should be noted that because of the constraint, $g_{m6}R_c = 1$, the loop gain in Eq. (E2-3) contains no finite frequency left half or right half plane zeros.

(3) The design of the two-stage op-amp under present consideration reflects a strategy wherein the compensation capacitance, C_c, is selected to ensure that its Miller multiplied time constant, $R_i(1 + g_{m6}R_o)C_c$, establishes the dominant pole of the open loop gain function, $A_{ol}(s)$, and hence, of the loop gain, $T_u(s)$. With p_1 symbolizing the frequency of this dominant pole, it follows from Eq. (E2-5) that

$$p_1 \approx \frac{1}{R_i(1 + g_{m6}R_o)C_c} \approx \frac{1}{g_{m6}R_i R_o C_c}, \quad \text{(E2-7)}$$

where the liberty of presuming $g_{m6}R_o \gg 1$ is much larger than one is taken. Recognizing that the parameter, b, in Eq. (E2-6) is the inverse product of dominant and nondominant pole frequencies, p_1 and p_2, respectively,

$$p_2 = \frac{1}{bp_1} \approx \left(\frac{C_c}{C_i + C_c}\right)\left(\frac{g_{m6}}{C_o + C_c}\right). \quad \text{(E2-8)}$$

To the extent that this pole frequency is larger than the radial unity gain frequency, ω_u,

$$\omega_u \approx A_{ol}(0)p_1 \approx \frac{g_{me}}{C_c}. \quad \text{(E2-9)}$$

It is useful to observe that the requirement, $p_2 > \omega_u$, implies, from Eqs. (E2-8) and (E2-9),

$$\frac{g_{m6}}{g_{me}} > \left(1 + \frac{C_i}{C_c}\right)\left(1 + \frac{C_o}{C_c}\right); \quad \text{(E2-10)}$$

in other words, the transconductance of the transistor used in the phase inverting second stage must exceed the transconductance indigenous to the transistors utilized in the differential first stage. In addition to this constraint, a second, possibly more demanding, requirement is imposed on g_{m6}. In particular, recall that $g_{m6}R_c$ must be one if no finite transmission zeros are to be incurred in the loop gain function. Moreover, $\omega_u R_c C_c$ must be substantially smaller than one if resistance R_c can be neglected in comparison to the impedance associated with capacitor C_c over the useful frequency range of amplifier operation. Accordingly

$$\omega_u R_c C_c \approx \frac{g_{me}}{g_{m6}} \ll 1. \quad \text{(E2-11)}$$

(4) Equation (E2-3) can now be cast in the form,

$$T_u(s) = \frac{A_{ol}(0)}{\left(1 + \dfrac{s}{p_1}\right)\left(1 + \dfrac{s}{p_2}\right)} = \frac{A_{ol}(0)}{\left(1 + \dfrac{A_{ol}(0)s}{\omega_u}\right)\left(1 + \dfrac{s}{k_u \omega_u}\right)}, \quad \text{(E2-12)}$$

where

$$k_u = \frac{p_2}{\omega_u} \approx \left(\frac{C_c}{C_i + C_c}\right)\left(\frac{C_c}{C_o + C_c}\right)\left(\frac{g_{m6}}{g_{me}}\right) \quad \text{(E2-13)}$$

is the approximate factor by which the frequency of the nondominant amplifier pole exceeds the open loop unity gain frequency of the op-amp. It is worthwhile noting that the absence of any finite frequency transmission zeros and the presence of only two finite frequency poles in the loop gain function of Eq. (E2-12) precludes closed loop instability since the phase response of this loop gain only approaches $-180°$ in the limit as the signal frequency tends toward infinity. Thus, a phase margin specification in the present case fails to convey a meaningful sense of relative stability. However, it does bracket the settling time to transient step input excitations, as discussed in Chapter 4.

The phase response, $\phi_u(\omega_u)$, associated with $T_u(s)$ at frequency ω_u is

$$\varphi_u(\omega_u) = -\tan^{-1}[A_{ol}(0)] - \tan^{-1}\left(\frac{1}{k_u}\right) = \tan^{-1}\left[\frac{k_u A_{ol}(0) + 1}{A_{ol}(0) - k_u}\right]. \quad \text{(E2-14)}$$

Since the phase margin is this angle added to 180° and in general, $\tan(x + 180) = \tan(x)$, the phase margin, ϕ_{pm}, of interest is

$$\varphi_{pm} = \tan^{-1}\left[\frac{k_u A_{ol}(0) + 1}{A_{ol}(0) - k_u}\right] \approx \tan^{-1}(k_u), \quad \text{(E2-15)}$$

where the indicated approximation exploits the reasonable assumption, $A_{ol}(0) \gg 1$. It follows that a phase margin in the range of 65°, which generally results in reasonable settling times to transient inputs, requires a nondominant pole frequency that is larger than the amplifier unity gain frequency by a factor slightly larger than 2.1.

8.6.0. Slew Rate Limitations

All of the high frequency discourse in this chapter and in preceding chapters is limited to small signal response characteristics. In particular, it has always been presumed that regardless of the amplitudes of applied input signal excitations, the active circuits embedded in the signal flow path coupling the input port to the output port of considered feedback systems and circuits continue to operate in their linear regimes. This linearity presumption is obviously fallacious when large input signals are applied to active feedback structures. In the case of open loop operation of an op-amp, which typically exudes very large open loop gain at low signal frequencies, the "large input" commensurate with nonlinearity onset can actually be quite small. For example, if the open loop gain of an op-amp is 1,000 volts/volt, or 60 dB, and if the largest achievable output voltage is 5 volts, as is usually prescribed by the utilized power supply voltages, any input above $5/1000 = 5\,\text{mV}$ necessarily constrains the open loop op-amp to a nonlinear mode of operation. In actual practice, op-amp nonlinearity is likely to be incurred at signal input levels that are considerably smaller than inputs deriving from a trivial division of the maximum achievable output voltage by the low frequency gain. As is demonstrated herewith, the onset of nonlinearity in a properly compensated op-amp is only indirectly affected by supply line voltages and more likely to be defined by the rate at which the utilized compensation capacitance can be charged. In turn, this maximum charging rate, or *slew rate*, is intimately related to the biasing levels imposed on the first stage of amplification.

8.6.1. Fundamentals of Slew Rate Issues

In order to gain an insightfully clear understanding of the design-oriented issues pervasive of slew rate limitations in active feedback networks, return to the compensated two-stage op-amp in Fig. 8.20, whose small signal, and therefore linear, equivalent circuit is the topology offered in Fig. 8.21. The open loop voltage gain implied by the model in question is the somewhat messy relationship of Eq. (8-47), where the open loop gain and the frequencies of the right half plane zero and both poles can be determined from Eqs. (8-48) through (8-51). Ideally, frequency compensation of the amplifier undergoing investigation entails choosing the compensation capacitance, C_c, and hence the net effective feedback capacitance, C_f, around the second stage to render the transmission zero and the nondominant pole inconsequential throughout the passband of interest. If the invoked design scenario converges toward this idealized goal, Eq. (8-47) can be supplanted by

$$V_o(s) \approx -\frac{A_{ol}(0)V_{di}(s)}{1+\dfrac{s}{p_1}} \approx -\frac{g_{me}V_{di}(s)}{\dfrac{1}{g_{m6}R_oR_i}+sC_f}, \qquad (8\text{-}90)$$

which gives rise to the elegantly simple first order op-amp macromodel submitted in Fig. 8.33.

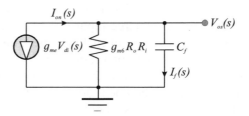

Figure 8.33. An approximate, small signal, frequency domain model of the compensated op-amp shown in Fig. 8.20. The branch element, C_f, which is the net effective feedback capacitance imposed around the phase inverting second stage, is presumed to establish a dominant pole open loop response.

Let the differential input voltage, V_{di}, applied to the op-amp in Fig. 8.20 be a step function of amplitude $-V_m$, whereby $V_{di}(s)$ in Eq. (8-90) becomes $-V_m/s$. Assuming that the capacitance, C_f, terminating the output port in the model of Fig. 8.33 is initially uncharged, the resultant time domain response follows as

$$v_o(t) = A_{ol}(0) V_m \left(1 - e^{-t/g_{m6} R_o R_i C_f}\right). \tag{8-91}$$

As expected in light of the subject model, capacitance C_f charges exponentially from its initial value of 0 volts to a steady state value of $A_{ol}(0) V_m$. The time constant of this charging characteristic is $g_{m6} R_o R_i C_f$, which is the inverse of the dominant pole frequency. The rate at which said capacitor charges, which is the impulse response of the linearized op-amp, is

$$\frac{dv_o(t)}{dt} = \left[\frac{A_{ol}(0) V_m}{g_{m6} R_o R_i C_f}\right] e^{-t/g_{m6} R_o R_i C_f}. \tag{8-92}$$

This charging rate is maximal at time $t = 0$; that is, maximal capacitive charging rate, which is indeed the slew rate of the op-amp (under the tacit presumption of linearity), occurs at the instant of step input application. Denoting this slew rate as SR_L, where subscript "L" reminds of presumed op-amp linearity for the given input voltage amplitude of V_m,

$$SR_L = \left|\frac{dv_o(t)}{dt}\right|_{max} = \frac{A_{ol}(0) V_m}{g_{m6} R_o R_i C_f} = \frac{g_{me} V_m}{C_f}. \tag{8-93}$$

Several interpretations of the slew rate defined by the preceding expression are potentially enlightening. Foremost among these is that a large slew rate is inherently desirable in that the op-amp should be capable of responding as rapidly as possible to even the most abrupt of applied input excitations. This desirability is akin to tromping on the accelerator of a sports car and relishing its sudden and agile acceleration. More to the electronics engineering point, the op-amp should achieve its steady state response, which in this case is $A_{ol}(0)V_m$ when the amplitude of the differential input voltage is V_m, as quickly as possible. A measure of this response dexterity is maximal capacitive charging rate, or slew rate, much like the rise and settling times discussed in previous chapters are also measures of response speed. Second, the combination of Eqs. (8-93) and (8-55) asserts

$$SR_L = \left| \frac{dv_o(t)}{dt} \right|_{max} = \frac{g_{me} V_m}{C_f} = \omega_u V_m, \qquad (8\text{-}94)$$

which stipulates the critical importance of the unity gain frequency in achieving large slew rate. Large V_m also conduces large SR_L, which is reasonable in that the response of a linear network is directly proportional to its input energy. In the sports car example, one naturally expects quicker acceleration with a progressively more aggressive right foot depressing the gas pedal. Finally, it is important to note that for $V_{di}(s) = -V_m/s$, the product, $g_{me}V_m$, in Eq. (8-93) or (8-94) is the maximum time domain value of the Norton current transform, $I_{on}(s)$, indicated in Fig. 8.33. Since the time domain current, $i_f(t)$, flowing through capacitance C_f in the same figure is $C_f dv_o(t)/dt$, this Norton current, say I_m, is also the maximum current conducted by C_f. The fact that the maximum capacitive current is identical to the maximum Norton current, I_m, is hardly surprising. In particular, maximum capacitor charging rate is evidenced at time $t = 0$, where the time domain capacitor voltage, $v_o(0)$, is zero, thereby precluding any initial current flow through the shunting resistance, $g_{m6}R_oR_i$, in Fig. 8.33. In summary,

$$SR_L = \left| \frac{dv_o(t)}{dt} \right|_{max} = \frac{g_{me}V_m}{C_f} = \omega_u V_m = \frac{I_m}{C_f}. \qquad (8\text{-}95)$$

It is prudent to underscore once again that Eq. (8-95) and its accompanying discourse are predicated on the presumed linearity of the op-amp, independent of the value of the input voltage amplitude, V_m. It is also worth interjecting that the preceding results are predicated on a negative input differential voltage solely for the purpose of conveniently establishing a positive output response in the time domain. Identical slew rate results derive from a positive input differential voltage, which generates a negative output response, since the slew rate metric merely defines the magnitude of time domain rate of output response change.

8.6.2. Slew Rate Limiting due to Nonlinearity

Return now to the op-amp depicted in Fig. 8.20 to reconsider the effects of a suddenly applied, large differential voltage. In particular, let $V_{di} = -V_m$ for time $t \geq 0$. If V_m approaches either of the two utilized power supply voltages or is otherwise large, the source to gate voltage of transistor $M2$ falls below the $M2$ threshold potential, thereby forcing this transistor into cutoff. On the other hand, the source to gate voltage of transistor $M1$ rises considerably above the threshold level of $M1$. But assuming an insignificant amount of channel length modulation in transistor $M5$, the maximum possible current that $M1$ can conduct, regardless of the value of input voltage V_m, is the quiescent current, previously noted to be I_{d5q}, supplied by the current source transistor, $M5$. This current flows through transistor $M3$ and because $M3$ and $M4$ function as a current mirror, the drain of transistor $M4$ conducts I_{d5q} as well. The relatively large current conducted by $M4$ drives the gate voltage of transistor $M6$ downward toward the negative supply bus, $-V_{ss}$, thereby holding $M6$ in cutoff immediately after the application of the large differential input voltage. Moreover, the current mirror nature of the $M5$–$M7$ pair forces $M7$ to conduct current I_{d5q}. Since $M6$ is turned off, a conduction path carrying current I_{d5q} is established from the $+V_{dd}$ supply line through capacitor C_c (or the net effective feedback capacitance, C_f) and thence through transistor $M4$ to the $-V_{ss}$ supply bus, or signal ground. As time progresses, $M2$ ultimately begins to conduct, thereby diminishing the current flowing through transistor $M4$. As the current flowing through $M4$ diminishes, the gate voltage of $M6$ rises to turn on this device. The actions of turning on $M6$ and incurring reduced current levels through $M4$ combine to diminish the charging current available to the feedback capacitance, which reduces its voltage charging rate. Evidently, however, the maximum charging current available to the net feedback capacitance occurs when transistors $M2$ and $M6$ are cutoff, and this maximum current is roughly I_{d5q}.

Since the net, presumably initially uncharged, feedback capacitance, C_f, appearing between the drain and gate terminals of transistor $M6$ conducts a nominally constant current of I_{d5q} immediately after the application of large differential input voltage,

$$v_o(t) = \left(\frac{I_{d5q}}{C_f}\right) t. \tag{8-96}$$

As argued in the preceding paragraph, this voltage expression remains valid until transistors $M2$ and/or $M6$ begin to conduct current. Regardless of when such conduction commences, the time rate of output voltage change inferred by Eq. (8-96) is the maximum possible rate of response change since the charging current decreases from its initial I_{d5q} value when $M2$ or $M6$ come out of cutoff. Thus, the slew rate, say SR, of the two-stage op-amp is simply[9]

$$SR = \left| \frac{dv_o(t)}{dt} \right|_{max} = \frac{I_{d5q}}{C_f}. \qquad (8\text{-}97)$$

A cursory inspection of Eq. (8-97) with Eq. (8-95) inappropriately suggests that current I_m in Eq. (8-95) has merely been replaced by current I_{d5q}. While this suggestion is algebraically valid, it incorrectly asserts that network linearity prevails and specifically, that $I_{d5q} = g_{me}V_m$. During actual and pragmatic slew rate testing, input voltage V_m is likely to be large enough to render $g_{me}V_m$ larger than I_{d5q} by a significant factor. Equivalently, SR_L, as defined by Eq. (8-95) is larger, and possibly substantially larger, than the observable slew rate, SR, given by Eq. (8-97). Stated in yet another way, note by Eq. (8-96) that the output voltage in the face of op-amp linearity charges linearly with time in the immediate neighborhood of $time\ t = 0$. On the other hand, op-amp linearity infers from Eq. (8-91) that the output voltage charges exponentially with time, which is considerably faster than that afforded by a linear voltage versus time dependence.

If Eq. (8-55) is inserted into Eq. (8-97), the slew rate can be expressed in terms of the approximate unity gain frequency of the op-amp; namely,

$$SR = \frac{I_{d5q}}{C_f} = \left(\frac{I_{d5q}}{g_{me}} \right) \omega_u. \qquad (8\text{-}98)$$

From Eq. (8-19), the effective transconductance, g_{me}, of the first stage of amplification is essentially equal to g_{m1}, the forward transconductance of transistor $M1$ (and of transistor $M2$, which is matched to $M1$). In turn, g_{m1} is determined by the quiescent current, I_{d1q}, conducted by $M1$. This bias current is one-half the current sourced by transistor $M5$; that is, $I_{d1q} = I_{d5q}/2$. Ignoring channel length modulation in $M1$, Eq. (6-29) establishes

$$g_{me} \approx g_{m1} = \sqrt{K_{n1} \left(\frac{W_1}{L_1} \right) I_{d5q}}, \qquad (8\text{-}99)$$

where K_{n1} is recalled as the $M1$ transconductance coefficient, which is the product of channel carrier mobility and oxide capacitance density. It follows that

$$SR = \frac{I_{d5q}}{C_f} \approx \omega_u \sqrt{\frac{I_{d5q}}{K_{n1}(W_1/L_1)}}, \qquad (8\text{-}100)$$

which suggests that for a given or a desired unity gain frequency, the slew rate of the two-stage op-amp rises in proportion to the square root of the quiescent current supplied to the differential first stage by transistor $M5$ in Fig. 8.20.

8.6.3. Full Power Bandwidth

The finite slew rate of an op-amp, or indeed of any type of electronic amplifier, limits the ability of the considered amplifier to respond faithfully in the steady state to sinusoidal excitations. To illustrate this contention, consider a differential input signal, V_{di}, in the circuit of Fig. 8.20 that is the sinusoid,

$$V_{di} = V_m \cos(\omega t), \qquad (8\text{-}101)$$

which effectively "slews" at a rate of

$$\frac{dV_{di}}{dt} = -\omega V_m \sin(\omega t). \qquad (8\text{-}102)$$

It is clear that the maximum time rate of change of the considered sinusoidal input is ωV_m. If this maximum rate of change is smaller than the amplifier slew rate, SR, the amplifier is capable of tracking the applied input signal, provided, of course, that the implemented amplifier biasing ensures an amplified input signal amplitude that is compliant with output port linearity. On the other hand, $\omega V_m > SR$ inhibits accurate tracking of the input sinusoid, since SR effectively bounds the maximum possible rate of amplifier response. The resultant ramification is potentially severe harmonic distortion, despite the fact that the amplified input amplitude lies well within the linear range of the amplifier output port.

A figure of merit serving to quantify the foregoing limitations is the amplifier *full power bandwidth*, say B_p. This parameter is defined to be the frequency of an applied input sinusoid of prescribed amplitude V_m at which the maximum rate of input signal change is identical to the amplifier slew rate. Accordingly,

$$B_p \triangleq \frac{SR}{V_m}, \qquad (8\text{-}103)$$

and by Eq. (8-100),

$$\frac{B_p}{\omega_u} = \sqrt{\frac{I_{d5q}}{K_{n1}(W_1/L_1)V_m^2}}. \qquad (8\text{-}104)$$

8.7.0. Biasing Subcircuits

The compensated two-stage op-amp of Fig. 8.31 requires, in addition to the $+V_{dd}$ and $-V_{ss}$ bus lines, two bias voltage ports. One of these ports, to which the static voltage, V_{bias}, is applied, is incident with the gates of the P-channel transistors, $M5$, $M7$, and $M9$. These three transistors respectively establish the current sources for the input stage differential amplifier, the second stage phase inverting unit, and the source follower used in the capacitive feedback compensation loop. The second port, to which the voltage, V_{comp}, is applied, is the gate of transistor $M10$, which operates in its triode regime to forge an electronically adjustable resistance in series with the compensation capacitance, C_c. Both of the requisite biasing voltages, V_{bias} and V_{comp}, can derive from simple resistive division of the applied bus voltages. But this approach can lead to undesirably large power dissipation in simple subcircuits whose only purpose is the realization of reliable sources of voltage for circuit ports conducting virtually zero static current. Moreover, it renders the derived biases vulnerable to power line voltage fluctuations that, for example, might be incurred because of electrical noise coupled to these lines.

In view of the foregoing observations, it is appropriate to consider at least a few of the biasing circuits that are exploited commonly in MOS technology operational amplifiers, as well as in other MOS circuit applications. The biasing treatment delivered herewith is not intended to be exhaustive. Rather, it is offered merely to provide the reader a cursory overview of biasing networks and a glimpse as to how these circuits can be designed and implemented in MOS electronics. In the process of discussing these circuits, groundwork is forged for understanding the functionality, attributes, and limitations of more complex biasing networks. The discussion also serves to underpin creative new approaches for the design of bias circuits delivering reliable and predictable performance in the face of temperature variations and fluctuations in power line voltages.

8.7.1. Active Divider

Figure 8.34(a) depicts the schematic diagram of a commonly used active voltage divider realized in NMOS technology. This circuit is the electronic analog of a simple resistive divider in that transistors *MA* and *MB* are in series with one another, and the desired output voltage, V_Q, is extracted with respect to ground at the node to which *MA* and *MB* are connected to form the series topology. Since both transistors have their gate and drain terminals connected together, both devices operate in saturation. Ignoring threshold voltage modulation induced by bulk-source voltages, channel length modulation, and assuming that the two transistors are identical, save

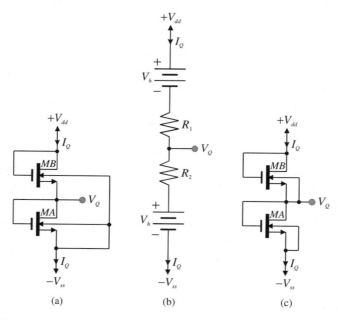

Figure 8.34. (a) An active divider circuit for generating a desired static voltage, V_Q, from two available power line buses. (b) The equivalent passive model of the circuit in (a). (c) An alternative to the topology of (a). This structure mitigates bulk-induced modulation of transistor threshold potentials.

for possible differences in their respective gate aspect ratios, η_A and η_B, the indicated current, I_Q, is given by

$$I_Q = \frac{K_n}{2}\eta_A \left(V_Q + V_{ss} - V_h\right)^2 = \frac{K_n}{2}\eta_B \left(V_{dd} - V_Q - V_h\right)^2, \quad (8\text{-}105)$$

where V_h symbolizes the gate-source threshold voltage of each transistor. Equation (8-105) gives rise to

$$V_Q = \left(\frac{1}{1 + \sqrt{\eta_A/\eta_B}}\right)(V_{dd} - V_h) - \left(\frac{\sqrt{\eta_A/\eta_B}}{1 + \sqrt{\eta_A/\eta_B}}\right)(V_{ss} - V_h),$$

(8-106)

which suggests the resistive macromodel submitted as Fig. 8.34(b). In the model at hand,

$$\frac{R_1}{R_2} = \sqrt{\frac{\eta_A}{\eta_B}} = \frac{V_{dd} - V_Q - V_h}{V_Q + V_{ss} - V_h}. \quad (8\text{-}107)$$

Clearly, an increase in the ratio, η_A/η_B, results in decreased quiescent output voltage, V_Q.

Although Eq. (8-106) and the companion result, Eq. (8-107), are simple expressions that allow the ratio of transistor gate aspect ratios to be stipulated in terms of desired static output voltage and available power line voltages, several transistor issues mandate that care be exercised when utilizing them in actual design environments. The first of these issues derives from bulk-induced modulation of threshold potential, which renders the threshold voltage of transistor *MB* slightly larger than that of transistor *MA*. The difference between the threshold voltages is exacerbated by progressively larger values of V_Q and the use of the negative supply line voltage, since the bulk-source voltage of *MA* is null, while for *MB* it is $-(V_Q + V_{ss})$. The computational error incurred by the tacit neglect of threshold voltage modulation is minimized if each bulk terminal is connected to its respective source terminal, as suggested in Fig. 8.34(c). When adopting this alternative architecture, design precautions must be taken to preclude measurable current conduction through the intrinsic PN junction diodes associated with the transistor bulk and the drain and source implants of transistor *MB*.

A second problem compromising the accuracy of Eq. (8-106) is channel length modulation, which increases the basic square law estimate of static drain current by a factor that is linearly related to transistor drain-source voltage. The resultant error in the prediction of static drain current is exacerbated for deep submicron channel lengths. For transistor *MA*, the drain-source voltage is $(V_Q + V_{ss})$, while for *MB*, it is $(V_{dd} - V_Q)$. Given a channel length modulation voltage, V_λ, Eq. (8-105) is accordingly modified to

$$I_Q = \frac{K_n}{2}\eta_A \left(V_{dd} - V_Q - V_h\right)^2 \left(1 + \frac{V_{dd} - V_Q}{V_\lambda}\right)$$

$$= \frac{K_n}{2}\eta_B \left(V_Q + V_{ss} - V_h\right)^2 \left(1 + \frac{V_Q + V_{ss}}{V_\lambda}\right), \quad (8\text{-}108)$$

for which the desired voltage, V_Q, derives as a solution to a cumbersome cubic algebraic equation. Fortunately, the problem at hand is rendered inconsequential through appropriate selection of the channel lengths for the two utilized transistors. In particular, by choosing the channel lengths of transistors *MA* and *MB* suitably long, parameter V_λ becomes large in comparison to both $(V_{dd} - V_Q)$ and $(V_Q + V_{ss})$. It should be understood that the adoption of relatively long channel lengths incurs increased carrier transit times and, for a fixed gate aspect ratio, increased device capacitances, which combine to diminish frequency response capabilities. But these effects are not shortfalls in a subcircuit whose functionality is limited to static applications and indeed, they may actually be desirable in that they mitigate the effects of high frequency noise that may superimpose with the static bus line voltages.

The computational errors in Eq. (8-106) are further propounded by its tacit neglect of degraded channel carrier mobility, which reduces the drain current predicted by the basic volt-ampere expression,

$$I_d = \frac{K_n}{2}\left(\frac{W}{L}\right)\left(V_{gs} - V_h\right)^2 \left(1 + \frac{V_{ds}}{V_\lambda}\right). \quad (8\text{-}109)$$

In particular, the transconductance parameter, K_n, is proportional to the carrier mobility that prevails under zero, or at least low, channel electric field. An added complication is the fact that two somewhat independent phenomena contribute to this mobility degradation. The first of these phenomena is the lateral electric field manifested in the nondepleted portion of the conduction channel by applied drain-source voltage. A first order account of the impact of lateral fields entails reducing parameter K_n by the factor, f_c, such that [10]

$$f_c = 1 + \frac{V_{gs} - V_h}{V_c}, \quad (8\text{-}110)$$

where
$$V_c = LE_c, \tag{8-111}$$

L is the actual, or drawn, channel length, and E_c is the lateral field strength for which the carrier velocity in the channel is one-half of the saturated value of carrier velocity. In silicon technologies, E_c is in the range of 5 volts/micron-to-7 volts/micron.

The second phenomena contributing to mobility degradation is the vertical electric field induced through the gate oxide and into the conducting channel by applied gate-source voltage. The corresponding additional reduction in parameter K_n in Eq. (8-109) is the factor,[11]

$$f_x = 1 + \frac{V_{gs} - V_h}{V_x}, \tag{8-112}$$

where the parameter, V_x, is termed the vertical field mobility degradation voltage. To first order, the subject voltage parameter relates to the thickness, T_{ox}, of the gate oxide in accordance with

$$V_x = k_{ox} T_{ox}, \tag{8-113}$$

where k_{ox} is typically 0.5 volts-per-nanometer.

The combined impact of Eqs. (8-110) and (8-112) on the basic volt-ampere characteristic of Eq. (8-109) is

$$I_d = \frac{K_n}{2 f_c f_x} \left(\frac{W}{L}\right) (V_{gs} - V_h)^2, \tag{8-114}$$

where the low field value of the transconductance parameter, K_n, is seen to be reduced by the voltage dependent factor,

$$f_c f_x = \left(1 + \frac{V_{gs} - V_h}{V_c}\right) \left(1 + \frac{V_{gs} - V_h}{V_x}\right). \tag{8-115}$$

It follows that the dominant effects of field-induced carrier mobility degradation can be absorbed into the square law MOSFET relationship in Eq. (8-109) through replacement of the transconductance parameter, K_n, by an effective transconductance parameter value, K_{ne}, such that

$$K_{ne} = \frac{K_n}{f_c f_x} = \frac{K_n}{\left(1 + \dfrac{V_{gs} - V_h}{V_c}\right)\left(1 + \dfrac{V_{gs} - V_h}{V_x}\right)}. \tag{8-116}$$

The indicated reduction factor is clearly a second order function of the excess gate voltage, $(V_{gs} - V_h)$. A reasonable first order estimate of the transconductance degradation factor in short channel devices is 30%-to-40%.

Alternatively, and under the presumption that channel length modulation phenomena are mitigated through appropriate selection of transistor channel lengths, the result in Eq. (8-114) can be expressed in the form,

$$I_d = \frac{K_n}{2}\left(\frac{W}{L}\right)V_e^2, \tag{8-117}$$

which preserves the original value of the transconductance parameter, K_n. However, observe that the actual excess gate-source voltage, $(V_{gs} - V_h)$, is now supplanted by an effective voltage difference, V_e, in accordance with

$$V_e = \frac{V_{gs} - V_h}{\sqrt{f_c f_x}} = \frac{V_{gs} - V_h}{\sqrt{\left(1 + \frac{V_{gs} - V_h}{V_c}\right)\left(1 + \frac{V_{gs} - V_h}{V_x}\right)}}. \tag{8-118}$$

The immediate upshot of the revised volt-ampere relationship in Eq. (8-117) is that Eq. (8-107) becomes

$$\frac{\eta_A}{\eta_B} = \left(\frac{V_{eB}}{V_{eA}}\right)^2. \tag{8-119}$$

Using Eq. (8-118) and Fig. 8.34, V_{eA} in Eq. (8-119) is

$$V_{eA} = \frac{V_Q + V_{ss} - V_h}{\sqrt{\left(1 + \frac{V_Q + V_{ss} - V_h}{V_c}\right)\left(1 + \frac{V_Q + V_{ss} - V_h}{V_x}\right)}}, \tag{8-120}$$

while

$$V_{eB} = \frac{V_{dd} - V_Q - V_h}{\sqrt{\left(1 + \frac{V_{dd} - V_Q - V_h}{V_c}\right)\left(1 + \frac{V_{dd} - V_Q - V_h}{V_x}\right)}}. \tag{8-121}$$

In an attempt to highlight, as a function of the desired quiescent output voltage, V_Q, the impact that mobility degradation has on the design of the biasing circuit in Fig. 8.34(c), Fig. 8.35 is submitted to compare the gate geometry computation deduced from Eq. (8-107) to that gleaned by way of Eq. (8-119). In this comparison, both transistors MA and MB have channel lengths (L) of 1.5 microns, a critical channel electric field (E_c) of 6 volts/micron, an oxide thickness (T_{ox}) of 9 nanometers, $k_{ox} = 0.5$ volts/nanometer, and a zero bias threshold voltage (V_h) of

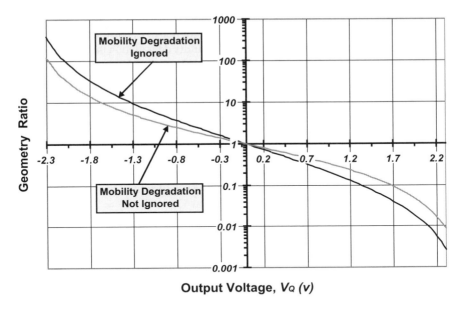

Figure 8.35. Semi-logarithmic plot of the transistor geometry ratio, η_A/η_B, as a function of the quiescent output voltage, V_Q, for the simple divider circuit in Fig. 8.34(c).

0.75 volts. The subject plot invokes $V_{dd} = V_{ss} = 3.3$ volts. Observe in Fig. 8.35 that significant differences prevail in the two computations when voltage V_Q approaches 25% of either of the two power supply voltages.

Example 8.3. Design the biasing network shown in Fig. 8.34(c) for a static output voltage, V_Q, of 1.0 volts when $V_{dd} = 2.7$ volts and $V_{ss} = 0$. The internal resistance, say R_p, indigenous to the positive supply line is 2 Ω. The bias circuit is to consume less than 700 μwatts of static power when it drives a capacitive load. Each transistor has a critical channel field, E_c, of 6 volts/μm, an oxide thickness of 10 nm, a zero bias threshold voltage (V_h) of 0.68 volts, and an oxide voltage parameter, k_{ox}, of 0.5 volt/nm. Moreover, each transistor has a transconductance coefficient, K_n, of 160 μmho/volt.

Solution 8.3.

(1) The net static power consumed by the circuit in Fig. 8.34(c) is $(V_{dd} + V_{ss})I_Q$. Since $(V_{dd} + V_{ss}) = 2.7$ volts, and the maximum allowable power dissipation is 700 μwatts, $I_Q \leq 259.3$ μA. Let the drain currents conducted by transistors MA and MB be restricted to $I_{dA} = I_{dB} = I_Q = 250$ μA.

(2) From Eq. (8-120), it is desirable to make $V_c = V_{cA} \gg (V_Q + V_{ss} - V_h)$ so that the effective gate-source voltage, V_{eA}, applied to transistor MA is rendered nominally independent of the effects of mobility degradation incurred by lateral channel fields. Since $(V_Q + V_{ss} - V_h) = 320\,\text{mV}$, choose $V_{cA} \geq 3.2$ volts, which is 10-times the 320 mV critical value. Recalling Eq. (8-111), $V_{cA} \geq 3.2$ volts and $E_c = 6\,\text{volts}/\mu\text{m}$ deliver $L_A \geq 0.533\,\mu\text{m}$.

(3) For the circuit of Fig. 8.34(c), $(V_{dd} - V_Q - V_h) = 1.02$ volts. It is therefore desirable to have $V_c = V_{cB} \geq 10.2$ volts, or $L_B \geq 1.70\,\mu\text{m}$. Recall that design constraints associated with transistor MA call for $L_A \geq 0.533\,\mu\text{m}$. From a circuit fabrication perspective, it is convenient to use transistors of equal channel lengths, each larger than the maximum of dimensions L_B and L_A. Accordingly, pick $L_A = L_B = 2.0\,\mu\text{m}$, which serves to satisfy the field mitigation requirements in both devices. Note then that $V_{cA} = L_A E_C = L_B E_C = V_{cB} = 12.0$ volts, while for both transistors, $V_x = k_{ox} T_{ox} = 5.0$ volts.

(4) As a peripheral calculation, it may be enlightening to evaluate the product, $f_c f_x$, for both transistors. For transistor MA, Eq. (8-115) and Fig. 8.34(c) give, with $V_{ss} = 0$,

$$f_{cA} f_{xA} = \left(1 + \frac{V_Q - V_h}{V_{cA}}\right)\left(1 + \frac{V_Q - V_h}{V_x}\right) = 1.092, \quad \text{(E3-1)}$$

while for transistor MB,

$$f_{cB} f_{xB} = \left(1 + \frac{V_{dd} - V_Q - V_h}{V_{cB}}\right)\left(1 + \frac{V_{dd} - V_Q - V_h}{V_x}\right) = 1.306.$$

$$\text{(E3-2)}$$

(5) With $V_{dd} = 2.7$ volts, $V_{ss} = 0$, $V_Q = 1.0$ volts, $V_h = 680\,\text{mV}$, $V_{cA} = V_{cB} = 12.0$ volts, and $V_x = 5$ volts, Eqs. (8-120) and (8-121) give $V_{eA} = 306.2\,\text{mV}$ and $V_{eB} = 892.4\,\text{mV}$. It follows from Eq. (8-119) that $\eta_A/\eta_B = 8.496$.

(6) Given that $I_{dA} = I_{dB} = 250\,\mu\text{A}$, $K_n = 160\,\mu\text{mho/volt}$, and $V_{eB} = 892.4\,\text{mV}$, Eq. (8-117) yields a gate aspect ratio for transistor MB of $\eta_B = W_B/L_B = 3.924$, which means that $W_B = 7.85\,\mu\text{m}$. Since $\eta_A/\eta_B = 8.496$, $\eta_A = W_A/L_A = 33.34$, thereby implying that $W_A = 66.7\,\mu\text{m}$.

(7) To complete the design, it is generally prudent (although not always necessary) to incorporate a capacitance, say C_p, across the biasing output port, as depicted in Fig. 8.36(a). This capacitance, which is invariably large enough to mandate its incorporation as an off chip element, protects the biased circuitry from sudden power line noise and the transients incurred by routine turn on and turn off of the applied voltage, V_{dd}. Fig. 8.36(b) offers the pertinent small signal model, where ΔV_{dd} symbolizes the transient component of the power line voltage, and R_p is the effective internal resistance of the power source. Moreover, with g_{mA} denoting the forward transconductance of transistor MA and g_{mB} signifying the forward transconductance of MB, $1/g_{mA}$ and $1/g_{mB}$

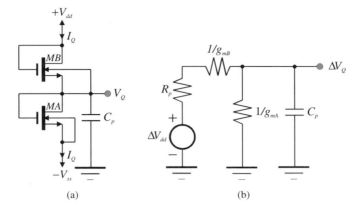

Figure 8.36. (a) The active divider circuit of Fig. 8.34(c) with capacitive decoupling incorporated at the output port. (b) Low frequency, approximate small signal equivalent circuit of the network in (a).

are respectively the small signal resistances presented to the circuit by the diode-connected transistors, MA and MB.

The time constant, T_p, manifested by capacitance C_p is

$$T_p = \frac{C_p}{g_{mA} + \dfrac{g_{mB}}{1 + g_{mB} R_p}}. \tag{E3-3}$$

Capacitance C_p should be chosen to ensure that this time constant is larger than the time length of any transient noise pulses that might reasonably superimpose with the power line voltage. In particular, a sufficiently large time constant, T_p, ensures a correspondingly large rise time in the bias voltage response to transient power line fluctuations, thereby rendering the biased circuitry nominally insensitive to such perturbation.

Ignoring channel length modulation phenomena, Eq. (6-29) produces

$$\left. \begin{array}{l} g_{mA} = \sqrt{\dfrac{2K_n \left(W_A/L_A\right) I_Q}{f_{cA} f_{xA}}} = 1.563 \text{ mmho} \\[2ex] g_{mB} = \sqrt{\dfrac{2K_n \left(W_B/L_B\right) I_Q}{f_{cB} f_{xB}}} = 490.3 \text{ } \mu\text{mho} \end{array} \right\}. \tag{E3-4}$$

With $R_p = 2\,\Omega$ and setting $T_p \geq 1$ mSEC for the purposes of this example, capacitance C_p is determined from Eq. (E3-3) to be $C_p = 2.053\,\mu\text{F}$. Accordingly, set $C_p = 5\,\mu\text{F}$.

Comments. The proof of the proverbial pudding lies in the simulated response of the completed circuit. To this end, HSPICE Level 3 model parameters for a representative 0.5-micron NMOS transistor reflective of the stipulated critical field, oxide thickness, threshold, oxide voltage parameter, and transconductance coefficient are given herewith. For future reference, PMOS parameters in a complementary 0.5-micron process are also submitted.

NMOS: .MODEL NM1 NMOS(LEVEL = 3 PHI = 0.66 TOX = 10N XJ = 0.15U TPG = 1 + VTO = 0.68 DELTA = 0.9 LD = 40N KP = 160U UO = 480 THETA = 0.22 + RSH = 1.3 GAMMA = 0.40 NSUB = 1.4E + 17 NFS = 7.1E + 11 VMAX = 180K + ETA = 21.27 M KAPPA = .1 CGDO = 330P CGSO = 330P CGBO = 420P + CJ = 520U CJSW = 270P MJ = .62 MJSW = 0.35 PB = 0.99)

PMOS: .MODEL PM1 PMOS(LEVEL = 3 PHI = 0.7 TOX = 10N XJ = 0.2U TPG = − 1 + VTO = − 0.75 DELTA = 0.25 LD = 67N KP = 44.5U UO = 240 THETA = 0.18 + RSH = 2.7 GAMMA = 0.45 NSUB = 9.8E + 16 NFS = 6.5E + 11 VMAX = 310K + ETA = 18M KAPPA = 6.3 CGDO = 370P CGSO = 370P CGBO = 430P + CJ = 930U CJSW = 150P MJSW = 0.3 PB = 0.95)

In addition to the foregoing model parameters, six other parameters must be incorporated into the net list entries that define the utilized transistors. In particular, the channel length, L, (in meters) and the gate width, W, (in meters) must be given. Moreover, the source and drain areas, A_S and A_D, respectively, (both in units of square meters) which are used in the calculation of relevant transistor capacitances, must be given. To first order, these areas may be calculated as

$$A_S \approx A_D \approx \frac{5LW}{2}. \tag{E3-5}$$

Finally, device perimeter dimensions, P_S and P_D, which are, like parameters A_S and A_D, exploited in transistor capacitance calculations, must be given. These may be estimated in units of meters as

$$P_S \approx P_D \approx W + 5L. \tag{E3-6}$$

For the purposes of this example and related forthcoming discussions in this text, it is not essential that the engineering significance of the model and device geometrical parameters be understood. Nevertheless, Table 8.1 offers tacit explanations.

Figure 8.37 gives the simulated static transfer curve of the bias circuit under present consideration. The curve confirms that for V_{dd} = 2.7 volts, V_Q = 0.991 volts, which is only 9 mV below the design target of 1.0 volts. This excellent accuracy with respect to voltage V_Q stems from the fact that the output response derives from a consideration of *ratios* of excess gate-source voltages. In contrast, the net power dissipation of 624.7 μwatts at V_{dd} = 2.7 volts is 10.8% lower than the computed power drain of 700 μwatts. Moreover, the simulated transistor drain currents of 231.4 μA are 7.4% smaller than their corresponding design estimates.

Table 1. Brief description of MOS technology transistor parameters used in HSPICE.

PARAMETER	DESCRIPTION	UNITS
PHI	Surface potential in strong inversion	volts
TOX	Oxide thickness	meters
XJ	Source & drain junction depths	meters
TPG	Gate material	—
VTO	Zero bias threshold voltage	volts
DELTA	Threshold dependence factor on width	—
LD	Source & drain lateral diffusion	meters
KP	Transconductance coefficient	mhos/volt
THETA	Vertical field mobility degrade factor	—
UO	Low field channel carrier mobility	meters^2/volt-sec
GAMMA	Body effect coefficient	volts^{-1}
RSH	Source & drain diffusion sheet resistance	ohms/square
NSUB	Effective substrate doping concentration	cm^{-3}
NFS	Fast surface state density	cm^{-3}
VMAX	Saturated limited carrier velocity	meters/sec
ETA	Lateral field mobility degrade factor	—
KAPPA	Channel length modulation factor	volts^{-1}
CGDO	Drain-gate overlap capacitance density	farads/meter
CGSO	Source-gate overlap capacitance density	farads/meter
CGBO	Bulk-gate overlap capacitance density	farads/meter
CJSW	Source & drain sidewall capacitance density	farads/meter
CJ	Bulk-source/drain capacitance density	farads/meter^2
MJ	Bulk-source/drain junction grading coefficient	—
MJSW	Source & drain sidewall grading coefficient	—
PB	Bulk-source/drain built-in potential	volts

Although these power dissipation and current results are acceptable from an engineering design perspective, they do reflect the inherent inaccuracies of the simple square law model for the saturation domain volt-ampere characteristics of a MOSFET.

Figure 8.38 pictures the simulated transient response of the circuit. For the indicated 2.7 volt power supply pulse train, observe an absence of perceptible overshoot and undershoot in the resultant output waveform. Note further that even

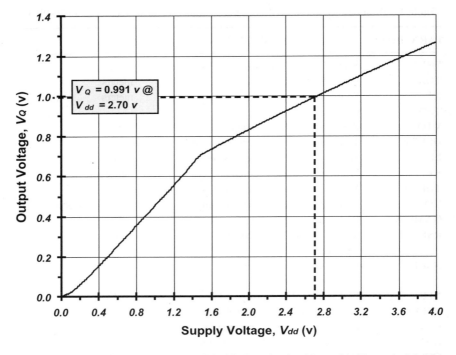

Figure 8.37. Static transfer response of the biasing circuit addressed in Example 8.3. The simulated response shows that the output voltage, V_Q, is only 0.9% below the design target of 1.0 Volts for a power supply voltage of $V_{dd} = 2.7$ Volts.

when the supply waveform is null, the output voltage decays to only about 50% of its steady state value owing to the capacitance, C_p, incorporated across the bias circuit output port. This is to say that the subject capacitance charges fully during those periodic time intervals when V_{dd} rises to 2.7 volts, but it does not discharge completely at times when V_{dd} is zero. An explanation of this observation is reserved for a problem exercise at the conclusion of this chapter.

8.7.2. Supply-Independent Biasing

The engineering price paid for the simplicity of the simple biasing divider in Fig. 8.34(c) is a static output voltage dependent on the applied power line voltage. This performance shortfall is clearly apparent in the simulated static response offered in Fig. 8.37. In order to mollify an output response dependence on power supply voltage, the output port of the biasing network must be isolated from the power line. With reference to the circuit

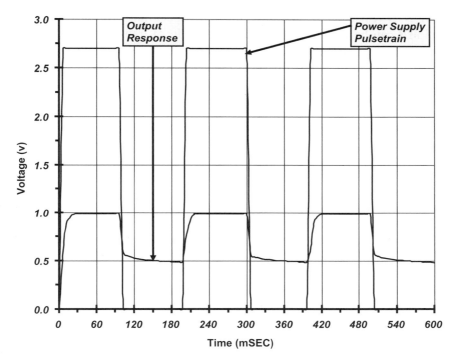

Figure 8.38. The simulated transient response of the biasing circuit addressed in Example 8.3. Observe the absence of observable overshoot and undershoot in the output response waveform.

in Fig. 8.34(c), such isolation can be accomplished by coupling the drain terminal of transistor *MB* to the supply voltage, V_{dd}, through a constant current source. Of course, the current, I_Q, supplied by this intervening current source must be precisely the current level that establishes gate-source voltages on transistors *MA* and *MB* that are commensurate with the desired output voltage, V_Q. To the extent that I_Q is independent of V_{dd}, response, V_Q, is likewise rendered independent of V_{dd}.

A practical example of a supply-independent biasing network is depicted in Fig. 8.39.[12] Observe that the series interconnection of transistors *MA* and *MB* is identical to the biasing configuration shown in Fig. 8.34(c). This series subcircuit is fed by a static current, I_Q, that is supplied by the current mirror comprised of the matched *P*-channel devices, *M*1 and *M*2, which are presumed to have identical gate aspect ratios. In the absence of conductive loads that might be connected to the ports where output responses V_{QA} and V_{QB} are extracted, current I_Q is resultantly conducted by all transistors, as well as by resistance R. Although the gate aspect ratios of transistors $M3$ and MB can be identical, the gate aspect ratio of transistor $M4$ must be larger than that of MA because the gate-source voltage, V_{gsA}, developed on transistor MA must supply the sum of the gate-source voltage, V_{gs4}, on $M4$ and the potential drop, $I_Q R$, developed across resistance R.

Figure 8.39 confirms

$$V_{gsA} = V_{gs4} + I_Q R, \qquad (8\text{-}122)$$

where the approximate square law characteristic of a MOSFET gives

$$I_Q = \frac{K_n}{2}\eta_A \left(V_{gsA} - V_{hn}\right)^2 = \frac{K_n}{2}\eta_4 \left(V_{gs4} - V_{hn}\right)^2. \qquad (8\text{-}123)$$

In the last expression, η_A and η_4 respectively denote the gate aspect ratios of transistors MA and $M4$, V_{hn} is presumed to be the threshold voltage of all N-channel transistors, and K_n is the transconductance coefficient of all N-channel units. If Eq. (8-123) is inserted into Eq. (8-122), resistance R is seen as satisfying the constraint,

$$R = \frac{2}{\sqrt{2K_n \eta_A I_Q}} \left(1 - \sqrt{\frac{\eta_A}{\eta_4}}\right). \qquad (8\text{-}124)$$

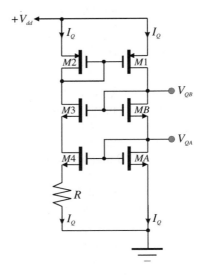

Figure 8.39. An example of a biasing network capable of realizing static output voltages, V_{QB} and V_{QA}, that are nominally independent of the supply voltage, V_{dd}. The substrate terminals of the P-channel transistors are connected to the V_{dd} voltage line, while the substrate terminals of all N-channel devices are returned to ground.

Recalling Eq. (6-29),

$$g_{mA} = \sqrt{2K_n \eta_A I_Q} \qquad (8\text{-}125)$$

is the small signal transconductance associated with transistor *MA*. Accordingly, Eq. (8-124) implies

$$g_{mA} = \frac{2}{R}\left(1 - \sqrt{\frac{\eta_A}{\eta_4}}\right), \qquad (8\text{-}126)$$

which in addition to providing a designable relationship for resistance R, portrays g_{mA} as a parameter that is independent of bias current and dependent only on resistance R and the ratio, η_A/η_4, of transistor gate aspect ratios. Clearly, η_4 must exceed η_A to ensure a physically realizable resistance, R. If $\eta_A/\eta_4 = 1/4$, there results the particularly convenient design relationship,

$$g_{mA}|_{\eta_A/\eta_4=1/4} = \frac{1}{R}. \qquad (8\text{-}127)$$

Returning to Fig. 8.39, the output voltage, V_{QA}, is given by

$$V_{QA} = V_{gsA} = V_{hn} + \sqrt{\frac{2I_Q}{K_n \eta_A}}, \qquad (8\text{-}128)$$

which, when combined with Eqs. (8-125) and (8-127), yields

$$V_{QA} = V_{hn} + 2I_Q R. \qquad (8\text{-}129)$$

Analogously,

$$V_{QB} = V_{gsB} + V_{QA} = V_{hn} + \sqrt{\frac{2I_Q}{K_n \eta_B}} + V_{QA}, \qquad (8\text{-}130)$$

and using Eq. (8-129),

$$V_{QB} = 2\left[V_{hn} + I_Q R\left(1 + \sqrt{\frac{\eta_A}{\eta_B}}\right)\right]. \qquad (8\text{-}131)$$

Equations (8-129) and (8-131) underscore the fact that if current I_Q is indeed a constant, independent of the power line voltage, V_{dd}, each of the two static output voltage responses, V_{QA} and V_{QB}, is independent of V_{dd}. The requirement of constant I_Q mandates that the *P*-channel transistor, $M1$, operate in its saturation domain and be characterized by an acceptably large channel length modulation voltage. In turn, the negligible channel length modulation implied by a large modulation voltage parameter requires a suitably large channel length. If large channel length transistors are unavailable or are otherwise impractical to implement, the current mirror comprised of transistors $M1$ and $M2$ can be supplanted by alternative current source structures that are compensated to deliver high output impedance.[13,14] A transparent shortfall of the present biasing scheme is the indicated dependence of V_{QA} and V_{QB} on the N-channel threshold voltage, V_{hn}, which gives rise to a slight sensitivity to operating temperature of the static output voltage responses.

Although the circuit in Fig. 8.39 satisfies its fundamental operational objective of delivering static output responses that are independent of the power supply voltage, utilization care must be exercised from two perspectives. First, the circuit at hand is a positive feedback structure. To confirm this contention, observe in Eqs. (8-129) and (8-131) that an increase in either V_{QA} or V_{QB}, precipitated by any phenomenological event, is necessarily accompanied by a commensurate increase in current I_Q. Because of the current mirror nature of the subject topology, an increase in I_Q causes the source-gate voltage, V_{sg2}, on transistor $M2$ to increase. In turn, increases in V_{sg2} precipitate increases in the source-gate voltage, V_{sg1}, developed at the source to gate terminals of transistor $M1$. There accordingly results a further increase in I_Q, which superimposes on the initial parasitic increase in said current; that is, positive feedback is evidenced. Despite this positive feedback, the circuit remains fortuitously stable as long as the gate aspect ratio, η_4, for transistor $M4$ is larger than the gate aspect ratio, η_A, for transistor MA. In particular, and as the reader is invited to confirm in a problem exercise at the conclusion of this chapter, the loop gain of the circuit remains smaller than unity, hence ensuring circuit stability, if $\eta_4/\eta_A > 1$.

The second problem derives from the inherent nonlinearity of the biasing network at hand. Specifically, two stable operating states for the static output responses are possible. The first of these states is the desired solution couched by Eqs. (8-129) and (8-131). The second stable solution is $V_{QA} = V_{hn}$ and $V_{QB} = 2V_{hn}$. From Eqs. (8-129) and (8-131), thus undesirable voltage solution implies $I_Q = 0$. In Eq. (8-123), $I_Q = 0$ gives rise to $V_{gsA} = V_{gsB} = V_{hn}$, which indeed satisfies Eq. (8-122) for $I_Q = 0$. In other words, the circuit in Fig. 8.39 may not converge at startup to the desired operating state associated with the appropriate nonzero value of current I_Q. Ordinarily, startup of the biasing configuration is not a serious issue because of unavoidable, albeit small, disparities among like parameters of presumably matched transistors. Startup problems may also be mitigated by the transient effects of a rapid application of power line voltage.

If startup problems persist, relatively simple startup subcircuits affecting biasing responses if and only if all transistor currents are latched null can be appended to the topology in Fig. 8.39.[15]

Example 8.4. Design the biasing network shown in Fig. 8.39 for static output voltages of $V_{QB} = 1.7$ volts and $V_{QA} = 900$ millivolts when $V_{dd} = 2.7$ volts. The internal resistance, say R_p, indigenous to the positive supply line is 1.5 Ω. The bias circuit is to consume less than 700 μW of static power when it drives a capacitive load. Each N-channel transistor has a critical channel field, E_c, of 6 volts/μm, an oxide thickness of 10 nm, a zero bias threshold voltage (V_{hn}) of 0.68 volts, an oxide voltage parameter, k_{ox}, of 0.5 volt/nm, and a transconductance coefficient, K_n, of 160 μmho/volt. On the other hand, each P-channel transistor has a critical channel field, E_c, of 5 volts/μm, an oxide thickness of 10 nm, a zero bias threshold voltage (V_{hp}) of 0.75 volts, an oxide voltage parameter, k_{ox}, of 0.5 volt/nm, and a transconductance coefficient, K_p, of 44.5 μmho/volt. Detailed Level 3 HSPICE parameters for both the N-channel and the P-channel devices appear in the preceding example.

Solution 8.4.

(1) The total static power consumed by the circuit in Fig. 8.9 is $2V_{dd}I_Q$. With $V_{dd} = 2.7$ volts and a maximum allowable power dissipation of 700 μW, $I_Q \leq 129.6$ μA. Let the indicated quiescent current, I_Q, conducted by all transistors be $I_Q = 125$ μA.

(2) Since $V_{QA} = 900$ mV, the gate-source voltage, V_{gsA}, applied to transistor MA must also be 900 mV. Recalling Eq. (8-115), $V_{cA} \gg (V_{QA} - V_{hn})$ ensures an effective gate-source voltage, V_{eA}, applied to transistor MA rendered nominally independent of the effects of mobility degradation incurred by lateral electric fields in the source to drain channel. Since $(V_{QA} - V_{hn}) = 220$ mV, choose $V_{cA} \geq 2.2$ volts, which is 10-times the 220 mV critical value. From Eq. (8-111), $V_{cA} \geq 2.2$ volts and $E_c = 6$ volts/μm deliver $L_A \geq 0.368$ μm.

(3) With $V_{QB} = 1.7$ volts and $V_{QA} = 0.9$ volt, V_{gsB}, the gate-source voltage established on transistor MB is $V_{gsB} = 0.8$ volt. Accordingly, $(V_{gsB} - V_{hn}) = 120$ mV. It is therefore desirable to have $V_c = V_{cB} \geq 1.2$ volts, or $L_B \geq 0.20$ μm. The minimal channel length restrictions for both transistors MA and MB are each smaller than the minimum allowable channel length of 0.50 μm. Because circuit response speed is not a design issue herewith, channel lengths considerably larger than the calculated minimum requirements can be selected. To this end, a channel length of 1.5 μm is reasonable for all N-channel transistors. Thus, $L_A = L_B = L_3 = L_4 = 1.5$ μm. Note then

that $V_{cA} = L_A E_C = L_B E_C = V_{cB} = 9.0$ volts, while for all transistors, $V_x = k_{ox} T_{ox} = 5.0$ volts.

(4) For transistor MA, Eq. (8-115) and Fig. 8.39 give

$$f_{cA} f_{xA} = \left(1 + \frac{V_{QA} - V_{hn}}{V_{cA}}\right)\left(1 + \frac{V_{QA} - V_{hn}}{V_x}\right) = 1.070, \quad \text{(E4-1)}$$

and for transistor MB,

$$f_{cB} f_{xB} = \left(1 + \frac{V_{QB} - V_{QA} - V_{hn}}{V_{cB}}\right)\left(1 + \frac{V_{QB} - V_{QA} - V_{hn}}{V_x}\right) = 1.038.$$

$$\text{(E4-2)}$$

Because these correction factors are barely greater than unity, taking $f_{c4} f_{x4}$ to be $f_{cA} f_{xA}$ incurs negligible computational error when evaluating the volt-ampere properties of transistor $M4$, which, save for differences in gate aspect ratios, is identical to transistor MA.

(5) The application of Eq. (8-114) to transistor MA in Fig. 8.39 results in $\eta_A = W_A/L_A = 34.53$, whence $W_A = 51.79\,\mu\text{m}$. Since $\eta_A/\eta_4 = 1/4$ is a design objective, $W_4/L_4 = 4W_A/L_A = 138.1$, or $W_4 = 207.2\,\mu\text{m}$.

(6) With $I_Q = 125\,\mu\text{A}$, $W_4/L_4 = 138.1$, and $f_{c4} f_{x4} \approx f_{cA} f_{xA} = 1.070$, the voltage, V_{gs4}, that must be established across the gate-source terminals of transistor $M4$ is $V_{gs4} = 0.790\,\text{V}$. Given that $V_{gsA} = 900\,\text{mV}$, it follows from Eq. (8-122) that the required value of circuit resistance, R, is $R = 880\,\Omega$.

(7) The gate-source voltage, V_{gsB}, of transistor MB is $V_{gsB} = V_{QB} - V_{QA} = 800\,\text{mV}$. For $f_{cB} f_{xB} = 1.038$ and $I_Q = 125\,\mu\text{A}$, Eq. (8-114) provides $\eta_B = W_B/L_B = 112.6$, which implies $W_B = 168.9\,\mu\text{m}$. The nominal symmetry between the two conductive branches of the circuit at hand allows this gate width to be stipulated as well for transistor $M3$; that is, $W_3 = W_B = 168\,\mu\text{m}$. Because of potentially significant bulk-induced modulation of the threshold voltage of transistor MB, it may prove expedient to connect the bulk terminal of MB directly to its source.

(8) The current source nature of transistor $M1$ in Fig. 8.39 demands that its dynamic channel resistance be large. In order to ensure large channel resistance, choose a channel length for both P-channel transistors that is two or three times larger than the channel lengths invoked for the N-channel devices. To this end, let $L_1 = L_2 = 4\,\mu\text{m}$, which means from Eq. (8-111) that $V_{c1} = V_{c2} = 20\,\text{V}$. The voltage parameters, V_{x1} and V_{x2}, remain 5 V. In order to assure the saturation regime operation of transistor $M1$, $(V_{dd} - V_{QB}) = 1\,\text{V} \geq (V_{sg1} - V_{hp})$. Since $V_{hp} = 750\,\text{mV}$, choose $V_{sg1} = 1\,\text{V}$. Then, by Eq. (8-115),

$$f_{c1} f_{x1} = \left(1 + \frac{V_{sg1} - V_{hp}}{V_{c1}}\right)\left(1 + \frac{V_{sg1} - V_{hp}}{V_{x1}}\right) = 1.063. \quad \text{(E4-3)}$$

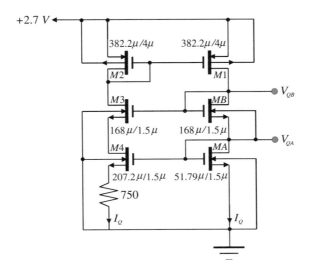

Figure 8.40. Schematic portrayal of the biasing network designed in Example 8.4.

and by Eq. (8-114), $\eta_1 = W_1/L_1 = 95.56$. Since transistors $M1$ and $M2$ function as a current mirror, it follows that $W_1 = W_2 = 382.2\,\mu\text{m}$.

Comments. The finalized circuit schematic diagram germinated by the foregoing design strategy is offered in Fig. 8.40. The 880-ohm calculation of resistance R is modified to 750 ohms because in the course of simulating the static transfer characteristics of the circuit, the currents conducted by all transistors were found to be about 30% lower than the design goal of 125 μA. This modification produces a steady state output voltage, V_{QB}, of 1.691 volts and a secondary steady state output voltage, V_{QA}, of 0.899 volts.

In Fig. 8.41, the simulated static transfer curves with respect to the power supply voltage, V_{dd}, shows that the design objective of nominal insensitivity of responses with respect to V_{dd} are met. Indeed, the curves confirm that the two measured output voltage responses are almost perfectly constant for $1.92\,\text{V} \leq V_{dd} \leq 3.5\,\text{V}$.

The academic perfectionist may experience heartburn over the apparent need to modify the originally calculated feedback resistance, R. Such modifications are commonplace in MOS technology circuit design and are even somewhat understandable in view of the analytical foundation upon which the design calculations are premised. In particular, the device models exploited are relatively simple in that they do not account for channel length modulation phenomena, nor do they embrace bulk-induced modulation of transistor threshold voltages, ohmic losses in the device drain and source leads, and several other second order effects. Moreover, it must be remembered that the fundamental purpose of mathematical circuit

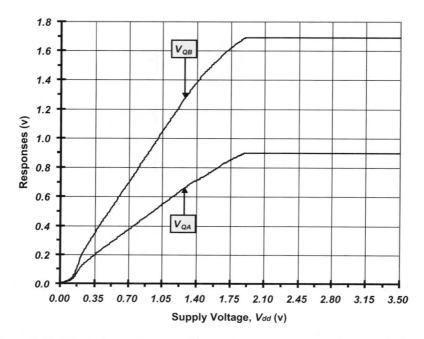

Figure 8.41. The static transfer curves of voltages V_{QB} and V_{QA} versus the power supply voltage, V_{dd}, for the biasing network whose final schematic diagram appears in Fig. 8.40.

analyses is not to arrive at precise circuit solutions. Rather, their purpose is to incur an insightful understanding of circuit dynamics that enables an expedient modification and optimization of circuit performance premised on either experimental observations or the fruits of computer assisted analyses. Stated in another fashion, one can question the need for computer-assisted design tools if all design-oriented calculations were guaranteed to result in precise delineations of designable network variables and device parameters.

References

1. A. B. Grebene, *Bipolar and MOS Analog Integrated Circuit Design* (Wiley Interscience, New York 1984), pp. 323–325.
2. J. Choma and J. Trujillo, Canonic cells of linear bipolar technology, *The Circuits and Filters Handbook*, (2nd Edn.), ed. W.-K. Chen (CRC Press, Boca Raton, Florida 2002), pp. 1639–1653.
3. S. Witherspoon and J. Choma, The analysis of balanced linear differential amplifiers, *IEEE Trans. on Education* **38** (1995) 40–50.

4. P. E. Allen and D. R. Holberg, *CMOS Analog Circuit Design* (Oxford University Press, New York 2002), pp. 180–199.
5. J. E. Solomon, The monolithic op-amp: A tutorial study, *IEEE J. Solid-State Circuits* **SC-9** (1974) 314–332.
6. Y. P. Tsividis and P. R. Gray, An integrated NMOS operational amplifier with internal compensation, *IEEE J. Solid-State Circuits* **SC-11** (1976) 748–753.
7. Y. P. Tsividis, Design considerations in single-channel MOS analog integrated circuits, *IEEE J. Solid-State Circuits* **SC-13** (1978) 383–391.
8. G. Palumbo and J. Choma, Jr., An overview of single and dual loop analog feedback; Part II: design examples, *J. of Analog Integrated Circuits And Signal Processing* **17** (1998) 195–219.
9. P. R. Gray and R. G. Meyer, *Analysis and Design of Analog Integrated Circuits* (John Wiley and Sons, New York, 1977), pp. 541–551.
10. J. Choma, Circuit level models of CMOS technology transistors, Univ. of Southern California, EE 533a Class Notes, 2002.
11. T. H. Lee, *The Design of CMOS Radio-Frequency Integrated Circuits* (Cambridge University Press, Cambridge, United Kingdom 2004), pp. 189–190.
12. J. M. Steininger, Understanding wide-band MOS transistors, *IEEE Circuits and Devices* **6** (1990) 26–31.
13. E. Säckinger and W. Guggenbühl, A high-swing, high-impedance MOS cascode circuit, *IEEE J. Solid-State Circuits* **25** (1990) 289–298.
14. K. Bult and G. J. G. M. Geelen, A fast-settling CMOS opamp for SC circuits with 90-dB DC gain, *IEEE J. Solid-State Circuits* **25** (1990) 1379–1384.
15. D. Johns and K. Martin, *Analog Integrated Circuit Design* (John Wiley & Sons, Inc., New York 1997), pp. 259–260.

Exercises

Problem 8.1
The differential amplifier studied in Example 8.1 is modified by appending capacitances, C, between ground and each transistor drain terminal, as shown in Fig. P8.1. Assume that capacitors, each of which is symbolized as C, are the dominant energy storage elements in relevant small signal models of the network. The circuit and device parameters, as well as the simplifying analytical assumptions, invoked in the aforementioned example, apply equally well to the problem considered herewith.

(a) What value of capacitance C delivers a 3-dB bandwidth of 400 MHz when a single ended output voltage response is extracted at the drain terminal of transistor $M2$?

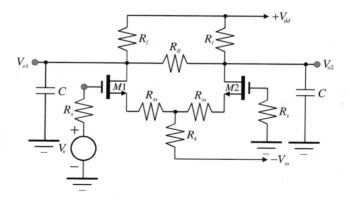

Figure P8.1.

(b) What fundamental purpose is served by the capacitor attached to the drain node of transistor $M1$?

Problem 8.2

The differential amplifier studied in Example 8.1 is modified by appending a capacitance, C, across the resistance, R_{ll}, as shown in Fig. P8.2. The circuit and device parameters, as well as the simplifying analytical assumptions, invoked in the aforementioned example, apply equally well to the problem considered herewith. What value of capacitance C delivers a 3-dB bandwidth of 400 MHz when a single ended output voltage response is extracted at the drain terminal of transistor $M2$?

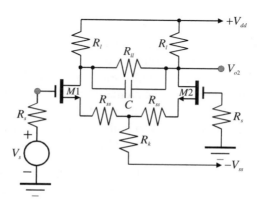

Figure P8.2.

Problem 8.3

The transistors in the balanced differential amplifier of Fig. P8.3 can be presumed to have infinitely large drain-source channel resistances and negligibly small bulk transconductance factors. These devices are identical and are biased identically to operate in their saturation regimes. The output response of interest is the small signal component of voltage V_{o1}. Only low signal frequencies, for which all intrinsic device capacitances can be tacitly ignored, are of interest in this exercise.

(a) Derive a general expression for the differential mode gain, A_d, of the amplifier.
(b) Derive a general expression for the common mode gain, A_c, of the amplifier.
(c) Give an expression for the Thévenin voltage gain, $A_{vt} = V_{o1s}/V_s$.
(d) Give an expression for the common mode rejection ratio, ρ.
(e) Derive a relationship for the differential mode output resistance, R_{do}.
(f) Derive a relationship for the common mode output resistance, R_{co}.
(g) What is the Thévenin resistance with respect to the output port at the drain terminal of transistor $M1$?

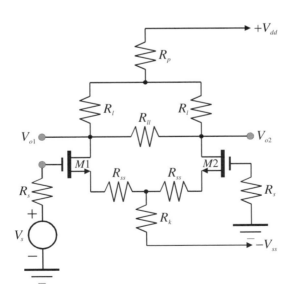

Figure P8.3.

Problem 8.4

Channel length modulation and bulk-induced threshold modulation phenomena can be tacitly ignored in all transistors embedded within the balanced differential amplifier of Fig. P8.4. All transistors are biased in their saturation regimes, and the small signal parameters of transistors $M1$, $M3$, and $M5$ are respectively identical to those of $M2$, $M4$, and $M6$. The output response of interest is the low frequency, small signal component of voltage V_{o2}.

(a) Derive a general expression for the differential mode gain, A_d, of the amplifier.
(b) Derive a general expression for the common mode gain, A_c, of the amplifier.
(c) Give a relationship for the Thévenin voltage gain, $A_{vt} = V_{o2s}/V_s$.
(d) Give a relationship for the common mode rejection ratio, ρ.
(e) Derive an expression for the differential mode output resistance, R_{do}.
(f) Derive an expression for the common mode output resistance, R_{co}.
(g) What is the Thévenin resistance with respect to the output port at the drain terminal of transistor $M2$?

Problem 8.5

Channel length modulation and bulk-induced threshold voltage modulation can be tacitly ignored in all MOSFETs embedded in the amplifier

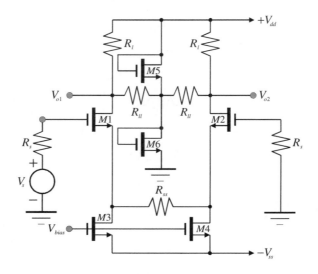

Figure P8.4.

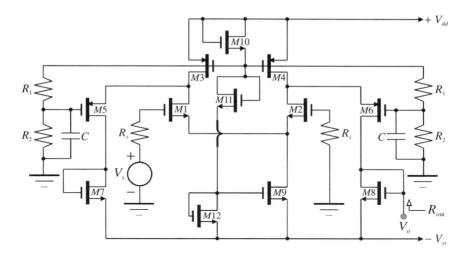

Figure P8.5.

of Fig. P8.5. All transistors operate in saturation. The transistor pairs, $M1-M2$, $M3-M4$, $M5-M6$, and $M7-M8$, are matched and biased identically. Transistors $M9$, $M10$, $M11$, and $M12$ are biased to yield respectively identical small signal parameters. Capacitance C is very large and therefore behaves as a short circuit over the signal frequencies of interest.

(a) Draw the differential mode, half circuit schematic diagram.
(b) What is the common mode rejection ratio, ρ, of the amplifier? Briefly rationalize your answer without explicitly solving for this metric.
(c) Derive an expression for the small signal voltage gain, $A_v = V_{os}/V_s$, of the amplifier.
(d) Give an expression for the small signal output resistance, R_{out}.
(e) Under differential mode operating conditions, what is the time constant, τ, associated with capacitance C?
(f) Is the circuit best suited for voltage, transimpedance, or transconductance amplification applications?
(g) Assuming differential operating conditions and that the source resistance, R_s, is 50 Ω, is stray capacitance appearing at the input port, the output port, the gate node of transistor $M5$, or the drain node of transistor $M2$ likely to establish the dominant amplifier pole?

(h) Assuming that the output port drives a balanced, purely capacitive load of capacitance C_l and that no other circuit or device capacitances are significant, what is the amplifier 3-dB bandwidth?

Problem 8.6

In the circuit of Fig. P8.6 all N-channel transistors have their bulk terminals incident with the negative supply bus, are biased in saturation, have infinitely large drain-source channel resistances, and have negligible bulk transconductances; that is, all $\lambda_b = 0$. Additionally, all transistors are identical, save for possibly different gate aspect ratios. The capacitance, C, which is obviously an open circuit at low signal frequencies, is inserted between ground and the indicated node to establish a predictable 3-dB bandwidth for the amplifier. The gate aspect relationships that are relevant to this problem are itemized herewith.

The gate aspect ratio of $M1$ is identical to that of $M1a$.
The gate aspect ratio of $M2$ is identical to that of $M2a$.
The gate aspect ratio of $M3$ is identical to that of $M3a$.
The gate aspect ratio of $M4$ is identical to that of $M4a$.
The gate aspect ratio of $M5$ is 4-times greater than that of $M7$.
The gate aspect ratio of $M1$ is 100-times greater than that of $M2$.
The gate aspect ratio of $M5$ is 9-times greater than that of $M6$.

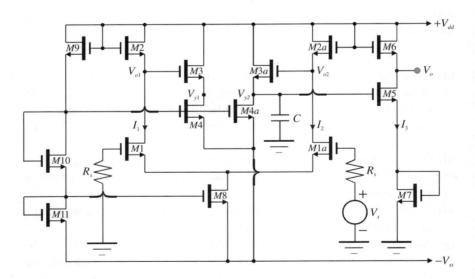

Figure P8.6.

(a) The amplifier at hand is technically unbalanced. What transistor gate terminal must be disconnected to render the configuration balanced at low signal frequencies?

(b) For the balanced subcircuit gleaned from the disconnection addressed in Part (a), what are the differential mode, low frequency, small signal components of currents I_1 and I_2? Express these currents in terms of the forward transconductance, g_{m1}, of transistor $M1$ and the indicated signal voltage, V_s.

(c) Using the results of Part (b), give the differential mode, low frequency, small signal components of the voltages, V_{o1}, V_{o2}, V_{y1}, and V_{y2}. By exploiting the stipulated gate aspect relationships for the transistors, individual answers can be expressed as a simple numerical constant multiplied by the signal voltage, V_s.

(d) What are the common mode, low frequency, small signal components of the voltages, V_{o1}, V_{o2}, V_{y1}, and V_{y2}?

(e) Using the results of Parts (c) and (d), give a simple expression for the small signal, low frequency current, I_3, conducted by the source terminal of transistor $M5$. Express your answer as a function of the forward transconductance, g_{m5}, of $M5$ and the signal voltage, V_s.

(f) Using the result of Part (e), give a simple expression for the small signal, low frequency voltage gain, V_o/V_s. Using previously stipulated gate aspect ratio relationships, the answer can be expressed as either a positive or a negative numerical value.

(g) Give an expression for the 3-dB bandwidth, B, of the amplifier.

(h) Would the gate aspect ratios of transistors $M3$ and $M3a$ need to be large or small in light of a large 3-dB circuit bandwidth objective?

(i) What principle function is served by the MOSFETs, $M9$, $M10$, and $M11$?

Problem 8.7

The CMOS circuit depicted in Fig. P8.7, which is used widely in systems compelled to operate from low supply voltages, is a so-called folded current mirror, differential amplifier. The substrate terminals of all NMOS devices are returned to ground potential, while the substrates of the PMOS transistors are connected to the positive voltage supply line. All NMOS transistors and all PMOS devices operate in their saturation regimes and are respectively matched, inclusive of gate aspect ratios. Observe that despite the preceding stipulation, all transistors do not necessarily conduct the same static drain currents, nor do all devices necessarily support the same drain-source

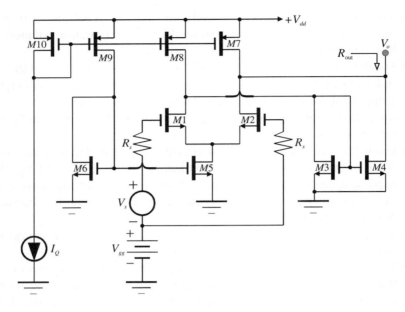

Figure P8.7.

quiescent potentials. Accordingly, the small signal forward transconductances of pertinent devices are not necessarily the same, and the small signal channel resistances cannot be presumed equal. Finally, voltage V_{gg} and current I_Q are biasing sources, while V_s represents a small signal input voltage. Note that there is no common mode input signal. Instead, only common mode input biasing for $M1$ and $M2$ is applied in the form of V_{gg}.

(a) Is the $M1 - M2$ differential pair balanced for all signal frequencies? If it is not balanced, what engineering strategy must be invoked to allow for a half circuit analysis of this pair?
(b) Determine the low frequency Thévenin equivalent voltages, say V_{to}, and the Thévenin equivalent resistances, say R_{to}, established at the drain terminals of transistors $M1$ and $M2$. *Do not* ignore relevant device channel resistances.
(c) Use the results of Part (b) to derive an expression for the low frequency voltage gain, $A_v = V_{os}/V_s$. Once again, do not ignore relevant channel resistances but in the interest of analytical simplicity, assume that the transconductance, g_{m3}, is large in comparison to the sum of the channel conductance of $M3$ and the Thévenin conductance, $1/R_{to}$.

(d) Use the results of Part (b) to derive an expression for the indicated output resistance, R_{out}. Do not ignore relevant device channel resistances.

(e) In terms of the constant current, I_Q, determine the static currents flowing:

   (i) in the drain of transistor $M6$;
   (ii) in the drain of transistor $M5$;
   (iii) in the drains of transistors $M1$ and $M2$;
   (iv) in the drains of transistors $M7$ and $M8$;
   (v) in the drains of transistors $M3$ and $M4$.

(f) In terms of current I_Q, derive an expression for the quiescent voltage established with respect to ground at the drains of transistors $M1$ and $M2$.

(g) In light of the results of Part (a), what can be said about the following small signal parameters?

   (i) The transconductances of transistors $M1$ and $M2$;
   (ii) The transconductances of transistors $M3$ and $M4$;
   (iii) The channel resistances of transistors $M1$ and $M2$;
   (iv) The channel resistances of transistors $M3$ and $M4$;
   (v) The channel resistances of transistors $M7$ and $M8$.

(h) Why is it inappropriate to ignore the effects of channel resistances in transistors $M1$, $M2$, $M7$, and $M8$?

(i) Why might it be appropriate to presume tacitly that the channel resistance of transistor $M3$ is infinitely large?

(j) What dominant device capacitances prevail at the amplifier output port?

Problem 8.8
Derive Eq. (8-62) through use of the basic stability concepts proffered in Chapter 4.

Problem 8.9
Derive Eq. (8-64) through use of the basic stability concepts proffered in Chapter 4.

Problem 8.10
Clearly explain the reason underlying the contention that unity closed loop gain in an electronic system poses the most significant network stabilization challenge.

Problem 8.11
Prove that a second order network whose phase margin is $\tan^{-1}(2)$ delivers a closed loop frequency response exuding maximally flat magnitude characteristics.

Problem 8.12
In the two-transistor subcircuit of a linear amplifier shown in Fig. P8.12(a), subscript "Q" in the indicated voltage and current variables denotes quiescent components, while subscript "s" symbolizes small signal components. Both transistors operate in saturation and are not necessarily characterized by negligible channel length modulation effects. Save for possible differences in the gate aspect ratios, the two transistors are otherwise identical. Show that the small signal model of the subcircuit in Fig. P8.12(a) is the equivalent circuit appearing in Fig. P8.12(b). In the latter topology, g_{mi} and g_{oi} respectively represent the forward transconductance and drain-source channel conductance of the ith transistor.

Problem 8.13
The topology depicted in Fig. P8.13, which utilizes the subcircuit shown in Fig. 8.12(a) is known as a Wilson current amplifier. All three transistors operate in saturation and are not necessarily characterized by negligible channel length modulation effects. Moreover, the transistors at hand may have significant bulk-induced modulation of threshold voltages. In general, the ith transistor postures a small signal transconductance of g_{mi}, a small signal channel conductance of g_{oi}, and a bulk transconductance factor of λ_{bi}. Transistors $M1$ and $M2$ have identical gate aspect ratios, while the gate aspect ratio of $M2$ is larger than that of $M3$ by a factor of k. The current,

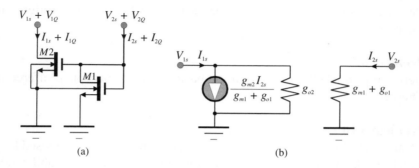

Figure P8.12.

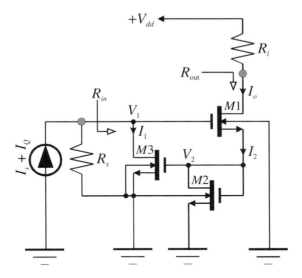

Figure P8.13.

I_s, is a small signal input current, I_Q is a bias current, and the variables, V_1, V_2, I_1, I_2, and I_o are each the superposition of quiescent and small signal components. The response of interest to input current signal I_s is the low frequency, small signal output current, I_{os}, flowing in the drain of transistor $M1$. In addressing the following inquiries, use the pertinent results of the preceding exercise.

(a) The amplifier in Fig. P8.13 is a global feedback current amplifier. Give an expression for the global feedback factor, say f.

(b) Using the feedback analyses strategies of preceding chapters, show that the open loop current gain, say A_{io}, is

$$A_{io} = \frac{g_{m1} R_s / (1 + g_{o3} R_s)}{1 + g_{o1} R_l + \dfrac{g_{o1} + (1 + \lambda_{b1}) g_{m1}}{g_{m2} + g_{o2}}}.$$

(c) Confirm that the loop gain, say T, is given approximately by

$$T \approx \left(\frac{g_{m3}}{g_{m2} + g_{o2}} \right) \left(\frac{g_{m1} R_s}{2} \right).$$

(d) Show that the closed loop, low frequency, small signal current gain, say A_i, is given approximately by

$$A_i \approx k.$$

(e) Confirm that the closed loop, driving point input resistance, R_{in}, satisfies

$$R_{in} \approx \frac{2k}{g_{m1}}.$$

(f) Show that the closed loop, driving point output resistance, R_{out}, can be approximated as

$$R_{out} \approx \frac{2 + g_{m1} R_s}{g_{o1}}.$$

Problem 8.14

Use the results of the preceding problem to deduce an approximate expression for the low frequency, small signal voltage gain, V_{os}/V_s, of the amplifier depicted in Fig. P8.14. All four transistors operate in saturation and have negligible channel length modulation and negligible bulk transconductance. In general, let the ith transistor have a small signal transconductance of g_{mi}. Transistors $M1$ and $M2$ have identical gate aspect ratios, while the gate aspect ratio of $M2$ is larger than that of $M3$ by a factor of k.

(a) Compare the functionality of the $M1$–$M2$–$M3$ subcircuit to that of a common gate cell in a conventional common source-common gate cascode configuration.

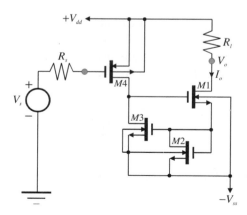

Figure P8.14.

(b) What advantage does the circuit at hand have over a conventional common source-common gate cascode?

Problem 8.15
Use a P-channel MOSFET and a capacitor to convert the amplifier in Fig. P8.14 into an integrator. Upon such conversion, what is the approximate 3-dB bandwidth of the resultant small signal voltage gain?

Problem 8.16
Design the amplifier in Fig. P8.14 to satisfy the following specifications. Use the HSPICE parameters given in the "Comments" accompanying Example 8.3.
  Maximum static power dissipation: 5 mW, with ± 2.7 volt supplies;
  Static output voltage: 0 volts;
  Small signal voltage gain: 5 volts/volt.

(a) Simulate the design realization on HSPICE or equivalent computer-sided design software. Modify the computed design results in an attempt to satisfy all specifications as close as possible.
(b) What is the simulated 3-dB bandwidth of the circuit?
(c) Does the amplifier exude a dominant pole response? Explain why or why not.

Problem 8.17
The static transfer characteristic offered in Fig. 8.37 clearly shows two prevailing slopes with respect to the applied power line voltage, V_{dd}. Analyze the circuit in question to explain these two distinct slopes.

Problem 8.18
Analyze the circuit in Fig. 8.40 to explain in Fig. 8.41 the nominally linear static transfer curves a low power supply voltage, V_{dd}.

Problem 8.19
Use the simple square law model for the volt-ampere characteristic of a MOSFET to deduce general expressions for the static voltages, V_{Q1} and V_{Q2}, in the three-transistor divider circuit of Fig. P8.19.

(a) What advantage and disadvantage pervades the connection of each bulk substrate terminal to the corresponding transistor source terminal, as shown in the diagram?

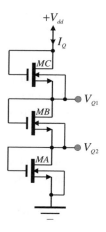

Figure P8.19.

(b) Design the biasing network in Fig. 8.19 to satisfy the following specifications. Use the HSPICE parameters given in the "Comments" accompanying Example 8.3.

Maximum static power dissipation: 200 μW, with $V_{dd} = 3.3$ volts;
Static output voltages: $V_{Q1} = 2.2$ volts & $V_{Q2} = 1.1$ volts;

(c) Simulate the static transfer characteristics of the design realization on HSPICE or equivalent computer-sided design software. Modify the computed design results in an attempt to satisfy all specifications as closely as possible.

(d) Let V_{dd} be represented by a suitable pulse train whose amplitudes range from 0 to 3.3 volts. Simulate on HSPICE the resultant transient responses evidenced with respect to V_{Q1} and V_{Q2} for the cases of no shunt capacitances incident with either output port and 5 μF of capacitance terminating each of the output ports to ground.

Problem 8.20

Analyze the biasing network addressed in Example 8.3 from the perspective of explaining, as accurately as reasonable approximations allow, the transient response plotted in Fig. 8.38. Of particular interest is the response evidenced when $V_{dd} = 0$.

Problem 8.21

Derive an expression for the small signal loop gain of the biasing configuration displayed in Fig. 8.39. Express results as a function of the gate aspect ratios, η_4 and η_A, for transistors $M4$ and MA, respectively. Give the necessary condition underlying the unconditional dynamic stability of the circuit.

Problem 8.22

The circuit diagram in Fig. P8.22 is an example of a biasing network that produces a static output response, V_Q, which is nominally independent of the supply voltage, V_{dd}. All transistors operate in their respective saturation regimes and to first order, channel length modulation and bulk-induced threshold modulation phenomena can be ignored in all devices. The substrate terminals of all N-channel devices are returned to circuit ground, and the substrate terminals of all P-channel transistors are connected to the positive supply line.
Using the simple square law model of the static volt-ampere characteristics for all transistors, verify that the indicated static output voltage, V_Q, is

$$V_Q = \left(\frac{2R_L}{K_n R^2 \eta_1}\right)\left(1 - \frac{1}{\sqrt{K_n}}\right)^2,$$

where K_n is the transconductance coefficient of all N-channel transistors.

Problem 8.23

Design the biasing network in Fig. P8.22 to satisfy the following specifications. Use the HSPICE parameters given in the "Comments" accompanying Example 8.3.

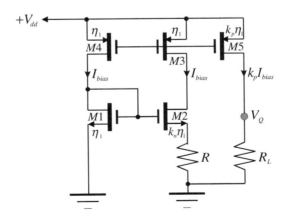

Figure P8.22.

Maximum static power dissipation: 5 mW, with a 2.7 volt supply;
Static output voltage: 1.5 volts.

(a) Simulate the static transfer characteristics of the design realization on HSPICE or equivalent computer-sided design software. Modify the computed design results in an attempt to satisfy all specifications as closely as possible.

(b) Let V_{dd} be represented by a suitable pulse train whose amplitudes range from 0 to 2.7 volts. Simulate on HSPICE the resultant transient responses evidenced with respect to V_Q and for the cases of no shunt capacitances incident at the output port and 5 μF of capacitance terminating the output port to ground.

(c) Discuss and explain the transient response deduced in Part (b) of this problem.

Problem 8.24

In the low voltage biasing circuit of Fig. P8.24, transistors $M1$ and $M2$ are identical except for the fact that while the gate aspect ratio of $M1$ is η, the gate aspect ratio of $M2$ is ηk^2, with k denoting a numerical constant. For the purposes of this problem, assume that the simple square law relationship for the volt-ampere characteristic of both transistors applies. The current, I_{bias}, is supplied by an ideal constant current source. The voltage, V_{bias}, is chosen to ensure that transistor $M2$ operates at the boundary of its triode and saturation regimes. The substrate terminals of both transistors are routed to circuit ground.

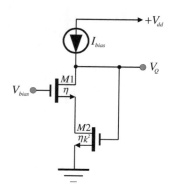

Figure P8.24.

(a) In terms of the constant, k, and the N-channel threshold voltage, V_{hn}, derive an expression that stipulates the maximum possible value of the voltage, V_{bias}.

(b) For the value of V_{bias} determined in Part (a), determine the static output voltage, V_Q.

(c) In terms of the small signal transistor transconductances, g_{m1} and g_{m2}, the small signal transistor channel conductances, g_{o1} and g_{o2}, and the bulk transconductance factor, λ_{b1}, for transistor $M1$, derive an expression for the small signal Thévenin resistance, say R_{out}, associated with the network output port.

Problem 8.25

The biasing cell in Fig. 8.24 is incorporated into the network offered in Fig. P8.25. Using the same assumptions invoked in the preceding problem, what $M3$ gate aspect ratio achieves the requisite value of V_{bias} deduced in the preceding problem?

Problem 8.26

In the circuit of Fig. P8.25, the current sources, I_{bias}, can be realized with P-channel transistors of appropriate geometries. Design the resultant network to satisfy the following specifications. Use the HSPICE parameters given in the "Comments" accompanying Example 8.3.

Maximum static power dissipation: 3 mW, with a 2.7 volt supply;
Static output voltage: 1.0 volts.

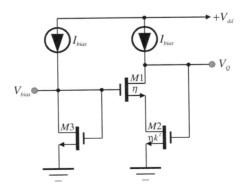

Figure P8.25.

(a) Simulate the static transfer characteristics of the design realization on HSPICE or equivalent computer-sided design software. Modify the computed design results in an attempt to satisfy all specifications as closely as possible.

(b) Let V_{dd} be represented by a suitable pulse train whose amplitudes range from 0 to 2.7 volts. Simulate on HSPICE the resultant transient responses evidenced with respect to V_Q and for the cases of no shunt capacitances incident at the output port and 5 μF of capacitance terminating the output port to ground.

(c) Discuss and explain the transient response deduced in Part (b) of this problem.

Chapter 9

Broadband and Radio Frequency MOS Technology Amplifiers

9.1.0. Introduction

Rendering an amplifier functional at progressively higher signal frequencies has been an underscored focus of analog circuit research and development activities since the day that DeForest introduced the technical community to his vacuum tube. Propelled by the omnipresent need to communicate, quickly and faithfully, large amounts of information and data in commercial, military, and space system applications, broadbanding interests continue to burgeon in the present state of the electronic systems art. It is therefore fitting to explore the fundamental circuit concepts and theories that underlie the design of broadbanded amplifiers and narrowband amplifiers tuned to high radio frequencies. To this end, the feedback network and signal flow theories that clarified the analysis and design of operational amplifiers in the preceding chapter can be gainfully exploited herewith to formulate practical high frequency design strategies. Although MOS technology is of primary interest in the following sections of material, the concepts disclosed apply equally well to virtually all monolithic device technologies, inclusive of bipolar, silicon-germanium (SiGe) heterostructure bipolar, and III-V compound transistors.

The topological foundations of a number of classic broadband analog circuit architectures are studied in this chapter. All of these configurations exploit one or both of two fundamental observations addressed, albeit peripherally, in previous chapters. The first of these observations is that negative feedback applied around a dominant pole open loop produces a closed loop exuding improved frequency response. Indeed, the closed loop

3-dB bandwidth of such a feedback arrangement is larger than the 3-dB bandwidth of the open loop by a factor of one plus the zero frequency value of the loop gain. Unfortunately, single pole open loop amplifiers are indigenous only to the rarified academic climate of a classroom. As the damping factor associated with the interaction of the dominant pole and the unavoidable higher frequency poles of an active network diminishes, the bandwidth improvement afforded by feedback becomes progressively less pronounced. Moreover, the bandwidth enhancement that is realized may be compromised by undesirable frequency response peaking within the passband and even potential circuit instability. Pragmatic limits to the bandwidth benefits incurred by negative feedback are accordingly suggested. In the pages that follow, these limits are studied, and methods to counteract them are developed and exemplified.

The second observation derives from the fact that the 3-dB bandwidth of a dominant pole amplifier having no finite frequency zeros is approximately the inverse sum of the open circuit time constants established by the energy storage elements within the amplifier. Although the open circuit time constant method of estimating bandwidth is a straightforwardly applied and even insightfully useful tool for high frequency circuit assessment, its utility as a broadbanding design vehicle is limited from several perspectives. First, the method suggests that broadbanding a dominant pole network reduces to the problems of determining the designable parameters that determine the frequency of the dominant pole and then adjusting an appropriate subset of these variables to incur an increased pole frequency. To some extent, this engineering strategy is viable. But the frequencies of the presumably nondominant network poles are invariably influenced by any parametric perturbations. Even if these other frequencies are magically unaffected, progressive successes recorded with respect to increasing the dominant pole frequency result ultimately in a nondominant pole circuit for which a bandwidth estimate predicated on pole dominance is erroneous. In short, moving the dominant pole to higher frequencies means that the originally dominant pole is displaced to a region of the complex frequency plane that is populated by higher order poles.

A second limitation, which is particularly evident in a network whose only energy storage elements are capacitors, is that a displacement of a dominant pole to a higher frequency entails reducing the effective open circuit resistance "seen" by one or more of the network capacitors. Such a design tack invariably entails either a reduction of shunt *I/O* port resistances or/and an increase in device transconductances, both of which mandate increases in circuit biasing currents. The increased circuit power dissipation typified by these types of broadbanding strategies forebodes battery longevity problems in portable electronic systems.

The third issue is more the result of an analytical restriction than it is an actual limitation. In particular, the open circuit time constant method of predicting bandwidth is premised on the nonexistence of finite frequency zeros, in addition to the presence of a dominant pole. Consider compensating an amplifier whose dominant pole has a radial frequency of p by introducing a zero at radial frequency z. If z is within a factor of the square root of two of p, the high frequency amplifier response never attenuates to 3-dB below its zero frequency gain. Accordingly, the compensated amplifier boasts infinitely large 3-dB bandwidth, despite the fact that the strict application of the open circuit time constant method predicts a finite 3-dB bandwidth of p. The inference of the foregoing elementary disclosure is that the uncompensated bandwidth of p might be appreciably extended, not by increasing p but rather, by introducing a finite zero into the amplifier transfer function to mitigate the effects of the dominant pole. In the sections that follow, broadband compensation via the introduction of zeros receives considerable attention.

9.2.0. Cascade of Dominant Pole Amplifiers

The majority of broadbanding schemes address the individual gain stages of an electronic network. A design-oriented concentration on individual stages is logical in the sense that the response of an overall electronic network can be no faster than is its slowest constituent. The fact that the gain requirements in most high frequency applications are rarely satisfied by a single amplification stage bodes the necessity of a cascade of presumably broadbanded individual network stages. Questions therefore arise as to the effects exerted by cascading on the resultantly observable 3-dB bandwidth of the system.

To the foregoing end, assume an electronic network comprised of a cascade of N stages, each of which boasts an nth order transfer function of

$$H(s) = \frac{H(0)}{\left(1 + \dfrac{s}{p_1}\right)\left(1 + \dfrac{s}{p_2}\right)\left(1 + \dfrac{s}{p_3}\right)\cdots\left(1 + \dfrac{s}{p_n}\right)}, \qquad (9\text{-}1)$$

where $H(0)$ represents the zero frequency, or "DC," gain of the stage, p_1 through p_n are the radial frequencies of the network poles, and no finite frequency zeros are evidenced. It is worthwhile interjecting that the method of open circuit time constants delivers a radial bandwidth estimate, B, deriving from

$$\frac{1}{B} \approx \sum_{i=1}^{n} \frac{1}{p_i}, \qquad (9\text{-}2)$$

which represents little more than the s-term coefficient in the algebraic expansion of the denominator on the right hand side of Eq. (9-1). It is also worth noting for future reference that while Eq. (9-2) is only an approximation of the actual 3-dB bandwidth, the right hand side of Eq. (9-2) is precisely the zero frequency value of the amplifier envelope delay. This metric is the time delay of the output response, measured with respect to the input excitation, in the sinusoidal steady state.

Let the stage in question be designed to ensure that the pole at $s = -p_1$ is dominant. This assumption implies that p_1 is a real number and that the product, $|H(0)|p_1$, closely approximates the unity gain frequency of the stage. It also implies that $|H(0)|p_1$ is smaller than the magnitude of any of the remaining pole frequencies, p_2 through p_n. The upshot of these stipulations is that (9-1) collapses to

$$H(s) \approx \frac{H(0)}{\left(1 + \dfrac{s}{p_1}\right)}, \qquad (9\text{-}3)$$

and, by Eq. (9-2), the 3-dB bandwidth becomes $B \approx p_1$. Note that the aforementioned product, $|H(0)|p_1$, is the gain-bandwidth product, GBP, of the stage at hand and that this product converges toward the unity gain radial frequency, ω_u, of the stage; that is,

$$GBP \triangleq |H(0)|B \approx |H(0)|p_1 \approx \omega_u. \qquad (9\text{-}4)$$

9.2.1. Bandwidth of *N*-Stage Cascade

A cascade of N stages, each characterized by the transfer relationship in Eq. (9-3) produces an overall transfer function, $H_T(s)$, given by

$$H_T(s) = H^N(s) = \frac{H^N(0)}{\left(1 + \dfrac{s}{p_1}\right)^N}, \qquad (9\text{-}5)$$

assuming that the input port of any one stage does not appreciably load the output port of its predecessor stage. One way of precluding interstage loading effects entails a design wherein the magnitude of the driving point input impedance of each stage is much larger than that of the stage output port within the signal passband of interest. A more practical way of mitigating loading effects at high signal frequencies is to design each stage for matched and constant driving point input and output resistances. More information is tendered later on broadbanding via constant resistances.

In the sinusoidal steady state, Eq. (9-5) becomes

$$H_T(j\omega) = H^N(j\omega) = \frac{H^N(0)}{\left(1 + \frac{j\omega}{B}\right)^N}, \qquad (9\text{-}6)$$

where use is made of the fact that the 3-dB frequency of a dominant pole circuit is the frequency of said pole. The cascaded 3-dB bandwidth resultantly derives from the requirement,

$$\left[1 + \left(\frac{B_T}{B}\right)^2\right]^{N/2} = \sqrt{2}, \qquad (9\text{-}7)$$

whence

$$\frac{B_T}{B} = \sqrt{2^{1/N} - 1}, \qquad (9\text{-}8)$$

which suggests bandwidth compression in a cascade configuration. For example, a cascade of two identical stages produces a bandwidth that is 35.6% smaller than the 3-dB bandwidth of any single stage.

A more utilitarian form of Eq. (9-8) derives from the observation that

$$2^{1/N} = e^{\ln(2^{1/N})} = e^{\ln 2/N}. \qquad (9\text{-}9)$$

But for large N,

$$2^{1/N} = e^{\ln 2/N} \approx 1 + \frac{\ln 2}{N}, \qquad (9\text{-}10)$$

which, when combined with Eq. (9-8), delivers

$$\frac{B_T}{B} = \sqrt{2^{1/N} - 1} \approx \sqrt{\frac{\ln 2}{N}}. \qquad (9\text{-}11)$$

For $N \geq 2$, Eq. (9-11) predicts a normalized bandwidth that is low by no more than 8.53%, which is close enough for government work. A slight computational error notwithstanding, the advantage of Eq. (9-11) is that it suggests that the overall bandwidth of a cascade diminishes as roughly the square root of the number of cascaded stages. This observation is interesting

734   Feedback Networks: Theory and Circuit Applications

in that Eq. (9-6) confirms that the decibel value of cascaded zero frequency gain rises linearly with the number of stages. Accordingly, the decibel gain value rises with N faster than the overall system bandwidth degrades, which suggests that acceptably high amplifier bandwidths can be achieved through prudent design measures, despite system requirements that drive a need for additional stages.

9.2.2. Optimized Bandwidth of a Cascade

The dependence of gain and bandwidth on the number of cascaded identical stages warrants further exploration. To this end, assume an N stage cascade whose overall gain magnitude at zero signal frequency is a given value, say K. It follows that each identical stage of the amplifier must deliver a gain magnitude at zero frequency of $K^{1/N}$, which implies an individual stage gain-bandwidth product of $GBP = K^{1/N} B$. Using Eq. (9-11), the overall bandwidth, B_T, of the cascade configuration is expressible as

$$B_T \approx B \sqrt{\frac{\ln 2}{N}} = \frac{GBP}{K^{1/N}} \sqrt{\frac{\ln 2}{N}}. \qquad (9\text{-}12)$$

Figure 9.1 plots the normalized overall bandwidth, B_T/GBP, as a function of the number of stages, N, for various values of the overall gain, K. For a fixed gain-bandwidth product, GBP, in each of the identical stages and a given overall gain magnitude specification, K, this 3-dB bandwidth expression is clearly a nonmonotonic function of N. It displays a maximum value, say B_{To}, at a particular value of N, which can be denoted as N_o. Note that the actual maximum overall bandwidth is somewhat soft in the sense that B_T is not an especially sensitive function of N for relatively large values of N and K. The number, N_o, can be determined by equating the derivative of B_T with respect to N in Eq. (9-12) to zero. The result of this unpleasant task is

$$N_o = \ln(K^2), \qquad (9\text{-}13)$$

which defines the optimum number of stages commensurate with a given cascade gain specification. The corresponding zero frequency gain of each stage in the cascade is relatively small and is, in particular,

$$H(0) = K^{1/N_o} = \sqrt{e}, \qquad (9\text{-}14)$$

where e is the base of natural logarithms. Upon insertion of Eq. (9-13) into

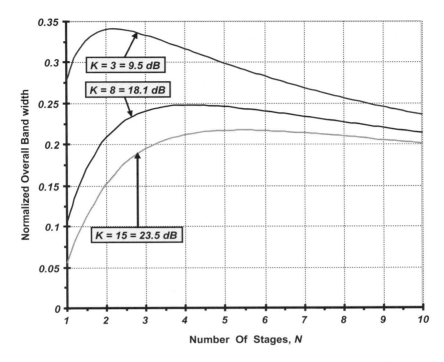

Figure 9.1. The overall bandwidth of a cascade of N identical stages as a function of the number of stages, N, for three values of the overall DC gain, K. The overall bandwidth on the vertical scale of this plot is normalized to the gain-bandwidth product of a single amplification stage.

Eq. (9-12), the maximum possible 3-dB bandwidth is found to be

$$B_T = GBP\sqrt{\frac{\ln 2}{2e \ln K}}. \tag{9-15}$$

Example 9.1. A low frequency gain magnitude of at least 6 is required of a lowpass amplifier that is to be designed to deliver a 3-dB frequency of 1.5 GHz. If this amplifier is to be realized as a cascade of a number of identical stages, determine the optimum number of stages and the corresponding 3-dB bandwidth to which each stage must be designed if maximum overall bandwidth is to be achieved.

Solution 9.1.

(1) From Eq. (9-13), an overall gain of 6 commands that the number of stages be $N_o = 3.584$. Since the realization of a cascade comprised of a non-integral number of stages is a daunting challenge, choose $N_o = 4$, which implies a revised gain magnitude design goal of $K = e^{4/2} = 7.389 = 17.4$ dB.

(2) With $B_T = 1.5$ GHz and $K = 7.389$, Eq. (9-15) stipulates a gain-bandwidth product for each stage of $GBP = 5.941$ GHz.

(3) Since the optimized gain-per-stage is the square root of the base of natural logarithms, or 1.649, the requisite 3-dB bandwidth of each dominant pole stage is $B = GBP/1.649 = 3.60$ GHz.

Comments. The simplicity of these computations warrants no further explanation, but the numerical results are enlightening in that they paint a somewhat foreboding design scenario. While a single example hardly posits generalized and reliable design guidelines, it appears that bandwidth optimization in cascaded, identical dominant pole stages has dubious broadbanding merit. In particular, observe that the bandwidth requirement of each stage in this example is 140% of the system bandwidth goal of 1.5 GHz. This stage bandwidth is doubtlessly attainable with state of the art monolithic processes, particularly since the gain magnitude of each stage is only 1.649, or 4.34 dB. However, four stages are required to produce a gain magnitude of only 7.389. The extant state of the art boasts monolithic transistors having unity gain frequencies in the mid-tens of gigahertz. It may therefore be more prudent, at least from a power dissipation perspective, to attempt a realization of the performance specifications with either one amplification stage or perhaps two, non-identical stages.

An additional concern is that the transfer function of the final circuit realization effectively projects a fourth order pole at nominally 3.60 GHz. In view of the parasitic feedback invariably confronted in the course of physically implementing the amplifier cascade, this multi-order pole contributes to poor phase and gain margins and possibly, outright instability.

9.3.0. Degenerative RC Broadbanding

In Chapter 6, the frequency response of the common source amplifier is definitively investigated, primarily because the common source stage is the workhorse of MOS technology systems requiring significant I/O gain. Fig. 9.2(a) offers a slight variation to the common source topology in that a degeneration resistance, R_f, is inserted into the source lead of the transistor. As might be expected and as is to be confirmed, this resistance acts as a negative feedback element to extend the 3-dB bandwidth of the basic common source stage at the expense of reducing the stage zero frequency gain. It also allows for the introduction of capacitive compensation that extends the degenerated bandwidth without additional gain penalty. Assuming transistor operation in its saturation regime, Fig. 9.2(b) is the pertinent small signal

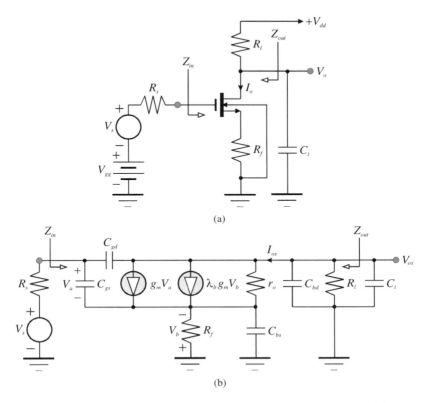

Figure 9.2. (a) Basic schematic diagram of a common source amplifier in which a source degeneration resistance, R_f, is used to establish negative feedback. (b) Small signal, high frequency model of the amplifier in (a). The transistor is presumed biased in saturation.

model, where resistance R_l is the drain load resistance, C_l is the net capacitance loading the amplifier output port, and R_s is the Thévenin resistance of the signal source. The transistor is represented by its traditional small signal parameters. In particular, g_m is the forward transconductance, λ_b is the bulk transconductance factor, r_o is the small signal channel resistance, C_{gs} is the gate-source capacitance, C_{gd} is the gate-drain capacitance, C_{bd} is the bulk-drain capacitance, and finally, C_{bs} symbolizes the bulk source capacitance.

9.3.1. Gain and Dominant Pole

Figure 9.3(a) diagrams the small signal low frequency version of the model in Fig. 9.2(b). In this low frequency equivalent circuit, the feedback

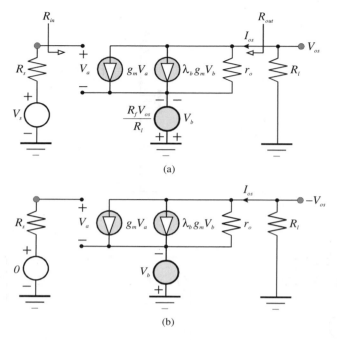

Figure 9.3. (a) The low frequency version of the small signal equivalent circuit offered in Fig. 9.2(b). (b) The model of (a) configures for the computation of the normalized return ratio with respect to the feedback factor, R_f/R_l.

resistance, R_f, is supplanted by a voltage controlled voltage source, $-R_f V_{os}/R_l$, where use is made of the fact that the small signal load current, $I_{os} = -V_{os}/R_l$, flows through the resistance, R_f, which is embedded in the source lead of the transistor. This replacement of a resistance by a controlled voltage source makes clear that the feedback factor at low signal frequencies is R_f/R_l, and that the low frequency, open loop (in the sense of $R_f = 0$) voltage gain, $A_o(0)$, is

$$A_o(0) = \left.\frac{V_{os}}{V_s}\right|_{R_f=0} = -g_m\left(r_o \| R_l\right) \approx -g_m R_l, \qquad (9\text{-}16)$$

where the approximation reflects the tacit, but generally reasonable, assumption that $r_o \gg R_l$.

Figure 9.3(b) reconfigures the model in Fig. 9.3(a) for the express purpose of evaluating the low frequency normalized return ratio, $Q_s(0)$, with respect to the feedback factor. A straightforward circuit analysis confirms that

$$Q_s(0) = \left.\frac{V_{os}}{V_b}\right|_{V_s=0} = (1+\lambda_b)g_m(r_o \| R_l) + \frac{R_l}{R_l + r_o}. \qquad (9\text{-}17)$$

Since the feedback generator is controlled by the output voltage response variable, the null return ratio with respect to the feedback parameter is itself null. Accordingly, the low frequency loop gain, $T(0)$, evolves as

$$T_s(0) = \left(\frac{R_f}{R_l}\right)Q_s(0) = \left(\frac{R_f}{R_l + r_o}\right)[1 + (1+\lambda_b)g_m r_o]. \qquad (9\text{-}18)$$

If, in addition to $r_o \gg R_l$, $g_m r_o \gg 1$ and $\lambda_b \ll 1$, this expression simplifies to

$$T_s(0) = \left(\frac{R_f}{R_l}\right)Q_s(0) \approx g_m R_f. \qquad (9\text{-}19)$$

It follows that the low frequency closed loop voltage gain of the amplifier at hand is

$$A(0) = \frac{V_{os}}{V_s} = \frac{A_o(0)}{1 + T_s(0)} \approx -\frac{g_m R_l}{1 + g_m R_f} \approx -\frac{R_l}{R_f}, \qquad (9\text{-}20)$$

which, assuming $g_m R_f \gg 1$, is essentially independent of transistor parameters. The approximations leading to Eqs. (9-19) and (9-20) require a suitably large MOSFET transconductance, which may command a relatively large gate aspect ratio and/or an appreciable drain bias current. However, care must be exercised to preclude a large drain current from jeopardizing the presumption that the channel resistance, r_o, is substantially larger than the drain load resistance, R_l.

Note that the immediate impact of the degeneration resistance is an attenuation of the common source low frequency gain by a factor of nominally, $(1 + g_m R_f)$. Gain attenuation is to be expected since the applied signal source in the circuit of Fig. 9.2(a) is not impressed across the gate-source terminals of the transistor, as it is in a traditional common source stage. Instead, the signal is applied from the gate terminal to ground, an electrical path that includes the series embodiment of the feedback degeneration resistance, R_f. If the input signal is small enough to cause an insignificant perturbation of the gate-source quiescent voltage, most of the signal, V_s, is developed across resistance R_f. This action results in a signal current, V_s/R_f, flowing in the transistor source lead and since the gate current is zero at low frequencies, the same signal current flows into the transistor drain terminal and through the drain load resistance, R_l. Consequently, an output signal voltage of $-R_l V_s/R_f$ is developed at the amplifier output port, whence the approximate gain given by Eq. (9-20). It might therefore be asserted that the validity of the approximation in Eq. (9-20) is premised on negligible signal swing induced by signal V_s in the gate-source biasing voltage. This signal swing is tantamount to the error signal observed in a generalized feedback system model. Recall from preceding chapters that this error signal is indeed small in negative feedback systems when the loop gain is sufficiently large.

A first order investigation of the frequency response of the degenerated common source amplifier derives from an examination of the open circuit time constants associated with each capacitance in the model of Fig. 9.2(a). The cumbersome topology of this model renders this task algebraically daunting unless the circuit implications of the approximations noted above are exploited a priori. In particular, a negligibly small transconductance factor, λ_b, means that the controlled source, $\lambda_b g_m V_b$, can be vanquished. On the other hand, a large channel resistance, r_o, renders feasible the removal of the shunt loading imposed by resistance r_o. To these ends, it can be shown that the open circuit time constant, τ_{gs}, associated with the transistor gate-source capacitance, C_{gs}, is (with the implicit understanding that all other capacitances in the model are open circuited)

$$\tau_{gs} = \frac{(R_s + R_f) C_{gs}}{1 + T_s(0)} \approx \frac{(R_s + R_f) C_{gs}}{1 + g_m R_f}, \qquad (9\text{-}21)$$

where use is made of the approximation projected by Eq. (9-19). It should be noted that in the absence of a source circuit degeneration resistance, R_f, this gate-source time constant is simply $R_s C_{gs}$. An inspection of Eq. (9-21) suggests that a potentially significant reduction in the value of this nondegenerated time constant is manifested by the utilized feedback. In particular and despite the fact that resistance R_f superimposes with the Thévenin source resistance, R_s, in the numerator of the subject expression, the gate-source time constant is diminished by virtue of the fact that the actual gate-source capacitance is effectively reduced by a factor of one plus the zero frequency value of the amplifier loop gain.

An analysis similar to that executed for the gate-source capacitive time constant yields expressions for the open circuit time constants associated with the remaining capacitances in the model. For the gate-drain capacitance, C_{gd}, the time constant, τ_{gd}, is

$$\tau_{gd} = \left[R_l + \left(1 + \frac{g_m R_l}{1 + g_m R_f}\right) R_s\right] C_{gd}. \tag{9-22}$$

The incorporated feedback resistance is seen to limit the Miller multiplication of the gate-drain capacitance, but it has no effect on the time constant component attributed to the interaction of the load resistance with this capacitance. Thus, feedback is seen to reduce the time constant associated with the gate-drain capacitance to varying degrees, depending on the value of the Thévenin source resistance.

For the bulk-source capacitance, C_{bs}, the open circuit time constant, τ_{bs}, is

$$\tau_{bs} = \frac{R_f C_{bs}}{1 + T_s(0)} \approx \frac{R_f C_{bs}}{1 + g_m R_f}, \tag{9-23}$$

which is again rendered small because of the effective reduction of capacitance C_{bs} by a factor of one plus the zero frequency loop gain. Finally, the net load capacitance, which is the sum of the transistor bulk-drain capacitance, C_{bd}, and the actual load capacitance, C_l, generates an open circuit time constant, τ_l, of

$$\tau_l = R_l (C_{bd} + C_l). \tag{9-24}$$

Feedback has no effect on the latter time constant, at least within the constraints imposed by the invoked analytical approximations. The reason underlying this discovery derives from the tacit presumption of a very large transistor channel resistance, r_o. In particular, large r_o means that the load resistance is driven by a controlled current source whose inherently high terminal resistance effectively isolates the feedback resistance in the source circuit from the load termination.

Assuming that a dominant pole prevails in the amplifier of Fig. 9.2(a), the 3-dB bandwidth, say B_d, of the degenerated amplifier can be approximated as the inverse sum of the four time constants evaluated above. Ignoring those time constant components that are inherently small by virtue of their respective inverse dependence on the zero frequency loop gain,

$$B_d \approx \frac{1}{\tau_{gs} + \tau_{gd} + \tau_{bs} + \tau_l} \approx \frac{1}{R_l (C_{gd} + C_{bd} + C_l) + R_s C_{gd}}. \tag{9-25}$$

In an active RC network, a necessary condition for pole dominance is that one, and only one, capacitive time constant be dominant in comparison to all other computed open circuit time constants. It is hardly a leap of

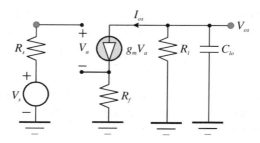

Figure 9.4. Simplified high frequency equivalent circuit of the amplifier in Fig. 9.2(a).

faith to argue that the time constant associated with the output port in the amplifier of Fig. 9.2(a) is the likely root of pole dominance. This contention is particularly germane if the load capacitance, C_l, is purposefully appended to set a particular 3-dB bandwidth that is within the frequency capability of the utilized transistor. Thus, $R_l(C_{gd} + C_{bd} + C_l) \gg R_s C_{gd}$ in Eq. (9-25) is the requirement that forges a dominant pole frequency response. The stipulated inequality is distinctly possible for two reasons. First, the signal source resistance, R_s, is rarely larger than 50 Ω to 75 Ω in broadband amplifiers. Second, self-aligning transistor gate technology and transistor saturation combine to render C_{gd} inherently small. It follows that Eq. (9-25) collapses to the simple form,

$$B_d \approx \frac{1}{R_l C_{lo}}, \qquad (9\text{-}26)$$

where C_{lo} represents the effective load port capacitance "seen" by resistance R_l; namely,

$$C_{lo} = C_{gd} + C_{bd} + C_l. \qquad (9\text{-}27)$$

Equations (9-26) and (9-20) combine to establish the simplified high frequency model depicted in Fig. 9.4, for which the closed loop voltage gain as a function of complex frequency s is

$$A(s) = \frac{V_{os}}{V_s} = \frac{A_o(s)}{1 + T_s(s)} \approx -\frac{g_m R_l}{\left(1 + g_m R_f\right)\left(1 + \dfrac{s}{B_d}\right)}. \qquad (9\text{-}28)$$

9.3.2. Broadband Compensation

As expected, Eq. (9-28) confirms that the net load capacitance in the amplifier modeled by Fig. 9.4 incurs a progressively reduced voltage gain with

Broadband and Radio Frequency MOS Technology Amplifiers

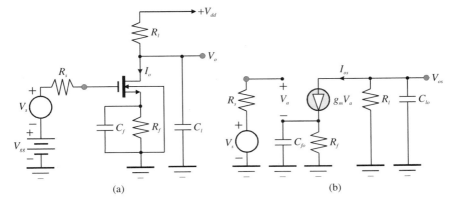

Figure 9.5. (a) Schematic diagram of compensated degenerative amplifier. (b) Small signal, high frequency approximate equivalent circuit of the amplifier in (a).

increasing signal frequency. A possible broadbanding clue derives from the observation that the load capacitance reduces the net load impedance from its zero frequency value of R_l to a value that approaches zero at very high frequencies. Since the gain is nominally proportional to the ratio of the load impedance to resistance R_f, it is logical to consider replacing R_f in the amplifier by a shunt interconnection of R_f and an appropriate capacitance, say C_f, as shown in the schematic diagram of the proposed compensated amplifier in Fig. 9.5(a). To the extent that the gain is directly dependent on the ratio of load impedance to feedback impedance, at least at low signal frequencies, a feedback impedance time constant matched to the time constant of the load impedance allows the feedback impedance to track the load impedance over frequency, whence a broadbanded amplifier frequency response.

The simplified small signal model for the structure in Fig. 9.5(a) appears in Fig. 9.5(b). Since the introduced capacitance, C_f, appears in shunt with the transistor bulk-source capacitance, C_{bs}, capacitance C_{fo} appears in the model, where it is understood that

$$C_{fo} = C_f + C_{bs}. \tag{9-29}$$

If resistance R_f in Eq. (9-28) is now replaced by the parallel combination of R_f and C_{fo}, the transfer function of the compensated amplifier becomes

$$A_c(s) = \frac{V_{os}}{V_s} = \frac{A(0)\left(1 + sR_f C_{fo}\right)}{\left(1 + \dfrac{sR_f C_{fo}}{1 + g_m R_f}\right)\left(1 + \dfrac{s}{B_d}\right)}, \tag{9-30}$$

where $A(0)$ is given by Eq. (9-20) and bandwidth B_d of the uncompensated amplifier is defined by Eq. (9-26). If the time constant, $R_f C_{fo}$, established by the net compensation capacitance, C_{fo}, is selected so that it equates to the inverse frequency, $1/B_d$, of the uncompensated dominant pole, the resultant compensated bandwidth, B_c, is

$$B_c = \frac{1 + g_m R_f}{R_f C_{fo}} = (1 + g_m R_f) B_d \approx [1 + T_s(0)] B_d, \qquad (9\text{-}31)$$

where Eq. (9-19) is exploited. The result indicates that the simple action of appending a capacitance across the feedback resistance in the source lead of the transistor can extend the uncompensated bandwidth by as much as one plus the zero frequency value of the loop gain.

Unfortunately, engineering care must be exercised when interpreting the foregoing result because the pole-zero cancellation on which it is premised is imperfect. In particular, Eq. (9-30) derives from Eq. (9-26), which presumes that pole dominance prevails in the original uncompensated amplifier. Imperfections also abound with respect to equating time constant $R_f C_{fo}$ to inverse B_d. For example, capacitance C_{fo} is a function of transistor bulk-source capacitance C_{bs}, which is rarely known accurately. Moreover, bandwidth B_d relies on capacitance C_{lo}, which in turn is functionally dependent on transistor bulk-drain capacitance C_{bd}, whose precise numerical value, like that of C_{bs}, is elusive. Accordingly, it is at least analytically prudent to investigate the alternative design constraint,

$$R_f C_{fo} = \frac{k}{B_d}, \qquad (9\text{-}32)$$

with the understanding that the positive number, k, reflects an uncertainty implicit to the desired time constant match. For example, $k = 0.75$ suggests that $R_f C_{fo}$ is 25% smaller than the desired value of $1/B_d$, while $k = 1.25$ implies a 25% larger than desired time constant value. Armed with Eq. (9-32), Eq. (9-30) can be written in the form,

$$\frac{A(s)}{A(0)} = \frac{1 + \dfrac{sk}{B_d}}{\left\{1 + \dfrac{sk}{[1 + T_s(0)] B_d}\right\} \left(1 + \dfrac{s}{B_d}\right)}. \qquad (9\text{-}33)$$

The magnitude response implicit to Eq. (9-33) is plotted in Fig. 9.6 for four values of the constant, k, and a zero frequency loop gain, $T_s(0)$, equal to 10, or 20 dB. The curve for $k = 0$ corresponds to the uncompensated amplifier case and as expected, it projects unity normalized bandwidth (the

Broadband and Radio Frequency MOS Technology Amplifiers 745

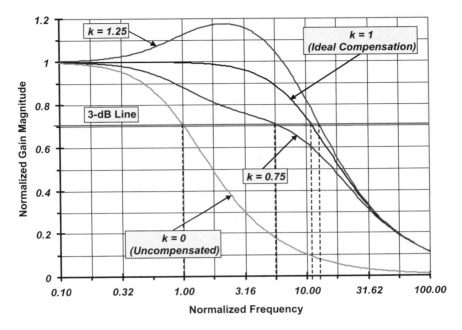

Figure 9.6. Frequency response of the compensated amplifier in Fig. 9.5(a) for a zero frequency loop gain of 10, and various values of k, as defined by Eq. (9.32). The signal frequency is normalized to the bandwidth, B_d, of the uncompensated, degenerated amplifier, while the gain is normalized to the zero frequency value, $A(0)$, of small signal voltage gain.

frequency scale in the plot is normalized to B_d). The $k = 1$ case reflects idealized pole-zero cancellation and shows a normalized bandwidth of 11, which is also expected in view of Eq. (9-31) and the fact that $T_s(0) = 10$. For $k = 1.25$, the normalized compensated bandwidth improves to 12.8, but at the expense of an invariably unacceptable gain peaking of almost 20%. The peaking results because the introduced zero lies at a frequency that is smaller than the frequency of the pole intended for cancellation. Finally, $k = 0.75$ results in a compensated bandwidth that is almost 50% smaller than the bandwidth corresponding to $k = 1$. Note also that the pole-zero doublet resulting from imperfectly cancelled critical frequencies is plainly evident in the normalized frequency range of nominally 1 to 5. The upshot of the matter is that the compensated frequency response is sensitive to the degree by which pole-zero cancellation is actually achieved. A pragmatic mitigation of this sensitivity may entail incorporating a capacitance adjustment capability in either capacitance C_l or C_f to fine-tune the ultimately realized frequency response.

Example 9.2. When suitably biased in saturation, the transistor in the amplifier of Fig. 9.5(a) delivers the following small signal parameters.

Forward Transconductance (g_m):	34 mS
Bulk Transconductance Factor (λ_b):	0.09
Drain-Source Channel Resistance (r_o):	15.8 KΩ
Gate-Source Capacitance (C_{gs}):	173 fF
Gate-Drain Capacitance (C_{gd}):	15 fF
Bulk-Drain Capacitance (C_{bd}):	12 fF
Bulk-Source Capacitance (C_{bs}):	10 fF

Prior to incorporating the feedback resistance, R_f, and its shunting capacitance, C_f, in the circuit, which is driven by a 50 Ω (R_s) signal source, the measured 3-dB bandwidth is 175 MHz, while the measured gain-bandwidth product is determined to be 7 GHz. The desired closed loop bandwidth is 1.5 GHz. Design the circuit by determining the appropriate values of resistance R_f, resistance R_l, capacitance C_f, and capacitance C_l. Using the small signal model provided in Fig. 9.2(b), simulate the small signal frequency response of the finalized design on HSPICE. Examine these responses for the case in which the compensation capacitance, C_f, is 20% smaller than the value required for precise cancellation of the zero forged by the feedback subcircuit and the dominant pole of the uncompensated version of the amplifier.

Solution 9.2.

(1) If the uncompensated amplifier is characterized by a dominant pole response, a gain-bandwidth product of 7 GHz and a 3-dB bandwidth of 175 MHz implies a zero frequency voltage gain of $|A(0)| = GBP/B_d = 40$. Using Eq. (9-16) with $g_m = 34$ mS and $r_o = 15.8$ KΩ, the requisite drain circuit resistance is $R_l = 1{,}271\ \Omega$.

(2) Recalling Eq. (9-25), $B_d = 2\pi(175\text{ MHz})$, $R_l = 1{,}271\ \Omega$, $R_s = 50\ \Omega$, $C_{gd} = 15$ fF, and $C_{bd} = 12$ fF deliver a terminating load capacitance of $C_l = 687.9$ fF.

(3) The factor by which the uncompensated bandwidth is to be enhanced is 1.5 GHz/175 MHz, or 8.571. To first order, this means that $g_m R_f$, by Eq. (9-31), is 7.571, whence $R_f = 7.571/3$ mS $= 222.7\ \Omega$. It should be noted that a bandwidth improvement by a factor of 8.571 implies gain degradation by nominally the same amount. Accordingly, the magnitude of voltage gain for the compensated circuit is $|A_c(0)| = |A(0)|/8.571 = 4.667$.

(4) Recalling Eq. (9-32), $R_f = 222.7\ \Omega$, $B_d = 2\pi(175\text{ MHz})$, and $k = 1$ for precise pole-zero cancellation give $C_{fo} = 4.084$ pF. It follows from Eq. (9-29) that with $C_{bs} = 10$ fF, $C_f = 4.074$ pF. A 20% smaller capacitance implies $C_f = 3.259$ fF.

(5) Figure 9.7 shows the small signal model of the finalized design for the case of precise pole-zero cancellation. The simulated magnitude and phase responses for $k = 1$ appear in Fig. 9.8.

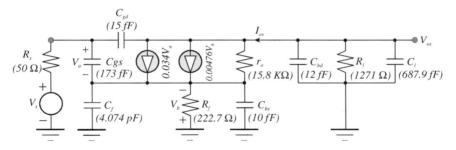

Figure 9.7. Small signal, high frequency equivalent circuit of the amplifier designed in Example 9.2.

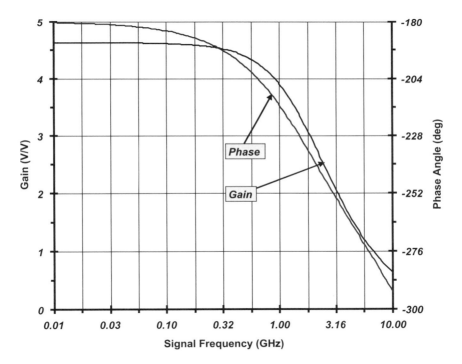

Figure 9.8. The simulated (HSPICE) magnitude and phase responses for the network whose small signal model appears in Fig. 9.7.

Comments. Although the simulated responses given in Fig. 9.8 track well with the foregoing computations, a few minor errors are apparent. These are to be expected in light of the approximations tacitly invoked in the foundational theoretical disclosures. For example, the simulated low frequency voltage gain is 4.623 volts/volt, which is a scant 0.94% lower than the calculated value of 4.667 volts/volt. Additionally, the simulated bandwidth is 1.589 GHz, which is larger than the 1.5 GHz objective by 5.93%. When C_f is replaced by its 20% smaller value, the 3-dB bandwidth falls to under 1.1 GHz, and a clear pole-zero doublet is evidenced in the magnitude response.

It is worthwhile interjecting that the three-place precision invoked on the model element values has dubious engineering merit in light of the tolerances routinely encountered in manufacturing processes. This precision is invoked herewith merely to demonstrate the engineering propriety of the theoretic background for the computations documented above.

9.4.0. Shunt Peaked Compensation

The preceding section of material dramatizes the positive impact exerted on the frequency response of a common source amplifier by a suitable left half plane zero introduced into the amplifier transfer function. In the foregoing discussion, the requisite zero is forged through shunt RC impedance degeneration inserted into the transistor source terminal. Unfortunately, several costs accompany the laudable bandwidth enhancement afforded by source RC degeneration. The first of these costs is the increased power dissipation resulting from a drain biasing current that necessarily flows through the resistance inserted in the transistor source lead. A second cost is decreased voltage gain. Recall that the closed loop gain is smaller than the amplifier open loop gain (gain with zero resistance in the source terminal) by an amount that roughly equals the bandwidth improvement factor afforded by the incorporated feedback subcircuit. In other words, the gain-bandwidth products of the uncompensated and the compensated amplifiers remain nominally the same, thereby implying that gain is traded for enhanced bandwidth. Finally, resistance in the transistor source lead degrades the noise characteristics of the amplifier. Although a discussion of electronic noise phenomena is beyond the scope of this textbook, suffice it to assert presently that a resistance in the source lead of a common source amplifier increases the equivalent input noise voltage. This noise voltage is a critical amplifier parameter in that it defines the minimum input signal level that can be detected reliably and processed faithfully by the network.

Fortunately, a few alternative means of incorporating transmission zeros into the transfer function of a common source amplifier mitigate most of the foregoing shortfalls. One such method, known as *shunt peaking*, installs the requisite zero through use of an inductor inserted into the transistor drain circuit, as depicted in Fig. 9.9. To be sure, shunt peaking does not result in

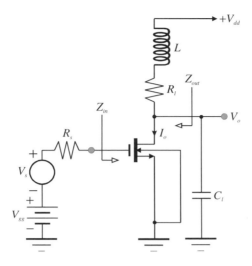

Figure 9.9. Basic schematic diagram of a shunt peaked common source amplifier. The transistor operates in its saturation regime.

a bandwidth enhancement that is as dramatic as that boasted by appropriate source degeneration. But the bandwidth improvement is nonetheless substantial and is afforded without compromising the zero frequency gain of the amplifier. It is therefore notable that shunt peaking delivers increased bandwidth by actually increasing the gain-bandwidth product of the uncompensated, common source configuration. A further advantage of shunt peaked compensation is, ignoring the small winding resistance implicit to the realization of the inductor, that the introduced inductance incurs no increase in static circuit power dissipation. Although inductor winding resistance is detrimental to circuit power dissipation, the quality factor of the utilized inductor, which can be poor in silicon monolithic realizations, is inconsequential to the small signal dynamics of the shunt peaked amplifier because the inductance is inserted in series with the drain load resistance.

9.4.1. Common Source Stage Revisited

Shunt peaking, as diagrammed in Fig. 9.9, is an effective broadbanding strategy, provided that the net capacitance incident at the output port, which includes the indicated load capacitance, C_l, and appropriate transistor capacitances, establishes the dominant time constant in the uncompensated (meaning $L = 0$) amplifier. To this end, a tacit re-investigation

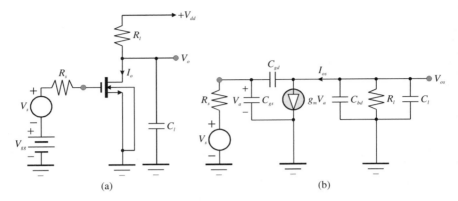

Figure 9.10. (a) Schematic diagram of a simple common source amplifier. The transistor is biased in saturation. (b) Approximate small signal, high frequency model of the network in (a). Drain-source channel resistance and bulk-induced threshold voltage modulation are tacitly ignored.

of the frequency response of the simple common source stage depicted in Fig. 9.10(a) is warranted. The pertinent equivalent circuit shown in Fig. 9.10(b) tacitly neglects the drain-source channel resistance of the transistor.

The zero frequency gain projected by the model in Fig. 9.10(b) is clearly identical to the expression postured by Eq. (9-16); namely, $A(0) = V_{os}/V_s = -g_m R_l$. The time constant analyses executed in the preceding section of material can be gainfully exploited to arrive at the individual open circuit time constants for the model at hand. In particular, with R_f set to zero, Eqs. (9-21), (9-22) and (9-24) give for the respective time constants associated with capacitances C_{gs}, C_{gd}, and C_l,

$$\tau_{gs} = R_s C_{gs}, \qquad (9\text{-}34)$$

$$\tau_{gd} = [R_l + (1 + g_m R_l) R_s] C_{gd}, \qquad (9\text{-}35)$$

and

$$\tau_l = R_l (C_{bd} + C_l). \qquad (9\text{-}36)$$

If pole dominance prevails, the bandwidth, say B_{cs}, of the common source amplifier derives from

$$\frac{1}{B_{cs}} \approx R_s \left[C_{gs} + (1 + g_m R_l) C_{gd} \right] + R_l \left(C_{gd} + C_{bd} + C_l \right). \qquad (9\text{-}37)$$

This result clearly suggests "Miller time" if the gain magnitude, $g_m R_l$, is large and/or the gate-drain capacitance is insufficiently small to warrant its neglect. Therefore, the load port time constant can be postured as dominating the 3-dB bandwidth estimate if and only if the signal source resistance, R_s, is small, $g_m R_l$ is small, and/or transistor gate self alignment renders capacitance C_{gd} inconsequential. If the load port indeed establishes the dominant time constant, the relevant small signal model reduces to the structure proffered in Fig. 9.11, where capacitance C_{lo} remains given by Eq. (9-27). Accordingly, the 3-dB bandwidth, B_{cs}, of the uncompensated (meaning $L = 0$) common source amplifier is given by the approximate relationship,

$$B_{cs} \approx \frac{1}{R_l C_{lo}}. \tag{9-38}$$

When the dreaded Miller multiplication of the gate-drain capacitance in the common source transistor observably influences the 3-dB bandwidth of the uncompensated circuit, the common source-common gate cascode configuration diagrammed in Fig. 9.12 may prove advantageous, albeit at the expense of an increase in requisite static power dissipation. Since the effective load resistance seen by the common source transistor, $M1$, is nominally the inverse of the transconductance, g_{m2} of the common gate device, $M2$, the Miller multiplier in Eq. (9-37) becomes $(1 + g_{m1}/g_{m2})$ for the stage at hand. If $1/g_{m2} \ll R_l$, which requires that transistor $M2$ be realized with a suitably large gate aspect ratio, the first term on the right hand

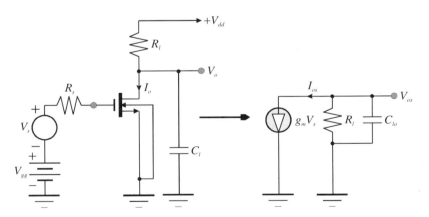

Figure 9.11. Simplified high frequency model of the common source amplifier. The simplification relies on presumed pole dominance established at the output port of the amplifier. The capacitance, C_{lo}, is given by $(C_l + C_{bd} + C_{gd})$.

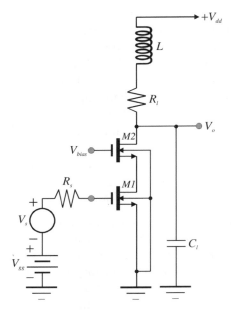

Figure 9.12. A shunt peaked amplifier utilizing a common base cascode, $M2$, to mitigate the effects of Miller multiplication of the gate-drain capacitance in the common source transistor, $M1$.

side of Eq. (9-37) is significantly diminished, thereby allowing the amplifier dominant pole to be established at the load port of the amplifier. To be sure, the inclusion of transistor $M2$ manifests additional time constants that are not embodied by Eq. (9-37). For example, the bulk-drain capacitance of $M1$, as well as the bulk-source and gate-source capacitances of $M2$, faces an effective resistance of roughly $1/g_{m2}$, which, as noted above, is presumably small. It should be observed that the capacitances, C_{gd} and C_{bd}, embedded in the second term on the right hand side of Eq. (9-37) now correspond respectively to the gate-drain and bulk-drain capacitances of transistor $M2$. Because the form factor of $M2$ is invariably larger than that of transistor $M1$, thereby giving rise to increased device capacitances, the second term on the right hand side of Eq. (9-37) is likely to be slightly larger than the value evidenced in the simple common source amplifier of Fig. 9.9. Nonetheless, the common gate current buffer likely improves the overall bandwidth because of the potentially dramatic reduction of Miller multiplication of the gate-drain capacitance in the common source driver.

9.4.2. Shunt Peaked Amplifier Response

In the shunt peaked stage in Fig. 9.9 or the cascode version shown in Fig. 9.12, the amplifier driving the load comprised of capacitance C_l in shunt with the series interconnection of resistance R_l and inductance L emulates a transconductor. If the resistively terminated transconductor functions as a dominant pole amplifier whose dominant pole frequency is determined by the net capacitance, C_{lo}, incident at the load port, the pertinent small signal equivalent circuit is the structure suggested in Fig. 9.13. An inspection of

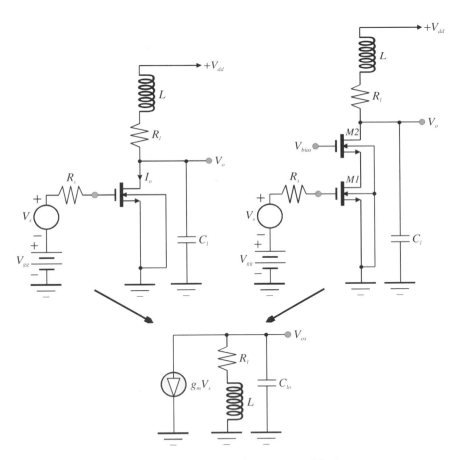

Figure 9.13. Approximate small signal, high frequency model of a common source or a common source-common gate cascode shunt peaked amplifier. The indicated model presumes pole dominance established at the output port when inductance L is zero.

the subject diagram reveals a voltage gain, $A(s)$, of

$$A(s) = \frac{V_{os}}{V_s} = -\frac{g_m R_l \left(1 + \dfrac{sL}{R_l}\right)}{1 + sR_l C_{lo} + s^2 L C_{lo}}. \qquad (9\text{-}39)$$

Not surprisingly, the zero frequency voltage gain is $A(0) = -g_m R_l$. If gain $A(s)$ is normalized to this zero frequency value, Eq. (9-39) can be cast in the form,

$$A_n(s) \triangleq \frac{A(s)}{A(0)} = \frac{1 + \dfrac{Qs}{\omega_n}}{1 + \dfrac{s}{Q\omega_n} + \left(\dfrac{s}{\omega_n}\right)^2}, \qquad (9\text{-}40)$$

where $A_n(s)$ denotes the normalized voltage gain function,

$$\omega_n = \frac{1}{\sqrt{LC_{lo}}} \qquad (9\text{-}41)$$

is the undamped natural frequency of the circuit, and, recalling Eq. (9-38),

$$Q = \frac{1}{\omega_n R_l C_{lo}} = \frac{B_{cs}}{\omega_n} = \frac{\sqrt{L/C_{lo}}}{R_l} \qquad (9\text{-}42)$$

is the circuit quality factor. It should be understood that the undamped natural frequency, ω_n, is the frequency at which inductance L resonates with capacitance C_{lo} under the zero damping circumstance of a null circuit resistance, R_l. Moreover, the quality factor, Q, is the quality factor of the circuit inductance at frequency ω_n; that is, $Q \equiv \omega_n L / R_l$. Equations (9-40) through (9-42) confirm that the insertion of the inductance in the drain load circuit of the uncompensated topology establishes a left half plane zero that can be exploited for broadbanding purposes. Unfortunately, Eq. (9-40) also suggests that the compensated structure may exhibit undue frequency response peaking. The obvious source of this peaking is the left half plane zero, which can lie at too low a frequency within the circuit passband if Q is too large and/or ω_n is too small. The second, and more subtle, cause of unacceptable peaking is that complex circuit poles, which arise for $Q > 1/2$, conduce underdamped responses.

The foregoing and related other issues are best examined by expressing the normalized gain in Eq. (9-40) as an explicit function of a complex frequency, say p, that is normalized to the bandwidth, B_{cs}, of the uncompensated amplifier; that is, the amplifier prior to insertion of the peaking coil. Thus, with

$$p \triangleq s/B_{cs}, \qquad (9\text{-}43)$$

Eq. (9-40) becomes

$$A_n(p) \triangleq \frac{A(p)}{A(0)} = \frac{1 + Q^2 p}{1 + p + Q^2 p^2}. \qquad (9\text{-}44)$$

In the sinusoidal steady state, the complex frequency, p, can be replaced by jy, with the understanding that y represents radial signal frequency, ω, normalized to bandwidth B_{cs}. With $y \triangleq \omega/B_{cs}$,

$$A_n(jy) \triangleq \frac{A(jy)}{A(0)} = \frac{1 + jQ^2 y}{1 - (Qy)^2 + jy}. \qquad (9\text{-}45)$$

Figure 9.14 displays a plot of the magnitude response implied by Eq. (9-45) for three values of the circuit quality factor. Observe that for $Q = 1$, almost 50% peaking is evidenced, while the amount of peaking diminishes rapidly

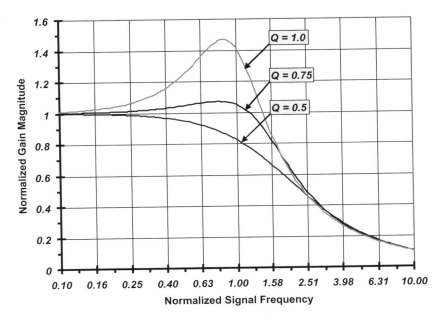

Figure 9.14. Magnitude response, plotted for various values of the circuit quality factor, Q, of the shunt peaked amplifier modeled in Fig. 9.13.

with progressive decreases in Q. Note further that the 3-dB bandwidth, which is the normalized frequency yielding a gain magnitude of the inverse of root two, increases with increasing Q.

9.4.2.1. Maximally Flat Magnitude Response

Figure 9.14 teaches the apparent existence of an optimum quality factor such that maximal bandwidth is attained within the constraint of a flat response exhibiting no peaking within the amplifier passband. This optimality is tantamount to the condition of a maximally flat magnitude (MFM) frequency response, for which the first $(N-1)$ frequency derivatives of the magnitude response for an Nth order system are identically null at zero frequency. For such a system, it can be shown that the normalized squared magnitude response, $|H_N(jy)|^2$, must mirror the general form,[1]

$$|H_N(jy)|^2 = \frac{P(y^2)}{P(y^2) + a_N y^{2N}}, \qquad (9\text{-}46)$$

where $P(y^2)$ is an even order polynomial in the squared normalized frequency variable, y^2, and a_N is a constant. Observe that this general expression suggests a squared magnitude response whose form is the ratio of a polynomial divided by the identical polynomial plus one additional, frequency dependent term. From Eq. (9-45),

$$|A_n(jy)|^2 = \frac{1 + Q^4 y^2}{1 + (1 - 2Q^2) y^2 + Q^4 y^4}, \qquad (9\text{-}47)$$

which mirrors the MFM form of Eq. (9-46), provided that the quality factor, Q, satisfies

$$Q^4 = 1 - 2Q^2. \qquad (9\text{-}48)$$

The positive solution of Eq. (9-48), which stipulates the specific quality factor, say Q_m, commensurate with maximal flatness of shunt peaked amplifier response, is

$$Q_m = \sqrt{\sqrt{2} - 1} = 0.6436. \qquad (9\text{-}49)$$

While Eq. (9-49) stipulates an optimized quality factor in the sense of achieving a MFM response, the bandwidth achieved by this optimal condition is not immediately apparent. There is some room for concern in this regard since maximal flatness attained with a compensated 3-dB bandwidth that is less than its uncompensated value has dubious merit. To this end, return to Eq. (9-47) to set the squared magnitude transfer function equal to $1/2$ at a normalized frequency, say y_b, where y_b represents the ratio of compensated bandwidth, say B_l, to the uncompensated bandwidth, B_{cs}. The fruit of an hour or two of depressing algebra is

$$y_b = \frac{B_l}{B_{cs}} = \sqrt{\frac{(2Q^4 + 2Q^2 - 1) + \sqrt{(2Q^4 + 2Q^2 - 1)^2 + 4Q^4}}{2Q^4}}. \quad (9\text{-}50)$$

For the quality factor given by Eq. (9-49), Eq. (9-50) delivers $y_b = 1.722$, which infers a laudable 72.2% increase in the uncompensated circuit bandwidth when the shunt peaking inductor is selected to ensure a maximally flat magnitude response.

It is essential to underscore the fact that Eq. (9-49) does not deliver maximum possible bandwidth in the compensated structure. Instead, it yields maximum bandwidth within the constraint of no peaking of the observable frequency response. This contention is highlighted by Fig. 9.15, which plots y_b in Eq. (9-50) as a function of the circuit quality factor, Q. This

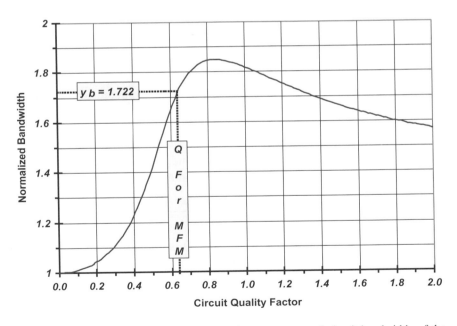

Figure 9.15. The bandwidth, normalized to the uncompensated circuit bandwidth, of the shunt peaked amplifier. A quality factor of $Q = 0.6436$ ensures a maximally flat magnitude response delivering a 72.2% improvement in circuit bandwidth.

figure infers an attainable bandwidth improvement approaching 85% in the neighborhood of $Q = 0.86$. Unfortunately, the price paid for this additional bandwidth enhancement is a peaked response, since, as Fig. 9.14 verifies, peaking becomes progressively more pronounced as the circuit quality factor increases.

In order to assess quantitatively the extent of the aforementioned peaking, the derivative, with respect to normalized frequency y, of the squared magnitude function in Eq. (9-47) can be equated to zero to ascertain the value of y, say y_m, where maximal squared gain is evidenced. The requisite mathematics are hardly enjoyable, but a closed form solution for y_m can be determined; namely, the nonzero value of y_m is

$$y_m = \frac{\sqrt{Q\sqrt{Q^2 + 2} - 1}}{Q^2}. \tag{9-51}$$

Since y_m is a particular frequency normalized to the 3-dB bandwidth of the uncompensated circuit, it must obviously be a real number. The relationship at hand projects real y_m if and only if

$$Q\sqrt{Q^2 + 2} \geq 1, \tag{9-52}$$

which requires $Q \geq 0.6436 \equiv Q_m$. Thus, frequency response peaking is evidenced only for circuit quality factors that exceed the quality factor commensurate with the realization of a maximally flat magnitude response. For $Q = Q_m$, y_m in Eq. (9-51) is zero, which is as expected, since a maximally flat magnitude frequency response implies the existence of a response maximum at only zero frequency.

Yet another immensely enjoyable exercise entails substituting Eq. (9-51) into Eq. (9-47) to discern the maximum value, say M, of the normalized magnitude response. The result, which is meaningful only when $Q \geq Q_m$, is

$$M|_{Q \geq Q_m} = |A_n(jy_m)|\|_{Q \geq Q_m} = \frac{Q^2}{\sqrt{2Q\sqrt{Q^2 + 2} - (2Q^2 + 1)}}. \tag{9-53}$$

For $Q < Q_m$, M is understood to be one. This understanding reflects the fact that for small Q, no peaking in excess of the zero frequency gain magnitude arises at nonzero frequencies, thereby implying that maximum gain, whose normalized value is one, is observed at only zero frequency. The percentage overshoot, obtained simply by multiplying $(M - 1)$ by 100, is plotted in Fig. 9.16. It is a bit distressing to note that this overshoot is a somewhat sensitive function of quality factor. To wit, zero percent overshoot is ensured for $Q = Q_m = 0.6436$, but at $Q = 0.775$, almost 10%

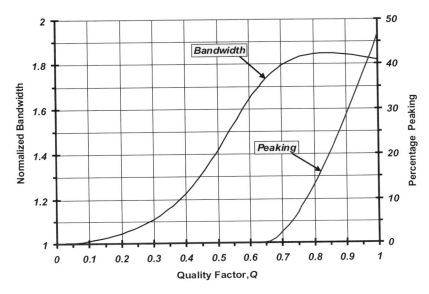

Figure 9.16. The bandwidth, normalized to the uncompensated circuit bandwidth, and the percentage peaking in the frequency response of the shunt peaked amplifier in Fig. 9.13.

overshoot prevails. In contrast, the normalized bandwidth increases only slightly over this Q-interval, from 1.722 at $Q = 0.6436$ to 1.838 at $Q = 0.775$.

Example 9.3. The inductor in the amplifier of Fig. 9.9 is to be selected to ensure a maximally flat magnitude frequency response. This response is to be characterized by a zero frequency gain of 18 dB and a 3-dB bandwidth of at least 1.6 GHz when the amplifier is driven by a 50 Ω signal source. The utilized transistor is the same as the device exploited in Example 9.2; its small signal model parameters are repeated herewith for reader convenience.

Forward Transconductance (g_m):	34 mS
Bulk Transconductance Factor (λ_b):	0.09
Drain-Source Channel Resistance (r_o):	15.8 KΩ
Gate-Source Capacitance (C_{gs}):	173 fF
Gate-Drain Capacitance (C_{gd}):	15 fF
Bulk-Drain Capacitance (C_{bd}):	12 fF
Bulk-Source Capacitance (C_{bs}):	10 fF

Design the circuit by determining the appropriate values of the peaking inductance, L, the drain load resistance, R_l, and the requisite load capacitance, C_l. Using the small signal model provided in Fig. 9.2(b), simulate the small signal frequency response of the finalized design on HSPICE. Compare the frequency response of the compensated amplifier with that of the uncompensated ($L = 0$) version.

Solution 9.3.

(1) The bandwidth enhancement afforded by a maximally flat, shunt peaked amplifier response is 72.2%. Thus, if the compensated bandwidth is to be at least 1.6 GHz, the bandwidth of the stage without inductive compensation must be no smaller than 1.6 GHz/1.722 = 929.1 MHz. To allow for the various approximations invoked in the course of shunt peaked response analysis, it is prudent to design the uncompensated stage for a 3-dB bandwidth of 950 MHz; that is, $B_{cs} = 2\pi(950\,\text{MHz})$.

(2) A voltage gain of 18 dB is equivalent to a numerical gain of 7.943. Since the zero frequency magnitude of voltage gain is $g_m(r_o \| R_l)$, the required drain load resistance is $R_l = 237.1\,\Omega$.

(3) Equation (9-38) is an applicable expression for the 3-dB bandwidth of the uncompensated amplifier. A slightly more accurate relationship is

$$B_{cs} = \frac{1}{(r_o \| R_l) C_{lo}}, \quad \text{(E3-1)}$$

whence, $C_{lo} = 717.1\,\text{fF}$. Recalling Eq. (9-27) and the gate-drain and bulk-drain capacitances given above, the load capacitance terminating the output port of the amplifier follows as $C_l = 690.1\,\text{fF}$.

(4) Using Eq. (9-42), the required inductance value of the shunt peaking coil is

$$L = (QR_l)^2 C_{lo}. \quad \text{(E3-2)}$$

Recalling that $Q = 0.6436$ for MFM response, $L = 16.70\,\text{nH}$.

(5) Figure 9.17 delineates the resultant small signal models for both the uncompensated and the shunt peaked compensated amplifiers. The simulated magnitude responses of both of these structures are depicted in Fig. 9.18.

Comments. The HSPICE results pictured in Fig. 9.18 confirm that the zero frequency gain magnitude of both the uncompensated common source amplifier and the shunt peaked common source unit is precisely the design target of nominally 7.94 volts/volt. The simulated frequency response of the uncompensated network shows a 3-dB bandwidth of 913.3 MHz, which is 3.86% lower than the design target of 950 MHz. Correspondingly, the simulated 3-dB bandwidth of the compensated structure is 1.55 GHz, which is 3.13% smaller than the 1.6 GHz design objective. These small errors can be attributed to the numerous approximations invoked to arrive at tractable, design-oriented results.

The quoted differences between simulation results and manual computations are actually smaller than the errors typically experienced in the course of broadband

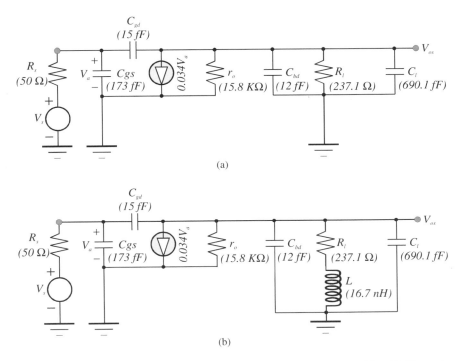

Figure 9.17. (a) Small signal model of the uncompensated amplifier considered in Example 9.3. (b) Small signal model of the shunt peaked common source amplifier designed for maximally flat magnitude response in Example 9.3.

amplifier design in MOSFET technologies. Designable circuit elements can be adjusted to correct for the observed errors. For example, the load capacitance can be decreased to incur a reduction in net capacitance C_{lo} by 3.13%, provided that inductance L is likewise reduced by the same factor to preserve the quality factor required for MFM response. However, these adjustments may comprise an exercise in engineering futility, given routinely encountered manufacturing tolerances with respect to electronic circuit components. In other words, response errors accruing from manufacturing tolerances are likely to be comparable to, or even larger than, those observed herewith.

A final point worthy of mention is that the requisite inductance of 16.7 nH may be too large for practical monolithic implementation. Inductances of at most 4 nH to 6 nH are generally an implicit requirement of monolithic circuit layout. In the present case, the specifications may need to be altered and/or different topological structures may need to be adopted if the ultimately designed circuit is to be realized pragmatically as an analog integrated circuit.

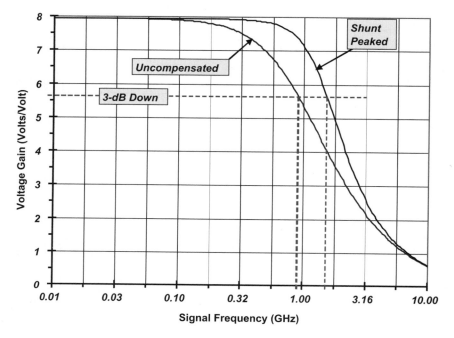

Figure 9.18. The simulated (HSPICE) magnitude responses for the two small signal amplifier models offered in Fig. 9.17.

9.4.2.2. *Maximally Flat Delay Response*

The realization of a maximally flat, or at least reasonably flat, magnitude response is arguably the most common objective of a circuit broadbanding strategy. But numerous electronic system applications require, in addition to a sufficiently broad passband, the ability to process input signals with nominally constant time delay, independent of the frequency spectrum of these signals. Examples of such applications include high speed electronic instrumentation, video systems, and digital communication systems.[2] Accordingly, it is prudent to investigate the shunt peaked amplifier from the perspective of its amenability to provide nominally constant I/O signal delay over suitably wide frequency passbands.

Of particular engineering interest is the network time delay achieved in the steady state. This steady state delay is commonly referenced as the *envelope delay*, $D(\omega)$, which relates to the I/O phase response, $\phi(\omega)$ as

$$D(\omega) = -\frac{d\phi(\omega)}{d\omega}. \qquad (9\text{-}54)$$

If the network transfer function, and thus its phase response, is cast explicitly as a function of a normalized frequency variable, y, as it is in Eq. (9-45) with $y = \omega/B_{cs}$,

$$-\frac{d\varphi(\omega)}{d\omega} = -\frac{1}{B_{cs}} \frac{d\varphi(y)|_{y=\omega/B_{cs}}}{dy} = \frac{D(y)}{B_{cs}}, \quad (9\text{-}55)$$

so that

$$D(y) \triangleq -\frac{d\varphi(y)}{dy} = B_{cs} D(\omega)|_{\omega=B_{cs} y} \quad (9\text{-}56)$$

is identified as the envelope delay, normalized to a delay value of $1/B_{cs}$.

Returning to Eq. (9-45), the phase response of the compensated shunt peaked network is

$$\varphi(y) = \tan^{-1}(Q^2 y) - \tan^{-1}\left[\frac{y}{1 - (Qy)^2}\right]. \quad (9\text{-}57)$$

Using Eq. (9-56), the normalized delay function is found to be

$$D(y) = \frac{1 + (Qy)^2}{1 + (1 - 2Q^2) y^2 + Q^4 y^4} - \frac{Q^2}{1 + Q^4 y^2}. \quad (9\text{-}58)$$

Observe a zero frequency normalized delay of $D(0) = (1 - Q^2)$, which is predictable via an inspection of Eq. (9-45). The delay response defined by Eq. (9-58) and plotted in Fig. 9.19 for various values of the circuit quality factor reveals an apparent optimal value of Q, say Q_d, such that the delay is rendered nominally constant over a suitably broad frequency passband.

The MFM concepts addressed previously can be applied to the delay function in the hope of achieving a maximally flat delay (MFD) response. To this end, the criterion underlying MFD response realization derives from an attempt to couch the right hand side of Eq. (9-58) into the form of the right hand side of Eq. (9-46). If Eq. (9-58) is expressed as a ratio of polynomials in the squared normalized frequency variable, y^2,

$$D(y) = (1 - Q^2)$$
$$\times \left[\frac{1 + \left(\frac{3Q^4}{1 - Q^2}\right) y^2}{1 + (Q^4 - 2Q^2 + 1) y^2 + 2Q^4(1 - Q^2) y^4 + Q^8 y^6}\right]. \quad (9\text{-}59)$$

Because the numerator on the right hand side in this expression is a first order polynomial in y^2, while the denominator is a third order polynomial

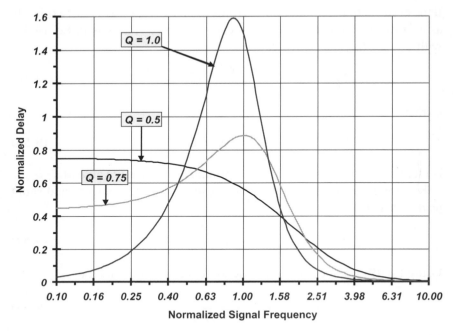

Figure 9.19. The normalized delay response of the shunt peaked amplifier shown in Fig. 9.13.

in y^2, $D(y)$ cannot be rendered a strictly maximally flat function of y in the sense of the form projected by the right hand side of Eq. (9-46). But MFD can nonetheless be approximated by equating the y^2 coefficients in the numerator and the denominator. Specifically, if Q_d signifies the optimal quality factor commensurate with an approximate MFD response,

$$\frac{3Q_d^4}{1 - Q_d^2} = Q_d^4 - 2Q_d^2 + 1, \tag{9-60}$$

which gives rise to the constraint,

$$Q_d^6 + 3Q_d^2 - 1 = 0. \tag{9-61}$$

Believe it or not, this equation can be solved precisely for Q_d by first changing variables to transform the sixth order relationship to an equivalent cubic.[3] The positive real root of Eq. (9-61) is

$$Q_d = \sqrt{\sqrt[3]{\left(\frac{\sqrt{5}+1}{2}\right)} - \sqrt[3]{\left(\frac{\sqrt{5}-1}{2}\right)}} = 0.5676. \tag{9-62}$$

At least three interesting observations with respect to this result can be proffered. First, no peaking in the magnitude response results from designing for $Q = Q_d$, since peaking is manifested only for $Q > Q_m = 0.6436$. Second, a consideration of the calculations leading to the bandwidth plot in Fig. 9.16 confirms a bandwidth improvement with respect to the uncompensated amplifier of almost 57%. Although this improvement is smaller than the bandwidth enhancement observed under the MFM condition, it is nonetheless laudable in view of the fact that this enhancement derives from the addition of only one, nominally lossless, component to the basic amplifier configuration. Finally, for $y = 1.57$ and $Q = Q_d = 0.5676$, the last term in the denominator on the right hand side of Eq. (9-59) evaluates numerically as 0.1614, which is 6.2-times smaller than unity, the zero frequency value of this denominator polynomial. Of course, the other terms in the subject denominator are themselves positive and hence add to the unity term, thereby making the final term comparatively smaller still. The reasonable engineering conclusion herewith is that MFD over a wide signal passband is well approximated by $Q = Q_d$ in the shunt peaked amplifier.

9.5.0. Series Peaked Compensation

In a shunt peaked amplifier, as diagrammed in Fig. 9.9, the inductive branch is placed in parallel with the amplifier output port to which the net capacitance serving to establish the dominant pole of the uncompensated amplifier is presumed to be incident. In contrast, a series peaked amplifier places the inductive branch in series with the load capacitance, C_l, that the amplifier is compelled to drive. As can be seen in the pertinent schematic diagram of Fig. 9.20(a), the compensating inductance, L, separates the load capacitance from the amplifier output port capacitance, C_o, which is fundamentally the sum of the bulk-drain and gate-drain capacitances associated with the common gate cascode transistor, $M2$. As is demonstrated shortly, the additional capacitance, C_x, appended to the amplifier output port proves indispensable with respect to achieving a broadbanded, maximally flat magnitude response in the compensated structure. The pertinent small signal model, given in Fig. 9.20(b), is premised on two presumptions. First, the drain-source channel resistance of both transistors is sufficiently large to warrant its tacit neglect. Second, with $L = 0$, pole dominance attributed to the net output port capacitance is presumed. Accordingly, the uncompensated

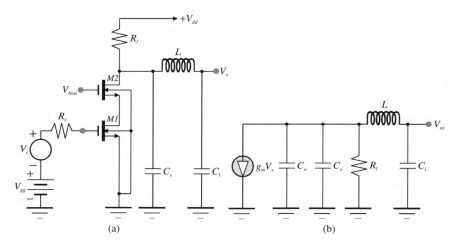

Figure 9.20. (a) Basic schematic diagram of series peaked amplifier. (b) Approximate small signal, high frequency model of the circuit in (a). Both transistors $M1$ and $M2$ are presumed to operate in saturation, and pole dominance attributed to the net output port capacitance is presumed when $L = 0$.

radial 3-dB bandwidth, B_u, is

$$B_u \approx \frac{1}{R_l(C_o + C_l)}. \tag{9-63}$$

Note that capacitance C_x is not included in this expression, for C_x is a requirement of only the compensated, series peaked topology.

A straightforward analysis of the model in Fig. 9.20(b) sets forth a voltage transfer function of

$$A_v(s) = \frac{V_{os}}{V_s}$$

$$= -\frac{g_m R_l}{1 + sR_l(C_o + C_x + C_l) + s^2 LC_l + s^3 LC_l R_l(C_o + C_x)}, \tag{9-64}$$

where, $-g_m R_l$ is confirmed as the zero frequency value, $A_v(0)$, of the voltage gain. Two special cases are of immediate interest: a compensated second order and a compensated third order response.

9.5.1. Second Order Compensated Response

The series peaking case traditionally visited by much of the textbook literature is the condition in which the capacitance sum, $(C_o + C_x)$ is negligible, whence the normalized gain deriving from Eq. (9-64) reduces to

$$A_n(s) = \frac{A_v(s)}{A_v(0)} = \frac{A_v(s)}{-g_m R_l} = \frac{1}{1 + s R_l C_l + s^2 L C_l}. \qquad (9\text{-}65)$$

This second order transfer relationship has no finite zeros to enhance its bandwidth attributes. Nonetheless, it produces a maximally flat magnitude response, known as a Butterworth response, when the normalized transfer function abides by the functional form,[4]

$$A_n(s) = \frac{1}{1 + \dfrac{\sqrt{2}s}{B_c} + \left(\dfrac{s}{B_c}\right)^2}, \qquad (9\text{-}66)$$

where B_c symbolizes the compensated 3-dB bandwidth. By a comparison of this expression with the right hand side of Eq. (9-65), a Butterworth MFM response materializes if

$$\left. \begin{aligned} \frac{\sqrt{2}}{B_c} &= R_l C_l \\ \frac{1}{B_c^2} &= L C_l \end{aligned} \right\}. \qquad (9\text{-}67)$$

Recalling Eq. (9-63), the time constant, $R_l C_l$, is the inverse of the uncompensated circuit bandwidth, B_u, if C_o is indeed a negligible capacitance. Accordingly, the first of the expressions in Eq. (9-67) suggests immediately that MFM series peaking in the face of negligible capacitance at the amplifier output port improves bandwidth by a factor of the square root of two; that is,

$$B_c = \sqrt{2} B_u. \qquad (9\text{-}68)$$

Additionally, this result and Eq. (9-67) combine to stipulate a required inductance value of

$$L = \frac{R_l}{2 B_u}, \qquad (9\text{-}69)$$

which highlights a requisite inductance such that its reactance at the uncompensated circuit bandwidth is precisely one-half of the drain load resistance. Equivalently, it can be stated that the quality factor of the shunt peaking coil at a frequency equaling the uncompensated amplifier bandwidth is precisely $1/2$.

Series peaking endures bad press in comparison to the exaltations showered on shunt peaking largely because of Eq. (9-68), which extols a slightly better than 41% enhancement in uncompensated circuit bandwidth. In contrast, shunt peaking designed for MFM response enhances bandwidth by better than 72%. Since broadbanding by even small amounts is generally a painful design experience, a 41% bandwidth improvement, gleaned without significant static power dissipation penalty, is nonetheless laudable. Series peaking does not deserve the aforementioned negative press. But the approximations underpinning the second order response in Eq. (9-65), which entail the tacit neglect of all capacitances incident at the output port of the uncompensated amplifier, are fair game for vigorous criticism.

9.5.2. Third Order Compensated Response

A third order circuit providing a maximally flat, Butterworth response requires a normalized transfer characteristic of the form,

$$A_n(s) = \frac{1}{1 + 2\left(\frac{s}{B_c}\right) + 2\left(\frac{s}{B_c}\right)^2 + \left(\frac{s}{B_c}\right)^3}, \quad (9\text{-}70)$$

where, as in the case in the second order circuit addressed in the preceding subsection, B_c represents the radial 3-dB bandwidth of the compensated amplifier. A term-by-term comparison of this expression with the right hand side of Eq. (9-64) leads to

$$\left.\begin{array}{l} \dfrac{2}{B_c} = R_l(C_o + C_x + C_l) \\[6pt] \dfrac{2}{B_c^2} = LC_l \\[6pt] \dfrac{1}{B_c^3} = LC_l R_l(C_o + C_x) \end{array}\right\}. \quad (9\text{-}71)$$

When combined with Eq. (9-63), the first of these expressions produces

$$\frac{B_c}{B_u} = \frac{2}{1 + \dfrac{C_x}{C_o + C_l}}, \quad (9\text{-}72)$$

which, assuming $C_x \ll (C_o + C_l)$, portends almost a doubling of the uncompensated circuit bandwidth. From the second expression in Eq. (9-71), the required inductance value is

$$L = \frac{2}{B_c^2 C_l}. \qquad (9\text{-}73)$$

Finally, the combination of all three expressions in Eq. (9-71) produces the design requirement,

$$C_o + C_x = \frac{C_l}{3}, \qquad (9\text{-}74)$$

whereupon Eq. (9-72) can be equivalently expressed as

$$\frac{B_c}{B_u} = \frac{3}{2}\left(1 + \frac{C_o}{C_l}\right). \qquad (9\text{-}75)$$

Obviously, C_o in Eq. (9-75) can be no smaller than zero, while from Eq. (9-74), C_o can be no larger than $C_l/3$. Accordingly, the factor by which the uncompensated circuit bandwidth can be improved via third order, Butterworth MFM compensation is between 1.5 and 2.0. In other words, third order, series peaked bandwidth compensation is guaranteed to be slightly better than that afforded by its second order counterpart and conceivably, it can exceed the bandwidth compensation factor implicit to maximally flat shunt peaked compensation.

Example 9.4. When driven by a 50 Ω signal source, a series peaked amplifier is to deliver a maximally flat response characterized by a zero frequency voltage gain of 14 dB and a 3-dB frequency of 2.2 GHz. The utilized transistor is the same as the device exploited in Example 9.3. Design the circuit by determining the appropriate values of the series peaking inductance, L, the drain load resistance, R_l, the requisite load capacitance, C_l, and the compensation capacitance, C_x, required at the amplifier output port. Using the small signal model provided in Fig. 9.2(b), simulate the small signal frequency response of the finalized design on HSPICE. Compare the frequency response of the compensated amplifier with that of its uncompensated version.

Solution 9.4.

(1) A voltage gain of 14 dB is equivalent to a numerical gain of 5.012. Since the zero frequency magnitude of voltage gain is $g_m(r_o \| R_l)$, the required drain load resistance is $R_l = 148.8 \, \Omega$.

(2) Using Eqs. (9-71) and (9-74),

$$B_c = \frac{3}{2R_l C_l}; \qquad (\text{E4-1})$$

With $B_c = 2\pi(2.2\,\text{GHz})$ and $R_l = 148.8\,\Omega$, the required load capacitance follows as $C_l = 729.3\,\text{fF}$.

(3) For $C_o = C_{gd} + C_{bd} = 27\,\text{fF}$, Eq. (9-74) yields $C_x = 216.1\,\text{fF}$.

(4) From Eq. (9-73), the required inductance is $L = 14.35\,\text{nH}$.

(5) From Eq. (9-75), the bandwidth improvement factor is $B_c/B_u = 1.556$, which means that the bandwidth of the uncompensated amplifier is $B_u = 2\pi(1.41\,\text{GHz})$. It is to be understood that in the present context, "uncompensated" means $L = 0$ and $C_x = 0$.

(6) Figure 9.21 diagrams the small signal models for both the uncompensated and the series peaked compensated amplifiers. The simulated magnitude responses of both of these structures are provided in Fig. 9.22.

Comments. The HSPICE results portrayed in Fig. 9.22 show a zero frequency gain magnitude for both amplifiers that is nearly precisely equal to the design goal. The simulated 3-dB bandwidth of the compensated amplifier is 2.23 GHz

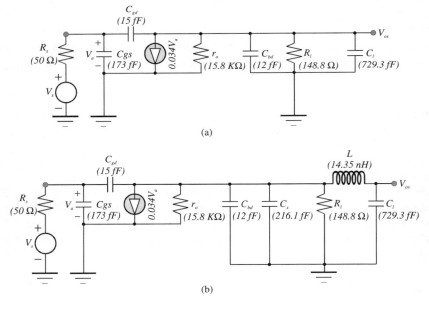

Figure 9.21. (a) Small signal model of the uncompensated common source amplifier addressed in Example 9.4. (b) Small signal model of the series peaked common source amplifier designed for maximally flat magnitude response in Example 9.4.

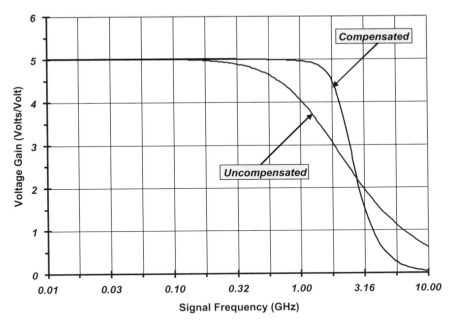

Figure 9.22. The simulated magnitude responses for the small signal models of the uncompensated and the compensated series peaked amplifier shown in Fig. (8.21).

and is once again nearly identical to the design objective. On the other hand, the uncompensated structure shows a 3-dB bandwidth of 1.37 GHz. Thus, the factor by which bandwidth improves as a result of series peaking in this example is $2.23/1.37 = 1.63$, which is higher than the calculated enhancement factor by about 4.5%.

The example at hand conveys no drama with respect to calculated and simulated amplifier performance. It does serve to show that maximally flat magnitude series peaking can be nominally as effective as the more traditionally invoked shunt peaking compensation strategy.

9.6.0. Series-Shunt Peaked Compensation

If shunt peaking and series peaking comprise effecting circuit broadbanding strategies, engineers are naturally inclined to suspect that the marriage of these two techniques may result in even more dramatic frequency response improvements. In order to examine the propriety of this suspicion, consider Fig. 9.23(a), which depicts the schematic diagram of a representative MOS technology series-shunt peaked amplifier driving a load of pure

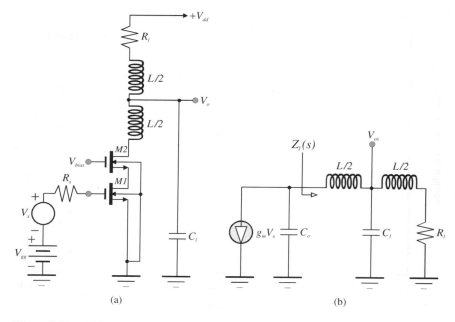

Figure 9.23. (a) Schematic diagram of a series-shunt peaked amplifier driving a capacitive load. (b) Approximate small signal model of the amplifier in (a). As usual, the transistors are biased in their saturation regimes.

capacitance, C_l. The $L/2$ inductance incident with the drain terminal of the cascode transistor, $M2$, serves as a series peaking element, while the inductance in series with resistance R_l functions as the shunt peaking element. Fig. 9.23(b) is the corresponding small signal model, which reflects the recurring themes of large transistor channel resistances, negligible bulk-induced modulation of threshold voltage, and an effective output capacitance, C_o, which is nominally the sum of the bulk-drain and gate-drain capacitances of transistor $M2$.

9.6.1. Design Criteria

In principle, it is possible to deduce the transfer function, V_{os}/V_s, for the model in Fig. 9.23(b), but the fourth order nature of the circuit renders a useful quantification of bandwidth enhancement a daunting analytical ordeal. A more pragmatic tack derives from the observation that the model at hand is the topology of a lumped approximation to a distributed transmission line.[5]–[7] In particular, it is possible to choose the drain load resistance,

R_l, in a manner that renders $Z_l(s)$ equal to R_l, independent of indicated inductances and capacitances, over a broad range of signal frequencies. The advantage of constant $Z_l(s)$ is a simplification of the pertinent transfer function and hence, a presumably tractable and designable bandwidth relationship.

An inspection of the model in Fig. 9.23(b) reveals

$$Z_l(s) = \frac{sL}{2} + \frac{R_l + \frac{sL}{2}}{1 + sR_lC_l + \frac{s^2LC_l}{2}}. \tag{9-76}$$

It is a straightforward matter to show that $Z_l(s) \equiv R_l$ if R_l is selected to satisfy

$$R_l = \sqrt{\left(\frac{L}{C_l}\right)\left[1 + \left(\frac{s}{\omega_h}\right)^2\right]}, \tag{9-77}$$

where

$$\omega_h = \frac{2}{\sqrt{LC_l}}. \tag{9-78}$$

Obviously, a frequency dependent, passive resistance of the form propounded by Eq. (9-77) is physically unrealizable. But if the frequency parameter, ω_h, is suitably large in comparison to the outer reach of the amplifier passband,

$$R_l \approx \sqrt{\frac{L}{C_l}} \tag{9-79}$$

is very much realizable, whereupon

$$\omega_h = \frac{2}{\sqrt{LC_l}} \approx \frac{2}{R_lC_l}. \tag{9-80}$$

Armed with the preceding two expressions and the tacit (and not completely satisfying) presumption that capacitance C_o can be ignored in the model of Fig. 9.23(b), the applicable transfer function can be shown to be

$$A_v(s) = \frac{V_{os}}{V_s} \approx -\frac{g_mR_l\left(1 + \frac{s}{\omega_h}\right)}{1 + 2\left(\frac{s}{\omega_h}\right) + 2\left(\frac{s}{\omega_h}\right)^2}. \tag{9-81}$$

Following straightforward analytical procedures invoked previously, the compensated 3-dB bandwidth, B_c, can be shown to be

$$B_c = 0.8995\omega_h = \frac{1.80}{R_l C_l}, \qquad (9\text{-}82)$$

where Eq. (9-80) is used. If capacitance C_o is indeed negligible, the inverse of the time constant, $R_l C_l$, is the amplifier bandwidth prior to series-shunt peaked compensation and thus, Eq. (9-82) suggests an outstanding 80% improvement in bandwidth. This upgraded performance comes with a slight price, which can be shown to be 2.91% of magnitude response peaking occurring at a frequency of $0.687/R_l C_l$.

9.6.2. A Design Problem

Although the tacit neglect of the amplifier output capacitance, C_o, encourages a tractably straightforward frequency response analysis of the series-shunt peaked amplifier in Fig. 9.23, such neglect also obscures a potentially catastrophic design problem. In an attempt to dramatize the nature of this problem, assume that the amplifier in question utilizes the transistors exploited in Example 9.2 and is to be designed for a voltage gain of 15 dB and a 3-dB bandwidth of 3 GHz. From Eq. (9-82), the required value of ω_h is $2\pi(3.335\,\text{GHz})$. Since 15 dB is a gain magnitude of 5.623, $g_m = 34$ mmho yields a required drain circuit load resistance of $R_l = 165.4\,\Omega$. It follows from Eq. (9-80) that the load capacitance delivering the stipulated bandwidth must be $C_l = 577.0\,\text{f}F$. Finally, Eq. (9-79) yields $L = 15.78\,\text{nH}$, whence $L/2 = 7.89\,\text{nH}$.

The HSPICE simulation of the magnitude response of the model in Fig. 9.23(b), given the foregoing computed parameters and the presumption that $C_o = 0$, is depicted in Fig. 9.24. The simulated 3-dB bandwidth is 3.0 GHz, which is precisely the bandwidth objective. Moreover, the simulated maximum value of the magnitude response is 5.787, or 2.86% of peaking, which doubtlessly differs from the theoretic percentage peaking of 2.91% only because of restrictions implicit to the quantized frequency scale adopted for the circuit simulation. On the other hand, Fig. 9.25 compares the foregoing $C_o = 0$ simulation of the magnitude response to its $C_o = 27\,\text{fF}$ (the sum of C_{gd} and C_{bd} for transistor $M2$) counterpart. Despite the close match between the two simulated frequency responses for frequencies extending through the 3-dB bandwidth, even the most sedentary

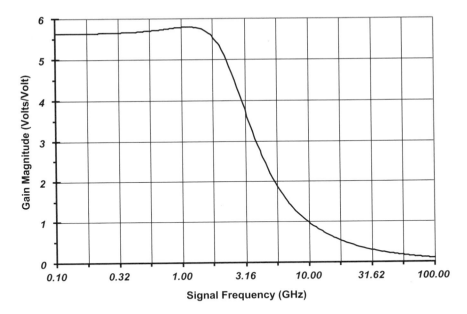

Figure 9.24. Magnitude response of the series-shunt peaked amplifier modeled in Fig. 9.23(b). The model parameters are $g_m = 34\,\text{mmho}$, $L/2 = 7.89\,\text{nH}$, $C_l = 577\,\text{fF}$, $R_l = 165.4\,\Omega$, and notably, $C_o = 0$.

of circuit designers is likely to become hypertensive at the sight of the apparent resonance, in this case at 10.96 GHz.

The cause of the indicated resonance in the $C_o = 27\,\text{fF}$ response curve depicted in Fig. 9.25 is somewhat transparent. In particular, at high signal frequencies, the relatively large load capacitance, C_l, begins to emulate a signal short circuit. At these frequencies, the inductance, $L/2$, to the left of capacitance C_l in Fig. 9.23(b) is resultantly placed in virtual shunt with the amplifier output port capacitance, C_o. It follows that the observed resonant frequency derives from the effective shunt interconnection of these two elements, whence, the radial resonant frequency, say ω_o, is

$$\omega_o \approx \frac{1}{\sqrt{(L/2)\,C_o}} = 2\pi(10.90\,\text{GHz}). \tag{9-83}$$

In order to mitigate the foregoing dilemma, steps must be taken to ensure that the parasitic output capacitance, C_o, of the transconductor cell does not face an inductive impedance, $Z_l(s)$, seen looking into the tee configuration

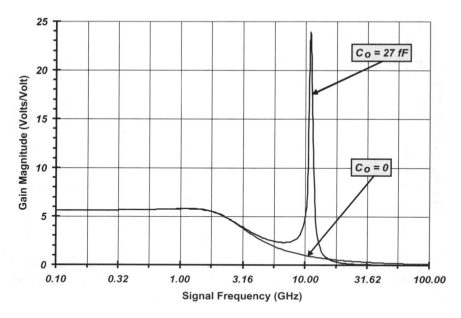

Figure 9.25. Magnitude responses of the series-shunt peaked amplifier modeled in Fig. 9.23(b). The model parameters are identical to those that underlie the simulated response drawn in Fig. 9.24, except for the indicated differences in the amplifier output port capacitance, C_o.

comprised of the two inductances, $L/2$, and the load capacitance, C_l, in Fig. 9.23(b). Since a capacitive tee input impedance is also undesirable because it foretells a possible degradation in circuit 3-dB bandwidth, it follows that a constant, frequency invariant $Z_l(s)$ is a necessary condition for design optimality in the senses of broadbanded and reasonably flat frequency responses. In other words, the subject tee network must be modified so that the frequency dependent $Z_l(s)$ that results from approximating the load requirement of Eq. (9-77) by Eq. (9-79) becomes frequency invariant for a stipulated drain load resistance, R_l.

9.6.2.1. *Capacitance Bridged, Coupled Inductor Load*

An intriguing broadband circuit architecture satisfying the foregoing design requirements is the circuit shown in Fig. 9.26.[8] This configuration, which is an outgrowth of classic shunt and double series peaking strategies, modifies the circuit of Fig. 9.23(a) by appending a bridging capacitance, C_f, across the two symmetrical coils in the hope that this element can mitigate

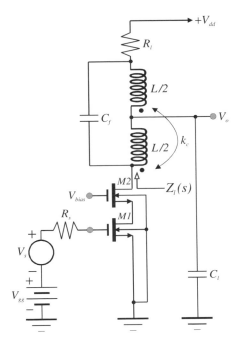

Figure 9.26. Modified version of the broadband peaking circuit in Fig. 9.23(a). Observe the use of magnetically coupled inductors and the incorporation of a bridging capacitance across the tee circuit.

the inductive component of the tee input impedance without appreciable bandwidth degradation. The mere fact that C_f can be selected to achieve a purely resistive $Z_l(s)$ does not necessarily guarantee an acceptably flat frequency response for the small signal voltage gain function, V_{os}/V_s. To this end, an additional design degree of freedom is forged by purposefully laying out the two coils to incur a presumably controllable mutual inductance, M. This mutual inductance is characterized classically by the so-called coupling coefficient, k_c, between the two coils, such that

$$k_c = \frac{M}{L/2}. \tag{9-84}$$

For $k_c = 0$, the inductors operate independently of one another (meaning that the electromagnetic fields surrounding one inductor do not envelope any portion of the coils implicit to the other inductor), while for $k_c = 1$, the inductors are maximally coupled, as in an ideal transformer. In silicon

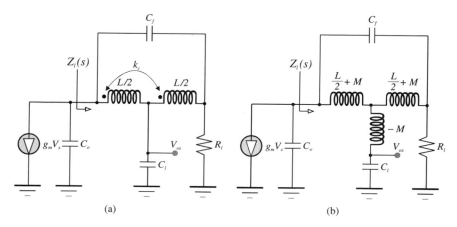

Figure 9.27. (a) Small signal, high frequency equivalent circuit of the broadband amplifier shown in Fig. 9.26. (b) Alternative form of the model in (a).

integrated circuits, spiral inductors can be laid out to yield coupling coefficients in the range of $k_c \leq 0.5$. On the other hand proximately located bond wires can be positioned to achieve inductive coupling coefficients of $k_c \leq 0.7$.

The small signal, high frequency model of the amplifier in Fig. 9.26 is provided in Fig. 9.27(a). In order to facilitate circuit analysis, and as the reader is invited to demonstrate in Problem 9.9 at the conclusion of this chapter, the two coupled inductances can be represented electrically as the uncoupled tee inductor two-port network drawn in Fig. 9.27(b). It is interesting to note that the second model suggests that the immediate effect of inductive coupling is to incur additional series peaking of the circuit load capacitance, C_l, thereby arguably leading to the possibility of improving the bandwidth established by conventional series-shunt peaking.

9.6.2.2. Constant Resistance Criteria

In order to arrive at the criteria ensuring constant $Z_l(s)$, and in particular, $Z_l(s) \equiv R_l$ for all signal frequencies, it is expedient to view the load network in Fig. 9.27(b) as a shunt interconnection of the bridging capacitance, C_f, and the inductive tee network, as is exemplified in Fig. 9.28. Such a view promotes the notion that the short circuit admittance, or y-parameters, of the overall network are the sum of the corresponding y-parameters of each of the two networks. To this end, the short circuit input admittance, y_{11a}, of

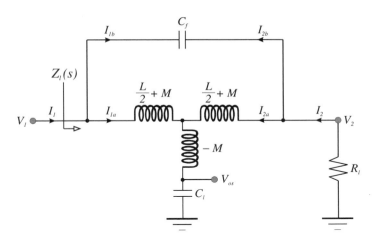

Figure 9.28. The load circuit in Fig. 9.27(b) drawn to underscore the fact that the structure is a shunt-shunt interconnection of two linear two-port networks.

the inductive tee configuration is

$$y_{11a} = \left.\frac{I_{1a}}{V_1}\right|_{V_2=0} = \frac{1 + \dfrac{s^2 L C_l}{2}}{s(1+k_c)L\left[1 + \dfrac{s^2(1-k_c)LC_l}{4}\right]}, \quad (9\text{-}85)$$

where Eq. (9-84) has been used. The transadmittance, y_{12a}, can be shown to be

$$y_{12a} = \left.\frac{I_{1a}}{V_2}\right|_{V_1=0} = -\frac{1 - \dfrac{s^2 k_c L C_l}{2}}{s(1+k_c)L\left[1 + \dfrac{s^2(1-k_c)LC_l}{4}\right]}. \quad (9\text{-}86)$$

Because of the symmetrical nature of the passive network at hand,

$$\left.\begin{array}{l} y_{11a} = \left.\dfrac{I_{1a}}{V_1}\right|_{V_2=0} \equiv y_{22a} = \left.\dfrac{I_{2a}}{V_2}\right|_{V_1=0} \\[2mm] y_{12a} = \left.\dfrac{I_{1a}}{V_2}\right|_{V_1=0} \equiv y_{21a} = \left.\dfrac{I_{2a}}{V_1}\right|_{V_2=0} \end{array}\right\rangle. \quad (9\text{-}87)$$

For the bridging capacitance two port, it is obvious that

$$y_{11b} = \left.\frac{I_{1b}}{V_1}\right|_{V_2=0} = sC_f \equiv \left.\frac{I_{2b}}{V_2}\right|_{V_1=0} = y_{22b}, \quad (9\text{-}87)$$

and

$$y_{12b} = \left.\frac{I_{1b}}{V_2}\right|_{V_1=0} = -sC_f \equiv \left.\frac{I_{2b}}{V_1}\right|_{V_2=0} = y_{21b}. \qquad (9\text{-}88)$$

It follows that the overall network is characterized by the y-parameter functions,

$$y_{11} \equiv y_{22} = y_{11a} + y_{11b} = \frac{1 + \dfrac{s^2 L C_l}{2}}{s(1+k_c)L\left[1 + \dfrac{s^2(1-k_c)LC_l}{4}\right]} + sC_f, \qquad (9\text{-}89)$$

and

$$y_{12} \equiv y_{21} = y_{12a} + y_{12b}$$

$$= -\frac{1 - \dfrac{s^2 k_c L C_l}{2}}{s(1+k_c)L\left[1 + \dfrac{s^2(1-k_c)LC_l}{4}\right]} - sC_f, \qquad (9\text{-}90)$$

A review of Sections 2.3.3 and 2.4.3 of this text confirms that the driving point impedance, $Z_l(s)$, highlighted in Fig. 9.28 derives from the expression,

$$\frac{1}{Z_l(s)} = y_{11} - \frac{y_{12}^2}{y_{11} + \dfrac{1}{R_l}}. \qquad (9\text{-}91)$$

Since $Z_l(s) \equiv R_l$ is the design objective, Eq. (9-91) gives rise to the constraint,

$$\frac{1}{R_l} = \sqrt{(y_{11} + y_{12})(y_{11} - y_{12})}. \qquad (9\text{-}92)$$

Upon substitution of Eqs. (9-89) and (9-90) into Eq. (9-92),

$$\frac{1}{R_l} = \sqrt{\frac{C_l}{(1+k_c)L}\left[\frac{1+s^2(1+k_c)LC_f}{1+\dfrac{s^2(1-k_c)LC_l}{4}}\right]}. \qquad (9\text{-}93)$$

Within the radical on the right hand side of this expression, note the presence of conjugate imaginary zero and pole pairs. If capacitance C_f is chosen as

$$C_f = \frac{C_l}{4}\left(\frac{1-k_c}{1+k_c}\right), \qquad (9\text{-}94)$$

the aforementioned imaginary zeros and poles cancel, thereby producing the frequency independent expression,

$$\frac{1}{R_l} = \sqrt{\frac{C_l}{(1+k_c)L}}, \quad (9\text{-}95)$$

for the resistive load termination, R_l, that produces a driving point input impedance that is identical to said load resistance. For a given value of R_l, Eq. (9-95) stipulates a requisite inductance, L, of

$$L = \frac{R_l^2 C_l}{1+k_c}. \quad (9\text{-}96)$$

Equations (9-94) and (9-96) comprise necessary and sufficient conditions underlying the realization of an input impedance, $Z_I(s)$, that is purely resistive and equal to the load resistance, R_l, for all signal frequencies. Note that this constant resistance characteristic can be achieved even when the two utilized inductors are uncoupled; that is, when $k_c = 0$. When the inductors are perfectly coupled in the sense of $k_c = 1$, Eq. (9-94) shows that no bridging capacitance is required for constant input resistance behavior of the network at hand.

9.6.2.3. Transfer Relationship

Armed with Eqs. (9-94) and (9-96), the small signal voltage transfer function, $A_v(s) = V_{os}/V_s$, can be cast, albeit somewhat painfully, in a useful closed form format. This analysis is also facilitated by the design constraint, $Z_I(s) \equiv R_l$ in Fig. 9.27(b), which implies that the time constant associated with the transconductor output capacitance, C_o, is simply $R_l C_o$. If the inverse of this time constant is significantly larger than the desired, compensated radial bandwidth, B_c, so that $B_c R_l C_o \ll 1$, capacitance C_o can be ignored tacitly to garner a modicum of simplified circuit analysis. Then, applying the same gain normalization as that introduced in Eq. (9-65), the normalized transfer relationship that derives because of Eqs. (9-94) and (9-96) is

$$A_n(s) = \frac{1}{1 + \dfrac{s}{Q_c \omega_x} + \left(\dfrac{s}{\omega_x}\right)^2}, \quad (9\text{-}97)$$

where the quality factor, Q_c, is

$$Q_c = \sqrt{\frac{1-k_c}{1+k_c}}, \quad (9\text{-}98)$$

and the frequency parameter, ω_x, is given by

$$\omega_x = \frac{2}{\sqrt{LC_l(1-k_c)}} = \frac{2}{Q_c R_l C_l}, \qquad (9\text{-}99)$$

where Eqs. (9-96) and (9-98) have been exploited. Since the uncompensated bandwidth, B_{cs}, indigenous to $L = M = 0$ and $C_f = 0$ is approximately the inverse of the time constant, $R_l(C_o + C_l)$, observe that ω_x in Eq. (9-99) is also expressible as

$$\omega_x = \frac{2}{\sqrt{LC_l(1-k_c)}} = \frac{2}{Q_c R_l C_l} \approx \frac{2B_{cs}}{Q_c}\left(1 + \frac{C_o}{C_l}\right). \qquad (9\text{-}100)$$

A comparison of Eq. (9-97) with Eq. (9-66) suggests that if quality factor Q_c is set to the inverse of root two, the peaking circuit at hand delivers a second order, Butterworth, maximally flat magnitude frequency response whose radial 3-dB bandwidth is precisely ω_x. From Eq. (9-98), $Q_c = 1/\sqrt{2}$ is tantamount to the requirement, $k_c = 1/3$, which lies well within the coupling coefficient domain of monolithic inductor pairs. For this design constraint, note from Eq. (9-100) that the resultant bandwidth, B_c, of the compensated structure, optimized for maximally flat magnitude response, is

$$\omega_x = B_c \approx 2\sqrt{2}\left(1 + \frac{C_o}{C_l}\right)B_{cs}, \qquad (9\text{-}101)$$

which suggests an impressive bandwidth enhancement factor approaching three.

It is important to stress that ω_x is the 3-dB bandwidth of the network represented by the transfer function in Eq. (9-97) if and only if Q_c is chosen to achieve a maximally flat magnitude response. In general, the compensated bandwidth, B_c, can be determined by studying the network at hand for steady state sinusoidal signal excitation, wherein the applicable transfer relationship is

$$A_n(j\omega) = \frac{1}{1 + \dfrac{j\omega}{Q_c \omega_x} - \left(\dfrac{\omega}{\omega_x}\right)^2}. \qquad (9\text{-}102)$$

Accordingly, bandwidth B_c derives from

$$\left[1 - \left(\frac{B_c}{\omega_x}\right)^2\right]^2 + \left(\frac{B_c}{Q_c \omega_x}\right)^2 = 2, \qquad (9\text{-}103)$$

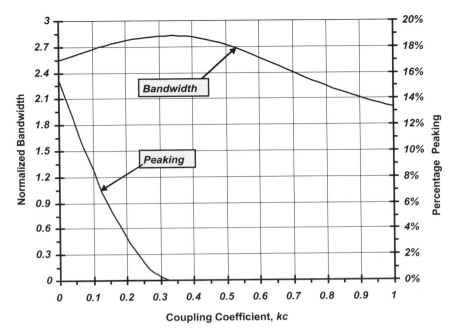

Figure 9.29. The normalized bandwidth and the percentage of magnitude response peaking for the broadbanded amplifier modeled in Fig. 9.27. The bandwidth curve is normalized to the uncompensated bandwidth, B_{cs}, of the amplifier and exploits the presumption of a transconductor output capacitance much smaller than the load capacitance driven by the amplifier.

whose solution, using Eq. (9-100), can be shown to be

$$\frac{B_c}{B_{cs}} = \left(1 + \frac{C_o}{C_l}\right)\frac{2}{Q_c}\sqrt{\sqrt{\left[\left(\frac{1}{2Q_c^2} - 1\right)^2 + 1\right]} - \left(\frac{1}{2Q_c^2} - 1\right)}.$$

(9-104)

Of course, this normalized bandwidth can be expressed explicitly in terms of the inductor coupling coefficient, k_c, by combining Eq. (9-98) with Eq. (9-104). To this end, Fig. 9.29 graphically depicts the dependence of the normalized bandwidth on the coupling coefficient for the case in which the transconductor output capacitance, C_o, is much smaller than the load capacitance, C_l. Observe that $k_c = 1/3$, which corresponds to $Q_c = 1/\sqrt{2}$, produces a bandwidth that, in addition to agreeing with Eq. (9-101), is the maximum achievable bandwidth of the subject network. Even more interesting, as well as appealing from an integrated circuit design and layout perspective, is the observation that the normalized bandwidth is not a particularly sensitive function of the coupling coefficient, whose precise numerical value is vulnerable to routinely encountered process vagaries and parametric uncertainties. For example, over the coefficient range, $0.1 \le k_c \le 0.5$, B_c/B_{cs} varies from only 2.67 to its peak of 2.82.

Although B_c/B_{cs} is not a sensitive function of k_c, a magnitude response peaking penalty is paid when $k_c \ne 1/3$. The frequency, say ω_p, at which the magnitude response is maximized can be deduced by equating the frequency derivative of the magnitude function in Eq. (9-102) to zero. The result is

$$\omega_p = \omega_x \sqrt{1 - \frac{1}{2Q_c^2}}. \tag{9-105}$$

Since ω_p is an imaginary number for $Q_c < 1/\sqrt{2}$, response peaking is manifested if and only if $Q_c > 1/\sqrt{2}$, or equivalently, for $k_c < 1/3$. Note further that for $Q_c = 1/\sqrt{2}$, $\omega_p = 0$, which suggests that the peak value of the transfer function occurs only at zero frequency, which is as it should be in a lowpass, maximally flat magnitude network. It is a simple matter to show that the actual magnitude peak, say M_p, corresponding to $\omega = \omega_p$ is

$$M_p = \frac{Q_c}{\sqrt{1 - \frac{1}{4Q_c^2}}}. \tag{9-106}$$

It is to be understood that in concert with the disclosures underpinning Eq. (9-105), Eq. (9-106) is valid only for $k_c \le 1/3$; otherwise, $M_p = 1$. Superimposed with the normalized bandwidth curve in Fig. 9.29 is a plot of the percentage by which the magnitude response peak, M_p, exceeds the zero frequency, unity gain value of the network transfer characteristic. This percentage peaking is somewhat sensitive to k_c but nonetheless, the maximum observable peaking, which occurs for the uncoupled inductor case evidenced by $k_c = 0$, is only about 15.5%.

Example 9.5. When driven by a 50 Ω signal source, the amplifier in Fig. 9.26 is to be designed to deliver a maximally flat response characterized by a zero frequency voltage gain of 16 dB and a 3-dB frequency of 10 GHz. The utilized transistor is the same as the device exploited in Example 9.3. Design the circuit by determining the appropriate values of the coupled inductances, the coupling coefficient, k_c, of the inductors, the drain load resistance, R_l, the requisite load capacitance, C_l, and the bridging capacitance, C_f, required at the amplifier output port. Using the small signal model in Fig. 9.27(a), simulate the small signal frequency response, V_{os}/V_s, of the finalized design on HSPICE. Compare the frequency response of the compensated amplifier with that of its uncompensated counterpart.

Solution 9.5.

(1) A voltage gain of 16 dB is equivalent to a numerical gain of 6.310. In a cascode arrangement of a common source amplifier, the effects of drain-source channel resistance are entirely negligible. Thus, the voltage gain magnitude is $g_m R_l$, whence a required drain load resistance of $R_l = 185.6\,\Omega$.

(2) Since the uncompensated common source amplifier bandwidth is $B_{cs} = 1/R_l(C_o + C_l)$, Eq. (9-101) yields

$$B_c = \frac{2\sqrt{2}}{R_l C_l}. \tag{E5-1}$$

With $B_c = 2\pi(10\,\text{GHz})$ and $R_l = 185.6\,\Omega$, the required load capacitance is seen to be $C_l = 242.6\,\text{fF}$. The net output capacitance, C_o, of the transconductor driver is the sum, $(C_{gd} + C_{bd})$, of gate-drain and bulk-drain capacitances of the cascode transistor, $M2$, in the circuit of Fig. 9.26. Since $C_o = C_{gd} + C_{bd} = 27\,\text{fF}$ is only about nine times smaller than the required value of load capacitance C_l, and since C_o is tacitly ignored in all analyses executed on the broadband configuration, engineering prudence dictates a reduction of the computed value of C_l by nominally five percent. Accordingly, and in the hope of mitigating the deleterious consequences of an output port capacitance, C_o, that is not entirely negligible in comparison to the load capacitance, select $C_l = 230\,\text{fF}$.

(3) The required coupling coefficient for a Butterworth, maximally flat magnitude response is $k_c = 1/3$. Using Eq. (9-96) with $R_l = 185.6\,\Omega$ and $C_l = 230\,\text{fF}$, the requisite half inductance value is $L/2 = 2.970\,\text{nH}$.

(4) From Eq. (9-94), the bridging capacitance with $C_l = 230\,\text{fF}$ and $k_c = 1/3$ is $C_f = 28.75\,\text{fF}$.

(5) Figure 9.30 depicts the small signal models for both the uncompensated and the compensated cascode amplifiers. The simulated magnitude responses of both of these structures appear in Fig. 9.31. Observe that both responses are flat over the frequency passband of interest.

Comments. An assiduous inspection of the HSPICE simulation results reveals a 3-dB bandwidth for the compensated amplifier of 10.02 GHz, which clearly satisfies the bandwidth design target. On the other hand, the uncompensated unit shows a bandwidth of 3.33 GHz, which, as previously conjectured, is less than the bandwidth of the compensated cell by a factor of nominally three-fold.

By any engineering measure, the bandwidth compensation strategy entailing a bridging capacitance in shunt with two coupled inductances comprises a satisfying broadbanding methodology. To be sure, a verification of the broadbanding scheme

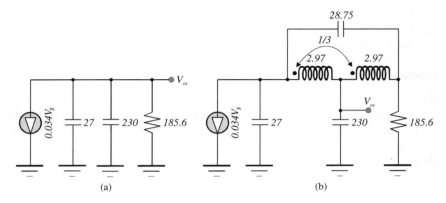

Figure 9.30. (a) Small signal model of the uncompensated amplifier addressed in Example 9.5. (b) Small signal model of the compensated amplifier designed for maximally flat magnitude response in Example 9.5. All capacitance values are in femtofarads, resistances are in units of ohms, and inductance values are in units of nanohenries.

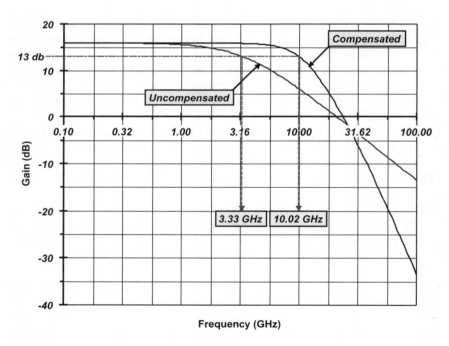

Figure 9.31. Frequency responses of the amplifiers addressed in Example 9.5. The "uncompensated" curve pertains to the small signal model in Fig. 9.30(a), while the "compensated" curve refers to the model of Fig. 9.30(b).

is provided by means of small signal analyses and simulations executed on only pertinent small signal models. This verification procedure is defendable from two points of view. First, the purpose of this textbook is not electronic circuit and system analysis and design. Instead, a principle purpose of this text is to demonstrate that the insightful engineering understanding deriving from an application of classic circuit theoretic tools and mathematically sound circuit analyses conduces innovative circuit design solutions. Second, it should be understood that the ability to glean acceptable responses in a relevant small signal model is a necessary condition for achieving satisfactory performance in the electronic amplifier ultimately configured for integrated circuit layout and processing.

9.7.0. Broadbanding via Feedback

The preceding sections of material underscore open loop broadbanding methods, wherein transmission zeros are introduced to mitigate the deleterious effects exerted on open loop bandwidth by dominant network poles. As alluded to in the introductory section of this chapter, negative feedback constitutes a viable alternative to open loop broadbanding, provided that problems surrounding potentially underdamped open loop responses are circumvented without incurring substantive penalties on achievable closed loop bandwidth. To this end, the dual loop, shunt-series feedback amplifier, whose basic schematic form is the structure in Fig. 9.32(a), is a commonly used broadband topology. In this amplifier, for which the approximate high frequency model is the diagram of Fig. 9.32(b), R_s represents the Thévenin signal source resistance, R_l is the load resistance driven by the circuit, while resistances R_i and R_o are incorporated for biasing purposes. The resistance, R_{ss}, establishes series feedback around the open loop, while resistance R_f realizes shunt feedback. The small signal equivalent circuit in Fig. 9.32(b) tacitly ignores the channel resistance and bulk-induced threshold modulation in the transistor. Moreover, C_{gs} and C_{gd} are the traditional gate-source and gate-drain capacitances of the MOSFET, while capacitance C_l accounts for the bulk-drain capacitance of the transistor, as well as for any load capacitance associated with the terminated output port.

Compared with the open loop topologies studied earlier, the advantages of the feedback amplifier in Fig. 9.32(a) include improved linearity and diminished sensitivity of forward gain with respect to active element parameters. Additionally, its dual loop nature provides controllable and predictable driving point input and output impedances, Z_{in} and Z_{out}, at least for signal frequencies lying within the amplifier passband. A clue as to the controllable nature of Z_{in} and Z_{out} derives from observing that if resistance

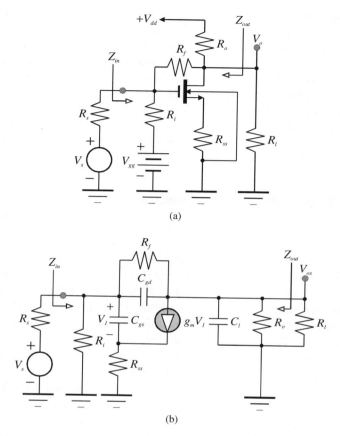

Figure 9.32. (a) Basic schematic diagram of a shunt-series feedback amplifier. The transistor is biased in saturation. (b) Approximate small signal, high frequency model of the amplifier in (a).

R_f is supplanted by an open circuit, the feedback resistance, R_{ss}, confers a transadmittance amplifier featuring extremely high driving point input and output impedances. On the other hand, a short circuited R_{ss} leaves feedback resistance R_f to forge a transimpedance amplifier characterized by very low input and output impedances. It is therefore reasonable to surmise the likelihood that the presence of both R_{ss} and R_f, which is tantamount to coalescing transadmittance operation with transimpedance signal processing, leads to finite and predictable Z_{in} and Z_{out}. The analyses that follow confirm the propriety of this expectation. Indeed, Z_{in} and Z_{out} are not only predictable, their low frequency constituents, R_{in} and R_{out}, respectively, can be equated if the source resistance, R_s, is matched to the load resistance, R_l. Moreover, it is possible to realize the subject dual loop amplifier in such a way that it delivers the so-called *match terminated* case of $R_s = R_{in} = R_{out} = R_l$. At least three engineering attributes accompany the match terminated circumstance. First, a cascade of stages of the form depicted in Fig. 9.32(a) can be implemented without the interstage loading effects that cause gain attenuation and even outright uncertainty in the cascaded value of gain. Second, maximum power transfer is engaged between the output port of an amplifier and the amplifier input port to which said output port is incident. Finally, if $R_s = R_l$ is a relatively small resistance (50 Ω or less), bandwidth degradation owing to parasitic I/O port capacitances is minimal. To the latter end, the achievable 3-dB bandwidth is more limited by active device characteristics and in particular, the unity gain frequency of the transistor, than it is by parasitic capacitances at either the source or load ports.

9.7.1. Low Frequency Characteristics

The low frequency version of the small signal model shown in Fig. 9.32(b) is given in Fig. 9.33(a). Using signal flow analyses, the low frequency voltage gain, A_{vo}, of the shunt-series feedback amplifier is expressible as

$$A_{vo} = \frac{V_{os}}{V_s} = A_{os}\left[\frac{1 + R_{To}(R_s, R_l)/R_f}{1 + R_T(R_s, R_l)/R_f}\right], \qquad (9\text{-}107)$$

where A_{os} is the null parameter voltage gain evaluated under the condition that the conductance, say G_f, associated with feedback resistance R_f is set to zero. The function, $R_T(R_s, R_l)$, is the normalized return ratio with respect to conductance G_f, while $R_{To}(R_s, R_l)$ is the normalized null return ratio with respect to G_f. On the other hand, the driving point input resistance, R_{in}, is expressible as

$$R_{in} = R_{ino}\left[\frac{1 + R_T(0, R_l)/R_f}{1 + R_T(\infty, R_l)/R_f}\right], \qquad (9\text{-}108)$$

while the driving point output resistance, R_{out}, is given by

$$R_{out} = R_{outo}\left[\frac{1 + R_T(R_s, 0)/R_f}{1 + R_T(R_s, \infty)/R_f}\right]. \qquad (9\text{-}109)$$

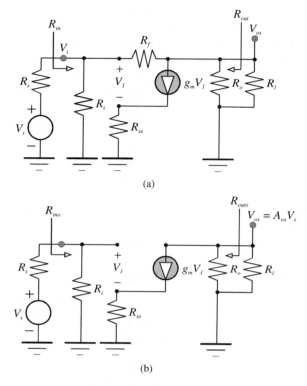

Figure 9.33. (a) Approximate low frequency model of the shunt-series feedback amplifier in Fig. 9.32(a). (b) Amplifier model appropriate for the evaluation of the low frequency values of the open loop voltage gain, A_{os}, the driving point input resistance, R_{ino}, and the driving point output resistance, R_{outo}, for the null feedback parameter case of $1/R_f = 0$.

Figure 9.33(b) is the model in Fig. 9.33(a), with resistance R_f open circuited; that is, the model at hand is pertinent to an open loop, with respect to the shunt-shunt feedback element, at low frequency operating conditions. An inspection of this model reveals

$$A_{os} = \left.\frac{V_{os}}{V_s}\right|_{R_f=\infty} = -g_{me}\left(\frac{R_i}{R_i + R_s}\right)(R_o \| R_l), \qquad (9\text{-}110)$$

where the effective forward transconductance, g_{me}, of the amplifier is

$$g_{me} = \frac{g_m}{1 + g_m R_{ss}}. \qquad (9\text{-}111)$$

A further inspection of the model at hand shows that the input resistance, R_{ino}, under the condition of $R_f = \infty$, is

$$R_{ino} = R_i, \qquad (9\text{-}112)$$

and the corresponding null feedback parameter output resistance, R_{outo}, is

$$R_{outo} = R_o. \qquad (9\text{-}113)$$

The equivalent circuit relative to the computation of the normalized return ratio, $R_T(R_s, R_l)$, appears in Fig. 9.34(a), while the normalized null return ratio, $R_{To}(R_s, R_l)$, derives from an analysis of the structure shown in Fig. 9.34(b). It is a straightforward task to show that in Fig. 9.34(a),

$$R_T(R_s, R_l) = \left.\frac{V_x}{I_x}\right|_{V_s=0} = (R_o \| R_l) + [1 + g_{me}(R_o \| R_l)](R_i \| R_s). \qquad (9\text{-}114)$$

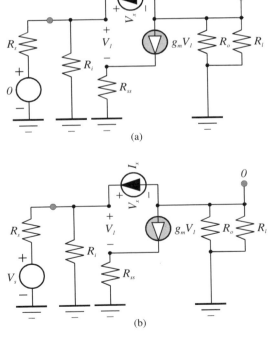

Figure 9.34. (a) Low frequency model of the shunt-series feedback amplifier used in the computation of the normalized return ratio, $R_T(R_s, R_l)$. (b) Low frequency model exploited to compute the null normalized return ratio, $R_{To}(R_s, R_l)$.

On the other hand, Fig. 9.34(b) confirms

$$R_{To}(R_s, R_l) = \left.\frac{V_x}{I_x}\right|_{V_{os}=0} = -\frac{1}{g_m}\left[1 + (1 + g_m R_i)\left(\frac{R_{ss}}{R_i}\right)\right]. \quad (9\text{-}115)$$

If Eqs. (9-114) and (9-112) are combined with Eq. (9-108), the driving point input resistance is found to be

$$R_{in} = R_i \left\| \left[\frac{R_f + (R_o \| R_l)}{1 + g_{me}(R_o \| R_l)}\right]. \quad (9\text{-}116)$$

For the output resistance, Eqs. (9-114), (9-113), and (9-109) coalesce to deliver

$$R_{out} = R_o \left\| \left[\frac{R_f + (R_i \| R_s)}{1 + g_{me}(R_i \| R_s)}\right]. \quad (9\text{-}117)$$

It is interesting to note that if $R_i = R_o$ and $R_s = R_l$, $R_{in} \equiv R_{out}$. If $R_s = R_l \triangleq R$, the match terminated constraint of $R_s = R_l = R_{in} = R_{out} \equiv R$ requires

$$R_f = [1 + g_{me}(R_i \| R)]\left(\frac{R_i R}{R_i - R}\right) - (R_i \| R). \quad (9\text{-}118)$$

Since resistance R must be small to support broadband design objectives, resistances R_i and R_o can be chosen to ensure that both R_i and R_o are significantly larger than R. Under this design guideline, Eq. (9-118) collapses to the requirement,

$$R_f \approx g_{me} R^2 = \frac{g_m R^2}{1 + g_m R_{ss}}. \quad (9\text{-}119)$$

A determination of the low frequency closed loop gain corresponding to the foregoing match terminated condition requires the substitution of Eqs. (9-110), (9-114), and (9-115) into Eq. (9-107). In view of Eq. (9-119) and its underlying approximations, this voltage gain is

$$A_{vo} = A_{os}\left[\frac{1 + R_{To}(R_s, R_l)/R_f}{1 + R_T(R_s, R_l)/R_f}\right]$$

$$\approx -\left(\frac{g_{me} R - 1}{2}\right) \approx -\left[\frac{(R_f/R) - 1}{2}\right]. \quad (9\text{-}120)$$

Observe that a gain magnitude exceeding unity requires $g_{me} R > 3$ or equivalently, $R_f > 3R$. Recalling Eq. (9-111), this constraint invariably mandates that the transistor in the circuit diagram of Fig. 9.32(a) have sufficiently large gate aspect ratio and/or that it be biased at sufficiently large drain current.

It is also interesting to note that since the driving point input resistance, R_in, predicted by the model in Fig. 9.33(a) is R, provided that Eq. (9-119) and its companion approximations are satisfied, the ratio, V_i/V_s, of input port signal voltage to Thévenin signal source voltage is precisely $1/2$. Since the low frequency closed loop voltage gain, A_{vo}, can be written as $A_{vo} = V_{os}/V_s = (V_i/V_s)(V_{os}/V_i)$, Eq. (9-120) confirms a port voltage gain, say A_{io}, of

$$A_{io} = \frac{V_{os}}{V_i} \approx -(g_{me}R - 1). \tag{9-121}$$

The importance of this port gain is underscored in a subsequent subsection addressing the driving point input impedance of the subject amplifier.

9.7.2. Amplifier Bandwidth

If the feedback amplifier in Fig. 9.32(a) postures a dominant pole frequency response, its small signal voltage transfer function, $A_v(s)$, can be approximated at high signal frequencies by the single pole expression,

$$A_v(s) \approx \frac{A_{vo}}{1 + \dfrac{s}{B}}, \tag{9-122}$$

where, assuming match terminated design conditions, A_{vo} is the low frequency voltage gain stipulated by Eq. (9-120), and B, the approximate radial 3-dB bandwidth of the amplifier, is precisely the sum of the open circuit time constants associated with the three individual capacitances embedded in the model of Fig. 9.32(b). Specifically, parameter B derives from an expression of the form,

$$\frac{1}{B} = R_{gs}C_{gs} + R_{gd}C_{gd} + R_{cl}C_l. \tag{9-123}$$

With the signal source voltage, V_s, set to zero, R_{gs} is the resistance that effectively faces capacitance C_{gs} when $C_{gd} = C_l = 0$, R_{gd} is the effective resistance seen by C_{gd} when $C_{gs} = C_l = 0$, and finally, R_{cl} is the resistance effectively in shunt with capacitance C_l, under the condition of $C_{gs} = C_{gd} = 0$. Assuming that resistances R_i and R_o in the equivalent circuit are very large and that the feedback resistance, R_f, is chosen in accordance with the match termination constraint in Eq. (9-119), the dividends of a bit of circuit analyses are

$$R_{gs} = \frac{1}{2}\left[\frac{R + R_{ss}}{1 + g_m R_{ss}} + \left(\frac{1}{1 + g_{me}R}\right)\left(\frac{R_{ss}}{1 + g_m R_{ss}}\right)\right], \quad (9\text{-}124)$$

$$R_{gd} = \left(\frac{2 + g_{me}R}{1 + g_{me}R}\right)\left(\frac{g_{me}R^2}{2}\right), \quad (9\text{-}125)$$

and

$$R_{cl} = \frac{R}{2}. \quad (9\text{-}126)$$

In the last expression, use is made of the fact that the load capacitance, C_l, shunts a terminating load resistance of R, as well as a low frequency driving point output resistance that is identical to R because of the design constraint imposed by Eq. (9-119). Assuming further that the design goal embraces a relatively large voltage gain magnitude at low frequencies, which mandates $g_{me}R \gg 1$, the combination of the preceding four expressions produces

$$B \approx \frac{2}{R(C_{gm} + C_{dm} + C_l) + R_{ss}C_{gm}}, \quad (9\text{-}127)$$

where

$$C_{gm} \triangleq \frac{C_{gs}}{1 + g_m R_{ss}} \quad (9\text{-}128)$$

and

$$C_{dm} \triangleq g_{me}RC_{gd}. \quad (9\text{-}129)$$

Observe that the effect of the transconductance feedback element, R_{ss}, is to reduce the effective value of transistor gate-source capacitance, C_{gs}. On the other hand, the transistor gate-drain capacitance, C_{gd}, is multiplied by the factor, $g_{me}R$, to which the transimpedance feedback element, R_f, is directly proportional and on which the low frequency voltage gain is linearly dependent. As is shown shortly, this factor is the Miller multiplier of capacitance C_{gd}; that is, C_{dm} is approximately the value of input port capacitance when capacitance C_{gd} is referred to the input port. In most practical situations, $C_l \ll (C_{gm} + C_{dm})$. Accordingly, Eq. (9-127) reduces to

$$B \approx \frac{2}{R(C_{gm} + C_{dm}) + R_{ss}C_{gm}}. \quad (9\text{-}130)$$

9.7.3. Input Impedance

An expression for the driving point input impedance, Z_{in}, of the shunt-series feedback amplifier can be derived by applying the trustworthy mathematical ohmmeter method to the model shown in Fig. 9.32(b). Although this derivation is a straightforward exercise, it produces results whose algebraic complexity masks crucial design-oriented implications. A more insightful, but admittedly approximate, analytical strategy examines the loading imposed on the applied signal source by the two capacitances, C_{gs} and C_{gd}, in Fig. 9.32(b). To this end, the considered model is redrawn as Fig. 9.35(a) for the design objective of match terminated operation. The input port resistance, R_i, and its counterpart output port resistance, R_o, are presumed sufficiently large to justify their tacit neglect. Additionally, the small, match terminated open circuit time constant associated with the net load capacitance, C_l, encourages ignoring this capacitance.

In terms of the input and output port signal voltages, V_i and V_{os}, respectively, the current, I_{gd}, conducted by capacitance C_{gd}, which must be supplied by the Thévenin input voltage, is

$$I_{gd} = sC_{gd}(V_i - V_{os}) = sC_{gd}V_i\left(1 - \frac{V_{os}}{V_i}\right). \tag{9-131}$$

But since capacitance C_l is ignored, the ratio, V_{os}/V_i, is precisely the low frequency port gain given by Eq. (9-121), whence

$$I_{gd} = sC_{gd}V_i\left(1 - \frac{V_{os}}{V_i}\right) = sC_{gd}V_i(1 - A_{io})$$

$$= sg_{me}RC_{gd}V_i = sC_{dm}V, \tag{9-132}$$

where Eq. (9-129) is recalled. Thus, the signal current supplied to C_{gd} by the signal voltage, V_s, is equivalent to the current conducted by an effective capacitance, C_{dm}, which is incident between the input port and ground. The alternative model depicted in Fig. 9.35(b) reflects this observation.

An analytical strategy similar to that just executed with respect to C_{gd} can be applied to capacitance C_{gs} in the model of Fig. 9.35(a). In particular, the input port voltage, V_i, is

$$V_i = V_1 + R_{ss}(g_m V_1 + sC_{gs}V_1),$$

which portends

$$V_1 = \frac{V_i}{1 + g_m R_{ss} + sR_{ss}C_{gs}}. \tag{9-133}$$

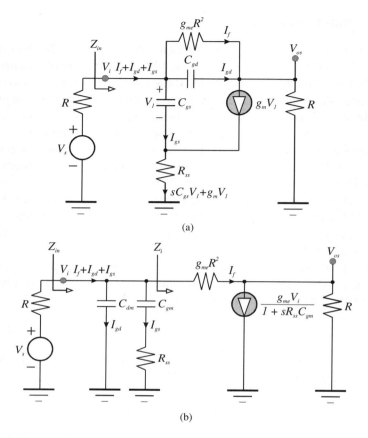

Figure 9.35. (a) Approximate high frequency model of the feedback amplifier in Fig. 9.32(a). (b) Approximated version of the model in (a). The model is valid only for the first order determination of forward transmission characteristics, including the driving point input impedance, Z_{in}.

Recalling Eq. (9-128), it follows that the signal current, I_{gs}, flowing through capacitance C_{gs} is

$$I_{gs} = sC_{gs}V_1 = \frac{sC_{gs}V_i}{1 + g_m R_{ss} + sR_{ss}C_{gs}} = \frac{V_i}{R_{ss} + \dfrac{1}{sC_{gm}}}. \qquad (9\text{-}134)$$

The last result suggests that the original current, I_{gs}, flowing through capacitance C_{gs} in the network of Fig. 9.35(a) is equivalent to the current flowing through a branch, incident with the amplifier input port, comprised of a series interconnection of resistance R_{ss} and the effective gate-source capacitance, C_{gm}. In addition to this capacitive current representation, it should be noted that Eq. (9-133) allows the controlled current, $g_m V_1$, to be rendered proportional to the input port voltage, V_i, in accordance with

$$g_m V_1 = \frac{g_m V_i}{1 + g_m R_{ss} + s R_{ss} C_{gs}} = \frac{g_{me} V_i}{1 + s R_{ss} C_{gm}}, \qquad (9\text{-}135)$$

where Eqs. (9-111) and (9-128) have been invoked. The electrical implications of Eqs. (9-134) and (9-135) are mirrored topologically by the alternative high frequency model in Fig. 9.35(b).

The model just derived clearly suggests that the input impedance, Z_{in}, of the shunt-series feedback amplifier is capacitive. An approximate expression for this impedance function can be straightforwardly deduced if the radial signal frequencies, ω, of interest are constrained to satisfy the inequality, $\omega R_{ss} C_{gm} \ll 1$. Under this circumstance, the impedance indicated as Z_i in Fig. 9.35(b) reduces to the low frequency value of Z_{in}, which has already been shown to be R for the match terminated case under investigation. Accordingly, Z_{in} is little more than the impedance of the parallel combination of R, capacitance C_{dm}, and the series interconnection of resistance R_{ss} and capacitance C_{gm}; that is,

$$Z_{in} \approx \frac{R\left(1 + s R_{ss} C_{gm}\right)}{1 + s\left[R_{ss} C_{gm} + R\left(C_{gm} + C_{dm}\right)\right] + s^2 R R_{ss} C_{gm} C_{dm}}. \qquad (9\text{-}136)$$

Note that for small R_{ss}, this result implies an effective input port capacitance, C_{in}, of

$$C_{in} \approx C_{gm} + C_{dm} = \frac{C_{gs}}{1 + g_m R_{ss}} + g_{me} R C_{gd}. \qquad (9\text{-}137)$$

9.7.4. Input Impedance Compensation

The input impedance degradation with frequency of the match terminated shunt-series feedback amplifier confirms the logical suspicion that impedance matching prevails at only low signal frequencies. At higher frequencies, the effect of a reduced input impedance magnitude is an amplifier input port voltage that is progressively smaller for a given Thévenin source voltage and is therefore in danger of falling below the detectable level determined by the equivalent input noise voltage, or *noise floor*, of the amplifier.[9] The remedy is a matching filter interposed, as suggested by Fig. 9.36(a), between the signal source and the input port of the amplifier. In an ideal world, this filter minimally satisfies two operational constraints. First, the driving point impedance, Z_a, seen looking into its input terminals is constant at a value equal to the match terminated Thévenin source

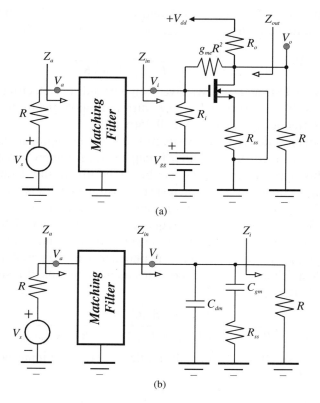

Figure 9.36. (a) The shunt-series feedback amplifier of Fig. 9.32(a), which incorporates an input port matching filter to compensate for the frequency dependence of the input impedance, Z_{in}. (b) The approximate input port model of the amplifier, wherein the results underlying the topology of Fig. 9.35(b) have been exploited.

resistance, R, for all signal frequencies of interest. Second, the filter should deliver a port voltage transfer function, V_i/V_a, of unity, or at least close to one, so that minimal loss of gain is incurred between the applied signal excitation and the amplifier input terminals. If both these performance objectives are met, maximum signal power transfer is assured between the signal source and the filter input port, which supports a signal level nominally mirroring that observed at the amplifier input port; that is, $V_i \approx V_a$. Using the results deduced in the preceding subsection, a viable input port model is the structure appearing in Fig. 9.36(b). The latter model is approximate in that $Z_i \approx R$ only if $\omega R_{ss} C_{gm} \ll 1$.

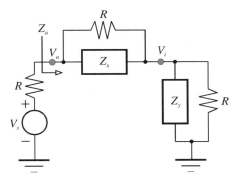

Figure 9.37. A plausible matching filter for the input port of the shunt-series amplifier of Fig. 9.36(a).

The technical literature is sprinkled with a number of filters that satisfy the foregoing proscribed requirements.[10] For the application at hand, a leading filter candidate is the architecture shown in Fig. 9.37, for which the indicated input impedance, Z_a, is identical to the resistance, R, provided that impedances Z_x and Z_y are the scaled duos implied by the constraint,

$$Z_x Z_y = R^2. \tag{9-138}$$

It is a simple matter to show that Eq. (9-138) stipulates a filter voltage transfer function, V_i/V_a, of

$$\frac{V_i}{V_a} = \frac{1}{1 + \dfrac{R}{Z_y}} \equiv \frac{1}{1 + \dfrac{Z_x}{R}}. \tag{9-139}$$

When adapted to the input port model of Fig. 9.36(b), the compensated configuration becomes the network offered in Fig. 9.38(a). The interesting feature of this adaptation is that impedance Z_y is not a designable element and is, in fact, the impedance associated with the parallel interconnection of capacitance C_{dm} and the series combination of resistance R_{ss} and capacitance C_{gm}. Specifically,

$$Z_y = \frac{\left(\dfrac{1}{sC_{dm}}\right)\left(R_{ss} + \dfrac{1}{sC_{gm}}\right)}{\dfrac{1}{sC_{dm}} + R_{ss} + \dfrac{1}{sC_{gm}}} = \frac{1 + sR_{ss}C_{gm}}{s(C_{gm} + C_{dm}) + s^2 R_{ss} C_{gm} C_{dm}}.$$

$$\tag{9-140}$$

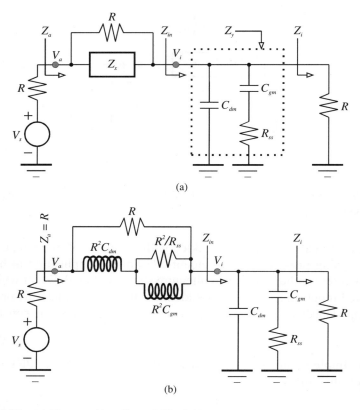

Figure 9.38. (a) The matching filter of Fig. 9.37 adapted for the input port of the shunt-series feedback amplifier. (b) The realization of impedance Z_x in the filter of (a).

It follows from Eq. (9-138) that the impedance, Z_x, required for matched compensation is

$$Z_x = \frac{R^2\left[s\left(C_{gm} + C_{dm}\right) + s^2 R_{ss} C_{gm} C_{dm}\right]}{1 + sR_{ss}C_{gm}}$$

$$= sR^2 C_{dm} + \frac{1}{\dfrac{R_{ss}}{R^2} + \dfrac{1}{sR^2 C_{gm}}}, \qquad (9\text{-}141)$$

which suggests the realization delineated in Fig. 9.38(b). Figure 9.39 shows the basic schematic diagram of the shunt-series feedback amplifier that results when the input port matching filter is incorporated. Also incorporated into the compensated structure is a suitably large blocking capacitor,

Broadband and Radio Frequency MOS Technology Amplifiers 801

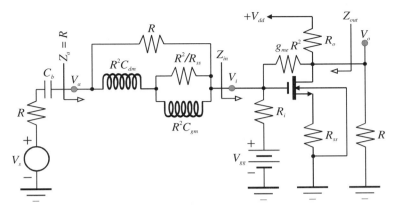

Figure 9.39. The shunt-series feedback amplifier of Fig. 9.32(a) with an input matching filter incorporated to compensate for the frequency dependence of the indicated amplifier input port impedance, Z_{in}.

C_b, to prevent static current flow through the applied signal source. Observe that the inductance, $R^2 C_{dm}$, can be implemented as a bond wire of suitable length and cross section area.

It should be noted in Fig. 9.38(a) that the input impedance, Z_{in}, seen looking into the gate of the transistor in the shunt-series feedback amplifier is little more than the parallel combination of the impedance, Z_y, and the resistance R, which is to say that

$$Z_y = \frac{R Z_{in}}{R - Z_{in}}. \tag{9-142}$$

If this relationship is substituted into Eq. (9-139), the voltage transfer function from the filter input port to the amplifier input port is simply

$$\frac{V_i}{V_a} = \frac{Z_{in}}{R}. \tag{9-143}$$

Since the impedance indicated as Z_a in Fig. 9.38(b) is R, $V_a/V_s = 1/2$, independent of signal frequency. With the load capacitance, C_l, tacitly ignored and with the forward characteristics of the amplifier at hand modeled by the approximate equivalent circuit shown in Fig. 9.35(b), the port gain, V_{os}/V_i, of the amplifier is given by Eq. (9-121), assuming that $\omega R_{ss} C_{gm} \ll 1$.

It follows that the overall I/O gain of the amplifier is

$$A_v(s) = \frac{V_{os}}{V_s} = \frac{V_{os}}{V_i} \times \frac{V_i}{V_a} \times \frac{V_a}{V_s} \approx -\left(\frac{g_{me}R-1}{2}\right)$$

$$\times \left\{\frac{1 + sR_{ss}C_{gm}}{1 + s\left[R_{ss}C_{gm} + R(C_{gm} + C_{dm})\right] + s^2 R R_{ss} C_{gm} C_{dm}}\right\}.$$

(9-144)

Equation (9-144) confirms that the incorporation of the matching filter at the amplifier input port does not degrade the zero frequency closed loop gain, A_{vo}, stipulated earlier as Eq. (9-120). It also confirms that because of the left half plane zero, the magnitude response rolls off with high signal frequencies at a rate of only 20 dB/decade, despite the inherently second order nature of the circuit used to model the active network. Recall that this frequency characteristic is desirable from the perspective of ensuring adequate gain and phase margins in the closed loop response. Unfortunately, however, Eq. (9-144) underscores a drawback to the use of the input port filter. In particular, Eqs. (9-120) and (9-130) allow Eq. (9-144) to be written in the form,

$$A_v(s) \approx \frac{A_{vo}(1 + sR_{ss}C_{gm})}{1 + \frac{2s}{B} + s^2 R R_{ss} C_{gm} C_{dm}}, \quad (9\text{-}145)$$

whose first order term in the indicated characteristic polynomial suggests, assuming that time constant $R_{ss}C_{gm}$ is small, an approximate radial bandwidth of $B/2$. Accordingly, a deleterious impact of the utilized matching filter is that the approximate bandwidth, B, achieved without the matching filter, is reduced by a factor of two.

Example 9.6. The shunt-series feedback amplifier depicted in Fig. 9.32 is to be designed to deliver a low frequency voltage gain magnitude of five (14 dB) when operated match terminated into 50 ohms. The transistor used has a short circuit unity gain frequency (f_T) of 35 GHz, a drain-source channel resistance (r_o) of 25 KΩ, and a bulk-induced threshold modulation factor (λ_b) of 0.025. For the given semiconductor process, the gate-drain capacitance is known to be roughly 15% of the gate-source capacitance. An estimate of the parasitic load capacitance (C_l) (inclusive of bulk-drain device capacitance) is 300 fF. Design the basic cell by calculating an appropriate value of the transconductance feedback resistance, R_{ss}, the required value of the transresistance feedback element, R_f, and the values of gate-source and gate-drain capacitances, C_{gs} and C_{gd}, respectively, appropriate to the geometry and biasing levels that produce the device transconductance,

g_m, required of the amplifier. Additionally, estimate the 3-dB bandwidth of the amplifier, divorced of any matching filter implemented at the input signal port. Submit two designs: one without an input impedance matching filter and one with an appropriate input filter embedded. Compare the small signal performance of the two designs through HSPICE simulations.

Solution 9.6.

(1) With a gain magnitude, $|A_{vo}|$, of 5 and for $R = 50\,\Omega$, Eq. (9-120) gives a required effective transconductance of $g_{me} = 220$ mmhos. Because of Eq. (9-111), the source lead resistance, R_{ss}, must be smaller than $1/g_{me}$, which is $4.545\,\Omega$ in this case. Select $R_{ss} = 2.5\,\Omega$, which requires, again by Eq. (9-111),

$$g_m = \frac{g_{me}}{1 - g_{me} R_{ss}} = 488.9 \text{ mmhos.} \quad \text{(E6-1)}$$

Since the analyses undertaken as a prelude to this example ignore drain-source channel resistance, bulk-induced threshold voltage modulation, and several other second order factors, it is prudent to increase the computed device transconductance value by 5% or so. In view of this strategy, let $g_m = 510$ mmhos.

(2) From Eq. (9-119) and using $g_m = 510$ mmhos, the requisite voltage feedback resistance is $R_f = 560.4\,\Omega$.

(3) Recall that for the purpose of this demonstration, $C_{gd} = 0.15\,C_{gs}$. Then, the unity gain radial frequency, $\omega_T = 2\pi f_T$, is stipulated by

$$\omega_T = \frac{g_m}{C_{gs} + C_{gd}} = \frac{g_m}{1.15 C_{gs}}. \quad \text{(E6-2)}$$

Given $g_m = 510$ mmhos and $\omega_T = 2\pi(35\,\text{GHz})$, C_{gs} must be $C_{gs} = 2.017\,\text{pF}$, which means that $C_{gd} = 0.15 C_{gs} = 302.5\,\text{fF}$.

(4) Using Eqs. (9-128) and (9-129), $C_{gm} = 886.4\,\text{fF}$ and $C_{dm} = 3.391\,\text{pF}$.

(5) With respect to the elements of the input matching filter drawn in Fig. 9.39, $R^2 C_{dm} = 8.319\,\text{nH}$, $R^2 C_{gm} = 2.216\,\text{nH}$, and $R^2/R_{ss} = 1\,\text{K}\Omega$.

(6) From Eq. (9-127), the estimated 3-dB bandwidth of the amplifier operated without the input port filter is $B = 2\pi(1.378\,\text{GHz})$.

(7) The foregoing design calculations precipitate Fig. 9.40, which shows the resultant small signal model schematic of the shunt-series amplifier. In order to facilitate future analyses and assessments, the output impedance, Z_{out}, is herewith redefined as the impedance effectively facing the shunt interconnection of the terminating load resistance, R, and the net load capacitance, C_l. With switch SW closed in this diagram, the amplifier operates without benefit of broadband input impedance matching. On the other hand, SW open activates the input port matching filter.

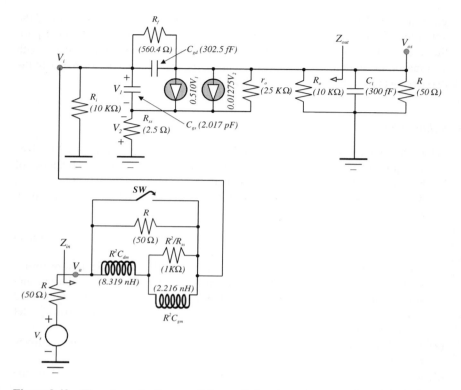

Figure 9.40. The schematic diagram of the small signal model of the shunt-series feedback amplifier designed in Example 9.6. With switch SW open, the amplifier operates with an input port impedance matching filter incorporated. A closed switch constrains the amplifier to operate without an input port filter.

Comments. Figure 9.41 displays the pertinent magnitude responses for the cases of input impedance matching and no matching filter implemented. Both responses show a low frequency gain of 14.04 dB, or 5.04 volts/volt, which is higher than the design goal of 5.0 volts/volt by only 0.80%. Obviously, the tack of increasing the originally computed device transconductance by 5% has proven minutely over zealous. For the case of no input port impedance matching, the simulated 3-dB bandwidth is 1.374 GHz, a scant 4 MHz lower than the design estimate of 1.378 GHz. This miniscule computational error confirms that the considered amplifier behaves as a dominant pole structure, despite its inherent third order nature and the presence of a finite frequency left half plane zero. With input port matching, the simulated bandwidth is 717.6 MHz, which is smaller than the non-matched simulated bandwidth by a factor of 1.92. In theory, the subject bandwidth ratio should be two. This slight disparity between simulated and computed bandwidth ratio can doubtlessly be attributed to the "Miller time" method invoked with respect to

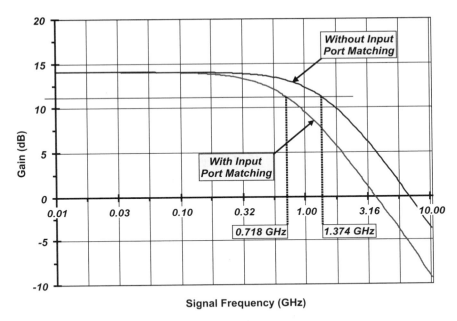

Figure 9.41. The frequency responses of the shunt-series feedback amplifier whose small signal equivalent circuit is diagrammed in Fig. 9.40. The curve labeled, "without input port matching," corresponds to switch *SW* closed in Fig. 9.40. The curve indicated "with input port matching" pertains to the case of switch *SW* open circuited in the aforementioned modeling diagram.

representing the transistor gate-drain capacitance as an effective input port capacitance. Also, recall that several approximations (neglect of channel resistance, neglect of bulk-induced threshold voltage modulation, etc.) were exploited in the course of all theoretic analyses.

The magnitude of the driving point input impedances, both with and without input port matching, are exhibited in Fig. 9.42. In both cases, the simulated low frequency values of the input impedance are 50.63 Ω, which differs from the design objective by only 1.26%. Without impedance matching, the input impedance magnitude plummets to roughly 25 Ω at the 3-dB bandwidth of the amplifier. A detailed analysis of the simulated data for the nonmatched case reveals a capacitive input impedance over all frequencies of interest. This capacitance characteristic is depicted in Fig. 9.43. Observe a low frequency input capacitance of about 3.68 pF, which is slightly larger than capacitance C_{dm}, but smaller than the capacitance sum, $(C_{dm} + C_{gm})$. This observation synergizes with the input port model postulated in Fig. 9.38. With input port matching, the input impedance magnitude remains at its low frequency value to within about 5 Ω through the entire range of considered

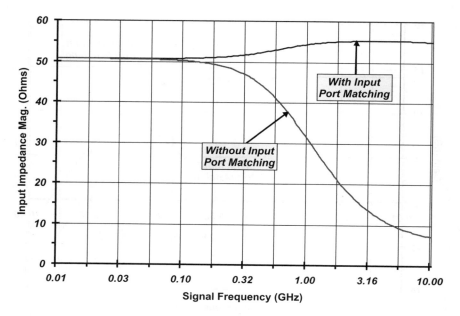

Figure 9.42. The magnitude of the input impedance for the shunt-series feedback amplifier modeled in Fig. 9.40. The curve labeled, "without input port matching," corresponds to switch *SW* closed in the subject model. The curve indicated "with input port matching" pertains to the case of switch *SW* open circuited.

signal frequencies. In contrast to the nonmatched circumstance, the matched input impedance is slightly inductive through, and substantially beyond, the amplifier 3-dB bandwidth. In view of the clearly inductive impedance that appears in series with the input port in Fig. 9.38, this state of affairs is hardly surprising.

9.7.5. Output Impedance

The output impedance, Z_{out}, seen by the terminating load resistance, R, and the net load capacitance, C_l, in the shunt-series feedback amplifier under consideration is simply the ratio, V_x/I_x, in the model of Fig. 9.44(a). If this impedance, whose low frequency value has already been established to be the matching resistance, R, is approximated as a first order frequency function,

$$Z_{out} \approx R\left(\frac{1+as}{1+bs}\right), \qquad (9\text{-}146)$$

where parameter "a" is the sum of the open circuit time constants associated with the impedance zeros, while parameter "b" is the sum of the open circuit

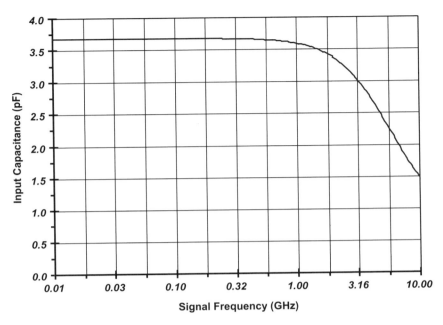

Figure 9.43. The input port capacitance of the shunt series modeled by the circuit structure in Fig. 9.40. For this capacitance simulation, switch SW in the aforementioned diagram remains closed; that is, no input port matching filter is utilized.

time constants associated with the poles of the output impedance.[11] With reference to the equivalent circuit in Fig. 9.44(b), parameters "a" and "b" are given precisely by

$$a = C_{gs}\left(\frac{V_1}{I_1}\bigg|_{V_x=0}\right) + C_{gd}\left(\frac{V_2}{I_2}\bigg|_{V_x=0}\right),$$
$$b = C_{gs}\left(\frac{V_1}{I_1}\bigg|_{I_x=0}\right) + C_{gd}\left(\frac{V_2}{I_2}\bigg|_{I_x=0}\right)$$
(9-147)

where each of the parenthesized voltage to current ratios is recognized as an effective resistance facing the corresponding multiplicative capacitance under the condition of either a short circuited ($V_x = 0$) or an open circuited ($I_x = 0$) output port. Assuming large R_i and R_o, the execution of analyses commanded by Eq. (9-147) results in

$$a = \left(\frac{g_{me}R^2}{1+g_{me}R}\right)C_{gm} + \left(\frac{R}{1+g_{me}R}\right)C_{dm}$$
(9-148)

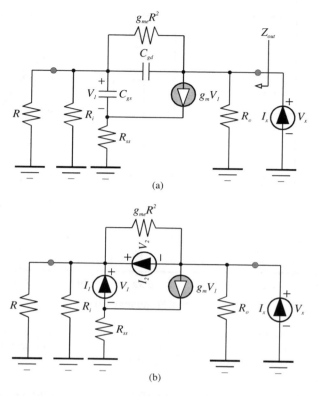

Figure 9.44. (a) Model used to evaluate the driving point output impedance, Z_{out}, of the shunt-series feedback amplifier. (b) Model exploited to calculate the time constants of the pole and zero associated with a first order frequency domain representation of the output impedance.

and

$$b = \left(\frac{R + R_{ss}}{1 + g_{me}R}\right) C_{gm} + RC_{dm}, \qquad (9\text{-}149)$$

where the effective transconductance and capacitance values, g_{me}, C_{gm}, and C_{dm}, are given respectively by Eqs. (9-111), (9-128), and (9-129).

A tacit inspection of Eq. (9-146) reveals that the driving point output impedance is inductive for $a > b$, purely resistive for $a = b$, and capacitive for $a < b$. From Eqs. (9-148) and (9-149), the inductive condition of $a > b$ requires

$$\frac{C_{gm} - C_{dm}}{C_{gm}} = \frac{C_{gs} - g_{me}R(1 + g_m R_{ss})C_{gd}}{C_{gs}} > \frac{R + R_{ss}}{g_{me}R^2}, \qquad (9\text{-}150)$$

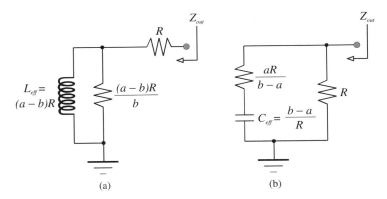

Figure 9.45. (a) Passive representation of an inductive driving point output impedance for the shunt-series feedback amplifier. (b) Passive representation of a capacitive driving point output impedance for the shunt-series feedback amplifier.

for which a necessary condition of satisfaction is

$$\frac{C_{gs}}{C_{gd}} > g_{me} R(1 + g_m R_{ss}). \tag{9-151}$$

Note that an inductive output impedance in the shunt-series feedback amplifier is likely to derive from the sufficiently small gate-drain capacitance, C_{gd}, achieved with self-aligned gate processing. Obviously, a capacitive output impedance follows from the dissatisfaction of Eq. (9-150).

For the inductive case of $a > b$, a meaningful expansion of Eq. (9-146) is

$$Z_{out} \approx R\left(\frac{1+as}{1+bs}\right) = R + \cfrac{1}{\cfrac{1}{R}\left(\cfrac{b}{a-b}\right) + \cfrac{1}{R(a-b)s}}, \tag{9-152}$$

whose passive realization is given in Fig. 9.45(a). Observe an effective inductance, L_{eff}, of

$$L_{eff} = R(a-b) = \left(\frac{g_{me}R}{1+g_{me}R}\right)R^2$$

$$\times \left[(C_{gm} - C_{dm}) - \left(1 + \frac{R_{ss}}{R}\right)\left(\frac{C_{gm}}{g_{me}R}\right)\right]. \tag{9-153}$$

For large $g_{me}R$, which is tantamount to stipulating a relatively large voltage gain at low signal frequencies,

$$L_{eff} \approx R^2(C_{gm} - C_{dm}). \tag{9-154}$$

In the capacitive case of $a < b$, a continued fraction expansion of Eq. (9-146) is

$$Z_{out} \approx R\left(\frac{1+as}{1+bs}\right) = \frac{1}{\dfrac{1}{R} + \dfrac{1}{aR + \dfrac{R}{b-a} + \dfrac{1}{(b-a)s}}}, \qquad (9\text{-}155)$$

whose passive representation appears in Fig. 9.45(b). The effective capacitance, C_{eff}, associated with this passive model is

$$C_{eff} = \frac{b-a}{R} = \left(\frac{g_{me}R}{1+g_{me}R}\right)$$

$$\times \left[(C_{dm} - C_{gm}) + \left(1 + \frac{R_{ss}}{R}\right)\left(\frac{C_{gm}}{g_{me}R}\right)\right]. \qquad (9\text{-}156)$$

For the large gain circumstance implied by large $g_{me}R$,

$$C_{eff} \approx C_{dm} - C_{gm}. \qquad (9\text{-}157)$$

Example 9.7. In the shunt-series feedback amplifier depicted in Fig. 9.32 and analyzed in Example 9.6, examine the resultant driving point output impedance. In particular, determine whether this output impedance is inductive or capacitive for the case of no broadband impedance matching implemented at the amplifier input port. Using the small signal model shown in Fig. 9.40, simulate the output impedance with HSPICE, and compare these simulated results with simulations executed on the appropriate macromodel offered in Fig. 9.45. For ease of reference, the pertinent results gleaned in the preceding example are listed herewith.

$$\begin{aligned}
g_m &= 510\,\text{mmhos} & g_{me} &= 224.18\,\text{mmhos} \\
C_{gs} &= 2.017\,\text{pF} & C_{gd} &= 302.5\,\text{fF} \\
C_{gm} &= 886.4\,\text{fF} & C_{dm} &= 3.391\,\text{pF} \\
R_f &= 560.4\,\text{ohms} & R_{ss} &= 2.5\,\text{ohms} \\
\lambda_b &= 0.025 & r_o &= 25\,\text{Kohms} \\
R_i &= 10\,\text{Kohms} & R_o &= 10\,\text{Kohms}
\end{aligned}$$

Solution 9.7.

(1) An assessment of the nature of the driving point output impedance requires only that parameters "a" and "b" in Eqs. (9-148) and (9-149) be compared. To this end, $a = 54.577$ pSEC and $b = 173.34$ pSEC. Since $a < b$, the output impedance is capacitive.

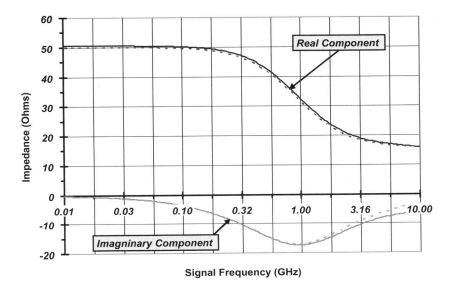

Figure 9.46. The simulated real part and imaginary part components of the driving point output impedance of the shunt-series feedback amplifier addressed in Examples 9.6 and 9.7. The solid curves correspond to the physically sound, small signal model of the amplifier shown in Fig. 9.40. The dashed curves reflect simulations executed on the impedance macromodel of Fig. 9.45(b).

(2) Since the output impedance is capacitive, the pertinent output port representation is the schematic diagram of Fig. 9.45(b). In this diagram,

$$\left. \begin{array}{l} R = 50\,\Omega \\ \dfrac{aR}{b-a} = 22.977\,\Omega \\ C_{\mathit{eff}} = \dfrac{b-a}{R} = 2.375\,\text{pF} \end{array} \right\}.$$

(3) The HSPICE simulations of the real and imaginary parts of the output impedance are shown in Fig. 9.46 for both the amplifier model of Fig. 9.40 and the impedance macromodel of Fig. 9.45(b).

Comments. The simulated results shown in Fig. 9.46 confirm the capacitive nature of the driving point output impedance for the shunt-series feedback amplifier in that the imaginary component of the subject impedance is negative for all signal frequencies of interest. The excellent agreement between the simulated results gleaned from the actual small signal equivalent circuit and the impedance macromodel is cause for celebration. Specifically, this agreement verifies the propriety of the dominant pole/dominant zero strategy adopted to assess the output impedance characteristics over frequency.

9.8.0. The f_T-Doubler

In most open loop amplifiers, such as the series peaked and series-shunt peaked configurations studied earlier, the achievable 3-dB bandwidth is limited by the effective load capacitance incident at the amplifier output port. On the other hand, the bandwidth of many closed loop amplifiers, for which a notable example is the dual loop feedback network examined in the preceding section, is invariably determined by the intrinsic capacitances of the utilized transistors. In these latter topologies, an f_T-doubler, which is essentially a Darlington configuration designed to deliver nominally twice the unity gain frequency afforded by the single transistor it replaces, can be gainfully exploited for broadbanding purposes. The principle underlying the operation of these doublers is elementary. In particular, the short circuit, unity gain frequency, ω_T, of a MOSFET characterized by a forward transconductance of g_m, a net gate-source capacitance of C_{gs}, and a net gate-drain capacitance of C_{gd} is

$$\omega_T = 2\pi f_T = \frac{g_m}{C_{gs} + C_{gd}}. \qquad (9\text{-}158)$$

If a circuit is contrived to double f_T while maintaining constant g_m, which is determined by geometry and biasing considerations, capacitances C_{gs} and C_{gd} are necessarily halved. To the extent that circuit bandwidth is linearly related to the gate-source and gate-drain capacitances of the transistor supplanted by the f_T-doubler, the observable 3-dB bandwidth is consequently doubled.

Figure 9.47 shows the basic schematic diagram of the f_T-doubler. A clue as to the alleged effectiveness of the circuit derives from observing that the small signal, high frequency impedance seen looking into the gate terminal of transistor $M1$ is rendered much larger than the gate to source impedance of $M1$ alone by the small signal series interconnection of the $M1$ source terminal with the $M2$ gate terminal. Since the high frequency gate-source impedance is dictated by gate-source capacitance and Miller-multiplied gate-drain capacitance, the effective values of these device capacitances are necessarily reduced by the aforementioned series topology. Maximal effectiveness of the circuit requires that transistors $M1$ and $M2$ be identical active devices having identical gate aspect ratios. When these transistors are biased in their saturation regimes at the same current, I_K, their indicated interconnection behaves as an effective single MOSFET, Me, whose unity gain

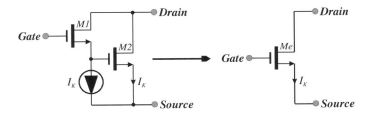

Figure 9.47. Basic schematic diagram of an f_T-doubler. Transistors $M1$ and $M2$ are identical devices biased in saturation at identical currents. The substrate terminals of both transistors are connected to the most negative potential afforded by the circuit in which the doubler is immersed. The doubler on the left behaves as an effective transistor, Me, whose unity gain frequency is nominally twice as large as the unity gain frequency of either transistors $M1$ or $M2$.

frequency, ω_e, is roughly twice ω_T, the unity gain frequency of $M1$ and $M2$, individually. In actual practice, ω_e is slightly less than $2\omega_T$, but approaches ω_T if the gate-drain capacitance of $M1$ (or of $M2$) is significantly smaller than the gate-source capacitance, C_{gs}, of each of these two devices.

9.8.1. Small Signal Analysis

A determination of the effective unity gain frequency, ω_e, of the structure given in Fig. 9.47 demands that its short circuit current gain, $A_{sc}(s) = I_{sc}/I_s$, be investigated as a function of signal frequency. To this end, the pertinent schematic diagram, divorced of biasing complexities, is provided in Fig. 9.48(a), for which the corresponding small signal model is the network in Fig. 9.48(b). The model in question takes the liberties of ignoring channel length modulation, bulk-induced threshold voltage modulation, transistor bulk-drain capacitances, and transistor bulk-source capacitances.

A straightforward analysis of the model in Fig. 9.48(b) results in the somewhat formidable transfer function,

$$A_{sc}(s) = \frac{I_{sc}}{I_s} = \left(\frac{\omega_T}{s}\right)^2 \left\{ \frac{1 + \dfrac{s}{\omega_T}\left(\dfrac{2C_{gs} - C_{gd}}{C_{gs} + C_{gd}}\right) - \left(\dfrac{s}{\omega_T}\right)^2 \left[\dfrac{C_{gd}\left(3C_{gs} + C_{gd}\right)}{(C_{gs} + C_{gd})^2}\right]}{1 + \dfrac{C_{gd}C_{gs}}{(C_{gs} + C_{gd})^2} + \left(\dfrac{\omega_T}{s}\right)\left(\dfrac{C_{gd}}{C_{gs} + C_{gd}}\right)} \right\},$$

(9-159)

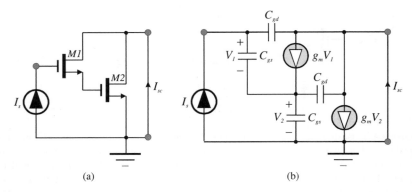

Figure 9.48. (a) The f_T-doubler circuit of Fig. 9.47 configured for an evaluation of the small signal, short circuit current gain, I_{sc}/I_s. (b) The approximate equivalent circuit for the network in (a).

where ω_T is given by Eq. (9-158) and is understood to be the unity gain frequency of either transistor $M1$ or $M2$ in the schematic diagram of Fig. 9.47. Assuming $C_{gd} \ll C_{gs}$ and given that interest herewith focuses on frequencies beyond ω_T, Eq. (9-159) can be approximated as

$$A_{sc}(s) \approx \left(\frac{\omega_T}{s}\right)^2 + \frac{2\omega_T}{s}, \tag{9-160}$$

which for sinusoidal excitation in the steady state becomes

$$A_{sc}(j\omega) \approx -\left(\frac{\omega_T}{\omega}\right)^2 - j2\left(\frac{\omega_T}{\omega}\right). \tag{9-161}$$

The resultant unity gain frequency, ω_e, is such that

$$|A_{sc}(j\omega_e)| = \sqrt{\left(\frac{\omega_T}{\omega_e}\right)^4 + 4\left(\frac{\omega_T}{\omega_e}\right)^2} \equiv 1, \tag{9-162}$$

for which the pertinent solution is

$$\frac{\omega_e}{\omega_T} = 2.058; \tag{9-163}$$

that is, the unity gain frequency of the doubler circuit is nominally twice that of the individual transistors embedded in the doubler. The exact solution for the doubler unity gain frequency, derived from a steady state sinusoidal analysis of Eq. (9-159), reveals that the frequency ratio, ω_e/ω_T, is a non-monotonic, but not especially sensitive, function of the capacitance ratio,

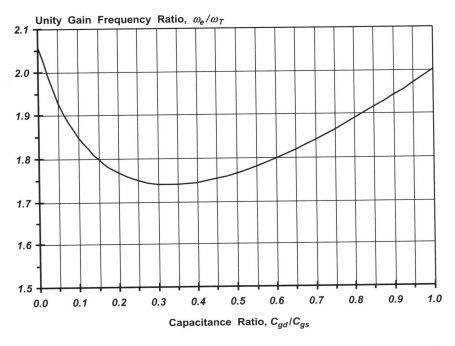

Figure 9.49. The unity gain frequency of the f_T-doubler in Fig. 9.47, normalized to the unity gain frequency of transistor $M1$ (or $M2$), viewed as a function of the gate-drain to gate-source capacitance ratio, C_{gd}/C_{gs}. The results reflect computer-aided analyses of the short circuit current transfer function of a doubler circuit that utilizes transistors deriving from a 28 GHz monolithic process.

C_{gd}/C_{gs}. Figure 9.49 is a graphical depiction of this functional dependence for the case of a doubler circuit realized with transistors having individual unity gain frequencies of 28 GHz.

9.8.2. Realization of the f_T-Doubler

A critical requirement underlying the optimal operation of the f_T-doubler in Fig. 9.47 is that transistors $M1$ and $M2$ conduct the same quiescent currents. This constraint stems primarily from the need to establish matched transconductances in the two utilized transistors. The traditional solution to the problem that no mechanism in the subject conceptual diagram assures requisite current mirroring is the modified doubler network depicted in Fig. 9.50.[12] Note therein that any quiescent current, I_K, conducted by transistor $M1$, and hence by transistor $M3$ as well, is mirrored by transistor

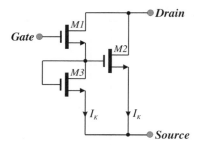

Figure 9.50. Doubler realization used to ensure that transistors $M1$ and $M2$ conduct nominally identical quiescent currents.

$M2$ if $M2$ and $M1$ are identical devices having matched gate aspect ratios. A drawback of the proposed compensation scheme is that the gate-source capacitance of transistor $M3$ is placed directly in shunt with the gate-source capacitance of $M2$, thereby incurring an effective doubling of the effective gate-source capacitance of the latter transistor. This shortcoming results in an f_T-doubler that actually delivers a unity gain frequency that is nominally only *1.5-times* larger than the unity gain frequency of either $M1$ or $M2$. A minor shortfall is the resistive load, whose approximate value is $1/g_m$, imposed on the gate-source terminals of $M2$ by the diode interconnection of transistor $M3$. In comparison to the native doubler of Fig. 9.47, this additional load serves to diminish, albeit slightly, the available short circuit current gain of the modified configuration.

9.9.0. Bandpass Feedback Amplifier

In contrast to the broadband architectures considered to this juncture, *bandpass, or narrowband, amplifiers* deliver gain only over a restricted signal passband that is geometrically apportioned about a *tuned, or center, frequency*. These bandpass structures are commonplace in a diversity of analog and digital communication systems. The receiver in these systems generally embodies an input, or *front-end*, linear amplifier to process audio, video, or digitally encoded information that is transmitted in the form of a modulated, fixed frequency carrier signal. This front-end amplifier, which is also typically referenced as a *radio frequency (RF) amplifier*, is invariably a bandpass structure satisfying several operational requirements.

In order to understand the salient aspects of the requisite properties of an RF amplifier, consider the abstraction posted in Fig. 9.51(a). The indicated

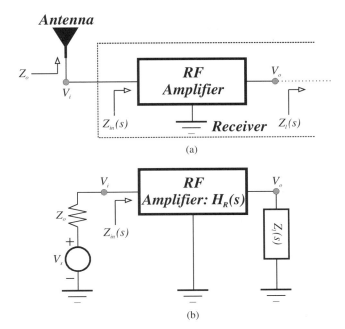

Figure 9.51. (a) Abstraction of the RF amplifier embedded in the front-end of a communications receiver. (b) Simplified electrical model of the system in (a).

RF amplifier is driven by an antenna that captures the modulated carrier signal. In narrowband communication systems, the frequency spectrum of this modulated carrier can be taken as embracing the frequency range, $(\omega_c \pm \omega_m/2)$, where ω_c is the radial value of the fixed carrier frequency, and ω_m is the radial value of the information passband. Since the antenna is a nominally linear entity, it can be represented as the Thévenin equivalent signal source depicted in Fig. 9.51(b). In the latter representation, V_s is the open circuit antenna voltage, and Z_o is the generally resistive characteristic impedance of the antenna, combined with the coupling medium that connects said antenna to the input port of the RF amplifier. The RF amplifier responds to the antenna excitation by delivering an output voltage, V_o, which can be related to V_s in the sinusoidal steady state through the transfer function,

$$H_R(j\omega) = \frac{V_o}{V_s} = \frac{H_o}{1 + jQ_o\left(\dfrac{\omega}{\omega_o} - \dfrac{\omega_o}{\omega}\right)}. \qquad (9\text{-}164)$$

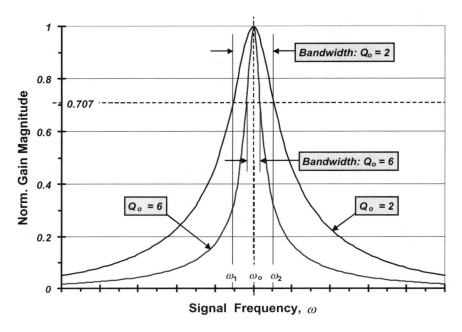

Figure 9.52. The generalized frequency response of an RF bandpass amplifier. The first order transfer function is given by Eq. (9-164). The 3-dB frequencies, ω_1 and ω_2, are delineated for only the case of $Q_o = 2$.

In this expression, which embodies the net load impedance, Z_l, that terminates the amplifier output port, ω_o is the center frequency to which the RF amplifier is tuned, H_o, a constant, symbolizes the voltage gain observed at the tuned center frequency, and Q_o is the quality factor of the amplifier. Observe, as is highlighted by the normalized frequency response plot in Fig. 9.52, that for signal frequencies, ω, that are either far below or substantially above the center frequency, ω_o, the gain magnitude is considerably smaller than H_o. Moreover, the rate of gain attenuation increases dramatically with increases in the circuit quality factor, Q_o. These observations suggest that the amplifier center frequency, ω_o, should be tuned, or adjusted, to the transmitted carrier frequency, ω_c, to reduce the likelihood of amplifying potentially interfering signals transmitted by other communication channels whose carrier frequencies lie in the neighborhood of ω_c. Additionally, designs achieving relatively high Q_o are motivated, since progressively higher quality factors afford the RF amplifier the capability of effectively rejecting undesirable carrier frequencies that happen to lie in the immediate neighborhood of ω_c. These contentions merely reflect the obvious desire that a radio tuned to a particular station or channel is expected to receive and process only the information emanating from the tuned station or channel, and not any information or data indigenous to another channel whose carrier frequency is proximate to that of the tuned station.

Equation (9-164) and its companion plots of Fig. 9.52 warrant further exploration with respect to the achievable bandwidth. Note the obvious fact that for any value of quality factor, the gain magnitude attenuates to 3-dB below the center frequency gain at precisely two signal frequencies. For the specific case of $Q_o = 2$ in Fig. 9.52, these frequencies are delineated as ω_1 and ω_2. At either of these two frequencies, the magnitude of the j-term coefficient in the denominator on the right hand side of Eq. (9-164) must be one. Since $\omega_1 < \omega_o$, it follows that

$$Q_o \left(\frac{\omega_1}{\omega_o} - \frac{\omega_o}{\omega_1} \right) = -1, \qquad (9\text{-}165)$$

while, since $\omega_2 > \omega_o$,

$$Q_o \left(\frac{\omega_2}{\omega_o} - \frac{\omega_o}{\omega_2} \right) = 1. \qquad (9\text{-}166)$$

The respective solutions for the 3-dB frequencies in these two expressions are

$$\left. \begin{array}{l} \omega_1 = -\dfrac{\omega_o}{2Q_o} + \dfrac{\omega_o}{2Q_o}\sqrt{1+4Q_o^2} \\ \omega_2 = +\dfrac{\omega_o}{2Q_o} + \dfrac{\omega_o}{2Q_o}\sqrt{1+4Q_o^2} \end{array} \right\}, \qquad (9\text{-}167)$$

where it is understood that $|H_R(j\omega_1)| = |H_R(j\omega_2)| \equiv H_o/\sqrt{2}$. The 3-dB bandwidth, say, B_o, of the bandpass architecture is simply the difference between the foregoing 3-dB frequencies, whence

$$B_o = \omega_2 - \omega_1 = \frac{\omega_o}{Q_o}. \qquad (9\text{-}168)$$

Since this bandwidth, which should be made at least as large as the information passband, ω_m, is inversely related to the circuit quality factor, Q_o, a progressively larger quality factor gives rise to an RF amplifier more sharply tuned to the center frequency, and thus an amplifier fully capable of discriminating between desired and most unwanted carriers. The signal discrimination attribute of large Q_o is also advantageous from an electrical noise perspective. In particular, the total output noise precipitated by shot, flicker, and thermal processes generated within the amplifier is proportional to the amplifier bandwidth. It follows that design care should be exercised insofar as designing for an unnecessarily large value of 3-dB bandwidth.

A second noteworthy point is the observation from Eq. (9-167) that

$$\omega_o \equiv \sqrt{\omega_1 \omega_2}; \tag{9-169}$$

that is, the amplifier center frequency is the geometric mean of its two 3-dB frequencies. But for $2Q_o \gg 1$, which mathematically defines the *narrowband* case, Eq. (9-167) reduces, with the help of Eq. (9-168), to

$$\left. \begin{array}{l} \omega_1 \approx \omega_o - \dfrac{B_o}{2} \\ \omega_2 \approx \omega_o + \dfrac{B_o}{2} \end{array} \right\}, \tag{9-170}$$

thereby inferring that for high Q_o, the amplifier center frequency approximates the arithmetic mean of its two 3-dB frequencies.

It is also important to appreciate the problems surrounding the fact that the Thévenin antenna voltage, V_s, is small in virtually all communication systems. Accordingly, the RF amplifier must satisfy three other fundamental design objectives, in addition to the basic requirement of sufficiently high gain at the tuned center frequency. The first of these is the desirability of ensuring that the amplifier input impedance, $Z_{in}(s)$, be matched to the antenna characteristic impedance, Z_o. This matching assures maximal signal power transfer between the antenna and the amplifier input port, thereby precluding unnecessary losses of unavoidably anemic signal power levels. But even with prevailing maximum power transfer conditions, there is a danger that the signal voltage associated with the actual power level observed at the amplifier input port may be so small as to be masked by the equivalent input noise voltage of the amplifier. To ensure sufficiently low equivalent input noise, design care must therefore be exercised with respect to the choice of amplifier topology, biasing of the active elements implicit to the selected topology, and the prudent use of inherently noisy passive resistances in the critical signal flow paths of the amplifier. Finally, the age of portable electronics demands that the RF amplifier be capable of operating optimally with low standby power requirements. Unfortunately, low power design objectives often run counter to the requirements of impedance matching, low noise, and high center frequency gain.[13]

9.9.1. Common Source RF Amplifier

Although the literature is rife with viable RF amplifiers, the topology currently favored in the state of the art is the low power, low voltage configuration depicted schematically in Fig. 9.53(a). In this structure, transadmittance feedback in the form of the source lead inductance, L_{ss}, is exploited as an ideally lossless vehicle for achieving the desired matching between

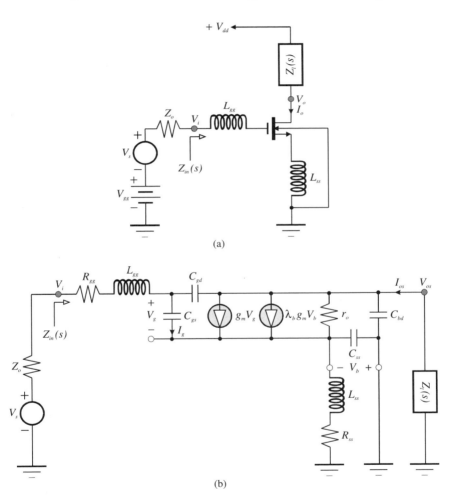

Figure 9.53. (a) The basic schematic diagram of a low voltage, low power RF amplifier realized in MOS technology. The transistor is biased in saturation. (b) High frequency, small signal equivalent circuit of the amplifier in (a).

the driving point input impedance, $Z_{in}(s)$, and the Thévenin signal source resistance, Z_o, at the tuned center frequency, ω_o, of the amplifier. Specifically, L_{ss} can be chosen to ensure $Z_{in}(j\omega_o) = Z_o$, where frequency ω_o is determined by the resonant interaction of the net gate lead inductance and the net effective input capacitance of the amplifier stage. Because only inductive impedances are exploited in both the source and the gate leads of the transistor, low power, low voltage biasing is facilitated, since within the ivy-cloaked halls of academe, inductors are lossless circuit elements. As is demonstrated shortly, the indicated load impedance, $Z_l(s)$, like the source lead impedance, must be dominantly inductive over the RF amplifier passband.

Recalling the modeling disclosures in *Chapter 7*, the pertinent small signal model of the RF amplifier is the network in Fig. 9.53(b). In this model, R_{ss} symbolizes the resistance implicit to the integrated circuit metallization spiral that forges inductance L_{ss}, while capacitance C_{ss} accounts for transistor bulk-source capacitance, the capacitance precipitated between the spiral metallization and inductor interconnect wiring crossing under the individual spiral segments, and the capacitance evidenced between signal ground and the top layer metallization winding.[14] Since the gate lead inductance, L_{gg}, can be synthesized with bond wire, no effort is expended to model the parasitic capacitance associated with this element. Although the quality factor of bond wire inductances is typically large, a resistance, R_{gg}, is nonetheless included in series connection with L_{gg}. The remaining modeling elements in the figure are those that are traditionally considered in high frequency representations of the small signal behavior of MOS technology devices.

Before attempting a definitive analysis of the subject RF amplifier, the model in Fig. 9.53(b) can be sanitized in concert with a few reasonable approximations and basic circuit theoretic principles. To wit, bulk-induced threshold voltage modulation can be ignored, assuming that at the amplifier center frequency, the controlled current, $\lambda_b g_m V_b$ is much smaller than the primary controlled current, $g_m V_g$. Accordingly, λ_b, which is a measure of said threshold phenomena, is set to zero. Second, the channel resistance, r_o, is presumed large enough to justify its neglect. In truth, r_o may be only a few thousand ohms in deep submicron MOS technology devices, but it is nevertheless likely that r_o substantially exceeds the magnitude, $|Z_l(j\omega_o)|$, of the load impedance at the amplifier center frequency. Third, the bulk-drain capacitance, C_{bd}, of the transistor can be conveniently absorbed into the terminating load impedance, which may include capacitive phenomena associated with inductances embedded in the drain circuit. Fourth, the transistor gate-drain capacitance, C_{gd}, is presumed small, but not necessarily negligible, particularly with respect to the delineation of the amplifier driving point input impedance. The presumption of small gate-drain capacitance requires the use of self-aligned gate technology processing and/or the incorporation of a common gate cascode between the load impedance and the drain of the degenerated common source amplifier. Finally, because considerable interest focuses on the evaluation of the driving point input impedance, which entails that the effective impedance in the transistor source terminal be referred to its gate lead, it is convenient to transform the voltage controlled current source, $g_m V_g$, into an equivalent current controlled element. To the latter end, Fig. 9.53(b) projects

$$g_m V_g = g_m \left(\frac{I_g}{sC_{gs}} \right) = \left(\frac{k_g \omega_T}{s} \right) I_g, \qquad (9\text{-}171)$$

where Eq. (9-158) is used, I_g is the indicated signal current conducted by the transistor gate-source capacitance, C_{gs}, and

$$k_g \triangleq 1 + \frac{C_{gd}}{C_{gs}}. \qquad (9\text{-}172)$$

The resultantly simplified model is shown in Fig. 9.54.

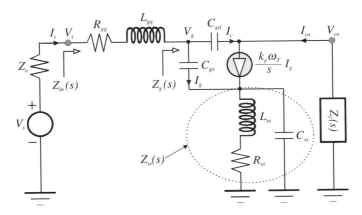

Figure 9.54. Simplified small signal, high frequency model of the RF amplifier shown in Fig. 9.53(a).

9.9.2. Impedance and Transfer Characteristics

It is a simple matter to show that the source circuit impedance indicated as $Z_{ss}(s)$ in Fig. 9.54 is given analytically by

$$Z_{ss}(s) = \frac{R_{ss}\left(1 + \dfrac{Q_s s}{\omega_s}\right)}{1 + \dfrac{s}{Q_s \omega_s} + \left(\dfrac{s}{\omega_s}\right)^2}, \qquad (9\text{-}173)$$

where

$$\omega_s = \frac{1}{\sqrt{L_{ss} C_{ss}}} \qquad (9\text{-}174)$$

is the undamped natural, or self-resonant, frequency of the inductance inserted into the source lead of the transistor. Moreover,

$$Q_s = \frac{\omega_s L_{ss}}{R_{ss}} = \frac{1}{R_{ss}} \sqrt{\frac{L_{ss}}{C_{ss}}} \qquad (9\text{-}175)$$

defines the quality factor of the source lead inductance at the aforementioned self-resonant frequency. It is important to underscore that ω_s is *not* the tuned center frequency, ω_o, of the amplifier undergoing investigation. Indeed, ω_s must be significantly larger than ω_o if the amplifier is to function properly in the sense of an *I/O* performance that is nominally insensitive to the net capacitance incident between signal ground and the source lead of the transistor. Ideally, $\omega_s = \infty$, which for finite inductance L_{ss}, demands the physically unachievable circumstance of $C_{ss} = 0$. In lieu of $\omega_s = \infty$, large ω_s is manifested by assiduously laying out the on chip spiral inductor to minimize its parasitic capacitances.[15,16] Additionally, a minimal geometry transistor can be used to avoid unduly large bulk-source capacitance and/or consideration can be given to connecting together bulk and source terminals. As usual, care must be exercised to preclude substrate current conduction when the bulk is connected to the source, particularly since in the circuit at hand, the inductive impedance in the transistor source lead can be large at high signal frequencies. In the limit of very large ω_s, Eq. (9-175) reduces Eq. (9-173) to the expected result,

$$Z_{ss}(s) \approx R_{ss} + s L_{ss}. \qquad (9\text{-}176)$$

The driving point input impedance, $Z_{in}(s)$, of the RF amplifier modeled in Fig. 9.54 is obviously

$$Z_{in}(s) = \frac{V_i}{I_i} = R_{gg} + s L_{gg} + Z_g(s), \qquad (9\text{-}177)$$

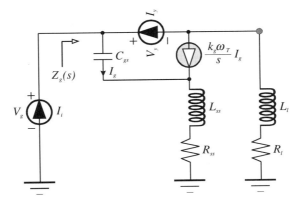

Figure 9.55. Small signal equivalent circuit for applying signal flow analysis techniques to the evaluation of the impedance, $Z_g(s)$, in Fig. 9.54.

where, in terms of the current, I_g, and the voltage, V_g, delineated in the diagram, the impedance, $Z_g(s)$, seen looking directly into the gate terminal of the transistor utilized in the amplifier is

$$Z_g(s) = \frac{V_g}{I_i}. \tag{9-178}$$

9.9.2.1. Gate Impedance

The gate terminal impedance identified by Eq. (9-178) is best evaluated with the signal flow analytical techniques developed in a preceding chapter. Unfortunately, the formulation of a useful expression for this impedance in terms of any general load termination, $Z_l(s)$, is a challenging, if not impossible, task. The determination of the subject gate impedance begins by assuming for the moment that $Z_l(s)$ is the simple inductive impedance,

$$Z_l(s) = R_l + sL_l, \tag{9-179}$$

where R_l is the presumably small resistance associated with the coil that forges inductance L_l. In advance of a discussion focused on the logistics of this particular selection of load impedance termination, Fig. 9.55 delineates the small signal model pertinent to a signal flow evaluation of the gate impedance of interest. In particular, $Z_g(s)$ can be expressed in the form,

$$Z_g(s) = \frac{V_g}{I_i} = Z_{go}(s) \left(\frac{1 + sC_{gd}Z_{qr}}{1 + sC_{gd}Z_{qs}} \right), \tag{9-180}$$

where $Z_{go}(s)$ is the input impedance when C_{gd} is tacitly ignored or equivalently, when $I_y = 0$; that is,

$$Z_{go}(s) = \left. \frac{V_g}{I_i} \right|_{I_y=0}. \quad (9\text{-}181)$$

By inspection of the model at hand, $Z_{go}(0)$ is noted to be infinitely large, thereby constraining the validity of Eq. (9-180) to only nonzero signal frequencies. This disclaimer hardly comprises a practical restriction since attention to the RF amplifier undergoing study is directed to signal frequencies lying in the neighborhood of the generally high tuned center frequency of the circuit. The parameter, Z_{qs}, in Eq. (9-180) is the normalized return ratio with respect to the capacitive admittance, sC_{gd}. It is literally the impedance facing capacitor C_{gd} under the condition of an open circuited gate port, so that

$$Z_{qs} = \left. \frac{V_y}{I_y} \right|_{I_i=0}. \quad (9\text{-}181)$$

On the other hand, the impedance function, Z_{qr}, is the normalized null return ratio with respect to admittance sC_{gd}. It is evaluated under the condition of a short circuited gate port in accordance with

$$Z_{qr} = \left. \frac{V_y}{I_y} \right|_{V_g=0}. \quad (9\text{-}182)$$

An inspection of the model in Fig. 9.55 reveals that when current I_y is set to zero,

$$V_g = \left(\frac{1}{sC_{gs}}\right) I_i + (R_{ss} + sL_{ss})\left(1 + \frac{k_g \omega_T}{s}\right) I_i. \quad (9\text{-}183)$$

It follows that $Z_{go}(s)$ is the impedance of a series RLC circuit in that

$$Z_{go}(s) = \left. \frac{V_g}{I_i} \right|_{I_y=0} = R_g + sL_{ss} + \frac{1}{sC_g}, \quad (9\text{-}184)$$

where

$$R_g = R_{ss} + k_g \omega_T L_{ss} \approx k_g \omega_T L_{ss} \quad (9\text{-}185)$$

and, recalling Eqs. (9-158) and (9-172),

$$C_g = \frac{C_{gs}}{1 + k_g \omega_T R_{ss} C_{gs}} = \frac{C_{gs}}{1 + g_m R_{ss}} \approx C_{gs}. \quad (9\text{-}186)$$

The approximations invoked in the preceding two relationships reflect the fact that the resistance, R_{ss}, associated with the commonly small source degeneration inductance, L_{ss}, is typically very small.

Observe that if current I_y is zero, which is closely approximated if C_{gd} is a negligibly small device capacitance, $Z_g(s)$ in Eq. (9-180) is precisely given by the right hand side of Eq. (9-184). It follows that when this V_g/I_i ratio is inserted into the input impedance expression of Eq. (9-177), an impedance match to a purely resistive source termination can be accomplished by choosing inductance L_{ss} correctly. The resultant series combination of this inductance and the gate lead inductance, L_{gg}, can then be made to resonate with capacitance C_g at the tuned center frequency of the amplifier. The significance of these observations is that the input resistance commensurate with maximum power transfer at the RF amplifier input port is realized at the amplifier center frequency in a nominally lossless fashion through the inductance, L_{ss}. In contrast, a desired input resistance realized straightforwardly as a resistive shunt across the amplifier input port incurs potentially unacceptable power dissipation.

Yet another advantage accrues from tuning the input port of the RF amplifier. In particular, only those signal currents whose frequencies lie in the immediate neighborhood of the tuned center frequency are predominantly conducted to the gate terminal of the transistor. On the other hand, currents whose frequencies lie appreciably outside the amplifier passband are substantially attenuated, as noted earlier in the general discussion of tuned circuits. It follows that any electrical noise detected as interfering energy at the antenna site is largely precluded from contaminating the signal processing capabilities of the transistor, and hence the amplifier response, provided that the power spectral density of the detected noise lies outside the amplifier passband. In short, the total output noise of an amplifier is nominally proportional to the circuit bandwidth. Accordingly, narrowbanding the input port shields the transistor from contaminating noise energy and serves to reduce the total output noise of the RF amplifier.

Continuing with the analysis, it can be demonstrated that

$$Z_{qs} = \left.\frac{V_y}{I_y}\right|_{I_i=0} = Z_{go}(s) + R_l + k_g \omega_T L_l + s L_l + \frac{k_g \omega_T R_l}{s}, \quad (9\text{-}187)$$

and

$$Z_{qr} = \left.\frac{V_y}{I_y}\right|_{V_g=0} = R_l + s L_l. \quad (9\text{-}188)$$

The substitution of Eqs. (9-187) and (9-188) into Eq. (9-180) produces

$$Z_g(s) = Z_{go}(s) \left(\frac{1 + sR_lC_{gd} + s^2L_lC_{gd}}{\begin{array}{c}1 + k_g\omega_T L_lC_{gd} + sZ_{go}(s)C_{gd} \\ + s(R_l + k_g\omega_T L_l)C_{gd} + s^2L_lC_{gd}\end{array}} \right). \quad (9\text{-}189)$$

It is more convenient to deal with the admittance, $Y_g(s)$, corresponding to impedance $Z_g(s)$, whereupon Eq. (9-189) is seen to imply

$$Y_g(s) = Y_{go}(s)$$

$$\times \left\{ 1 + \frac{k_g\omega_T R_lC_{gd} + s[k_g\omega_T L_lC_{gd} + Z_{go}(s)C_{gd}]}{1 + sR_lC_{gd} + s^2L_lC_{gd}} \right\}, \quad (9\text{-}190)$$

where, of course,

$$Y_{go}(s) = \frac{1}{Z_{go}(s)} = \frac{sC_g}{1 + sR_gC_g + s^2L_{ss}C_g}. \quad (9\text{-}191)$$

The algebraic complexity of Eq. (9-190) encourages a loudly tolling bell for suitable approximations. To this end, recall that for the ideal circumstance of $C_{gd} = 0$, the impedance seen looking directly into the gate terminal is simply $Z_{go}(s)$ or equivalently, the gate terminal admittance is $Y_{go}(s)$. While null C_{gd} is a dubious presumption for high frequency signal processing, small C_{gd} is a prudent analytical tack, particularly if the utilized transistor derives from a self-aligned gate process. If, in addition to small C_{gd}, R_l assumes its expected small value, the magnitudes of the critical frequencies associated with the terms in sR_lC_{gd} and $s^2L_lC_{gd}$ in the denominator of the second term on the right hand side of Eq. (9-190) can be presumed to be far larger than the tuned center frequency, ω_o, of the amplifier. Accordingly, Eq. (9-190) reduces to

$$Y_g(s) \approx Y_{go}(s) + sC_{gd} + \left[\frac{sk_g\omega_T R_lC_{gd}C_g + s^2k_g\omega_T L_lC_{gd}C_g}{1 + sR_gC_g + s^2L_{ss}C_g} \right]. \quad (9\text{-}192)$$

The last expression suggests that the admittance seen looking into the gate terminal of the transistor in the RF amplifier under consideration is comprised of a shunt interconnection of three distinct electrical branches. Unfortunately, the bracketed term on the far right hand side of Eq. (9-192) defies a positive real circuit interpretation. Indeed, the presence of the s^2-term in the numerator of this bracketed quantity even gives rise to a possible negative branch element resistance in the extreme (and impractical) situation of large gate-drain capacitance and/or large drain circuit inductance. In an attempt to facilitate a pragmatic design strategy, the term in $s^2 L_{ss} C_g$ can be neglected for most RF amplifier designs. Moreover, the frequency associated with the time constant, $R_g C_g$, is invariably much larger than the tuned center frequency of the amplifier, if said center frequency is smaller than the unity gain frequency of the utilized transistor by a factor of at least 20. Using Eqs. (9-158), (9-172), and (9-186), the foregoing observations allow Eq. (9-192) to be collapsed further to the first order result,

$$Y_g(s) \approx Y_{go}(s) + sC_m, \qquad (9\text{-}193)$$

where "Miller time" is apparent through the defining capacitance,

$$C_m \triangleq \left(1 + \frac{g_m R_l}{1 + g_m R_{ss}}\right) C_{gd}. \qquad (9\text{-}194)$$

The Miller multiplier in Eq. (9-196) differs from the traditionally encountered factor, $(1 + g_m R_l)$, because at low signal frequencies, the forward transconductance, g_m, of the transistor is degenerated by an amount of $(1 + g_m R_{ss})$ due to the source terminal resistance, R_{ss}. At this juncture, Eqs. (9-184) and (9-193) combine to forge the high frequency modeling approximation advanced in Fig. 9.56(a).

Two additional modeling tasks remain. The first involves rendering the model in Fig. 9.56(a) more analytically convenient by couching a dependence of the indicated current controlled current source on the net input current, I_i, as opposed to a direct dependence on the gate-source capacitance current, I_g. To this end, a simple current divider gives rise to

$$\frac{I_g}{I_i} = \frac{sC_g}{sC_g + sC_m\left(1 + sR_g C_g + s^2 L_{ss} C_g\right)} \approx \frac{C_g}{C_g + C_m}, \qquad (9\text{-}195)$$

where the approximation reflects the recurring presumption that the critical frequencies, $1/R_g C_g$ and $1/\sqrt{L_{ss} C_g}$, are significantly larger than the tuned center frequency of the amplifier. The second of the aforementioned two modeling tasks is the reduction of the input port circuit in Fig. 9.56(a) to the simple series impedance interconnection promoted by Fig. 9.56(b). To this end, Eq. (9-193), the critical frequency assumptions invoked to arrive at Eq. (9-195), and Fig. 9.56(a) combine to deliver an effective series input

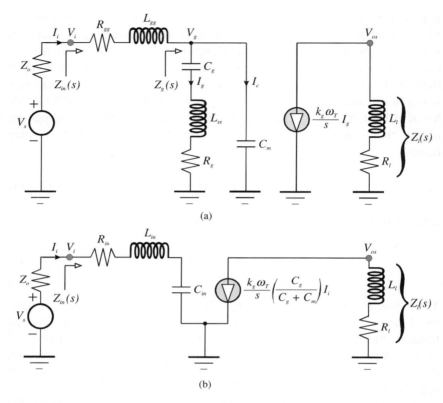

Figure 9.56. (a) The approximate high frequency model of the RF amplifier considered in Fig. 9.53(a). (b) Alternative form of the equivalent circuit in (a).

resistance, R_{in}, of

$$R_{in} = R_{gg} + \left(\frac{C_g}{C_g + C_m}\right) R_g \approx R_{gg} + \left\{\frac{C_{gs}}{C_{gs} + [1 + g_m(R_{ss} + R_l)]C_{gd}}\right\}$$
$$\times (R_{ss} + k_g \omega_T L_{ss}), \tag{9-196}$$

an effective series input port inductance, L_{in}, of

$$L_{in} = L_{gg} + \left(\frac{C_g}{C_g + C_m}\right) L_{ss}$$

$$\approx L_{gg} + \left\{\frac{C_{gs}}{C_{gs} + [1 + g_m(R_{ss} + R_l)]C_{gd}}\right\} L_{ss}, \tag{9-197}$$

and an effective series input port capacitance, C_{in}, of

$$C_{in} = C_g + C_m \approx \frac{C_{gs}}{1 + g_m R_{ss}} + \left(1 + \frac{g_m R_l}{1 + g_m R_{ss}}\right) C_{gd}. \quad (9\text{-}198)$$

An inspection of Eq. (9-196) reveals that impedance matching to the Thévenin source resistance, Z_o, shown in Fig. 9.56(b) can be effected at the tuned center frequency, ω_o, if inductance L_{in} resonates with capacitance C_{in} at radial frequency ω_o and if L_{ss} is selected to deliver $R_{in} = Z_o$. Thus,

$$\omega_o = \frac{1}{\sqrt{L_{in} C_{in}}}, \quad (9\text{-}199)$$

and

$$R_{in} = Z_o = R_{gg} + \left\{\frac{C_{gs}}{C_{gs} + [1 + g_m(R_{ss} + R_l)]C_{gd}}\right\} R_g, \quad (9\text{-}200)$$

where R_g remains given by Eq. (9-185). Equations (9-199) and (9-200) now allow the driving point input impedance to be formulated as

$$Z_{in}(j\omega) \approx Z_o + j\omega_o L_{in}\left(\frac{\omega}{\omega_o} - \frac{\omega_o}{\omega}\right), \quad (9\text{-}201)$$

which analytically substantiates $Z_{in}(j\omega_o) \approx Z_o$.

Before proceeding further, it is wise to interject a word of engineering caution as regards the propriety of Eqs. (9-195) through (9-198) and (9-200). In particular, all of these expressions are predicated on crude, first order approximations invoked on Eq. (9-192). The errors potentially resulting from these approximations are best explained by returning to the basic impedance relationship of Eq. (9-180). In this relationship, it should be understood that capacitance C_{gd} is inherently very small in any MOSFET deemed viable for incorporation into a high performance RF amplifier. In the neighborhood of the tuned center frequency, it follows that the magnitudes of the numerator and denominator within the parenthesized term on the right hand side of Eq. (9-180) are respectively very nearly unity. In addition to near unity magnitudes, the phasor angles of these numerator and denominator quantities are invariably very small near the amplifier center frequency. Thus, the real part of impedance $Z_g(j\omega_o)$, to which the real part of the amplifier driving point input impedance is intimately related, is dependent on the cosine of a difference of two very small angles, which begets potentially large computational errors unless the individual numerator and denominator angles are accurately determined. This inherent problem of analytical accuracy is exacerbated by the fact that in general,

$Z_{go}(j\omega_o)$ is strongly capacitive, which portends an impedance angle in the range $-80°$ to $-90°$. Since the real part of the input impedance is linearly related to the cosine of an angle that equals the sum of the angle of $Z_{go}(j\omega_o)$ and the aforementioned difference angle between the numerator and denominator terms in Eq. (9-180), a potentially large computational error can be rightfully anticipated. For example, there is a difference of almost 25% between the cosine of $(-82°)$ and the cosine of $(-84°)$. It is entirely inappropriate to presume tacitly that the approximations invoked herewith imply an impedance angle accuracy of $\pm 10°$, yet alone a $\pm 2°$ precision.

The foregoing disclaimer does not infer that the equations noted above have zero engineering value. Indeed, the relationships are worthy of the paper that espouses them, for they offer the designer a means of computing first order values for relevant circuit elements. Armed with these first order parameter estimations, a design optimization can be executed on MATLAB, EXCEL, or suitable other software to discern final values for all designable circuit elements. Implicit to this design optimization tack is the fact that $Z_g(j\omega)$ in Eq. (9-180), and hence $Z_{in}(j\omega)$ in Eq. (9-177), can be programmed as a function of the radial signal frequency, ω, and all resistive, capacitive, and inductive elements of the amplifier model. A methodically manual, or even semi-automated, adjustment in the inductances, L_{ss}, L_{gg}, and L_l, can then be pursued to arrive at the optimal design solution.

9.9.2.2. Voltage Transfer Function

The analyses and associated approximations that deliver the simplified high frequency equivalent circuit offered in Fig. 9.56(b) produce an input port signal current, I_i, of

$$I_i(j\omega) = \frac{V_s(j\omega)}{Z_o + Z_{in}(j\omega)} = \frac{V_s(j\omega)}{Z_o + Z_o + j\omega L_{in} + \dfrac{1}{j\omega C_{in}}}, \quad (9\text{-}202)$$

or

$$I_i(j\omega) = \frac{V_s(j\omega)/2Z_o}{1 + jQ_o\left(\dfrac{\omega}{\omega_o} - \dfrac{\omega_o}{\omega}\right)}, \quad (9\text{-}203)$$

where in terms of center frequency ω_o and 3-dB bandwidth B_o,

$$Q_o = \frac{\omega_o}{B_o} = \frac{\omega_o L_{in}}{2Z_o} = \frac{1}{2Z_o}\sqrt{\frac{L_{in}}{C_{in}}}. \quad (9\text{-}204)$$

The voltage transfer function, $H_R(j\omega)$, of the RF amplifier now follows directly from Fig. 9.56(b) and Eq. (9-203). In particular,

$$H_R(j\omega) = \frac{V_{os}(j\omega)}{V_s(j\omega)} = \frac{V_{os}(j\omega)}{I_i(j\omega)} \times \frac{I_i(j\omega)}{V_s(j\omega)}$$

$$= -\frac{\left(\dfrac{k_g \omega_T}{j\omega}\right)\left(\dfrac{C_g}{C_g + C_m}\right)\left[\dfrac{Z_l(j\omega)}{2Z_o}\right]}{1 + jQ_o\left(\dfrac{\omega}{\omega_o} - \dfrac{\omega_o}{\omega}\right)}. \quad (9\text{-}205)$$

This result is similar to the desired transfer function form of Eq. (9-164), with the notable exception that the numerator on the right hand side of Eq. (9-205) is not frequency invariant, as is the constant signified by parameter H_o in Eq. (9-164). The problem at hand is circumvented if the terminating load is the impedance, $Z_l(j\omega) = j\omega L_l$, of an ideal inductance, L_l. In such a circumstance, Eq. (9-205) emulates Eq. (9-164) in the sense that H_o is rendered frequency invariant. Specifically,

$$H_R(j\omega_o) \triangleq H_o = -k_g \left(\frac{C_g}{C_g + C_m}\right)\left(\frac{\omega_T L_l}{2Z_o}\right)$$

$$= -k_g \left(\frac{C_g}{C_g + C_m}\right) Q_o \left(\frac{\omega_T}{\omega_o}\right)\left(\frac{L_l}{L_{in}}\right), \quad (9\text{-}206)$$

where Eq. (9-204) is exploited.

Unfortunately, the realization of the drain circuit load impedance, $Z_l(s)$, as an ideal inductance, L_l, comprises engineering fantasy. Any attempt to incorporate an inductor into this drain circuit unavoidably incurs parasitic series resistance and shunt capacitance and results in an inductive subcircuit that is topologically identical to the subcircuit associated with the source terminal inductance, L_{ss}. This contention is highlighted in Fig. 9.57, where L_l is understood to be the desired drain circuit inductance, R_l is the series resistance associated with said inductance, and C_l embodies parasitic load capacitance at the output port, transistor bulk-drain capacitance, and the capacitance associated with the realization of inductance L_l. Because the network defining $Z_l(s)$ in Fig. 9.57 is topologically identical to that which defines impedance $Z_{ss}(s)$ in the same figure, Eqs. (9-173) through (9-175) can be adapted to write for $Z_l(s)$,

$$Z_l(s) = \frac{R_l\left(1 + \dfrac{sQ_l}{\omega_o}\right)}{1 + \dfrac{sx}{Q_l\omega_o} + \left(\dfrac{s}{\omega_o}\right)^2 x} = \frac{sL_l\left(1 + \dfrac{\omega_o}{Q_l s}\right)}{1 + \dfrac{sx}{Q_l\omega_o} + \left(\dfrac{s}{\omega_o}\right)^2 x}, \quad (9\text{-}207)$$

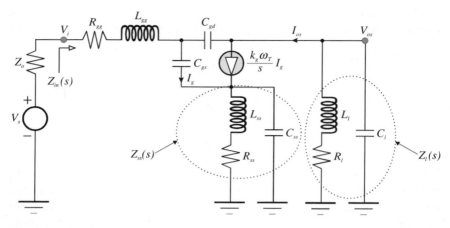

Figure 9.57. The high frequency model of the RF amplifier in Fig. 9.53(a) with the drain circuit load impedance realized as a practical circuit integrated circuit inductor.

where ω_o is the amplifier center frequency defined by Eq. (9-199),

$$Q_l = \frac{\omega_o L_l}{R_l}, \tag{9-208}$$

is the quality factor of the load circuit inductance measured at the amplifier center frequency, and parameter x is

$$x = \omega_o^2 L_l C_l. \tag{9-209}$$

Note that x represents little more than the inverse square of the self-resonant frequency of the load circuit inductance, normalized to the tuned center frequency. Clearly, if $x \ll 1$ and $Q_l \gg 1$, the impedance given as Eq. (9-207) approximates the desired ideal load circuit inductance, at least in the neighborhood of frequency ω_o. The requirement, $x \ll 1$, imposes a limit on the allowable load circuit capacitance, while the objective of a very large load quality factor can be met easily if the drain circuit inductance is realized as a bond wire connecting the transistor drain terminal to the positive circuit supply voltage.

Example 9.8. When operated from a 50 ohm signal source, the bandpass amplifier of Fig. 9.53(a) is to be designed to deliver a voltage gain magnitude of five (14 dB) at a tuned center frequency of 1900 MHz and a 3-dB bandwidth of 360 MHz. The transistor utilized in the amplifier is appropriately biased in saturation where it has a short circuit unity gain frequency of 42 GHz, a drain-source channel resistance of 10 KΩ, and a bulk-induced threshold modulation factor of 0.03. For the given MOSFET process, the gate-drain capacitance is known to be roughly 2.5% of the gate-source capacitance. Assume that the source and drain circuit inductances are realized as on chip spirals having a nominal quality factor of no more than 9 at 1900 MHz. On the other hand, the gate circuit inductance is to be implemented either with bond wire or as an off chip element and has a nominal quality factor of 40 at the tuned center frequency of the amplifier. Design the network by calculating all requisite inductances and stipulating the maximum allowable capacitive loading at both the source and drain terminals of the transistor. Use HSPICE to simulate the small signal response of the designed amplifier, using the small signal model shown in Fig. 9.53(b).

Solution 9.8.

(1) From Eq. (9-168), a 3-dB bandwidth of $B_o = 2\pi(360\,\text{MHz})$ and a center frequency of $\omega_o = 2\pi(1900\,\text{MHz})$ requires a circuit quality factor of $Q_o = 5.278$. Since $Z_o = 50\,\Omega$, Eq. (9-204) stipulates a net input circuit inductance of $L_{in} = 44.21\,\text{nH}$. In Eq. (9-199), the requisite net input capacitance corresponding to $\omega_o = 2\pi(1900\,\text{MHz})$ is $C_{in} = 158.7\,\text{fF}$. Thus, an effective input inductance of $L_{in} = 44.21\,\text{nH}$ establishes the circuit quality factor ($Q_o = 5.278$) appropriate to the bandwidth design target of $B_o = 360\,\text{MHz}$. In tandem with this inductance requirement, a net effective input capacitance of $C_{in} = 158.7\,\text{fF}$ sets the amplifier center frequency, about which the aforementioned bandwidth is centered, to the design goal of $\omega_o = 2\pi(1900\,\text{MHz})$.

(2) In an attempt to maximize the accuracy of relevant design computations, it is necessary to estimate values for the parasitic resistances, R_{gg}, R_{ss}, and R_l, associated respectively with circuit inductances L_{gg}, L_{ss}, and L_l. One way of accomplishing this feat is to predicate first order values of all circuit inductances on the tacit assumption of negligible parasitic resistances. Once these first order inductance values are available, the subject resistances can be estimated from the stipulated quality factors of the inductances. Inductances can then be reevaluated to arrive at their respective initial design values by accounting for the estimated resistances in the appropriate design equations. This procedure obviously results in finalized inductance quality factors that differ somewhat from their nominal specifications. This variance is acceptable in light of the fact that quality factors are not mere constants, but are actually intricate functions of frequency and numerous physical parameters. Accordingly, the quality factors provided in the problem statement should be viewed as only reasonable estimates of actual performance metrics related to inductor losses.

Since $C_{gd} = 0.025\,C_{gs}$, k_g in Eq. (9-172) is $k_g = 1.025$. With parasitic resistances ignored, Eq. (9-200) yields

$$Z_o \approx \left(\frac{C_{gs}}{C_{gs} + C_{gd}}\right) k_g \omega_T L_{ss} = \left(\frac{1}{k_g}\right) k_g \omega_T L_{ss} = \omega_T L_{ss}, \qquad \text{(E8-1)}$$

where Eq. (9-172) has been applied. For $Z_o = 50\,\Omega$ and $\omega_T = 2\pi(42\,\text{GHz})$, the preliminary value of source circuit inductance L_{ss} is 189.5 pH. Given a nominal inductor quality factor of 9 at the amplifier center frequency, $R_{ss} = \omega_o L_{ss}/9 = 0.2513\,\Omega$. Conservatism is arguably prudent in bandpass high frequency amplifier designs and thus, R_{ss} might be increased by some 10% to a value of $R_{ss} = 0.275\,\Omega$.

Continuing with the tacit neglect of inductor resistances, Eq. (9-197) provides

$$L_{\text{in}} \approx L_{gg} + \frac{L_{ss}}{k_g}, \tag{E8-2}$$

and with $L_{\text{in}} = 44.21\,\text{nH}$, $L_{ss} = 189.5\,\text{pH}$, and $k_g = 1.025$, $L_{gg} = 44.02\,\text{nH}$. Since the quality factor of the gate inductance is roughly 40 at $\omega = \omega_o$, R_{gg} follows as $R_{gg} = \omega_o L_{gg}/40 = 13.14\,\Omega$. As with the adjustment to R_{ss} above, increase this resistance value by approximately 10% to $R_{gg} = 14.5\,\Omega$.

Using Eq. (9-198), the input capacitance can be approximated as

$$C_{\text{in}} \approx C_{gs} + C_{gd} = k_g C_{gs}. \tag{E8-3}$$

Since $C_{\text{in}} = 158.7\,\text{fF}$ and $k_g = 1.025$, $C_{gs} = 154.8\,\text{fF}$, whence $C_{gd} = 0.025 C_{gs} = 3.871\,\text{fF}$. Accordingly, $\omega_T = 2\pi(42\,\text{GHz})$ in Eq. (9-158) requires transistor biasing and geometry commensurate with a forward transconductance of $g_m = 41.88\,\text{mmho}$.

Continuing with the approximations invoked thus far, the ratio, $[C_g/(C_g+C_m)]$, approximates $1/k_g$ so that the center frequency voltage gain relationship in Eq. (9-206) becomes

$$H_o \approx Q_o \left(\frac{\omega_T}{\omega_o}\right)\left(\frac{L_l}{L_{\text{in}}}\right). \tag{E8-4}$$

For $Q_o = 5.278$, $\omega_T = 2\pi(42\,\text{GHz})$, $\omega_o = 2\pi(1900\,\text{MHz})$, and $L_{\text{in}} = 44.21\,\text{nH}$, a gain magnitude of $|H_o| = 5$, is seen to require a drain circuit inductance of $L_l = 1.895\,\text{nH}$. With an inductance quality factor of 9, this inductance implies a series parasitic resistance of $R_l = 2.513\,\Omega$. Design conservatism beckons an increase of this estimated inductor resistance to $R_l = 2.75\,\Omega$.

(3) Having deduced estimates of all parasitic resistances associated with the required circuit inductors, the foregoing preliminary computations of inductances, capacitances, and transistor transconductance must now be updated in accordance with the computer-aided optimization guidelines delineated in the preceding subsection of text. The procedure starts by discerning the source terminal inductance, L_{ss}, commensurate with the requisite 50 ohm driving point input impedance at the tuned center frequency of the amplifier. The gate lead inductance, L_{gg}, can then be determined in terms of the revised value of L_{ss} and the originally computed value of the net input circuit inductance, L_{in}. Updated values of the gate-source and gate-drain capacitances, C_{gs} and C_{gd}, respectively, can now be determined in terms of the computed value of net

Broadband and Radio Frequency MOS Technology Amplifiers

input capacitance, C_{in}. Armed with the new values of C_{gs} and C_{gd}, as well as with the stipulated value of the transistor unity gain frequency ω_T, the transistor transconductance, g_m, can be updated. Finally, the load circuit inductance commensurate with the desired center frequency gain magnitude of 5 volts/volt can be determined. The small signal model of the subject amplifier can now be simulated to ascertain the propriety of the revised parameter estimates. If necessary, the optimization procedure can be revisited to arrive at an acceptable finalized design result.

In the problem at hand, the optimized circuit parameters are as itemized below. Interestingly enough, the optimized values of inductances, capacitances, and transistor transconductance differ only minimally from their respective, originally computed first order estimates.

$$R_{ss} = 0.275 \, \Omega$$
$$R_{gg} = 14.5 \, \Omega$$
$$R_l = 2.75 \, \Omega$$
$$Q_o = 5.278$$
$$k_g = 1.025$$
$$L_{in} = 44.02 \, \text{nH}$$
$$C_{in} = 158.7 \, \text{fF}$$
$$L_{ss} = 184.96 \, \text{pH}$$
$$L_{gg} = 44.08 \, \text{nH}$$
$$L_l = 1.835 \, \text{nH}$$
$$C_{gs} = 156.1 \, \text{fF}$$
$$C_{gd} = 3.903 \, \text{fF}$$
$$g_m = 42.33 \, \text{mmho}$$
$$\lambda_b g_m = 1.267 \, \text{mmho}$$

(4) In order to guarantee that the parasitic capacitances in both the source and drain terminals of the transistor exert negligible influence on the amplifier frequency response, the self-resonant frequencies of both the source and drain inductances should be substantially larger than the tuned center frequency by at least a factor of root ten. Recalling Eqs. (9-174) and (9-209), this restriction means that $C_{ss} \leq 5.134 \, \text{pF}$ and $C_l \leq 361.8 \, \text{fF}$. Fortunately, the upper limit value of capacitance C_{ss} is far larger than typical bulk-source capacitances. Likewise, the upper bound for capacitance C_l is much larger than representative bulk-drain capacitances. Generally, these device capacitances are of the order of 40% of the gate-source capacitance value. For this design, it is reasonable to presume $C_{ss} = 60 \, \text{fF}$ and $C_l = 100 \, \text{fF}$, the latter accounting to first order for parasitic capacitance associated with any extrinsic load incident with the amplifier output port.

(5) The foregoing design computations are effectively summarized by the small signal schematic diagram of the bandpass amplifier offered in Fig. 9.58. The

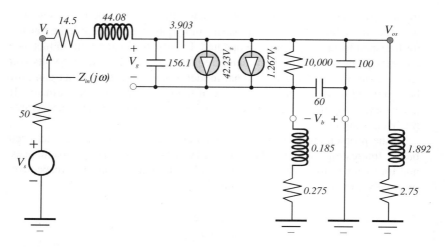

Figure 9.58. Small signal model of the bandpass amplifier designed in Example 9.8. In the diagram, all resistances are in ohms, all inductances are in units of nanohenries, and all capacitances are in femtofarads. Moreover, the conductance multipliers of the two controlled sources are dimensioned in millimhos.

transfer and input impedance characteristics implied by this model were simulated with HSPICE.

Figure 9.59 plots the resistive and reactive components of the driving point input impedance. The reactive portion of this graph is especially interesting, for it serves to pinpoint the resonant frequency of the amplifier. In particular, resonance prevails at the frequency where the reactive input impedance vanishes. A detailed inspection of the simulated data reveals resonance to occur at 1.901 GHz, a scant 0.04% higher than the design target. At 1.901 GHz, the data pertinent to the resistive component of the driving point input impedance offers an input resistance of 50.081 ohms, which is 0.16% above the desired 50 ohm value.

Figure 9.60 displays the overall and input port frequency responses of the amplifier modeled in Fig. 9.58. At the simulated resonant frequency of 1.901 GHz, the data confirms a center frequency I/O gain magnitude of 4.98, which is 0.39% below the design target gain magnitude of five. A degradation of this center frequency gain by 3-dB corresponds to an I/O gain magnitude of 3.552, which is realized at signal frequencies of 1.737 GHz and 2.093 GHz. The simulated 3-dB bandwidth follows as (2.093 GHz − 1.737 GHz), or 356.3 MHz, which is only 1.02% smaller than the design goal of 360 MHz.

The input port frequency response depicted in Fig. 9.60 shows gain magnitudes of nominally unity at both very low and very high frequencies. This result aligns with expectations because the effective series input capacitance, C_{in},

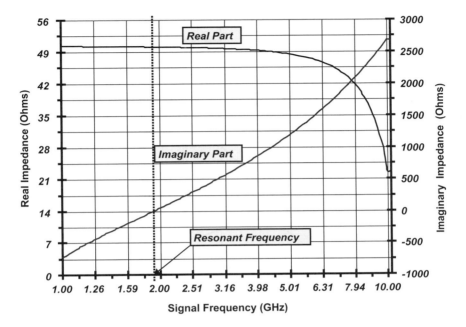

Figure 9.59. Simulated resistive and reactive components of the driving point input impedance, $Z_{in}(j\omega)$, implied by the model in Fig. 9.58.

produces a very large input impedance reactance at low signal frequencies, while the effective series input inductance, L_{in}, likewise yields high reactance at high signal frequencies. At the amplifier center frequency, the input port gain magnitude should be $1/2$ because the amplifier input port is presumably matched to the signal source impedance at the tuned center frequency. A detailed inspection of the simulated data shows that at the simulated center frequency of 1.901 GHz, the I/O gain magnitude is 0.504, which is 0.80% larger than its expected value. This slight discrepancy synergizes with simulated input resistance, which is 0.16% larger than 50 ohms at the center frequency.

Comments. In this example, extraordinarily good agreement is obtained between design predictions of amplifier performance and simulated responses. This good agreement is even more remarkable in light of the fact that the first order design calculations of circuit element values differ only minimally from their optimized counterparts. In order to avoid being lulled into a proverbial false feeling of security, however, it is important to interject that the observed success of this design venture is strongly aided by the fact that the gate-drain capacitance of the utilized transistor is only 1/40th of the transistor gate-source capacitance. Progressively

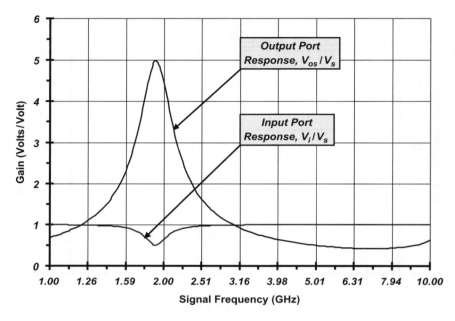

Figure 9.60. Simulated small signal frequency responses of the RF amplifier designed in Example 9.8.

larger gate-drain capacitances can be expected to incur corresponding larger disparages between design objectives and simulated performance. In extreme cases, large gate-drain capacitance may even preclude an acceptable optimization with respect to I/O gain requirements, center frequency goals, and input port impedance matching at the tuned center frequency.

Appreciable gate-drain capacitance causes two, albeit somewhat related, design problems. The first and most obvious of these problems is that the gate-drain capacitance modifies the driving point input impedance by incurring a frequency dependent load on the impedance presented to the input port by the gate to source high frequency current path. Specifically, the latter impedance is $Z_{go}(s)$, as incorporated into Eq. (9-180), which also confirms the modification of $Z_{go}(s)$ to an impedance, $Z_g(s)$, by a frequency variant factor (the parenthesized term in said equation) that depends on the gate-drain capacitance. The second problem, which is highlighted by this frequency variant factor, is that C_{gd} causes an input port impedance that is perturbed in accordance with a somewhat intricate function of the frequency dependent load termination imposed on the output port of the amplifier. Thus, for example, adjustments made to the drain load inductance, L_l, to achieve a desired center frequency gain are likely to de-tune the input impedance and specifically, the conditions that effect an input port matching to the signal source impedance.

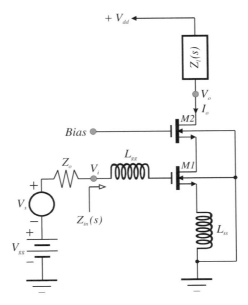

Figure 9.61. The RF amplifier of Fig. 9.53 modified by incorporation of a common gate cascode stage to mitigate the deleterious effects of gate-drain capacitance in the common source driver, $M1$.

The lesson to be learned herewith is that gate-drain capacitance incurs analytical and engineering pain in the design of an inductively degenerated common source RF amplifier. This pain can be treated by inserting a common gate cascode between the drain of the common source amplifier and the inductive load port, as suggested in Fig. 9.61. The principle effect of the cascode transistor, $M2$, is to isolate the inductive load impedance, $Z_l(s)$, from the amplifier input port. If transistors $M1$ and $M2$ have identical geometries, the Miller multiplier of the $M1$ gate-drain capacitance is nominally only a factor of two. To first order, therefore, the gate-source impedance of $M1$ is simply shunted by a capacitance of $2C_{gd}$. Detailed analytical considerations are left as an exercise for the reader.

References

1. J. Choma, Jr., *Electrical Networks: Theory and Analysis* (Wiley Interscience, New York, 1985), pp. 679–681.
2. B. P. Lathi, *Modern Digital and Analog Communication Systems* (Oxford University Press, New York, 1998), pp. 102–106.
3. C.R.C. *Standard Mathematical Tables*, 12th edn. (The Chemical Rubber Publishing Company, Cleveland, Ohio, 1959), p. 358.

4. W.-K. Chen, *Passive and Active Filters: Theory and Implementations* (John Wiley & Sons, New York, 1986), pp. 50–58.
5. E. L. Ginzton, W. R. Hewlett, J. H. Jasberg and J. D. Noe, Distributed amplification, *Proc. IRE* **36** (1948) 956–969.
6. W.-K. Chen, Theory and design of distributed amplifiers, *IEEE J. Solid-State Circuits* **SC-3** (1968) 165–179.
7. B. M. Ballweber, R. Gupta and D. J. Allstot, A fully integrated 0.5-5.5 GHz CMOS distributed amplifier, *IEEE J. Solid-State Circuits* **35** (2000) 231–239.
8. T. H. Lee, *The Design of CMOS Radio-Frequency Integrated Circuits* (Cambridge University Press, Cambridge, United Kingdom, 2004), pp. 279–282.
9. D. Johns and K. Martin, *Analog Integrated Circuit Design* (John Wiley and Sons, Inc., New York, 1997), Chapter 4.
10. N. Balabanian and T. A. Bickart, *Electrical Network Theory* (John Wiley and Sons, Inc., New York, 1969), pp. 406–414.
11. J. Choma, Jr. and S. A. Witherspoon, Computationally efficient estimation of frequency response and driving point impedance in wideband analog amplifiers, *IEEE Trans. on Circuits and Systems* **CAS-37** (1990) 720–728.
12. C. Battjes, Monolithic wideband amplifier, United States Patent, no. 4,236,119, November 25, 1980.
13. T. H. Lee, *Op. Cit.*, Chapter 12.
14. S. S. Mohan, Modeling, design, and optimization of on-chip inductors and transformers, Ph. D. Dissertation, Stanford University, 1999.
15. C. P. Yue and S. S. Wong, On-chip spiral inductors with patterned ground shields for Si-based RF IC's, *IEEE J. Solid-State Circuits* **33** (1998) 743–752.
16. A. Zolfaghari, A. Chan and B. Razavi, Stacked inductors and 1 to 2 transformers in CMOS technology, *Proc. 2000 Custom Integrated Circuits Conf.*, May 2000.

Exercises

Problem 9.1

A low frequency gain of at least 20 dB is required of a lowpass amplifier that is to be designed to deliver a 3-dB frequency of 1.0 GHz.

(a) Determine the optimum number of cascaded identical stages and the corresponding 3-dB bandwidth of each stage if maximum overall bandwidth is to be achieved.

(b) If the resultant circuit comprises the open loop of a feedback amplifier, plot the magnitude and phase responses of the loop gain for the case

Broadband and Radio Frequency MOS Technology Amplifiers   843

of unity closed loop gain. Comment on the closed loop stability of the amplifier.

Problem 9.2
The approximate transfer function for the compensated amplifier diagrammed in Fig. 9.4(a) is given by Eq. (9-33). Assume that the zero frequency value, $T_s(0)$, of the loop gain is 10.

(a) Derive an expression for the phase response of the amplifier. Plot this response for $k = 0$, $k = 0.75$, $k = 1$, and $k = 1.25$. Give an engineering assessment of the plotted responses.
(b) Derive an expression for the transient step response of the amplifier; use an input step amplitude of $1/A(0)$. Plot this response for $k = 0$, $k = 0.75$, $k = 1$, and $k = 1.25$. Give an engineering assessment of the plotted responses. Specifically, discuss the impact exerted on the transient response by the pole-zero doublet.
(c) Derive an expression for the transient unit impulse response of the amplifier. Plot this response for $k = 0$, $k = 0.75$, $k = 1$, and $k = 1.25$. Give an engineering assessment of the plotted responses.

Problem 9.3
The balanced differential amplifier depicted in Fig. P9.3 is designed for a maximally flat magnitude response and maximal 3-dB bandwidth. All transistors are biased in their saturation domains, all have their bulk terminals connected to their respective source terminals, and all are characterized by infinitely large drain-source channel resistance. Moreover, each transistor features a dominant pole response, with the dominant capacitance established at respective drain terminals.

(a) Develop a general expression for the differential voltage gain, $A_d(s) = V_o/V_s$.
(b) Give design criteria that support the objective of a maximally flat, optimally broadbanded, differential frequency response.
(c) Are common mode input signals a problem in light of the approximations invoked in this exercise?

Problem 9.4
In the cascode configuration of Fig. 9.12, set the inductance, L, to zero and determine all open circuit time constants. Assume that the two transistors are biased in their saturation regimes, have large drain-source channel resistances, and nonmatched small signal parameters. Assuming that pole dominance prevails, deduce the criteria that must be satisfied if the common

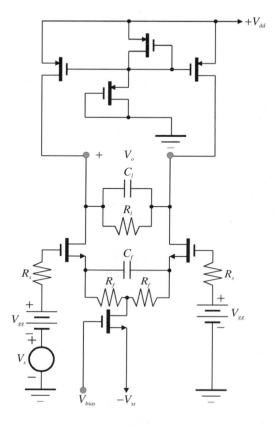

Figure P9.3.

gate cascode is to deliver a 3-dB bandwidth estimate that is larger than that afforded by a conventional common source amplifier.

Problem 9.5

Repeat Example 9.3, but now design for a low frequency voltage gain of 18 dB and a low frequency envelope delay of 200 pSEC. The delay response should emulate, as closely as possible, maximally flat delay. Use the small signal model parameters provided in conjunction with the aforementioned example.

(a) Provide a finalized small signal model of the completed design realization.

(b) Use HSPICE or equivalent computer-aided analysis software to simulate the following performance characteristics:
   i. the small signal magnitude response of the amplifier;
   ii. the small signal phase response of the amplifier;
   iii. the small signal delay response of the amplifier;
   iv. the small signal, unit step response of the amplifier.
(c) Discuss the executed simulations, focusing specifically on any disparages observed between theoretically predicted and simulated responses.

Problem 9.6
A third order Bessel filter, which realizes maximally flat delay without benefit of either left half or right half plane zeros has a normalized transfer function of

$$A_n(s) = \frac{1}{1 + (T_{do}s) + \frac{2}{5}(T_{do}s)^2 + \frac{1}{15}(T_{do}s)^3},$$

where T_{do} represents the zero frequency envelope delay between amplifier input and output ports.

(a) Design a maximally flat delay, series peaked amplifier that supports a voltage gain of 0 dB and a zero frequency delay of 30 pSEC. Use the transistor whose small signal parameters are supplied in conjunction with Example 9.2.
(b) Use HSPICE or equivalent computer-aided design software to simulate the magnitude response and the envelope delay response.

Problem 9.7
Return to the schematic diagram of the series peaked amplifier shown in Fig. 9.20. Assume that the net capacitance, $(C_o + C_x)$, is negligibly small.

(a) Derive the condition commensurate with a maximally flat delay (MFD) response.
(b) What bandwidth enhancement is afforded by selecting inductance L in concert with a maximally flat delay response?
(c) Derive an expression for the magnitude peaking as a function of circuit quality factor.
(d) Submit a plot depicting the dependencies of both the magnitude peaking and the 3-dB bandwidth on circuit quality factor.

Problem 9.8
Equation (9-81) defines the approximate voltage transfer function of the series-shunt peaked amplifier of Fig. 9.23 subject to neglect of capacitance C_o and the condition and definition imposed respectively by Eqs. (9-79) and (9-80).

(a) Derive the bandwidth expression given in Eq. (9-82).
(b) Show that the frequency at which peaking in the magnitude response is observed is $0.687 B_u$, where B_u is the bandwidth of the uncompensated amplifier ($L = 0$) if capacitance C_o is tacitly ignored. Verify that the percentage peaking, measured with respect to the zero frequency value of voltage gain, is just slightly under 3%.

Problem 9.9
Show that from the perspective of the terminal currents, I_1 and I_2, and the terminal voltages, V_1 and V_2, the coupled inductor circuit of Fig. P9.9(a) is electrically identical to the structure in Fig. P9.9(b). The parameter, M, signifies the mutual inductance between the two coils of inductance L.

Problem 9.10
Repeat Problem 9.9 for the case in which the two terminals of the inductive coil on the right in Fig. P9.9(a) are interchanged.

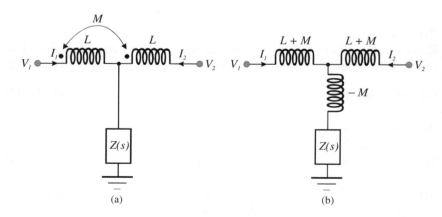

Figure P9.9.

Problem 9.11

Lossless constant resistance networks have been propounded in the technical literature as an effective vehicle for achieving broadband frequency responses in active transconductor amplifiers. One such alternative to the networks addressed in the text is the filter shown in Fig. P9.11, where the terminating load impedance, $Z_o(s)$, is selected to ensure that the driving point input impedance, $Z_{in}(s) \equiv Z_o(s)$. The indicated two inductors are uncoupled.

(a) Show that the required value of load impedance $Z_o(s)$ is expressible as

$$Z_o(s) = R_o \sqrt{1 + \left(\frac{s}{\omega_h}\right)^2},$$

where resistance parameter R_o and frequency parameter ω_h are given by

$$R_o = \sqrt{L/C}$$

and

$$\omega_h = 2/\sqrt{LC}.$$

(b) Clearly, a constant driving point input resistance is manifested for a strictly resistive load only for signal frequencies, ω, that satisfy the inequality, $\omega \ll \omega_h$, whereupon, $Z_o(s) \approx R_o$. For this resistive load constraint, demonstrate that the voltage transfer function, $H_o(s) = V_o/V_s$, mirrors that of a Butterworth filter.

(c) What is the 3-dB bandwidth associated with the transfer function, $H_o(s)$, deduced in Part (b)?

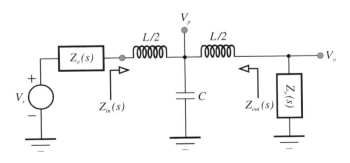

Figure P9.11.

(d) For $Z_o(s) = R_o$, derive an expression for the transfer function, $H_p(s) = V_p/V_s$. Is this transfer characteristic representative of a Butterworth filter?

(e) What is the zero frequency group delay associated with the transfer function, $H_o(s)$?

Problem 9.12

Yet another lossless, constant resistance structure is shown in Fig. P9.12, where the terminating load impedance, $Z_o(s)$, is to be selected to ensure that the driving point input impedance, $Z_{in}(s) \equiv Z_o(s)$.

(a) Show that the required value of load impedance $Z_o(s)$ is expressible as

$$Z_o(s) = \frac{R_o}{\sqrt{1 + \left(\frac{s}{\omega_h}\right)^2}},$$

where

$$R_o = \sqrt{L/C}$$

and

$$\omega_h = 2/\sqrt{LC}.$$

(b) Clearly, a constant driving point input resistance is manifested for a strictly resistive load only for signal frequencies, ω, that satisfy the inequality, $\omega \ll \omega_h$, whereupon, $Z_o(s) \approx R_o$. For this resistive load constraint, demonstrate that the voltage transfer function, $H_o(s) = V_o/V_s$, mirrors that of a Butterworth filter.

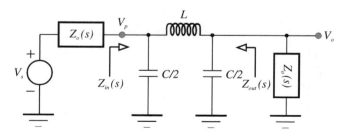

Figure P9.12.

(c) What is the 3-dB bandwidth associated with the transfer function, $H_o(s)$, deduced in Part (b)?
(d) For $Z_o(s) = R_o$, derive an expression for the transfer function, $H_p(s) = V_p/V_s$. Is this transfer characteristic representative of a Butterworth filter?
(e) What is the zero frequency group delay associated with the transfer function, $H_o(s)$?

Problem 9.13
For the amplifier modeled in Fig. 9.27(b), what inductor coupling coefficient yields a maximally flat group delay response for the transfer function, V_{os}/V_s? What 3-dB bandwidth is manifested by this maximally flat delay constraint?

Problem 9.14
Repeat all parts of Example 9.5 but this time, design for a maximally flat group delay response that delivers a zero frequency group delay of 20 pSEC.

Problem 9.15
An f_T-doubler can be realized straightforwardly as a balanced differential pair, as suggested by the simplified schematic diagram in Fig. P9.15 in which both transistors are identical devices that are biased identically.

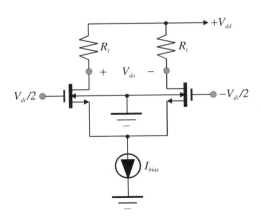

Figure P9.15.

(a) Give an engineering argument that rationalizes the presumed doubling of the unity gain frequency of each transistor when the circuit is operated as a differential amplifier.
(b) Execute a small signal analysis of the circuit to confirm the presumed doubling of transistor f_T.
(c) Does this problem teach the advisability of realizing all amplifiers designed for broadbanded frequency responses as balanced differential pairs? What engineering circumstances must prevail if differential circuit technology is to prove effective as a broadband topological tool?

Problem 9.16
Repeat Example 9.6 in the text for the case in which the gate-drain capacitance is 5% of the transistor gate-source capacitance, and the desired gain is 20 dB.

Problem 9.17
Repeat Example 9.7 in the text for case in which input port matching is implemented.

Problem 9.18
Generate a curve, similar to that supplied in the text as Fig. 9.49, for the case in which transistors exuding a 45 GHz unity gain frequency are exploited in the f_T-doubler of Fig. 9.47.

Problem 9.19
Repeat Example 9.6, but let the transistor embedded in the dual loop feedback amplifier shown in Fig. 9.32 be replaced by the f_T-doubler configuration offered in Fig. 9.47. Assume that the transistors in the doubler have a unity gain frequency of 28 GHz.

Problem 9.20
Perform a small signal analysis of the short circuit current gain response of the Battjes f_T-doubler, whose basic circuit schematic diagram appears in Fig. 9.50. Assume that the transistors utilized in the Battjes circuit have negligible gate-drain capacitance, are geometrically identical, and are biased identically.

Problem 9.21
In the cascode RF amplifier of Fig. 9.61, assume identical transistors that are biased identically and that are characterized by very large drain-source

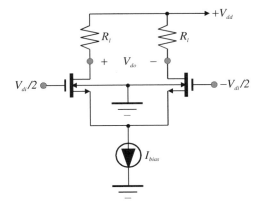

Figure 9.21.

channel resistances and negligibly small bulk-induced threshold voltage modulation factors. Assume further that the source terminal inductance, L_{ss}, as well as the drain load impedance, $Z_l(s)$, are realized as monolithic inductances. Analyze the circuit in a fashion similar to the work executed in Section 9.9.2 of the text so that appropriate responses can be generated to the following inquiries.

(a) Give an expression for the net effective series input resistance, R_{in}.
(b) Derive an expression for the net effective input port inductance, L_{in}.
(c) Derive an expression for the net effective input port capacitance, C_{in}.
(d) Deduce a relationship for the tuned center frequency, ω_o, of the RF amplifier.
(e) Deduce an expression for the 3-dB bandwidth of the amplifier.
(f) How must inductance L_{ss} be chosen to achieve an input port impedance match to a resistive source impedance at the tuned center frequency of the amplifier?
(g) Assuming input port matching, what is the amplifier voltage gain, H_o, at the tuned center frequency?
(h) What is the time constant established at the source terminal of transistor $M2$? What condition must be satisfied by this time constant if proper amplifier operation is to be ensured?
(i) What is the net capacitance that prevails at the amplifier output port? What condition must be satisfied by this capacitance if proper amplifier operation is to be ensured?

Problem 9.22

Using the results of the preceding exercise, repeat Example 9.8, but use the cascode configuration offered in Fig. 9.61. Comment as to the overall engineering quality of this amplifier, as compared to the amplifier specifically addressed in Example 9.8.

Problem 9.23

The transistor used in the source follower depicted in Fig. P9.23 is biased in its saturation regime. For the purposes of this problem, assume that the transistor is characterized by an infinitely large drain-source channel resistance and a negligibly small bulk-induced threshold modulation factor. Observe that the follower drives a capacitive load of capacitance C_l, and since the circuit is intended for use in high frequency signal processing applications, the gate-source, bulk-source, bulk-drain, and gate-drain capacitances of the device cannot be ignored.

(a) Derive an expression for the small signal, driving point input impedance, $Z_{in}(s)$.
(b) In the sinusoidal steady state, what is the reactive component of the driving point input impedance found in Part (a)? Is the input port of the follower capacitive or inductive at high signal frequencies?
(c) In the sinusoidal steady state, what is the real part component of the driving point input impedance found in Part (a)? Determine the load capacitance conditions that deliver a negative real input impedance.
(d) What are the engineering implications, with respect to the transient pulse response of a source follower, of a negative real part input impedance?

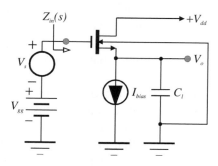

Figure P9.23.

(e) Compensate the circuit to alleviate the high frequency negative resistance problem to which the preceding part of this problem alludes. Provide design guidelines or formulae for any compensation elements added to the circuit.

Problem 9.24

The transistor used in the source follower depicted in Fig. P9.24 is biased in its saturation regime. For the purposes of this problem, assume that the transistor is characterized by an infinitely large drain-source channel resistance and a negligibly small bulk-induced threshold modulation factor. Observe that the follower drives an inductive load of inductance, L_l, and since the circuit is intended for use in high frequency signal processing applications, the gate-source, bulk-source, bulk-drain, and gate-drain capacitances of the device cannot be ignored. The load inductance is realized monolithically, so that an account of series resistance and shunt parasitic capacitance must be made.

(a) Derive an expression for the small signal, driving point input impedance, $Z_{in}(s)$.
(b) In the sinusoidal steady state, what is the reactive component of the driving point input impedance found in Part (a)? Is the input port of the follower capacitive or inductive at high signal frequencies?
(c) In the sinusoidal steady state, what is the real part component of the driving point input impedance found in Part (a)? Determine the load inductance conditions that deliver a negative real input impedance.
(d) Compensate the circuit to alleviate any high frequency negative resistance problems to which the preceding part of this problem alludes. Provide design guidelines or formulae for any compensation elements added to the circuit.

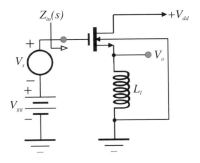

Figure P9.24.

Problem 9.25

A common gate circuit used as a cascode in conjunction with a common source amplifier engenders potential engineering design problems because of parasitic layout inductance manifested in the gate lead of the circuit. To this end, consider the circuit shown in Fig. P9.25, in which the transistor is characterized by an infinitely large drain-source channel resistance and a negligibly small bulk-induced threshold modulation factor. All relevant device capacitances should not be ignored so that due account can be made of relevant high frequency phenomena. The inductance, L_g, in the gate lead can be presumed to be an ideal element.

(a) Derive an expression for the indicated small signal, source input impedance, $Z_{in}(s)$.
(b) In the sinusoidal steady state, what is the reactive component of the source input impedance found in Part (a)? Is the input port of the common gate stage capacitive or inductive at high signal frequencies?
(c) In the sinusoidal steady state, what is the real part component of the driving point impedance found in Part (a)? Determine the gate inductance conditions that deliver a negative real source input impedance.
(d) Compensate the circuit to alleviate any high frequency negative resistance problems to which the preceding part of this problem alludes. Provide design guidelines or formulae for any compensation elements added to the circuit.

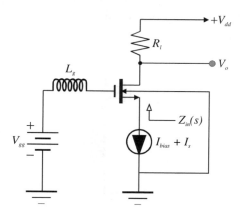

Figure P9.25.

Index

N stage cascade, 732, 734
S-Parameters, 250
S-parameter characterization, 226
ω_{max}, 204
π-type, 148
π-type network model, 137
c-parameter, 151, 153
c-parameter matrix, 154
g-parameter modeling, 128
h-parameter, 118, 119
h-parameter equivalent circuit, 119
h-parameter model, 119
m-port multiport, 260
y-parameters, 135, 469
z-parameters, 146
"scattered" current, 227
3-dB bandwidth, 66, 286, 289, 318, 342, 345, 536, 731
60-cycle "hum", 58

active divider, 692
adjustable resistance, 679
amplifier bandwidth, 793
available load power, 195
available power, 194
available power gain, 191
average power, 258

balanced circuit, 636
balanced differential amplifiers, 206, 651
balanced differential configuration, 207
balanced differential realization, 207
bandpass, 265

bandpass amplifiers, 51
bandpass architecture, 819
bandpass feedback amplifier, 816
bandpass frequency responses, 51
bandpass units, 58
bandwidth, 66
Barkhausen criterion, 292
Battjes f_T-doubler, 850
Bessel, 261
biasing, 115, 116
bilateral, 122
bilateral circuit element, 670
bilateral linear network, 252
bilateral two-port network, 137
bilinear coefficient, 230
bilinear transformation, 230
biquadratic filters, 57, 540, 591
Blackman's formula, 457, 486
Blackman's impedance formula, 476, 484
Blecher's procedure, 469
Boltzmann voltage, 507
Boltzmann's constant, 506
bond wire, 778, 801
bridging capacitance, 776
broadbanded dominant pole, 301
buffered capacitive compensation, 676
buffered capacitive feedback, 670
buffers, 32
bulk-induced modulation of threshold voltage, 655
bulk-source capacitance, 737
bulk transconductance factor, 737
bulk-drain capacitance, 737
Butterworth, 261
Butterworth filter, 847
Butterworth MFM, 769

856   Feedback Networks: Theory and Circuit Applications

Butterworth MFM response, 767
Butterworth response, 767

capacitance density of the oxide-channel, 655
carrier frequency, 817
cascade interconnection, 154
cascaded 3-dB bandwidth, 733
center frequency, 51, 816
chain matrix, 155
chain parameters, 151
channel length, 502, 655
channel length modulation voltage, 510, 655
channel pinch off, 512
channel resistance, 737
characteristic impedance, 241, 271, 817
characteristic polynomial, 282
circuit instability, 292
circuit partitioning, 337
circuit quality factor, 754
circuit stability, 293
closed loop, 181
closed loop bandwidth, 286
closed loop characteristic polynomial, 298
closed loop current gain, 188, 361
closed loop damping factor, 290, 297, 311
closed loop gain, 161, 169, 176, 178, 279, 337
closed loop input impedance, 190
closed loop instability, 291
closed loop output impedance, 191
closed loop pole, 286
closed loop stability, 300
closed loop system, 178
closed loop transadmittance, 171, 190
closed loop transfer function, 286
closed loop transimpedance, 170, 188
closed loop undamped natural frequency, 290
closed loop voltage gain, 332
CMOS amplifier, 550
Colpitts oscillator, 316
column vectors, 259
common drain amplifier, 498, 552
common gate amplifier, 498, 499, 568

common gate cascode, 841
common gate I/O characteristics, 571
common mode excitation, 640, 645
common mode gain, 713
common mode input signal, 634
common mode model, 641
common mode Norton transconductance, 641
common mode output resistance, 642, 713
common mode rejection ratio, 637, 642, 647, 650, 713
common mode signal rejection, 636
common mode transconductance, 636
common mode voltage gain, 646
common source amplifier, 498, 524, 535
common source RF amplifier, 821
common source-common gate cascode, 569, 575, 753
compensated operational amplifier, 673
compensated source follower, 617
compensation, 300
compensation pole, 303
complex conjugate poles, 76
constant reactance circles, 233
constant resistance criteria, 778
constant resistance networks, 847
continued fraction expansion, 261
controlled vector, 477
controlling parameter, 448
controlling vector, 477
coupled inductor load, 776
coupling coefficient, 777
Cramer's rule, 446
critical damping, 78, 85
critical feedback parameter, 331, 409
critical frequencies, 75, 289
critical frequency parameter, 294, 295
critical parameter, 324, 326, 338, 341, 448
crucial parameter, 448
current amplifier, 30, 122, 348
current buffer, 34, 40
current buffering, 576
current controlled voltage source, 149
current extracting node, 447
current feedback amplifier, 360
current gain, 448, 456

current injecting node, 447
current mirror, 652
cutoff, 505

damping factor, 62, 83, 288, 298, 309
Darlington configuration, 812
dBm, 51
dBm Power Measure, 50
decibel, 15
decibel value, 64
decoupling capacitance, 525, 621
degenerative RC broadbanding, 736
delay, 69
depletion mode, 500
desensitization, 279
dielectric constant, 507
differential input signal, 631, 634
differential mode, 644
differential mode excitation, 645
differential mode gain, 713
differential mode half circuit, 640
differential mode Norton transconductance, 641
differential mode output resistance, 713
differential mode transconductance, 636
differential mode voltage gain, 646
differential mode, half circuit schematic, 715
differential to single ended conversion, 634
differential to single ended converter, 638, 642
diode-connected transistor, 651
distributed transmission line, 231, 772
dominant, 77
dominant energy storage element, 106
dominant open circuit time constant, 577
dominant pole, 78, 528, 680
dominant pole amplifiers, 731
dominant pole open loop, 287
dominant pole response, 84, 533, 681
dominant pole, approximation, 285
double series peaking, 776
drain saturation current, 512
drain saturation voltage, 512
driving point, 118

driving point function matrix, 477
driving point impedance, 445, 446, 457, 473
driving point input impedance, 19, 120, 123, 132, 194, 337–339, 342
driving point input resistance, 343, 361, 789
driving point null output resistance, 343
driving point output impedance, 123, 132, 195, 340–342
driving point output resistance, 345, 361, 789
dual loop, 787
dual loop feedback, 385, 387, 401

Early resistance, 381
effective forward transconductance, 174
effective loop gain, 248
electrical noise, 55
electron mobility, 504
electronic potentiometer, 511
emitter degeneration resistance, 209, 380
energy incidence, 241, 242
enhanced common gate cell, 581
enhancement mode, 500
envelope delay, 72, 273, 732, 762, 763
equicofactor matrix, 443, 446
equivalent input noise voltage, 797
error signal, 278, 326

f_T-doubler, 812
feedback, 246, 285, 456
feedback admittance, 206
feedback branch admittance, 366
feedback branch impedance, 377
feedback current amplifier, 357
feedback factor, 136, 278, 284, 286, 314, 324, 394
feedback network, 177
feedback parameter, 177, 327, 332
feedback signal flow path, 188
feedback transimpedance, 147
feedback voltage amplifier, 354
feedback voltage gain, 244
Fermi potential, 505
final value theorem, 309

first order cofactor, 443
flatband potential, 505
folded current mirror, 717
forward current gain, 122
forward gain, 120, 136, 147, 162, 250
forward network gain, 339
forward short circuit current gain, 120
forward transadmittance, 249
forward transconductance, 516, 737
forward transimpedance, 162, 526
forward, signal path, 187
frequency response, 16, 63, 285
frequency response transformation, 265
front-end, 816
full power bandwidth, 690
fundamental matrix feedback flow graph, 478, 492

gain bandwidth product, 90, 287
gain margin, 292, 295, 297, 299, 300
gain-bandwidth product, 732
gate aspect ratio, 509, 654
gate impedance, 825
gate oxide overlap, 518
gate-drain capacitance, 518, 737
gate-source capacitance, 518, 737
gate-source threshold voltage, 655
global architecture, 178
global feedback, 178, 277, 348
gyrator, 267

h-parameters, 250, 252
half circuit analysis, 637
Hartley oscillator, 317
headroom, 295
highpass, 58, 265
hole mobility, 504
hybrid, 119
hybrid g-parameters, 128
hybrid h-parameters, 118

ideal current buffer, 40
ideal current source, 123
ideal feedback model, 464
ideal transadmittance amplifier, 27
ideal transconductor, 141

ideal transimpedance amplifier, 29
ideal transresistor, 149
ideal voltage amplification, 181, 182
ideal voltage amplifier, 24, 132, 181
ideal voltage buffer, 34
ideal voltage source, 149
identity matrix, 260, 479
immittance, 176, 449
impedance measurements, 472
impulse response, 79
impulsive source, 79
incident component of load current, 229
incident component of load voltage, 229
incident current, 227, 228
incident energy variable, 258
incident energy wave, 240
incident power, 229
incident voltage, 228
indefinite admittance matrix, 142, 144, 442, 449, 453, 455
indirect measurement of return difference, 486
inductive load port, 841
information passband, 817
input admittance, 136
input impedance, 147, 176, 448
input noise voltage, 748
input port reflection coefficient, 241, 245, 246
instability, 164, 200
integrator, 591, 592
intermodulation, 255
intrinsic carrier concentration, 506
I/O transfer admittance, 448

Kron's network partitioning theorem, 322

lateral field, 695
linear feedback circuit, 326
linear phase, 71
linear phase response, 71
linear transconductor, 26
linear two-port network, 239
linear two-port systems, 226
linearized network, 116
Llewellyn constraint, 199

Llewellyn stability factor, 199, 202
loop gain, 160–163, 168, 170, 171, 176, 188, 279, 284–286, 289, 314, 361, 527, 667
loop gain phase response, 539
loop transmission, 452
loop transmission matrix, 481, 491
lossless two-port filter, 261
lossless two-port network, 255
lossless, passive two-port network, 260
lowpass network, 14
lowpass response, 58

magnitude response peaking, 783
match terminated, 49, 195, 377, 789, 793
match terminated I/O ports, 376
match terminated condition, 792
match terminated design condition, 50
match termination, 172
matching filter, 797
matrix signal flow graph, 478
matrix singularity, 131
matrix transposition, 258
maximal flatness, 756
maximally flat delay (MFD), 74, 77, 763, 765, 845
maximally flat delay response, 762
maximally flat magnitude (MFM), 66, 77, 756, 765
maximally flat magnitude frequency, 759
maximally flat magnitude frequency response, 289, 782
maximally flat magnitude network, 784
maximally flat magnitude response, 756, 757, 761, 767, 782
maximally flat response, 784
maximum load power, 49
maximum possible transducer power gain, 202
maximum power, 195
maximum power transfer, 48
maximum signal source power, 230
maximum transducer gain, 201, 203
memoryless, 15
microwave amplifiers, 226
Miller capacitance multiplication, 633

Miller effect, 530, 561, 633, 661
Miller multiplication, 577, 598, 741, 751, 752
Miller multiplier, 794, 829, 841
Miller time, 750, 804, 829
Miller-limited frequency response, 532
mixed signal, 498
mixed signal integrated circuits, 630
mobility degradation, 695, 696
mobility of electrons, 510
model, 119
monic polynomials, 301
multi-parameter sensitivity, 492
multiloop feedback, 475
multiloop feedback circuits and systems, 322
multiparameter sensitivity function, 493
multiple feedback paths, 322
mutual inductance, 777
multiple feedback theory, 476
multiple loop feedback amplifiers, 476

N-channel, 500
N-channel driver transconductor, 634
narrow banding, 306
narrowband, 820
narrowband, amplifiers, 816
natural frequencies, 467
negative, 285
negative feedback, 164, 282, 286, 394
network I/O gain metric, 137
network block diagram, 325
network functions, 456
network power gain, 194
network stability, 286
network time delay, 762
NMOS, 500
noise, 256
noise floor, 255, 797
nondominant network pole, 78
nondominant open loop pole, 299
nondominant pole, 301, 680
nondominant pole frequency, 298
normalized loop gain, 386

normalized null return ratio, 325, 327, 331, 334, 336–341, 346, 347, 386, 789, 791, 826
normalized return ratio, 325, 327, 328, 331, 334, 335, 337–339, 341, 342, 346, 347, 349, 789, 791, 826
normalizing inductance, 263
Norton current, 5, 9
Norton current gain, 10, 152
Norton equivalent network, 470
Norton forward transadmittance, 152
Norton impedance, 5
Norton transadmittance, 10
Norton's theorem, 3, 4
notch, 265
notch filter, 58, 97, 210
notch frequency, 97, 210
null driving point input resistance, 343
null feedback parameter, 790
null forward gain, 339
null forward transadmittance, 354
null impedance, 340
null input impedance, 338, 339
null output impedance, 341, 342
null output response, 340
null parameter gain, 325, 327, 386
null parameter voltage gain, 334, 356, 789
null return difference, 454, 482
null return difference matrix, 476, 482, 483
null return ratio, 325, 331, 344, 454, 609
null return ratio matrix, 483
null signal flow metric, 327
null Thévenin admittance, 380
null Thévenin impedance, 369
null Thévenin resistance, 374
null transadmittance, 352
null transimpedance, 348

ohmic regime, 509, 513
ohmmeter method, 8
one-port linear network, 239
one-stage operational amplifier, 631
op-amp, 629
open circuit, 122, 151
open circuit forward voltage gain, 128

open circuit impedance parameters, 146
open circuit impedances, 146
open circuit input impedance, 147, 253
open circuit input port admittance, 128
open circuit output impedance, 147
open circuit time, 531
open circuit time constant, 529, 531, 557, 732
open circuit transimpedance, 10, 147
open circuit voltage gain, 10
open circuit z-parameters, 146
open circuit, open loop voltage gain, 632
open loop, 284, 287, 526
open loop current gain, 168, 361, 363, 609
open loop dominant pole, 299
open loop gain, 35, 134, 160, 161, 176, 248, 278, 314, 356
open loop gain function, 318
open loop input impedance, 165
open loop network transimpedance, 350
open loop pole dominance, 293
open loop response, 680
open loop self-resonant frequency, 298
open loop transadmittance, 352
open loop transfer function, 286
open loop transimpedance, 170
open loop voltage gain, 410, 790
open-circuited capacitive time constants, 371
operational amplifier, 281, 629
operational transconductance amplifier-capacitor (OTA-C) integrator, 540
operational transconductor amplifier, 25
oscillation, 196
oscillator, 282
OTA, 25
output admittance, 122
output impedance, 4, 129, 176
output port reflection coefficient, 245
output port time constant, 540
overdamped, 77
overdamped network, 83
overshoot, 86
oxide capacitance, 510

Index

P-channel, 500
P-channel driver transconductor, 634
P-channel transconductor, 635, 638
passive highpass feedback, 670
passive highpass frequency compensation, 677, 678
peak overshoot, 308
phase, 69
phase distortion, 71
phase margin, 292, 294, 297, 299, 311, 539, 680
phase response, 762, 763
phasor formats, 257
pi topology, 519
PMOS, 500
PMOS load, 548
pole splitting capacitance, 661
poles, 75
polysilicon, 502
port current vector, 258
port voltage gain, 132, 154
port voltage vector, 258
positive feedback, 163, 282
positive loop gain, 164
positive real functions, 196, 197
potential instability, 164, 200
potential stability, 196
potentially unstable, 164, 197, 282
power busses, 525
power gain, 191
power scattering, 229
prototype lowpass filter, 265

quality factor, 53, 62, 288, 749, 818
quiescent, 113
quiescent operating conditions, 515

radio frequency (RF) amplifier, 816
reference impedance, 227, 241, 243
reference impedances, 239
reflected current, 227
reflected energy variable, 258
reflected energy wave, 240
reflected load current, 229
reflected load voltage, 229
reflected power, 229
reflected voltage, 228

reflection coefficient, 227, 228, 230, 239, 243, 246
reflection coefficient plane, 231, 232
reflection plane, 230, 232
response peaking, 64–66
response sensitivity, 448
return difference, 290, 291, 448, 449, 452, 457, 459, 461, 464, 468, 472, 475
return difference function, 487
return difference matrix, 476, 480, 481, 485, 491
return ratio, 325, 331, 343, 349, 451, 609
return ratio matrix, 481, 491
returned signal, 482
returned voltage, 459
reverse gain, 162
reverse transadmittance, 162
reverse voltage gain, 121, 122, 243, 252
root mean square (RMS), 46, 257

Säckinger circuit, 426
saturated domain, 513
saturation, 514
saturation regime, 512
scalar sensitivity function, 492
scattering, 226
scattering analysis, 244
scattering parameter, 226, 228, 239, 240, 252
second order circuits, 57
second order closed loop network, 309
second order cofactor, 444, 454
second order compensated response, 767
second order filters, 58
second order model, 285
second order network, 319
self-resonant, frequency, 824
sensitivity, 463
sensitivity function, 463, 491
sensitivity matrix, 488, 490
sensitivity metric, 464
series peaked amplifier, 765
series peaked common source amplifier, 770
series peaked compensation, 765
series peaking, 768

series peaking element, 772
series peaking inductance, 769
series-series architecture, 399
series-series feedback, 178, 189, 380
series-series/shunt-shunt, 388
series-series/shunt-shunt dual loop feedback, 405
series-series/shunt-shunt feedback, 388
series-series/shunt-shunt feedback pair, 398
series-shunt feedback, 178, 405, 460
series-shunt feedback amplifier, 452, 455, 462, 481
series-shunt peaked amplifier, 771
series-shunt peaked compensation, 771
series-shunt/shunt-series dual loop feedback amplifier, 411
series-shunt/shunt-series feedback, 405
settling time, 82, 308, 309
sgn, 444
short circuit, 136
short circuit admittance parameters, 135
short circuit current, 9
short circuit current gain, 10, 123, 140, 253, 813
short circuit input admittance, 136, 137
short circuit input impedance, 137
short circuit output admittance, 136
short circuit output impedance, 129
short circuit transadmittance, 136
short circuit transconductance, 631
short circuit transfer admittance, 10
short circuit unity gain frequency, 204
short circuit, common base current gain, 362
shunt peaked amplifier, 753, 755
shunt peaked compensation, 748
shunt peaking, 748
shunt peaking element, 772
shunt peaking inductor, 757
shunt-antiphase shunt, 204, 207
shunt-antiphase shunt compensation, 204–206
shunt-series feedback, 178, 360, 405
shunt-series feedback amplifier, 186, 787, 790

shunt-series topology, 187
shunt-series/series-shunt, 388
shunt-shunt feedback, 178, 188
shunt-shunt global feedback, 427
shunt-shunt loop, 399
signal flow parameters, 326
signal flow theory, 322
signal incidence, 241
signal reflection, 241
silicon dioxide, 502
single ended output voltage, 631
single loop feedback amplifier, 457
single stage op-amp, 630, 632
sinusoidal oscillator, 164
slew rate, 685–687
slew rate limitations, 685
small signal drain-source conductance, 511
small signal model, 518
small signal MOS equivalent circuit, 517
Smith chart, 231
source follower, 499, 552
source follower I/O impedances, 560
source follower transfer function, 555
spiral inductor, 778, 824
spiral metallization, 822
stability, 62, 164, 195
stability headroom, 293
stable network, 467
standby, 113
step response, 82
subcircuits, 498
substrate, 502
subthreshold regime, 608
superposition theory, 8
supply independent biasing, 702, 703

Tchebyschev, 261
tee-type, 148
tee-type network model, 148
Thévenin equivalent model, 345
Thévenin impedance, 4
Thévenin input impedance, 19
Thévenin transimpedance, 10
Thévenin voltage, 4, 9
Thévenin voltage gain, 10, 151, 195

Index

Thévenin's theorem, 3
Thévenin, forward transimpedance, 151
thermal gradient, 639
third order compensated response, 768
threshold voltage, 505
time constant, 14, 528, 536
time domain response, 306
transadmittance, 171
transadmittance amplifier, 20, 189, 278, 348
transadmittance feedback amplifier, 352
transconductance amplifier, 141
transconductance coefficient, 509
transconductance parameter, 654
transconductor, 20, 25, 141, 381
transducer power gain, 191, 194, 196, 243, 260
transfer admittance, 448
transfer function matrix, 476, 477, 479, 480, 484
transfer impedance, 445
transient response, 79
transimpedance, 188, 445
transimpedance amplifier, 20, 27, 278, 348
transimpedance feedback amplifier, 348
transimpedance network, 526
transition capacitance, 519
transmission matrix, 155
transmission parameters, 151
transresistance amplifier, 149
transresistor, 20, 27, 149
transverse cutoff frequency, 613
tuned amplifiers, 51
tuned center frequency, 818, 824
tuned frequency, 816
two-port networks, 111
two-port parameters, 112
two-port scattering parameters, 239
two-stage op-amp, 632
two-stage operational amplifier, 633

unconditional stability, 197–199
unconditionally stable, 188, 197, 202, 282
undamped natural frequency, 81, 288, 297, 754

undamped natural frequency of oscillation, 62
undamped self-resonant frequency, 62
underdamped, 76
underdamped network, 86
underdamping, 77
undershoot, 86
unilateral, 206
unilateral amplifier, 204
unilateral capacitive feedback, 670
unilateral feedback, 670
unilateral network, 122
unilateral two-port system, 205
unilateralization, 204
unit step response, 306
unity gain frequency, 299, 305, 310, 311, 318, 522, 523, 534, 535, 680, 732, 813, 815
unity loop gain frequency, 293
unity power gain frequency, 204
unstable, 282

vertical field mobility degradation, 695
virtual dominant pole, 304
virtual short circuit, 281
voltage amplifier, 20, 21, 185, 348
voltage buffer, 34
voltage controlled current source, 141
voltage controlled voltage source, 132
voltage gain, 446, 448, 453
voltage measurement node, 447
voltage reference node, 447
voltage scattering, 228
voltage transfer function, 154, 247
voltage-series feedback amplifier, 483, 494

Wien bridge oscillator, 316
Wilson amplifier, 619
Wilson current amplifier, 618, 720
winding resistance, 749

zero feedback parameter gain, 325
zero frequency gain, 528